Free-Convective Heat Transfer

Oleg G. Martynenko   Pavel P. Khramtsov

Free-Convective Heat Transfer

With Many Photographs of Flows and Heat Exchange

Springer

Professor Oleg G. Martynenko
Belarus Academy of Sciences
Heat and Mass Transfer Institute
P. Brovka str. 15
220072 Minsk
Belarus

Dr. Pavel P. Khramtsov
International Center of Excellence
for Research Eng. and Technology (ICERET)
Auf dem Gossberg
55471 Wüschheim
Germany
Email: iceret@list.ru

Belarus Academy of Sciences
Heat and Mass Transfer Institute
Physical and Chemical
Hydrodynamics Laboratory
P. Brovka str. 15
220072 Minsk
Belarus

Library of Congress Control Number: 2005921210

ISBN-10 3-540-25001-8 Springer Berlin Heidelberg New York
ISBN-13 978-3-540-25001-2 Springer Berlin Heidelberg New York

Springer is a part of Springer Science+Business Media
springeronline.com

Printed in The Netherlands

Typesetting: by the authors and TechBooks using a Springer LaTeX macro package

Cover design: Cover design: *design & production* GmbH, Heidelberg

Printed on acid-free paper SPIN: 11370413 62/TechBooks 5 4 3 2 1 0

The authors express gratitude to Irina A. Shikh and Tatyana A. Baranova for their help in preparing and editing the text of the book and also for advice concerning its style and presentation. Thanks also to Natalya K. Shveeva and Greta R. Maljavskaya for their help in editing the English version of the book. The authors acknowledge the help from Sergey V. Volkov and Victor S. Burak in carrying out the research work.

Oleg G. Martynenko
Pavel P. Khramtsov

Contents

1

Basic Statements and Equations of Free Convection

A body which is brought into a fluid having another temperature is a source of disturbance of the equiprobable state of the medium. The elements of the fluid bordering on the body surface assume its temperature, and the process of heat distribution in the fluid by molecular thermal conductivity – the process of conduction – begins. For a small difference of temperatures, this is the basic mechanism of heat transfer. The arising temperature nonuniformity which is connected with the nonuniformity of density $\Delta\rho$ leads to the occurrence of upward (downward) flows or convection which transfers heat from the object.

In general, natural-convection heat transfer occurs in a nonuniform field of mass forces:

$$\Delta \vec{F} = \Delta(\rho \vec{g_t}) = \Delta\rho \vec{g_t} + \rho\Delta \vec{g_t} \ .$$

If the density nonuniformity $\Delta\rho$ is due to the temperature nonuniformity, then the occuring motion is referred to as thermal gravitational convection. A change in the density can also be due to nonuniform distribution of the concentration of any mixture component or to chemical reactions (in this case we speak of concentration diffusion, or convection), to the presence of phases with different densities or to surface tension forces at the phase interface, etc.

Natural-convection flows can be induced by both gravitational and other mass forces (centrifugal, Coriolis, electromagnetic, etc.). For example, in rotating gas-filled channels the nonuniform mass force field is caused not only by density difference, but also by a nonuniform acceleration field.

Motion and heat exchange occurring in an infinite space are called free convection. The pressure in the field of thermal nonuniformity and in the zone of convective flow can be considered constant. Motion and heat exchange in a bounded volume is called natural convection.

Natural-convection flows can be laminar and turbulent. Experimental data show that in free convection the basic area of thermal and hydrodynamic disturbances is concentrated in a rather thin boundary layer of fluid near the heat transfer surface. For example, at the bottom of a heated vertical plate a laminar boundary layer is formed. With increase in the height of the plate the boundary layer thickness increases and thus heat transfer decreases. At a certain height, the laminar motion is disturbed and becomes turbulent. In this region, the flow represents random motion of the masses of the fluid whose characteristics are described by stochastic functions of space and time variables. For a part of the heated surface, where the characteristics of thermal turbulence become statistically identical, the heat transfer coefficient is independent of the body dimensions.

For free convection one cannot consider separately thermal and hydrodynamic boundary layers, since the motion of fluid is fully determined by the process of heat transfer.

1.1 Equations and Uniqueness Conditions

To describe free-convective motion and heat transfer, the laws of momentum, mass, and energy conservation in the fluid moving under the action of mass, surface, and inertial forces are used. In a rectangular coordinate system the conservation equations have the form [1.1]

$$\rho\left(\frac{\partial u_i}{\partial \tau}+u_j\frac{\partial u_i}{\partial x_j}\right)=\rho F_i+\frac{\partial}{\partial x_j}\left[\mu\left(\frac{\partial u_i}{\partial x_j}+\frac{\partial u_j}{\partial x_i}\right)\right]-\frac{2}{3}\frac{\partial}{\partial x_j}\left(\mu\frac{\partial u_j}{\partial x_j}\right)-\frac{\partial p}{\partial x_i}\,, \tag{1.1.1}$$

$$\frac{\partial \rho}{\partial \tau}+\frac{\partial}{\partial x_j}(\rho u_j)=0\,, \tag{1.1.2}$$

$$\rho c_p\left(\frac{\partial T}{\partial \tau}+u_j\frac{\partial T}{\partial x_j}\right)=Q_\nu+\beta T\left(\frac{\partial p}{\partial \tau}+u_j\frac{\partial p}{\partial x_j}\right)+\frac{\partial}{\partial x_j}\left(\lambda\frac{\partial T}{\partial x_j}\right)+\mu\left[\frac{1}{2}\left(\frac{\partial u_i}{\partial x_j}+\frac{\partial u_j}{\partial x_i}\right)^2-\frac{2}{3}\left(\frac{\partial u_j}{\partial x_j}\right)^2\right]. \tag{1.1.3}$$

Closure of system (1.1.1)–(1.1.3) is achieved through the thermodynamic equation of state

$$\rho=\rho(p,T) \tag{1.1.4}$$

and the equations relating the coefficients of viscosity, heat capacity, heat conduction, and volumetric expansion to pressure and temperature.

The system of differential equations should be augmented with uniqueness conditions to single out the considered process from the whole class of the phenomena described by the system of differential equations (1.1.1)–(1.1.3).

The geometric conditions specify the form and the linear dimensions of the body in which the process proceeds.

Initial conditions are necessary in the problems of nonstationary free convection. They represent the distribution of velocities and temperatures at a certain initial moment of the time $\tau = \tau_0$:

$$\begin{aligned} u_i &= u_{i0}(x_i) \,, \\ T &= T_0(x_i) \,. \end{aligned} \tag{1.1.5}$$

Boundary conditions specify the values of the required functions at the boundaries of the region considered and can be described in a number of ways.

For a solid body in a viscous fluid flow, when the free path of molecules in the fluid is mach smaller than the characteristic size of the body, the velocity of the particles of the fluid on a fixed surface is equal to zero, whereas on a moving one it coincides with the velocity of the points of the surface (no-slip condition):

$$u_{i\mathrm{w}}\left(\tau, x_{i\mathrm{w}}\right) = 0 \,. \tag{1.1.6}$$

In weakly rarefied gases the sleep velocity on a solid surface is proportional to the derivative of the tangent velocity component with respect to the normal to the surface. If there is mass transfer on the surface, the normal velocity component is determined by the rate of absorption (release) of substance by the wall. The boundary conditions also include setting specification of the velocity far from the body immersed in a flow. A large variety of boundary conditions exist for temperature.

The boundary condition of the 1st kind consists in specification of temperature distribution over the heating surface at any instant of time:

$$T = T_{\mathrm{w}}\left(\tau, x_{i\mathrm{w}}\right) \,. \tag{1.1.7}$$

The boundary condition of the 2nd kind specifies the heat flux density for each point of the body surface as a function of time:

$$-\lambda_{j\mathrm{w}}\left(\frac{\partial T}{\partial n_j}\right) = q_{j\mathrm{w}}\left(\tau, x_{i\mathrm{w}}\right) \,, \tag{1.1.8}$$

where j is the number of continuous boundary surfaces of the body.

The simplest boundary conditions are the constancy of temperature or of the heat flux density on the body surface:

$$\begin{aligned} T = T_{\mathrm{w}} &= \mathrm{const} \,, \\ -\lambda_{j\mathrm{w}}\left(\frac{\partial T}{\partial n_j}\right)_{\mathrm{w}} &= q_{\mathrm{w}} = \mathrm{const} \,. \end{aligned} \tag{1.1.9}$$

The boundary condition of the 3rd kind characterizes the law of convective heat exchange between the body surface and environment. In this case, the heat flux density is directly proportional to the difference of temperatures between the body surface and the environment:

$$-\lambda_{j\mathrm{w}}\left(\frac{\partial T}{\partial n_j}\right)_{\mathrm{w}}=\alpha_j\left(T_{j\mathrm{w}}-T_\infty\right)\,. \tag{1.1.10}$$

Relation (1.1.10) holds only for constant $T_{j\mathrm{w}}$. In the majority of cases $T_{j\mathrm{w}}$ changes along the surface depending on the body properties. Therefore, for free-convective heat transfer relation (1.1.10) can be accepted as a first-approximation boundary condition.

The boundary condition of the 4th kind corresponds to heat exchange of the body surface with the surrounding medium or to heat exchange between adjoining bodies when the temperature on the boundary of these bodies is the same. When a solid body is immersed in a flow of fluid, in addition to the equality of the temperatures:

$$T_{j\mathrm{w}}(x_{i\mathrm{w}},\tau)=\left[T_\infty(x_i,\tau)\right]_{j\mathrm{w}} \tag{1.1.11}$$

the equality of heat fluxes also holds:

$$-\lambda_{j\mathrm{w}}\left(\frac{\partial T}{\partial n_j}\right)_{\mathrm{w}}=-\lambda_{j\infty}\left(\frac{\partial T_\infty}{\partial n_j}\right)_{\mathrm{w}}\,. \tag{1.1.12}$$

The boundary condition of the 4th kind leads to conjugate problems of free-convective heat transfer.

1.2 Boussinesq Approximation

When pressure and temperature differences in a flow are small, then it follows from the thermodynamic equation of state (1.1.4) that a change in the density is also small:

$$\begin{aligned}\frac{\rho(p,T)}{\rho_{\mathrm{av}}}&=1+\left(\frac{1}{\rho}\frac{\partial\rho}{\partial T}\right)_{p_{\mathrm{av}}}(T-T_{\mathrm{av}})+\left(\frac{1}{\rho}\frac{\partial\rho}{\partial p}\right)_{T_{\mathrm{av}}}(p-p_{\mathrm{av}})\\&\quad+\frac{1}{2}\left(\frac{1}{\rho}\frac{\partial^2\rho}{\partial T^2}\right)_{\mathrm{av}}(T-T_{\mathrm{av}})^2+\frac{1}{2}\left(\frac{1}{\rho}\frac{\partial^2\rho}{\partial p^2}\right)_{\mathrm{av}}(p-p_{\mathrm{av}})^2\\&\quad+\frac{1}{2}\left(\frac{1}{\rho}\frac{\partial^2\rho}{\partial T\partial p}\right)_{\mathrm{av}}(T-T_{\mathrm{av}})(p-p_{\mathrm{av}})+\dots\end{aligned} \tag{1.2.1}$$

Here, the coefficients of volumetric expansion and isothermal compression precede the temperature and pressure differences. For the majority of working bodies the numerical values of $\varepsilon_1=\beta\theta_0$ and $\varepsilon_2=\beta_p\Delta p$ are rather small (for ***water*** $\varepsilon_1=1.5\cdot10^{-4}\theta_0$ and $\varepsilon_2=4.8\cdot10^{-8}l$; for ***air*** $\varepsilon_1=3.5\cdot10^{-3}\theta_0$ and $\varepsilon_2=1.2\cdot10^{-6}l$). This allows one to neglect the quadratic terms in the expansion

$$\rho=\rho_{\mathrm{av}}\left(1-\beta\theta+\beta_p\Delta p\right)\,. \tag{1.2.2}$$

The smallness of the ratio $\varepsilon_2/\varepsilon_1=(10^{-3}-10^{-4})l/\theta_0$ permits the assertion that for the overwhelming majority of real geometrical sizes and temperature differences a change in the pressure in free motion of the fluid does not exert a substantial influence on the change in the density. In view of the above, we get

$$\rho = \rho_{\mathrm{av}} \left(1 - \beta\theta\right) \ . \tag{1.2.3}$$

The motion driven by the temperature difference, subject to (1.2.3), is the elementary model of thermal convection.

The system of free-convective motion and heat transfer equations is [1.2, 1.3]

$$\rho_{\mathrm{av}} \left(\frac{\partial u_i}{\partial \tau} + u_j \frac{\partial u_i}{\partial x_j} \right) = -\beta \rho \theta_{\mathrm{av}} F_i + \frac{\partial}{\partial x_j} \left[\mu \left(\frac{\partial u_i}{\partial x_j} + \frac{\partial u_j}{\partial x_i} \right) \right] - \frac{\partial p}{\partial x_i} \ , \tag{1.2.4}$$

$$\frac{\partial u_j}{\partial x_j} = 0 \ , \tag{1.2.5}$$

$$\rho_{\mathrm{av}} c_p \left[\frac{\partial \theta}{\partial \tau} + u_j \frac{\partial \theta}{\partial x_j} + u_j \left(\frac{\partial T_{\mathrm{av}}}{\partial x_j} - \beta T \frac{F_j}{c_p} \right) \right] = \Delta Q_v + \beta T \frac{\partial p}{\partial \tau}$$
$$+ \beta T u_j \frac{\partial p}{\partial x_j} + \frac{\partial}{\partial x_j} \left(\lambda \frac{\partial \theta}{\partial x_j} \right) + \frac{\mu}{2} \left(\frac{\partial u_i}{\partial x_j} + \frac{\partial u_j}{\partial x_i} \right)^2 \ . \tag{1.2.6}$$

In (1.2.4) the component of the buoyancy $O(\varepsilon_1)$ is preserved. It is shown in [1.4] that this kind of approximation for the momentum equation is valid if the acceleration $\rho_{\mathrm{av}} \partial u_i / \partial \tau)$ is small in comparison with the lifting force $|F_i|$ (equal to the Earth gravity force).

The system of equations of free-convective heat transfer in the form (1.2.4)–(1.2.6) represents the Boussinesq approximation. The comparison of the solutions in this approximation with experimental data shows that this system correctly reflects the basic special features of thermal convection. We note that the Boussinesq approximation does not impose restrictions on changes in thermophysical characteristics and on the effect exerted by the work of compression and dissipation of energy on the flow and heat transfer.

The state equation in the form of (1.2.3) was first applied for research of free convection in the atmosphere by Oberbeck in 1879 [1.5].

The Boussinesq approximation can also be applied to complex thermodynamic systems with the state equation

$$\rho = \rho \left(p, T, a_i\right) \tag{1.2.7}$$

for small changes in the of parameters a_i, pressure, and temperature:

$$\rho = \rho_{\mathrm{av}} \left(1 - \beta\theta + \beta_p \Delta p + \sum \beta_i \Delta a_i + \ldots \right) \ . \tag{1.2.8}$$

As an example, we can mention homogeneous multicomponent systems or solutions with a small concentration of the corresponding component [1.6].

For dropping liquids it is necessary to take into account the quadratic dependence of density on temperature difference in (1.2.3) [1.2, 1.7–1.9].

1.3 Method of Generalized Variables

A complex system of nonlinear partial differential equations is used in investigation of the processes of motion and free-convective heat transfer. For example, in Boussinesq's approximation it includes five equations for determining the unknown variables u_i, p, T and supplementary equations for the dependences of the thermophysical parameters ρ, β, c_p, and λ on temperature and pressure.

To impart a generalized form to the results of a numerical or experimental solution and also to reduce the number of the parameters of a problem, a method of generalized variables is used [1.10, 1.11]. It rests on the replacement of individual parameters of the problem by their certain combinations representing generalized variables. The structure of the parameters-complexes depends on the form of the differential operators used in equations.

The efficiency of the method of generalized variables can be demonstrated on the equations in the Boussinesq approximation (1.2.4)–(1.2.6). Consider free-convective motion in the field of the gravity forces $F_i = -g_i$ near a surface with a given temperature T_{w}. The thermophysical characteristics are assumed to be constant and independent of temperature. The free convection equations here have the form

$$\begin{aligned}
&\frac{\partial u_i}{\partial \tau} + u_j \frac{\partial u_i}{\partial x_j} = -g_i \beta \theta + \nu \frac{\partial^2 u_i}{\partial x_j^2} - \frac{1}{\rho}\frac{\partial p}{\partial x_i}\,, \\
&\frac{\partial u_j}{\partial x_j} = 0\,, \\
&\frac{\partial \theta}{\partial \tau} + u_j \frac{\partial \theta}{\partial x_j} + u_j\left(\frac{\partial T_\infty}{\partial x_j} - \frac{g_j \beta T}{c_p}\right) = \frac{\Delta Q_\nu}{\rho c_p} + \frac{\beta T}{\rho c_p}\frac{\partial p}{\partial \tau} \\
&\qquad + \frac{\beta T u_j}{\rho c_p}\frac{\partial p}{\partial x_j} + a\frac{\partial^2 \theta}{\partial x_j^2} \\
&\qquad + \frac{\nu}{2c_p}\left(\frac{\partial u_i}{\partial x_j} + \frac{\partial u_j}{\partial x_i}\right)^2 .
\end{aligned} \tag{1.3.1}$$

We reduce (1.3.1) to a dimensionless form. For this purpose, we introduce characteristic scales whose role can be conveniently played by the values entering into the uniqueness conditions. For a linear scale, we select any characteristic dimension of the body l, for the velocity u_0, and for temperature, pressure, time, gravity force, and volumetric heat generation θ_0, p_0, τ_0, g, $Q_{\nu 0}$, respectively. After making the system of (1.3.1) dimensionless, we get

$$\begin{aligned}
&\frac{l}{\tau_0 u_0}\frac{\partial \overline{u}_i}{\partial \overline{\tau}} + \overline{u}_j \frac{\partial \overline{u}_i}{\partial \overline{x}_j} = \frac{g\beta\theta_0 l}{u_0^2}\overline{g}_i \theta + \frac{\nu}{u_0 l}\frac{\partial^2 \overline{u}_i}{\partial \overline{x}_j^2} - \frac{p_0}{\rho u_0^2}\frac{\partial \overline{p}}{\partial \overline{x}_i}\,, \\
&\frac{\partial \overline{u}_j}{\partial \overline{x}_j} = 0\,,
\end{aligned} \tag{1.3.2}$$

$$\frac{l}{\tau_0 u_0}\frac{\partial \theta}{\partial \overline{\tau}} + \overline{u}_j\left(\frac{\partial \theta}{\partial \overline{x}_j} + \frac{\partial \theta_\infty}{\partial \overline{x}_j}\right) + \frac{\beta g l}{c_p}\overline{u}_i\overline{g}_i(\theta+\theta_\infty) = \frac{Q_{v0} l}{\rho c_p u_0 \vartheta_0}\Delta\overline{Q}_v$$
$$+ \frac{p_0\beta}{\rho c_p}(\theta+\theta_\infty)\left(\frac{l}{\tau_0 u_0}\frac{\partial \overline{p}}{\partial \overline{\tau}} + \overline{u}_j\frac{\partial \overline{p}}{\partial \overline{x}_j}\right) + \frac{a}{u_0 l}\frac{\partial^2\theta}{\partial x_j^2}$$
$$+ \frac{1}{2}\frac{\nu u_0}{c_p\theta_0 l}\left(\frac{\partial \overline{u}_i}{\partial \overline{x}_j} + \frac{\partial \overline{u}_j}{\partial \overline{x}_i}\right)^2 .$$

Equation (1.3.2) involve the parameters and scales, which cannot be found from the uniqueness conditions and can be selected proceeding from the physical nature of the problem. First of all, this concerns the characteristic velocity u_0. We refer to (1.3.2) and assume the coefficient at the volumetric force in the momentum equation to be equal to unity:

$$u_0 = \sqrt{g\beta\theta_0 l}\ . \tag{1.3.3}$$

The characteristic velocity can also be obtained by comparing the volumetric and viscosity forces ($u_0 = g\beta v_0 l^2/\nu$) or the inertia forces in the momentum equation ($u_0 = l/\tau_0$). Moreover, it is possible to introduce the characteristic velocity from the equation of energy.

The pressure scale p_0 is usually selected to be equal to the doubled value of the dynamic pressure ρu_0^2.

To define the characteristic velocity we make use of formula (1.3.3) and rewrite the system (1.3.2) in the form

$$\frac{l}{\mathrm{Zh}\,\mathrm{Gr}^{1/2}}\frac{\partial \overline{u}_i}{\partial \overline{\tau}} + \overline{u}_j\frac{\partial \overline{u}_i}{\partial \overline{x}_j} = \overline{g}_i\theta + \mathrm{Gr}^{-1/2}\frac{\partial^2 \overline{u}_i}{\partial \overline{x}_j^2} - \frac{\partial \overline{p}}{\partial \overline{x}_i}\,,$$

$$\frac{\partial \overline{u}_j}{\partial \overline{x}_j} = 0\,,$$

$$\frac{1}{\mathrm{Zh}\,\mathrm{Gr}^{1/2}}\frac{\partial \theta}{\partial \overline{\tau}} + \overline{u}_j\left(\frac{\partial \theta}{\partial \overline{x}_j} + \frac{\partial \theta_\infty}{\partial \overline{x}_j}\right) + \mathrm{Ec}\,\overline{u}_i\overline{g}_i(\theta+\theta_\infty) \tag{1.3.4}$$
$$= \frac{\mathrm{Os}}{\mathrm{Pr}\,\mathrm{Gr}^{1/2}}\Delta\overline{Q}_\nu + \mathrm{Ec}\,\varepsilon_1(\theta+\theta_\infty)\left(\frac{\mathrm{Gr}^{1/2}}{\mathrm{Zh}}\frac{\partial \overline{p}}{\partial \overline{\tau}} + \overline{u}_j\frac{\partial \overline{p}}{\partial \overline{x}_j}\right)$$
$$+ \frac{1}{\mathrm{Pr}\,\mathrm{Gr}^{1/2}}\frac{\partial^2\theta}{\partial x_j^2} + \frac{1}{2}\frac{\mathrm{Ec}}{\mathrm{Gr}^{1/2}}\left(\frac{\partial \overline{u}_i}{\partial \overline{x}_j} + \frac{\partial \overline{u}_j}{\partial \overline{x}_i}\right)^2 .$$

The solution of system of (1.3.4) is determined by six dimensionless complexes: Zh, Gr, Os, Ec, Pr and ε_1. We note that the choice of characteristic scales determines the system of similarity numbers which governs the solution of the problem [1.34]. Using the above numbers, it is possible to construct another system of dimensionless complexes, but their number will not exceed six.

Generally, the expression for velocity, temperature, and pressure are

$$\begin{aligned} u_i &= f_i(x_i,\ \tau,\ \mathrm{Zh},\ \mathrm{Gr}\ ,\mathrm{Pr},\ \mathrm{Os},\ \mathrm{Ec},\ \varepsilon_1;\ \mathrm{P}_k)\ , \\ \theta &= f_4(x_i,\ \tau,\ \mathrm{Zh},\ \mathrm{Gr},\ \mathrm{Pr},\ \mathrm{Os},\ \mathrm{Ec},\ \varepsilon_1;\ \mathrm{P}_k)\ , \\ \Delta p &= f_5(x_i,\ \tau,\ \mathrm{Zh},\ \mathrm{Gr},\ \mathrm{Pr},\ \mathrm{Os},\ \mathrm{Ec},\ \varepsilon_1;\ \mathrm{P}_k)\ . \end{aligned} \tag{1.3.5}$$

The group of dimensionless numbers P_k includes the ratios that determine the geometric properties of the system, the scale values of the excess temperature and of the same kind of constants that characterize the properties of the dissimilar parts of the system.

To calculate heat transfer, we avail ourselves of the boundary conditions on a separation surface:

$$\alpha\theta_{\mathrm{w}} = -\lambda\left(\frac{\partial\theta}{\partial x_i}\right)_{\mathrm{w}} .$$

After making the above equation dimensionless,

$$\mathrm{Nu}\ \theta_{\mathrm{w}} = -\left(\frac{\partial\theta}{\partial x_i}\right)_{\mathrm{w}} \tag{1.3.6}$$

we obtain the new complex $\mathrm{Nu} = \alpha l/\lambda$.

The form of the generalized equation for the dimensionless heat-transfer coefficient can be determined from (1.3.5):

$$\mathrm{Nu} = f_6(x_i,\ \tau\ ,\mathrm{Zh},\ \mathrm{Gr},\ \mathrm{Pr},\ \mathrm{Os},\ \mathrm{Ec},\ \varepsilon_1;\ \mathrm{P}_k)\ . \tag{1.3.7}$$

In determining the mean coefficient of heat transfer, the coordinates and time drop out of the arguments:

$$\mathrm{Nu} = f_7(\mathrm{Zh},\ \mathrm{Gr},\ \mathrm{Pr},\ \mathrm{Os},\ \mathrm{Ec},\ \varepsilon_1;\ \mathrm{P}_k)\ . \tag{1.3.8}$$

The quantity of the governing dimensionless numbers can be reduced if we divide the process of heat transfer into the groups differing in the uniqueness conditions. For stationary processes, the numbers containing the time scale τ_0 (i.e., the Zhukowski homochromism criterion Zh) drop out of the governing ones. The Gr number contains the parametric criterion ε_1. In free convection, heat generation due to Joule dissipation and work of compression exert a little influence on heat transfer. Moreover, for heat transfer in a medium without internal heat sources (1.3.8) it is possible to write

$$\mathrm{Nu} = f_8(\mathrm{Gr}, \mathrm{Pr}; \mathrm{P}_k)\ . \tag{1.3.9}$$

The dependence (1.3.9) can be simplified further by using the similarity theory methods [1.12]. In the case of slow motion of relatively ***viscous fluids*** the inertia forces in the motion equations are small in comparison with other forces acting in the fluid, and they can be neglected. If we select the value

$u_0 = g\beta v_0 l^2/\nu$ as the characteristic velocity, then (1.3.4) in the simplest case of stationary motions transforms to

$$\begin{gathered}\overline{g}_i\theta + \frac{\partial^2 \overline{u}_i}{\partial \overline{x}_j^2} - \frac{\partial \overline{p}}{\partial \overline{x}_i} = 0\,,\\ \frac{\partial \overline{u}_j}{\partial \overline{x}_j} = 0\,,\\ \overline{u}_j\left(\frac{\partial \theta}{\partial \overline{x}_j} + \frac{\partial \theta_\infty}{\partial \overline{x}_j}\right) = \frac{1}{\mathrm{Ra}}\frac{\partial^2 \theta}{\partial x_j^2}\,.\end{gathered} \tag{1.3.10}$$

As (1.3.10) contain only the number Ra, dependence (1.3.9) can be presented in the form

$$\mathrm{Nu} = f_9(\mathrm{Ra};\, \mathrm{P}_k)\,. \tag{1.3.11}$$

This motion occurs, when $\mathrm{Pr} \gg 1$.

Another limiting case is a model of an ideal fluid when the internal friction forces in the equations of motion are negligibly small in comparison with the volumetric and inertial forces. The motion and heat transfer (1.3.4) for an ideal fluids are

$$\begin{gathered}\frac{\partial \overline{u}_i}{\partial \overline{\tau}} + \overline{u}_j\frac{\partial \overline{u}_i}{\partial \overline{x}_j} = \overline{g}_i\theta - \frac{\partial \overline{p}}{\partial \overline{x}_i}\,,\\ \frac{\partial \overline{u}_j}{\partial \overline{x}_j} = 0\,,\end{gathered} \tag{1.3.12}$$

and heat transfer can be defined as

$$\mathrm{Nu} = f_{10}(\mathrm{Gr}\,\mathrm{Pr}^2;\, \mathrm{P}_k)\,. \tag{1.3.13}$$

The dependence (1.3.13) describes convection processes in liquid metals ($\mathrm{Pr} \ll 1$).

Nusselt was the first to formulate the physical similarity laws for the processes of free-convective heat transfer [1.13]. Thereafter, the similarity theory in free-convective heat transfer were extended in [1.14, 1.15]. At the suggestion of Greber, in 1931 three basic similarity criteria in free convection Gr, Pr and Nu were given the names of the famous scientists [1.16]. The designations of other criteria were introduced as they appeared. The basic natural convection similarity criteria and their standard designations are given in Appendix 1.11.

1.4 Dimensional Analysis

In studying the transfer processes the mathematical description of which does not exist, an effective tool for investigation is the experiment. It is expedient to represent experimental results in a generalized form. Similarity numbers can be found by the dimensional analysis method, but for this it is necessary to define a list of the physical quantities which are essential for the process considered.

All the physical quantities can conventionally be divided into primary and secondary ones.

For heat transfer processes it is convenient to select the length L, mass M, time T, the amount of heat Q, and the temperature Θ as the primary quantities.

The secondary quantities are expressed in terms of the primary ones according to definitions or physical laws. The dimensional formulas for the secondary quantities φ have the form of exponential monomials:

$$[\varphi] = L^{n1} M^{n2} T^{n3} Q^{n4} \Theta^{n5} . \tag{1.4.1}$$

The dimensionality of the secondary quantity relative to the given primary one is determined by the value of the exponent n_i at this primary quantity.

Let us compose a product of dimensional formulas, raised to certain powers, for physical quantities, which are essential of the process. We assume that the dimensionality of the exponential monomial is equal to zero. Then the exponential monomial can be represented as a product of dimensionless complexes of dimensional quantities.

Let us consider the application of the dimensional analysis to the problems of free convection. For free-convective heat transfer the following quantities (the dimensionality is specified in parentheses) are essential:

$$l(L),\ g(LT^{-2}),\ \beta(\Theta^{-1}),\ \mu(MT^{-1}L^{-1}),\ \rho(ML^{-3}), \lambda(QL^{-1}T^{-1}\Theta^{-1}),\ \theta(\Theta),$$
$$c_p(QM^{-1}\Theta^{-1}),\ \alpha(QL^{-2}T^{-1}\Theta^{-1}),\ \tau(T),\ Q_v(QL^{-3}),\ Jg(L^2MT^{-2}Q^{-1}) .$$

Dimensionless variables must have the form

$$t^a g^b \beta^c \mu^d \rho^e \lambda^f \theta^h c_p^k \alpha^l \tau^m Q_v^n Jg^p , \tag{1.4.2}$$

where $a, b, c, \ldots, p$ denote exponents. If we substitute the dimensionality of each of the quantities into (1.4.2), then for the dimensionality of the variable we get

$$L^a(LT^{-2})^b\Theta^{-c}(MT^{-1}L^{-1})^d(ML^{-3})^e(QL^{-1}T^{-1}\Theta^{-1})^f\Theta^h(QM^{-1}\Theta^{-1})^k$$
$$(QL^{-2}T^{-1}\Theta^{-1})T^m(QL^{-3})^n(L^2MT^{-2}Q^{-1})^p .$$

The equality to zero of the sum of the exponents at each of the symbols of the primary quantities is the condition for the entire expression to be dimensionless. According to the number of the primary quantities of the system of units selected we obtain five equations for determining 12 exponents. The number of the exponents, for which the values can be selected arbitrarily is equal to $12 - 5 = 7$. Then the remaining exponents are defined in terms of the first ones. After transformations we obtain a system of seven dimensionless complexes: Nu, Zh, Gr, Pr, Os, Ec, and ε_1. The required functional dependence connects these seven complexes:

$$F(\text{Nu}, \text{Zh}, \text{Gr}, \text{Pr}, \text{Os}, \text{Ec}, \varepsilon_1) = 0 ,$$

or

$$\text{Nu} = f_7(\text{Zh}, \text{Gr}, \text{Pr}, \text{Os}, \text{Ec}, \varepsilon_1)$$

and it has already the known form of (1.3.8).

In going over to the dimensionless quantities the number of the variables decreased from 12 to 7. This corresponds to the π-theorem: the number of dimensionless complexes is equal to the number of physical quantities essential for the given process minus the number of primary quantities.

We may analyze special cases from the previous paragraph by the dimensional method. Using the assumptions made in deriving formula (1.3.9), the essential physical quantities must include l, $g\beta$, ρ, μ, λ, θ, c_p and α. The required dependence must connect here three $(8 - 5 = 3)$ dimensionless complexes Nu, Gr, and Pr.

When the viscosity is small, the coefficient μ must be excluded from the essential physical quantities. The number of dimensional physical quantities decreases to 7, and with the same five primary quantities we obtain only two dimensionless exponential products. This result coincides with dependence (1.3.13).

The inertial forces acting in a medium are determined in the dimensional analysis by the density ρ. However, even of the case of small inertial forces the density cannot be excluded completely from consideration because it is contained in the mass force and in the enthalpy of the system. Then the essential physical quantities must include l, μ, ρc_p, λ, θ, $\rho g\beta$, and α. We obtain two dimensionless complexes Nu and Ra and dimensionless dependence (1.3.11).

The latter two cases correspond to intermediate asymptotics in the limiting cases of the number Pr [1.17]. In a Cartesian coordinate system we can take into account the direction (L_x, L_y, L_z) and the direction sign of the velocity vector (L_v and L_{-v}) [1.18]. In multiphase media the vector character of the linear dimensions must be taken into account separately for each phase. This makes it possible to investigate the conjugate problems of free-convective heat transfer using the dimensional analysis [1.19].

As an example we consider stationary free convection on a vertical plane surface in laminar fluid flow. We assume that there is no volumetric heat generation and the work of the forces of compression and energy dissipation are negligibly small. We select a rectangular coordinate system so that the axis x is directed along the plate in the direction of the volumetric force and the axis y along the normal to the surface. Seven quantities are essential for the given process:

$$l(L_x), g\beta(L_xT^{-2}\Theta), \mu(L_x^{-1}L_yMT^{-1}), \rho(L_x^{-1}L_y^{-1}M),$$
$$\lambda(L_x^{-1}\,L_y\,QT^{-1}\Theta^{-1}), \theta(\Theta), c_p(QM^{-1}\Theta^{-1}), \alpha(L_x^{-1}QT^{-1}\Theta^{-1}) .$$

There are two $(8-6=2)$ independent dimensionless complexes in the given problem: $\mathrm{Nu}/\mathrm{Gr}^{1/4}$ and Pr, between which we can establish the dependence

$$\mathrm{Nu}/\mathrm{Gr}^{1/4} = f_{11}(\mathrm{Pr}) \tag{1.4.3}$$

instead of the generalized formula (1.3.9).

For $\mathrm{Pr} \ll 1$

$$\mathrm{Nu} = C_1(\mathrm{Gr}\,\mathrm{Pr}^2)^{1/4} , \tag{1.4.4}$$

for $\mathrm{Pr} \gg 1$

$$\mathrm{Nu} = C_2\mathrm{Ra}^{1/4} . \tag{1.4.5}$$

Dependences (1.4.3)–(1.4.5) are more convenient for processing experimental results than similar formulas (1.3.9), (1.3.11) and (1.3.13).

Replacing the heat transfer coefficient α by the heat flux q, we transform the simple dependence (1.3.9) to [1.20]

$$\begin{aligned} \mathrm{Gr}^* &= f_{12}(\mathrm{Gr},\, \mathrm{Pr}) , \\ \mathrm{Gr}^* &= \mathrm{Nu}\,\mathrm{Gr} . \end{aligned} \tag{1.4.6}$$

If we exclude from consideration the heat transfer coefficient α, then (1.4.6) will relate to the boundary conditions of the 1st kind. For the boundary conditions of the 2nd kind, on the surface [1.21]

$$\mathrm{Gr} = f_{13}(\mathrm{Gr}^*, \mathrm{Pr}) . \tag{1.4.7}$$

1.5 Free-Convective Boundary Layer

The main ideas of the boundary layer approximation for natural and forced convections are similar [1.22]. The main difference is that the pressure outside a boundary layer is not determined by the main stream conditions and is hydrostatic, and the velocity beyond the boundary layer is equal to zero [1.23]. And it is assumed that free-convective flow and mass and energy transfers by this flow are concentrated in the main in a thin layer near the surface. Outside this layer, the fluid is assumed to be immobile, which is confirmed by numerous experimental studies. This entails the assumption that the gradients along the surface are much smaller than those along the normal to it. The scales of the boundary layer are determined by some characteristic quantities δ and δ_T, where δ is the thickness of a hydrodynamic boundary layer equal to the distance over which the main change in the velocity occurs and δ_T is the thickness of a thermal boundary layer equal to the distance over which the temperature of the flow changes from the wall temperature T_0 to the ambient temperature T_∞. In general, $\delta \neq \delta_T$. If $\delta_T < \delta$, then the motion outside the thermal layer, where the buoyancy is absent, is determined by viscous interaction between the moving fluid layers. When the thickness of the viscous layer δ_ν is smaller than that of the thermal one ($\delta_\nu < \delta_T$), the motion outside the viscous boundary layer is potential and differs from zero within the limits of the thermal layer δ_T [1.24]. For a free-convective flow, the motion in the boundary layer is primary, whereas the main stream is secondary. For free convection we cannot consider thermal and hydrodynamic boundary layers separately, since fluid motion is fully determined by the process of heat transfer.

The concept of the boundary layer in free convection in comparison with the forced one is extended due to the substantial influence of buoyancy and forces of compression in addition to viscosity and inertia.

In deriving approximate boundary layer equations use is mainly made of the concept of smallness of the thermal and dynamic boundary layer thicknesses in comparison with the characteristic size of the system at large Grashoff numbers [1.2, 1.23, 1.25–1.27]. Let us consider the system of Boussinesq equations which describes the simplest vertical free-convective flows in which the thermophysical parameters of transfer $\mu(T)$ and $\lambda(T)$ do not change appreciably in the region of flow. In practice, this occurs when μ and λ are weak functions of temperature or at small temperature differences:

$$\frac{\partial u}{\partial x} + \frac{\partial v}{\partial y} = 0 , \tag{1.5.1}$$

$$u\frac{\partial u}{\partial x} + \nu\frac{\partial u}{\partial y} = \nu\left(\frac{\partial^2 u}{\partial x^2} + \frac{\partial^2 u}{\partial y^2}\right) - \frac{1}{\rho}\frac{\partial p}{\partial x} + g\beta\left(T - T_\infty\right) , \tag{1.5.2}$$

$$u\frac{\partial v}{\partial x} + \nu\frac{\partial v}{\partial y} = \nu\left(\frac{\partial^2 v}{\partial x^2} + \frac{\partial^2 v}{\partial y^2}\right) - \frac{1}{\rho}\frac{\partial p}{\partial y} , \tag{1.5.3}$$

$$\begin{aligned} u\frac{\partial T}{\partial x} + \nu\frac{\partial T}{\partial y} &= a\left(\frac{\partial^2 T}{\partial x^2} + \frac{\partial^2 T}{\partial y^2}\right) + \frac{\beta T}{\rho c_p}\left(u\frac{\partial p}{\partial x} + \nu\frac{\partial p}{\partial y}\right) \\ &\quad + \frac{\mu}{\rho c_p}\Phi + \frac{q}{\rho c_p} . \end{aligned} \tag{1.5.4}$$

A detailed analysis of the relative influence of different physical processes in this system [1.28] allows one to obtain the equations that describe a two-dimension stationary free-convective vertical flow in a boundary layer:

$$\begin{gathered} u\frac{\partial u}{\partial x} + \nu\frac{\partial u}{\partial y} = \nu\frac{\partial^2 u}{\partial y^2} + g\beta\theta , \\ \frac{\partial u}{\partial x} + \frac{\partial v}{\partial y} = 0 , \\ u\frac{\partial \theta}{\partial x} + \nu\frac{\partial \theta}{\partial y} = a\frac{\partial^2 \theta}{\partial y^2} + \frac{q_v}{\rho c_p} + \frac{\nu}{c_p}\left(\frac{\partial u}{\partial y}\right)^2 + \frac{\beta T}{\rho c_p}u\frac{\partial p}{\partial x} . \end{gathered} \tag{1.5.5}$$

The boundary and initial conditions depending on the character of heat transfer on the surface can be written as

$$\theta = \theta_{\mathrm{w}} \quad \text{or} \quad \frac{\partial \theta}{\partial y} = -\frac{q_{\mathrm{w}}}{\lambda}\,; \quad \text{or} \quad \frac{\partial \theta}{\partial y} = \frac{\lambda_b}{\lambda}\frac{\partial \theta_{\mathrm{b}}}{\partial y} \quad \text{for} \quad y = 0\,;$$
$$u = 0,\ \theta = 0 \quad \text{or}\ y \to \infty;\ u = u_0(y),\ \theta = \theta_0(y) \quad \text{for} \quad x = x_0\,. \tag{1.5.6}$$

The dimensionless system of the equations of free-convective heat transfer near a vertical plate with a constant temperature T_{w} for a stationary case in the absence of volumetric heat sources is [1.2]

$$u\frac{\partial u}{\partial x} + v\frac{\partial u}{\partial y} = \theta + \mathrm{Gr}^{-1/2}\left(\frac{\partial^2 u}{\partial x^2} + \frac{\partial^2 u}{\partial y^2}\right) - \frac{\partial \Delta p}{\partial x}\,,$$
$$u\frac{\partial v}{\partial x} + v\frac{\partial v}{\partial y} = \mathrm{Gr}^{-1/2}\left(\frac{\partial^2 v}{\partial x^2} + \frac{\partial^2 v}{\partial y^2}\right) - \frac{\partial \Delta p}{\partial y},$$
$$\frac{\partial u}{\partial x} + \frac{\partial v}{\partial y} = 0\,, \tag{1.5.7}$$

$$u\frac{\partial \theta}{\partial x} + \nu\frac{\partial \theta}{\partial y} + u\frac{\partial \theta_\infty}{\partial x} + \mathrm{Ec}\, u\,(\theta + \theta_\infty)$$
$$= \mathrm{Ec}\,\varepsilon_1\,(\theta + \theta_\infty)\left(u\frac{\partial \Delta p}{\partial x} + \nu\frac{\partial \Delta p}{\partial y}\right) + \mathrm{Pr}^{-1}\mathrm{Gr}^{-1/2}\left(\frac{\partial^2 \theta}{\partial x^2} + \frac{\partial^2 \theta}{\partial y^2}\right)$$
$$+ \mathrm{Ec}\,\mathrm{Gr}^{-1/2} \times \left[\left(\frac{\partial u}{\partial x}\right)^2 + \left(\frac{\partial \nu}{\partial y}\right)^2 + \left(\frac{\partial u}{\partial y} + \frac{\partial \nu}{\partial x}\right)^2\right].$$

Using as a basis the assumption about the finite nature of increments for temperature and longitudinal velocity in a boundary layer of thickness δ and measuring the longitudinal coordinate in the scale $y = \delta Y$, from the continuity equation we get

$$\nu = \delta V\,. \tag{1.5.8}$$

Substituting (1.5.8) into the system of (1.5.7) and preserving the values of the same order of magnitude lead to the equation [1.2]

$$u\frac{\partial u}{\partial x} + V\frac{\partial u}{\partial Y} = \theta + \delta_\nu^{-2}\,\mathrm{Gr}^{-1/2}\frac{\partial^2 u}{\partial Y^2} - \frac{\partial \Delta p}{\partial x}\,,$$
$$\frac{\partial \Delta p}{\partial Y} = 0\,,$$
$$\frac{\partial u}{\partial x} + \frac{\partial V}{\partial Y} = 0\,, \tag{1.5.9}$$

$$u\frac{\partial \theta}{\partial x} + \delta_\nu\delta_T^{-1}V\frac{\partial \theta}{\partial Y} + u\frac{\partial \theta_\infty}{\partial x} + \mathrm{Ec}\,u\,(\theta + \theta_\infty) = \mathrm{Ec}\,\varepsilon_1\,(\theta + \theta_\infty)\,u\frac{\partial \Delta p}{\partial x}$$
$$+\mathrm{Pr}^{-1}\mathrm{Gr}^{-1/2}\delta_T^{-2}\frac{\partial^2 \theta}{\partial Y^2} + \mathrm{Ec}\,\mathrm{Gr}^{-1/2}\delta_\nu^{-2}\left(\frac{\partial u}{\partial Y}\right)^2.$$

The second equation of motion points to the constancy of the pressure in the section of the boundary layer normal to the body surface. The motion is determined by three dimensionless criteria: Gr, Pr, and Ec. Depending on the order of magnitude of the similarity numbers, various relationships are possible between the thicknesses of the viscous δ_ν and thermal δ_T boundary layers. In the majority of cases, the work of compression and viscous dissipation of energy exert a weak influence on the process of free-convective heat transfer, and the corresponding components are usually neglected. If $\mathrm{Pr} = O(1)$, then under the boundary conditions of the 1st kind the thicknesses of the thermal and dynamic boundary layers coincide and are the values of the order of $\mathrm{Gr}^{-1/4}$. Then (1.5.9) take the form:

$$
\begin{gathered}
u\frac{\partial u}{\partial x} + V\frac{\partial u}{\partial Y} = \theta + \frac{\partial^2 u}{\partial Y^2} - \frac{\partial \Delta p}{\partial x} \,, \\
\frac{\partial u}{\partial x} + \frac{\partial V}{\partial Y} = 0 \,, \\
u\frac{\partial \theta}{\partial x} + V\frac{\partial \theta}{\partial Y} + u\frac{\partial \theta_\infty}{\partial x} = \mathrm{Pr}^{-1}\frac{\partial^2 \theta}{\partial Y^2} \,.
\end{gathered}
\tag{1.5.10}
$$

The heat transfer is determined by the dimensionless relation

$$
\mathrm{Nu}/\mathrm{Gr}^{1/4} = f\,(\mathrm{Pr}) \,. \tag{1.5.11}
$$

For small Prandtl numbers the thickness of the viscous boundary layer is much smaller than that of the thermal one ($\delta_\nu/\delta_T = \mathrm{Pr}^{1/2}$). The thermal boundary layer in this case consists of a thin viscous near-wall layer and a zone of inviscid flow.

The motion is induced by the temperature difference, which is observed over the entire thermal layer. The thickness of the hydrodynamic layer δ coincides then with the thickness of thermal layer δ_T and is determined from the heat balance equation

$$
\delta_T = \delta = \mathrm{Pr}^{-1}\ \mathrm{Gr}^{-1/2} \,. \tag{1.5.12}
$$

The condition of the applicability of the boundary-layer approximation in this case is $\mathrm{Gr}\ \mathrm{Pr}^2 \gg 1$. The system of boundary layer (1.5.1)–(1.5.4) in the limit $\mathrm{Pr} \to 0$ is

$$
\begin{gathered}
u\frac{\partial u}{\partial x} + \nu\frac{\partial u}{\partial y} = g\beta\theta \,, \\
\frac{\partial u}{\partial x} + \frac{\partial \nu}{\partial y} = 0 \,, \\
u\frac{\partial \theta}{\partial x} + \nu\frac{\partial \theta}{\partial y} = a\frac{\partial^2 \theta}{\partial y^2}
\end{gathered}
\tag{1.5.13}
$$

and the system of dimensionless (1.5.10) is transformed to

$$
\begin{gathered}
u\frac{\partial u}{\partial x} + V\frac{\partial u}{\partial Y} = \theta - \frac{\partial \Delta p}{\partial x} \,, \\
\frac{\partial u}{\partial x} + \frac{\partial V}{\partial Y} = 0 \,, \\
u\frac{\partial \theta}{\partial x} + V\frac{\partial \theta}{\partial Y} + u\frac{\partial \theta_\infty}{\partial x} = \frac{\partial^2 \theta}{\partial Y^2} \,.
\end{gathered}
\tag{1.5.14}
$$

The dimensionless relation for heat transfer is

$$\frac{\mathrm{Nu}}{\left(\mathrm{Gr}\,\mathrm{Pr}^2\right)^{1/4}} = \text{const}\ . \tag{1.5.15}$$

This model corresponds to the thermal boundary layer of an ideal fluid near the surface.

At large Prandtl numbers, the thermal boundary layer is thin compared with the viscous one ($\delta_\nu/\delta_T = \mathrm{Pr}^{1/2}$). The buoyancy in this case acts only within the limits of the thermal layer. In the viscous layer $\theta \equiv 0$, and the flow in it occurs due to the entrainment of the fluid by friction. The equations of boundary layer (1.5.5) for this case are

$$\begin{aligned} \nu\frac{\partial^2 u}{\partial y^2} + g\beta\theta = 0\ , \\ \frac{\partial u}{\partial x} + \frac{\partial v}{\partial y} = 0\ , \\ u\frac{\partial \theta}{\partial x} + v\frac{\partial \theta}{\partial y} = a\frac{\partial^2 \theta}{\partial y^2} + \frac{\nu}{c_p}\left(\frac{\partial u}{\partial y}\right)^2 . \end{aligned} \tag{1.5.16}$$

Neglect of the convective terms in the equation of motion leads to singular disturbances of the system (1.5.16), since the condition $u = 0$ for $y \to \infty$ is not satisfied. It is replaced by the condition of velocity limitation, which gives $\partial u/\partial y = 0$ for $y \to \infty$. The viscous flow outside the boundary layer is determined by the equations [1.25]

$$\begin{aligned} u\frac{\partial u}{\partial x} + \nu\frac{\partial u}{\partial y} = \nu\frac{\partial^2 u}{\partial y^2}\ , \\ \frac{\partial u}{\partial x} + \frac{\partial \nu}{\partial y} = 0\ , \end{aligned} \tag{1.5.17}$$

which satisfy the boundary condition $u = 0$ for $y \to \infty$. The boundary condition on the wall for a longitudinal velocity is its equality to longitudinal velocity for the internal thermal layer.

Selecting the expression

$$u_0 = \frac{g\beta\theta_0 L^2}{\nu} \tag{1.5.18}$$

as the characteristic scale of velocity and forming, on its basis, the "Reynolds number" for free convection, the square of which is equal to the Grashof number

$$\mathrm{Gr}_L = \frac{u_0^2 L^2}{\nu^2} = \frac{g\beta L^3 (T_\mathrm{W} - T_\infty)}{\nu^2}\ ,$$

from (1.5.10) we obtain the dimensionless equations for a thermal layer [1.2]:

$$\theta + \frac{\partial^2 u}{\partial Y^2} - \frac{\partial \Delta p}{\partial x} = 0 ,$$
$$u\frac{\partial \theta}{\partial x} + V\frac{\partial \theta}{\partial Y} + u\frac{\partial \theta_\infty}{\partial x} = \frac{\partial^2 \theta}{\partial Y^2} , \tag{1.5.19}$$
$$\frac{\partial u}{\partial x} + \frac{\partial V}{\partial Y} = 0 .$$

The condition for the applicability of the approximation is Gr Pr $\gg 1$. This condition can be satisfied due to large values of Pr at small Gr numbers. For example, for an aqueous solution of glycerine the Prandtl number can reach the values $\sim 10^6$. The considered boundary layer has essential features connected with the thermophysical characteristics of the object investigated. The dimensionless relation for heat transfer is [1.29]

$$\frac{\mathrm{Nu}}{\mathrm{Ra}^{1/4}} = \mathrm{const} , \tag{1.5.20}$$

and the viscous flow outside the thermal boundary layer is described then by the equations

$$u\frac{\partial u}{\partial x} + V\frac{\partial u}{\partial Y} = \frac{\partial^2 u}{\partial Y^2} ,$$
$$\frac{\partial u}{\partial x} + \frac{\partial V}{\partial Y} = 0 . \tag{1.5.21}$$

Equations (1.5.21) satisfy the boundary condition $u = 0$ for $Y \to \infty$.

The system of boundary-layer equations for large Eckert numbers Ec is constructed similarly. The quantity Ec characterizes the amount of heat evolved due to friction and compression work. The large values of the Ec number occur in atmospheric phenomena and astrophysics at small temperature differences and large linear dimensions of the system. In the limiting case Ec $\to \infty$, the equations of motion and heat transfer (1.5.1)–(1.5.4) are converted to the form:

$$\nu\frac{\partial^2 u}{\partial y^2} + g\beta\theta = 0 ,$$
$$\frac{\partial u}{\partial x} + \frac{\partial v}{\partial y} = 0 , \tag{1.5.22}$$
$$a\frac{\partial^2 \theta}{\partial y^2} + \frac{\nu}{c_p}\left(\frac{\partial u}{\partial y}\right)^2 = 0 .$$

Outside the thermal boundary layer there is a viscous boundary layer of thickness $\delta_\nu = \delta_T(\mathrm{Pr\ Ec})^{1/2}$, where the flow is independent of buoyancy and in the limit Ec $\to \infty$ is determined by (1.5.17).

The boundary-layer thickness and the characteristic scale of velocity at large Ec numbers are evaluated from the conditions

$$\delta = (\mathrm{Gr\ Pr\ Ec})^{-1/4}, \quad u_0 = \mathrm{Pr}^{-1/2}\mathrm{Ec}^{-1/2} \tag{1.5.23}$$

and the dimensionless system of the boundary-layer equations in this case is

$$\begin{gathered} \theta + \frac{\partial^2 u}{\partial Y^2} - \frac{\partial \Delta p}{\partial x} = 0\,, \\ u\,(\theta + \theta_\infty) = \varepsilon_1\,(\theta + \theta_\infty)\,u\frac{\partial \Delta p}{\partial x} + \frac{\partial^2 \theta}{\partial Y^2} + \left(\frac{\partial u}{\partial Y}\right)^2\,, \\ \frac{\partial u}{\partial x} + \frac{\partial V}{\partial Y} = 0\,. \end{gathered} \tag{1.5.24}$$

The dimensionless relation for calculating the heat transfer coefficient is

$$\frac{\mathrm{Nu}}{(\mathrm{Pr\ Gr\ Ec})^{1/4}} = \mathrm{const}\,. \tag{1.5.25}$$

The conditions for the applicability of the approximation of such a boundary-layer model are Gr Pr Ec $\gg 1$ and Ec $\gg 1$ without any additional restrictions on the values of Gr and Pr.

The boundary-layer equations in application to natural convection can also be obtained as the zero approximation of the method of small disturbances. The dimensionless stream function $f = \Psi/(\nu G)$ and the temperature θ can be represented as an expansion in terms of the small parameter ε:

$$\begin{aligned} f &= f_0(\eta) + \varepsilon f_1(\eta) + \varepsilon^2 f_2(\eta) + \dots\,, \\ \theta &= \theta_0(\eta) + \varepsilon\theta_1(\eta) + \varepsilon^2\theta_2(\eta) + \dots\,, \end{aligned}$$

where $\varepsilon(x) = 1/G$, $\eta = \frac{y}{x}\sqrt[4]{\mathrm{Gr}_x}$, and $G = 4\left[g\beta x^3(T - T_\infty)/4\nu^2\right]^{1/4}$.

Substituting these expansions into full partial-differential equations and isolating the terms containing ε^0, we obtain the system of boundary-layer equations for a free-convective flow [1.23].

Similarly we can write the corresponding equations for axisymmetric flows, for example, for a flow near a vertical cylinder or a flow in a wake above a point heat source and also for a number of other cases of flow. If the boundary-layer thickness δ is not a small value in comparison with x, the boundary-layer approximation becomes inappropriate. These conditions exist in the regions near the leading edge of a vertical surface and also at small surface temperatures and at some particular values of the physical parameters of a fluid [1.23].

1.6 Integral Methods

One of the commonest approximate methods used for calculating free-convective heat transfer is the integral method allowing one to determine the behavior of the velocity and temperature fields near the body surface; this behavior characterizes heat transfer between the body and the medium. The application of the method requires satisfaction of the summarized relations obtained from the differential equations of boundary layer, without determining velocity and temperature at each point. Moreover, the satisfactions the boundary conditions are also required.

In the majority of cases, the main integrated equations are the Karman momentum and the heat balance equations. There are two ways of obtaining these equations: one is based on the momentum principle in the Euler form and on the heat balance on the surface; the second is analytical consisting of the transformation of the boundary-layer equations. In the first case, a certain control surface is isolated near the plate, and the balance of various characteristics on it is considered. The momentum flux through such an immobile surface is equal to the sum of the integral of volumetric forces and the friction resistance of a part of the plate from its leading edge to the section with the coordinate x. The change in the total amount of heat in the boundary layer is determined by heat transfer due to thermal conductivity through the immobile fluid layer near the body surface.

We refer to the boundary-layer equations for the case of viscous developed motion in the absence of volumetric heat generation, negligibly small energy dissipation, and constant ambient temperature [1.30]:

$$\begin{gathered}\frac{\partial u}{\partial x}+\frac{\partial v}{\partial y}=0\,,\\ u\frac{\partial u}{\partial x}+\nu\frac{\partial u}{\partial y}=g\beta\theta+\nu\frac{\partial^2 u}{\partial y^2}\,,\\ u\frac{\partial \theta}{\partial x}+\nu\frac{\partial \theta}{\partial y}=a\frac{\partial^2 \theta}{\partial y^2}\,,\end{gathered}\tag{1.6.1}$$

with the boundary conditions

$$\begin{gathered}u=0,\ v=0,\ \theta=\theta_{\mathrm{w}}\quad\text{or}\quad q_{\mathrm{w}}=-\lambda\frac{\partial\theta}{\partial y}\quad\text{for}\quad y=0\,,\\ u=0,\quad \theta=0\quad\text{for}\quad y\to\infty\,.\end{gathered}\tag{1.6.2}$$

Using the continuity equation, we rewrite the second and the third equations of the system (1.6.1) as

$$\begin{gathered}\frac{\partial u^2}{\partial x}+\frac{\partial u\nu}{\partial y}=g\beta\theta+\nu\frac{\partial^2 u}{\partial y^2}\,,\\ \frac{\partial u\theta}{\partial x}+\frac{\partial \nu\theta}{\partial y}=a\frac{\partial^2 \theta}{\partial y^2}\,.\end{gathered}\tag{1.6.3}$$

We integrate the obtained equations over y from 0 to ∞ or to a certain finite boundary-layer thickness $\delta_{\min}$. In the latter case, the asymptotic boundary conditions are replaced by approximated ones:

$$u = 0, \ \frac{\partial u}{\partial y} = 0, \ \theta = 0, \ \frac{\partial \theta}{\partial y} = 0 \ \text{at} \ y = \delta_m \, . \tag{1.6.4}$$

The integration yields

$$\begin{aligned}
&\int\limits_0^{\delta_{\min},\infty} \frac{\partial u^2}{\partial x} dy + (u\nu)\Big|_0^{\delta_{\min},\infty} = g\beta \int\limits_0^{\delta_{\min},\infty} \theta dy + \nu \left(\frac{\partial u}{\partial y} \right)\Bigg|_0^{\delta_{\min},\infty} , \\
&\int\limits_0^{\delta_{\min},\infty} \frac{\partial u\theta}{\partial x} dy + (\nu\theta)\Big|_0^{\delta_{\min},\infty} = a \left(\frac{\partial \theta}{\partial y} \right)\Bigg|_0^{\delta_{\min},\infty} .
\end{aligned} \tag{1.6.5}$$

Using boundary conditions (1.6.4) and assuming it possible to change the order of differentiation and integration, we get

$$\begin{aligned}
&\frac{d}{dx} \int\limits_0^{\delta_{\min},\infty} u^2 dy = g\beta \int\limits_0^{\delta_{\min},\infty} \theta dy + \nu \left(\frac{\partial u}{\partial y} \right)_0 , \\
&\frac{d}{dx} \int\limits_0^{\delta_{\min},\infty} u\theta dy = -a \left(\frac{\partial \theta}{\partial y} \right)_0 .
\end{aligned} \tag{1.6.6}$$

The quantity $\delta_{\min}$ represents the least of the boundary-layer thicknesses δ_v and δ_T. Equations (1.6.6) are the main integral equations of the free-convective boundary layer [1.31].

The essence of the method used to investigate boundary-layer equations, based on application of integral equations, rests on approximation of an unknown velocity or temperature profile by some function satisfying the boundary conditions and containing a free parameter (for example, the boundary-layer thickness) which thereafter is determined from the momentum equation (heat balance equation).

In free convection, the main results were obtained by the Karman-Pohlhausen integral method at equal thermal and hydrodynamic boundary-layer thicknesses and when low degree polynomials are used as approximations of the velocity and temperature profiles [1.31].

It is necessary to note the special features of the application of the integral methods for two different cases: $\Pr > 1$ and $\Pr < 1$, since on transition through the values $\Pr \sim 1$ the structure of the boundary layer changes substantially. For the case $\Pr > 1$ the following functions can be used as temperature and velocity approximations [1.32]:

$$T - T_\infty = \Delta T \exp(-x/\delta_T) \, , \tag{1.6.7}$$

$$\nu = V \exp(-x/\delta) \left[1 - \exp(-x/\delta_T)\right] \, , \tag{1.6.8}$$

where V, δ, and δ_T are the unknown functions of the coordinate y and $\Delta T = T_0 - T_\infty = \text{const}$. Substituting the selected profiles as integrated functions into the momentum and energy integrals:

$$\frac{d}{dy}\int_0^X v^2 dx = -\nu\left(\frac{\partial v}{\partial x}\right)_{x=0} + g\beta\int_0^X (T - T_\infty)\, dx \,, \tag{1.6.9}$$

$$\frac{d}{dy}\int_0^X v\,(T_\infty - T)\, dx = a\left(\frac{\partial T}{\partial x}\right)_{x=0} , \tag{1.6.10}$$

and assuming $X \to \infty$, we obtain

$$\frac{d}{dy}\left[\frac{V^2\delta q^2}{2\,(2+q)\,(1+q)}\right] = -\frac{\nu V q}{\delta} + g\beta\Delta T\frac{\delta}{q} \,, \tag{1.6.11}$$

$$\frac{d}{dy}\left[\frac{V\delta}{(1+q)\,(1+2q)}\right] = \frac{a}{\delta} \,, \tag{1.6.12}$$

where q is a function of Pr:

$$q(\mathrm{Pr}) = \frac{\delta}{\delta_T} \,. \tag{1.6.13}$$

In (1.6.11) and (1.6.12) three functions are unknown: $V(y)$, $\delta(y)$, and $q(\mathrm{Pr})$. The third equation can be selected in various ways. Sometimes, the condition $\delta = \delta_T$ $(q = 1)$ is used as a closing equation [1.33]. However, in this case a large number of problems dealing with a situation with variable δ/δ_T drops out of consideration; therefore, it is more expedient to assume that $\delta \neq \delta_T$ and to use as a closing equation other conditions following from the physical statement of the problem. It is necessary to bear in mind that (1.6.11)–(1.6.12) are approximate, and the initial equations can be satisfied at various approximations of integrated functions. This affords some freedom in applying the integral method. Directly near the surface, under the non-slip condition the inertia terms in the Navier-Stokes equation can be neglected in comparison with buoyancy and viscous forces irrespective of the Prandtl number, and then the equation

$$0 = \nu\frac{\partial^2 v}{\partial x^2} + g\beta\,(T_0 - T_\infty) \tag{1.6.14}$$

can be used for closing the system (1.6.11)–(1.6.12). The solution of the three equations obtained leads to the following dependence $q(\mathrm{Pr})$ [1.32]:

$$\mathrm{Pr} = \frac{5}{6}q^2\frac{q + 1/2}{q+2} \,. \tag{1.6.15}$$

Here, we have the following formula for Nu:

$$\mathrm{Nu} = \left[\frac{3}{8}\frac{q^3}{(q+1)\,(q+1/2)\,(q+2)}\right]^{1/4} \mathrm{Ra}_y^{1/4} \,. \tag{1.6.16}$$

In the limit $\mathrm{Pr} \to \infty$

$$\frac{\delta}{\delta_T} = \left(\frac{6}{5}\,\mathrm{Pr}\right)^{1/2} \quad \text{and} \quad \mathrm{Nu} = 0.783\ \mathrm{Ra}_y^{1/4}\ . \tag{1.6.17}$$

In case where Pr < 1, with an exponentially decreasing temperature profile (1.6.7), the other velocity profile can be selected as an integrated function:

$$\nu = V_1 \exp\left(-x/\delta_T\right)\left[1 - \exp\left(-x/\delta_\nu\right)\right]\ , \tag{1.6.18}$$

where V_1, δ_T, and δ_ν are unknown functions of y. In this case, the solution of the system of (1.6.9)–(1.6.10) and (1.6.14) leads to other dependences [1.32]:

$$\mathrm{Pr} = \frac{5}{3}\left(\frac{q_1}{1+q_1}\right)^2, \quad q_1 = \frac{\delta_T}{\delta_\nu}\ , \tag{1.6.19}$$

$$\mathrm{Nu} = \left(\frac{3}{8}\right)^{1/4}\left(\frac{q_1}{2q_1+1}\right)^{1/2} \mathrm{Ra}_y^{1/4}\ . \tag{1.6.20}$$

In the limit $\mathrm{Pr} \to \infty$

$$\frac{\delta_\nu}{\delta_T} = \left(\frac{3}{5}\,\mathrm{Pr}\right)^{1/2}\ ,$$

$$\mathrm{Nu} = 0.689\left(\mathrm{Pr}\,\mathrm{Ra}_y\right)^{1/4}\ . \tag{1.6.21}$$

The calculation of the Nusselt number depends to some extent on a choice of analytical functions for velocity and temperature profiles, which must correspond to the physically acceptable form of the profile and at the same time lead to the least complex analytical transformations. In the foregoing analysis the choice of the exponential of form approximation for the temperature and velocity profiles implies rather simple transformations. The integral analysis made in [1.33] and based on polynomial distribution for the corresponding profiles at $\delta = \delta_T$ also predicts correct Nusselt numbers for a wide range of Pr, although the condition $\delta = \delta_T$ restricts this range by the values near 1 ($\mathrm{Pr} \sim 1$).

The integral form of the momentum equation can be obtained by multiplying it by u^k with subsequent integration ($k \geq 1$) [1.25]:

$$\frac{1}{k+1}\frac{d}{dx}\int_0^{\delta,\infty} u^{k+2}dy = g\beta^3 \int_0^{\delta_{\min},\infty} \theta u^k dy - \nu k \int_0^{\delta,\infty} u^{k-1}\left(\frac{\partial u}{\partial y}\right)^2 dy\ . \tag{1.6.22}$$

When $k = 1$, (1.6.22) represents an integral Leibenson's relation of mechanical energy balance for free convection [1.34].

Similarly, from the heat balance equation we have

$$\frac{1}{k+1}\frac{d}{dx}\int_0^{\delta,\infty} u\theta^{k+1}dy = a \int_0^{\delta_T,\infty} \theta^k \frac{\partial^2\theta}{\partial y^2}dy\ . \tag{1.6.23}$$

Integral relations are widely applied in numerical methods to check the accuracy of calculation and the convergence of the results obtained. Using the integral equations, one can find averaged characteristics which determine the scales of initial quantities. In a number of cases, in assigning the distribution of the unknown parameters from experiments or theoretical calculations the integral equations allow one to determine the averaged characteristics of the process irrespective of the internal features of the structure of the flow and of heat transfer in the volume investigated. Most frequently, the integral relations are used in approximate practical calculations. These methods made it possible to solve many problems in free-convective heat transfer, however, there are no proofs of convergence or error estimations for the general case. Recently various velocity and temperature distributions are applied to increase the accuracy of the integral methods. Their accuracy is evaluated by comparing them with analytical solutions or with asymptotic dependences. However, the very choice of suitable velocity and temperature profiles remains arbitrary, and one cannot claim with certainty which one leads to the most accurate results.

1.7 Loss of Stability and Transition to Turbulence

For the majority of free-convective flows, the intensity of transfer processes depends strongly on the mode of flow, therefore investigation of the boundary-layer stability and conditions of transition from a laminar to a turbulent flow determines the accuracy of calculation of transfer characteristics. The transition to a turbulent regime occurs as a result of the effect of external disturbances leading to the development of flow instability. The sources of disturbances can be, for example, vibration of buildings and equipment, fluctuations in heat supply to a heated surface, instability in the stratification of the environment, etc. Depending on the conditions of their occurrence, buoyancy magnitude, physical and geometrical characteristics of the process considered, the amplitude of these disturbances can increase on mutual effect of pressure, buoyancy, and viscosity.

The first stage in the occurrence and development of instability, for many flows is the initial growth of small disturbances [1.23, 1.28, 1.35]. It has been established that the disturbance developing in a natural-convection flow often has the form of a periodic wave moving downstream. If during mutual the effect of hydrodynamic and gravitational forces these waves receive additional energy, their amplitude grows, and the motion becomes unstable. However, if the disturbance has a fixed frequency, this occurs at different distances from the leading edge depending on the Grashof number Gr_x. The disturbance of any given frequency is unstable here only in a certain range of the Grashof numbers and, consequently, of the corresponding values of the longitudinal distance [1.28].

For example, experimental investigation of the process of transition from a laminar to a turbulent regime in natural convection along a vertical plane surface with a constant heat flux was carried out in [1.36]. The flow visualization in water was carried out using liquid crystals and a fluorescent dye solution. There were horseshoe-shaped vortical flows the time and spatial scales of which are independent of the heat flux magnitude and of the arrangement of these vortices. There are also W-shaped flows near the wall with an unstable three-dimensional structure playing a significant role in transition from a laminar to a turbulent regime.

The experimental results for the local Nu number in natural convection ***along a vertical plane surface with a constant heat flux*** are adequately described by empirical relations:

for a laminar regime [1.37]

$$\mathrm{Nu} = \left(\frac{\mathrm{Pr}}{4 + 9\sqrt{\mathrm{Pr}} + 10\,\mathrm{Pr}}\right)^{1/5} \left(\frac{\nu_\infty}{\nu_\mathrm{w}}\right)^{0.17} (\mathrm{Ra}^*)^{1/5} , \tag{1.7.1}$$

for a turbulent regime [1.38]

$$\mathrm{Nu} = 0.568\,(\mathrm{Ra}^*)^{0.22} , \tag{1.7.2}$$

where the modified Rayleigh number $\mathrm{Ra}^* = \mathrm{Gr}^*\,\mathrm{Pr}$ changes within the range $3 \cdot 10^{12} \leq \mathrm{Ra}^* \leq 3.5 \cdot 10^{13}$, and $\nu_\mathrm{w}, \nu_\infty$ are the kinematic viscosity values at the wall and ambient temperatures, respectively.

On the basis of the linear theory of stability and analysis of the Orr-Sommerfeld equations for perturbations of the functions of temperature and streamlines the analysis of transition to turbulence in the case of one-dimensional free-convective water flow ($\mathrm{Pr} = 6.7$) near a vertical plane surface with a stepwise change in the heat flux from 0 to a certain constant value q_w on it was carried out in [1.39]. Unlike the majority of works analyzing transient flow induced by a sudden change in the temperature or heat flux on the investigated surface in the initial instant of time, work [1.40] analyzes the instability of a natural-convection flow near a vertical plate placed into a stratified medium for a physically more real problem in which the boundary conditions for the surface temperature are described as $T_\mathrm{w} = T_{\mathrm{w}0} + \Delta T_0 \phi\,(t)$, where $T_{\mathrm{w}0}$ is the basic stable distribution of the surface temperature and $\phi\,(t)$ describes temperature fluctuations dependent on time and characterizing the perturbations of the temperature field. The results of the investigation show strong dependence of the temperature and velocity profiles on the character of change in T_w and on the stratification level of the medium. The local heat transfer coefficient calculated on the basis of the initial temperature difference decreases with increase in the stratification. As T_w decreases, the influence of the stratification becomes more significant. These effects are felt more a strongly for large Pr numbers.

Experimental study of transition to turbulence in a free-convective flow along an isothermal vertical surface, carried out in [1.41], shows that transition to turbulence is initiated by interaction between a zone with vertical stratification of the temperature outside the boundary layer and the boundary layer near the vertical surface. The times of the start of transition which are in good agreement with the experimental data on flow visualization with the aid of interferograms, are determined on the basis of the dimensional analysis with a count for the vertical temperature gradient and the analogy between free and forced convections.

A two-layer turbulence model [1.42] is used for a detailed numerical analysis of transient free-convective flows in various geometrical configurations (vertical slot, heated plane vertical plate, rectangular cavities with various aspect ratios).

The averaged velocity and temperature distributions in a turbulent free-convective flow along a heated horizontal surface [1.43] obtained on the basis of the dependence $\mathrm{Nu} = N_0 \mathrm{Pr}^{1/7} \mathrm{Ra}^{2/7}$, which differs from the classical one $\mathrm{Nu} = N_0 \mathrm{Ra}^{1/3}$, lead to the dependences: $\sim x^{-1/2}$ for the averaged temperature and $\sim \log x$ for the averaged vertical velocity component, where x is the distance from the surface and N_0 is an empirical constant. The results are confirmed by experimental data for both the Rayleigh convection ($\mathrm{Nu} = 0.23\ \mathrm{Ra}^{0.282}$) and unstable free-convective flow (for the cases of miscible and immiscible flows).

Figure 1.7.1(a, b) presents the of the interferograms of the disturbances of the laminar boundary layer on a vertical plate of constant temperature created by a horizontal plate oscillating harmonically. For the disturbance frequency

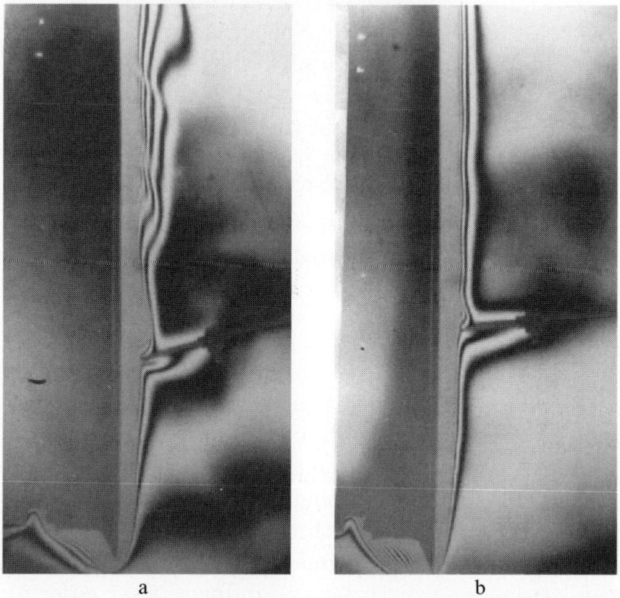

a b

Fig. 1.7.1. Laminar boundary-layer disturbances on a vertical plate of constant temperature: (**a**) disturbance frequency 0.5 Hz, the disturbances are created by a horizontal plate oscillating harmonically, (**b**) disturbance frequency 2 Hz

$f = 0.5\,\mathrm{Hz}$ (Fig. 1.7.1a) small initial disturbances increase downstream, for $f = 2\,\mathrm{Hz}$ (Fig. 1.7.1b) the initial disturbances damp. Figure 1.7.2 shows an interferogram of the disturbances of the laminar boundary layer created by periodic pulse heating of a horizontal wire. The period of heating was equal to about 2 sec. It is seen from the interferogram that originating the disturbance propagates downstream virtually not changing its form.

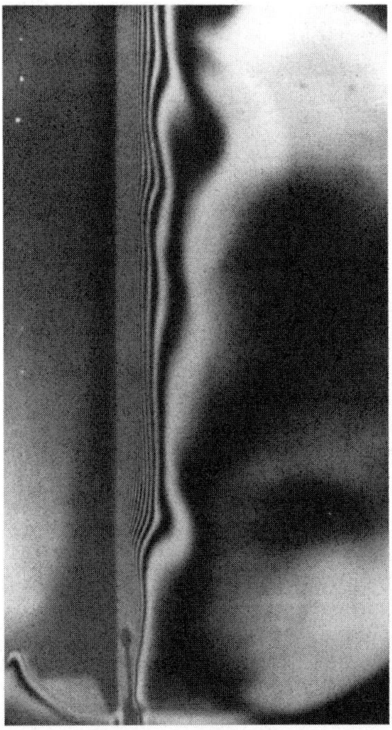

Fig. 1.7.2. Laminar boundary-layer disturbances on a vertical plate of constant temperature: the disturbances are created by pulsed heating of a horizontal wire (heating-up period ~2 sec)

The linear stability theory allows one to study initial development of relatively weak disturbances, which, as is shown in [1.44] in application to natural convection, can be considered two-dimensional. But as the amplitude of the disturbance increases downstream, the two-dimensional linear theory of stability becomes unsuitable for investigating main mechanisms, because nonlinear phenomena and the occurrence of three-dimensional disturbances begin to exert a substantial effect on the formation of flow. Nonlinear interaction of longitudinal and transverse disturbances induces a secondary averaged flow in the boundary layer which causes appreciable transfer of momentum in the transverse direction and represents a combination of the pairs of longitudinal vortices. In contrast to forced convection, where the vortices arranged in a line in the inner region of the boundary layer are observed [1.45], in free-convective flows vortices appear also in the outer part of the boundary layer and in the region joint to immobile fluid, which can strongly deform the profile of the longitudinal mean velocity component [1.28].

The development of disturbances with account for nonlinear effects was investigated theoretically by the method of successive approximations of the linear stability theory [1.28, 1.46]. The calculations carried out in studying the development of longitudinal disturbance, modulated by transverse disturbance in the form of a stable wave showed the existence of conditions for origination of a double line of longitudinal vortices, which was experimentally confirmed by detailed investigations of the mechanisms of nonlinear amplification of both regulated and naturally occurring disturbances in a free-convective flow [1.47–1.51]. Secondary averaged flows lead to periodic increase and decrease in the boundary-layer thickness due to the redistribution of momentum in the transverse direction. In some sections, external vortices transfer the fluid with a higher velocity into outer regions, in others – from the immobile surrounding medium to the surface. The internal vortices exert the opposite effect. As a result, a periodic change in the steepness of the longitudinal velocity profile occurs near the inflection point. The local increase in the steepness corresponds to the increase in disturbances with this increase becoming more intensive downstream. These data agree well with the theoretical results of [1.46].

The process of transition from a laminar to a turbulent mode of flow nears a vertical surface for free and forced convection is similar. However, there are also characteristic distinctive features of the development of this process.

In the forced convection flows the regions with high flow shear represent a source of secondary instability and cause high-frequency oscillations in the boundary layer leading to the formation of turbulent patches. The transition to turbulence for natural convection is characterized by the interdependence of the velocity and temperature fields, as the result of which the Pr number becomes an additional parameter essential for the transient process.

For example, the main features of transient processes in natural convection for water (Pr = 6.7) and air (Pr = 0.71) are the same, however, there are also characteristic features associated with different Prandtl numbers. One of the most interesting features consists in the preservation of the dominant frequencies of the laminar flow, on which, according to the stability theory, the greatest amplification of disturbance occurs. These frequencies lie in a very narrow range and are also observed in the laminar parts of the disturbed flow in its downstream sections. At the beginning of the transition the turbulent velocity and temperature oscillations appear at considerably higher frequencies than in the laminar region of the flow. This occurs near the maximum of the velocity profile and probably is explained by the effect of the shear layer induced by the nonlinear effects during the increase of disturbances [1.23]. When Pr = 6.7, the isolated turbulent patch appears first in the dynamic boundary layer, and then in the thermal one. The intensity of the disturbance increases, the fluid from the thermal boundary layer starts to diffuse into the dynamic boundary layer, and this in turn causes a change in the average temperature profile. There occur mixing and increase in the thickness of the dynamic and thermal boundary layers [1.28]. The mechanisms and the characteristics features of the transition process for natural convection of water were investigated in detail in [1.50,1.52,1.53]. On completion of the transition to turbulence in the water flow, a further expansion of the spectrum velocity oscillations and decrease in the intensity of temperature oscillations occur.

Experimental investigations of transition to turbulence in natural-convection gas flows carried out in [1.54–1.57] showed that the special feature of transition for $\mathrm{Pr} \sim 1$ is the same order of magnitude of thermal and dynamic boundary-layer thicknesses resulting in the earlier effect of velocity fluctuations on the temperature field, and the transient processes occur in both boundary layers almost simultaneously. The difference between the flows of gas and water seems to be due to the effect of the Prandtl number, because in the case of water the initial delay in the development of temperature disturbances influences substantially the further entire process of transition [1.28].

To determine the boundaries of the transition region, different criteria can be used. But the simplest and most reliable is considerate to be the moment of the deviation of the mean temperature profile from the laminar one [1.23,1.28]. It is shown experimentally that the start of transition is determined not only by the Grashof number, but also by the distance along the flow x and the coefficient of kinematic viscosity ν. The appearance of noticeable velocity and temperature fluctuations and also the deviation of the average temperature profile from the laminar one, which characterize the beginning of transition to turbulence, are well described by the parameter which characterizes the flux of kinetic energy in a boundary layer $E = G^*(\nu^2/gx^3)^{2/15}$. For gases it is possible also to use another correlation parameter of the start of transition, namely, $\mathcal{Q} = q(gx^3/\nu^2)^{2/15}$, where q is the fifth degree root of the local dimensionless density of the heat flux supplied from the surrounding medium [1.28].

The most essential, from the practical point of view, factor which characterizes the process of transition to turbulence, is the increase in the intensity of heat transfer. The increase of the local heat transfer coefficient in comparison with a laminar regime is accompanied by a decrease in the local temperature of the surface.

The end of transition to turbulence cannot also be determined only by the magnitude of the Grashof number. On the basis of numerous experimental data the following value is offered as the correlation parameter for determining the end of the region of transition: $\mathcal{Q}_{ET} = G^* \,\mathrm{Pr}^{1/5}(\nu^2/gx^3)^{1/6}$ [1.28].

Turbulent transfer in a developing flow. The results of experimental [1.50,1.58–1.60] and theoretical [1.61,1.62] investigations of the characteristics of turbulent transfer in the field of the flow developing near a vertical wall allow one to describe more precisely the mechanisms of transition to a developed turbulence. The data obtained in [1.50, 1.58] for the intensity of heat transfer of the transient processes in water made it possible to reveal the existence of the additional stages of flow relaxation after the complete destruction of the laminar flow with an increasing of the local heat transfer coefficient. The results of an experimental study of the spectral characteristics [1.60] show that the effect noted result from the earlier concentration of the energy by large-scale vortices.

As a result of the integral analysis applied for the first time to flows induced by buoyancy, in [1.63] at the Prandtl numbers close to unity: there where received the following relations were obtained for the mean Nusselt number:

$$\mathrm{Nu}_x = 0.0246\,\mathrm{Gr}^{2/5}\,\mathrm{Pr}^{7/5}\left[1 + 0.494\,\mathrm{Pr}^{2/3}\right]^{-2/5} . \tag{1.7.3}$$

The extension of the integral method to the case of natural convection ***in liquid metals*** [1.64] led to the following relation valid in the range of the Grashof numbers $10^{10} \le \mathrm{Gr} \le 10^{15}$:

$$\mathrm{Nu} = 0.08\,\mathrm{Gr}^{1/4} . \tag{1.7.4}$$

Among the experimental investigations of natural-convection turbulent boundary layers the greatest number of works concerns simple geometrical configurations, in particular, flows near a vertical surface. In one of the first of such works [1.65] measurements of the flow velocity profiles near a plane vertical plate were made. Investigation of natural convection ***in water and mercury*** [1.66] yielded the following correlation for calculating Nu for $\mathrm{Ra} > 10^{10}$:

$$\mathrm{Nu} = 0.17\,\mathrm{Ra}^{1/3} , \tag{1.7.5}$$

which fits well the experimental data on heat transfer in a turbulent ***air flow near an isothermal surface*** [1.56].

The empirical dependence [1.53, 1.58]

$$\mathrm{Nu}_x = 0.568(\mathrm{Gr}_x^*\,\mathrm{Pr})^{0.22} \tag{1.7.6}$$

describes the results of investigation of natural convection ***in water*** ($3.6 < \mathrm{Pr} < 10.5$) ***at a constant density of a heat flux on a wall*** in the range $10^{13} < \mathrm{Gr}_x^* < 10^{16}$. The turbulent heat transfer ***in air*** [1.67] is characterized by the following dependence:

$$\mathrm{Nu}_x = 0.17(\mathrm{Gr}_x^*\,\mathrm{Pr})^{0.25} . \tag{1.7.7}$$

The calculation ***of heat fluxes at constant surface temperature and constant heat flux density for laminar and turbulent regimes*** can be carried out on the basis of empirical relations valid for all the values of the Prandtl number [1.68]:

$$\mathrm{Nu}^{1/2} = 0.825 + \frac{0.387\,\mathrm{Ra}^{1/6}}{\left[1 + (0.492/\,\mathrm{Pr})^{9/16}\right]^{8/27}} , \tag{1.7.8}$$

$$\mathrm{Nu}^{1/2} = 0.825 + \frac{0.387\,(\mathrm{Gr\ Pr})^{1/6}}{\left[1 + (0.437/\,\mathrm{Pr})^{9/16}\right]^{8/27}} \,. \tag{1.7.9}$$

In calculating a constant heat flux directed upward ***from a heated inclined surface***, relation (1.7.7) can be used [1.67]. If a flow is directed downwards, Gr_x^* should be replaced by $\mathrm{Gr}_x^* \cos^2\theta$. Relations (1.7.8) for an isothermal surface and relations (1.7.9) for the constant heat flux density on the wall are modified similarly for an inclined surface. The empirical dependence for calculating an upward heat flux ***from a horizontal heated surface***, that is,

$$\begin{gathered} 8 \cdot 10^6 \le \mathrm{Ra}_L \le 1.6 \cdot 10^9 \,, \\ \mathrm{Nu} = 0.15\ \mathrm{Ra}_L^{1/3} \,, \end{gathered} \tag{1.7.10}$$

is valid for horizontal surfaces of different shapes, if the area-to-perimeter ratio is used as the characteristic length for calculating the number Ra_L [1.69].

In investigation of heat transfer ***from a horizontal cylinder*** the following correlation for the Nusselt number was obtained, which is valid for the Rayleigh numbers in the range $10^9 < \mathrm{Ra} < 10^{12}$ [1.70]

$$\mathrm{Nu} = 0.13\ \mathrm{Ra}^{1/3} \,. \tag{1.7.11}$$

The equation obtained in [1.71] describes heat transfer ***from circular horizontal cylinders for both laminar and turbulent flow regimes:***

$$\mathrm{Nu} = 0.6 + 0.387 \left\{ \mathrm{Ra} / \left[1 + (0.559/\,\mathrm{Pr})^{9/16}\right]^{16/9} \right\}^{1/6} \,. \tag{1.7.12}$$

The following empirical relation, obtained from (1.7.12), is used to calculate a turbulent heat flux ***from spheres*** [1.71]

$$\mathrm{Nu} = 0.589\ \mathrm{Ra}^{1/4} / \left[1 + (0.469/\,\mathrm{Pr})^{9/16}\right]^{4/9} \,. \tag{1.7.13}$$

Using the correlation which takes into account the fact that when $\mathrm{Ra} \to 0$, the Nusselt number ***for a sphere*** tends to the value 2.0 corresponding to the Nusselt number for the heat flux due to heat conduction ($10^{\circ} \le \mathrm{Gr} \le 10^5$),

$$\mathrm{Nu} = 2 + 0.45\ \mathrm{Gr}^{1/4}\ \mathrm{Pr}^{1/3} \tag{1.7.14}$$

one can generalize well the experimental data on heat and mass transfer at the Prandtl numbers near unity [1.72]. The results of investigation of turbulent heat transfer ***from spheres*** for $\mathrm{Ra} \cong 10^8$ deviate from dependence (1.7.13) but are well generalized by the relation [1.73]

$$\mathrm{Nu} - 2 = \frac{0.589\ \mathrm{Ra}^{1/4}}{\left[1 + (0.469/\mathrm{Pr})^{9/16}\right]^{4/9}} \left\{ 1 + \frac{7.44 \cdot 10^{-8}\ \mathrm{Ra}}{\left[1 + (0.469/\mathrm{Pr})^{9/16}\right]^{16/9}} \right\}^{1/12} . \tag{1.7.15}$$

Turbulence in nonstationary convection. The analysis of the instability of nonstationary flows is much more complex than in the case of a steady-stable flow, since due to the change in time of the free stream parameters the simple increase in disturbances cannot be considered as a sufficient criterion of instability. There are different approaches to the study of the stability of nonstationary flows [1.28]. If the disturbances develop faster than the free stream, then it can be considered quasi-stationary, and the equations of the linear stability theory can be used. There is no need then to specify initial conditions. Another approach consists in solving the Caushy problem with initial conditions. The numerical solutions of linearized equations agree well with the experimental data.

In [1.74], a nonstationary transition in natural convection near a plane vertical surface heated by a constant heat flux is investigated experimentally. The visualization of the temperature field with the aid of an interferometer demonstrated a complex picture of the nonstationary transition. When the heat flux density is low the flow regime is turbulent during transition, but then again becomes laminar. At large heat loads reversal to laminarization does not occurs. For intermediate heat fluxes the region of transition moves gradually to some stationary location.

The results of the investigation of the stability characteristics for a wide class of nonstationary accelerating and decelerating flows for several Grashof numbers between two vertical plates are present in [1.75]. It is shown that accelerating flows are more stable in comparison with stationary ones, whereas decelerating ones are, on the contrary, less stable.

The Monte Carlo method and the method of moments for different Rayleigh numbers where used to solve the problem of the Rayleigh-Benard convection induced by random disturbances [1.76, 1.77].

The basic equations for turbulent free-convective flow are similar to the equations for a laminar flow, and they can be solved if the coefficients of turbulent transfer are known. Since these coefficients depend strongly on the character of flow, they are often determined from experiments or selected so as to obtain agreement with the corresponding measured results.

Compared with extensive amount of theoretical and experimental data on forced turbulent flows, the natural-convection turbulence has been investigated insufficiently, though their main mechanisms are similar. The main difference is that in natural-convection flows the levels of averaged velocities are smaller, and the levels of disturbances are higher compared to forced turbulence. Moreover, due to interdependence of velocity and temperature fields, they are to be analyzed simultaneously, which significantly complicates both theoretical approaches and experimental investigations [1.23].

The theoretical studies of the turbulent natural convection are aimed mainly at determination of the Nusselt number. The application of the dimension analysis on assumption that in turbulent flow the heat transfer coefficient is independent of x showed that $\mathrm{Nu}(x)$ must be proportional to $\mathrm{Gr}_x^{1/3}$ and the neglect of the effect of viscosity lead to the dependence of $\mathrm{Nu}(x)$ on $\mathrm{Gr}_x \mathrm{Pr}^2$ [1.23]. Using the results of the previous calculations for forced convection and modifying them, many authors obtained correlations applicable to the free-convective turbulence.

In particular, for large values of Gr the following formula is valid [1.63]:

$$\mathrm{Nu} = 0.0246\ \mathrm{Gr}^{2/5}\ \mathrm{Pr}^{7/15} \left(1 + 0.494\ \mathrm{Pr}^{2/3}\right)^{-2/5} . \tag{1.7.16}$$

Using the integral method for small Pr and $10^{10} < \mathrm{Gr} < 10^{15}$, the following correlation was obtained [1.78]:

$$\mathrm{Nu} = 0.08\ \mathrm{Gr}^{1/4} . \tag{1.7.17}$$

Based on the analogy between the Reynolds and Grashof numbers, the relations for the Nusselt number at large Gr were obtained [1.79]:

$$\mathrm{Nu} = \frac{\mathrm{Pr}\ \mathrm{Gr}^{1/2}}{8\left\{\mathrm{Pr} - 1 + \ln \frac{\mathrm{Nu}(5\,\mathrm{Pr}+1)}{3\,\mathrm{Pr}}\right\}^2} . \tag{1.7.18}$$

As a result of experimental investigations in a transient regime the velocity profiles in a flow were measured [1.65] and the data on heat transfer [1.80] near a plane vertical surface were obtained. It is shown that in a turbulent flow the generalized velocity and temperature profiles virtually do not change in the direction of flow. High (up to 30%) turbulence levels characterized by the ratio $\left\langle \nu'^2 \right\rangle^{1/2} / \langle \nu \rangle_{\max}$ and the expansion of the velocity profile on reduction of its dimensionless value were observed [1.73]. In the experimental work [1.81] the velocity and temperature profiles for vertical surfaces in water were measured and also high (up to 50%) turbulence levels were obtained. The general laws characterizing velocity and temperature profiles are similar for a transient and a developed turbulent flow regimes.

Flow stability loss and transition to turbulence in plumes. Free-convective flows in plumes differ substantially from flows near a surface, the effect of which leads to attenuation of disturbances. The existence of free boundaries in a two-dimensional plume exerts a direct effect on the mechanism of the increase of disturbances and leads to lesser stability of flow in a plume compared with a flow near a surface at the same Grashof numbers. The temperature in the mid plane of the plume decreases downstream as $x^{-3/5}$, and the velocity increases as $x^{1/5}$. The modified Grashof number Gr is selected as the local determining parameter [1.28].

Figure 1.7.3 presents the interferograms and the results of track visualization of a laminar free-convective plume above a horizontal cylinder with a

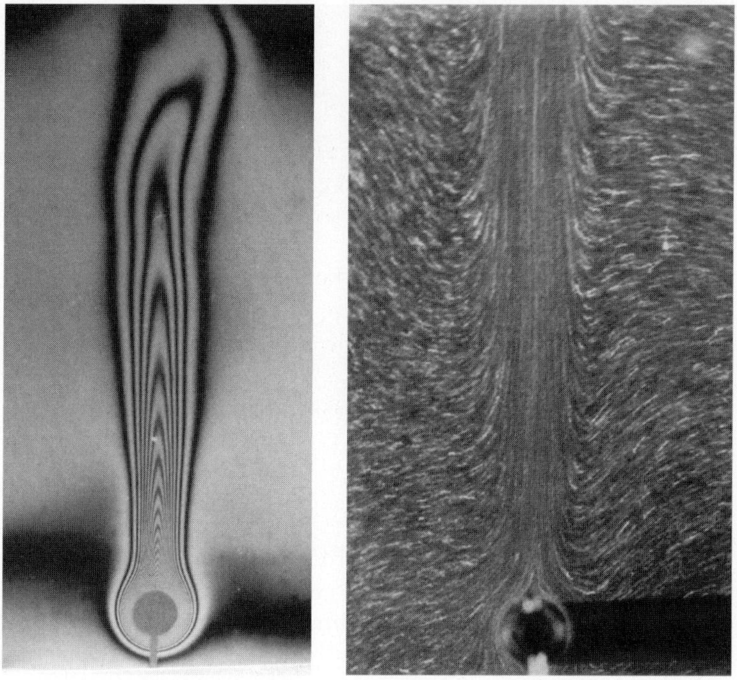

Fig. 1.7.3. Interferogram and the results of track visualization of a laminar free-convective plume above a horizontal cylinder with a constant heat flux on the wall

constant heat flux on the wall. A laminar free-convective plume above a thin electrically heated wire is shown in Fig. 1.7.4.

The structure of the thermal boundary layer of a laminar plume is represented in Fig. 1.7.4a as an interferogram, whose lines can be interpreted as the lines of equal temperature. The structure of a dynamic boundary layer is obtained by the method of track visualization (Fig. 1.7.5). Figure 1.7.4b presents an interferogram of the characteristic instability of a free-convective plume due to the motion of the environment.

An interferogram and results of track visualization of the disturbances of a free-convective plume above a horizontal cylinder with a constant heat flux on the wall are presented in Fig. 1.7.6. The disturbances are created by a harmonically oscillating horizontal plate. The frequency of the disturbances is $f = 1.6\,\mathrm{Hz}$. An interferogram and results of track visualization of the free-convective plume disturbances with the frequency $f = 4\,\mathrm{Hz}$ are given in Fig. 1.7.7. It is seen from Figs. 1.7.5 and 1.7.7 that in the regions of bending of the thermal boundary layer stable two-dimensional vortices are formed. The more significant increase in the disturbances with the frequency $f = 4\,\mathrm{Hz}$ compared to the disturbances with the frequency $f = 1.6\,\mathrm{Hz}$ is apparently associated with the nonlinear mechanism of the growth of disturbances.

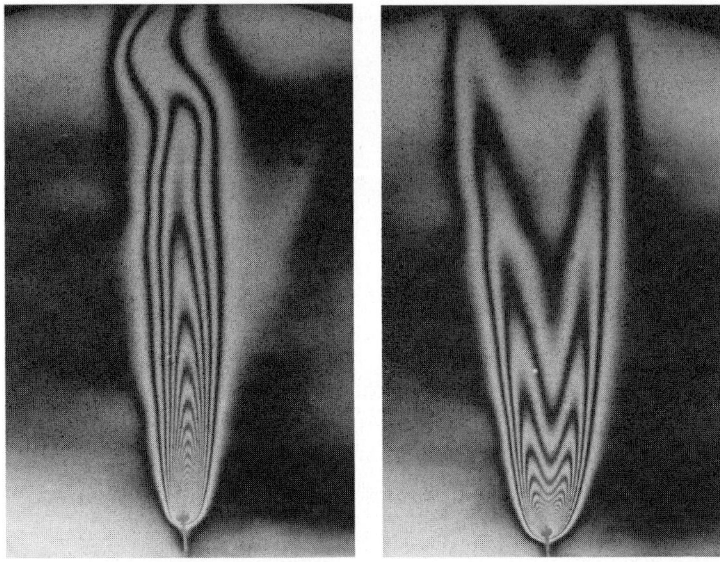

Fig. 1.7.4. Laminar free-convective plume above a thin electrically heated wire: (**a**) interferogram of the plume, (**b**) instability of the free-convective plume caused by motion of the environment

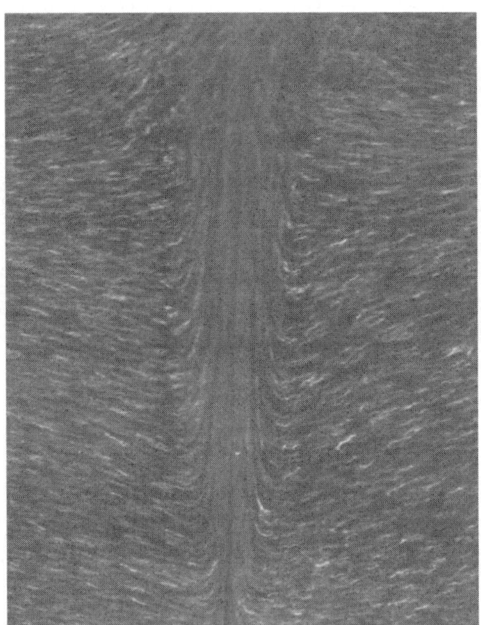

Fig. 1.7.5. Results of track visualization of a laminar free-convective plume above a thin electrically heated wire

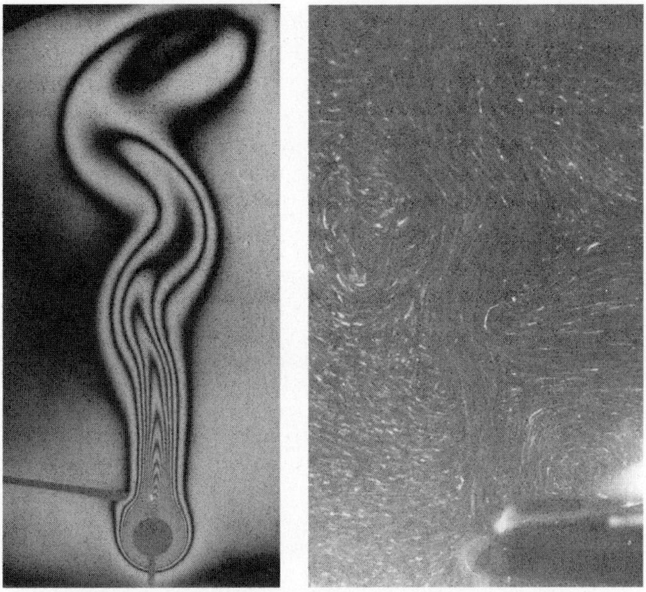

Fig. 1.7.6. Disturbances of a free-convective plume above a horizontal cylinder with a constant heat flux on the wall. Diameter of the cylinder $d = 10\,\mathrm{mm}$. Disturbance frequency 1.6 Hz. The disturbances are created by a harmonically oscillating horizontal plate

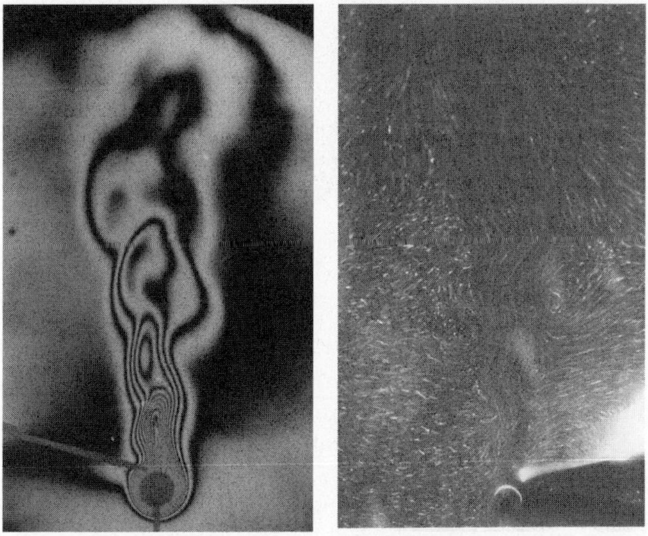

Fig. 1.7.7. Disturbances of a free-convective plume above a horizontal cylinder with a constant heat flux on the wall. Diameter of the cylinder $d = 10\,\mathrm{mm}$. Disturbance frequency is 4 Hz. The disturbances are created by a harmonically oscillating horizontal plate

An interferogram of a turbulent free-convective plume above a horizontal thin electrically heated wire is given in Fig. 1.7.8. The region of flow can be conventionally divided into three subregions: the subregion of a laminar flow located just near the wire, further there is a transient subregion, and finally the subregion of a turbulent flow.

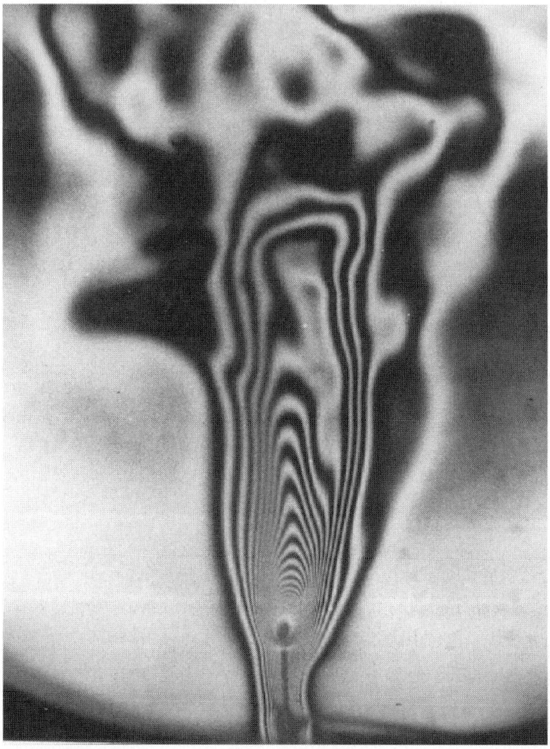

Fig. 1.7.8. Interferogram of a turbulent free-convective plume above a horizontal thin electrically heated wire

The estimations of the location of the boundaries for the transition region are given by the Grashof numbers corresponding to the beginning of transition determined from the interferograms, $\mathrm{Gr}_x = 5 \cdot 10^8$ [1.82], and also from the results of the measurement of temperature [1.83], $\mathrm{Gr}_x = 11.2 \cdot 10^6$. The transition ends respectively at $\mathrm{Gr}_x = 7.9 \cdot 10^8$ [1.82] and $\mathrm{Gr}_x = 5 \cdot 10^9$. In the latter case, the end of the transition region is determined by the expansion of the temperature profile. After the full destruction of the free boundary layer the flow parameters approach the values of a turbulent regime. The temperature on the plume axial line becomes lower than for that a laminar regime. Further, a decrease in the intensity of turbulent fluctuations and in the sizes of vortices occurs [1.28].

1.8 Outer and Inner Flow Regions

In free convection, the nonuniformities of the velocity and temperature fields are localized in the wall region. If the heat conduction and viscosity coefficients are small, strong temperature and velocity gradients are formed in the thin layer near the body surface in the direction perpendicular to the surface. The flow in the region far from the surface virtually remains isothermal, it is referred to as the outer flow and it is well described by the theory of the inviscid fluid. The division of the entire region of flow into a boundary layer and an outer flow makes the study of the flow characteristics on the whole easier, since for both the boundary layer and the outer flow simpler equations than those in system (1.5.1–1.5.4) are valid.

The large difference between the boundary layer flow and the outer flow does not make it possible to find the solution on the basis of simplified equations which could have been valid for the whole region. This difficulty can be overcome by constructing solutions being in a certain way compatible in each of the regions according to the method of the matched asymptotic expansions [1.2]. In application to the problems of free convection the idea of the method can be illustrated as follows: the flow in the boundary layer near the heated surface gives a major contribution to the solution, while the outer flow induced by the boundary layer, being secondary, provides a higher-order correction. The outer flow in turn induces a secondary boundary layer, that is, "the boundary layer of the boundary layer", causing the outer flow with the correction of higher order of smallness, etc.

The summation of the corrections gives asymptotic representation in each of the regions. The boundary conditions are selected by matching the external and internal solutions in the region where they overlap. The matching is efficient if the behavior of the internal solution on the outer edge of the boundary layer can be presented in an analytical form.

The equations of the boundary layer are parabolic, while the initial system of (1.5.1–1.5.4) is elliptic. Therefore, in the vicinity of the leading and trailing edges singularities appear nonremovable within the framework of the method of matched asymptotic expansions. The use of the method of deformed coordinates makes it possible to shift these singularities in the direction of their actual position and thus to take into account the leading edge effect. The vicinity of the trailing edge is investigated on the basis of a two-layer boundary layer model in this region.

The velocity and temperature distribution in a boundary layer on a vertical plane surface with a constant temperature is defined by the expressions [1.2]

$$\begin{aligned} \theta &= r_0 + r_1 \mathrm{Gr}_x^{-1/4} + r_{4/3} \mathrm{Gr}_x^{-1/3} + r_2 \mathrm{Gr}_x^{-1/2} + O\left(\mathrm{Gr}_x^{-3/4}\right), \\ ux/(\nu \mathrm{Gr}_x^{1/2}) &= p_0 + p_1 \mathrm{Gr}_x^{-1/4} + p_{4/3} \mathrm{Gr}_x^{-1/3}\ p_2 \mathrm{Gr}_x^{-1/2} + O\left(\mathrm{Gr}_x^{-3/4}\right), \end{aligned} \tag{1.8.1}$$

where $r_i = r_i(\eta)$, $p_i = p_i(\eta)$, and $\eta = yx^{--1}\mathrm{Gr}_x^{1/4}$. In the expansions, the first term corresponds to the boundary layer theory, the second to the first approximation, the third to the eigen solution, and the fourth corresponds to the second approximation.

1.9 Experimental Methods in Free Convection

The processes of free-convective flow and heat transfer have a number of specific features distinguishing them from other types of flow, therefore, to carry out for experimental investigation of free convection often special methods are to be used which take into account these features. Special attention is to be given to the interrelationship between the fields of the parameters of flow and temperature or concentration with account for the smallness, in the general case, of the measured velocity and heat flux compared to the corresponding values for forced convection.

Like in the majority of experimental works, in studying free convection it is necessary to consider mainly two basic points relating to conducting experiments. The first concerns the measurement technique and measuring equipment and the other the experimental equipment and the ways of the implementation of the specific type of flow and of heat transfer.

The main purpose of the experimental investigation free convection is determination of the parameters of flow and heat transfer, which leads to the necessity of measuring the velocity, temperature, and heat flux fields. Due to the presence of the density gradients and other flow and heat transfer parameters occurring in the nonuniform temperature field, various methods of visualization, both qualitative and quantitative, can be applied for studying the heat transfer characteristics.

Velocity measurement. The study of the velocity fields of free-convective flow has a number of essential features compared to similar cases of forced convection. The main difficulty which considerably complicates velocity measurement in a free-convective flow is assotiated with the presence of the temperature or concentration fields which caused the appearance of buoyancy and became a source of the flow considered. Moreover, due to the interrelationship between the velocity and temperature fields, the introduction of a probe into a convective flow for velocity measurement creates disturbances in both fields. Certain difficulties also arise due to the smaller, compared to forced convection, characteristic values of the velocities of free-convective flows. Moreover, the wave numbers of the velocity disturbances in an unstable, transient or turbulent mode of free convection lie in a low-frequency range.

To measure the velocity of a laminar flow or the averaged velocity of a turbulent flow, wide use is made of a combined Pitot–Prandtl tube, which is a cylindrical tube with a hemispherical nose, with the tube axis being set up along the flow. The total pressure P is measured through the central hole in the hemisphere (critical point); another hole (or a number of holes) are made in the side wall of the tube far enough from the nose and from the holder and serves for measurement of the static pressure P_0. The geometric shape of the Pitot–Prandtl tube, the shape of the holes, and the distances from them to the tube nose are selected so that the pressure in the side holes could differ as slightly as possible from the static pressure at the studied point of the flow. A slight discrepancy between the pressures is allowed for by the correction factor ξ, which is found from individual calibrations. Knowing P and P_0, one can calculate the flow velocity ν from the Bernoulli equation. For an incompressible fluid

$$\nu = \sqrt{2\xi\,(P - P_0)\,/\rho}\,,$$

where the density ρ can be found from the Clapeyron equation. The readings of the tube are rather accurate (the error being less than 1%) provided that the velocity vector makes an angle not larger than $10 - 25°$ with the tube axis. The absolute dimensions of the tube must be such that one could neglect the pressure (velocity) gradients within the distance from the nose to the sidewall. At small velocities of the flow ($\nu < 6\,\mathrm{m/sec}$) or under high vacuum a sharp increase in the coefficient ξ is observed.To measure the direction of a flow, cylindrical and spherical attachments are used which represent a combination of three inclined Pitot tubes and the readings of which are very sensitive to the flow direction.

The velocity fields in the boundary layer near a solid wall are investigated with the aid of the Stanton tube which measures the velocity head in a flow with a large vertical velocity gradient; the tube is set up directly on the surface of the solid body in a flow and is moved vertically by means of a micrometer screw. The velocity is calculated from the difference between the total pressure measured by the tube and the static pressure on the channel wall. To measure the velocity, use is also made of the Brabe, Preston, Venturi tubes and others, which are modifications of the Pitot–Prandtl tube.

Laser Doppler anemometer. A device with a very short response time (about 1 μces) and spatial resolution determined by velocity field averaging over a very small volume (usually of about $1\,\mathrm{mm}^3$) is required to investigate transient or turbulent flows. For this purpose, the thermoanemometers were constructed that had the needed spatial and time resolution. However, despite the efficiency with which there used to study the convective flow structure, there are a number of purely technical problems of fundamental nature.

The main problem is associated with the disturbances of a flow by introduced probes, resulting in the distortion of the experimental data obtained. This is the common problem for all the standard velocity meters. The perturbation of the flow by a probe can result in substantial errors, especially in small-scale experiments or when several probes are used. A very large problem is that of calibration. Since all the wire probes differ somewhat from one another, they are to be calibrated using a standard device, i.e., the voltage on the probe is to be compared with the appropriate flow velocity fluctuations. Pitot tubes are usually used as these calibration standards. Thus, only relative rather than absolute measurements can be carried out using wire thermoanemometers.

The signal at the hot wire thermoanemometer outlet is not a linear function of velocity, since the output voltage changes approximately as a fourth degree root of velocity. This considerably complicates the analysis of experimental data, and therefore the accuracy with which the turbulence parameters are measured is not high. These problems limit the use of thermoanemometers, which inevitably makes researchers to look for new, more precise methods of measurement.

The flow visualization by means of photographs or direct visual observation of the trajectories of certain particles or of the dye introduced into the flow represents one of the earliest and most reliable ways of flow measurement. The main difficulty of these methods is the quantitative analysis of the results; because for this purpose it is necessary to trace thousands of particles in the flow, which undoubtedly presents a difficulty even for modern computers. To restore the three-dimensional structure of the flow attempts were to analyze 2D photos, however, this still increases more greatly the amount of data to be processed [1.84].

The advantage of the optical methods of measuring the flow parameters such as flow visualization is that they allow one to avoid introduction of probes into the region of measurement and, hence, to avoid the corresponding disturbances. However, photographic methods, seem to be most useful for studying large-scale flow pictures. In observing small-scale turbulent fluctuations it is necessary to track microscopically small isolated particles which follow high-frequency velocity fluctuations. The invention of the laser made it possible to create modern optical systems that measure velocity at a given point. For the first time the laser Doppler system has been described in [1.85]. From that time both the optical scheme and the scheme of electronic processing of a signal have been considerably improved.

The principle of operation of the Laser Doppler anemometer is the following: if the object moving with the velocity u scatters a wave packet traveling with the velocity c and circular frequency ω, then the frequency of the scattered waves that reach a motionless observer changes by the value of the Doppler shift of frequency:

$$\Delta\omega = \left(2\omega|\ \vec{u}\ |\big/c\right)\ \sin\left(\varphi_1 - \varphi_2/2\right)\ \sin\varphi_2\ , \tag{1.9.1}$$

where φ_1 is the angle between the incident beam and the velocity vector $\vec{u}$, and φ_2 is the angle between the incident beam and the direction of the distribution of the waves that reach the observer. The scattered waves usually propagate in all the directions, though they do it nonuniformly, so that the angle φ_2 can be selected independently. The measured velocity component $(|\ \vec{u}\ |/c)\ \sin(\varphi_1 - \varphi_2/2)$ (Fig. 1.9.1) is the projection of the velocity onto a bisector of the obtuse angle between the incident and the scattered beams. To create, on this basis, real systems for measuring the velocity of the particles moving in a fluid, the width of the band of the frequencies of the incident radiation $\delta\omega$ must be small compared to the measured frequency shift. Otherwise, it is necessary to use interferometric methods. The output signal of typical continuous lasers with the wavelength about of $6 \cdot 10^{-7}$ m consists of several very narrow ($\sim$ 10 Hz) spectral lines divided by the frequency interval from 10^8 up to 10^9 Hz. The apparent relative width of the band $\delta\omega/\omega$ equals approximately $6 \cdot 10^{-6}$. However, due to the smallness of the Doppler frequency shift in comparison with $\delta\omega$, different lines can be considered separately as coherent sources, for which the Doppler shifts differ by the negligibly small value $\delta\omega/\omega$. Thus, if the Doppler frequency shift can be measured directly, there are no difficulties in measuring of the flow velocity. To obtain an output signal of frequency $\Delta\omega$, the incident and scattered beams are directed at the entrance to the photodetector responding to the radiation intensity (Fig. 1.9.1).

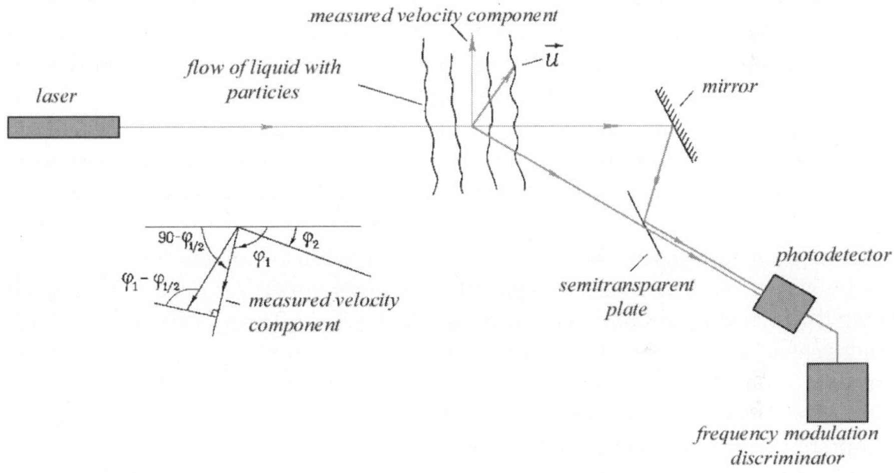

Fig. 1.9.1. Optical scheme of a laser Doppler velocimeter

The output signal from a photodetector is proportional then to the square of the sum of two fluctuations and thus contains the low-frequency component $\Delta\omega$, and the frequencies of the order of $\delta\omega$ and ω are smoothed out by a detector. The requirement that both beams are to be almost parallel to each other at the entrance to the photodetector, i.e., that the relative phase shift over the beam width be small enough is rather strict. This requirement is partially satisfied by focusing both beams on the photodetector surface, due to which their widths turn to be small enough.

The output photodetector signal of frequency $\Delta\omega$ (which for the velocity $u = 1$ m/sec is $\sim$1 MHz) is directed to the discriminator with frequency modulation, at the outlet of which the output voltage is proportional to the frequency.

In the simplest scheme of such a discriminator a short pulse of constant duration and amplitude corresponds to each cycle of the input signal. After smoothing in a filter, which transmits only certain frequencies, the output voltage of the discriminator becomes proportional to $\Delta\omega$ and, consequently, to $\vec{u}$. In measuring the flow velocity, one usually has to add smoke or fine powder to the gas flow or a dye to a liquid flow to obtain sufficient scattering of light. It is impossible to use molecular scattering even for sufficient light intensity, since then the frequency shift depends on the velocity of random motion of molecules, which usually considerably exceeds the average velocity of the liquid. The minimum size of the used smoke or dye particles is limited by the scale of Brownian movement, which can turn out to be appreciable. Moreover, in the voltage of the discriminator a modulation noise appears, because the signal from one scattering particle is admitted only during the time the particle passes through the light beam and, consequently, consists of the final number of wave packets. If a large enough number of particles arrives at the photo detector aperture, the broadening of Doppler frequency range caused by the finite length of the wave packets is independent of the number of particles, because the output signal is averaged. The only way to reduce this broadening and, consequently, to decrease the noise at the outlet is an increase in the number of oscillations in each wave packet by increasing the scattering volume or decreasing the wavelength. There exist few practical possibilities of decreasing the wavelength, and the scattering volume is to be reduced to a minimum one, so that the resolution of an anemometer could not be worse than that of a hot-wire thermoanemometer. On the other hand, a small scattering volume complicates the tuning of an optical system. The other reason for the broadening of the band is the finite beam aperture, which causes a change in the angle φ_1 when a particle passes through the beam. This effect can be reduced to minimum by appropriate selection of optical equipment and using lasers with a narrow output beam.

The above scheme of measuring the flow velocity using the Doppler effect is not the only one. At the present time, many different optical schemes of laser Doppler anemometers have been developed which can be found in special literature [1.86]. It should be noted that the laser Doppler anemometer has a number of indisputable advantages. First, it does not introduce disturbances into the test flow, because probing is carried out by the optical method. Second, there is no need to calibrate the device, because the flow velocity is measured directly. Third, the velocity component is registered in a certain singled-out direction, with the output signal being a linear function of the given velocity component. Due to these advantages, there occurs a rapid development of laser Doppler systems attracting intent attention of research workers. There are also other advantages of this method compared with the methods, for example, that use wire anemometers for measuring the flow parameters. For example, the optical method can be used for studying flows with a very high turbulence level and in the range of the measured velocities from several millimeters per second to supersonic ones and above. Moreover, this method is also applicable for simultaneous measurement of three flow velocity components.

Thermoanemometer. One of the most widespread flow velocity measurement devices is thermoanemometer, the operating principle of which is based on the dependence of the electrical resistance of a very thin and short metal wire heated by an electric current and cooled by the flow around it, on the rate of heat transfer from it. The sensitive element of the anemometer is included as a Witston bridge arm. The current flowing through the bridge heats the sensitive wire up to the needed temperature, and the output signal of the bridge is a measure of heat losses by the wire when it is cooled by the flow. There are two operation regimes of the thermoanemometer: a constant current regime and a constant temperature regime. In the first case, a constant current is supplied to the sensitive element from the constant current generator; in second, the sensitive element temperature is maintained constant using the feedback mechanism which adjusts the change in the temperature due to heat transfer. The second technique has become more popular, because it allows one to measure high-frequency temperature fluctuations with a strongly varied amplitude. The influence of the temperature dependence of the wire electrical resistance has the first order; this is the effect on which the application of the heated wire as an anemometer is based. The specified temperature dependence can be written as follows:

$$R_\omega = R_0 \sum_{k=0}^{\infty} b_k (T_\infty - T_0) \,, \tag{1.9.2}$$

where R_0 is the wire resistance at the characteristic temperature T_0 (usually $T_0 = 273°\mathrm{C}$) and b_k are the coefficients of the temperature dependence. Here $b_0 = 1$, and b_1 is the linear temperature coefficient of the specific electrical wire resistance. We should note that cooling by the flow is not the only factor which exerts its influence on the heat transfer of the wire in the thermoanemometer. One of such factors is also heat release through the elements of the probe and radiation losses of heat at rather high temperatures of the wire. If we neglect these factors, the heat losses per time unit from a wire of length l and diameter d with a uniformly distributed temperature to the ambient gas should be equal to the quantity of the heat evolved by the electric current, which flows through the wire. The heat balance equation in this case is

$$\alpha \pi d l (T_\mathrm{w} - T_g) = I^2 R_\mathrm{w} \,. \tag{1.9.3}$$

Wires of length 1–3 mm and thickness 2–10 microns are usually used in practical measurements, so that the aspect ratio $l/d \sim 200$. Even for such the wire length-to-thickness ratio the question of concerning the uniform distribution of the temperature along the wire remains open [1.87]. When the cooling of the wire by the free-convective flow induced by the wire itself can be neglected, to calculate the Nusselt number the dependence which yields good enough results for many liquids and gases can be used [1.88]:

$$\mathrm{Nu} = 0.42\,\mathrm{Pr}^{0.2} + 0.57\,\mathrm{Pr}^{0.33}\,\mathrm{Re}^{0.5}\,. \tag{1.9.4}$$

For air and diatomic gases relation (1.9.4) is valid in the range of the Reynolds numbers $10^{-2} < \mathrm{Re} < 10^{5}$.

Subject to the linear dependence of the electrical resistance of the wire on temperature, relations (1.9.3) and (1.9.4) lead to the equation

$$\frac{I^2 R_{\mathrm{w}}}{R_{\mathrm{w}} - R_g} = A + B\sqrt{U}\,, \tag{1.9.5}$$

where U is the velocity of the flow considered,

$$\begin{aligned} A &= 0.42\frac{\pi\lambda l}{bR_0}\,\mathrm{Pr}^{0.2}, \\ B &= 0.57\frac{\pi\lambda l}{bR}\,\mathrm{Pr}^{0.33}\left(\frac{pd}{\mu}\right)^{0.5}. \end{aligned} \tag{1.9.6}$$

In carrying out practical measurements the coefficients A and B are determined experimentally by calibrating the velocity probe. The calibration must be made in an undistorted gas flow, in which the turbulence level is as low as possible.

Two different methods are applied for measuring velocity fluctuations. In the first method, a constant electrical current is maintained, whereas the temperature and hence the electrical resistance change depending on the velocity. In the second method, the temperature and resistance remain constant, whereas the electric current is a fluctuating value. It should be noted that as concern the accuracy of measurement, the constant temperature method has a principal advantage over the constant current method. The temperature field, which for free-convective heat transfer is always nonuniform, considerably, complicates measurements, because heat transfer from the heated wire changes with the liquid temperature. When the heated wire crosses the boundary layer, the liquid temperature T changes, and this induces a signal, which is to be distinguished from a signal caused by a change in the flow velocity. A change in the liquid temperature causes a change in its physical properties and this change can also affect the probe signal. Much smaller wire overheating is used when measurements are carried out in dropping liquids compared with gases to prevent formation of bubbles and liquid boiling on the surface of the sensitive element. Therefore, the temperature of the sensitive element cannot be raised to reduce the error caused by a change in the liquid temperature.

It should be noted that the thermoanemometer could be applied to measure various velocity components using two and more sensitive elements. In order to measure two components lying in one plane a V-shaped configuration of two wires located in the same plane can be used. Then addition of the signals gives one component, while deduction gives the other [1.23]. The directed sensitivity of the wire is determined by the formula

$$V^2 = V_0^2(\cos^2\vartheta + k\sin^2\vartheta) , \tag{1.9.7}$$

where V_0 is the velocity directed at the angle ϑ to the sensitive element.

On the basis of the foregoing the inference can be made that the most exact method of measuring the velocity in natural convection is based on probe calibration in the considered temperature and velocity ranges, simultaneous measurement of temperature and velocity, and use of a calibrating curve at the given temperature of the liquid.

Temperature measurements. The methods of temperature measurement relate to the most important ones in experimental investigation of the processes of free-convective heat transfer. Temperature measurements in a convective flow, surrounding medium and on a heated surface near which natural convection is induced are necessary for studying the basic mechanisms of heat transfer and of the convective flow characteristics. The majority of temperature measurements are carried out only in the flow region, even through in some cases (for example, in a stratified medium) measurements outside this region are also required. Having determined the temperature gradient on the surface, one can connect them with heat fluxes. For a flow with free boundaries, integration of the measured velocity and temperature profiles across the flow allows one to determine the flow rate, the total heat release flux, and the flux of momentum. For a surface with a constant heat flux the temperature itself is a measure of the local heat transfer coefficient α, since the product $\alpha(T_w - T_0)$ is equal to the heat flux which remains constant. Therefore, the surface and ambient temperatures and the temperature gradient in the liquid determined near the surface are of special interest in experimental studies.

The general methods of measuring temperature in laminar and transient convective flows and the measuring devices used when these methods are applied are well known. These are thermometers, thermistors representing semiconductor devices, the resistance of which decreases with increase in temperature, resistance thermometers, the operating principle of which is based on a change in the resistance of the sensitive element with temperature, and thermocouples, whose operation is based on the Zeebeck thermoelectric effect. The thermistor is a very sensitive probe; it can be applied at low temperatures and in liquid media which do not interact with the material of the thermistor. Resistance thermometers are generally applied at relatively high temperatures and temperature gradients. Usually, a change in the temperature is measured by means of the bridge electrical circuit.

The most widespread temperature probes are thermocouples, whose operation is based on the existence of the difference of potentials between two conductors running from the junction of heterogeneous metals, depending on the junction temperature. Usually, two junctions are used, one of which is maintained at a certain characteristic temperature (equal, for example to the ice melting point temperature) and the other is placed at the point of temperature measurement. The output signal in this case is a measure of the difference of the temperatures. The following materials are usually used for production of thermocouples: copper-constantan, iron-constantan, chromel-alumel, chromel-constantan, platinum-platinorodium, tungsten-rhenium, and tungsten-molybdenum. In an appreciable range of measured temperatures, the dependence of the output voltage on difference of the temperatures of the cold and hot junctions is linear. To increase the accuracy of measurements and the range of measurements, the thermocouple calibrating curve is approximated by polynomial ordinarily not higher than of the third degree:

$$E = AT + BT^2 + CT^3 , \tag{1.9.8}$$

where the coefficients A, B, and C depend on the thermocouple material and probe sensitivity.

There are a number of important features relating to the measurement of the local surface temperature. The measuring device should not exert its influence on the local temperature due to the change in the heat transfer characteristics on the surface near the location of a probe. In the general case, the thermocouple fixed on the surface adjoining the surrounding fluid changes the local heat flux due to the heat release through the contacts and wires of the thermocouple. The contact of the thermocouple with the surface also brings a measurement error caused by the contact resistance between the surface and the junction and its finite dimensions. If the surface is heated by electrical current, then the thermocouple must be electrically insulated from the surface. The most frequently applied method for measuring the surface temperature is by imbedding thermocouples closest to the heated surface so that they could not violate the surface integrity. However, here it is necessary to provide a reliable thermal contact of the thermocouple with the surface and its electrical insulation from the source of electrical heating. When measuring the surface temperature with the aid of a thermocouple, special attention is to be given to the possibility of an error brought under nonstationary regimes by the finite heat capacity of the surface element. A change in the surface temperature on change in heat release is registered by the thermocouple output signal with some delay depending on the surface material and the position of the probe. For a periodic law of the change in the surface temperature the imbedded thermocouple will generally show the amplitude different from that on the surface. The transient characteristic of the thermocouple determines the highest disturbance frequency which can be measured.

The measurement of a temperature in natural convection in both the flow field and the surrounding medium is a rather complex process, since it depends not only on the distribution of heat fluxes but also on the flow character. These measurements carried out allow one to compare the results of theoretical and experimental investigations, to determine the limits of transient and turbulent flows, and to study the structure of the boundary layer. The results of these measurements are widely used in studies of free convection.

Another error in temperature measurements by a thermocouple can be brought in by thermal radiation of the heated surface. To reduce this error, the thermocouple surface is polished and sometimes is also shielded, which reduces the amount of the radiation absorbed. To minimize the error connected with heat release along the thermocouple wires, it is optimal to locate it parallel to isotherm, so that there is no temperature gradient along the wires, which is especially important for the problems of natural convection.

In the case of a turbulent free-convective flow, the most suitable device for measuring temperature is thermoanemometer. Temperature fluctuations and its mean value can be measured using a cold wire as a resistance thermometer, but a heated wire reacts both to the velocity and temperature fluctuation, therefore they must be separated even when only velocity fluctuations are to be determined. In investigating heat transfer in the case of turbulent natural convection, one usually needs mean values of the squared velocity and temperature fluctuations and also a velocity and temperature covariation. There are two main techniques for dividing velocity and temperature fluctuations. In the first one, the instantaneous signals of the cold wire working as a resistance thermometer, i.e., of the wire conducting a very low current, at which its heating is insignificant,

$$E_c = I_c R_c \alpha_c \theta \tag{1.9.9}$$

and the instantaneous signals of the neighboring heated wire

$$E_h = au + b\theta \tag{1.9.10}$$

are multiplied by the appropriate coefficients, to find the difference

$$bE_c - I_c R_c \alpha_c E_h \ , \tag{1.9.11}$$

which is proportional to the velocity u with the accuracy of the second order of smallness. The value of the temperature θ is measured by the cold wire directly, so that the required values $\langle u^2 \rangle$, $\langle \theta^2 \rangle$, and $\langle u\theta \rangle$ can be determined, if various calibrating coefficients are known [1.87–1.89]. It should be noted that using the combinations of heated and cold wires is the best way of carrying out measurements in flows with small velocities and small temperature gradients, since at large temperature differences the coefficient b changes considerably, and the instantaneous values of the velocity u determined by deduction of signals become less exact. For each probe the coefficients a and b must be determined during their calibration.

In the second technique, the same wire is used at different temperatures and, consequently, the coefficients a and b acquire different values. For the known values of these coefficients it is possible to determine three turbulence parameters by measuring triply the value

$$\langle E_h^2 \rangle = a^2 \langle u^2 \rangle + b^2 \langle \theta^2 \rangle + 2ab \langle u\theta \rangle \ . \tag{1.9.12}$$

It is evident that one of these measurements can be carried out with the aid of a cold wire, i.e., with the coefficient a equal to zero, but it is possible only in the constant current regime, when the thermal inertia of the cold wire can be compensated in the exactly the same way as in the case of the hot wire. The anemometers with a constant wire temperature cannot work at very small differences of the mean temperatures of the wire and flow (if the probe is not specially cooled), since no method of supplying of a negative electrical power is available and, moreover, the frequency characteristic of an amplifier becomes worse at a small output signal of the anemometer. The accuracy of calibrations by which a and b are determined depend on the intervals of measuring mean velocities and temperatures, on the required measurement accuracy, and, of course, on the possibilities of the experimental setup.

Interference methods. The interference methods are more often used for quantitative measurements than shadowgraph ones, in spite of large complexity, cost, and, as a rule, strict limitation on the range of measurements, since they give a more detailed information on the model considered and are more accurate. The designs and schemes of the two-beam interferometers usually used for measuring transparent objects differ greatly one another. The areas of their applicability differ substantially too.

The wave fields of a measuring and reference beams must interfere, and therefore they must be coherent, i.e., have a constant difference of phases. In interferometers this is achieved by splitting an initial beam from a light source into two, measuring and reference ones, which subsequently join again, because it is difficult to get separate physical light sources with identical frequencies and phases even with the aid of a laser.

Two-beam interferometers can be classified according to the method of light beam splitting. The complexity of an interferometer grows with increase in the light beam diameter and spatial shift between the measuring and the reference beams. In classical mirror interferometers, semi-transparent mirrors are applied as the elements which split a light beam. For conventional diameters of a measuring beam equal to 0.1–0.5 m, it is necessary to use large and expensive interferometer mirrors the accuracy of the treatment of which must be equal to the value of about $\lambda/20$, and the semi-transparent plates used must be strictly parallel. The splitting properties of semi-transparent mirrors with metal coating are independent of the radiation wavelength, but their drawback is the large absorption which lowers the light efficiency of the device and, hence, the quality of the interference picture obtained. Dielectric layers are more convenient for splitting the initial beam into two beams with identical amplitudes, however, their splitting properties depend on radiation length.

The Mach–Zehnder interferometer provides a very accurate way for investigating free-convective flows; it allows one to obtain quantitative data. Its general scheme is shown in Fig. 1.9.2. A collimatied beam from a small light source is incident on a semi-transparent mirror so that one part of it passes through the mirror and the other part is reflected. Further, one beam passes through the working part, and the other through the reference beam of the interferometer of the same optical length. The beams join at the second semi-transparent mirror and are focused to the screen or a photosensitive material. If there are density gradients in the working area the reference and measuring

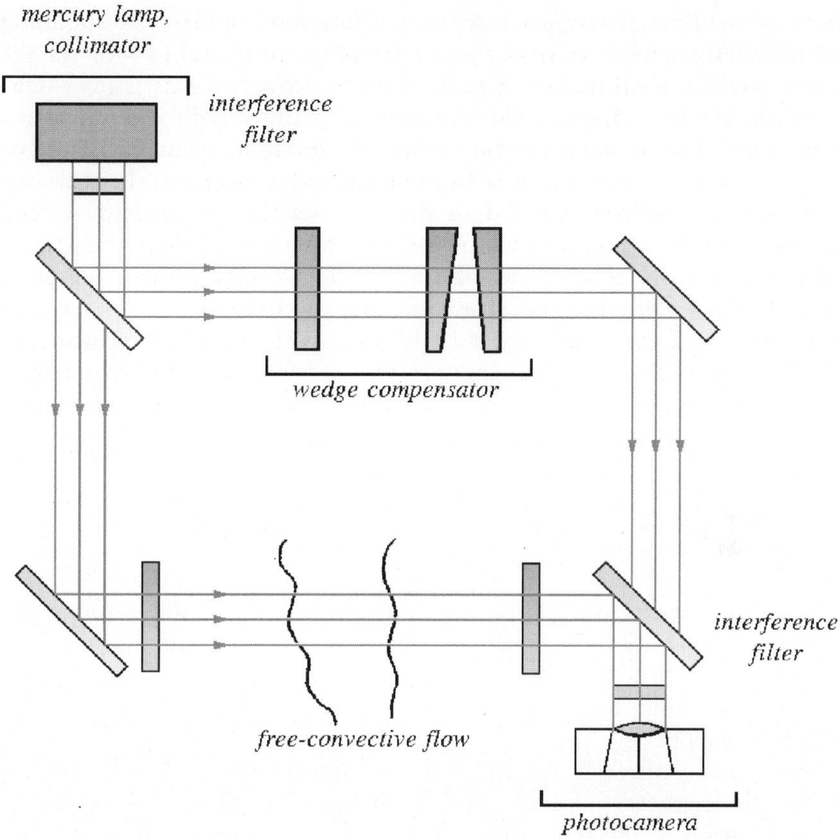

Fig. 1.9.2. Optical scheme of Mach–Zehnder interferometer

beams cover different optical lengths and then interfere due to the phase shift between them and this results in alternating dark and light regions on the screen called interference lines. The number of the bands obtained depends on the difference between the optical path lengths of the reference and measuring beams, with one band (dark or light) corresponding to the difference between the optical path lengths equal to half the radiation wavelength.

When a laser is used as a coherent light source, the precise equality of the optical path lengths is not required in the absence of changes in the density. However, when a white light is used as an initial beam (for example, radiation of a mercury lamp) the path lengths in the absence of density disturbances should be identical.

Experimental studies begin, as a rule, in the absence of bands. This condition is satisfied in the case of identical path lengths in white light over the entire working field, which practically is possible only in high-quality devices. Then, the appearance of bands is directly associated with a change in the density of the flow investigated, with each interference line corresponding to its own, strictly definite value of the density of the medium or to the refraction index associated with it. Sometimes, when a convective flux is investigated, a necessity arises to find the distribution of the phase shift within the limits of one band. This situation occurs when the quantity of interference bands is very small or when the flux is to be investigated in detail. In these cases, the interference picture obtained on the photosensitive material is subjected to digitization on a device which measures the darkening density of the photosensitive material D and the distribution of the phase shift between the reference and probing beams calculated from the formula

$$\delta\varphi = \arccos\left(\frac{2D}{D_{\max}} - 1\right) , \qquad (1.9.13)$$

where $D_{\max}$ is the darkening density corresponding to the maximum interference image intensity.

The Mach–Zehnder interferometer is widely applied for studying various free-convective processes [1.90–1.95]. The interference method is applicable in a wide range of change in the flow velocity, i.e., from the characteristic velocities for natural convection up to supersonic ones. The number of the interference bands obtained, which can be used for determining the surface temperature and the temperature gradient across the flow area, depend on the kind of the fluid, temperature difference across the flow, and on the width of the flow region.

The quantity directly measured by the interferometer is the phase shift between the reference and the probing beams; its value is associated with the distribution of the refraction index in the flow considered. However, the value of the refraction index is not the final goal in investigation of convective processes, therefore, of great interest is its correlation with other characteristics of the medium. In studying gas flows, it is necessary to know the basic quantities that determine the state of the gas: density, temperature, and pressure. Moreover, if in the system there are processes of diffusion of various mixture components, then the chemical composition of the mixture at each point of the considered object is an essential parameter from the point of view of density change.

The gas density can be determined from the values of the refraction index using the Gladston–Dail relation

$$\frac{n-1}{p} = K , \qquad (1.9.14)$$

where K is a constant for the given gas.

The given formula is based on the simplest theory which ignores a great variety of factors. It is not taken into account that the field of the light wave is variable and that the polarizability depends on the external field frequency; no question was raised about the interaction of the wave with a substance.

The application of a more rigorous molecular theory and consideration of the phenomenon from the quantum-mechanical point of view gives somewhat different results. The molecular refraction of gases appears to be dependent on temperature, which is explained by the excitation of the vibrational levels of the molecule. This phenomenon was considered in [1.96]. In the calculations it was assumed that the molecule can be considered as an unharmonic oscillator with a small oscillation energy. Moreover, it was assumed that the distribution of the molecules over the energy levels obeys the Boltzmann law. As a result of the calculation the following dependence is obtained:

$$\frac{n-1}{p} = K\left(1+\beta T\right) , \qquad (1.9.15)$$

where β is a constant for the given gas. This dependence is not usually used in practical calculations, because of the smallness of β ($\sim 10^{-9} - 10^{-7}$). The following relation is used most frequently in practice:

$$\frac{n-1}{p} = \frac{n_0 - 1}{p_0} , \qquad (1.9.16)$$

where p_0 and n_0 are the density and refraction index at normal conditions, respectively. The values of the refraction index n_0 for different gases are presented in Table 1.9.1 for light wavelength $\lambda = 5893$ Å. This table also presents the coefficients of the Cauchy dispersion formula

$$n - 1 = A(1 + \beta/\lambda^2) \qquad (1.9.17)$$

the use of which yields the refraction index for any other wavelength. All the data of Table 1.9.1 refer to dry gases. To determine the refraction index of humid gases the following relation is used:

$$n_{\mathrm{w}} = n - \frac{4.1 \cdot 10^{-5} m}{760} , \qquad (1.9.18)$$

where m is the partial pressure of water vapor in torrs [1.97].

In the case where natural convection occurs due to mixting of gases with different densities, it is necessary to measure the concentration of the mixture component in the flow. The gradient of the refraction index and the relevant phase shift and light deviation in passing optical inhomogeneity are caused in this case by the change in the concentration of the components in the flow. The possibility of determining the concentrations is based on additive refraction of the mixture components. The refraction index is expressed here in terms of the sum of the polarizability and concentration of the mixture components

$$\frac{n^2-1}{n^2+2} = \frac{4\pi}{3} \sum_i N_i a_i . \qquad (1.9.19)$$

In investigating the magnetohydrodynamic convection one have to take into account the fact that the polarization of the medium depends not only on

Table 1.9.1. The Values of the Refraction Index (n_0) for Different Gases

($P = 760\,\text{mm Hg}$, $T = 0°\text{C}$, $\lambda = 5893$ Å)

Gas	$(n_0 - 1) \cdot 10^4$	$A \cdot 10^5$	$B \cdot 10^{11}$
Air	2.92	28.71	5.67
O_2	2.71	26.63	5.07
N_2	2.98	29.06	7.7
H_2	1.39	13.58	7.52
Mercury vapor	9.33	87.8	22.65
Ne	0.67	6.66	2.4
Ar	2.81	27.92	56
He	0.35	3.48	2.3
Kr	4.27	41.89	6.97
Xe	7.02	68.23	10.14
NH_4	3.77	36.45	11.8
CO_2	4.50	45	65
NO	2.95	29.5	74
HCl	4.46	43.56	8.7
HBr	5.77	56	11
HJ	9.28	89	15
O_3	5.11	48	23
Cl	7.61	74	10
Water vapor	2.52	24.35	8.7

coupled electrons, but also on free ones since both factors act independently, then to determine the refraction of a partially ionized plasma it is necessary to consider the motion of a free electron in the field of the light wave and to add the calculated polarization value to the polarization caused by coupled electrons. If the absorption of the medium is small, then the value of the refraction index is determined by the expression

$$n^2 = 1 - \frac{e^2 N}{m\varepsilon_0\left(\omega^2 + \nu^2\right)} + \frac{e^2}{m} \sum_{i,k} N_i \frac{f_{ik}}{\omega_{ik}^2 - \omega^2}\,, \tag{1.9.20}$$

where N is the concentration of the free electrons, N_i is the concentration of each species of the particles available in the plasma (ions, atoms or molecules), e and m are the electron charge and mass, ω is light wave oscillation frequency, ω_{ik} is the frequency of the ith absorption band of the kth species of particles, f_{ik} is the oscillator strength for the transition $i \to k$, and ν is the frequency of collision of free electrons with heavy particles. We can assume that in the considered optical range of wavelengths the value of ω is very large. Free electrons make negative contribution to the refraction index, with each of them acting approximately $(\omega_{ik}/\omega)^2$ times more strongly than an atom or an ion. Their influence is virtually substantial already when the ionization level is equal to several percents. At large ionization degrees the refraction is determined only by free electrons, with the refraction index being less than unity. For this case we can write

$$n = 1 - \frac{\lambda^2 e^2 N}{2\,mc^2}\,, \tag{1.9.21}$$

where c is the speed of light in vacuum. The specifics of measurements in plasma is its strong emission. Therefore, when it is necessary to apply accurate photometric methods one has either to make the light source much brighter or to limit the plasma luminosity by installing light filters and special diaphragms. Experience shows that if the density of the plasma is small, the influence of its own luminosity is neutralized rather easily. The above reasoning concerning the relationship between the refraction index and the thermodynamic parameters of the medium investigated are applicable not only when using the Mach–Zehnder interferometer. At the present time there exist a significant number of various schemes of obtaining interference in a flow of liquid or gas. The precursor of the Mach–Zehnder interferometer is the Jamin interferometer (Fig. 1.9.3) consisting of two mirrors. The face side of each glass plate is covered with a semitransparent layer which either splits or joins the light beam. The backsides are covered with a fully reflecting layer.

These four layers are four mirrors in the Mach–Zehnder interferometer. The shift between the reference and the probing beams is not large and is determined by the thickness of the glass plate. The Jamin interferometer can be used as a differential interferometer.

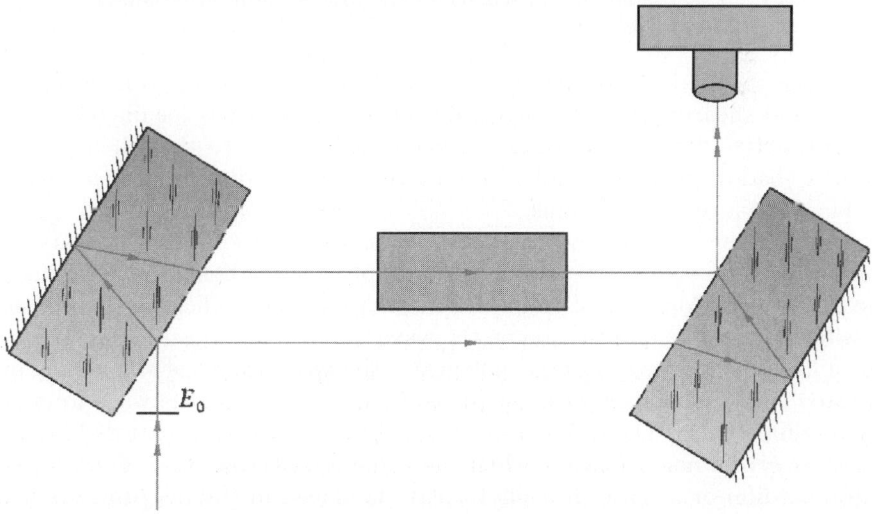

Fig. 1.9.3. The optical scheme of the Jamin interferometer

The most popular mirror interferometer is the Michelson interferometer (Fig. 1.9.4). However, it is not very convenient for measurements in transparent objects. The measuring beam crosses twice the test object following different trajectories because of the deviation of the beam due to the gradient of the refraction index in the object investigated. This circumstance makes the processing of the obtained interferograms difficult.

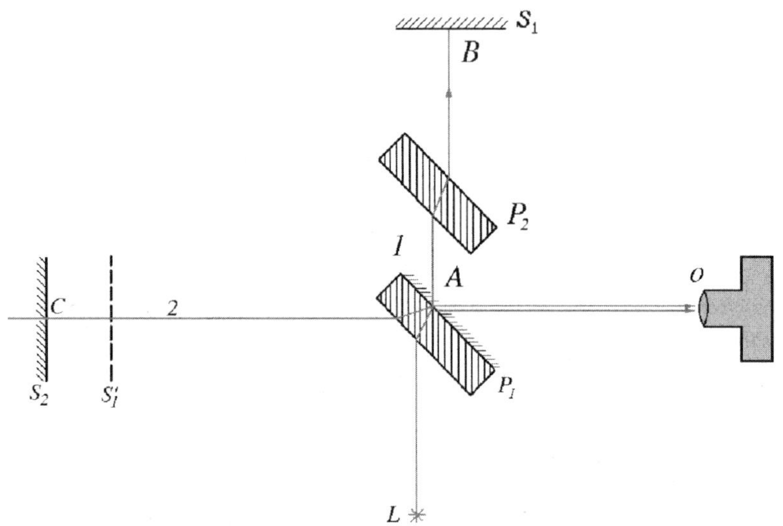

Fig. 1.9.4. The optical scheme of the Michelson interferometer

Mention should also be made of the use of the interferometers on Wollaston prisms and shearing interferometers on diffraction gratings for diagnostics of free-convective flows. Their optical schemes are similar to the scheme of the Toepler shadow graph and differ only by the way of radiation splitting into a probing and a reference beams.

Use of lasers as light sources in the optical schemes of the diagnostics of convective flows opens wide prospects for application of holography. Owing to the further development of this method for practical applications it became possible to preserve information on the wave fronts of an object in a hologram. To reproduce an image using a hologram, the wave fronts of the object are restored. This image is equivalent to the image created by the wave fronts of the original object. The object wave fronts illuminated by a scattered light of the laser give a normal image when the camera objective projects on a photographic film or a plate. If a photo plate is placed in the instrument's field of vision, then the information on the object contained in the distributions of phases and in the intensities of wave fronts of the object can be stored as microscopic interference patterns. These patterns appear on a photograph as a result of the interference of the wave fronts of the object with comparative wave fronts incident at a certain angle. The comparative beam is separated from the same laser beam which lights the object, for example, by means of a semitransparent mirror. When the image is restored from the developed holographic plate, it is illuminated at the same angle and under the same conditions which were used in exposure. The wave fronts of the light diffracted on the interference plates correspond to the wave fronts of the object. In the interference holography, two holograms, from a transparent and an opaque objects, are registered on one photo plate by the method of double exposure. In reproduction of this type of holograms, the phase differences of two exposures are observed as interference bands superposed onto the image of the object. Two superposed holograms making the interference hologram act as the comparative and measuring fronts just as in the method of two-beam interference. Carrying out measurements in transparent objects with the aid of an interference hologram, it is possible to obtain two-beam interference similar to the Mach–Zehnder interference if first a hologram without the object is obtained, and then a hologram with the object. To obtain a hologram of a nonstationary process, two holograms corresponding to the different time moments can be combined. This time differential interferometrics cannot be obtained on conventional interferometers.

Virtually each classical interferometer designed to measure optical transmission path length has its holographic analogue [1.98]. The classical interferometers are characterized not so much by the arrangement of optical elements, which can differ greatly depending on a specific scheme, as whether the interferometrically compared wave fronts are almost plane or spherical with a rather small phase deviations from an ideal wave front. For this reason, the optical elements used in the classical interferometer must be manufactured with a high degree of accuracy, so that they could not introduce spurious bands into the resulting interference pattern, whereas holography allows restoring wave fronts with an arbitrary phase change across the wave front, and therefore optical elements of lower quality can be used in manufacturing these interferometers. Holographic interferometric system can be constructed on scattering elements, which cannot be at all used in classical methods. Since in the classical interferometers the wave fronts are compared but not recorded, they operate in real time, and this requires high stability of the interferometer optical elements and to some extent also high stability of the phenomenon investigated. In a holographic interferometer the compared wave fronts are remembered, and this provides the experimenter with one more dimension, that is time. The presence of the time variable is a very essential part of holographic interferometry, which triggered its new numerous applications.

In classical interferometers with beam splitting, for example, in the Mach–Zehnder and Michelson interferometers, the plane wave is split into two, travelling in different directions. One of them passes through the area investigated, and the other acts as a reference wave. The second light splitter, or the second passage through the light splitter as in the case in the Michelson interferometer, joins the split beams and thus producing an interference pattern across the wave front. The lines of this interference pattern represent the lines corresponding to the constant difference of the optical path lengths across the wave front. Replacing the light splitter by a hologram we get a holographic interferometer. A holographic implementation of the Mach–Zehnder interferometer is shown in Fig. 1.9.5 and Fig. 1.9.6 presents schematically a holographic analogue of the Michelson interferometer. In both cases, the interferometer is formed by recording the hologram of the given sequence of optical elements. Restoration of the hologram in the same place where it was recorded leads to the formation of a reference wave front, which makes it possible to carry out interferometric measurements of subsequent changes in the system. Therefore, any hologram represents in principle an interferometer, in which the initial wave fronts, which generated this hologram, can interfere in restoration. In practice, a holographic interferometer is frequently used whose scheme is identical to that of the Toepler device (Fig. 1.9.7). As a rule, this type of devices is constructed so that both holographic interferometric investigations and shadow measurements could be made.

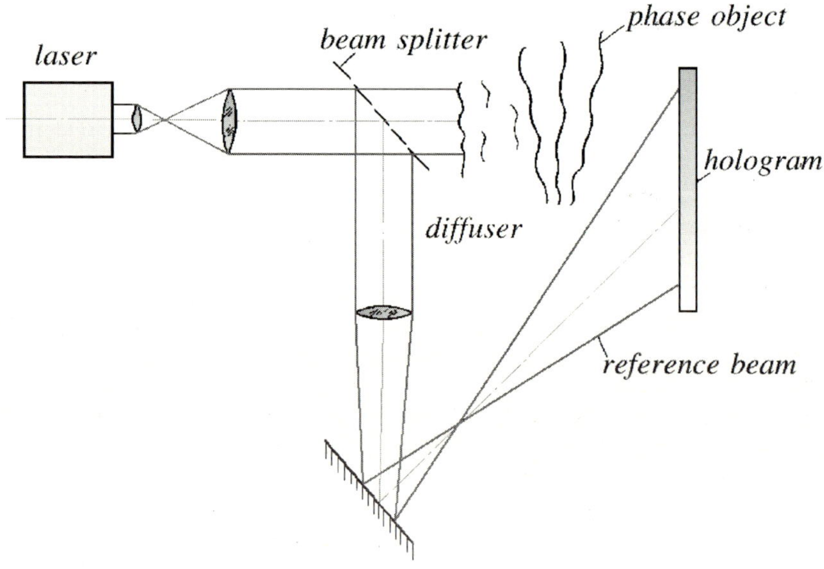

Fig. 1.9.5. Mach-Zander interferometer holographic realization

In other classical interferometers, light splitting and polarizing elements are used to shift the wave front in the plane perpendicular to the direction of wave propagation. The shearing interferometers compare different parts of the same wave front and therefore they are sensitive to a phase change across the front, but not to its absolute value at the given point. The holographic shearing interferometer is implemented on one or two wave front holograms. In any case,

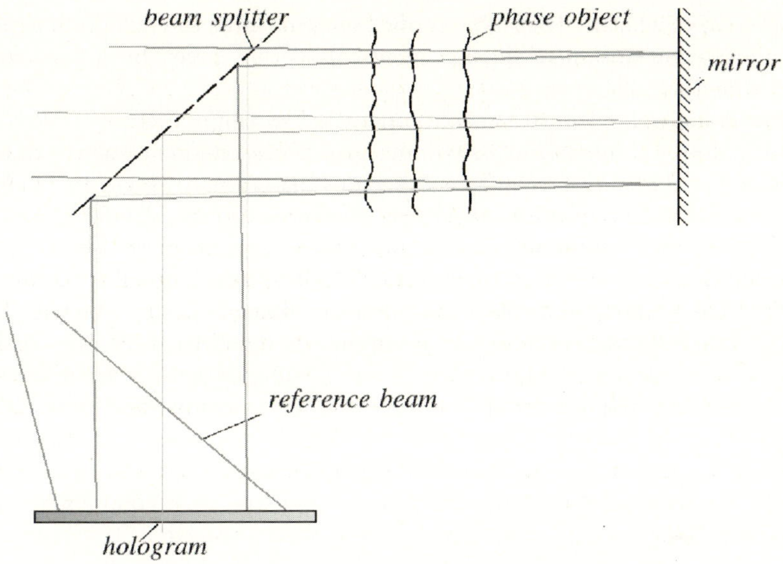

Fig. 1.9.6. Scheme of holographic implementation of the Michelson

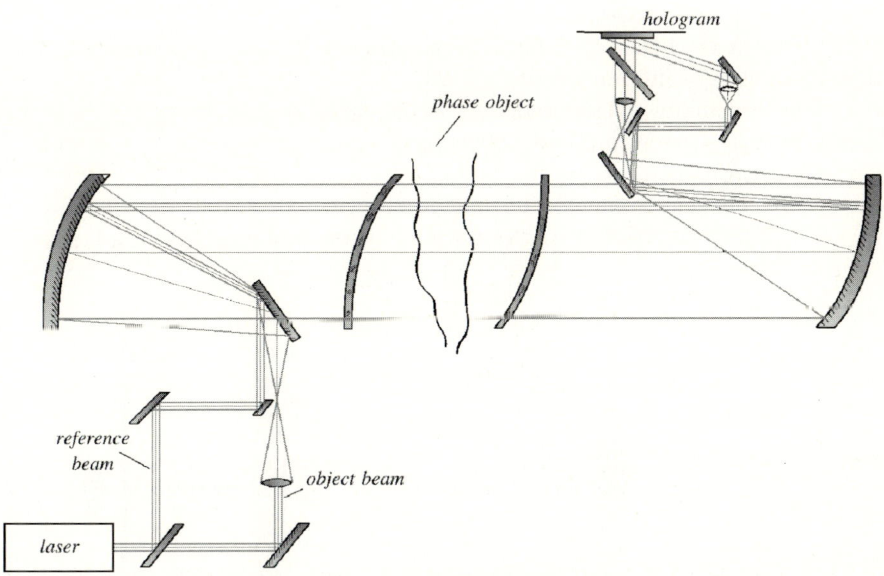

Fig. 1.9.7. The optical scheme of a holographic interferometer constructed on the basic of the Toepler device

irrespective of whether the shift is done before or after the hologram exposure the information obtained appears identical to that given by a conventional interferometer.

The holographic interferometers operating in the real time are less practical and stable than their classical analogues. The unique property of holography is the ability to record a multitude of images onto the same hologram, which in restoration interfere as independent wave fronts. Interferometry with double hologram exposure made it possible to significantly weaken the requirements to stability of exposure time. This ability to record wave fronts for different times and also the fact that now any wave fronts can be used in the interferometry made the holographic approach much more flexible than the classical one. Moreover, the holograms of moving objects contain the information on their motion, which is essentially new in comparison with classical methods.

Shadow methods. Shadow methods are widely used in investigation of natural convection, especially the flows which are very complex to be investigated theoretically or numerically and which received little study. This method is useful for studying the occurrence of turbulence, flow separation from a heated surface due to buoyancy force, flows in stratified media, penetrating flows, flows near curvilinear surfaces, and etc.

The shadow method was first suggested by the French astronomer Leon Foucault in 1858 to control the quality of large high-resolution astronomical objectives. In 1864 the German physicist August Toepler applied this method to investigate gas inhomogeneities [1.99].

There are many optical schemes which allow one to carry out measurements by the shadow method. Their large quantity is attributed to a large variety of problems and to the ambiquity in the requirements imposed on the results of measurements. The shadow devices in which the inhomogeneity investigated is located in the parallel beam have become most popular (Fig. 1.9.8).

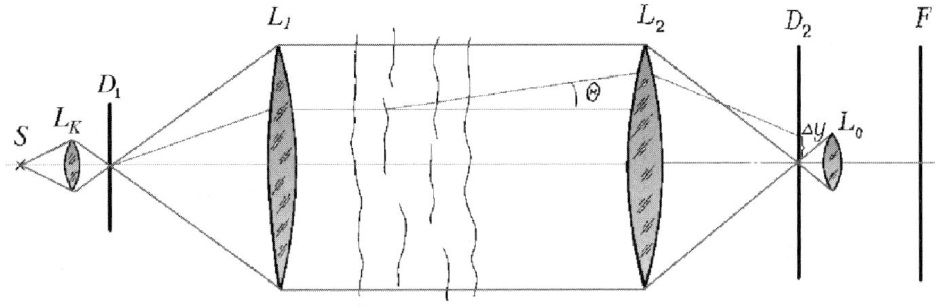

Fig. 1.9.8. Scheme of operation of the shadow device in diagnosting transparent objects in a parallel light beam

The inhomogeneity is visualized due to the introduction of a diaphragm D_2, resulting in phase and amplitude changes in a part of the light beam which cause the redistribution of the light intensity in the image plane F, which allows one to judge the character and some parameters of the considered inhomogeneity. This is the feature which distinguishes the shadow methods from the methods based on studying light emission or absorption where no diaphragms are used and also from interference and polarizing methods where phase and polarizing changes are brought into the whole light flux. The diaphragm D_2 bringing amplitude and phase changes into a part of the light beam is usually called a visualizing diaphragm. It may differ in shape: Foucault knife, phase knife, curvilinear diaphragm, groove grating, thin filament etc.

One of the most widespread variants of the shadow device developed by M. M. Maksutov is the z-shaped shadow device with a variable optical path length (Fig. 1.9.9).

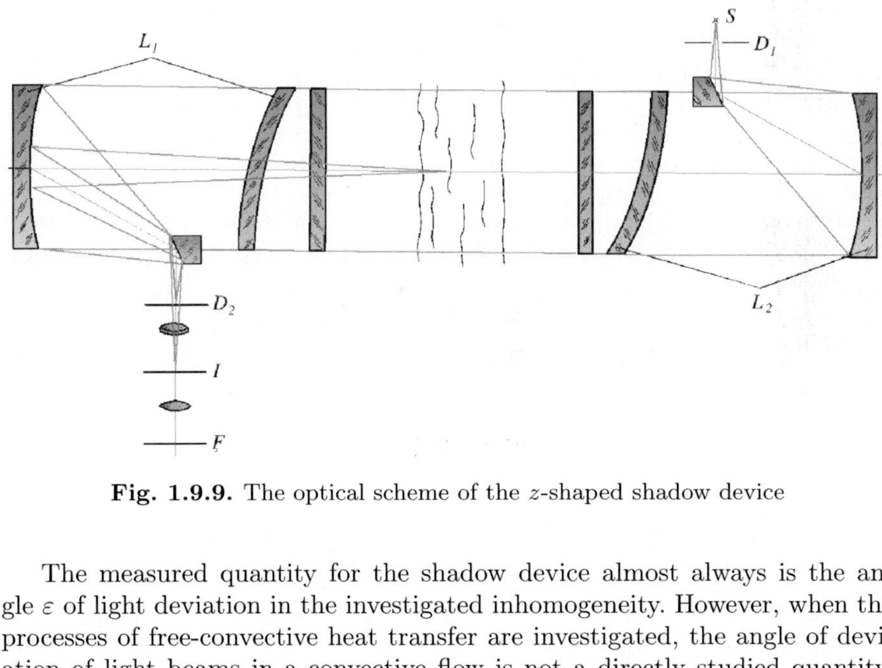

Fig. 1.9.9. The optical scheme of the z-shaped shadow device

The measured quantity for the shadow device almost always is the angle ε of light deviation in the investigated inhomogeneity. However, when the processes of free-convective heat transfer are investigated, the angle of deviation of light beams in a convective flow is not a directly studied quantity. To analyze this kind of problems it is necessary to consider light propagation in an isotropic transparent medium. For a nonmagnetic and no conducting isotropic medium the Maxwell equations can be written as

$$\begin{aligned} &\text{rot } H - \frac{\varepsilon}{c} E' = 0\,, \qquad \text{div} E = 0\,, \\ &\text{rot } E - \frac{1}{c} H' = 0\,, \qquad \text{div} H = 0\,, \end{aligned} \tag{1.9.22}$$

where E is the electrical field intensity vector, H is the magnetic field intensity vector, c is the speed of light, and ε is the dielectrical permeability of the medium considered. Excluding the electrical or magnetic field intensity vectors from (1.9.22), we can obtain the wave equation

$$\Delta E - \frac{\varepsilon}{c^2} E'' = 0 \,. \tag{1.9.23}$$

Substituting into (1.9.23) the electrical field intensity in the form

$$E = f(x, y, z) e^{ikct} \,,$$

we get the Helmholtz equation

$$\Delta f + k^2 n^2 f = 0 \,, \tag{1.9.24}$$

where $n = \sqrt{\varepsilon}$ is the refraction index of the medium.

To obtain the equation of the geometrical optics we consider the wave

$$f = U(x, y, z) e^{ikL(x,y,z)} \,, \tag{1.9.25}$$

where $L(x, y, z)$ is an almost linear function and $U(x, y, z)$ is a slowly changing function of the coordinates. Substituting (1.9.25) into (1.9.24), we get

$$k^2 U \left(n^2 - |\mathrm{grad}\ L|^2\right) + ik \left(U \Delta L + 2 \mathrm{grad}\ U\, \mathrm{grad}\ U\right) + \Delta U = 0 \,. \tag{1.9.26}$$

The relations of the geometrical optics are valid for large values of k; therefore, in (1.9.26) we may neglect all the terms except for the first one:

$$|\mathrm{grad}\ L|^2 = n^2 \,. \tag{1.9.27}$$

Introducing a unit vector S normal to the surface $L = \mathrm{const}$, we get

$$\int_{r_1}^{r_2} n dS = L(r_2) - L(r_1) \,. \tag{1.9.28}$$

The principle of the minimum optical path length of the propagating light (Fermat principle) follows from equality (1.9.28); it can be written mathematically as

$$\delta \int n dS = 0 \,. \tag{1.9.29}$$

The light beam trajectories corresponding to the Fermat principle (1.9.29) must satisfy the Euler equation

$$\frac{d}{dS}\left(n \frac{dx}{dS}\right) = \frac{dn}{dx} ; \quad \frac{d}{dS}\left(n \frac{dy}{dS}\right) = \frac{dn}{dy} ; \quad \frac{d}{dS}\left(n \frac{dz}{dS}\right) = \frac{dn}{dz} \,. \tag{1.9.30}$$

Let the probing radiation propagate in the shadow device along the z-axis. We assume that the light deviation angles in the convective flow investigated are small enough to neglect the curving of the path light beam in passing through the optical inhomogeneity and to consider that the beam travels rectilinearly, deviating very little from the path which it had covered in the absence of this inhomogeneity. In this case the Euler equations can be transformed to

$$\begin{aligned}\frac{d^2x}{dz^2} &= \left[1+\left(\frac{dx}{dz}\right)^2+\left(\frac{dy}{dz}\right)^2\right]\left[\frac{d(\ln n)}{dx}-\frac{\partial x}{\partial z}\frac{d(\ln n)}{dz}\right],\\ \frac{d^2y}{dz^2} &= \left[1+\left(\frac{dx}{dz}\right)^2+\left(\frac{dy}{dz}\right)^2\right]\left[\frac{d(\ln n)}{dy}-\frac{\partial y}{\partial z}\frac{d(\ln n)}{dz}\right].\end{aligned} \tag{1.9.31}$$

Using (1.9.31) we can obtain an expression which determines the additional difference of the phases and the angles of beam inclination to the coordinate axes which appeared as a result of the effect of the optical inhomogeneity [1.97]

$$\begin{aligned}\tau(x,y) &\approx \frac{1}{c}\int\limits_{z_1}^{z_2}\left[n\left(x,y,z\right)-n_0\right]dz\,,\\ \tan\varepsilon_x &\approx \int\limits_{z_1}^{z_2}\frac{d\left[\ln(x,y,z)\right]}{dx}dz\,,\\ \tan\varepsilon_y &\approx \int\limits_{z_1}^{z_2}\frac{d\left[\ln(x,y,z)\right]}{dy}dz\,,\end{aligned} \tag{1.9.32}$$

where z_1 and z_2 are the coordinates of the points of the input and output of probing radiation and of the exit of the probing radiation in passing through the flow studied.

In carrying out practical measurements, it is required to solve the reverse problem: find the distribution function of the refraction index from the measured values of the light deviation angles or from the additional difference of the optical path length. From this point of view, the simplest objects are the inhomogeneities in which the refraction index does not change along one of the directions. In practice, the device is usually located so that its optical axis coincides with this direction. Then the functions $\partial n/\partial x$ and $\partial n/\partial y$ do not depend on the z coordinate, and expression (1.9.32) takes the form

$$\begin{aligned}\varepsilon_x &= \frac{1}{n_0}\frac{\partial n}{\partial x}\left(z_2-z_1\right)\,,\\ \varepsilon_y &= \frac{1}{n_0}\frac{\partial n}{\partial y}\left(z_2-z_1\right).\end{aligned} \tag{1.9.33}$$

By means of a simple transformation (1.9.33) can yield expressions to determine the refraction index gradients:

$$\frac{\partial n}{\partial x}(x,y) = \frac{n_0 \varepsilon_x(x,y)}{L} ,$$
$$\frac{\partial n}{\partial y}(x,y) = \frac{n_0 \varepsilon_y(x,y)}{L} , \tag{1.9.34}$$

where L is light path length in the flow investigated.

Thus, in carrying out measurements in plane inhomogeneities, the field of the light deviation angles obtained by a shadow device can be transformed into the field of the refraction index gradients corresponding to this inhomogeneity.

The second class of the inhomogeneities for which the calculation techniques has been well developed which allow one to find the distribution of the refraction index inside the object investigated are the inhomogeneities possessing an axial symmetry. If the symmetry axis is perpendicular to the device optical axis and is directed along the x-axis, the light beam does not leave the section $x = \text{const}$ and, consequently, the dependence of the refraction index on x can be neglected, with each section $x = \text{const}$ being considered individually. On change from the rectangular coordinate system to the cylindrical one, (1.9.34) is transformed into the Abel-type integrated equations

$$(\varepsilon_y)_j = \frac{2}{n_0} \int_{r_j}^{R} \frac{\partial n}{\partial y} r_j \left(r^2 - r_j^2\right)^{-1/2} dr ,$$
$$(\varepsilon_x)_j = \frac{2}{n_0} \int_{r_j}^{R} \frac{\partial n}{\partial x} r_j \left(r^2 - r_j^2\right)^{-1/2} dr . \tag{1.9.35}$$

At the present time, there are some techniques of solving these equations [1.97, 1.100].

In carrying out practical measurements the methods have found widespread application which make it possible to isolate the curves of the same deviation angle. In the focal plane of the main objective of the receiving part L_2 (Fig. 1.9.8) a knife is set as the diaphragm D_2 parallel to the image of the illuminating slit D_1 at a certain distance from it. The regions that deviate the light beams so that the coming light is cut by the knife seem to be dark and the entire remaining field is light. The border between the light and shadow determines the locus of the image of the points which deviate the light beams by an angle corresponding to the distance between knife edge and the illuminating slit. Making photographs with different positions of the knife it is possible to determine the field of the light deviation angles over the whole object investigated. Frequently, a thin opaque wire is used instead of the Foucault knife. In this case, the shadow of the wire represents a blurred structure which is influenced by the finite width of the light source and by the diffraction effects, with the position of the shadow center corresponding to the locus of the points which deviate light by the distance equal to the stretch from the middle of the illuminating slit to the wire middle.

The ordinary shadow methods have been extended and developed to give color image methods [1.101]. The size and form of the color visualizing diaphragm and its separate parts can be very different. Most frequently, use is made of a set of color glass strips or round diaphragms with sectors of different colors. When a diaphragm in the form of a set of strips is used, the light source is in the form of a slit. The lines of the same color in a visualizing diaphragm are parallel straight lines, therefore the deviation of the slit image along a strip does not change the color of the image and, consequently, the method registers the change in the component of the light deviation angle only in the direction perpendicular to the lines of the same color of the diaphragm. A round diaphragm is used together with a round light source. In this case, displacement of the source image in any direction influences the change in the image color. Theoretically, using this diaphragm, it is possible to measure both the value and the direction of the deviation angle, applying, for example, spectral methods which allow one to determine quantitatively the correlation between the intensities of different monochromatic radiation components at each point of the image.

The methods of defocused diaphragms have become very popular in investigating free-convective heat transfer processes [1.102]. At the present time, these methods are used to measure the deviation angle by determining the displacement of shadows from the diaphragm which is a set of dark details against a transparent background. Among themselves the methods differ by the shape and location of diaphragms and by the techniques used to interpret the photographs obtained. Generally the shape and arrangement of the opaque details and light intervals in the diaphragm can be arbitrary. A system of concentric circles or ellipses, a fan of straight lines emanating from one point, a set of any holes and screens can be used as a diaphragm, but the diaphragms in the form of alternating dark and light strips have become most popular. In this case the magnitude of the light deviation angle can be calculated from the formula

$$\varepsilon = \frac{\delta_f P}{P_f f}\,, \tag{1.9.36}$$

where δ_f is the shift of the grating shadow on the photograph, P is the grating period, P_f is the grating period on the photograph, and f is the focal distance of the objective in the receiving part of the shadowgraph.

The main drawback of the methods of defocused diaphragms is that they allow one to measure light beam deviation angles only in the limited number of the lines corresponding to the middle parts of the diffraction pictures of grating grooves or light gaps. This is not enough for many physical investigations, since the parts of the object investigated which are of interest for observation can fall into the regions between opaque grooves and transparent gaps and thus become invisible. The method of absolute photo metering of knife and slit is to a large extent devoid of this drawback.

Let initially the knife masks the image of the slit so that only its part of width ξ_1 remains open. If the brightness of the slit image in the direction perpendicular to the knife is assumed to be constant, then the illuminance of any part of the image is equal to

$$I_1 = \xi_1 \int_{\eta_1}^{\eta_1+l} B(\eta)\, d\eta \,, \tag{1.9.37}$$

where η is the coordinate along the slit image, $B(\eta)$ is the function which describes the brightness of the source and absorption of the system, η_1 and $\eta_1 + l$ are the coordinates of the illuminating slit edges, and l is the length of the slit image. The optical density of the sensitive material blackening corresponding to this illumination is determined by the relation

$$D_1 = \gamma \log(I_1) - b \,, \tag{1.9.38}$$

where γ is the contrast range of the sensitive material, and b is a constant. If the investigated inhomogeneity deviates the light beam by the angle ε, the image of the slit shifts by the value $\delta = \varepsilon f$, and the illuminance of the same point of the inhomogeneity becomes equal to

$$I_2 = (\xi_1 + \delta) \int_{\eta_1}^{\eta_1+l} B(\eta)\, d\eta \,. \tag{1.9.39}$$

Combining the expressions for optical densities, we obtain

$$\varepsilon = \frac{\xi_1}{f}\left(10\frac{D_2 - D_1}{\gamma} - 1\right) . \tag{1.9.40}$$

It should be noted that the use of the photometric method is associated with a number of specific difficulties such as the necessity of determining the contrast range of the material used, controlling carefully all the stages of the chemical treatment of a photolayer, and carrying out density measurements.

Flow visualization. One of the most widespread and at the same time most informative methods for investigating the structure of the velocity field in natural convection is the method of track visualization of a flow. Photographs of the trajectories of particles or of the dye introduced continuously or at several fixed points into the flow give qualitative and even quantitative information on the velocity field, since the Lagrange description of the motion of a certain labeled particle is as valid as the Euler velocity description at a point, though the latter is more conveniently used in a theoretical analysis, and this is why it is preferred in experimental investigations. The Euler description of motion can be obtained by exposing a film during quite a long time interval, so that the particles could leave on it short strokes, whose length and direction allow one to determine the velocity vector at different points in the flow. This method is connected with the velocity measurement methods by the time of covering the known distance. The scheme of the set for implementation of the method considered is shown in Fig. 1.9.10. The general method of flow analysis consists either in photographing and subsequent digitization on any device which makes it possible to carry out automatical recording of the sensitive material blackening density by a computer, or in recording by a video camera and analyzing the voltage of the output brightness signal.

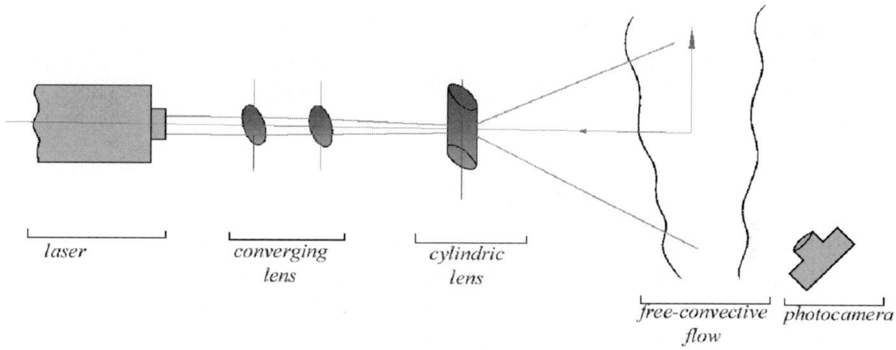

Fig. 1.9.10. Scheme of a set for convective flow track visualization

Computer analysis of this kind of information make it possible to obtain any data on the character of a free-convective flow.

The absorption of light by a dye or smoke can also be used for the analysis of the convective flow structure and concentration measurements. The calibration of this method causes certain difficulties, if the properties of an impurity are not carefully controlled. Measurement of the attenuation of radiation intensity over the entire length of the beam allows one to determine only a certain mean density of the impurity or the concentration of impurities; however, if the beam length is finite, the radiation intensity will be fluctuating. This method is of greatest interest for investigating free-convective flows which are considered to be two-dimensional. The block diagram of the experimental setup is shown in Fig. 1.9.11. A more a detailed description of the method in application to turbulent flows can be found in [1.103].

In the case of a laminar free-convective flow, the visualization of the velocity profile is of great interest. The velocity profile can be visualized using a smoke filament, when smoke or dying substance is generated as a result of some process on a wire located across the flow. In an inhomogenious field of the flow velocity different parts of the smoke filament are entrained to different distances, so that after a certain time the smoke filament takes the form of the flow velocity graph. The magnitude of the shift of each part of this filament is proportional to the flow velocity and to the time period which passed from the beginning of smoke generation. Figure 1.9.12 presents schematically a setup for visualizing the velocity profile in a boundary layer near a vertical surface of

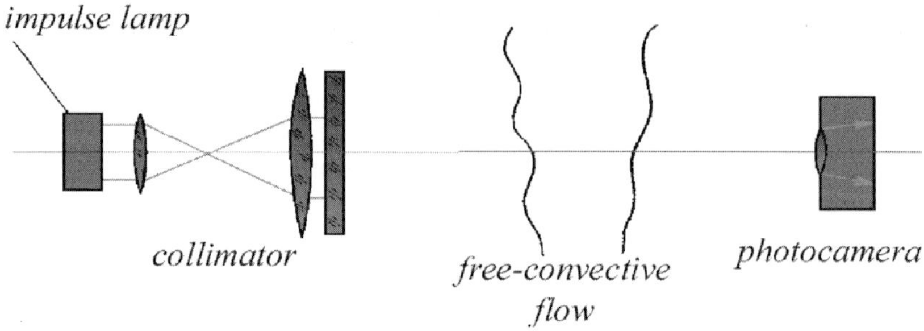

Fig. 1.9.11. Optical scheme of convective flow visualization due to principle based on light absorbtion by a dye or smoke

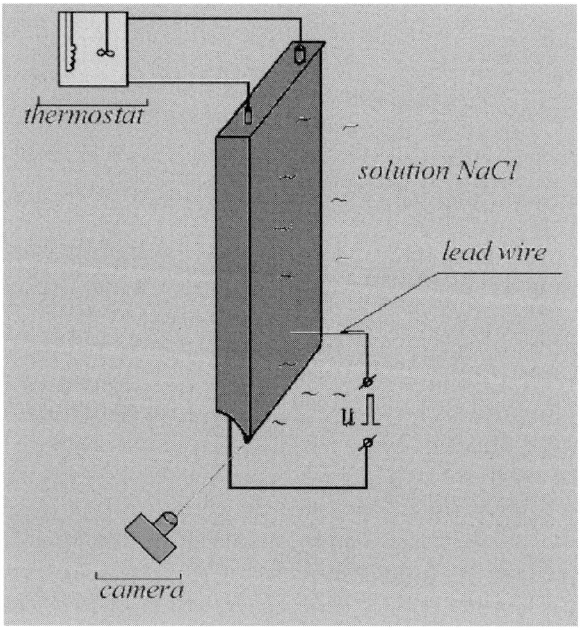

Fig. 1.9.12. A set for visualization the velocity profile in natural convection on a vertical surface

constant temperature. When an electric current is switched on, some amount of white $\mathbf{PbCl_2}$ is produced as a result of electrolysis in a solution of **NaCl** adjoining the lead wire. Entrained by the flow, $\mathbf{PbCl_2}$ forms a thin white filament of dye which visualizes the flow velocity profile.

A great number of various convective flow visualization methods is described in [1.104], however, by no means all of them are suitable for investigating of the process of natural convection.

1.10 Processing of Experimental and Calculated Data on Heat Transfer

The quantitative characteristics of heat transfer are determined from the dimensionless equation obtained by similarity theory or dimensional analysis methods. The most common method of processing predicted and experimental results is the use of a one-term power-law dependence between the similarity numbers

$$\mathrm{Nu} = C\,\mathrm{Gr}^{m}\,\mathrm{Pr}^{n}\mathrm{P}_{k}^{l_k}\,, \tag{1.10.1}$$

where C, n, m, and l_k are constant dimensionless numbers. The practical method of constructing the power-law dependence is based on the fact that in logarithmic coordinates all the points lie on a straight line. But if the experimental points lie on a curve, it is replaced by a broken line. For some of the parts of this curve the values of C, n, m, and l_k are different.

To expand the application of dependence (1.10.1) it is represented as a sum [1.105]:

$$\mathrm{Nu} = C_1\,\mathrm{Gr}^{m_1}\,\mathrm{Pr}^{n_1} + C_2\,\mathrm{Gr}^{m_2}\,\mathrm{Pr}^{n_2} + C_3\,\mathrm{Gr}^{m_3}\,\mathrm{Pr}^{n_3} + \ldots \tag{1.10.2}$$

The generalizing dependences are based on the equation [1.106]

$$y^{n}(z) = y_0^{n}(z) + y_\infty^{n}(z)\,, \tag{1.10.3}$$

where $y_0(z)$ and $y_\infty(z)$ correspond to the limiting values of the variable z. The functions $y_0(z)$ and $y_\infty(z)$ are determined from both theoretical calculations and experimental data.

For a thin cylindrical body the boundary layer of thickness δ_T is considered to be such a motionless layer, through which heat is transferred only by heat conduction. In this case [1.107]

$$\mathrm{Nu} = 2/\ln\left(1 + 2\delta_T/\delta\right)\,, \tag{1.10.4}$$

where the limiting value of the thickness δ_T can be determined from calculations of free convection on a plate.

The following formula is used to calculate heat transfer between parallel walls [1.108]:

$$\mathrm{Nu} = A\,\mathrm{Ra}\left\{1 - \exp\left[-B/\mathrm{Ra}^{0.75}\right]\right\}\,, \tag{1.10.5}$$

where A and B are the coefficients which describe the geometrical parameters of a heat generating surface.

If the quantities characterizing the physical properties of the system depend on temperature, then the equations determining the form of these dependences should be included into the number of these basic equations. The similarity numbers corresponding to these equations should be considered as the arguments of generalized dependences. The application of generalized analysis is impossible in this case and thus only approximate solutions can be used. In particular, in approximation of thermophysical characteristics as power-law dependences on temperature, the parametric criterion P_T is introduced as a ratio of environmental temperature T_∞ to the wall temperature T_{w} [1.10]. The physical parameters should be related to one of the two characteristic temperatures. This method is applicable to gases.

The dependence of heat transfer of dropping liquids on the heat flux direction and temperature difference can approximately be taken into account by introducing the additional factor $(\mathrm{Pr}_\infty/\mathrm{Pr}_{\mathrm{w}})$ into the similarity equation [1.109]. For a heated liquid $\mathrm{Pr}_\infty/\mathrm{Pr}_{\mathrm{w}} > 1$, for a cooled one $\mathrm{Pr}_\infty/\mathrm{Pr}_{\mathrm{w}} < 1$. The larger the temperature difference, the more the ratio $\mathrm{Pr}_\infty/\mathrm{Pr}_{\mathrm{w}}$ differs from unity. The dependence of the physical parameters on temperature can be taken into account by the parametric criteria P_k of the form $\lambda_\infty/\lambda_{\mathrm{w}}$, $\mu_\infty/\mu_{\mathrm{w}}$, $c_{p\infty}/_{\mathrm{pw}}$, $c_{p,\mathrm{av}}/n_{p\mathrm{w}}$, etc., and by introducing the temperature determining for this process. Nusselt suggested that the averaging of the physical parameters be made by the formula [1.13]

$$\varphi_{\mathrm{av}} = \frac{1}{\theta_{\mathrm{w}}} \int\limits_{T_\infty}^{T_{\mathrm{w}}} \varphi dT \;, \tag{1.10.6}$$

and the determining temperature be calculated as a mean logarithmic one:

$$T_{\mathrm{av}} = \theta_{\mathrm{w}} / \ln\left(1 + \theta_{\mathrm{w}}/T_\infty\right) \;. \tag{1.10.7}$$

For $\theta_{\mathrm{w}} \ll T_\infty$ we represent dependence (1.10.7) as a polynomial of $\theta_{\mathrm{w}}/T_\infty$. If we leave only the first term of the polynomial, then $T_{\mathrm{av}} = T_\infty$; if two, then $T_{\mathrm{av}} = (T_\infty + T_{\mathrm{w}})/2$. In free convection problems the determining temperature is often selected as a linear combination of the wall and environment temperatures:

$$T_{\mathrm{av}} = aT_{\mathrm{w}} + bT_\infty \;, \tag{1.10.8}$$

where a and b are the coefficients changing from 0 to 1; $a + b = 1$.

Usually, that linear dimension is taken to be determining l which mostly reflects the physical essence of the process (for example, height of a plate, diameter of a cylinder or a sphere, thickness of a gap, of a boundary layer, etc.). The remaining dimensions enter into the similarity equation as simplexes $\mathrm{P}_{lk} = l_k/l$ (width and thickness of a plate, height of a vertical cylinder or length of a horizontal cylinder, height of a gap). In some cases, a combination of dissimilar physical quantities entering into the uniqueness condition is taken as the determining linear dimension (length scale in the asymptotic theory $l/\mathrm{Gr}^{1/4}$ [1.2], linear dimension in jet free convection [1.110]). To make the computational relations universal and exclude parametric criteria, a single characteristic dimension is introduced. For example, $1/l = 1/a + 1/b$ for a horizontal plate, πd for a horizontal cylinder, $\pi d/2$ for a sphere, $d_{ef} = 4F/p_{ef}$ for a horizontal channel of any section [1.111].

Table 1.10.1.

Ra_l	C	N
10^{-3}–$5 \cdot 10^2$	1.18	1/8
$5 \cdot 10^2$–$2 \cdot 10^7$	0.54	1/4
$2 \cdot 10^7$–10^{13}	0.135	1/3

In applied problems for calculating a free-convective component of heat transfer from surfaces of arbitrary shape into an infinite space the following formula is recommended:

$$\mathrm{Nu} = C\, a_{l\infty}^n\, (\mathrm{Pr}_\infty \,/\, \mathrm{Pr}_\mathrm{w})^{1/4} \tag{1.10.9}$$

or in dimension form

$$\alpha_\mathrm{av} = A\left(\theta_\mathrm{w}/L^{1/n-3}\right)^n\,, \tag{1.10.10}$$

where the quantities C and n depend on the number Ra_l (Table 1.10.1). The mean boundary-layer temperature is considered to be the determining temperature. For inclined and horizontal surfaces correction coefficients are introduced.

1.11 Basic Similarity Criteria and Parameters of Free-Convective Heat Transfer

General Similarity Criteria

$$\mathrm{Gr}_l = g\beta l^3 \theta_\mathrm{w}/\nu^2$$

is the Grashof criterion; it is the main determining similarity criterion of heat exchange with the environment at a given wall temperature (the boundary conditions of the 1st kind). It is the measure of the ratio between the buoyancy forces caused by the difference of the fluid densities at individual points of the non-isothermal flow and the molecular viscosity forces. In calculating heat transfer in a limited space between surfaces of different temperatures T_w1 and T_w2, the Grashof criterion $\mathrm{Gr}_l = g\beta l^3(T_\mathrm{w1} - T_\mathrm{w2})/\nu^2$ characterizes also the motion of a liquid or gas along the heat transfer surface.

$$\mathrm{Gr}_l^* = g\beta l^4 q_\mathrm{w}/(\lambda\nu^2)$$

is the modified Grashof criterion; it is the main determining similarity criterion for a given heat flux on a wall (boundary conditions of the 2nd kind).

$$\mathrm{Pr} = \nu/a$$

is the Prandtl criterion. This is the criterion of the physical properties of liquids and gases for convective heat transfer, it is measure of the ratio between molecular momentum transfer and molecular heat transfer in a liquid or gas. It establishes the coupling and similarity between the velocity and temperature fields in free convection.

$$\mathrm{Ra}_l = \mathrm{Gr}_l\mathrm{Pr}$$

is the Rayleigh criterion. It characterizes the relationship between buoyancy and viscosity forces for media with small velocities of motion (the inertia forces are small) and a large viscosity (Pr > 1), and it is the main determining criterion of heat transfer by free convection under the boundary conditions of the 1st kind and at the numbers $\mathrm{Pr} \gg 1$.

$$\mathrm{Ra}_l^* = \mathrm{Gr}_l^*\mathrm{Pr}$$

is the modified Rayleigh criterion. It is applied under the boundary conditions of the 2nd kind on a surface and at the numbers $\mathrm{Pr} \gg 1$.

$$\mathrm{Gr}_{l\vartheta} = \mathrm{Gr}_l \cos\vartheta$$

is the Grashof criterion for a surface located at an angle ϑ to a vertical.

$$\mathrm{Gr}_{l\vartheta}^* = \mathrm{Gr}_l^* \cos\vartheta$$

is the modified Grashof criterion for a surface located at an angle ϑ to a vertical.

$$\mathrm{Nu}_l = \alpha l/\lambda$$

is the Nusselt criterion (dimensionless heat transfer coefficient); it is the main criterion to be determined in convective heat transfer. It characterizes the coupling between the heat transfer rate and the temperature field in the near-wall region. In the form $\alpha\theta/(\lambda\theta/l)$ it is a measure of the ratio between the density of a heat flux through a surface to the density of a heat flux by heat conduction in a liquid layer of thickness l. In the form $(l/\lambda)/(1/\alpha)$ it is a measure of the ratio between the thermal resistance by heat conduction to the thermal resistance by heat transfer. In heat transfer between walls, in a layer of thickness l filled with a thermal conductivity coefficient λ it represents a ratio of the effective thermal conductivity coefficient λ_{ef} to the real one λ and characterizes an increase in the heat transfer rate due to convection in comparison with pure heat conduction (in a motionless medium). If heat is transferred in a liquid only by heat conduction, the criterion Nu remains constant and dependent only on the geometrical form of the body, on the surface of which heat transfer occurs. If we define the Nusselt criterion without the heat transfer coefficient, then $\mathrm{Nu}_l = q_\mathrm{w} l/\lambda\theta_\mathrm{w} = \mathrm{Gr}^*/\mathrm{Gr}$.

$$\mathrm{P}_l = l_k/l$$

is the criterion of the body shape determining the effect of the characteristic geometric dimension along coordinate axes on heat transfer.

Variable Temperature of the Wall and of the Environment, Variable Heat Flux on the Wall, Volumetric Heat Sources

$$\mathrm{P}_{\vartheta} = \frac{x}{\theta_{\mathrm{w}}}\frac{d\theta_{\mathrm{w}}}{dx}, \quad \mathrm{P}_{\vartheta k} = \frac{x^k}{\theta_{\mathrm{w}}}\frac{d^k\theta_{\mathrm{w}}}{dx^k}, \quad \Lambda = 2\frac{x}{u_1(x)}\frac{du_1}{dx}$$

are the parametric similarity criteria of convective heat transfer which take into account the effect of variable wall temperature and of its gradients. They are used in parametric methods of calculating convective heat transfer in a boundary layer.

$$\mathrm{P}_q = \frac{x}{q_{\mathrm{w}}}\frac{dq_{\mathrm{w}}}{dx}$$

is the parametrical similarity criterion for a heat flux variable on a surface.

$$\mathrm{P}_s = \frac{x}{\theta_\infty}\frac{d\theta_\infty}{dx}, \quad \mathrm{S} = \frac{l}{\theta_{\mathrm{av}}}\frac{d\theta_\infty}{dx}$$

are the parametric criteria of the stratification of a liquid in the environment. For free convection of a cylinder, in a layer of the liquid of height H_1 the parametric criterion of stratification is $\mathrm{P}_s = \beta(T_{\mathrm{w}} - T_p)d/(H_1 + H_2)$.

$$\mathrm{Gr}_{ls} = \mathrm{Gr}_l P_s = \frac{g\beta l^4}{\nu^2}\frac{d\theta_\infty}{dx}$$

is the Grashof criterion for a stratified flow.

$$\mathrm{Gr}_v = g\beta l^5 Q_v/(\lambda\nu^2)$$

is the Grashof criterion; it is the main determining similarity criterion of free-convective heat transfer for the given volumetric heat sources in a medium.

$$\mathrm{Ra}_I = g\beta l^5 k_p I_0/(\lambda\nu a)$$

is the Rayleigh criterion determining the conditions of heat transfer in the presence of electromagnetic radiation in a medium.

$$\mathrm{O}s = Q_v l^2/\lambda\theta_{\mathrm{w}}$$

is the Ostrogradski criterion. It determines a measure of the ratio between heat fluxes due to volumetric heat generation ($Q_v l$) and heat conduction in a layer of thickness $l(\lambda\theta_0/l)$. The criterion Os characterizes the enhancement of heat transfer due to volumetric heat generation and establishes relationships between thermal resistances due to heat conduction through the layer of thickness l and due to volumetric heat generation.

Conjugation Parameters

$$\mathrm{P}_{\mathrm{w}p} = \frac{\lambda_\infty l}{\lambda_\mathrm{w} b}\left(\frac{\mathrm{Gr\ Pr}^2}{1+\mathrm{Pr}}\right)^{1/4}, \quad \mathrm{P}_{\mathrm{w}p1} = \frac{\lambda_\infty l}{\lambda_\mathrm{w} b}\mathrm{Gr}_l^{1/4}, \quad \mathrm{P'}_{\mathrm{w}p1} = \frac{\lambda_\mathrm{w} b}{\lambda_\infty l}$$

are the parametric criteria of thermal similarity, when the heat flux due to heat conduction is directed along the x-axis. They establish relationships between heat fluxes due to heat transfer in the boundary layer on a wall and due to heat conduction in a layer of thickness l.

$$\mathrm{P}_{\mathrm{w}p2} = \frac{\lambda_\infty b}{\lambda_\mathrm{w} l}\mathrm{Gr}_l^{1/4}, \quad \mathrm{P'}_{\mathrm{w}p2} = \frac{\lambda_\mathrm{w} l}{\lambda_\infty b}$$

are the parametric criteria of conjugation, when the main flux due to heat conduction is directed along the y-axis. They establish the relationship between heat fluxes due to convection from a surface of height l and due to heat conduction in a layer of thickness b.

$$\mathrm{k} = \frac{\lambda_{\infty 2}}{\lambda_{\infty 1}}\left(\frac{\mathrm{Ra}_{l2}}{\mathrm{Ra}_{l1}}\right)^{1/4}\left[\frac{\mathrm{Pr}_2\,(1+\mathrm{Pr}_1)}{\mathrm{Pr}_1\,(1+\mathrm{Pr}_2)}\right]^{1/4}$$

is the parametric criterion of conjugation of free convection on a vertical interface between the media with the temperatures $T_{\infty 1}$ and $T_{\infty 2}$. It establishes the relationship between thermal resistances in a boundary layer on the different sides of a vertical surface.

$$\mathrm{P}_{\mathrm{w}p3} = \frac{\lambda_\infty \rho_\infty c_{p\infty}}{\lambda_\mathrm{w} \rho_\mathrm{w} c_{p\mathrm{w}}}, \quad \mathrm{P'}_{\mathrm{w}p} = \frac{\rho_\mathrm{w} c_\mathrm{w} b}{\rho_\infty c_{p\infty} l}\mathrm{Gr}_l^{1/4}$$

are the conjugation parameters of non-stationary free convection.

$$\mathrm{P}_\lambda = \lambda_\mathrm{w}/\lambda_\infty$$

is the simplest parametric criterion of conjugation between a body and a liquid medium. It establishes the relationship between the heat fluxes due to heat conduction in two media of equal thicknesses.

$$\mathrm{Bi} = \alpha l/\lambda_T$$

is the Bio criterion. It characterizes the rate of heat transfer between a liquid (gas) and the body surface and establishes the relationship between the temperature field in the body and the conditions of heat transfer on its surface. For stationary heat transfer it may serve as an approximate measure of the ratio between the temperature difference over the thickness of the body and the temperature difference between the medium ambient and the body. If $\mathrm{Bi} > 100$, the thermal resistance of heat transfer on the body surface is small compared with the thermal resistance of the heat conduction of the body, and the temperature field in the body has a marked nonuniformity. If $\mathrm{Bi} < 0.1$, the temperature difference θ_w is large, and the temperature difference inside the body is small due to the low thermal resistance of the body, and the temperature field in the body is uniform.

Compressibility and Variable Properties of the Medium

$$\varepsilon_1 = \beta\theta_0$$

is the parametric criterion of the thermal expansion of the medium; it is the measure of the ratio of the actually occurring a change in the density to the characteristic value of the density of the medium; it determines the order of the magnitude of relative volumetric deformations. In the region of small values of ε_1 the gas compressibility does not manifest itself and the criterion ε_1 falls out of the number of quantitative characteristics essential for the given process (Boussinesq approximation). The change in the density in the buoyancy force is then taken into account in the Grashof criterion. For gases ε_1 coincides with the parametric criterion $\mathrm{P}_T = (T_\mathrm{w} - T_\infty)/T_\infty$ (temperature factor).

$$\varepsilon_2 = \beta_p \Delta p_0$$

is the parametric criterion of the expansion (compression) of the medium. Usually, $\varepsilon_2 \ll \varepsilon_1$, and the effect of ε_2 is taken into account only for critical and supercritical states of a substance.

$$\mathrm{Ec} = g\beta l/c_p$$

is the Eckert criterion in free convection. It characterizes heat generation due to friction and compression related to the heat flux on a surface. The criterion Ec can be defined as a ratio of the kinetic energy to heat transferred to the liquid. In the majority of technical sets the criterion Ec is small.

$$\mathrm{M} = gl/c_p\theta_\mathrm{w}$$

is the criterion taking into account the effect of compression forces on flow and heat transfer. The criterion M establishes the ratio of the work of compression to the magnitude of the heat flux transferred by the liquid moving with the velocity $u_0 = \sqrt{g\beta l\theta_0}$.

$$\mathrm{Ar}_l = \frac{\rho_\infty - \rho_\mathrm{w}}{\rho_\infty}\frac{gl^3}{\nu^2}$$

is the Archimed criterion; the modification of the Grashof criterion for a compressible flow; it establishes the measure of the ratio of the buoyancy force to the force of molecular friction in a nonisothermal compressible flow.

$$\mathrm{P}_\beta = \frac{\theta_\mathrm{w}}{\beta}\frac{d\beta}{d\theta_\mathrm{w}}$$

is the parametric criterion of the flow compressibility gradient.

$$\mathrm{K}_{ph} = r/c_p\theta_\mathrm{w}$$

is the criterion of phase change in physical and mechanical transformations; it is the measure of the ratio of heat flux spent for phase transformation to the overheating (overcooling) heat of one of the phases.

$$\mathrm{Gr}_{lm} = g\beta_s l^3 (T_\mathrm{w} - T_\infty)_q / \nu^2$$

is the Grashof criterion for water at a temperature of about 4°C in view of a nonlinear change in the volumetric expansion coefficient depending on temperature.

$$\mathrm{P}_{Ts} = (T_m - T_\infty)/(T_\mathrm{w} - T_\infty)$$

is the parametric criterion (temperature factor), taking into account a temperature change in water at a maximum density.

$$\mathrm{Kn} = l^*/l$$

is the Knudsen criterion; it characterizes the degree of gas rarefaction as a degree of its deviation from the state of continuum. It determines the conditions under which the effects caused by gas rarefaction can arise.

Nonstationary Processes

$$\mathrm{Zh} = \nu\tau_0/l^2$$

is the Zhukovski criterion (hydrodynamic homochronicity criterion). It characterizes the similarity condition of the velocity fields for nonstationary motion of a viscous incompressible fluid.

$$\mathrm{Fo} = a\tau_0/l^2$$

is the Fourier criterion. It is the homochronicity criterion of temperature fields; it characterizes the coupling between the rate of change in the temperature field, the physical properties and the dimensions of a body and establishes the conditions for temporal similarity of the temperature fields in bodies with molecular heat transfer in nonstationary heat conduction. It expresses the conformity between the rate of change in the conditions in the surrounding medium and the rate of rearrangement of the temperature field inside the body. The larger (smaller) the value of Fo, the more (less) uniformly the body temperature is distributed over the volume.

$$\mathrm{k}_{T1} = \frac{\partial\theta_\mathrm{w}}{\partial\tau}\sqrt{\frac{xT_\mathrm{av}}{g}}\theta_\mathrm{w}^{-3/2}, \quad \mathrm{k'}_{T1} = \frac{\partial\theta_\mathrm{w}}{\partial\tau}\frac{\delta^2}{\theta_\mathrm{w} a}, \quad \mathrm{k}_{T2} = \frac{\partial^2\theta_\mathrm{w}}{\partial\tau^2}\frac{xT_\mathrm{av}}{g\theta_\mathrm{w}}$$

are the parametric criteria characterizing nonstationary processes in heating (cooling) of bodies with a finite value of heat capacity.

$$\mathrm{Re}_{\Delta u} = \Delta u/\nu$$

is the amplitude Reynolds criterion.

$$\mathrm{Re}_\omega = \omega l^2/\nu$$

is the oscillation Reynolds criterion.

$$\mathrm{Re}_v = \Delta u^2/\omega\nu$$

is the Reynolds criterion of secondary flows.

$$\mathrm{M}_{\Delta u} = \Delta u/a$$

is the oscillatory Mach number.

$$\mathrm{J} = \Delta A\omega^{2/3}/\nu^2$$

is the criterion of vibrational acceleration.

$$\mathrm{Jg} = \omega^2 A/g$$

is the vibrational criterion of acceleration.

Mixed Convection

$$\mathrm{Lu} = D/a$$

is the Lewis criterion; it characterizes coupling of heat and mass transfer in gas mixtures and is the measure of the rate of change in the field of chemical potentials with respect to the temperature field in the mixture flow.

$$\mathrm{Sc} = \nu/D$$

is the Schmidt criterion; it characterizes the relationship between the substance and momentum transfer in process of heat transfer; it can be considered as the measure of similarity of the velocity and concentration fields in mass transfer.

$$\mathrm{Du} = Q_T R_\mathrm{w} M^2 T_\mathrm{w}/M_1 M_2 c_{p\mathrm{w}}(T_\mathrm{w} - T_\infty)$$

is the Dufour criterion; it characterizes the magnitude of diffusion heat conduction with respect to convective heat transfer in a gas mixture.

$$\mathrm{Ki} = \sigma l T_\infty^3/\lambda$$

is the Kirpichev criterion; it is the measure of the ratio of heat transfer by radiation to a heat flux due to molecular heat conduction.

$$\mathrm{R} = q_\mathrm{w}/\sigma\varepsilon T_\infty^4$$

is the criterion of radiation heat transfer; it characterizes the relationship between the heat flux from the wall and the heat flux due to radiation.

$$\mathrm{S} = 2\lambda/\left(r_0 \varepsilon \sigma T_\infty^3\right)$$

is the criterion of radiation heat transfer on a curved surface. It can be considered as a ratio of two length scales: the radiative one $2\lambda/\varepsilon\sigma T_\infty^3$ and of the cylinder radius r_0.

$$\mathrm{Re}_l = u_0 l/\nu$$

is the Reynolds criterion; it is the main criterion of hydrodynamic similarity in forced convection. It is the measure of the ratio of inertia force to molecular friction force in the flow.

$$\mathrm{Pe}_l = \mathrm{Re}_l \mathrm{Pr}$$

is the Peclet criterion; it is the measure of the ratio of the intensities of convective and molecular heat transfers in moving fluids.

$$\mathrm{Re}_\mathrm{w} = v_0 l/\nu$$

is the Reynolds criterion at injection or suction on permeable body surface.

$$\mathrm{Pe}_\mathrm{w} = \mathrm{Re}_\mathrm{w} \mathrm{Pr}$$

is the Peclet criterion on a permeable body surface.

$$\mathrm{Gr}_D = g\beta_\mathrm{w} l^3 \Delta c/\nu^2$$

is the diffusion Grashof criterion. It characterizes diffusion mass transfer due to the difference of the component concentrations on a wall and in the surrounding medium.

References

1. G. K. Batchelor, *An Introduction to Fluid Dynamics*, Cambridge Univ. Press, London (1970).
2. O. G. Martynenko, A. A. Berezovsky, and Yu. A. Sokovishin, *Asymptotic Methods in the Theory of Free-Convective Heat and Mass Transfer*, Nauka i Tekhnika Press, Minsk (1979).
3. D. D. Joseph, *Stability of Fluid Motions*, Vol. 2, Heidelberk, Berlin (1976).
4. G. Z. Gershuni and E. M. Zhuhovitsky, *Convective Stability of Noncompressible Liquids*, Nauka Press, Moscow (1972).
5. A. Oberbeck, *Ann. Phys. Chem.*, **7**, 271 (1879).
6. I. G. Shaposhnikov, *Uch. Zap. Permsk. Univ.*, **8**, 81 (1954).
7. F. H. Busse, *Rep. Progr. Phys.*, **41**, 1929 (1978).
8. V. V. Alekseev, N. S. Blokhina, A. M. Gusev, and E. K. Zhdanova, *Free Convection in the Athmosphere and Ocean*, MGU Press, Moscow (1979).
9. B. Gebhart, *Trans. ASME, J. Fluid Eng.*, **101**, 5 (1979).
10. A. A. Gukhman, *Introduction to Similarity Theory*, Vysshaya shkola Press, Moscow (1973).

11. L. I. Sedov, *Similarity and Dimensional Methods in Mechanics*, Nauka Press, Moscow (1977).
12. G. Greber, S. Erk, and U.Grigull, *Heat Transfer Fundamentals*, I. L. Press, Moscow (1958). (H. Gröber und S. Erk, Die Grundgesetze der Wärmeübertragung. 3. Völlig neu bearb. Aufl. von U. Grigull. Berlin u.a., Springer-Verl., 1955).
13. W. Nusselt, *Gesundh.-Ing.*, **38**, 477 (1915); **38**, 490 (1915).
14. A. A. Gukhman, *Trudy Len. Fiz.-Tekhn. Lab.*, No. 4, 43 (1926).
15. W. Okoto-Kutar, *Zesk. Nauk Politehn. Slaskiej*, No. 19, 35 (1959).
16. M. Jacob, *Forsch. Geb. Ing.*, **2**, 380 (1931).
17. G. I. Barenblatt, *Similarity, Self-Similarity, and Asymptotics. Theory and Applications in Geophysics and Hydrodynamics*, Gidrometeoizdat Press, Leningrad (1978).
18. H. E. Huntley, *Dimensional Analysis*, Dover Publ., Inc., New York (1967).
19. K. Chida and I. Katto, *Int. J. Heat Mass Transfer*, **19**, 453 (1976).
20. E. F. Adiutory, *The New Heat Transfer*, The Ventuno Press (1974).
21. O. G. Martynenko and Yu. A. Sokovishin, in: *Energy Conservation in Heating, Cooling and Ventilating of Buildings*, Vol. 1 (1978), pp. 231–234.
22. H. B. Squire, in: S. Goldstein (ed.), *Modern Developments in Fluid Dynamics*, Oxford Univ. Press, New York (1938).
23. Y. Jaluria, *Natural Convection Heat and Mass Transfer*, Pergamon, Oxford, U. K., (1980).
24. O. G. Martynenko, and Yu. A. Sokovishin, *Free-Convective Heat Transfer. Handbook*, Nauka i Tekhnika Press, Minsk (1982).
25. A. A. Zhukauskas and O. G. Martynenko (eds.), *Advances in Heat Transfer*, Mokslas Press, Vilnius (1989).
26. E. Schmidt and W. Beckmann, *Tech. Mech. Thermodyn.*, **1**, 341 (1930).
27. H. K. Kuiken, *J. Eng. Math.*, **2**, 95 (1968).
28. B. Gebhart, Y. Jaluria, R. L. Manajan, and B. Sammakia, *Buoyancy-Induced Flows and Transport*, Hemisphere Publ. Corp., Washington (1988).
29. J. D. Hellums and S. W. Churchill, in: *Chem. Eng. Progr. Symp. Ser.*, Vol. 57, No. 32 (1961), pp. 75–80.
30. O. G. Martynenko and Yu. A. Sokovishin, *Introduction to the Theory of Free-Convective Heat Transfer*, LGU Press, Leningrad (1982).
31. O. G. Martynenko and Yu. A. Sokovishin, *Free-Convective Heat Transfer on a Vertical Plate*, Nauka i Tekhnika Press, Minsk (1977).
32. A. Bejan, *Convection Heat Transfer*, A Wiley-Intersci. Publ., New York (1984).
33. H. B. Squire, in: S. Goldstein (ed.), *Modern Developments in Fluid Dynamics*, Vol. 2, Dover, New York (1965), pp. 641–643.
34. B. A. Tonkoshkurov and A. S. Asaturyan, *Tr. NII transp. i khran. neft. i nefteprod.*, No. 1, Moscow (1961), pp. 22–28.
35. E. R. G. Eckert and E. Soehngen, *Proc. Gen. Discuss. Heat Transfer*, London (1951), p. 21.
36. T. Inagaki and K. Komori, *Int. J. Heat Mass Transfer*, **38**, 3485 (1995)
37. *JSME Data Book: Heat Transfer*, 4th edn., Maruzen, Tokyo (1991).
38. G. C. Vliet and C. K. Liu, *Trans. ASME, Ser. C, J. Heat Transfer*, **91**, 517 (1965).
39. M. J. M. Krane and B. Gebhart, *Int. J. Heat Mass Transfer*, **36**, 977 (1993).
40. R. K. Tripathi, and G. Nath, *Int. J. Heat Mass Transfer*, **36**, 1125 (1993).

41. M. Jannot and T. Kunc, *Int. J. Heat Mass Transfer*, **41**, 4327 (1998).
42. W. Xu, Q. Chen, and F. T. M. Nieuwstadt, *Int. J. Heat Mass Transfer*, **41**, 3161 (1998).
43. R. J. Adrian, *Int. J. Heat Mass Transfer*, **39**, 2303 (1996).
44. C. P. Knowles and B. Gebhart, *J. Fluid Mech.*, **34**, 657 (1968).
45. P. S. Klebanoff, K. D. Tidstrom, and L. M. Sargent, *J. Fluid Mech.*, **12**, 1 (1962).
46. T. Audunson and B. Gebhart, *Int. J. Heat Mass Transfer*, **19**, 737 (1976).
47. Y. Jaluria, S. Saric, and A. A. Szewczyk, *ASME Heat Transfer Div.*, **54**, 1 (1985).
48. Y. Jaluria, *Int. J. Heat Mass Transfer*, **19**, 1057 (1976).
49. Y. Jaluria and B. Gebhart, *J. Fluid Mech.*, **61**, 337 (1973).
50. Y. Jaluria and B. Gebhart, *J. Fluid Mech.*, **66**, 309 (1974).
51. Y. Jaluria and B. Gebhart, *Int. J. Heat Mass Transfer*, **20**, 434 (1977).
52. F. Godaux and B. Gebhart, *Int. J. Heat Mass Transfer*, **17**, 93 (1974).
53. Z. M. Qureshi and B. Gebhart, *Int. J. Heat Mass Transfer*, **21**, 1967 (1978).
54. R. Chessewright, *Trans. ASME, Ser. C, J. Heat Transfer*, **90**, 1 (1968).
55. G. M. Regnier and C. Kaplan, *Proc. Heat Transfer Fluid Mech. Inst.*, Stanford Univ. Press, Stanford, Calif. (1963).
56. C. Y. Warner and S. S. Apraci, *Int. J. Heat Mass Transfer*, **11**, 397 (1968).
57. R. L. Mahajan and B. Gebhart, *J. Fluid Mech.*, **91**, 131 (1979).
58. G. C. Vliet, and C. K. Liu, *Trans. ASME, Ser. C, J. Heat Transfer*, **91**, 517 (1969).
59. D. D. Papailou and P. S. Lykoudis, *Int. J. Heat Mass Transfer*, **17**, 161 (1974).
60. R. G. Ir. Bill and B. Gebhart, *Int. J. Heat Mass Transfer*, **22**, 267 (1979).
61. J. L. Lumley, *J. Atmos. Sci.*, **21**, 99 (1965).
62. G. N. Shur, *AID Rept.* T-63-55, *Aerospace Information Division*, Library of Congress, Washington, D.C. (1962).
63. E. R. G. Eckert and T.W. Jackson, *NACA Tech. Note*, 2207 (1950).
64. H. Kato, *Int. J. Heat Mass Transfer*, **11**, 1117 (1968).
65. E. Griffiths and A. H. Davis, *DSIR Special Rept.* No. 9, British Food Investigation Board, London (1922).
66. O. A. Saunders, *Proc. Roy. Soc., (London), Ser. A*, **55**, 172 (1939).
67. G. C. Vliet and D. C. Ross, *Trans. ASME, Ser. C, J. Heat Transfer*, **97**, 549 (1975).
68. S. W. Churchill and H. H. S. Chu, *Int. J. Heat Mass Transfer*, **18**, 1323 (1975).
69. R. J. Goldstein, E. M. Sparrow, and D. C. Jones, *Int. J. Heat Mass Transfer*, **16**, 1025 (1973).
70. W. H. McAdams, *Heat Transmission*, 3rd edn., McGraw-Hill, New York (1957).
71. S. W. Churchill, in: E. Schlunder (ed.), *Heat Exchanger Design Handbook*, Pt. 2, Hemisphere, Washington, D.C. (1983).
72. G. Schutz, *Int. J. Heat Mass Transfer*, **6**, 873 (1963).
73. S. S. Kutateladze, *Principles of the Theory of Heat Transfer*, Atomizdat Press Moscow (1979).
74. J. C. Mollendorf and B. Gebhart, *Trans. ASME, Ser. C, J. Heat Transfer*, **92**, 628 (1970).
75. R. C. Gunness and B. Gebhart, *Phys. Fluids*, **12**, 1968 (1969).
76. B. Jhaveri and G. M. Homsy, *J. Fluid Mech.*, **98**, 39 (1980).
77. B. Jhaveri and G. M. Homsy, *J. Fluid Mech.*, **114**, 251 (1982).

78. F. J. Bayley, *Proc. Inst. Mech. Engrs.*, **169**, 361 (1955).
79. A. J. Ede, *Advances in Free Convection – Advances in Heat Transfer*, **4**, (1967).
80. O. A. Saunders, *Proc. Roy. Soc., (London), Ser. A*, **157**, 278 (1936).
81. G. S. H. Lock and F. J. D. Trotter, *Int. J. Heat Mass Transfer*, **11**, 1225 (1968).
82. R. G. Jr. Bill and B. Gebhart, *Int. J. Heat Mass Transfer*, **18**, 513 (1975).
83. R. J. Formstrom and E. M. Sparrow, *Int. J. Heat Mass Transfer*, **10**, 321 (1967).
84. J. Cristic, J. G. Burns, and M. A. S. Ross, in: *Electrooptic Systems in Flow Measurements*, Univ. of Southampton, 25–26 Sept., (1972).
85. J. Yeh and H. Cummins, *Appl. Phys. Lett.*, **4**, 176 (1964).
86. T. Durrany and K. Greyted, *Laser Systems in Hydrodynamic Measurements*, Energia Press, Moscow (1980).
87. P. Bradshaw, *An Introduction to Turbulence and its Measurement,* Pergamon Press (1971).
88. J. O. Hinze, *Turbulence. An Introduction to Its Mechanism and Theory*, McGraw-Hill, New York (1959).
89. P. J. Bourke and D. J. Pulling, *Int. J. Heat Mass Transfer*, **13**, 1331 (1970).
90. E. R.G. Eckert and R. M. Drake, *Analysis of Heat and Mass Transfer*, McGraw-Hill, New York (1972).
91. I. A. Vatutin, V. F. Vinokurov, O. G. Martynenko, P. P. Khramtsov, and I. A. Shikh, *Exper. Heat Transfer*, **6**, 69 (1993).
92. B. Gebhart and C. P. Knowles, *Rev. Sci. Instr.*, **37**, 12 (1966).
93. B. Gebhart and C. P. Knowles, *Prog. Heat Mass Transfer*, **2**, Pergamon Press (1969).
94. B. Gebhart, *Ann. Rev. Fluid Mech.*, **5**, 213 (1973).
95. R. P. Dring and B. Gebhart, *J. Fluid Mech.*, **36**, 447 (1969).
96. L. M. Levin, *Izv. Akad. Nauk SSSR, Ser. Fiz.*, **4**, 111 (1940).
97. A. A. Vasil'ev, *Shadow Methods*, Nauka Press, Moscow (1964).
98. H. G. Caulfield, *Handbook of Optical Holography*, Vol. **2**, Mir Press, Moscow (1982.)
99. A. Toepler, *Beobachtungen nach einer neuen optischen Methode*, Bonn (1864).
100. M. I. Skotnikov, *Shadow Quantitative Methods in Gas Dynamics*, Nauka Press, Moscow (1976).
101. G. D. Salamandra, *Photographic Investigation Methods High-Rate Developing Processes*, Nauka Press, Moscow (1974).
102. G. A. Ostroumov, *Free Convection under the Condition of Internal Problem*, Gostekhizdat Press, Moscow–Leningrad (1952), pp. 159–170.
103. M. S. Vasudevan and P. E. Janson, *Optical Methods of Measuring Turbulence*, Sauthampton Univ., A.A.S.U. Rep. (1969).
104. W. Merzkirch, *Flow Visualization*, 2nd edn., Acad. Press, Orlando (1987).
105. H. Hausen, *Allgem. Warmetechnik*, **9**, 75 (1959).
106. S. W. Churchill and R. Usagi, *IEC Fund.*, **13**, 39 (1974).
107. I. Langmuir, *Phys. Rev.*, **32**, 401 (1912).
108. W. Elenbaas, *Physica*, **9**, 1 (1942).
109. I. N. Mikheeva, in: *Heat Transfer and Heat Simulation*, Moscow (1959), pp. 226–238.
110. G. Turner, *Buoyancy Effect in Liquids*, Mir Press, Moscow (1977).
111. H. Borner, *VDI-Forschungsheft*, No. 512, 1 (1965).

2

Free Convection on a Plane

Laminar convection of an incompressible fluid on a plane surface inclined at an angle ϑ to a vertical under stationary conditions is described by the equations

$$
\begin{gathered}
u\frac{\partial u}{\partial x}+\nu\frac{\partial u}{\partial y}=g\beta\theta\cos\vartheta-\frac{1}{\rho}\frac{\partial p}{\partial x}+\nu\left(\frac{\partial^2 u}{\partial x^2}+\frac{\partial^2 u}{\partial y^2}\right),\\
u\frac{\partial \nu}{\partial x}+\nu\frac{\partial \nu}{\partial y}=g\beta\theta\sin\vartheta-\frac{1}{\rho}\frac{\partial p}{\partial x}+\nu\left(\frac{\partial^2 \nu}{\partial x^2}+\frac{\partial^2 \nu}{\partial y^2}\right),\\
\frac{\partial u}{\partial x}+\frac{\partial \nu}{\partial y}=0\,,
\end{gathered}
\tag{2.1}
$$

$$
\begin{aligned}
u\frac{\partial T}{\partial x}+\nu\frac{\partial T}{\partial y}=a\left(\frac{\partial^2 T}{\partial x^2}+\frac{\partial^2 T}{\partial y^2}\right)+\frac{1}{\rho c_p}\left(u\frac{\partial p}{\partial x}+\nu\frac{\partial p}{\partial y}\right)\\
+\frac{Q_\nu}{\rho c_p}+\frac{\nu}{c_p}\left[\left(\frac{\partial u}{\partial x}\right)^2+\left(\frac{\partial v}{\partial y}\right)^2+\left(\frac{\partial u}{\partial y}+\frac{\partial \nu}{\partial x}\right)^2\right]
\end{aligned}
$$

with the appropriate boundary conditions.

Free-convective heat transfer on a vertical plane surface. For mathematical description of a two-dimension free-convective flow near a vertical wall in a stationary regime with constant physical properties of the medium investigated the system of Navier–Stokes and energy balance equations is used:

$$\frac{\partial u}{\partial x}+\frac{\partial u}{\partial y}=0\,, \tag{2.2}$$

$$\rho\left(u\frac{\partial u}{\partial x}+\nu\frac{\partial u}{\partial y}\right)=-\frac{\partial p}{\partial x}+\mu\nabla^2 u-\rho g\,, \tag{2.3}$$

$$\rho\left(u\frac{\partial \nu}{\partial x}+\nu\frac{\partial \nu}{\partial y}\right)=-\frac{\partial p}{\partial y}+\mu\nabla^2 \nu\,, \tag{2.4}$$

$$\rho c_p\left(u\frac{\partial T}{\partial x}+\nu\frac{\partial T}{\partial y}\right)=\lambda\nabla^2 T\,. \tag{2.5}$$

Equations (2.2)–(2.5) can be substantially simplified if we limit ourselves to consideration of the boundary-layer region ($y \sim \delta_T$, $x \sim L$), when the wall-averaged heat transfer coefficient equal to $h = k/\delta_T$ and the thermal boundary-layer thickness δ_T are negligibly small compared with the wall height L ($\delta_T \ll L$) [2.1]. In particular, from (2.2) it follows that the pressure in the boundary layer depends only on the longitudinal coordinate

$$\frac{\partial P}{\partial x} = \frac{dP}{dx} = \frac{dP_\infty}{\partial x} , \tag{2.6}$$

and (2.4) and (2.5) take the form

$$\rho\left(u\frac{\partial u}{\partial x} + \nu\frac{\partial u}{\partial y}\right) = -\frac{dP_\infty}{dx} + \mu\frac{\partial^2 u}{\partial y^2} - \rho g , \tag{2.7}$$

$$\rho c_p\left(u\frac{\partial \nu}{\partial x} + \nu\frac{\partial \nu}{\partial y}\right) = \lambda\frac{\partial^2 T}{\partial y^2} . \tag{2.8}$$

Further, as the hydrostatic pressure gradient $dP_\infty/dx = -\rho_\infty g$, (2.7) is transformed to

$$\rho\left(u\frac{\partial u}{\partial x} + \nu\frac{\partial u}{\partial y}\right) = \mu\frac{\partial^2 u}{\partial y^2} + (\rho_\infty - \rho)g . \tag{2.9}$$

Equations (2.8) and (2.9) are closed by the equation of the state of the medium (for an ideal gas it is $P = \rho RT$).

In the Boussinesq approximation (2.9) is

$$u\frac{\partial u}{\partial x} + \nu\frac{\partial u}{\partial y} = \nu\frac{\partial^2 u}{\partial y^2} + g\beta(T_\infty - T) . \tag{2.10}$$

For complete description of the vertical free-convective boundary layer, (2.2), (2.8), and (2.10) should be supplemented with the boundary conditions:

(a) ***constant wall temperature***

$$\begin{aligned} &u(y,0) = \nu(y,0) = \nu(y, x \to \infty) = 0 , \\ &T(y,0) = T_0, \quad T(y, x \to \infty) = T_\infty , \end{aligned} \tag{2.11}$$

(b) ***constant heat flux density on a wall***

$$\begin{aligned} &u(y,0) = \nu(y,0) = \nu(y, x \to \infty) = 0 , \\ &-\lambda\frac{\partial T}{\partial x}(y,0) = q_0, \; T(y, x \to \infty) = T_\infty . \end{aligned} \tag{2.12}$$

Figure 2.1 presents the structure of the flow, of the thermal boundary layer, and also the results of flow visualization by a laser knife method on a vertical surface: (a) ***for a constant wall temperature***; and (b) ***for a constant heat flux*** on a wall.

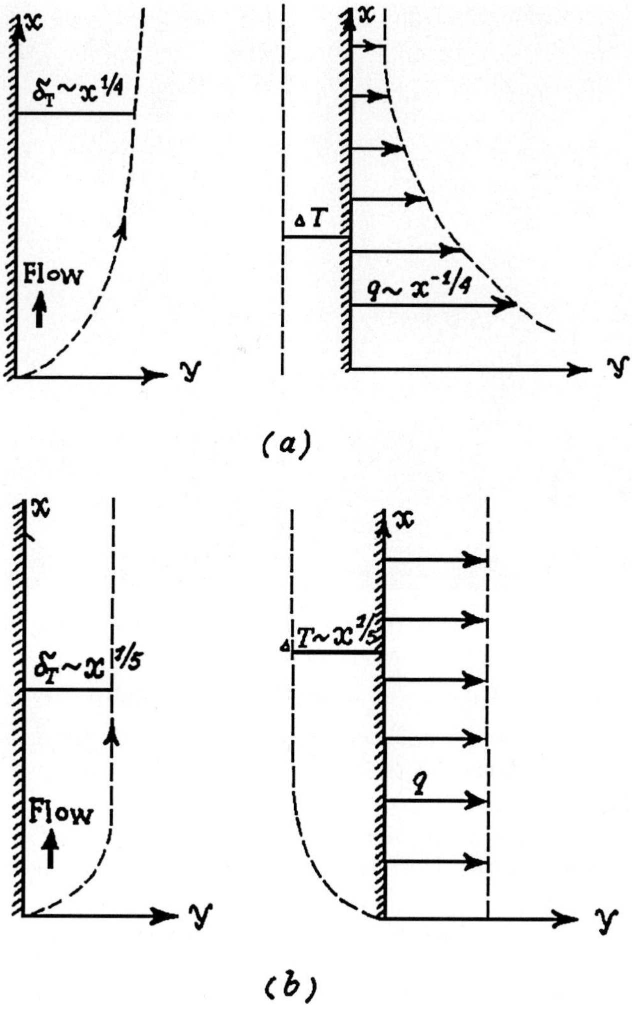

Fig. 2.1. Structure of a free-convective flow and a thermal boundary layer on a vertical surface (**a**) at a constant wall temperature and (**b**) for a constant heat flux on a wall

The flow velocity profile in free convection on a vertical surface can be visualized with the aid of various dying substances. Figure 2.2 shows the results of velocity profile visualization on a vertical plate ***of constant temperature*** immersed into a weak water solution of **NaCl**. The velocity profile was visualized due to the evolvement of $\mathbf{PbCl_2}$ on a lead wire stretched perpendicularly to the plate surface when an electric current was passed through it. The $\mathbf{PbCl_2}$ evolved was nonuniformly swept transferred by the flow in the boundary layer of the plate, with the displacement of the dying particles being

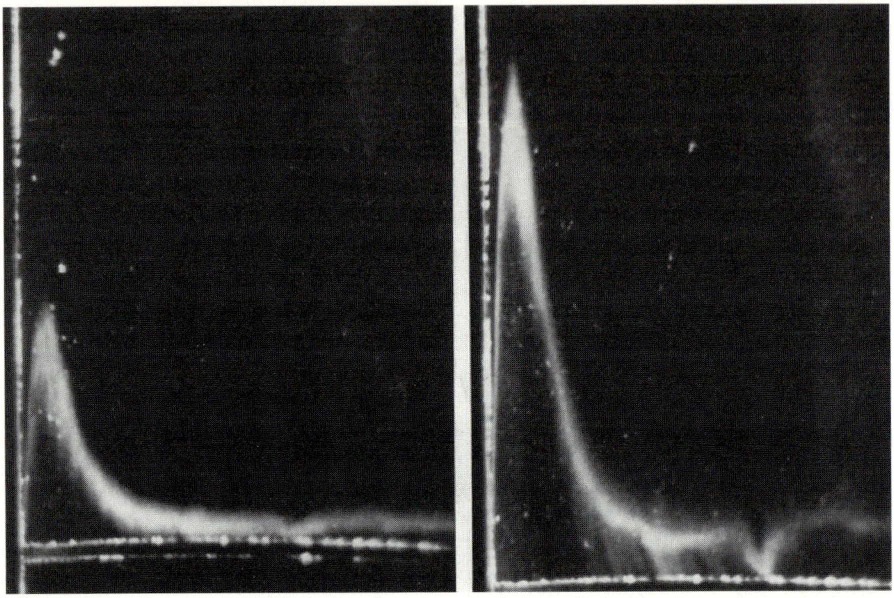

Fig. 2.2. Velocity profiles on a vertical plate of constant temperature, photographed 2 sec (**a**) and 4 sec (**b**) after the passage of an electric pulse

directly proportional to the flow velocity and time period from the moment of electric pulse passage through the wire. The dying substance was illuminated by a vertical laser knife.

Figure 2.3a presents an interferogram of a free-convective flow on a vertical surface ***with a constant wall temperature***. In this case, the interference lines are isotherms.

Based on the results of calculation of this interferogram the two-dimension temperature distribution in the boundary layer is constructed (Fig 2.3b).

The interference patterns of the process of development of a free-convective flow ***in air*** on a vertical plate ***at a constant heat flux on a wall***, $q_w = 9.1 \cdot 10^5$ W/m^2, are presented in Fig. 2.4. The distribution of the temperature in the boundary layer near the plate is presented in Fig. 2.5 for the time moment $t = 9$ sec from the beginning of plate heating.

The temperature distribution is obtained on the basis of calculation of the interferogram (Fig. 2.4). The flow near the plate kept its stability up to the wall temperature $T_w = 400$°C. Figure 2.6a gives the results of visualization of the flow described above.

The idea about the general picture of free-convective flow can be obtained as a result of visualization of streamlines. In the case of a vertical plate, the track method gives good results. The pattern of the streamlines of a free-convective flow on a vertical plate of constant temperature obtained by the laser knife method is given in Fig. 2.6b. Chalk particles suspended in water

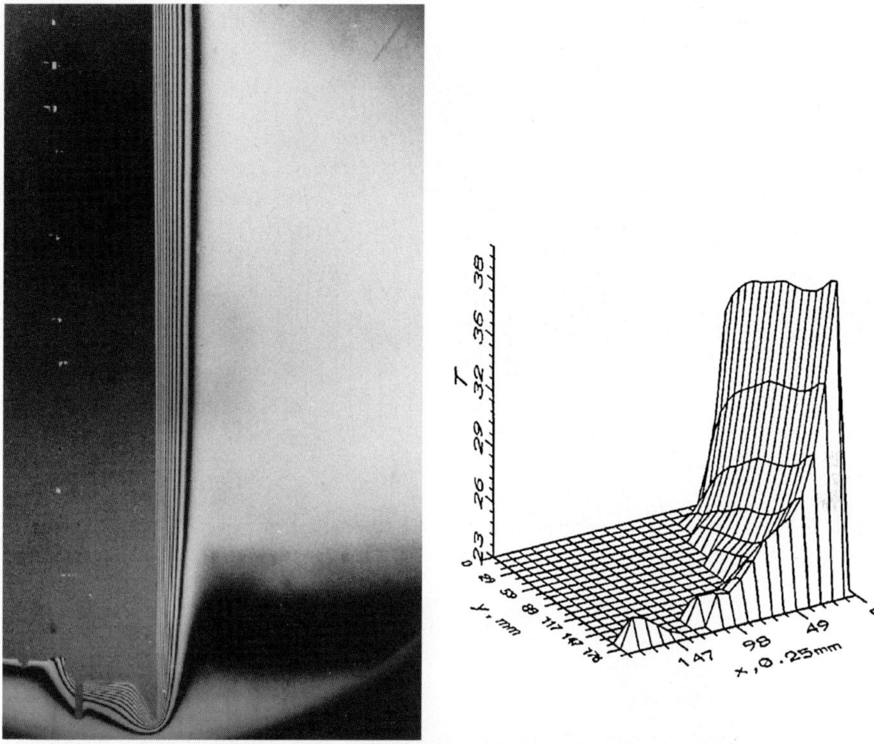

Fig. 2.3. Interferogram of a free-convective flow on a vertical surface with a constant wall temperature (**a**) and temperature distribution in a boundary layer near a plate (**b**)

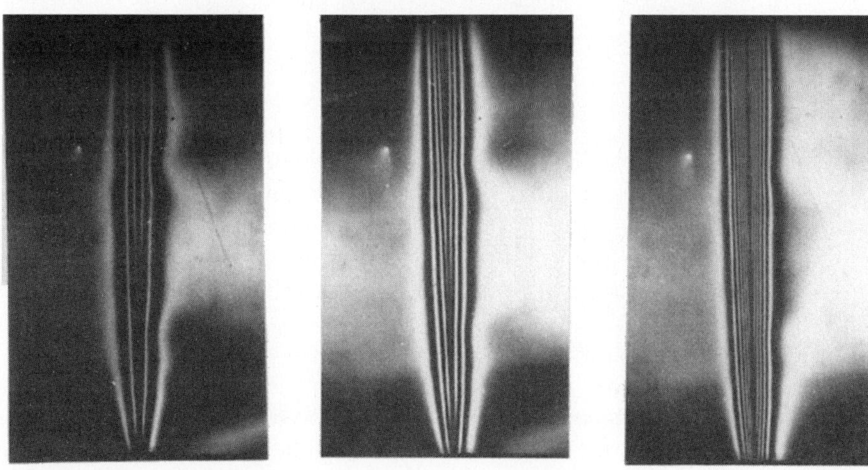

Fig. 2.4. Interferograms of a free-convective flow on a vertical plate with a constant heat flux obtained 3, 4.5, 9 sec after the beginning of plate heating

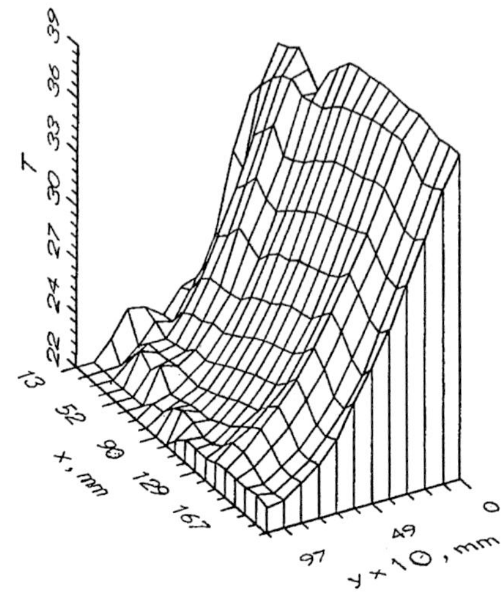

Fig. 2.5. The temperature distribution in a boundary layer of a vertical plate with a constant heat flux

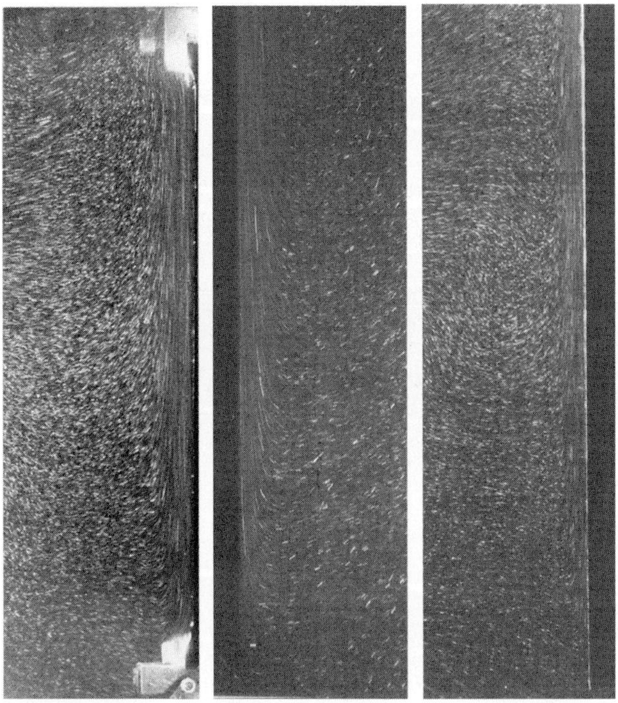

a b c

Fig. 2.6. Track visualization of a free-convective flow on a vertical plate: (**a**) in air for a constant heat flux, (**b**) in water at a constant temperature, (**c**) in air at a constant temperature

were used as a visualizing substance. The results of track visualization of free-convective motion of ***air*** near a vertical plate ***with a constant wall temperature*** are presented in Fig. 2.6c.

The numerical solution of the problem of laminar free convection ***of a non-Newtonian fluid*** on a vertical plate for the cases of a ***constant temperature and a heat flux*** [2.2] shows that the most essential factor which exerts its influence on heat transfer is the Pr number. As the modified Pr number increases, the heat transfer rate increases; the effect of the parameter n is less noticeable. The results of calculations of the mean Nu number for the modified Pr in the range from 1 to 2000 and of the parameter n from 0.5 to 1.5 are presented in Tables 2.1 and 2.2.

Table 2.1. Isothermal Plate

N	$\mathrm{Nu}/\mathrm{Ra}^{1/(3n+1)}$
0.5	0.6105
1.0	0.6701
1.5	0.7012

Table 2.2. Plate with a Constant Heat Flux

	$\mathrm{Nu}/\mathrm{Ra}^{*1/(3n+2)}$		
N	$\mathrm{Pr}_x^* = 2000$	$\mathrm{Pr}_x^* = 100$	$\mathrm{Pr}_x^* = 10$
0.5	0.7381	0.7336	0.6907
1	0.7883	0.7751	0.7445
1.5	0.8213	0.8058	0.7790

The modified Rayleigh and Prandtl numbers are determined from the formulas

for a constant surface temperature

$$\mathrm{Ra}_x = \rho g \beta \left(T_\mathrm{w} - T_\infty\right) x^{2n+1} / \left(K \alpha^n\right) ,$$

$$\mathrm{Pr}_x = \frac{1}{\alpha} \left(\frac{K}{\rho}\right)^{3/(n+1)} x^{(n-1)/2(n+1)} \left[g\beta \left(T_\mathrm{w} - T_\infty\right)\right]^{3(n-1)/2(n+1)} ,$$

for the constant heat flux density

$$\mathrm{Ra}_x^* = \rho g \beta q_\mathrm{w} x^{2(n+1)} / \left(K \alpha^n k\right) ,$$

$$\mathrm{Pr}_x^* = \frac{1}{\alpha} \left(\frac{K}{\rho}\right)^{5/(n+4)} x^{2(n-1)/(n+4)} \left[g\beta q_\mathrm{w}/k\right]^{3(n-1)/(n+4)} ,$$

where k and α are the thermal conductivity and thermal diffusivity of the fluid, and K is the proportionality coefficient in the power law; it depends on the species of the non-Newtonian fluid.

2.1 Vertical Flat Plate with a Constant Wall Temperature

From the analysis of numerous results and experimental data McAdams [2.3] suggested the general formula to calculate the length-averaged Nusselt number:

$$\mathrm{Nu}_L = C\mathrm{Ra}_L^m , \tag{2.1.1}$$

where C and m are the functions of the Rayleigh number Ra_L. For a laminar regime ($10^4 < \mathrm{Ra}_L < 10^9$) McAdams recommends the value $m = 1/4$, for a turbulent one ($10^9 < \mathrm{Ra}_L$), $m = 1/3$.

An increase in the Gr number, at a certain height from the bottom edge of the plate the motion becomes unstable and wavy. The criteria of turbulence occurrence are the Grashof and Rayleigh numbers. On a plane surface the critical value of the Ra_k number for various fluids changes in a wide range – from 10^7 to 10^{13} [2.4].

From the dimension analysis it follows that in a developed turbulent regime on the heat transfer surface the heat transfer coefficient ceases to depend on the linear dimension. The general equation

$$\mathrm{Nu} = \mathrm{Gr}^{1/3} f\,(\mathrm{Pr}) \tag{2.1.2}$$

is confirmed by experiments for arbitrarily shaped bodies. In a developed turbulent flow the effect of viscosity is insignificant.

If in the near-wall area there is turbulent heat transfer, we can neglect the molecular heat conduction. The resulting heat transfer law [2.5]

$$\mathrm{Nu} = C(\mathrm{Gr}\mathrm{Pr}^2)^{1/2} \tag{2.1.3}$$

is the limiting dependence.

In the case of turbulent natural convection near the heat transfer surface there exists a quasi-stable layer whose thickness is determined by the hydrodynamical stability parameter, when the gravity vector is parallel to the heat transfer surface, and by the thermal stability parameter, when the gravity vector is normal to this surface [2.6].

For a vertical free-convective boundary layer ***at a constant surface temperature*** the local Nusselt number can be written as [2.7]

$$\mathrm{Nu}(x) = F(\mathrm{Pr})(\mathrm{Gr}_x)^{1/4} \tag{2.1.4}$$

or

$$\mathrm{Nu}\,(x) = F_1\,(\mathrm{Pr})\,\mathrm{Ra}_x^{1/4}\,. \tag{2.1.5}$$

Churchill and Chu [2.8], on the basis of analysis of a large number of works, suggested the following correlations:

$0 < \mathrm{Ra}_L < 10^9$

$$\mathrm{Nu}_L = 0.68 + 0.67\ \mathrm{Ra}_L^{1/4}\left[1 + \left(\frac{0.492}{\mathrm{Pr}}\right)^{9/16}\right]^{-4/9}\,, \tag{2.1.6}$$

$\mathrm{Ra}_L > 10^9$

$$\mathrm{Nu}_L^{1/2} = 0.825 + 0.387\,\mathrm{Ra}_L^{1/6}\left[1 + \left(\frac{0.492}{\mathrm{Pr}}\right)^{9/16}\right]^{-8/27}\,. \tag{2.1.7}$$

Formulas (2.1.6) and (2.1.7) are valid for all Pr ($0 < \mathrm{Pr} < \infty$). Actually, the second of the two equations can be applied to all Ra_L values, however, the use of the first equation provides greater accuracy. The presence of the addend 0.68 allows one to obtain the non-zero Nusselt number for very low Ra_L, when the boundary-layer approximation is inapplicable, and the convective heat transfer becomes prevailing compared with the heat conduction in the surrounding medium.

A detailed analysis of various methods for solving the problems of flow and heat transfer near a vertical isothermal surface can be found in [2.7] and [2.18].

One of the first attempts to solve the differential equations for the given boundary conditions belongs to Oberbeck [2.9]. The further development in solving the equations was made by Lorenz [2.10], who obtained the following expression for the average Nusselt number:

$$\mathrm{Nu}_L = 0.548\,(\mathrm{Gr}_L\mathrm{Pr})^{1/4}\,, \tag{2.1.8}$$

where the constant 0.548 refers ***to air.*** The value of the constant ***for oil,*** equal to 0.555, was also obtained by Lorenz [2.10], and ***for mercury***, equal to 0.33, by Saunders [2.11]. The solution obtained by Schmidt and Beckmann by the similarity method [2.12] yields the following dimensionless equations:

$$\mathrm{Nu}(x) = 0.39\,(\mathrm{Gr}_x\mathrm{Pr})^{1/4}\,, \tag{2.1.9}$$

$$\mathrm{Nu}_L = 0.52\,(\mathrm{Gr}_L\mathrm{Pr})^{1/4}\,. \tag{2.1.10}$$

Using approximated methods, Saunders [2.11] and Schuh [2.13] obtained solutions for various Prandtl numbers. Ostrach [2.14], using a numerical method, obtained the following dependence for the local number Nu:

$$\mathrm{Nu}_x = f(\mathrm{Pr})\left(\frac{\mathrm{Gr}_x}{4}\right)^{1/4} \tag{2.1.11}$$

in the range of Prandtl numbers from 0.01 to 1000. Schuh [2.15] considered the case of large Pr numbers characteristic ***for oils***, and Sparrow and Gragg [2.16] obtained solutions for very small Pr numbers. Le Fevre [2.17] investigated the limiting cases $\mathrm{Pr} \to 0$ and $\mathrm{Pr} \to \infty$:

$$\mathrm{Nu}_L = 0.8\mathrm{Gr}_L^{1/4}\mathrm{Pr}^{1/2}\,, \quad \text{for } \mathrm{Pr} \to 0\,, \tag{2.1.12}$$

$$\mathrm{Nu}_L = 0.67\mathrm{Gr}_L^{1/4}\mathrm{Pr}^{1/4} \quad \text{for } \mathrm{Pr} \to \infty\,. \tag{2.1.13}$$

If we write the average Nusselt number as [2.18]

$$\mathrm{Nu}_L = \frac{4}{3} H\left(\mathrm{Pr}\right) \mathrm{Gr}_L^{1/4}\,, \tag{2.1.14}$$

then the values of $H(\mathrm{Pr})$ can be obtained from the numerical solution of the basic differential equations. Le Fevre also suggested a general form of empirical dependence for $H(\mathrm{Pr})$:

$$H\left(\mathrm{Pr}\right) = \frac{3}{4}\left[\frac{\mathrm{Pr}}{2.43478 + 4.884\,\mathrm{Pr}^{1/2} + 4.95283\,\mathrm{Pr}}\right]^{1/4}\,. \tag{2.1.15}$$

The integral method developed by Karman was first applied to the problems of natural convection by Squire [2.19], and ***in the case of an isothermal vertical surface*** he obtained the following dependences for the average and local Nusselt numbers:

$$\begin{aligned} \mathrm{Nu}\left(x\right) &= 0.508\,\mathrm{Pr}^{1/2}\left(0.952 + \mathrm{Pr}\right)^{-1/4}\mathrm{Gr}_x^{1/4}\,, \\ \mathrm{Nu}_L &= 0.68\,\mathrm{Pr}^{1/2}\left(0.952 + \mathrm{Pr}\right)^{-1/4}\mathrm{Gr}_L^{1/4}\,. \end{aligned} \tag{2.1.16}$$

For air this formula yields [2.18]

$$\mathrm{Nu}_L = 0.55\left(\mathrm{Gr}_L\,\mathrm{Pr}\right)^{1/4}\,. \tag{2.1.17}$$

Ede [2.20] analyzed the well-known experimental data in the range of *Ra* from 10^6 to 10^9 and showed, that they are rather well described by the generalized formula for the local Nusselt number:

$$\mathrm{Nu}_x = \frac{3}{4}\left[\frac{2\,\mathrm{Pr}}{5\left(1 + 2\,\mathrm{Pr}^{1/2} + 2\,\mathrm{Pr}\right)}\right]^{1/4}\left(\mathrm{Gr}_x\,\mathrm{Pr}\right)^{1/4}\,. \tag{2.1.18}$$

Based on the analysis of various experimental data, Churchill and Usagi [2.21] considered the limiting cases $\mathrm{Pr} \to 0$ and $\mathrm{Pr} \to \infty$ and suggested the following correlation:

$$\mathrm{Nu}_x = \frac{0.503\,\mathrm{Ra}_x^{1/4}}{\left[1 + \left(0.492/\,\mathrm{Pr}\right)^{9/16}\right]^{4/9}}\,, \tag{2.1.19}$$

which is valid for $0 < \Pr < \infty$ and $10^5 < \mathrm{Ra}_x < 10^9$. The average Nusselt number here is

$$\mathrm{Nu}_L = \frac{0.67\,\mathrm{Ra}_L^{1/4}}{\left[1 + (0.492/\Pr)^{9/16}\right]^{8/27}}\,. \tag{2.1.20}$$

A transition of the free-convective motion from laminar to turbulent occurs ***for air*** at $\mathrm{Ra}_k = 4\cdot 10^8$. The critical Rayleigh number increases with increase in Pr [2.4]. For instance, for $\Pr = 10^3–10^4 \mathrm{Ra}_k \cong 10^{13}$. The transient region in the case of free convection is wide enough, and this complicates exact determination of Ra_k.

For vertical plates in a urbulent regime ***with a constant wall temperature*** the following expressions obtained by solving the integral equations of motion and energy are recommended [2.22]:

$$\mathrm{Nu} = \begin{cases} 0.10(\mathrm{Gr}_x \Pr)^{1/2}, & \Pr = 0.72 \\ 0.060\mathrm{Gr}_x^{1/4}, & \Pr = 0.01 \end{cases}\,. \tag{2.1.21}$$

Proceeding from the integral equations and using the dimensional method, the analysis of different regimes of free-convective flow near ***an isothermal vertical plate due to the simultaneous action of temperature and concentration gradients*** was carried out in [2.23]. A wide range of parameters determining the development and interaction of the temperature and concentration boundary layers is considered. The range of the investigated parameters includes two asymptotic cases: (1) $1 \ll \Pr \ll \mathrm{Sc}$ (***solutions***) and (2) $\Pr \ll 1 \ll \mathrm{Sc}$ (***metal melts***).

The theoretical evaluations of various parameters that determine the regime of flow depending on the ratio between the buoyancy forces and the Le number, obtained on the basis of the dimensional analysis for the cases of ***thermal and diffusion*** free convection from a vertical surface ***in a porous medium saturated with a non-Newtonian liquid at constant temperature and concentration and also at constant heat and mass fluxes on the surface*** are given in [2.24]. A numerical solution is also obtained ***for an arbitrary in change temperature and concentration on the wall***. The results of the analytical investigation are presented in Tables 2.1.1 and 2.1.2 (n is an exponent index in the reological law of fluid flow, H is the wall height, G is the buoyancy ratio for constant temperature and concentration condition).

For a small temperature difference, when the thermophysical properties change insignificantly, the determining temperature T_{av} is introduced into the above-given dependences. ***For gases*** $T_{\mathrm{av}} = T_{\mathrm{w}} - 0.38(T_{\mathrm{w}} - T_\infty)$ in the range $T_{\mathrm{w}}/T_\infty = 0.5 - 3$; ***for mercury*** $T_{\mathrm{av}} = T_{\mathrm{w}} - 0.3(T_{\mathrm{w}} - T_\infty)$ [2.25]. In the range $0.25 \leq T_{\mathrm{w}}/T_\infty \leq 4.0$ the use of the determining temperature gives results differing only by 0.6% from those obtained for constant physical characteristics of the medium which correspond to the boundary layer temperature $T_{\mathrm{bl}} = (T_{\mathrm{w}} + T_\infty)/2$. This allows the use of the boundary layer temperature

Table 2.1.1. Scales of Various Quantities for Constant Temperature and Concentration on a Surface

Driving Potential	δ_T	δ_C	Nu	Sh
Temperature gradients, $\lvert G\rvert \ll 1$, Le $\gg 1$	$H\mathrm{Ra}^{-1/2n}$	$H\mathrm{Le}^{-1/2}(\mathrm{Ra})^{-1/2n}$	$\mathrm{Ra}^{1/2n}$	$\mathrm{Le}^{1/2}\mathrm{Ra}^{1/2n}$
Temperature gradients, $\lvert G\rvert \ll 1$, Le $\ll 1$	$H\mathrm{Ra}^{-1/2n}$	$H\mathrm{Le}^{-1}(\mathrm{Ra})^{-1/2n}$	$\mathrm{Ra}^{1/2n}$	$\mathrm{LeRa}^{1/2n}$
Concentration gradients, $\lvert G\rvert \gg 1$, Le $\ll 1$	$H(\mathrm{Ra}\lvert G\rvert)^{-1/2n}$	$H\mathrm{Le}^{-1/2}\times(\mathrm{Ra}\lvert G\rvert)^{-1/2n}$	$(\mathrm{Ra}\lvert G\rvert)^{1/2n}$	$\mathrm{Le}^{1/2}\times(\mathrm{Ra}\lvert G\rvert)^{1/2n}$
Concentration gradients, $\lvert G\rvert \gg 1$, Le $\gg 1$	$H\mathrm{Le}^{1/2}\times(\mathrm{Ra}\lvert G\rvert)^{-1/2n}$	$H\mathrm{Le}^{-1/2}\times(\mathrm{Ra}\lvert G\rvert)^{-1/2n}$	$\mathrm{Le}^{-1/2}\times(\mathrm{Ra}\lvert G\rvert)^{1/2n}$	$\mathrm{Le}^{1/2}\times(\mathrm{Ra}\lvert G\rvert)^{1/2n}$

Table 2.1.2. Scales of Various Quantities for Constant Heat and Mass Fluxes on a Surface

Driving Potential	δ_T	δ_C	Nu	Sh
Temperature gradients, $\lvert G^*\rvert \ll 1$, Le$\gg 1$	$H(\mathrm{Ra}^*)^{-1/(2n+1)}$	$H\mathrm{Le}^{-1/2}\times(\mathrm{Ra}^*)^{-1/(2n+1)}$	$(\mathrm{Ra}^*)^{1/(2n+1)}$	$\mathrm{Le}^{1/2}(\mathrm{Ra}^*)^{1/(2n+1)}$
Temperature gradients, $\lvert G^*\rvert \ll 1$, Le$\ll 1$	$H(\mathrm{Ra}^*)^{-1/(2n+1)}$	$H\mathrm{Le}^{-1}\times(\mathrm{Ra}^*)^{-1/(2n+1)}$	$(\mathrm{Ra}^*)^{1/(2n+1)}$	$\mathrm{Le}(\mathrm{Ra}^*)^{1/(2n+1)}$
Concentration gradients, $\lvert G^*\rvert \gg 1$, Le$\ll 1$	$H\mathrm{Le}^{1/2(2n+1)}\times(\mathrm{Ra}^*\lvert G^*\rvert)^{-1/(2n+1)}$	$H\mathrm{Le}^{-n/(2n+1)}\times(\mathrm{Ra}^*\lvert G^*\rvert)^{-1/(2n+1)}$	$\mathrm{Le}^{-1/2/(2n+1)}\times(\mathrm{Ra}^*\lvert G^*\rvert)^{-1/(2n+1)}$	$\mathrm{Le}^{n/(2n+1)}\times(\mathrm{Ra}^*\lvert G^*\rvert)^{-1/(2n+1)}$
Concentration gradients, $\lvert G^*\rvert \gg 1$, Le$\ll 1$	$H\mathrm{Le}^{(n+1)(2n+1)}\times(\mathrm{Ra}^*\lvert G^*\rvert)^{-1/(2n+1)}$	$H\mathrm{Le}^{-n/(2n+1)}\times(\mathrm{Ra}^*\lvert G^*\rvert)^{-1/(2n+1)}$	$\mathrm{Le}^{(n+1)(2n+1)}\times(\mathrm{Ra}^*\lvert G^*\rvert)^{-1/(2n+1)}$	$\mathrm{Le}^{n/(2n+1)}\times(\mathrm{Ra}^*\lvert G^*\rvert)^{-1/(2n+1)}$

in most cases, when the temperature difference in the flow is large. For gas flows a higher accuracy is achieved by using the relation for the determining temperature T_{av}.

Example 2.1. Determine the heat transfer coefficient for a vertical wall of height 1 m in air, if the average wall temperature is $t_{\mathrm{w}} = 80°\mathrm{C}$ and the air temperature far from the wall is $t_\infty = 20°\mathrm{C}$. The determining temperature is $t_{\mathrm{av}} = 80 - 0.38 \cdot 60 = 57.2°\mathrm{C}$. For the temperature t_{av} for air:

$$\beta = \frac{1}{273.16 + 57.4}\ 1/g, \quad \nu = 18.7 \cdot 10^{-6}\ \mathrm{m^2/sec}\ ,$$

$$\lambda = 2.88 \cdot 10^{-2}\ \mathrm{W/m \cdot deg}, \ \mathrm{Pr} = 0.697\ .$$

Hence,

$$\mathrm{Gr}_1 = \frac{9.81 \cdot 10^3 \cdot 10^{12} \cdot 60}{330.56 \cdot 18.7^2} = 5.092 \cdot 10^9\ ,$$

$$\mathrm{Ra}_1 = 3.549 \cdot 10^9\ .$$

The number $\mathrm{Ra}_1 > 4 \cdot 10^8$, and $\mathrm{Nu_{av}}$ is determined from the dimensionless relation (2.1.7):

$$\mathrm{Nu_{av}^{1/2}} = 0.825 + \frac{0.387 \cdot (3.549 \cdot 10^9)^{1/8}}{\left[1 + (0.492/0.697)^{9/16}\right]^{8/27}}\ ,$$

$$\mathrm{Nu_{av}} = 181.7\ .$$

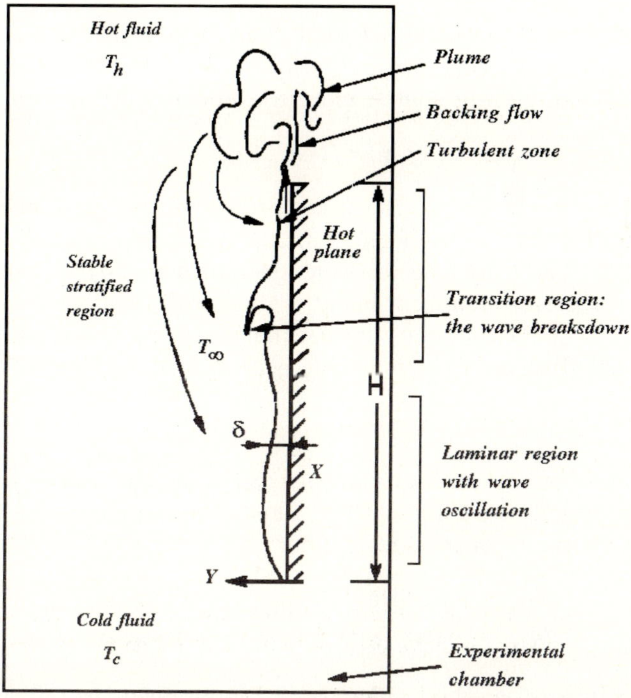

The coefficient of heat transfer from the surface is

$$\alpha_1 = \frac{\mathrm{Nu_{av}}\lambda}{1} = \frac{181.7 \cdot 2.88 \cdot 10^{-2}}{1} = 5.23\ \mathrm{W/m^2\,deg}\,.$$

2.2 Constant Heat Flow on a Vertical Plane Surface

The determining criterion of motion and heat transfer of liquid at a constant heat flux density on the surface is the modified Gr* or Ra* number. The conditions of steady heat transfer depending on the Pr number are characterized by the dependences of the form [2.26]

for $\Pr \ll 1$

$$\mathrm{Nu} = f_1\left(\mathrm{Gr}^*\,\Pr^2\right)\,,$$

for $\Pr = \mathrm{O}(1)$

$$\mathrm{Nu} = f_1\left(\mathrm{Gr}^*\,\Pr\right)\,, \tag{2.2.1}$$

for $\Pr \gg 1$

$$\mathrm{Nu} = f_3\left(\mathrm{Ra}^*\right)\,.$$

The following relationship is established between the Gr and Gr* numbers by the similarity theory methods:

$$\mathrm{Gr}^* = \mathrm{O}\left(\mathrm{Gr}^{5/4}\right)\,. \tag{2.2.2}$$

For problems ***with a constant heat flux on a wall*** the unknown quantity is the wall excess temperature θ_{w}. Since $\mathrm{Gr}^* = \mathrm{Nu}\mathrm{Gr}$, the dimensionless relation (2.2.1) can be represented more conveniently in the form

$$\mathrm{Gr} = f(\mathrm{Gr}^*, \Pr) \tag{2.2.3}$$

However, in the literature the main predicted and experimental results are usually represented in the form of (2.2.1) or, if using (2.2.2), in the form of (2.1.4–2.1.5). The heat transfer at a constant heat flux is usually 10–15% higher than the corresponding value at a given wall temperature.

The analysis of the boundary-layer equations near a vertical surface ***at a constant heat flux*** gives the following general formula for the local Nusselt number [2.7]:

$$\mathrm{Nu}_x = F(\Pr)\left(\mathrm{Gr}_x^*\,\Pr\right)^{1/5}\,. \tag{2.2.4}$$

On the basis of the similarity theory, Sparrow and Gragg [2.27] carried out investigations in the range $0.01 < \Pr < 100$ and suggested the limiting dependences for the local Nusselt number

$$\mathrm{Nu}_x = 0.616\,\mathrm{Ra}_x^{*1/5} \qquad \Pr \to \infty\,, \tag{2.2.5}$$

$$\mathrm{Nu}_x = 0.644\,\mathrm{Ra}_x^{*1/5}\,\Pr^{1/5} \qquad \Pr \to 0\,. \tag{2.2.6}$$

Using the integral method, Sparrow [2.28] obtained the following formula for the local Nusselt number making the same assumptions concerning the scales of the boundary layer as in [2.29]:

$$\mathrm{Nu}_x = \frac{2}{360^{1/5}}\left(\frac{\Pr}{4/5+\Pr}\right)^{1/5}\mathrm{Ra}_x^{*1/5}\,. \tag{2.2.7}$$

The application of the integral method ***in case of a constant heat flux on a surface*** also allowed one to find the temperature distribution corresponding to the given value of the heat flux [2.28]

$$T_{\mathrm{w}} - T_{\infty} = \frac{360^{1/5}}{2} \frac{qx}{k} \left(\mathrm{Gr}_x^*\right)^{1/5} \left[\frac{0.8 + \mathrm{Pr}}{\mathrm{Pr}^2}\right]^{1/5} . \tag{2.2.8}$$

Churchill and Ozoe [2.30] generalized the data of works [2.21, 2.27, 2.31, 2.32] and suggested the following formula which is valid for all Pr ***in a laminar mode of flow***:

$$\mathrm{Nu}_x = \frac{0.563\ \mathrm{Ra}_x^{1/4}}{\left[1 + (0.437/\,\mathrm{Pr})^{9/16}\right]^{4/9}} . \tag{2.2.9}$$

In [2.33] the solutions obtained by various researchers [2.27,2.30,2.34–2.37] were analyzed and the following formula was suggested to calculate $F(\mathrm{Pr})$:

$$F\left(\mathrm{Pr}\right) = \left(\frac{\mathrm{Pr}}{4 + 9\ \mathrm{Pr}^{1/2} + 10\,\mathrm{Pr}}\right)^{1/5} . \tag{2.2.10}$$

The formula is valid for all Pr and $10^5 < \mathrm{Gr}_x^* < 10^{12}$.

One fails to establish a distinct boundary of transition from a laminar to a turbulent flow. Some works point to the presence of an intermediate region with the heat transfer mechanism corresponding to a laminar regime [2.38]. ***For air*** the transition can occur at $\mathrm{Ra}_k^* = 10^{10}{-}10^{11}$ [2.38], ***for a liquid sodium*** at $\mathrm{Gr}_{xk}^* = 2 \cdot 10^{11}$ [2.39].

In [2.8] the results of [2.40–2.43] were generalized, and the following dependence was suggested to calculate the average Nusselt number in a wide range of GrPr ($1 < \mathrm{GrPr} < 10^{11}$) corresponding to both laminar and turbulent regimes of flow at any Pr values:

$$\mathrm{Nu}_L^{1/2} = 0.825 + \frac{0.387\left(\mathrm{Gr}_L\,\mathrm{Pr}\right)^{1/6}}{\left[1 + (0.437/\,\mathrm{Pr})^{9/16}\right]^{8/27}} . \tag{2.2.11}$$

By solving a self-similar problem for local and average Nusselt numbers, the following dimensionless relations are obtained [2.11, 2.18, 2.40]:

$$\mathrm{Nu}_x = \frac{G^*}{5\theta(0)} = \frac{1}{\theta(0)} \left(\frac{\mathrm{Gr}_x^*}{5}\right)^{1/5} , \tag{2.2.12}$$

$$\mathrm{Nu}_L = 6\left(\mathrm{Gr}_L^*\right)^{1/5} 5^{6/5} \theta\left(0\right) = 6G^*/25\theta\left(0\right) , \tag{2.2.13}$$

where

$$G^* = 5\left(\frac{\mathrm{Gr}_x^*}{5}\right)^{1/5} . \tag{2.2.14}$$

For vertical plates ***with the heat flux density constant over the surface*** in a turbulent regime the experimental data are approximated by the generalized formula [2.22]

$$\mathrm{Nu}_x = 0.568\,(\mathrm{Gr}_x^*\,\mathrm{Pr})^{0.22}\ . \tag{2.2.15}$$

Example 2.2. A copper rectangular bus bar of $l = 100\,\mathrm{mm}$ and $\delta = 3\,\mathrm{mm}$, is located on a rib in air with the temperature $t_\infty = 45°\mathrm{C}$. Under steady conditions the bus bar temperature must not exceed 80°C. Calculate the current value, if the specific resistance of the bus material is equal to $\rho_e = 2.18 \cdot 10^8\ \Omega{\cdot}\mathrm{m}$.

The determining air temperature is $t_{\mathrm{av}} = 80 - 0.38 \cdot 35 = 66.7°\mathrm{C}$, the characteristic dimension is l. Moreover,

$$\beta = \frac{1}{339.86} 1/\deg\ , \qquad \nu = 19.67 \cdot 10^{-6}\mathrm{m}^2/\sec\ ,$$
$$\lambda = 2.94 \cdot 10^{-2}\ \mathrm{W/m} \cdot \deg\ , \qquad \mathrm{Pr} = 0.694\ ,$$
$$\mathrm{Ra}_l = \frac{9.81 \cdot 10^{-3} \cdot 35 \cdot 0.694}{339.86 \cdot 19.67^2} \cdot 10^{12} = 1.812 \cdot 10^6 < \mathrm{Ra}_{lk}\ .$$

Using the dimensionless dependence (2.2.6), we obtain

$$\mathrm{Nu}_{\mathrm{av}} = 0.68 + \frac{0.67 \cdot (1.812 \cdot 10^6)^{1/4}}{\left[1 + (0.437/0.694)^{9/16}\right]^{4/9}} = 1975\ ,$$
$$\mathrm{Nu}_{\mathrm{av}} = \frac{ql}{\theta_{\mathrm{w}}\lambda}\ .$$

The heat $Q = I^2 R_e = \rho_e L/(\delta I)$ generated in the bus bar of length L is dissipated in the environment

$$Q = q\mathrm{Nu}_{\mathrm{av}} = 2\frac{ql}{\theta_{\mathrm{w}}\lambda} \cdot lL\ .$$

The heat balance condition

$$I^2 \frac{\rho_e L}{\delta I} = \frac{\mathrm{Nu}_{\mathrm{av}}\theta_{\mathrm{w}}\lambda}{l} 2lL\ ,$$

yields

$$I = \left(\frac{\mathrm{Nu}_{\mathrm{av}}\lambda\theta_{\mathrm{w}}2l\delta}{\rho_e}\right)^{1/2} = \left(\frac{19.75 \cdot 2.94 \cdot 10^{-2} \cdot 35 \cdot 2 \cdot 10^2 \cdot 3 \cdot 10^{-6}}{2.18 \cdot 10^{-8}}\right)^{1/2} = 237\ \mathrm{A}\ .$$

Note that according to (2.2.11) $I = 236$ A.

2.3 Plane Vertical Plate with a Variable Surface Temperature

In calculating the parameters of free-convective heat transfer on a surface having a variable wall temperature $T_w(x)$, one must take into account the dependence of the heat transfer coefficient α on plate height l, excess temperature θ_w, gravity g, the thermophysical characteristics of the flow β, ρ, μ, c_p and on the heat flux magnitude in the longitudinal direction $d\theta_w/dx$. The use of the dimensional analysis for free-convective heat transfer ***at a variable wall temperature*** leads to the following dimensionless dependence [2.44]:

$$\mathrm{Nu}/\mathrm{Gr}^{1/4} = f_1(\mathrm{Pr},\ \mathrm{Ec},\ \mathrm{P}_\vartheta) . \tag{2.3.1}$$

The wall temperature being variable, the number of determing quantities involves the parametrical criterion [2.26]

$$\mathrm{P}_\vartheta = \frac{x}{\theta_w}\frac{d\theta_w}{dx} , \tag{2.3.2}$$

which takes into account the effect of a change in the longitudinal heat flux $q_x = -\lambda(d\theta_w/dx)$ on heat transfer from the surface. For positive P_ϑ there corresponds free convection on the surface with an increasing temperature; for the negative P_ϑ, same with a decreasing temperature. The variety of wall temperature distributions can be taken into account by the criteria

$$\mathrm{P}_{\vartheta k} = \frac{x^k}{\theta_w}\frac{d\theta_w}{dx} . \tag{2.3.3}$$

Explicit relations for calculating heat transfer ***for an arbitrary change in the wall temperature*** were obtained by the approximate integral method [2.45]

$$\mathrm{Nu}_x = 0.614\, C_1\,(\mathrm{Pr})\, \frac{\theta_w^{0.95} x^{0.25} \left(\int_0^x \theta_w^{0.6} dx\right)^{0.5}}{\left[\int_0^x \theta_w^{2.2} \left(\int_0^x \theta_w dx\right)^{0.5} dx\right]^{0.5}} \mathrm{Ra}_x^{0.25} , \tag{2.3.4}$$

where $C_1 = 0.667/[1 + (0.49/\mathrm{Pr})^{9/16}]^{4/9}$.

Formula (2.3.4) can be used when there is a little change in the wall temperature.

Power-law surface temperature distribution $T_w - T_\infty = Nx^n$. This distribution of the wall temperature admits self-similar solutions. For this case, the system of ordinary differential equations obtained by the method of self-similar equations together with the boundary-layer approximation has the form [2.18]

$$\begin{gathered} f''' + (n+3)\, f'' - (2n+2)\, f'^2 + \theta = 0 , \\ \theta'' + \mathrm{Pr}\,[(n+3)\, f\theta' - 4nf'\theta] = 0 , \\ f(0) = f'(0) = 1 - \theta(0) = f'(\infty) = \theta\,(\infty) = 0 , \end{gathered} \tag{2.3.5}$$

where f and θ are the stream function and temperature in self-similar variables.

For the given power-law temperature distribution the heat flux at a given point with the coordinate x is written as

$$q(x) \sim x^{(5n-1)/4} . \tag{2.3.6}$$

Sparrow and Gragg [2.18] obtained the solution of system (2.3.5) for Pr = 0.7 and 1 for n varying from –0.8 to 3:

$$\mathrm{Nu}\,(x)\,/\mathrm{Gr}_x^{1/4} = [-\theta'\,(0)]\,/\sqrt{2} = H\,(\mathrm{Pr}, n) \ . \tag{2.3.7}$$

The difference from a similar expression for an isothermal surface ($n = 0$) is the value of $\theta'(0)$ and the appearance of the term $T_\mathrm{w} - T_\infty$ in the form of Nx^n in determining the heat transfer coefficient and the local Grashoff number. Now, the parameter H depends not only on Pr, but also on n. It is difficult to determine the general coefficient, since there is no characteristic temperature difference in this problem. The best way is to find the local heat transfer coefficient which describes the solution at a given value of n, and to determine Nu for this value of n, since it is impossible to exclude the dependence on n by a simple choice of the temperature difference [2.18].

Power-law change in the wall temperature $\theta_\mathrm{w} = Bx^m, \mathrm{P}_\vartheta = m$: [2.26]

$$\begin{aligned} \mathrm{Nu}_x &= 0.614 \left(\frac{2.5m + 1.5}{\sqrt{0.6m + 1}}\right)^{0.5} C_1(\mathrm{Pr})\,\mathrm{Ra}_x^{0.25} , \\ \mathrm{Nu}_\mathrm{av} &= \frac{4}{3+m}\,\mathrm{Nu}_1 ,\quad m > -0.6 \ . \end{aligned} \tag{2.3.8}$$

When $m \leq -0.6$, the self-similar problem corresponds to convection near a vertical plate with a given heat source at the bottom.

For a linearly increasing surface temperature $\theta_\mathrm{w} = \theta_0\,(x + x_0)$, if the flow begins to develop in the section $x = -x_0$, the self-similar solution of the Navier–Stokes and energy equations leads to the following dependence for the heat transfer coefficient [2.44]:

$$\frac{\mathrm{Nu}_x}{\mathrm{Ra}_x^{1/4}} = 0.895 \left(\frac{x}{x + x_0}\right)^{1/4} \left(\frac{\mathrm{Pr}}{1 + 1.58\,\mathrm{Pr}^{1/2} + 2.54\,\mathrm{Pr}}\right)^{1/4} . \tag{2.3.9}$$

Parabolic change in the wall temperature $\theta_\mathrm{w} = \theta_0\,[1 + (x/x_0)^m]$. When $x > x_0, m > 1$, the solution of this problem must tend rather rapidly to a self-similar one. For $m = 2$ and $x > 0.5$, the calculations carried out by finite-difference methods lead to the following dimensionless relation [2.46]:

$$\frac{\mathrm{Nu}_x}{\mathrm{Ra}_x^{1/4}} = 1.31 C_1 \left(\frac{x^2}{x^2 + x_0^2} \right)^{1/4} , \tag{2.3.10}$$

where $C_1 = 0.667/[1 + (0.49/\mathrm{Pr})^{9/16}]^{4/9}$.

Linear and parabolic decrease in the wall temperature $\theta_\mathrm{w} = \theta_0 \left[1 - (ax)^m\right]$. A one-parameter approximation of the universal equations is the exact solution for $m = 1$ [2.44]. The application of integral methods in the case of a decreasing wall temperature does not give sufficiently accurate results [2.28]. The velocity distribution in the boundary layer over plate height was experimentally investigated at $\mathrm{Pr} = 7.06$, $\theta_\mathrm{w} = 5{-}20°\mathrm{C}$, $m = 1$ [2.47].

Harmonic wall temperature distribution $\theta_\mathrm{w} = \theta_0 \sin(x/l)$. The approximation dependence for heat transfer with an error of 1.5% can be represented as [2.48] ($\mathrm{Pr} = 0.01 - 100$, $\bar{x} = 0 - 2.2$)

$$\frac{\mathrm{Nu}_x}{\mathrm{Ra}_x^{1/4}} = 0.062 \left[1 + \left(\frac{0.394}{\mathrm{Pr}} \right)^{0.63} \right]^{-0.397} \frac{(2.48 - \bar{x})(\bar{x} + 1.45)}{\pi - \bar{x}}$$
$$\times \left[1 + \frac{\mathrm{Pr}^{0.1} - 1}{7.7\,\mathrm{Pr} - 3.3} \exp\{4.3(x - 2)\} \right] . \tag{2.3.11}$$

Exponential wall temperature distribution $\theta_\mathrm{w} = \theta_0 \exp(mx)$. This problem was investigated in detail by Sparrow and Gragg [2.25] and also by Gebhart and Mollendorf [2.49]. All the criteria involve $1/m$ instead of x. It is shown that [2.18]

$$\mathrm{Nu}\,(m) = \frac{h\,(1/m)}{k} = \left[-\theta'\,(0)\right] (\mathrm{Gr}_m/4)^{1/4} , \tag{2.3.12}$$

where $\mathrm{Gr}_m = \left[g\beta\,(T_\mathrm{w} - T_\infty)\,(1/m)^3 \right] \nu^2$ and $\eta = ym\,(\mathrm{Gr}_m/4)^{1/4}$. Hence,

$$\mathrm{Nu}(x) = \frac{hx}{k} = (mx)^{1/4} \left[-\theta'(0)\right] (\mathrm{Gr}_x/4)^{1/4} , \tag{2.3.13}$$
$$\eta = (y/x) \left[(\mathrm{Gr}_x/4)mx\right]^{1/4} .$$

The parameter $[-\theta'\,(0)]$ depends only on Pr and does not depend on m. However, $\mathrm{Nu}(m)$ depends on m and so does $\mathrm{Nu}(x)$, due to the dependence of Gr_m on $(T_\mathrm{w} - T_\infty)$. The local Nusselt number $\mathrm{Nu}(x)$ changes with x as $\mathrm{Nu}(x) \sim x \exp(mx/4)$, and $\mathrm{Nu}(m)$ as $\sim\exp(mx/4)$. The heat flux $q(x)$ changes as $\sim\exp(5mx/4)$. Since the case $m = 0$ is inapplicable from the physical point of view, then it is impossible to obtain a problem with a constant heat flux from the analysis presented, like in the case of power-law distribution [2.18]. In the Boussinesq approximation, the analysis of the given problem yields nonzero values of the momentum and heat flux on the leading edge and an exponential drop in the temperature behind the leading edge at negative values of x. The velocity and temperature distributions obtained by numerical solution of the differential equations (2.3.5) with the respective boundary conditions depending on the Prandtle number [2.25, 2.49] are similar to those found for the power law. It is shown that the contribution of the leading edge can be neglected in comparison with the local momentum and heat transfer values, when $mL \gg 1$. But when $m > 0$, the boundary-layer thickness decreases with increase in x, and the physical applicability of this temperature distribution is limited to the problems where this behavior of a boundary layer is admissible [2.18].

Exponentially increasing wall temperature $\theta_\mathrm{w} = \theta_0 e^{mx}$, $\mathrm{P}_\vartheta = mx$:

$$\mathrm{Nu}_x = 1.103 C_1 (\mathrm{Pr}) \left(mx\mathrm{Ra}_x\right)^{0.25} . \tag{2.3.14}$$

The dependence (2.3.14) is applicable at some distance from the end face of a semi-infinite plate [2.46].

Piecewise continuous change in wall temperature: on the lower portion of a vertical surface the temperature $T_{\mathrm{w}1}$ is given, on the upper portion ($x > x_0$) the temperature $T_{\mathrm{w}2}$ and in the surrounding medium the temperature $T_\infty (T_{\mathrm{w}1} > T_\infty, T_{\mathrm{w}2} > T_\infty)$. By methods of the dimensional analysis it can be established that on the upper portion of the surface the intensity of heat transfer is determined from the equation [2.44]

$$\mathrm{Nu}/\mathrm{Gr}^{1/4} = f_2 \left(\mathrm{Pr}, \zeta, \theta_2\right), \zeta > 1 . \tag{2.3.15}$$

Near the point of the discontinuity of temperatures ($x = x_0$) the singularity has the form $(x - x_0)^{1/3}$ [2.50], and the intensity of heat transfer takes on infinite ($q_{\mathrm{w}1}/q_{\mathrm{w}2} \sim (\zeta - 1)^{-1/3}$ when $\zeta \to 1$).

To calculate the heat flux transferred from the surface the following dependences are suggested [2.51]:

$$\begin{aligned} \left(\frac{\mathrm{Nu}_x}{\mathrm{Gr}_x^{1/4}}\right)_2 \Bigg/ \left(\frac{\mathrm{Nu}_x}{\mathrm{Gr}_x^{1/4}}\right)_1 &= \theta_{\mathrm{W}2}^{-5/4} \left[\frac{\zeta}{(\zeta - 1)\,\theta_{\mathrm{w}2} + 1}\right]^{1/4} \\ &\times \left[1 + \left(\theta_{\mathrm{W}2}^{3/2} - 1\right) \{1 - \theta_{\mathrm{W}2} (\zeta - 1)\}^{-c}\right]^{-1/3} , \end{aligned} \tag{2.3.16}$$

where

$$c = 0.9016\, A(\theta_{\mathrm{w}2}^{-3/2} - 1)\theta_{\mathrm{w}2}^{3/2}(1 - \theta_{\mathrm{w}2})^{-1},$$

$$A = \left[\ \mathrm{Pr}^{1/2} \left(1 + 4\ \mathrm{Pr}^{1/2} + 3\ \mathrm{Pr}^{1/4}\right)\right] \left(1 - 1.4903\ \mathrm{Pr}^{9/16}\right)^{4/3} .$$

This problem was investigated in the main by numerical methods. Experiments were also carried out to study the temperature distribution in the boundary layer and heat transfer from a surface ***in air*** [2.52,2.53]. The largest divergence immediately after the discontinuity point is explained by the effect of finite thermal conductivity over the plate height [2.44].

Figure 2.3.1 presents numerical data on heat transfer ***at an exponentially decreasing wall temperature*** $\theta_w = \theta_0 \exp(-mx)$, Fig. 2.3.2 presents the same at $\theta_w = \theta_0(1 - (ax)^m)$ [2.54], and Table 2.3.1 lists the same at $\theta_w = B \sin(x/l)$ [2.46].

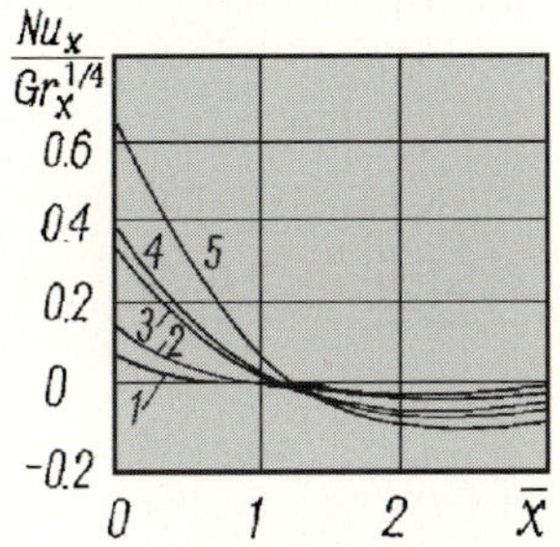

Fig. 2.3.1. Heat transfer from a vertical surface with an exponentially decreasing wall temperature: (1) Pr = 0.01; (2) 0.1; (3) 0.7; (4) 1; (5) 5

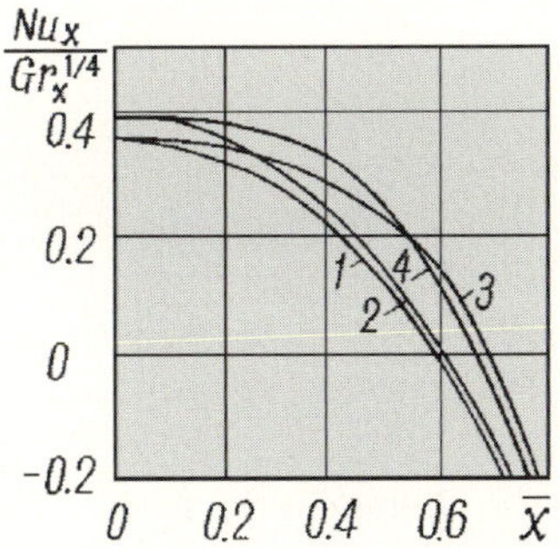

Fig. 2.3.2. Heat transfer of a vertical surface at $\theta_w = \theta_0[1 - (ax)^m]$: (1) Pr = $0.7; m = 1$; (2) $1, 1$; (3) $0.7, 2$; (4) $1, 2$

For a decreasing wall temperature the average heat flux on a semi-infinite plate, depending on its height, can adopt both positive and negative values.

In the general case, for calculating heat transfer we can recommend two one-parameter methods [2.26].

According to the universal one-parameter method, the heat transfer is calculated by the equation [2.55]:

$$\frac{\mathrm{Nu}_x}{\mathrm{Ra}_x^{1/4}} = -\frac{\theta'(\zeta, 0)}{\sqrt{2}} \left[\frac{x\theta_w(x)}{\int_0^x \theta_w dx} \right]^{0.25}, \tag{2.3.17}$$

Table 2.3.1. The Values of $\mathrm{Nu}_x/\mathrm{Gr}_x^{1/4}$

$\bar{x}$	Pr = 0.01	Pr = 0.1	Pr = 1.0	Pr = 10	Pr = 100
0	0.0927	0.5279	0.5951	1.1914	2.2039
0.2	0.0930	0.5260	0.5960	1.1933	2.2122
0.4	0.0914	0.5225	0.5886	1.1789	2.1778
0.6	0.0903	0.5161	0.5835	1.1708	2.1706
0.8	0.0879	0.5065	0.5713	1.1468	2.1198
1	0.0851	0.4932	0.5579	1.1236	2.0854
1.1	0.0830	–	0.5475	1.1028	2.0408
1.2	0.0809	0.4761	0.5377	1.0862	2.0177
1.3	0.0782	–	0.5247	1.0612	1.9660
1.4	0.0754	0.4520	0.5113	1.0376	1.9298
1.5	0.0719	–	0.4945	1.0057	1.8663
1.6	0.0680	0.4208	0.4761	0.9729	1.8129
1.7	0.0633	–	0.4536	0.9308	1.7318
1.8	0.0579	0.3773	0.4280	0.8848	1.6540
1.9	0.0513	–	0.3968	0.8271	1.5455
2.0	0.0435	0.3166	0.3602	0.7607	1.4300
2.1	0.0338	–	0.3147	0.6772	1.2760
2.2	0.0218	0.2256	0.2589	0.5760	1.0969

where

$$\zeta = \int_0^x \theta_{\mathrm{w}} dx \ , \quad \bar{\beta} = \frac{\int_0^x \theta_{\mathrm{w}} dx}{\theta_{\mathrm{w}}^2} \frac{d\theta_{\mathrm{w}}}{dx}, \quad \varepsilon(\bar{\beta}) = 4\zeta \frac{d\bar{\beta}}{d\zeta} \ ,$$

$$\overline{\beta} = \beta + \varepsilon(\overline{\beta})\Delta_0(\overline{\beta}), \ \varepsilon' = d\varepsilon/d\overline{\beta} \ .$$

In local self-similarity the change in the main quantities over the coordinate x is neglected in the initial equations ($\varepsilon = 0, \overline{\beta} = \beta$). With the temperature $\theta_{\mathrm{w}}(x)$ given, the quantity β is calculated and the quantity $\theta'(\beta, 0)$ is taken from Table 2.3.2.

Table 2.3.2.

β	$\theta'(\beta,0)$	β	$\theta'(\beta,0)$
1	−0.7344	−0.4	−0.3821
0.8	−0.6946	−0.6	−0.3188
0.6	−0.6511	−0.8	−0.2527
0.4	−0.6040	−1	−0.1838
0.2	−0.5534	−1.2	−0.1122
0	−0.4995	−1.4	−0.0380
−0.2	−0.4423	−1.5	0

In the method of local non-self-similarity the sequence of calculation is as follows: using the given $\theta_{\rm w}$ the quantities $\zeta, \overline{\beta}$, and ε' are calculated; the quantity β is determined from the graphs of Fig. 2.3.3 by successive iterations, and the quantity $\theta'(\beta, 0)$ is taken from Table 2.3.2. Using the known values of $\beta(x)$ and $\zeta(x)$, one finally determines $\theta(x, 0)$ in (2.3.17).

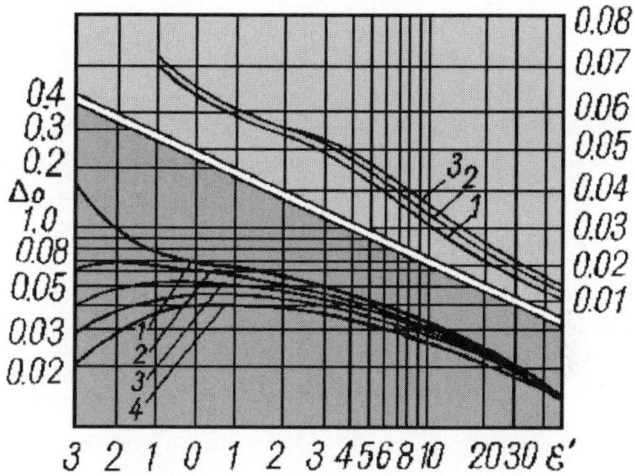

Fig. 2.3.3. Dependence of Δ_0 on ε' and β: (1) $\beta = 1$, (2) 0, (3) (-1), (4) $(-1, 5)$

The data of Table 2.3.2 and the curves of Fig. 2.3.3 are given for Pr = 0.7. In Fig. 2.3.4 the results obtained are compared with the known numerical data.

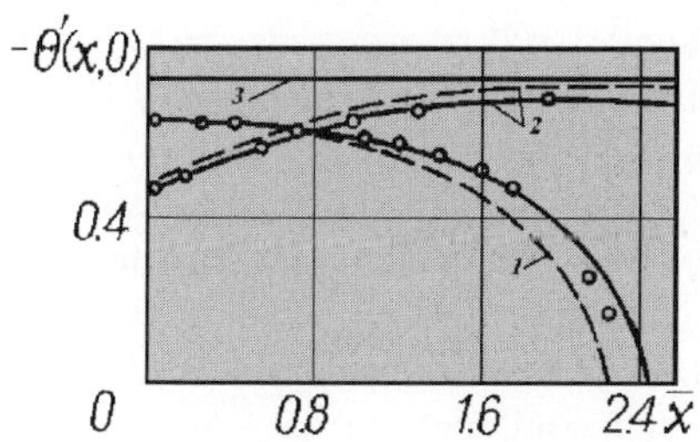

Fig. 2.3.4. Comparison of numerical calculations (*solid lines*) with the methods of local self-similarity (*dashed lines*) and local non-self-similarity (*points*): (1) $\theta_{\rm w} = \theta_0 \sin(\overline{x})$, (2) $\theta_{\rm w} = \theta_0 \exp(\overline{x})$, (3) self-similar solution in a large distance from the leading edge for $\theta_{\rm w} = \theta_0 \exp(\overline{x})$

More logically consecutive is the Kochin–Loytsyanskii method, although certain additional calculations are needed to obtain numerical results [2.56].

To calculate the local heat transfer value

$$\frac{{\rm Nu}_x}{{\rm Ra}_x}^{1/4} = \left(\frac{x}{p(x)f(x)}\frac{\theta'_{\rm w}}{\theta_{\rm w}}\right)^{1/4} Q(F) \tag{2.3.18}$$

we successively calculate $m_0 = \lim_{x \to 0} \mathrm{P}_\vartheta$; from tables or graphs we find f_0, and

$$\left(\frac{\mathrm{Nu}_x}{\mathrm{Gr}_x^{1/4}} \right)_{x=0} = Q\left(f_0\right) / [\gamma(f_0)\sqrt{2}] \, .$$

Further, from the equation

$$\left(f\theta_\mathrm{w}/\theta''_\mathrm{w}\right)' = F(f) + G(f) \tag{2.3.19}$$

by numerical integration we calculate $f(x)$ and, using specially compiled tables, we determine the remaining functions, and the heat transfer coefficient from (2.3.18). Table 2.3.3 lists the functions needed for calculations [2.56,2.57].

The liquid temperature $T(x, y)$ is a function of the wall temperature θ_w and parametric criteria $\mathrm{P}_{\vartheta k}$. The degree of the approach of the true temperature to the given $T_0(x, y)$ can be estimated with the aid of the functional

$$J(\theta_\mathrm{w}) = \int_0^{\delta_T} \left[T_0\left(x,y\right) - T\left(x,y\right)\right]^2 dy \, . \tag{2.3.20}$$

on the multitude $\{\theta_\mathrm{w}(x)\}$ [2.58]. The problem is to find such a wall temperature distribution $\theta_\mathrm{w}(x)$ which would provide the minimum of the functional (2.3.20). The solution of the problem can be obtained by the method of conventional gradient.

The analytical solution obtained by the method of small disturbances and valid for small values of the Ra number shows that in the case of steady two-dimension free-convective flow ***near an infinite vertical surface in a porous medium heated and cooled according to the harmonic law***, the structure of the flow in the range $0 \le \mathrm{Ra} \le 150$ represents a series of oppositely rotating cells adjacent to the vertical surface [2.59]. As Ra increases to values ~40, the cells separate from the surface. With the further increase in Ra the application of the dimensional method to the flow in the boundary layer near the vertical surface is more expedient. It allows one to calculate the average velocity and heat transfer coefficient.

2.4 Plane Vertical Plate with a Variable Heat Flux on a Surface

Application of the dimensional analysis to investigation of free-convective heat transfer for a variable heat flux on a surface carried out with account for the dependence of the heat transfer coefficient α on the plate height l, excess temperature θ_w, gravity g, and the thermophysical characteristics of the flow $\beta, \rho, \mu, \lambda$, and c_p leads to the appearance of the additional criterion P_q [2.26, 2.44]

Table 2.3.3.

M	γ	f	$F+G$	p	Q
		Pr = 0.7			
–0.6	1.0712	2.3334	2.2570	2.2570	0.0001
–0.55	0.9513	1.8543	1.7668	1.7668	0.0810
–0.5	0.8589	1.5175	1.4344	1.4344	0.1327
–0.45	0.7856	1.2712	1.1986	1.1986	0.1668
–0.4	0.7261	1.0853	1.0245	1.0245	0.1899
–0.35	0.6769	0.9410	0.8923	0.8923	0.2058
–0.3	0.6355	0.8266	0.7891	0.7891	0.2169
–0.2	0.5700	0.6585	0.6408	0.6408	0.2304
–0.1	0.5200	0.5420	0.5395	0.5395	0.2371
0	0.4809	0.4576	0.4676	0.4676	0.2402
0.2	0.4235	0.3450	0.3730	0.3730	0.2414
0.4	0.3832	0.2742	0.3146	0.3146	0.2400
0.6	0.3532	0.2262	0.2752	0.2752	0.2374
0.8	0.3300	0.1918	0.2473	0.2473	0.2348
1.0	0.3114	0.1661	0.2264	0.2264	0.2324
1.2	0.2961	0.1462	0.2103	0.2103	0.2301
1.4	0.2832	0.1304	0.1973	0.1973	0.2280
1.6	0.2723	0.1175	0.1871	0.1871	0.2261
2	0.2545	0.0981	0.1712	0.17.12	0.2229
2.4	0.2406	0.0839	0.1597	0.1597	0.2202
3	0.2246	0.0689	0.1264	0.1264	0.2170
5	0.1920	0.0430	0.1090	0.1090	0.2105
		Pr = 0.5			
–0.45	0.2430	0.11299	0.12349	0.12349	0.1039
–0.4	0.2204	0.09322	0.10124	0.10124	0.1149
–0.35	0.2024	0.07865	0.08529	0.08529	0.1217
–0.3	0.1877	0.06757	0.07345	0.07345	0.1257
–0.25	0.1755	0.05892	0.06438	0.06438	0.1282
–0.2	0.1652	0.05203	0.05796	0.05796	0.1295
–0.15	0.1564	0.04643	0.05155	0.05155	0.1302
–0.1	0.1488	0.04180	0.04600	0.04690	0.1304
–0.05	0.1421	0.03794	0.04304	0.04304	0.1303
0	0.1363	0.03466	0.03981	0.03981	0.1300
0.1	0.1264	0.02944	0.03470	0.03470	0.1289
0.2	0.1184	0.02548	0.03088	0.03088	0.1277
0.3	0.1118	0.02240	0.02792	0.02792	0.1264
0.4	0.1063	0.01993	0.02558	0.02558	0.1251
0.5	0.1015	0.01792	0.02368	0.02368	0.1238
0.6	0.0974	0.01626	0.02211	0.02211	0.1226
0.7	0.0938	0.01487	0.02080	0.02080	0.1215
0.8	0.0906	0.01368	0.01968	0.01968	0.1204
1	0.0852	0.01177	0.01790	0.01790	0.1185
1.2	0.0808	0.01031	0.01654	0.01654	0.1169
1.6	0.0741	0.00823	0.01461	0.01461	0.1143
2	0.0691	0.00683	0.01331	0.01331	0.1122
2.5	0.0643	0.00562	0.01218	0.01218	0.1103
3	0.0607	0.00477	0.01138	0.01138	0.1088
4	0.0554	0.00365	0.01034	0.01034	0.1066
5	0.0157	0.00295	0.00968	0.00968	0.1051

$$\mathrm{Nu}/\mathrm{Gr}^{*1/4} = f_3\left(\mathrm{Pr},\ \mathrm{Ec},\ \mathrm{P}_q\right)\ , \tag{2.4.1}$$

where $\mathrm{P}_q = \dfrac{x}{l}\dfrac{dq_{\mathrm{w}}}{dx}$.

The application of the integral method in the case of a variable heat flux on a wall $q = q(x)$ made it possible to obtain the following differential equations [2.18]

$$\frac{d}{dx}\left(\frac{V^2\delta}{105}\right) = \frac{\delta^2 q}{6q_0} - \frac{V}{\delta}\ ,$$
$$\frac{d}{dx}\left(\frac{V\delta^2 q}{30}\right) = \frac{2q}{\mathrm{Pr}}\ . \tag{2.4.2}$$

where q_0 is the heat flux for $x = 0$.

Power-law change in a heat flux. When $q = Ax^n$, the solution of (2.4.2) for V and δ yields [2.60]

$$\mathrm{Nu}\,(x) = \frac{\left(\frac{7}{24}\,\mathrm{Pr}\right)^{1/4(n+1)}\left(\mathrm{Gr}_x^{1/2}\,\mathrm{Pr}\right)^{1/4}}{\left[\left(\frac{35\,\mathrm{Pr}}{12} + 1\right)(n+1) + \frac{4}{3}\right]^{1/4}}\ . \tag{2.4.3}$$

The excess temperature changes in this case as $x^{(4n+1)/5}$.

The numerical calculations [2.61] were carried out at $n = -0.9 - 2; \mathrm{Pr} = 0.5 - 2$ and at $n = -0.4, 1;\ \mathrm{Pr} = 0.7, 7$ [2.62]. The wall temperature depends on the height as [2.44]

$$\frac{T_{\mathrm{w}} - T_\infty}{(T_{\mathrm{w}} - T_\infty)_0} = (x/l)^{(4n+1)/5}\ . \tag{2.4.4}$$

We note special cases: $n = -0.25$ ***corresponds to a constant wall temperature***, $n = 1$ ***to a linearly changing wall temperature.***

For a power-law change in the heat flux $q_{\mathrm{w}} = Nx, \mathrm{P}_q = n$ [2.63]

$$\mathrm{Nu}_x = \frac{1}{\theta\,(0)\,[5\,(n+1)]^{1/5}}\mathrm{Gr}_x^{*1/5}\ ,$$
$$\mathrm{Nu}_{\mathrm{av}} = \frac{4n+6}{5}\ \mathrm{Nu}_l\ , \tag{2.4.5}$$
$$\theta_{\mathrm{w}} \sim x^{(4n+1)/5},\quad n > -1\ .$$

When $n < -1$, a heat source is located in the initial section of the plate, and the liquid can move only in the direction opposite to the gravity direction.

Linear increase in the heat flux $q_{\mathrm{w}} = q_0(x + x_0)$. At a sufficient distance from the leading edge, the problem is reduced to a self-similar one, and to calculate the intensity of heat transfer it is possible to use the dependence [2.44]

$$\left(\frac{\mathrm{Nu}_x}{\mathrm{Ra}_x^{*1/5}}\right)_\infty = 0.7596\left[1+\left(\frac{0.3941}{\mathrm{Pr}}\right)\right]^{0.313} . \tag{2.4.6}$$

The numerical calculations of the boundary-layer equations are carried out for the entire region of the change of parameters. The following approximate dependence is recommended to calculate of the intensity of the heat transfer [2.64]

$$\begin{aligned}\frac{\mathrm{Nu}_x}{\mathrm{Ra}_x^{*1/5}} = \left(\frac{\mathrm{Nu}_x}{\mathrm{Ra}_x^{*1/5}}\right)_0 + \left[1+\left(1.8578\frac{x_0}{x}\right)^{1.1593}\right.\\ \times\left.\left[\left(\frac{\mathrm{Nu}_x}{\mathrm{Ra}_x^{*1/5}}\right)_0 - \left(\frac{\mathrm{Nu}_x}{\mathrm{Ra}_x^{*1/5}}\right)_\infty\right]\right] .\end{aligned} \tag{2.4.7}$$

The error of calculation from (2.4.7) does not exceed 3% for $x = 0$ and $\mathrm{Pr} = 0.02 - 100$.

For a linear change of the heat flux on the wall $q_\mathrm{w} = Nx$ [2.63]

$$\begin{aligned}\mathrm{Nu}_x &= [1.09C_l\,(\mathrm{Pr})]^{4/5}\,\mathrm{Gr}_x^{*1/5} ,\\ \mathrm{Nu}_\mathrm{av} &= \mathrm{Nu}_l .\end{aligned} \tag{2.4.8}$$

For a linear decrease in the heat flux density on the wall ($\mathrm{Pr} = 0.7$) the calculations of the wall temperature by the method of local self-similarity with the shifted parameter $x = x - x_0$ are applicable only for $x/x_0 < 0.3$ [2.55].

The interferograms of the heat transfer process on a vertical surface for a linear change in the heat flux on the wall $q_\mathrm{w} = q_0 + qx$, where $q_0 = 7382.2\,\mathrm{W/m^2}$ and $q = 28393.1\,\mathrm{W/m^3}$ are presented in Fig. 2.4.1 for

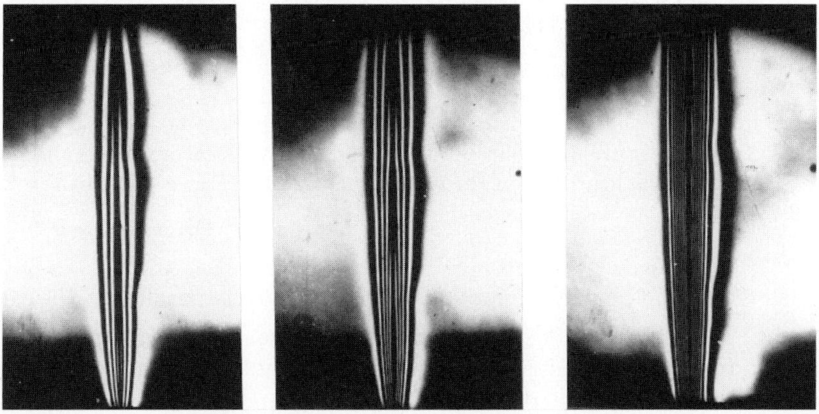

Fig. 2.4.1. Interferograms of a free-convective flow on a plate for a linear change in the heat flux on a wall taken at the time instants from the beginning of plate heating $t = 5; 8; 23$ sec (*from left to right*), respectively

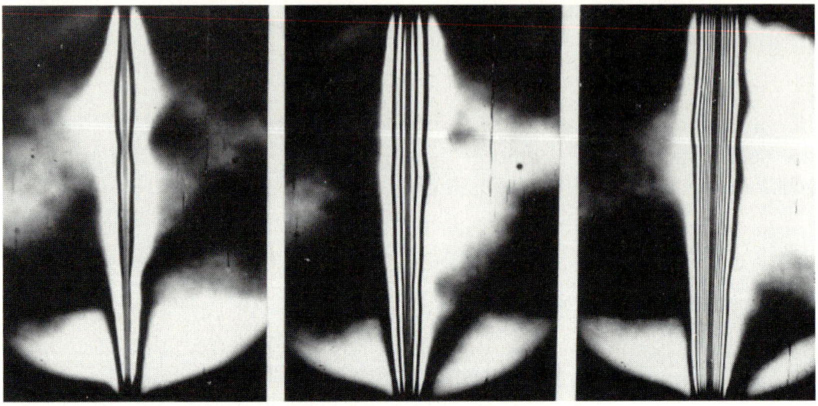

Fig. 2.4.2. Interference pictures of a free-convective flow on a plate with a linear change in the heat flux on the wall taken at the times $t = 7$; 11; 23 sec from the beginning of plate heating (*from left to right*), respectively. The heat flux increases along the plate from above downwards

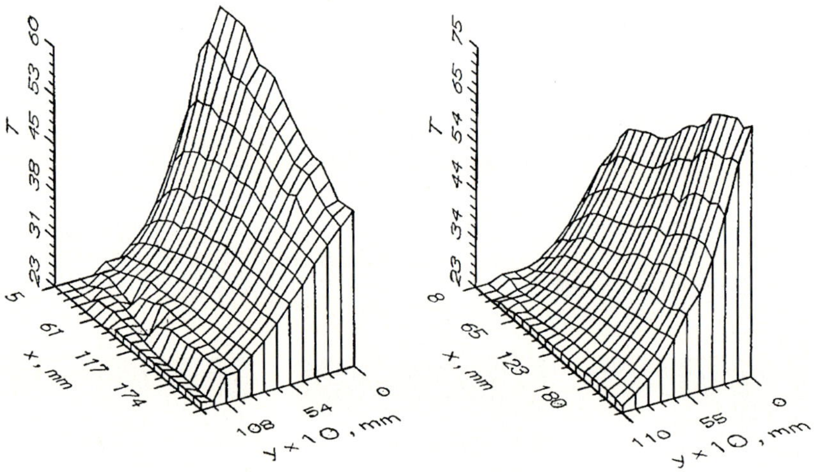

Fig. 2.4.3. Temperature distribution in a boundary layer near a vertical plate with a linear change in the heat flux on the wall: (**a**) $t = 23$ sec (Fig. 2.4.1), (**b**) $t = 23$ sec (Fig. 2.4.2)

the case where the heat flux gradient vector $\partial q_{\mathrm{w}}/\partial x$ is parallel to the vector $\vec{g}$ and in Fig. 2.4.2 when $\partial q_{\mathrm{w}}/\partial x$ is directed opposite to $\vec{g}$. In Fig. 2.4.3a and 2.4.3b the temperature distributions in a boundary layer near plates are presented which were calculated from the interferograms presented in Figs. 2.4.1 and 2.4.2.

The magnified image of the characteristic interferogram pattern of the temperature field in the boundary layer of a vertical plate with a linear change in the heat flux along the length is presented in Fig. 2.4.4a for the case where the gravity field vector has the same direction as the temperature gradient vector on the plate. Figure 2.4.4 (b, c) presents the results of track visualization of a free-convective flow on a vertical plate ***with a linear change in the heat flux along the length***: (b) $\partial q_w/\partial x$ is parallel to $\vec{g}$ and (c) $\partial q_w/\partial x$ is antiparallel to $\vec{g}$.

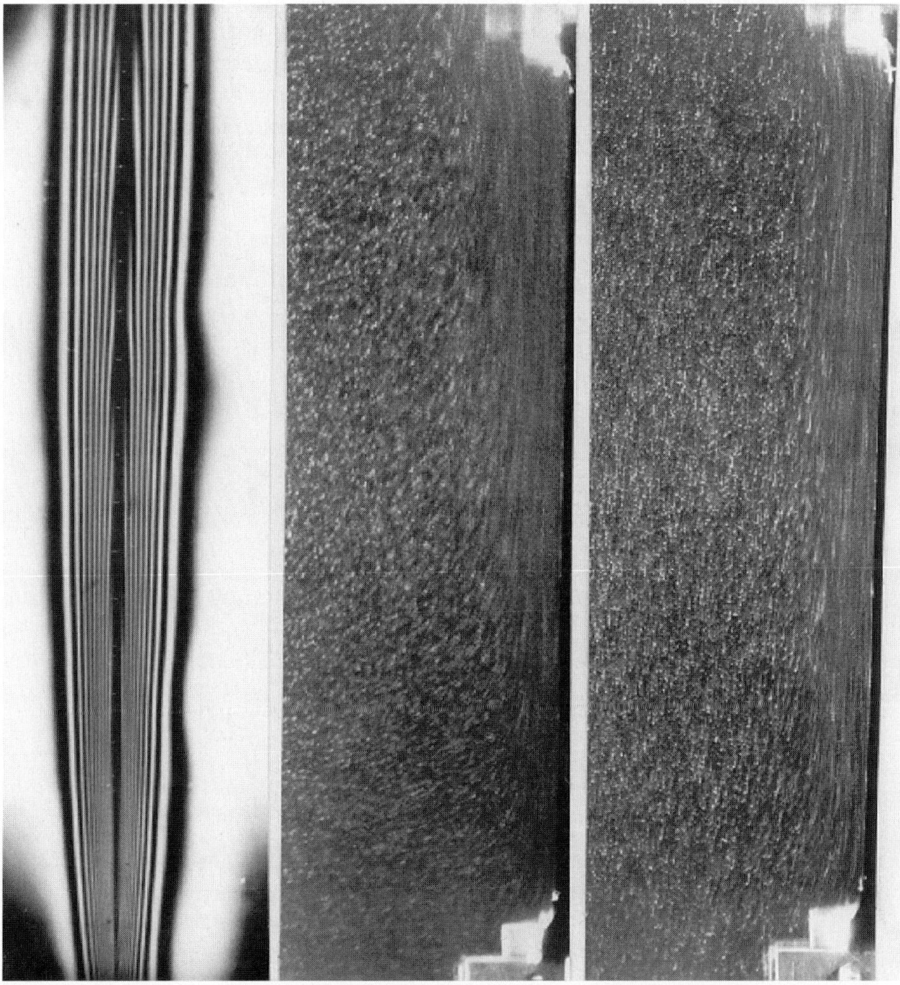

Fig. 2.4.4. Magnified images of the characteristic interferogram (**a**) and of the result of convective flow visualization by the method of photoregistration of particle tracks illuminated by a laser knife (**b**, **c**) in the case of a linear change in the heat flux along the vertical plate

Parabolic change in the heat flux. In [2.65, 2.66] for $q_w = q_0[1 + (x/x_0)^n]$ the integral method is applied to calculate the wall temperature distribution. For $Pr = O(1)$ a simplified formula is suggested:

$$\frac{\theta_w}{\theta_{w0}} = \frac{q_w}{q_0}\left[1 + \frac{1}{n+1}\left(\frac{x}{x_0}\right)^n\right]^{0.4} , \tag{2.4.9}$$

For $n = 0$–3 its error does not exceed 5% [2.26].

The integral methods for calculating heat transfer ***with a parabolic decrease in the heat flux*** $q_w = q_0[1 - (x/x_0)^n]$ give inaccurate results [2.44].

Exponentially increasing heat flux $q_w = q_0 e^{nx}$. Applying the similar transformations method to solving the self-similar problem, the value of Nu_x can be calculated from the formula [2.26]

$$\mathrm{Nu}_x = 1.099 C_l \left(mx\, \mathrm{Ra}_x\right)^{1/4} , \tag{2.4.10}$$

which with an error of 3% generalizes the experimental and calculated data on heat transfer at some distance from the end face of a semi-infinite plate.

Exponential decrease in the heat flux $q_w = q_0 e^{-mx}$. Numerical calculations of the boundary layer equations demonstrate the decrease in θ_w on increase in the parameter ml [2.44]. To calculate heat transfer rate with an error of 5% the following relation is suggested [2.26]

$$\frac{\mathrm{Nu}_x}{\mathrm{Ra}_x^{*1/5}} = \left(\frac{\mathrm{Pr}}{4 + 4\,\mathrm{Pr}^{1/2} + 10\,\mathrm{Pr}}\right)^{1/5} (mx)^{1/5} \left(e^{mx} - 1\right) . \tag{2.4.11}$$

Similar calculations allow determination the optimum height of cooling devices.

The effect of ***viscous dissipation for a variable heat flux on a surface*** can be taken into account by using integral methods [2.63]

$$\frac{\mathrm{Nu}_x}{\mathrm{Nu}_{x0}} = \exp\left(-0.2\mathrm{Ec}_x\right) \left\{\frac{q_w x}{\int_0^x q_w \exp\left(-\mathrm{Ec}_x\right) dx}\right\}^{0.2} . \tag{2.4.12}$$

The accuracy of the formula suggested is determined by the accuracy of calculations of heat transfer rate in the absence of dissipation Nu_{x0}.

The change in the excess temperature $\theta_w = (\mathrm{Gr}_x^*)^{1/5}(c/q_w x)$ for $q_w = q_0 \exp(-mx)$ and $\mathrm{Pr} = 0.699$ is shown in Fig. 2.4.5 [2.67].

The explicit dependences for heat transfer were obtained with the aid of the integral relations [2.63]

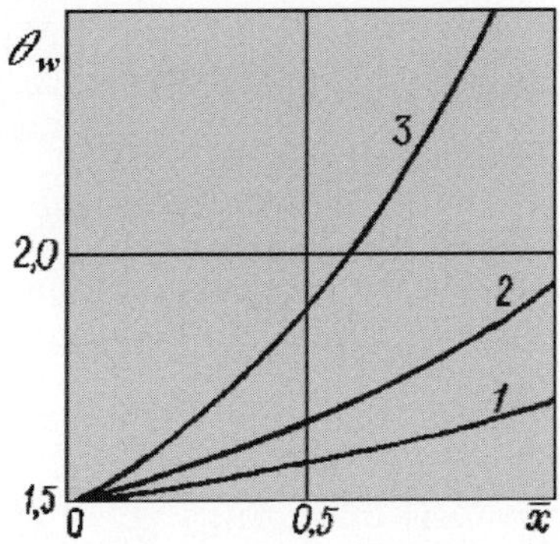

Fig. 2.4.5. The distribution of the excess temperature $\theta_w = (\mathrm{Gr}_x^*/5)^{0.2}\mathrm{Nu}_x$ over the plate height: (1) ml = 0.5, (2) 1; (3) 2

$$\frac{\mathrm{Nu}_x}{\left(\mathrm{Gr}_x^*\ \mathrm{Pr}^2\right)^{1/5}} = x^{1/5}\left(\frac{5I_2I_3^2b_1^2}{3I_1}\right)^{1/5} q_{\mathrm{w}}^{1/15}\left(\int\limits_0^x q_{\mathrm{w}}dx\right)^{-1/3\left(2+\frac{I_3}{I_1b_1}\mathrm{Pr}\right)}$$
$$\times \int\limits_0^x q_{\mathrm{w}}^{-1/3}\left(\int\limits_0^x q_{\mathrm{w}}dx\right)^{4/3+5/3\frac{I_3}{I_1b_1}\mathrm{Pr}} dx\ ,$$
$$\frac{\mathrm{Nu}_x}{\left(\mathrm{Gr}_x^*\ \mathrm{Pr}^2\right)^{1/5}} = x^{1/5}\left(\frac{2I_3^3b_1^2}{3I_4}\right)^{1/5} q_{\mathrm{w}}^{2/5}\left(\int\limits_0^x q_{\mathrm{w}}dx\right)^{-1/5\left(3+2\frac{I_3I_5}{I_4b_1}\mathrm{Pr}\right)}$$
$$\times \int\limits_0^x \left(\int\limits_0^x q_{\mathrm{w}}dx\right)^{1+2\frac{I_3I_5}{I_4b_1}\mathrm{Pr}} dx\ . \tag{2.4.13}$$

With the distribution of the velocities in the boundary layer given by the third order polynomial and the temperature profile given by the second order polynomial, we get $I_1 = 1/105, I_2 = 1/3, I_3 = 1/30, I_4 = 1/840, I_5 = 2/15, b_1 = 2$.

The approximate one-parameter method yields the dimensionless relation [2.55]

$$\frac{\mathrm{Nu}_x}{\mathrm{Gr}_x^{*1/5}} = \frac{1}{\theta(\beta,0)}\left(\frac{10}{3}\frac{x^{-1}}{q_{\mathrm{w}}^{2/3}}\int\limits_0^x q_{\mathrm{w}}^{2/3}dx\right)^{-1/5}, \tag{2.4.14}$$

where

$$\zeta = \int\limits_0^x F(x)dx; \quad \bar{\beta} = \frac{\int_0^x Fdx}{F^2}\frac{dF}{dx}; \quad \varepsilon(\bar{\beta}) = 4\zeta\frac{d\bar{\beta}}{d\zeta}; \quad \varepsilon' = \frac{d\varepsilon}{d\bar{\beta}}\ ;$$

$$\bar{\beta} = \beta + \varepsilon(\beta)\Delta_0(\beta); \quad F(x) = \frac{x}{\lambda} q_{\mathrm{w}} \mathrm{Gr}_x^{*1/5} \left(\frac{10}{3} \frac{1}{x q_{\mathrm{w}}^{2/3}} \int_0^x q_{\mathrm{w}}^{2/3} dx \right)^{1/5} .$$

In the local non-self-similarity method, first $\zeta, \overline{\beta}\varepsilon(\overline{\beta})$ and $\varepsilon'(\overline{\beta})$ are calculated. Values of the parameter β are calculated from the graphs of Fig. 2.4.6 for the corresponding values of ε' and Δ_0 by consecutive iterations. The values of $\theta(\beta,\,0)$ are listed in Table 2.4.1.

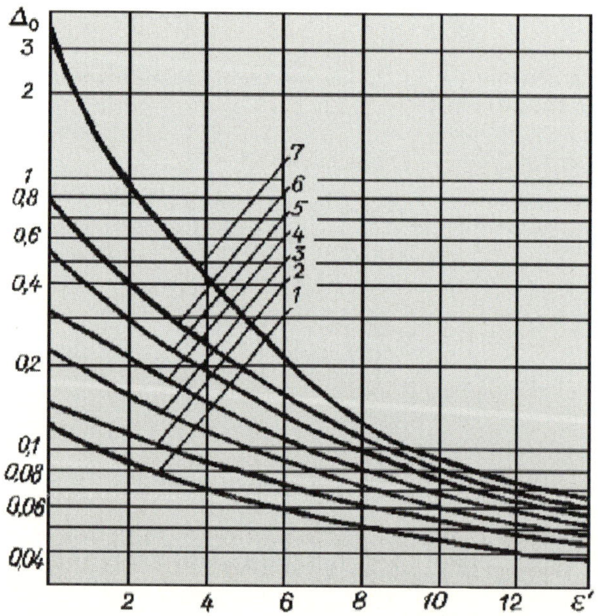

Fig. 2.4.6. Dependence of Δ_0 on ε' and β for a given heat flux on a surface: (1) $\beta = 1$; (2) 0.6; (3)0; (4)(−0.4); (5) (−0.8); (6) (−1); (7) (−1.4)

Table 2.4.1.

β	$\theta(\beta,0)$	β	$\theta(\beta,0)$
1	1.2802	−0.4	2.1591
0.8	1.3385	−0.6	2.6956
0.6	1.4096	−0.8	3.0058
0.4	1.4968	−1	3.8780
0.2	1.6053	−1.2	5.7556
0	1.7425	−1.4	13.6787
−0.2	1.9204	−1.5	∞

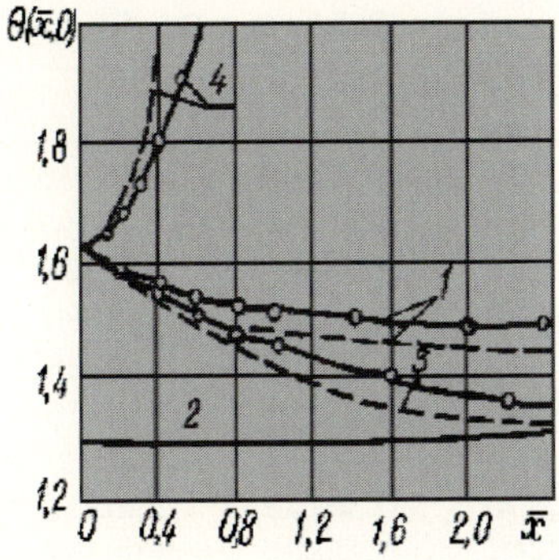

Fig. 2.4.7. Comparison of numerical calculations (*solid lines*) with the results obtained by the local self-similarity method (*dashed lines*) and by the local non-self-similarity method (*points*): (1) $q_w = q_0 \exp \overline{x}$, (2) self-similar solution for $q_w = q_0 \exp \overline{x}$, (3) $q_w = q_0(1 + \overline{x})$; (4) $q_w = q_0(1 - \overline{x})$

The results of numerical calculations and the comparison of the approximate methods for three cases of the change in the heat flux q_w are presented in Fig. 2.4.7.

In the Kochin–Loitsyanskii method the intensity of the heat transfer is calculated by the equation [2.57]

$$\frac{\mathrm{Nu}_x}{\mathrm{Gr}_x^{*1/5}} = \frac{1}{Q}\left(\frac{x}{pfq_w}\frac{dq_w}{dx}\right)^{1/5}\Bigg|_{f=f(x)} . \tag{2.4.15}$$

The calculation begins from determination of f_0 and heat transfer for $x = 0$

$$n = \lim_{x \to 0} \mathrm{P}_q, \quad f\Bigg|_{x=0} = \frac{an}{bn+1},$$

$$\frac{\mathrm{Nu}_x}{\mathrm{Gr}_x^{*1/5}}\Bigg|_{x=0} = \left(\frac{\gamma}{Q2^{2/5}}\right)_{f=f(n)} .$$

When $x > 0$, the equation

$$f = \frac{a}{q_w^{b+1}}\frac{dq_w}{dx}\int\limits_0^x q_w^b\, dx \tag{2.4.16}$$

yields $f(x)$. The values of the functions $p(x)$ and $Q(f)$ are presented in Table 2.4.2, and the intensity of the heat transfer can be determined by equation (2.4.16). For $\mathrm{Pr} = 0.7$ $a = 1.428$, $b = 1$.

Table 2.4.2.

n	γ	f	p	Q
–0.9	2.54285	–13.22991	28.92986	2.07671
–0.8	1.85804	–5.77449	12.27192	1.80337
–0.7	1.56045	–3.35196	7.73020	1.70417
–0.6	1.38486	–2.15004	5.68589	1.65701
–0.5	1.26603	–1.43159	4.54396	1.63209
–0.4	1.17892	–0.95354	3.82119	1.61840
–0.3	1.11165	–0.61264	3.32520	1.61097
–0.2	1.05774	–0.35722	2.96521	1.60721
–0.1	1.01331	–0.15871	2.69257	1.60571
0	0.97589	0	2.47939	1.60564
0.2	0.91595	0.23789	2.16807	1.60794
0.4	0.86965	0.40766	1.95220	1.61180
0.6	0.83250	0.53489	1.79418	1.61623
0.8	0.80183	0.63379	1.67350	1.62076
1.0	0.77596	0.71285	1.57854	1.62515
1.2	0.75373	0.77751	1.50177	1.62935
1.4	0.73435	0.83135	1.43853	1.63331
1.6	0.71726	0.87689	1.38556	1.63699
2.0	0.68835	0.94969	1.30182	1.64361
2.4	0.66464	1.00531	1.23706	1.64934
3.0	0.63588	1.06781	1.16833	1.65654
3.6	0.61278	1.11393	1.11693	1.66246
4.2	0.59366	1.14935	1.07781	1.66731
5.0	0.57257	1.18550	1.03810	1.67268

2.5 Free Convection on a Vertical Surface in Stratified Media

In nature and engineering we often encounter free-convective flows occurring in the media whose density changes over the height. This change in the density, known as ambient stratification, can exert a significant effect on the flow and heat transfer, and therefore it presents interest for theoretical and experimental investigation. Usually, the heat and mass transfer process itself causes stratification, which can be illustrated on an example of heat transfer between a heated vertical wall and an adjacent limited isothermal liquid volume (Fig. 2.5.1).

The ascending flow of a heated liquid, while moving along the vertical wall, reaches the top limiting surface and further moves horizontally to the opposite wall. The horizontal motion is due to the fact that the flow contains a warmer liquid in comparison with the remaining liquid in the limited volume. Then, being cooled, the flow moves downwards and returns to the heated wall, covering the horizontal stretch below the trajectory of the warmer top flow. The long-term effect of this mechanism leads to the flow under the conditions

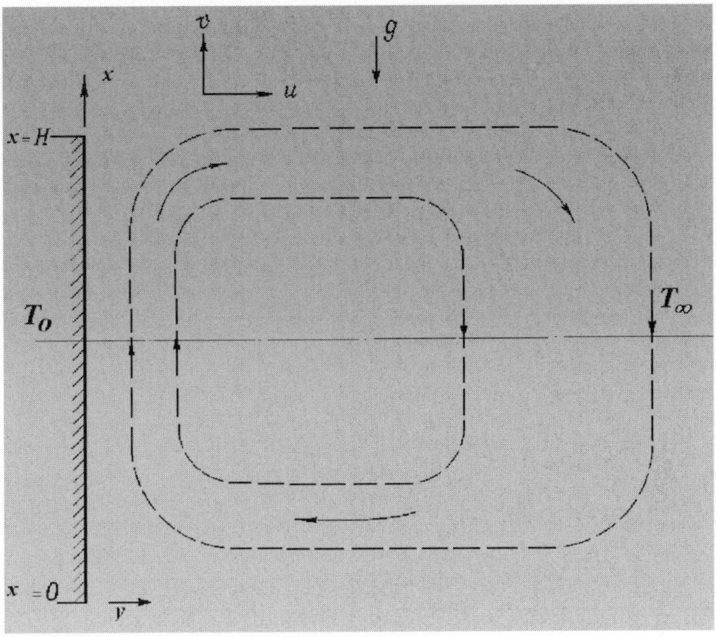

Fig. 2.5.1. Liquid stratification in a heat transfer process between a heated vertical wall and a limited isothermal liquid volume near it

of thermal stratification which in this case is characterized by the flow of warmer layers above the more cold ones [2.1]. An example of such closed flow in a stratified medium can be circulation of air in a heated room. On heat removal into an unlimited medium a similar process of the upstream development occurs. A steady stratification of the medium leading to temperature inversion can sometimes be observed near the Earth's surface. In the presence of inversion, a steady air layer arises above a certain place, outside which the temperature decreases with decrease in the surrounding medium density.

A condition of steady stratification consists in the necessity of decreasing the density over the height. If ρ is the liquid density, and x is the coordinate increasing with the height, this condition is written as $\partial\rho/\partial x < 0$. When the density depend only on temperature it follows from the stability condition that the liquid temperature must increase with the height for $\partial\rho/\partial T < 0$, which is satisfied for the majority of liquids, and in this case the stratification stability condition is represented by the inequality $\partial T/\partial x > 0$. If the density increases with the temperature (which in the case, for example, for water at a temperature from 0 to 4°C) this condition turns into $\partial T/\partial x < 0$ for $\partial\rho/\partial T > 0$.

In case where the density depends on temperature and pressure (compressible liquid) the condition of neutral equilibrium is the constancy of entropy over the vertical coordinate y. In an ascending vertical flow the liquid is cooled as a result of isoentropic expansion caused by a pressure decrease. If in this case the ambient temperature decreases by the same value by which the temperature of the liquid in the flow decreases, the latter temperature is in neutral equilibrium, and in this case the medium is called adiabatically stratified. The increase in the density caused by the temperature decrease is compensated by the density decrease due to the pressure decrease with the height [2.18].

The main condition of neutral equilibrium is that the density of the medium must not change with the height, i.e.,

$$\frac{\partial \rho}{\partial x} = \left(\frac{\partial \rho}{\partial s}\right)_p \frac{ds}{dx} . \tag{2.5.1}$$

At a constant pressure

$$T ds = c_p dT \quad \text{and} \quad \left(\frac{\partial T}{\partial s}\right)_p = \frac{T}{c_p} ,$$

hence

$$\left(\frac{\partial \rho}{\partial s}\right)_p = \left(\frac{\partial \rho}{\partial T}\right)_p \left(\frac{\partial T}{\partial s}\right)_p = \frac{T}{c_p} \left(\frac{\partial \rho}{\partial T}\right)_p .$$

Thus, (2.5.1) transforms into

$$\left(\frac{d\rho}{\partial x}\right) = \frac{T}{c_p} \left(\frac{\partial \rho}{\partial T}\right)_p \frac{ds}{dx} = 0 \tag{2.5.2}$$

and the condition of neutral equilibrium is in the entropy constancy over the height

$$\frac{ds}{dx} = 0 . \tag{2.5.3}$$

For steady stratification $\partial \rho / \partial x < 0$. Since, generally, for majority ofliquids $(\partial \rho / \partial T)_p < 0$, and T and c_p are always positive, the condition of steady stratification means a condition of entropy increase with the height:

$$\frac{ds}{dx} > 0 . \tag{2.5.4}$$

In an adiabatically stratified medium

$$\frac{dT}{dx} = \left(\frac{dp}{dx}\right) \left(\frac{dT}{dp}\right)_s \tag{2.5.5}$$

Taking into account the fact, that the pressure gradient is determined by the dependence

$$\frac{dp}{dx} = -\rho g , \tag{2.5.6}$$

equation (2.5.5) is transformed to

$$\frac{dT}{dx} = -\rho g \left(\frac{dT}{dp}\right)_s . \tag{2.5.7}$$

For dropping liquids (2.5.7) passes into the condition of isothermicity owing its origin to incompressibility:

$$\frac{dT}{dx} = 0 .$$

Generally, for gases the compressibility must not be neglected. For an ideal gas

$$\left(\frac{\partial T}{\partial p}\right)_s = \left(\frac{\partial V}{\partial s}\right)_p = \left[\left(\frac{\partial T}{\partial s}\right)\left(\frac{\partial V}{\partial T}\right)\right]_p = \frac{1}{T}\left(\frac{\partial T}{\partial s}\right)_p \frac{RT}{p} = \frac{1}{c_p}\frac{1}{\rho} , \tag{2.5.8}$$

where V is a specific volume and R is the universal gas constant.

Substituting (2.5.8) into (2.5.7), we get

$$\frac{dT}{dx} = -g/c_p . \tag{2.5.9}$$

Integrating (2.5.9) over the height leads to the dependence of the temperature on the vertical coordinate:

$$T_0 - T_r = -(g/c_p)x , \tag{2.5.10}$$

where T_r is the characteristic temperature equal to T_∞ for $x = 0$.

In the case of stable stratification, a decrease in the temperature of the medium decrease with increase in the height occurs more slowly, than in an adiabatically stratified medium, which is described by (2.5.10).

Thus, the use of the condition of stable stratification

$$\frac{dT}{dx} > -\rho g \left(\frac{dT}{dp}\right)_s \tag{2.5.11}$$

leads to the following dependences:

for an ideal gas

$$\frac{dT}{dx} > -g/c_p , \tag{2.5.12}$$

for dropping liquids

$$\frac{dT}{dx} > -\frac{g}{c_p}\beta T_0 , \tag{2.5.13}$$

since in this case

$$\left(\frac{\partial T}{\partial p}\right)_s = \left[\left(\frac{\partial T}{\partial s}\right)\left(\frac{\partial V}{\partial T}\right)\right]_p = \frac{T}{c_p}\left(\frac{\partial\,(1/\rho)}{\partial T}\right)_p = \frac{T}{c_p}\frac{\beta}{\rho} , \tag{2.5.14}$$

where $\beta = -(1/\rho)(\partial\rho/\partial T)_p$.

Equation (2.5.12) follows from (2.5.13) on substitution $\beta = 1/T$, which corresponds to the case of an ideal gas [2.18].

Thus, to determine the limits of stable stratification one has to find the temperature gradient corresponding to adiabatic stratification. For example, for ***a dry air*** at a temperature 300 K this gradient is equal to about $9.7 \cdot 10^{-3}$ K/m. For ***water*** under the same conditions the temperature gradient corresponding to adiabatic stratification is approximately equal to $1.26 \cdot 10^{-4}$ K/m, which is much less than for air. Therefore, the isothermicity condition is a good approximation to the condition of adiabatic stratification for dropping liquids.

In considering the applicability of self-similarity conditions to stratified media it was shown that in this case in the energy equation there appears an additional term which takes stratification into account. If the self-similar transformations of the variable and of the stream function are

$$\eta = b(x)y, \quad \Psi = \nu c(x) f(x, y),$$

then the self-similarity condition for the stratified medium can be written as [2.7, 2.18]

$$\frac{cj_x}{bd} = J \, , \tag{2.5.15}$$

where the ambient temperature changes as $T_0 - T_r = j(x)$, and the surface temperature is determined by the expression $T_w - T_0 = d(x)$. Then the value J, called the stratification parameter, must preserve its constant value over the height, as for the existence of the self-similar solution the dependence on x must be excluded. From the self-similarity condition for the stratified medium it follows that a change in the ambient temperature must be of the same form as a change in the surface temperature. The stratification parameter characterizes a relative value of temperature stratification, since it connects the change in the ambient temperature with the change in the surface temperature.

For a power-law change in the surface temperature, with the self-similarity condition (2.5.15) being satisfied, the system of basic equations and boundary conditions has the form

$$f''' + (n+3)\, f'' - 2\,(n+1)\, f' + \theta = 0 \, , \tag{2.5.16}$$

$$\frac{\theta'}{\mathrm{Pr}} + (n+3)\, f\theta - 4nf'\theta - Jf' = 0 \, , \tag{2.5.17}$$

$$f\,(0) = f'\,(0) = 1 - \vartheta\,(0) = f'\,(\infty) = \theta\,(\infty) = 0 \, .$$

The surface temperature is defined by the expression

$$T_w - T_0 = Nx^n = d\,(x) \, ,$$

and the ambient temperature by the expression

$$T_\infty - T_0 = N_0 x^n = j(x) \ .$$

The Grashof number Gr_x and the self-similar variable η are expressed in terms of $d(x)$ as

$$\mathrm{Gr}_x = \frac{g\beta x^3 d(x)}{\nu^2}, \quad \eta = \frac{\gamma}{x}(\mathrm{Gr}_x/4)^{1/4} \ .$$

In case of an exponential change in the temperature ($T_\mathrm{w} - T_0 = Me^{mx}$), the self-similarity is possible when

$$j_x = (JM/4)\, me^{mx}, \quad T_0 - T_r = (JM/4)\,(e^{mx} - 1) \ ,$$

where the characteristic temperature $T_r = T_0$ for $x = 0$.

The analysis of the stability condition for these problems shows that the stratification parameter J can be both positive and negative, with the positive values of J corresponding to the majority of cases of stable stratification [2.18].

Self-similar solutions of (2.5.16)–(2.5.17) ***for a vertical surface in a nonisothermal surrounding medium*** expressed in terms of f' and θ in dimensionless form and depending on Pr, n, and J were obtained in [2.68]. The local heat flux on the surface with the T_∞ changing exponentially, is defined by the formula

$$q(x) = \frac{[-\theta(0)]\, kNx^n}{x}\frac{G}{4} \sim x^{(5n-1)/4} \ , \tag{2.5.18}$$

where

$$G = 4(\mathrm{Gr}_x/4)^{1/4}, \quad \mathrm{Gr}_x = \left[g\beta x^3 (T_\mathrm{w} - T_0)\right]/\nu^2 \ . \tag{2.5.19}$$

Hence, just as in the case of a nonstratified medium, the value of $n = 1/5$ corresponds to the vertical surface with a constant heat flux. Then

$$N = \left(\frac{q}{k(-\theta'(0))}\right)^{4/5}\left(\frac{4\nu^2}{g\beta}\right)^{1/5} \ . \tag{2.5.20}$$

However, since in a stratified medium $-\theta'(0)$ depends on J, a change in the temperature of the surface exceeding the constant determining temperature T_r also depends on the stratification parameter [2.18]

$$T_\mathrm{w} - T_r = N'x^n \ , \quad \text{where} \quad N' = N(1 + J/4n) \ . \tag{2.5.21}$$

The value of N refers to a vertical surface with a constant heat flux in a nonstratified medium.

Thus, for the given distribution of the surface temperature $T_\mathrm{w} - T_0 = Nx^n$ such local characteristics as heat flux $q(x)$, heat transfer coefficient $h(x)$, and the Nusselt number $\mathrm{Nu}(x)$ for stratified media reach larger values in comparison with a similar situation in the absence of stratification because of

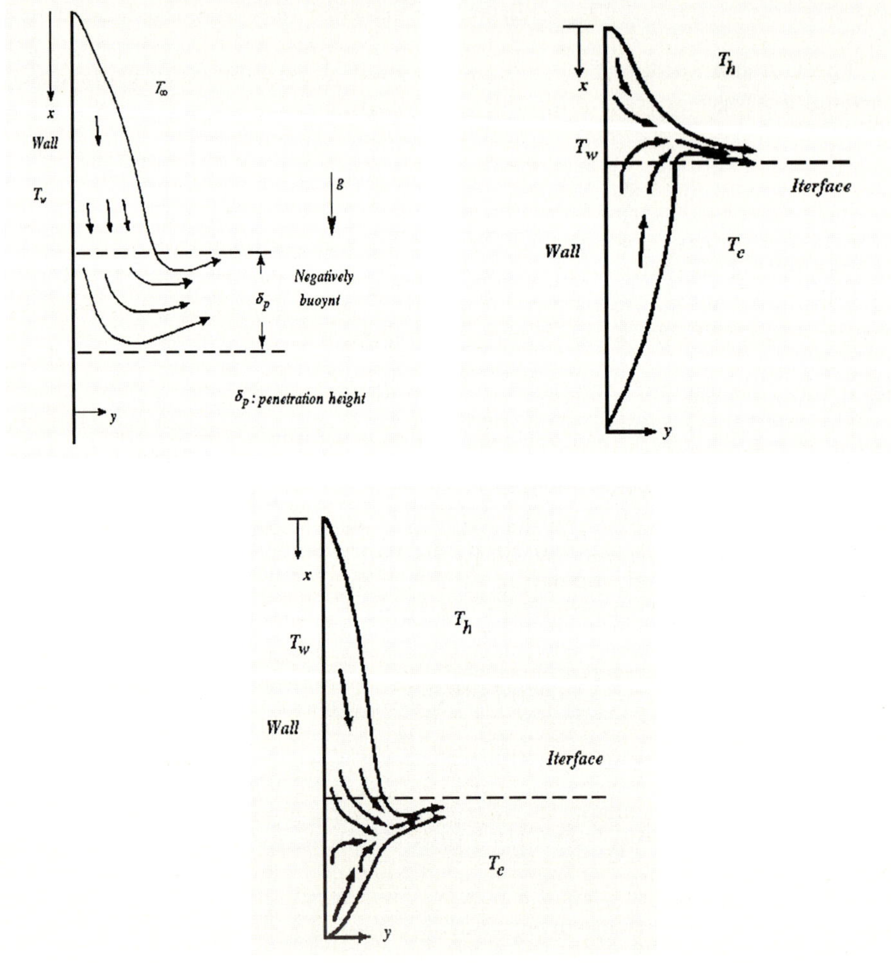

Fig. 2.5.2. The structure of free-convective flow near a vertical plane surface in a stratified medium [2.86]

the larger value of $-\theta'(0)$. This seems to be a consequence of the effect of the ascending liquid flow which has the temperature smaller than the local value of T_0. If a constant temperature is maintained on the surface or it is changed by a given law, the local temperature difference will be smaller in the case of the increase in the ambient temperature with x, which leads to a decrease in the heat flux even on increase on the heat transfer coefficient. Thus, for a constant surface temperature heat transfer from an isothermal surface placed in a stably stratified medium appears to be smaller than the heat transfer in an isothermal medium, since the effect of stratification leads to a decrease in the difference between the temperatures of the surface and the surrounding medium and to a decrease in the flow velocity [2.18].

The investigation of natural convection heat transfer from vertical surfaces on change in the ambient temperature according to

$$T_\infty = Bx^{-r} - T_\mathrm{w}$$

was carried out in [2.69]. The case $r < 0$ corresponds to the increase of temperature with x (stable stratification), and the case $r > 0$ represents the condition of instability. It is shown that when $r < 0$, the local number Nusselt is larger, and the temperature profile near the wall is sharper than in the absence of stratification. In the outer part of the boundary layer negative values of θ were noticed, but this local decrease in the temperature appeared to be insufficient for the flow direction inversion.

In [2.70] a detailed study of self-similar solutions was made for a vertical surface in a stratified medium for a wide range of variation of Pr, n, and J. The limit values of the Prandtl number are considered, and it is shown, that at these values the change in the flow parameters is independent of the thermal stratification of the medium. When $\mathrm{Pr} \to 0$, the friction stress on the wall and the heat flux change as $\mathrm{Pr}^{1/2}$, whereas for $\mathrm{Pr} \to \infty$ the heat flux increases as $\mathrm{Pr}^{1/4}$, whereas the friction stress on the wall changes as $\mathrm{Pr}^{-1/4}$. In case where the temperature increases rapidly with x, and the temperature excess in the boundary layer lags behind the increase in the ambient temperature, the flow rotates through 180° due to the negative values of θ, which is also noted in [2.68]. The regions of reversed flow and the regions with the temperature smaller than the ambient temperature are determined in [2.71].

In the majority of cases, thermal stratification decreases with time. The presence of a temperature difference in a stratified medium leads to an increase in heat transfer by heat conduction, and, moreover, heat is transferred from the region of stratification to the surrounding layers. These effects amplify the process of mixing, and eventually lead to gradual disappearance of the stratification. However in the majority of practically interesting problems the characteristic time of the decay of stratification greatly exceeds the characteristic times connected with transfer mechanisms, therefore the flow in stratified media can usually be considered as a quasi-stationary one.

The experimental investigations of quasi-stationary stratification [2.68, 2.72, 2.73] showed, that in the course of time the arbitrary stratification approaches that with linear temperature distribution which decays more slowly than other forms of stratification.

The investigations carried out in a stratified atmosphere [2.74] relate to such flow conditions which appear near an infinite inclined surface whose temperature is higher than that of the surrounding air. The excess temperature is supposed to be constant over the entire surface, and the medium is considered to be steadily and linearly stratified. The resulting flow is one-dimensional and parallel, and the velocity and temperature fields depend only on the distance from the surface.

If the infinite vertical surface is located in a stratified fluid with linear distribution of the temperature, then the constant vertical temperature gradient N_∞ is positive. The difference between the surface temperature and that of the surrounding medium is also maintained constant, i.e.,

$$T = N_\infty x \quad \text{for y} \to \infty , \tag{2.5.22}$$

$$T = \Delta T + N_\infty x \quad \text{for y} = 0 . \tag{2.5.23}$$

For such a flow there exists a simple solution of the form

$$T = \Delta T \theta (y) + N_\infty x , \tag{2.5.24}$$

$$V_x = V_x (y) , \tag{2.5.25}$$

where θ and V_x must satisfy the following boundary conditions:

$$\begin{aligned} \theta = 1 , \quad V_x = 0 \quad \text{for } y \to \infty , \\ \theta = 0 , \quad V_x = 0 \quad \text{for } y = 0 . \end{aligned} \tag{2.5.26}$$

The stratification in this vertical flow does not decay with time, since the temperature gradient is the same everywhere, and, consiquently, the heat flux supplied to each fluid element is equal to the heat flux from it. Here, the momentum and energy equations are [2.18]

$$\nu \frac{d^2 V_x}{dy^2} + g\beta \Delta T = 0 , \tag{2.5.27}$$

$$N_\infty V_x = a \Delta T \frac{d^2 \theta}{dy^2} . \tag{2.5.28}$$

These equations with boundary conditions (2.5.22)–(2.5.23) determine the solution in the form

$$\theta = e^{-y'} \cos y' , \tag{2.5.29}$$

$$V_x = \left(\frac{g\beta a}{\nu N_\infty} \right)^{1/2} \Delta T e^{-y'} \sin y' . \tag{2.5.30}$$

where $y' = y/l$, and l is the horizontal linear scale defined by the formula

$$l = \left(\frac{4\nu a}{g\beta N_\infty} \right)^{1/4} . \tag{2.5.31}$$

A similar result for the inclined surface was obtained by Prandtl [2.74] only as l the following scale is used:

$$l = \left(\frac{4\nu a}{g\beta N_\infty \sin^2 \gamma} \right)^{1/4} , \tag{2.5.32}$$

where γ is the angle of surface inclination to the horizontal.

Such solutions for a surface infinite in two directions correspond to stable stationary stratification. They also are obtained in studying the influence of instability of natural convection near inclined [2.75] and vertical [2.76] surfaces in a stably stratified medium. The analysis of the stability and transient processes in a free-convective vertical boundary layer in a stably stratified medium carried out in [2.68] shows, that the stable stratification improves the flow stability and delays transition.

The calculations of heat transfer ***of a vertical isothermal surface in a stratified medium with the temperature increasing linearly with height*** made on the basis of the theory of a thin laminar layer lead to the following dependences for the average Nusselt number [2.77, 2.78]:

$$\frac{\mathrm{Nu}}{\mathrm{Nu}_{iso}} = F\,(\mathrm{Pr}, S)\left(\frac{S+2}{2}\right)^{1/2} \qquad S \leq 2\,,$$
$$\frac{\mathrm{Nu}}{\mathrm{Nu}_{iso}} = F\,(\mathrm{Pr}, S)\,(2S)^{1/4} \qquad S \geq 2\,, \tag{2.5.33}$$

where $S = L/\Delta T_m(dT_\infty/dz)$ is the stratification parameter, $\Delta T_\mathrm{m} = T_\mathrm{w} - T_\infty$ is calculated in the middle boundary-layer section near the plate, L is the plate height, Nu_{iso} is the average Nusselt number calculated for a vertical plate in the case, when T_w and T_∞ are constant, and, consequently, the value of ΔT_m is also constant along the entire surface. The Nusselt number calculated on the basis of L and ΔT_m. The functions $F(\mathrm{Pr},\, S)$ are approximately described by the dependences

$$\begin{aligned} F &= 1 + 0.086(S/2) \quad \mathrm{Pr} = 0.71\,,\\ F &= 1 - 0.008(S/2) \quad \mathrm{Pr} \approx 6\,,\\ F &= 1 - 0.067(S/2) \quad \mathrm{Pr} \to \infty\,. \end{aligned} \tag{2.5.34}$$

For $0 \leq S \leq 2$ the value of $T_\mathrm{w} - T_\infty$ is positive within the entire boundary layer; for $S = 2, T_\mathrm{w} = T_\infty$ at $z = L$, and for $S \to \infty T_\mathrm{w} < T_\infty$ for the upper half of the plate and $T_\mathrm{w} > T_\infty$ for the lower one.

For the case of ***turbulent flow near a vertical surface in a stably stratified medium*** for $0 \leq S \leq 2$

$$\mathrm{Nu}_t = C_t\,\mathrm{Ra}^{1/3}\frac{3}{7S}\left[\left(\frac{2+S}{2}\right)^{7/3} - \left(\frac{2-S}{2}\right)^{7/3}\right], \tag{2.5.35}$$

where $\mathrm{Ra} = g\beta\Delta T_m L^3/\nu a$. For $S = 2$ the average value of heat transfer is only by 8% exceeds the values calculated on the assumption that the temperature difference is constant along the entire surface and is equal to ΔT_m.

Under the conditions of ***a linear change in the wall and ambient temperatures for a vertical surface in a stratified medium*** $\theta_\mathrm{w} = \theta_0 m x$ and $\theta_\infty = \theta_0(1+m)x$ [2.79]:

$$\mathrm{Nu}_x = \frac{1}{\sqrt{2}} \left(|m| \ \mathrm{Ra}_x\right)^{1/4} ,$$

$$\mathrm{Pr} = 0.01 - 100, \quad |m| \gg 1 . \tag{2.5.36}$$

The solution is unstable for $-1 < m < 0$.

For a power-law change in the wall and ambient temperatures $\theta_\mathrm{w} = \theta_0 m x^n$ and $\theta_\infty = \theta_0(1+m)x^n$ [2.70]:

$$\mathrm{Nu}_x = \frac{-\theta'(0)}{\sqrt{2}} \mathrm{Gr}_x^{1/4} . \tag{2.5.37}$$

The graphs of $\theta'(0)$ are presented in Fig. 2.5.3 for two values of Pr.

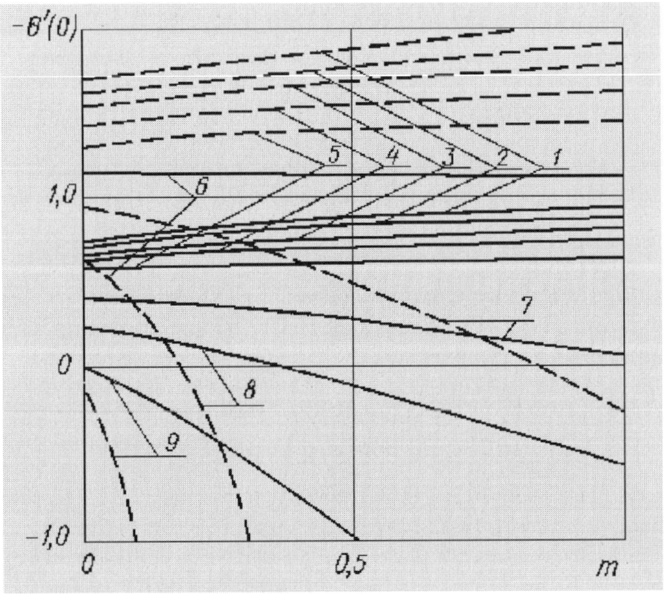

Fig. 2.5.3. The graph of $\theta'(0)$ for Pr = 0.72 (*solid lines*) and Pr = 10 (*dashed lines*): (1) $m = 1$, (2) 0.8, (3) 0.6, (4) 0.4, (5) 0.2, (6) 0, (7) −0.2, (8) −0.4, (9) −0.6

The results of the calculations carried out for different m and n [2.80–2.82] yield the following approximated dependences:

$$\mathrm{Nu}_x = C_s\left(\mathrm{Pr}, m\right) \ \mathrm{Ra}_x^{1/4} , \tag{2.5.38}$$

$$\mathrm{Nu}_x = 1/[5^{1/5}\theta\left(0\right)]\mathrm{Gr}_x^{*1/5} , \tag{2.5.39}$$

$$\mathrm{Nu}_x = \frac{3}{4} C \left[1 + \frac{5n}{3} + 4\left(k+1\right)\frac{mn}{3}\right]^{1/4} \ \mathrm{Ra}_x^{1/4} . \tag{2.5.40}$$

The values of the coefficients C_s, k, and $\theta(0)$ are given in Tables 2.5.1, 2.5.2, and 2.5.3.

Table 2.5.1. The values of $C_s(\Pr, m)$

m	$\Pr = 0.7$	$\Pr = 100$	$\Pr = \infty$
0	0.515	0.549	0.562
0.2	0.534	0.553	0.555
1	0.590	0.602	–

Table 2.5.2. The Values of $\theta(0)$

Pr	0.72	1	5	10
$\theta(0)$	1.438106	1.3085	0.8595256	0.7312685

Table 2.5.3. The values of k

k	2	2.5	3.5
Pr	10	5	0.7

In the case of ***a constant wall temperature and a variable ambient temperature*** ($\theta_\infty = \theta_0 x^{0.7}$) [2.78]

$$\frac{\mathrm{Nu}_x}{\mathrm{Gr}_{x0}^{1/4}} = \frac{1}{\sqrt{2}}(1.0315691 - 0.99213580\theta - 0.20880460\theta^2 + 0.22165380\theta^3) , \tag{2.5.41}$$

$$\Pr = 6.5, \mathrm{Gr}_{x0} = g\beta x^3(T_\mathrm{w} - T_0)/\nu^2, \theta = (T_\infty - T)_0(T_\mathrm{w} - T_0) .$$

If ***the wall temperature*** T_w ***is constant, and the temperature of the surrounding fluid grows from the leading edge as*** $T_\infty = T_0 + ax$, then the local temperature difference $\theta_\mathrm{w} = \theta_0(1 - ax)$ decreases linearly with increase in the coordinate x. The positive values of $a = dT_\infty/dx$ correspond to a hydrodynamically stable state for the majority of fluids [2.83]. The local and average values of the heat transfer coefficient for $ax \ll 1$ [2.71] are

$$\frac{\mathrm{Nu}_x}{\mathrm{Ra}_x^{1/4}} = \frac{3}{4}C\left[1 - ax\frac{\theta_1'(0)}{\theta_0'(0)} + (ax)^2\frac{\theta_2'(0)}{\theta_0'(0)}\right]\frac{1}{(1-ax)^{5/4}} , \tag{2.5.42}$$

$$\frac{\mathrm{Nu}_x}{\mathrm{Ra}_{x,\mathrm{av}}^{1/4}} = C\left[1 - \frac{3}{7}ax\frac{\theta_1'(0)}{\theta_0'(0)} + \frac{3}{11}(ax)^2\frac{\theta_2'(0)}{\theta_0'(0)}\right]\frac{1}{(1-\frac{1}{2}ax)^{5/4}} , \tag{2.5.43}$$

where $\mathrm{Ra}_{x,\mathrm{av}} = (\frac{g\beta x^3}{\nu^2})\theta_0(1 - \frac{1}{2}ax)$.

The values of the coefficients $\theta_i'(0)$ are presented in Table 2.5.4.

The correlation dependences for calculating the local and average values of the heat transfer coefficient for two Pr [2.78] are

Table 2.5.4. The Values of $\theta_i'(0)$

Pr	$-\theta_0'(0)$	$-\theta_1'(0)$	$-\theta_2'(0)$
100	2.191	4.790	16.06
30	1.589	2.171	2.048
10	1.169	1.161	0.4967
6	1.007	0.9069	0.2750
2	0.7165	0.5834	0.0645
1	0.5672	0.4580	0.0135
0.7	0.4995	0.4047	0.0010
0.3	0.3624	0.2974	–0.0098
0.1	0.2302	0.1921	–0.0096

for $S \le 2$

$$\frac{\mathrm{Nu}_x}{\mathrm{Ra}_{x,\mathrm{av}}^{1/4}} = \frac{3}{4} C \left[1 + c_1 ax + c_2 (ax)^2 + (ax)^3 + (ax)^4\right] , \tag{2.5.44}$$

$$\frac{\mathrm{Nu}_x}{\mathrm{Ra}_{x,\mathrm{av}}^{1/4}} = C \left[1 + \frac{3}{7} c_1 ax + \frac{3}{11} c_2 (ax)^2 + \frac{3}{15} (ax)^3 + \frac{3}{19} (ax)^4\right] , \tag{2.5.45}$$

for $S \ge 2$

$$\mathrm{Nu}_{\mathrm{av}} = C c_5 \left(S\, \mathrm{Ra}_{x0}\right)^{1/4} . \tag{2.5.46}$$

The values of the coefficients c_i are presented in Table 2.5.5.

Table 2.5.5. The Values of c_i

Pr	c_1	c_2	c_3	c_4	c_5
6.0	–0.8620	–0.1989	0.0865	–0.0256	1.179
0.7	–0.7572	–0.7485	2.2715	–1.1770	1.148

The following approximated dependences were also calculated [2.82]:

for $S \le 2$

$$\mathrm{Nu}_{\mathrm{av}} = C\, \mathrm{Ra}_{x,\mathrm{av}}^{1/4} \left(\frac{S+2}{S}\right)^{1/2} F(k, S) , \tag{2.5.47}$$

for $S \ge 2$

$$\mathrm{Nu}_{\mathrm{av}} = C\, \mathrm{Ra}_{x0}^{1/4} (2S)^{1/4} F(k, 2) . \tag{2.5.48}$$

The values of $F(k, s)$ are presented in Table 2.5.6.

The experiments for the fluids with the number $\mathrm{Pr} = 5.5 - 7.5$ [2.78] show that the deviation from the data calculated by (2.5.45)–(2.5.46) comes to $\pm 2.5\%$; for $S = 1$ ($ax = 2/3$) formula (2.5.44) gives the results underestimated by 9%.

Table 2.5.6. The Values of $F(k, s)$

Pr	S = 0	S = 0.25	S = 0.5	S = 0.75	S = 1.0	S = 1.25	S = 1.5	S = 1.75	S = 2.0
10	1	0.992	0.986	0.982	0.979	0.977	0.975	0.974	0.974
0.7	1.001	1.015	1.027	1.035	1.040	1.044	1.046	1.047	1.047

Heat transfer from a vertical plane wall with ***a constant heat flux on the surface and a linearly increasing ambient temperature in liquid*** **Na** is determined with an error of 1% from the equation [2.39]:

$$\mathrm{Nu}_x = 0.568 \left(\mathrm{Gr}_x^* \, \mathrm{Pr}^2\right)^{0.225} , \tag{2.5.49}$$

$$\mathrm{Pr} = 0.0068, \quad \mathrm{Gr}_x^* = 5 \cdot 10^7 - 10^{10}, \quad S = 0.4 .$$

In a vertical layer with insulated end faces and isothermal walls at the temperatures T_{w1} ***and*** T_{w2} in the case of a turbulent flow regime the parameter S in the middle part is constant and equal to 0.36 ± 0.04 [2.84]. The heat transfer ***from the vertical surface of constant temperature*** with a positive parameter S is determined from the formulas ($\mathrm{Ra}_x = g\beta x^3 (T_{\mathrm{w1}} - T_{\mathrm{w2}})/\nu^2$):

$$\mathrm{Pr} > 2, \quad \mathrm{Ra}_x = 5 \cdot 10^{10} - 2 \cdot 10^{11}, \quad S = 0.36 \pm 0.04,$$

$$\mathrm{Nu}_x = 0.108 \, \mathrm{Ra}_x^{1/3} . \tag{2.5.50}$$

The experimental data for the average temperature field and local heat transfer coefficient in a free-convective boundary layer ***near a vertical isothermal surface located in a thermally stratified medium*** are given in [2.85].

The results of numerical analysis presented in [2.86] for free-convective heat transfer from an ***isothermal vertical surface in a stably stratified medium saturated with a liquid having a low porosity which obeys the Darcy law*** can be generalized by the relation

$$\frac{\mathrm{Nu}}{\mathrm{Nu}_0} = 1 - 0.44S , \tag{2.5.51}$$

where $0 \le S \le 1.5$ is the parameter of thermal stratification and Nu_0 is the Nusselt number for $S = 0$.

The experimental investigation of free-convective flow and heat transfer from an ***isothermal vertical surface in a stably stratified medium consisting of two isothermal layers of the same fluid heated up to different temperatures*** was carried out in [2.87].

2.6 Conjugated Problems on Vertical Surface

In the majority of free-convective heat transfer investigations the distribution of the temperature or of a heat flux on a vertical surface is assumed to be given. However, under in real physical conditions these situations are encountered rather rarely. Much more often the real boundary conditions are determined by solving the problem of heat conduction in a solid body. In turn, the boundary conditions for the equation of heat conduction contain the distribution of the local heat transfer coefficient on the body surface. This leads to the necessity of solving a joint problem including free convection and heat conduction equations. These problems are called conjugated. The necessity of conjugated statement of the problem arises, for example, in analysis of cooling electronic equipment, calculation of thermal insulation of buildings, designing furnaces, in solidification in moulds.

Physically and mathematically the problem of calculation of free-convective heat transfer in conjugated statement resembles the problem of flow around a body with account for the effect of the boundary layer on the external flow. In both cases, this is the question of the effect of transfer processes in the boundary layer on the external flow. The solution of this problem requires application of iteration methods. In both cases the system containing elliptic and parabolic equations is solved jointly.

The general investigation of conjugated problems of convective heat transfer was carried out by Luikov [2.88], who offered the Brun number as the conjugation criterion for these problems:

$$\mathrm{Br} = \frac{\lambda\,\delta}{\lambda_\mathrm{b}\,x}\mathrm{Gr}_x^m\,\mathrm{Pr}^n\,. \tag{2.6.1}$$

In [2.88] the conditions were obtained, under which the traditional statement of the problem is possible, when the transfer processes in the body and in the surrounding medium can be considered separately. From the condition of the equality of temperatures and heat fluxes for the body and the surrounding medium

$$T = T_\mathrm{b}, \quad \lambda\frac{\partial T}{\partial n} = \lambda_\mathrm{b}\frac{\partial T_\mathrm{b}}{\partial n}\,, \tag{2.6.2}$$

one can get the relationship between the temperature differences corresponding to them:

$$\frac{\Delta T_\mathrm{b}}{\Delta T_\mathrm{w}} = \frac{\lambda}{\lambda_\mathrm{b}}\frac{\delta}{x}\,\mathrm{Nu}_x\,, \tag{2.6.3}$$

where the local Nusselt number depends on the local values of the Prandtl and Grashof numbers, i.e.,

$$\frac{\Delta T_\mathrm{b}}{\Delta T_\mathrm{w}} = C\frac{\lambda}{\lambda_\mathrm{b}}\frac{\delta}{x}\mathrm{Gr}_x^m\,\mathrm{Pr}^n\,. \tag{2.6.4}$$

From (2.6.4) it is seen, that at small values of the Brun number intense processes of heat conduction in a body lead to the fast equalizing of its temperature. In this case, the temperature difference is insignificant, which enables one to solve the problem of convective heat transfer without account for conjugation, using the wall temperature constancy as a boundary condition [2.44].

At the present time, there are different variants of mathematical models of two-dimensional and three-dimensional conjugated problems of natural convection, whose review is given in [2.89].

Figure 2.6.1 presents some of the most frequently encountered types of conjugated problems of free convection.

The application of parametrical methods to the solution of conjugated problems allows one to obtain analytical dependences for the surface temperature [2.90].

A two-dimensional problem of heat conduction for a vertical semi-infinite plate with heat sources distributed in it is solved by means of the Fourier transformation [2.91]. The boundary layer on the plate surface is calculated using the Goertler series [2.92]. The problem parameters are the ratio of thermal conductivities of the liquid and of the plate material and also a dimensionless function of heat generation. In [2.93]] the plate with a constant temperature or heat flux density on one side and with a free-convective heat transfer on the other is investigated. The temperature inside the plate is calculated by the methods of the heat conduction theory, the free-convective heat transfer is calculated by the method of local self-similarity in Goertler variables, with the dependence of the dimensionless derivative of temperature along the transverse coordinate on the wall on the problem parameter being approximated by a linear function. The plate surface temperature has the form of the fourth order polynomial along the longitudinal coordinate. The dimensionless parameter of the problem for cases of constant temperature and constant heat flux is determined differently. The experimental investigation ***for the case of a constant heat flux*** is carried out in [2.94]. The experimental and theoretical studies of a heat conductive plate with tape heaters on a surface (Fig. 2.6.1 h) is carried out in [2.95]. The system of the initial equations consists of the Laplace equation for stationary heat conduction in the plate and the equations of a free-convective boundary layer:

$$
\begin{gathered}
u\frac{du}{dx} + \nu\frac{du}{dy} = g\beta\theta + \nu\frac{\partial^2 u}{\partial y^2}\,, \\
\frac{du}{dx} + \frac{dv}{dy} = 0\,, \\
u\frac{d\theta}{dx} + \nu\frac{d\theta}{dy} = \alpha\left(\frac{\partial^2\theta}{\partial x^2} + \frac{\partial^2\theta}{\partial y^2}\right)
\end{gathered}
\tag{2.6.5}
$$

and the energy equation for stationary heat conduction in a solid body is

$$
\frac{\partial^2\theta}{\partial x^2} + \frac{\partial^2\theta}{\partial y^2} = 0\,. \tag{2.6.6}
$$

The longitudinal heat conduction in the energy equation for a boundary layer is caused by the presence of large vertical temperature gradients which

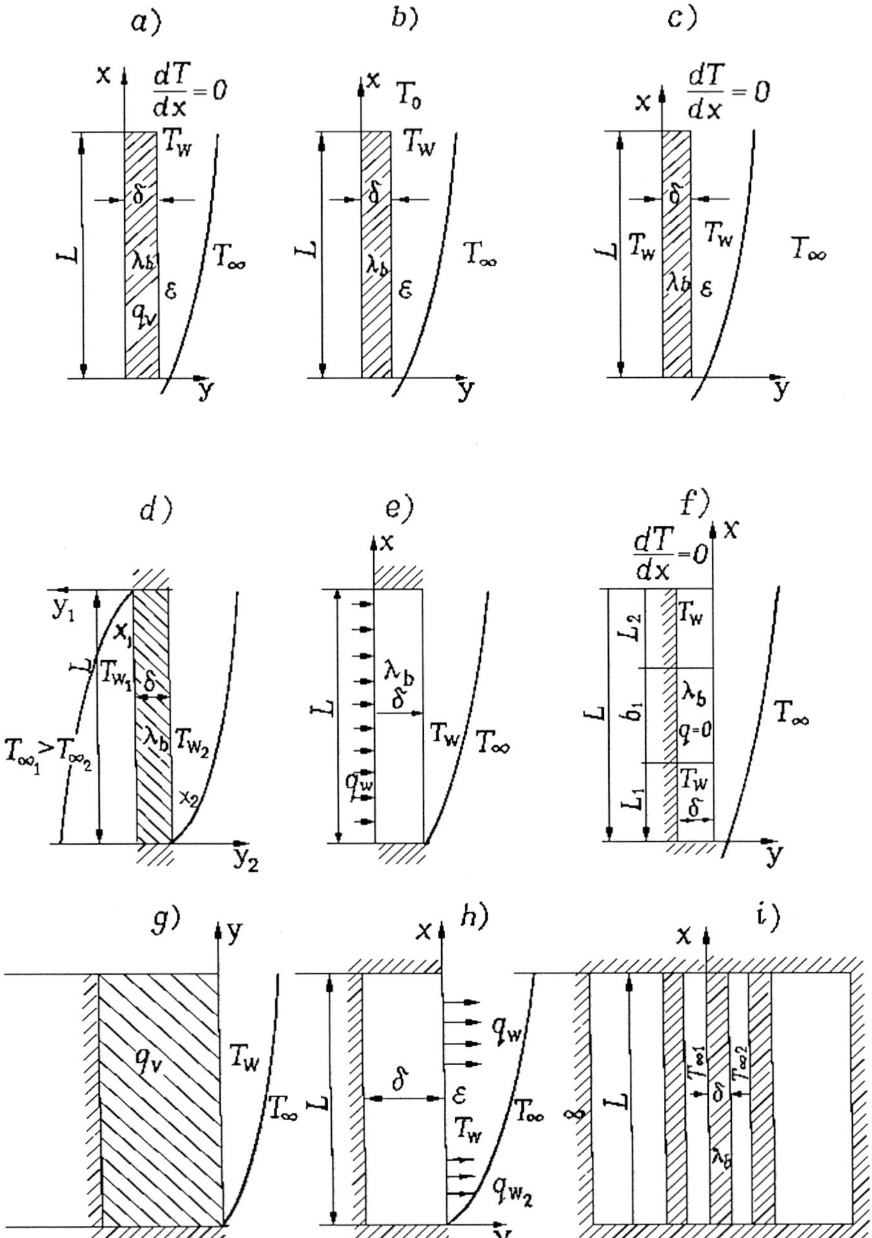

Fig. 2.6.1. Conjugated problems of free convection heat transfer on a vertical plane surface

appear in a fluid near the interface due to the action of local heat sources. Heat transfer by radiation is also taken into account. Flow visualization was carried out using a laser holographic interferometer. The local heat transfer coefficient changes appreciably along the plate and the greater the ratio of coefficients, the smaller the heat conduction. For a glass plate the values of the local Nusselt number in the upper part (for heating from below) become negative (inversion of the local heat flux occurs). This means that the heat transfer in the plate due to heat conduction is so small that separate parts thermally interact only through the boundary layer. The upper part of the plate is heated by an ascending flow. If the ratio of heat conduction coefficients has the order of 1, then the greater amount of heat generated by the heaters remains in their vicinity [2.90]. In the case of one source, the temperature gradient in the fluid on the interface decreases over the longitudinal coordinate downstream, which corresponds to the approximation to an adiabatic regime. In the presence of several heat sources the temperature gradient and the boundary layer thickness increase in the same direction [2.95].

The works of many researchers [2.44, 2.90, 2.96–2.103] are devoted to theoretical and experimental investigations of conjugated problems of free-convective heat transfer ***on a vertical plane nonisothermal surface under various boundary conditions***. We consider some characteristic cases presented in Fig. 2.6.1.

A thin vertical plate gives off heat by free convection and radiation from a side surface. The upper end of the plate is insulated, a constant temperature T_0 is maintained on the lower one. In general, volumetric heat generation of power q can occur (Fig. 2.6.1a).

A detailed analysis with the use of the dimensional analysis method for a vertical plate ***insulated from the upper and lower ends with a constant heat flux maintained on one of the side surfaces*** (Fig. 2.6.1e) and for ***a vertical rib heated at its base in the absence of volumetric heat generation*** (Fig. 2.6.1c) is given in [2.90]. If the plate length l considerably exceeds its thickness δ, and $\mathrm{Bi} \ll 1$, then the heat conduction process can be considered one-dimensional and it can be described by the heat conduction equation

$$\lambda\delta_{\mathrm{b}}\frac{d^2\theta_{\mathrm{w}}}{dx^2} = 2\alpha_x\theta_{\mathrm{w}} + 2\sigma\varepsilon\left[(\theta_{\mathrm{w}} + T_\infty)^4 - T_\infty^4\right] . \tag{2.6.7}$$

where ε is the emissivity of the plate investigated, and α_x is the heat transfer coefficient

$$\alpha_x = \frac{\lambda}{x}\mathrm{Gr}_x^{1/4} f\left(\mathrm{Pr}, \mathrm{P}_\vartheta\right) . \tag{2.6.8}$$

Nondimensionalization of (2.6.7) leads to the equation

$$\frac{d^2\Theta_{\mathrm{w}}}{d\zeta^2} = 2\frac{\lambda}{\lambda_{\mathrm{b}}}\frac{1}{\delta}\mathrm{Gr}_x^{1/4} f\left(\mathrm{Pr}, \mathrm{P}_\vartheta\right)\frac{1}{\zeta^{1/4}}\Theta_{\mathrm{w}}^{5/4} + 2\frac{\varepsilon\sigma T_\infty^4 l^2}{\delta\lambda_{\mathrm{b}}\theta_0}\left[\left(1 + \frac{\theta_0}{T_\infty}\Theta_{\mathrm{w}}\right)^4 - 1\right] , \tag{2.6.9}$$

$$\Theta_{\mathrm{w}} = \frac{\theta_{\mathrm{w}}}{\theta_0}\,, \qquad \zeta = \frac{x}{l}\,.$$

subject to the boundary conditions

$$\theta_{\mathrm{w}} = 1 \quad \text{for} \quad \xi = 0, \qquad \frac{d\theta_{\mathrm{w}}}{d\xi} = 0 \quad \text{for} \quad \xi = 1\,. \tag{2.6.10}$$

Free convection ***on a vertical plane nonisothermal surface*** is determined by the solution of the equations

$$\begin{gathered} u\frac{du}{dx} + \nu\frac{du}{dy} = g\beta\theta + \nu\frac{\partial^2 u}{\partial y^2}\,, \\ \frac{du}{dx} + \frac{dv}{dy} = 0\,, \\ u\frac{d\theta}{dx} + \nu\frac{d\theta}{dy} = \alpha\frac{\partial^2\theta}{\partial y^2} \end{gathered} \tag{2.6.11}$$

with the boundary conditions

$$\begin{gathered} u = 0, \quad \theta = \theta_{\mathrm{w}} \quad \text{or} \quad q_{\mathrm{w}} = -\lambda\frac{\partial\theta}{\partial y} \quad \text{for} \quad y = 0\,, \\ u = 0, \quad \theta = 0 \quad \text{for} \quad y \to \infty\,. \end{gathered} \tag{2.6.12}$$

An iteration procedure for solving the given nonlinear equations is suggested in [2.90]. The stability condition is determined by the sign of the total heat transfer in each section. Negative values of the excess temperature of the plate are associated with boundary layer separation [2.104]. It is established experimentally that a change in the direction of the total heat transfer leads to disturbance of the stationary regime and to appearance of wave oscillations along the plate length and in time.

The heat transfer coefficient $\alpha = -(\lambda_\infty/\theta_{\mathrm{w}})(\partial T/\partial y)_0$ is determined from the solution of the free-convective boundary layer equations. The conjugation criterion for the heat flux direction in the plate along the x axis is the quantity $\mathrm{P} = (\lambda_\infty/\theta_{\mathrm{w}})(l/b)\mathrm{Gr}_1^{1/4}$. In the limiting cases of the change of the Pr number, the modified conjugation criterion is suggested [2.26]:

$$P = (\lambda_\infty/\theta_{\mathrm{w}})\,(1/b)\left[\mathrm{Gr}_1\ \mathrm{Pr}^2/\,(1+\mathrm{Pr})\right]^{1/4}\,.$$

A similar problem for $\mathrm{Ki} = 0$ was investigated in [2.97, 2.98] using the integral method. The theoretical analysis by simultaneous solution of the boundary layer equations in self-similar variables and of the equation of heat conduction in the plate was carried out in [2.99–2.101].

In the case where ***the lower and upper ends of the plate are insulated, and the power of the internal heat sources is q*** (Fig. 2.6.1g), the heat transfer inside the plate obeys the equation

$$\frac{\delta \lambda_b \theta_0}{L^2} \frac{d^2\Theta}{d\zeta^2} = \alpha\theta_0\Theta - q\delta + \varepsilon\sigma T_\infty^4 \left[\left(1 + \frac{\theta_0}{T_\infty}\Theta \right)^4 - 1 \right]$$

with the boundary conditions

$$\frac{d\Theta}{d\zeta} = 0 \quad (\zeta = 0, \zeta = 1) \ .$$

This kind of problems appear in calculation of the heat transfer of ***the elements of electrical equipment with volumetric heat generation due to Joule dissipation*** $q = \rho I^2$, where ρ is the specific resistance and I is the current density.

The analysis of the distribution of the dimensionless heat transfer coefficient obtained numerically shows that at the conjugation parameter values exceeding 500 in the mid portion of the plate there is a region, where the heat transfer coefficient changes but little, and its value is close to 0.4, which corresponds to the power-law temperature distribution with an exponential index equal to 0.2, or, in other words, to distribution of the height-constant heat flux over the wall. The case of the height-constant heat flux is limiting for the given conjugated problem and corresponds to the conditions of complete absence of heat conduction [2.90].

The presence of intense radiative heat transfer leads to a decrease in the plate temperature and to a better its distribution over the length.

Experimental investigations were carried out ***on a steel plate*** of width 10 cm, thickness 1 cm, and height 45, 55, and 65 cm at the excess temperature at the base $\theta_w < 150°C$. On the radiating plate in a compressible fluid the heat transfer coefficient decreases over the height, and for $l > 0.5$ m it can become negative, but the resultant heat flux from the cooled plate remains positive, since in the upper part the radiating component has a significant effect. For the heated plate the heat transfer coefficient increases with height 2 – 2.5 times. The experimental data on surface temperature and velocity and temperature distributions in the boundary layer are given in [2.96,2.99,2.100], and on the total heat transfer of a conducting plate of width 2.5 cm, thickness 0.5 cm and height 10, 20, and 40 cm with the excess temperature at the base $\theta_w = 6 - 50°C$ – in [2.102].

The calculation results for the case of heat transfer ***to air*** (Pr = 0.7) in the absence of radiation show that at small values of the conjugation parameter the plate temperature changes little along the length and insignificantly depends on the way the base temperature is prescribed. The average heat transfer coefficient in this case is calculated by formula (2.1.17). Unlike an isothermal plate, the complex $\mathrm{Nu}_x/\mathrm{Gr}_x^{1/4}$ cannot be considered constant [2.44].

For a ***triangular rib with the given temperature at the base and zero heat flux at the apex*** the heat transfer coefficient increases ***in air*** with the conjugation parameter. The effect of the Pr number and of the rib profile on heat transfer is insignificant [2.103].

In a vertical heat conductive wall of thickness δ which separates fluids with different properties and temperatures T_{w1} and T_{w2} (Fig. 2.6.1d) the main heat flux is directed along the y axis. Here the conjugation criteria are the complexes

$$\kappa = \frac{\lambda_{\infty 1}}{\lambda_{\infty 2}} \left(\frac{\mathrm{Ra}_2}{\mathrm{Ra}_1} \right)^{1/4} \left[\frac{\mathrm{Pr}_2(1+\mathrm{Pr}_1)}{\mathrm{Pr}_1(1+\mathrm{Pr}_2)} \right]^{1/4} ,$$

$$P = \frac{\delta}{l} \frac{\lambda_{\infty 1}}{\lambda_{\infty 2}} \mathrm{Ra}_2^{1/4} \left(\frac{\mathrm{Pr}_2}{1+\mathrm{Pr}_2} \right)^{1/4} .$$

For $\mathrm{Pr}_1 = \mathrm{Pr}_2 = 0.7$ and $\kappa = 1$ [2.105]

$$\mathrm{Nu} = (0.223 - 0.0164P)\,\mathrm{Ra}_1^{1/4} . \quad (2.6.13)$$

An approximate temperature distribution over the height and an expression for the wall-average heat transfer for a ***plate with insulated ends faces*** are [2.106]

$$\frac{T_{\mathrm{w}} - T_{\infty 2}}{T_{\infty 1} - T_{\infty 2}} = \frac{1}{2} \left(1 + C\mathrm{Pr}_2\ \mathrm{Pr}^{1/4} \right) \left(\frac{x}{l/2} \right)^{1/5} , \quad (2.6.14)$$

$$\mathrm{Nu} = C\mathrm{Ra}_l^{1/4} , \quad (2.6.15)$$

where

$$C = 0.282 \left(1 + C\,\mathrm{P}\ \mathrm{Pr}^{1/4} \right)^{5/4} \Big/ \left[1 + (0.437/\,\mathrm{Pr})^{9/16} \right]^{4/9} ,$$

and the coordinate origin is selected at the center of the plate.

The Nu number decreases with increase in the plate thermal resistance. The effect of the conjugation parameter for ***the air-air system*** for P = 1.64 becomes insignificant [2.26].

The heat transfer from a side surface of a vertical heat conducting plate of thickness δ and height h ***with the given temperature or heat flux on the opposite side and insulated ends faces*** is investigated by a parametrical method [2.93].

Of great practical interest is interaction ***between two flows on both sides of a thin vertical partition, one side of which heated by an ascending and the other is cooled by a descending one***. This situation is often encountered in various ***isolation systems and heating systems of civil buildings***. As an example we can specify a usual window with glass dividing two gas volumes with different temperatures. The boundary layers are formed on both sides of the wall, but the distribution of temperature or of a heat flux on them is to be calculated in view of interaction of two flows on the interface.

Investigation of conjugated heat transfer in a similar system is carried out for ***a case of natural convection on the one side of the partition***

and forced convection on the other [2.107,2.108]. Two coupled flows are considered in which the spatial distributions of the convective heat transfer coefficients are different on both sides of the wall. Both flows are coupled by the conditions of continuity of temperature and heat flux on the interface, which considerably complicates numerical solution of the problem. The results of an analytical investigation on the basis of the Oseen linearization method for $\mathrm{Pr} \to \infty$ show that the heat transfer rate decreases with increase in the wall thermal resistance parameter ω defined as

$$\omega = \frac{t}{H}\frac{k}{k_{\mathrm{w}}}\,\mathrm{Ra}_H^{1/4}\,, \qquad (2.6.16)$$

where t, H, k, and k_{w} are the thickness and height of the wall and the heat conduction coefficients of the fluid and wall material, respectively. The dimensionless parameter ω can be considered as the ratio of the thermal resistance of the wall $t/(Hk_{\mathrm{w}})$ to the scale of the thermal resistance of the boundary layer $(H\mathrm{Ra}_H^{1/4})/(Hk)$. The Oseen linearization method consists of construction of the integral conservation equations for both surfaces of the wall taking into account that the wall temperature $T_0(y)$ is considered to be unknown function which is to be determined. For this purpose, use is made of the condition of heat flux continuity in the direction x from one surface of the wall to the other [2.1]. In [2.106] the situation is considered when ***on both sides of the partition there is the same fluid with different temperatures at some distances from both sides***. Dimensionless parameters of the problem are the Prandtl number and the conjugation criterion. As a result of application of the Oseen linearization method it is shown that the heat flux distribution on the wall between two conjugated free-convective boundary layers is quite adequately approximated by the model of a constant heat flux.

A general statement of the problem with ***different fluids in divided volumes*** is considered in [2.109]. The free-convective heat transfer was calculated by an approximate parametrical method [2.90].

The effect of thermal stratification on the conjugated heat transfer is investigated in [2.110]. It is shown that the coefficient of proportionality between Nu_H and Ra_H increases with the thermal stratification on both sides, which is different from a similar case of one isothermal wall. Such abnormal behavior can be explained by the fact that for the isothermal wall both Nu_H and Ra_H are calculated on the basis of the maximum temperature difference, whereas the same numbers for conjugated convection are calculated from the difference of the temperatures of two boundary layers at the midheight of the wall [2.1].

Numerical data for conjugated heat transfer ***along a thin horizontal wall dividing two regions of a porous medium saturated with a fluid at different temperatures*** are given in [2.111]. For a similar problem ***with a vertical dividing partition*** the results of analytical and numerical investigation are given in [2.112]. Data for the average Nusselt number depending on the ratio of the thermal resistances of the surrounding medium on both sides of the wall

$$\beta = \frac{k_1}{k_2}\left(\frac{\mathrm{Ra}_1}{\mathrm{Ra}_2}\right)^{1/2}$$

are given in Table 2.6.1.

Table 2.6.1. The Values of the Average Nusselt Number depending on β

β	1	0.75	0.5	0.25	0
$\mathrm{Nu}/\mathrm{Ra}_1^{1/2}$	0.371	0.421	0.502	0.608	0.888

Numerical data for conjugated heat transfer along a ***vertical wall dividing two fluids with different temperatures*** are presented in Table 2.6.2 [2.113].

Table 2.6.2. The Values of the Average Nusselt Number depending on the Ratio between the Thermal Resistances of the Surrounding Medium on both Sides of the Wall $\beta = (k_1/k_2)(\mathrm{Ra}_1/\mathrm{Ra}_2)^{1/4}$

β	1	0.667	0.5	0.4	0.333	0
$\mathrm{Nu}/\mathrm{Ra}_1^{1/4}$	0.242	0.291	0.324	0.349	0.367	0.535

The analytical investigation of the conjugated free-convective heat transfer on a vertical surface shows that the system of initial equations for this problem can be made dimensionless so that it contains only one determining parameter, viz., the Pr number [2.114].

The numerical investigation carried out in [2.115] for the effect of such parameters, as the conjugation parameter p, the material micropolarity parameter Δ, and the Pr number on the conjugated free-convective heat transfer of ***micropolar fluids along a vertical plane plate*** shows that the effect of the thermal conductivity of the wall on the local heat transfer rate is more significant for large Pr numbers or small values of the micropolarity parameter. The values of the average Nusselt number for different values of the Pr numbers and conjugation parameter P at $\Delta = 1.0$ are given in Table 2.6.3.

The numerical simulation of conjugated free-convective heat transfer ***on a vertical plate dividing two regions of different temperatures with account for the longitudinal and transverse thermal conductivity of the plate*** is carried out in [2.116]. The results of calculations of the average Nusselt number for the flow area with a lower temperature are given in Table 2.6.4 ($R_t = (k_1/k_\mathrm{w})(t/L)\mathrm{Ra}_1^{1/4}$, k_1, and k_w are heat conduction coefficients of the warmer flow area and of the wall, respectively; t and L are plate

Table 2.6.3. The Values of $Nu/Gr^{1/4}$ for Different Pr and P

	$Nu/Gr^{1/4}$	
Pr	P = 0.0	P = 0.1
20.0	1.27226	1.11226
6.0	0.87825	0.80117
2.0	0.62809	0.58340
0.7	0.44600	0.42147

Table 2.6.4. The Values of $Nu_2/Ra_2^{1/4}$ for $R_t = 0.1$ and $Pr_1 = 0.7$ depending on the Parameter A

	$Nu_2/Ra_2^{1/4}$		
Pr_2	$A = 10$	$A = 1$	$A = 0.1$
0.7	0.4427	0.2122	0.0422
2	0.4800	0.2222	0.0426
∞	0.5184	0.2309	0.0429

thickness and height, Ra_1 and Pr_1 are the Rayleigh and Prandtl numbers for the warmer flow area, $A = (k_1/k_2)(Ra_1/Ra_2)^{1/4})$.

The analytical and numerical investigation of conjugated free-convective heat transfer near a ***heated vertical plate of finite dimensions placed in a semi-infinite environment*** is carried out in [2.117]. Numerical data are presented for a wide range of change of the main parameters of the problem such as Ra, Pr, and k (ratio of the heat conduction coefficients of the plate and of the fluid), $\lambda = a/b$ ($a$and b are the thickness and length of the conducting plate).

Conjugated free-convective heat transfer from a ***vertical conducting plate*** was considered analytically and experimentally in [2.118]. The theoretical analysis based on the assumption of the existence of the vertically averaged interface temperature between the heated plate and the fluid shows that the average heat transfer coefficient for ***both laminar and turbulent flow regimes*** can be expressed as a simple function of the single dimensionless conjugation parameter σ which characterizes the given problem. The results of the analysis are presented in Table 2.6.5, where k is the ratio between the heat conduction coefficients of the plate and the fluid; θ is the dimensionless temperature, $\lambda = a/H$, a and H are the thickness and height of the plate, respectively.

The results of numerical solution of the problem of conjugated free convection near a ***vertical and a horizontal plate*** in the range of the Pr numbers from 0.001 to infinity are generalized by the dependences [2.119]:

Table 2.6.5. Correlation Dependences for the Conjugation Parameter and Average Nusselt Number for Different Flow Regimes

Pr	σ	C	Nu
Laminar, $\Pr \gg 1$	$\lambda C \mathrm{Ra}^{1/4}/k$	0.67	$C\mathrm{Ra}^{1/4}\theta^{5/4}$
Laminar, $\Pr \ll 1$	$\lambda C(\Pr\,\mathrm{Ra})^{1/4}/k$	0.8	$C(\Pr\,\mathrm{Ra})^{1/4}\theta^{5/4}$
Turbulent, $\Pr \gg 1$	$\lambda C \mathrm{Ra}^{1/3}/k$	0.15	$C\mathrm{Ra}^{1/3}\theta^{4/3}$
Turbulent, $\Pr \ll 1$	$\lambda C(\Pr\,\mathrm{Ra})^{1/3}/k$	0.19	$C(\Pr\,\mathrm{Ra})^{1/3}\theta^{4/3}$

local Nu ***for a vertical surface***

$$\frac{\mathrm{Nu}}{\lambda} = \xi\left[\frac{\mathrm{Nu}_t}{(\sigma \mathrm{Ra}_t)^{1/4}}\right] + (1-\xi)\left[\frac{\mathrm{Nu}_h}{(\sigma \mathrm{Ra}_h)^{1/5}}\right], \tag{2.6.17}$$

where Nu_t and Nu_h are the local values of the Nusselt number for a non-conjugated problem of free convection ***for a constant temperature and a constant heat flux***, ***respectively***:

$$\frac{\mathrm{Nu}_t}{(\sigma \mathrm{Ra}_t)^{1/4}} = 0.502\left[\frac{1+\Pr}{0.492 + 0.986\sqrt{\Pr} + \Pr}\right]^{1/4}, \tag{2.6.18}$$

$$\frac{\mathrm{Nu}_h}{(\sigma \mathrm{Ra}_h)^{1/5}} = 0.631\left[\frac{1+\Pr}{0.396 + 0.918\sqrt{\Pr} + \Pr}\right]^{1/5}, \tag{2.6.19}$$

local Nu ***for a horizontal surface***

$$\frac{\mathrm{Nu}}{\lambda} = \xi\left[\frac{\mathrm{Nu}_t}{(\sigma \mathrm{Ra}_t)^{1/5}}\right] + (1-\xi)\left[\frac{\mathrm{Nu}_h}{(\sigma \mathrm{Ra}_h)^{1/6}}\right], \tag{2.6.20}$$

where

$$\frac{\mathrm{Nu}_t}{(\sigma\ \mathrm{Ra}_t)^{1/5}} = 0.456\left[\frac{1+\Pr}{0.313 + 0.856\sqrt{\Pr} + \Pr}\right]^{1/5}, \tag{2.6.21}$$

$$\frac{\mathrm{Nu}_h}{(\sigma\ \mathrm{Ra}_h)^{1/6}} = \left[\frac{1+\Pr}{0.177 + 0.625\sqrt{\Pr} + \Pr}\right]^{1/6} \tag{2.6.22}$$

and $\sigma = \Pr/(1+\Pr)$, ξ is the dimensionless coordinate along the plate surface equal to:

for a vertical plate

$$\xi = \left[1 + \sigma\ \mathrm{Ra}_t/(\sigma\ \mathrm{Ra}_h)^{4/5}\right]^{-1},$$

where

$$b = \frac{9\,\theta_2^{-3/2}}{4\,b_1\,\mathrm{Pr}}\left[\frac{\left(\theta_2^{-3/2}-1\right)b_2}{1.1198\,(1-\theta_2)}\right]^3 ,$$

$\overline{x} = x/x_0, \theta_2 = (T_{\mathrm{w2}}-T_\infty)/(T_{\mathrm{w1}}-T_\infty)$, and b_1 and b_2 are given in Table 2.7.1. For $\overline{x} \to 1 (\mathrm{Nu}_x/\mathrm{Ra}_x^{1/4})_2 \to \infty$ due to the discontinuity of the temperatures near the section $x = x_0$, and for $\overline{x} \to \infty (\mathrm{Nu}_x/\mathrm{Ra}_x^{1/4})_2 = (\mathrm{Nu}_x/\ \mathrm{Ra}_x^{1/4})_1$, since at a large distance from the singular point its influence on the solution becomes insignificant. Dependence (2.7.2) is invalid at small values of the Prandtl number.

Table 2.7.1.

Pr	0.02	0.1	0.7	1	2	5	10	100	1000	∞
b_1	0.9284	0.8591	0.6789	0.6422	0.5713	0.4817	0.4192	0.2517	0.1449	$0.8245 \times \mathrm{Pr}^{-1/4}$
b_2	0.0784	0.2301	0.4995	0.5671	0.7165	0.9536	1.1693	2.1914	3.9651	$0.7110 \times \mathrm{Pr}^{1/4}$

This problem is investigated mainly by numerical methods. The experimental data for temperature distribution in the boundary layer and heat transfer from the surface ***in air*** are obtained [2.52, 2.53]. The largest divergence immediately downstream the discontinuity point is explained by the effect of finite heat conduction over the plate height [2.44].

In the case where the upper part of the plate is insulated, and the lower one is maintained at a constant temperature T_{w1}, to calculate heat transfer for $x \leq x_0$ one can use the equation

$$\mathrm{Nu}_x/\mathrm{Ra}_x^{1/4} = (3/4)\,C_l\,(\mathrm{Pr}) \;, \tag{2.7.3}$$

and the temperature distribution on the surface of the insulated part of the plate is described by the dependence

$$\theta_{\mathrm{w2}}/\theta_{\mathrm{w1}} = \theta_{\mathrm{w2,0}}/\theta_{\mathrm{w1,0}}\left(\frac{1-x_1/x_0}{\overline{x}-x_1/x_0}\right)^{3/5} ,$$

where the values of x_1/x_0 and $\theta_{\mathrm{w2,0}}/\theta_{\mathrm{w1,0}}$ are given in Table 2.7.2.

A problem of ***two vertical surfaces one above the other, a distance s apart***, is considered as one of the variants of boundary condition discontinuity. ***The plates of heights*** l_1 ***(upper) and*** l_2 ***(lower) are maintained at different temperatures*** T_{w1} and T_{w2}, respectively. Comparing the heat transfer of the upper plate in the presence (q_0) and in the absence ($q_{2,0}$) of the lower one shows that short vertical plates are more sensitive to the presence of the lower plate than long ones.

An increase in the distance s up to a certain value is accompanied by an increase in heat transfer even in the case $T_{\mathrm{w2}} < T_{\mathrm{w1}}$. Here, when the values

Table 2.7.2.

Pr	x_1/x_0	$\theta_{w2,0}/\theta_{w1,0}$
0.5	0.3592	0.6909
0.733	0.3572	0.6709
1	0.3643	0.6591
2	0.3713	0.6333
5	0.3789	0.6089
10	0.3835	0.5971

of s are small, the heat flux is directed from the fluid to the wall. When $T_{w2} > T_{w1}$, the effect of the lower plate on heat transfer is insignificant [2.26]. However, near the leading edge the local value of the heat flux is sensitive to the magnitude of the velocity and temperature of the fluid. The velocity and temperature profiles in the section of length s between the plates are dome-shaped due to the problem symmetry [2.44]. Experimental investigation of a free-convective boundary layer ***on a vertical surface with discontinuity in the wall temperature*** was carried out in [2.52]. Various cases of the relationship between temperatures of the lower ($T_{0,1}$) and upper ($T_{0,2}$) parts of the wall and the ambient temperature (T_∞) are considered. In particular, when $T_{0,2} < T_\infty$, which can occur in a thermally stratified medium and is often found in practical applications, a descending flow is formed near the upper part of the plate and ascending flows near the lower one. They meet and separate from the surface in the section whose position is determined by the relationship between the temperatures and heights of the upper and lower parts of the plate. Investigation of these flows requires the solution of complete governing equations due to the appearance of the oppositely directed flows and large gradients in the longitudinal direction [2.7]. The same requirement arises in all the problems with temperature discontinuity in the vicinity of the point of discontinuity. In the regions at farther distances from this point the investigation of free-convective flow is possible with the application of the boundary layer theory.

Piecewise continuous change in the heat flux. When the upper part of the plate is insulated ($x \leq x_0$), and the temperature of the lower one is fixed T_w, heat transfer from the part of the surface ($x < x_0$) is determined from the formulas for an isothermal plane wall. In the upper part, the flow can be represented as a jet on an insulated surface with the given value of heat content at $x = x_0$, which remains constant in any section. When $x > x_0$, the solution can be presented in a self-similar form with the coordinate origin $x = x_0$. Integral and numerical methods and the methods of matched asymptotic expansions are applied to calculate the temperature distribution on the insulated part of the surface. This distribution is approximated by the dependence [2.44]

$$\frac{\theta_{w2}}{\theta_{w1}} = \left(\frac{1 - x/x_0}{x/x_0 - x_1/x_0} \right)^{3/5} . \tag{2.7.4}$$

In the case of free-convective flow ***caused by several insulated heating elements of height*** h_1, ***located a distance*** l_1 ***from each other on a vertical adiabatic surface*** the ratio of Nu_B for the upper element to the corresponding value of Nu_H for the lower one changes in inverse proportion to the ratio of temperature differences, if both elements generate equal amounts of heat. Here, $\mathrm{Nu}_B / \mathrm{Nu}_H < 1$ for $h/l < 4$ and $\mathrm{Nu}_B/\mathrm{Nu}_H > 1$ for $h/l \geq 4$. For $Q_B/Q_H = 1$ the ratio of the Nusselt numbers changes from 0.9 to 1.05 with increase in h/l from 1 to 8. This ratio quickly decreases, when h/l becomes less than 1 [2.127].

Numerical calculations of heat transfer for the case ***where in the lower part of the surface*** ($x < x_0$) ***the heat flux is equal to*** q_{w1}, ***and in the upper one*** ($x > x_0$) ***to*** q_{w2}, show that the disturbances caused by heat flux discontinuity decay over the height [2.44].

Figure 2.7.1 presents an interferogram of a free-convective flow near a vertical plate with heat flux discontinuity on the surface (the greater value of the heat flux corresponds to the upper half), recorded in 50 seconds after the beginning of heating, and the constructed, on its basis, three-dimensional

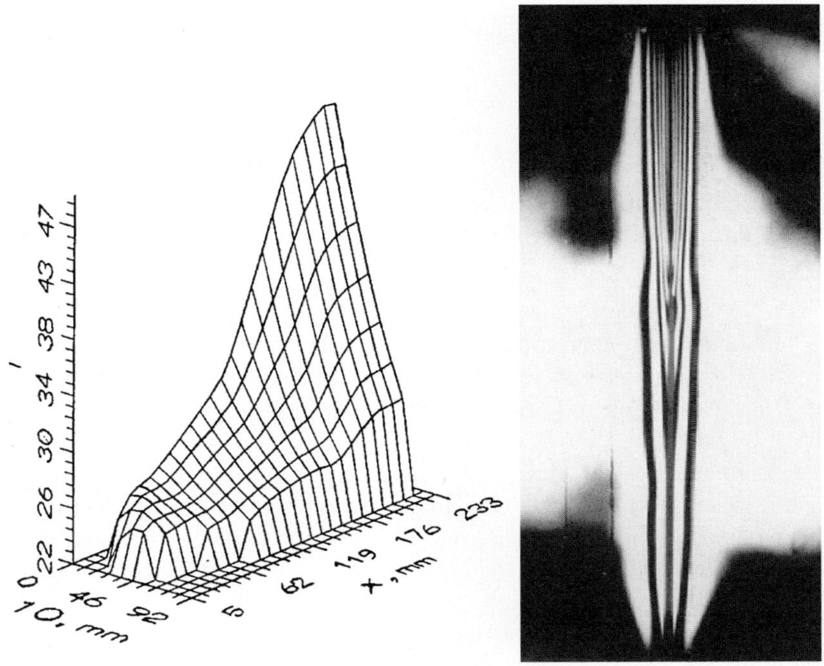

Fig. 2.7.1. Interferogram and the restored, on its basis, temperature distribution in a boundary layer near a vertical plate with discontinuity in the heat flux. The larger heat flux is above

diagram of temperature distribution in the boundary layer. The coordinate x is reckoned along the plate, the coordinate origin corresponds to the lower edge, the coordinate y is directed across the boundary layer.

The existence of the heat flux discontinuity on the surface leads to substantial change in the heat transfer character. Figure 2.7.2a presents the diagrams of the local heat transfer coefficient distribution along the plate length.

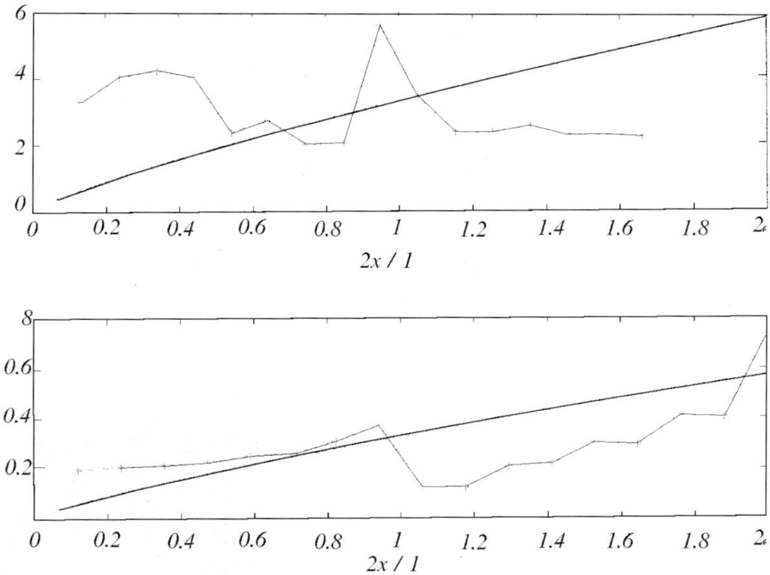

Fig. 2.7.2. Distribution of the local heat transfer coefficient over a vertical plate with heat flux discontinuity on the wall: (**a**) larger heat flux is above, (**b**) larger heat flux is below

The solid lines correspond to the interpolation dependence $\mathrm{Nu}_x = 0.55\ (\mathrm{Ra}_x^*)^{1/5}$ suggested in [2.128] for a vertical plate ***in air at a constant heat flux***. According to [2.65], this dependence differs from the self-similar solution by not more than 0.5% in the range $20 < \mathrm{Ra}_x^* < 5 \cdot 10^7$. The interpolation dependence is calculated for a heat flux equal to 220 W/m^2. It should be noted that the values of the local heat transfer coefficient on the lower and upper parts of the plate decrease according to the power law, which qualitatively agrees with theoretical [2.123] and experimental [2.52] results obtained for the temperature discontinuity on a vertical surface.

However, the local values of the heat transfer coefficient for the lower part of the plate are 20–30% higher compared with a plate of constant heat flux. Moreover, in the region lying immediately behind interface between heat fluxes there is a substantial (up to 40%) increase of the local heat transfer coefficient caused by a quick increase in the temperature gradient. Further downstream its decrease to the value making approximately 50% of the average value for the plate with a constant heat flux is observed. It can be due to the flow of a heated air from the lower part of the plate. However, the qualitative character of the temperature field profile remains the same, as in the case of a uniform heat flux.

Figure 2.7.3 presents an interferogram and a corresponding three-dimensional diagram of temperature distribution for surfaces with a large heat flux on the lower part of the plate. The recording was made in 75 seconds after the beginning of heating. The existence of three characteristic regions (Fig. 2.7.4) is noted in the temperature field distribution. Region I is formed near the lower half of the surface with a large heat flux. The temperature field has a power-law character, i.e., it is similar to the field on a plate with a uniform heat flux; the maximum of the temperature is located on the surface. Region II is a transient one; it is formed in a small region above the border dividing the surface into two parts. In this region the character of temperature decrease across the flow changes due to the fact that the gas, having a higher temperature, comes in contact with a colder part of the plate. As a result, the gas cools off, and a characteristic plateau appears in the temperature distribution profile near the surface.

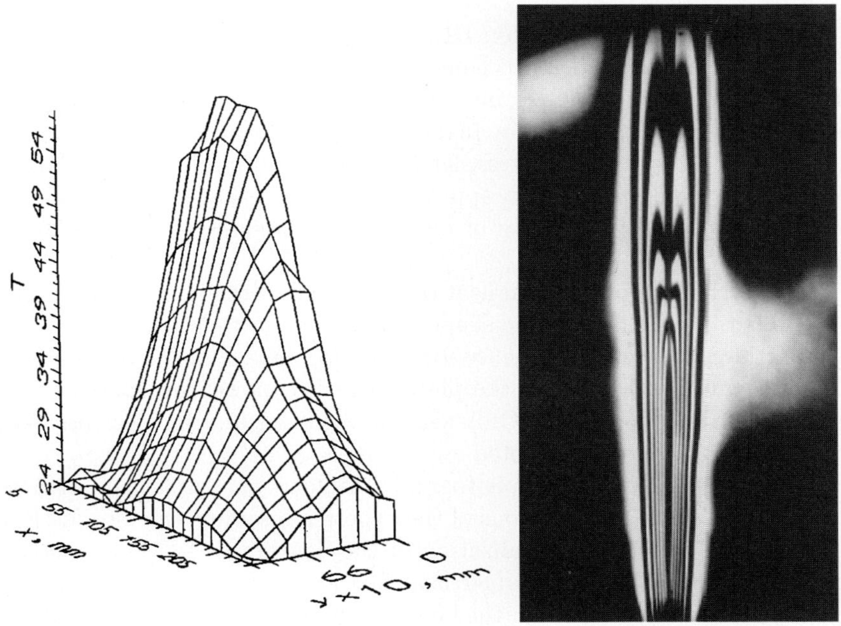

Fig. 2.7.3. Interferogram and the restored on its basis temperature distribution in a boundary layer near a vertical plate with heat flux discontinuity. The larger heat flux is below

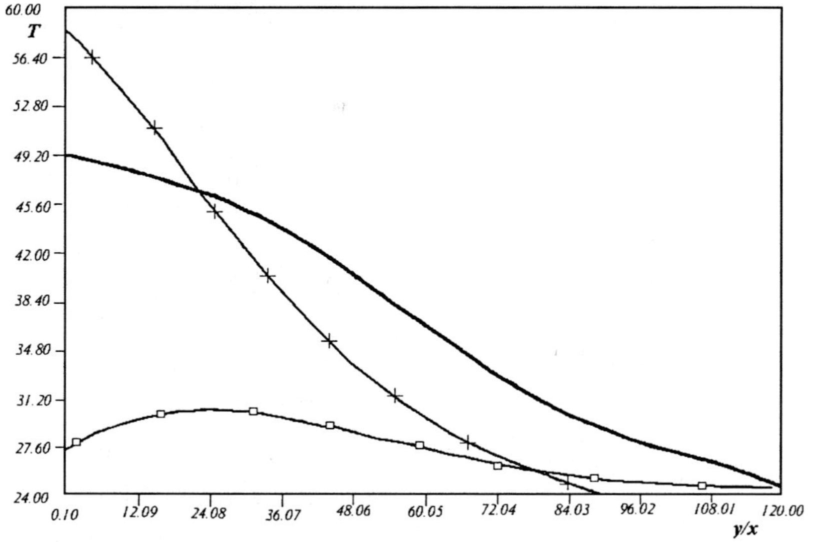

Fig. 2.7.4. Temperature distribution in a boundary layer near a vertical plate with heat flux discontinuity

Further along the flow region III is formed, where the near-wall gas layer is cooled as a result of heat exchange with the cooler surface, and the local temperature maximum moves into the boundary layer, forming a parabolic temperature profile here. For a plate of larger length, this maximum must disappear as a result of heat transfer between neiboring flow layers, and the temperature profile must again approach the profile near a uniformly heated surface. However, for the plate of investigated dimensions this was not observed.

The distribution of the local heat transfer coefficient along the plate length is presented in Fig. 2.7.2b. The shape of the curve in this case slightly differs from the interpolation dependence for a uniform heat flux [2.128], that is, on the upper and lower parts of the plate there is a power-law increase in the local heat transfer coefficient. Only near the interface its local decrease caused by the heat flux decrease is noted.

Figure 2.7.5 presents the photographs of flows near a surface with heat flux discontinuity. The conditions of the experiments when the interferograms were recorded and track visualization of the flow was made are exactly the same. For the surface with a larger heat flux above, a gradual increase in the hydrodynamic boundary layer thickness is observed upstream. For the plate with a larger heat flux on the lower part of the surface the boundary layer thickness grows gradually to the border between the heat fluxes, where it reaches its maximum value, and further upstream there is some decrease in its thickness.

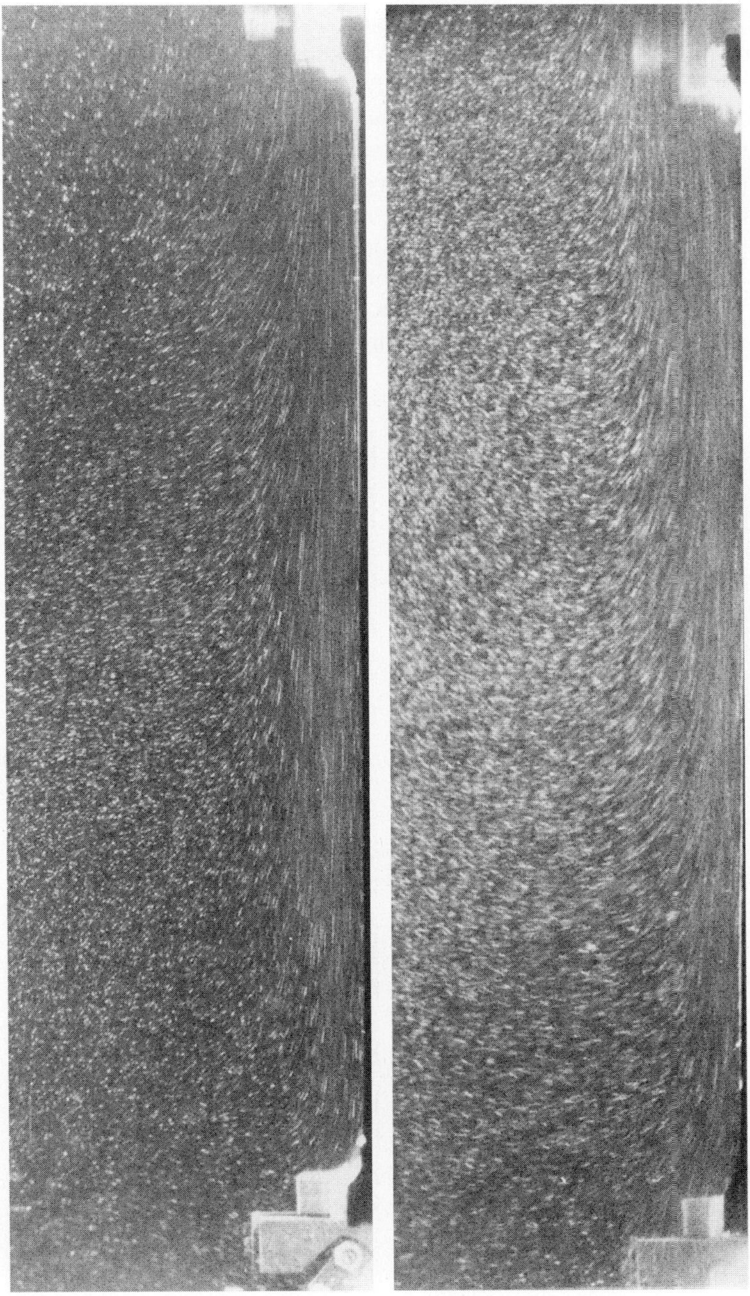

Fig. 2.7.5. Visualization of free-convective flow in a boundary layer near a vertical plate with heat flux discontinuity: (**a**) larger heat flux is above, (**b**) larger heat flux is below

Of certain interest is the case of fast heating of a plate by a larger heat flux on the lower part (density of heat generation from the upper and lower parts of the plate are equal respectively to 170 W/m^2 and 600 W/m^2). In this case, the flow on the lower part of the surface is formed much earlier, than on the upper one, as a result of faster temperature increase. The interaction of two regions leads to the occurrence of a two-dimensional vortex moving upwards. In Fig. 2.7.6a there is an interferogram of the originating circulation flow recorded in 11 seconds after the beginning of heating. From it is seen, in which way the warmer air from the lower part of the reaches the motionless, colder region, and separates it aside from the surface, not mixing with it. The process of retardation of the warm air leads to formation of a vortex whose

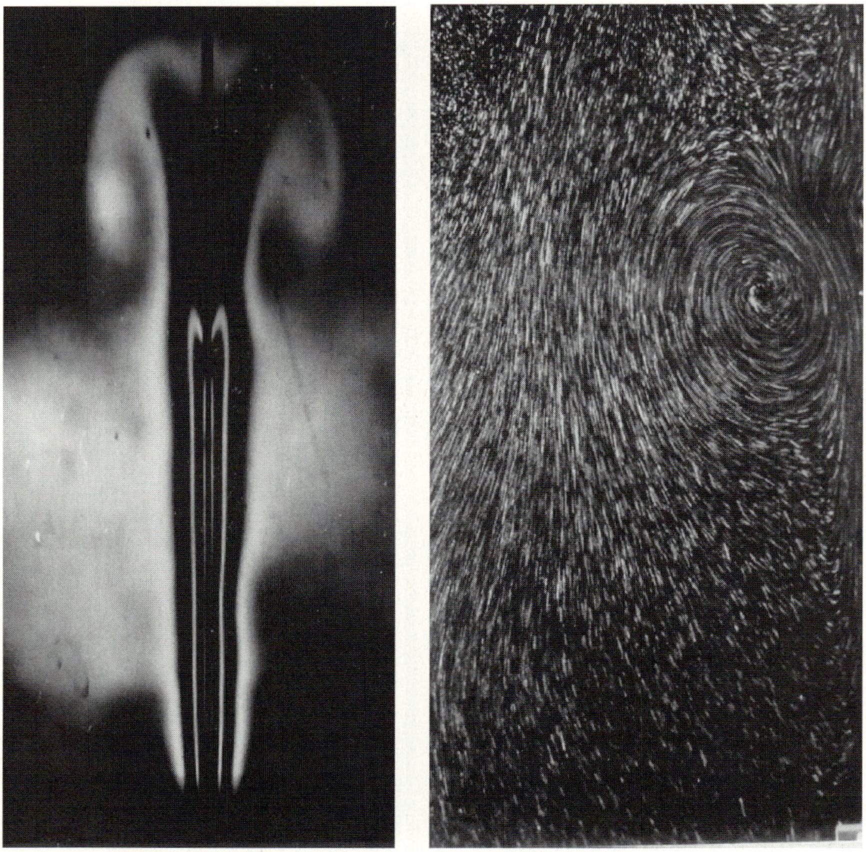

Fig. 2.7.6. Interferogram of the initial period in the development of a free-convective flow on a plate (large heat flux is supplied from below, heat generation density is 170 W/m^2 and 600 W/m^2) – (**a**); track visualization of the initial period of the free-convective flow development on a plate with heat flux discontinuity – (**b**)

core consists of a colder air. As soon as the vortex is formed, it moves upwards along the plate surface with the free-convective flow (Fig. 2.7.6b). Figure 2.7.7 presents the increased images of interferograms of the temperature field near a vertical plate with heat flux discontinuity for the cases, where the greater heat flux is located below (Fig. 2.7.7a) and where the greater heat flux is located above (Fig. 2.7.7b).

Figures 2.7.8 and 2.7.9 display a series of interferograms representing the development of a thermal boundary layer ***on a vertical plate with periodic heat flux discontinuities***. The scale of the periodic discontinuities is 3 cm (Fig. 2.7.8) and 1 cm (Fig. 2.7.9). At the initial moment of heating of the vertical plate with periodic heat flux discontinuities the dynamic boundary layer near the plate is formed as a sequence of two-dimensional vortical structures along the plate. The results of track visualization of this process are given in Figs. 2.7.10a and 2.7.11a. When the periodic heat flux discontinuities are equal to 3 cm near the plate, three two-dimensional vortices are formed. When the scale of the heat flux discontinuities is 1 cm, the boundary layer near the plate consists of 6 vortices. The plate being heated further, the vortical structures are destroyed, and the dynamic boundary layer thickness undergoes low-frequency periodic changes along the plate. The results of track visualization of the dynamic boundary layer near a vertical plate with periodic heat flux discontinuities for rather long times of heating are presented in Figs. 2.7.10b and 2.7.11b.

The free-convective heat transfer on a vertical plane surface with obstacles on it is used in many practical problems and technical devices. The presence of different obstacles is also one of the ways of heat transfer intensification.

The effect of obstacles on natural convection heat transfer in a vertical fluid layer ***with isothermal side walls*** at temperatures T_1 and $T_2 (T_1 > T_2)$ was investigated in [2.129]. One obstacle of height equal to the double distance from the wall to the region of maximum velocity mounted on a vertical wall has an insignificant effect on the velocity profile already at a distance of 3.5 heights from the obstacle. In the presence of the obstacle of height of the same order as the boundary layer thickness its influence on the temperature structure is insignificant already at the distance from the obstacle equal to its height, and the deviation of the local Nusselt number from the experimental dependence for a flow without obstacles does not exceed 13%. The experimental data on heat transfer are obtained for ***single obstacles of height $\Delta x = 10$ mm and width equal to 10 mm, and for regular obstacles with $\Delta x = 5$ mm and step equal to 20 mm*** [2.129].

The experimental and numerical investigation of natural convection ***on a two-dimensional vertical plate with rectangular horizontal grooves*** was carried out in [2.130].

The local values of the Nusselt number are found for each surface of a groove (inner, lower, outer and upper ones) by processing the interferograms obtained by the Mach–Zander interferometer. In some cases (for certain aspect ratios of the sizes of the groove and low Ra values) the overall heat transfer

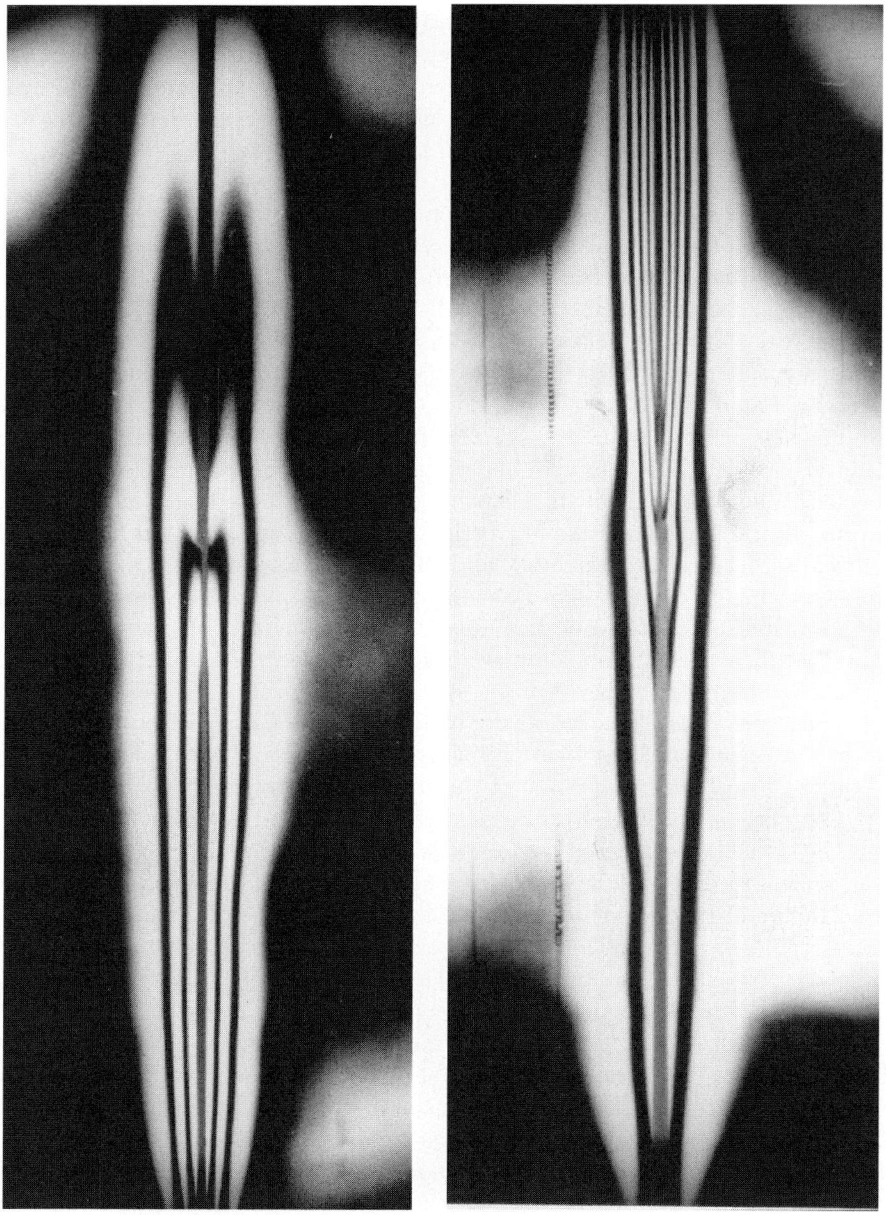

Fig. 2.7.7. Magnified images of interferograms of the temperature field near a vertical plate with heat flux discontinuity; (**a**) the larger heat flux is located below, (**b**) the larger heat flux is located above

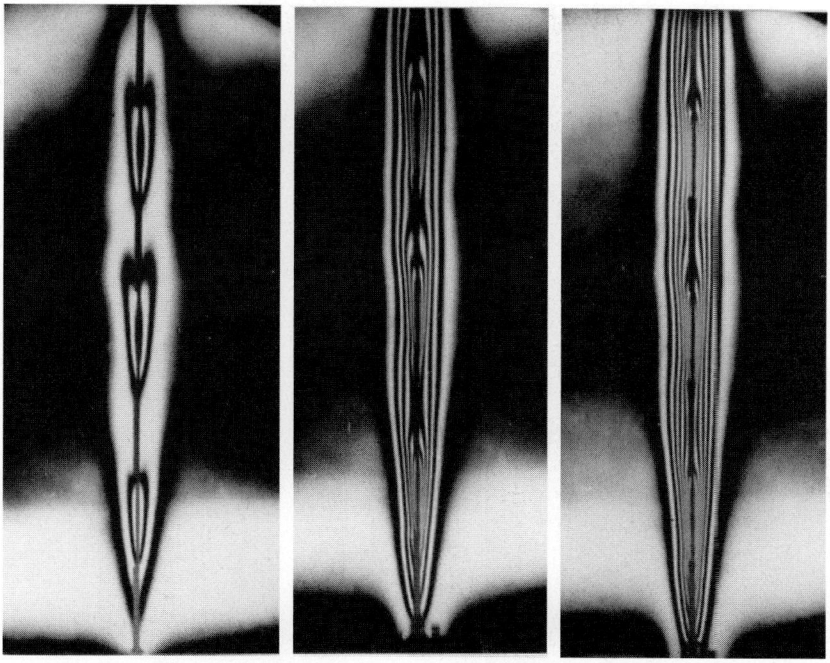

Fig. 2.7.8. A series of interferograms displaying the development of a thermal boundary layer on a vertical plate with periodic heat flux discontinuities. The discontinuity step is equal to 3 cm

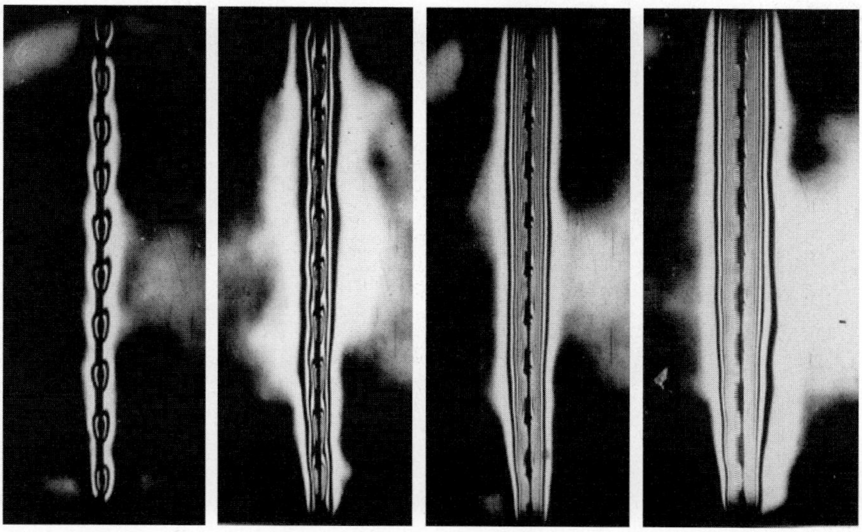

Fig. 2.7.9. Series of interferograms displaying development of a thermal boundary layer on a vertical plate with periodic heat flux discontinuities. Discontinuity step is 1 cm

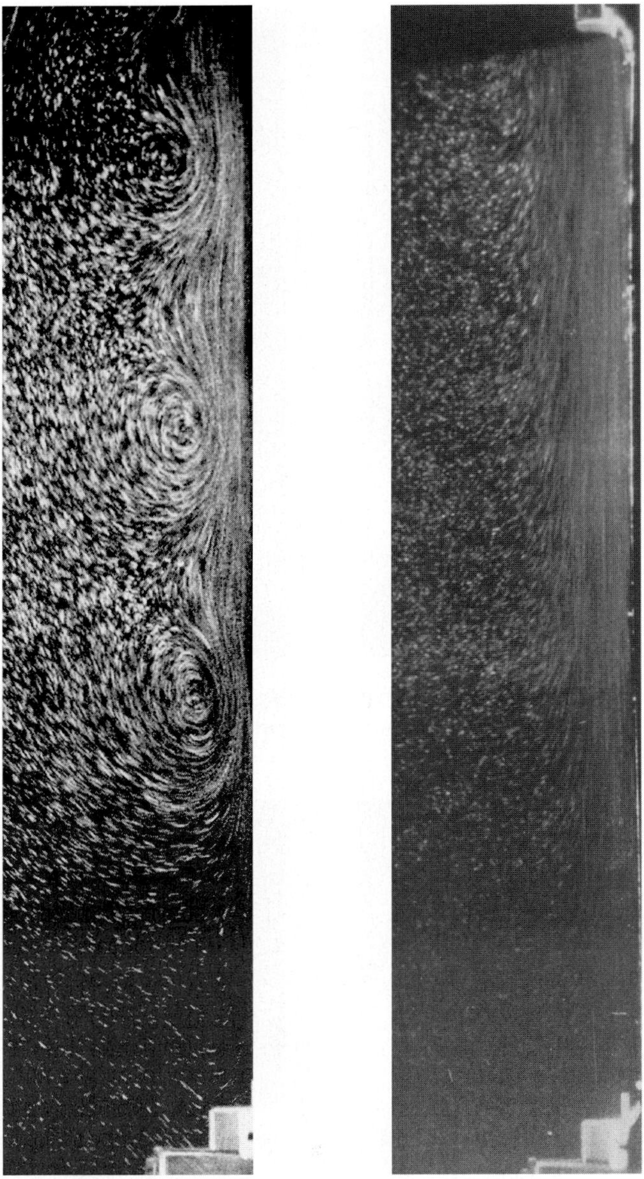

a) b)

Fig. 2.7.10. Track visualization of a dynamic boundary layer on a vertical plate with periodic heat flux discontinuities; (**a**) nonstationary heating regime, (**b**) regular heating regime

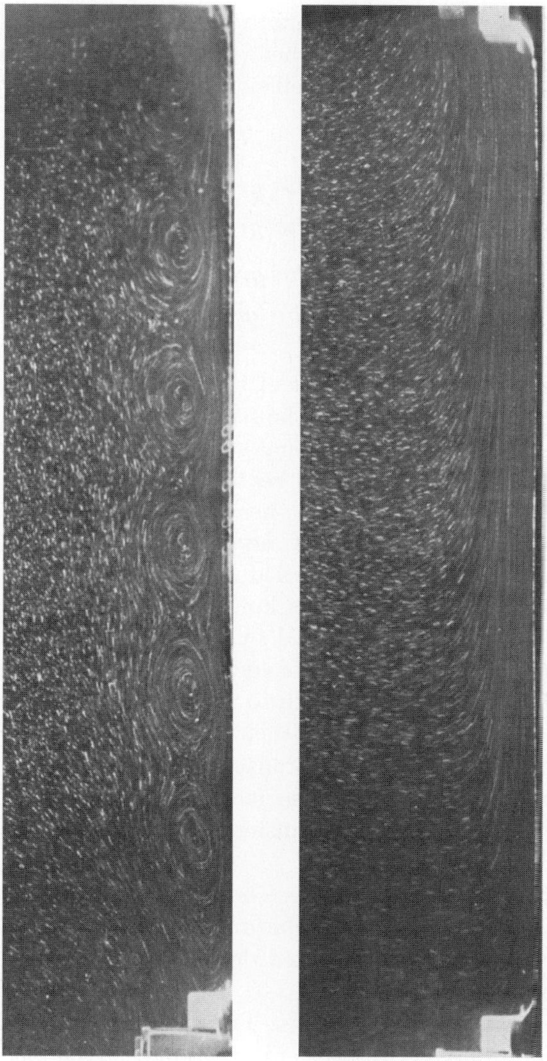

a) b)

Fig. 2.7.11. Track visualization of a dynamic boundary layer on a vertical plate with periodic heat flux discontinuities: (**a**) nonstationary heating regime, (**b**) regular heating regime. Discontinuity step is 1 cm

from the surface with grooves appears to be smaller than for a plane surface, despite the increase of the overall heat transfer surface. The experimental data for the average heat transfer coefficient are generalized by the correlation dependences:

$$\begin{aligned}
&\textit{for the inner surface of the groove} && \mathrm{Nu} = 0.45\ \mathrm{Ra}^{0.29}\ ,\\
&\textit{for the lower surface of the groove} && \mathrm{Nu} = 0.46\ \mathrm{Ra}^{0.28}\ ,\\
&\textit{for the outer surface of the groove} && \mathrm{Nu} = 0.45\ \mathrm{Ra}^{0.3}\ ,\\
&\textit{for the upper surface of the groove} && \mathrm{Nu} = 0.4\ \mathrm{Ra}^{0.3}\ ,
\end{aligned} \tag{2.7.5}$$

The numerical data on the effect of the dimensions of the rib and of the heat power distribution in them on the structure of the temperature field and free-convective flow in two-dimensional steady convection ***along a vertical plate with five discretely heated rectangular ribs*** are given in [2.131].

Experimental investigation of the heat transfer characteristics of laminar natural convection ***near a vertical surface with various roughness elements (steps and ribs)*** was carried out in [2.132]. The enhancement of heat transfer for the surfaces of this kind can be achieved by respective selection of the suitable dimensions and shapes of the roughness elements. The results of investigation of the effect of such factors, as the ratio of rib spacing to the rib height (or of the step width to its height) p/q, the inclination angle of the elements to the surface, and their thermal conductivity showed, that the maximum increase in the heat transfer rate achieved in experiments was 23% in comparison with a smooth surface of the same dimensions, and it is observed at $p/q = 16$. The heat transfer rate decreases with the parameter p/q.

Experimental investigation of free convection ***on a vertical surface of constant temperature with an obstacle facing upwards the flow*** was carried out in [2.133]. The generalized dependences for the heat transfer coefficient obtained for $\mathrm{Pr} = 0.7$ and $\mathrm{Gr}_x = 2 \cdot 10^5 - 8{\cdot}10^7$ describe the experimental data with an error of 20%. The heat transfer decreases up to the discontinuity point, has a zero value at it, then increases further along the flow, and at a sufficient distance attains its constant value, characteristic for a flat plate [2.26]

$$\begin{aligned}
\frac{\mathrm{Nu}_x}{\mathrm{Gr}_x^{1/4}} &= \left(0.657 - 0.83\frac{\Delta}{\delta_{b0}}\right)\left(1 - \frac{x}{x_0}\right)^{1/5}, && \frac{\Delta}{\delta_{b0}} = 0.25 \div 0.61\ ,\\
\frac{\mathrm{Nu}_x}{\mathrm{Gr}_x^{1/4}} &= \left(-0.289 + 0.721\frac{\Delta}{\delta_{b0}}\right)\left(1 - \frac{x}{x_0}\right)^{1/5}, && \frac{\Delta}{\delta_{b0}} = 0.61 \div 0.72\ ,\\
\frac{\mathrm{Nu}_x}{\mathrm{Gr}_x^{1/4}} &= \left(0.808 - 1.327\frac{\Delta}{\delta_{b0}}\right)\left(1 - \frac{x}{x_0}\right)^{1/5}, && \frac{\Delta}{\delta_{b0}} = 0.25 \div 0.48\ ,
\end{aligned} \tag{2.7.6}$$

$$\frac{\mathrm{Nu}_x}{\mathrm{Gr}_x^{1/4}} = \left(-0.707 + 1.832\frac{\Delta}{\delta_{b0}}\right)\left(1 - \frac{x}{x_0}\right)^{1/5}, \qquad \frac{\Delta}{\delta_{b0}} = 0.48 \div 0.61\,,$$

$$\frac{\mathrm{Nu}_x}{\mathrm{Gr}_x^{1/4}} = \left(1.446 - 1.672\frac{\Delta}{\delta_{b0}}\right)\left(1 - \frac{x}{x_0}\right)^{1/5}, \qquad \frac{\Delta}{\delta_{b0}} = 0.61 \div 0.72\,,$$

where Δ is the width of the horizontal area of the step, $\delta_{b0} = 5.4x_0(4/\mathrm{Gr}_{x0})^{1/4}$ is the boundary layer thickness for $x = x_0, x_0$ is the distance up to the leading edge of the plate.

The effect of the boundary layer separation on the heat transfer rate ***near a small step on a vertical flat plate*** was investigated numerically in [2.134]. The results of numerical calculations ***for air*** ($\mathrm{Pr} = 0.7$) ***and water*** ($\mathrm{Pr} = 8.0$) are presented for two kinds of thermal boundary conditions: (1) ***the temperature of the obstacle is equal to the temperature of the plate*** and (2) ***the temperature of the obstacle exceeds the temperature of the plate***. The data are presented for the steps of height $\sim L\varepsilon^{9/7}$ and length $\sim L\varepsilon^{6/7}$, where L is the distance from the plate leading edge to the obstacle, and $\varepsilon = \mathrm{Gr}^{-1/4}$.

The effect of ***four configurations of lateral ribs*** differing in the material and geometrical dimensions on heat transfer by free convection ***of a vertical isothermal copper plate*** was investigated experimentally in [2.135].

The empirical dependence which describes experimental data in the ranges of Pr numbers from 4.75 to 5.25 and Ra numbers from $8.5 \cdot 10^7$ to 10^9 has an error of ~1%. The universal dependence for heat transfer of ***single bodies of various shapes located on a vertical wall of a plane channel*** with free-convective motion of ***air*** was also obtained in [2.136]. ***Short cylinders and rectangular parallelepipeds made of steel and aluminum*** were used as investigated objects. The generalized dependence for the average heat transfer coefficient is valid in the range of the Grashof numbers $1.5 \cdot 10^3 \leq \mathrm{Gr} \leq 1.1 \cdot 10^6$.

Non-steady free-convective heat transfer ***in vertical parallel plates with a rectangular rib*** was investigated experimentally for the cases of transient processes with and without heating [2.137] and of the local steady-state regime under the conditions of asymmetric uniform heating [2.138]. It is shown that the heat transfer characteristics downstream behind the obstacle are similar to those of a turbulent flow, though the Rayleigh number in experiments changed from $3.75 \cdot 10^8$ to $1.93 \cdot 10^9$, which corresponds to a laminar regime for an equivalent heated flat plate. The investigations of a convective heat flux were carried out, and new correlations for the Nusselt numbers which describe the local characteristics of the rib in the downstream area were suggested.

Experimental and theoretical investigation of a laminar natural convection of ***air in a vertical channel with a single obstacle*** was carried out in [2.139]. The distributions of the temperature and heat flux for ***an isothermal wall*** and the visual flow patterns were obtained by optical methods using a Wollaston prisms-based interferometer and a Mach–Zander interferometer. The numerical calculations were performed ***for two cases of***

boundary conditions: constant temperature and constant heat flux on the channel walls.

The presence of the obstacle leads to essential change of the heat transfer rate compared with smooth walls. As the flow velocity increases near the obstacle, the local heat transfer coefficient increases up to some maximal value, then it decreases together with the velocity to the minimum, and then it increases again. This tendency is noticed for all investigated Ra values, which changed from 10^2 to 10^4. However the average heat transfer coefficient appears smaller compared with the smooth channel due to the flow velocity decrease and to the occurrence of circulating flows caused by the effect of the obstacle. Moreover, the average Nusselt number decreases by 5% at Ra $= 10^4$ and by 40% at Ra $= 10$ for the case of constant temperature on the channel walls.

The flow visualization and measurement of the temperature field ***in the region behind a backward-facing step*** in natural convection of ***water near a vertical plate with a rectangular step*** were carried out in [2.140]. The flow and heat transfer ***behind a two-dimensional obstacle in a flow of deairated water on a vertical flat surface*** were investigated experimentally in [2.141]. The obstacle surface and vertical heat exchange surfaces were heated by uniform and homogeneous heat fluxes. The range of the Grashof numbers at the inlet to the separation zone was from $4{\cdot}10^6$ to $2{\cdot}10^9$. For the length of the surface before the obstacle is equal to 30, 60, and 120 mm, the height of the step was varied from 0 up to 70 mm. The temperature fields were measured by chromel-alumel thermocouples. Dyes and heat-sensitive plates on liquid crystals were used for visualization of the flow and temperature field. When the height of the obstacle is small, a non-separating flow is observed, but as the height increases, the flow separates and transition to a turbulent flow occurs. The regimes of heat transfer by natural convection and the wall temperature distribution along the vertical flat plate ***in the presence of various obstacles*** were also investigated in detail in [2.142].

Investigation of temperature fields on a vertical surface ***with a constant heat flux was carried out for a plate with one, two, and three rectangular obstacles*** spaced 30 mm apart. The value of the heat flux on the wall changed within $q_w = 0 - 10^3$ W/m^2. An increase in the heat generation intensity leads to an increase in the plate heating rate and heat transfer intensification, but does not qualitatively change the interference flow pattern observed.

Figure 2.7.12 presents interferograms of the temperature field in the boundary layer near a vertical plate ***with one (a) two (b), and three (c) obstacles at a constant heat flux on the wall*** $q_w = 325$ W/m^2. The recording was made with an exposure time of 1/15 sec after 45 sec from the beginning of plate heating. The results of longitudinal visualization showed that the free-convective flow near the plate in the given experiments could be considered as two-dimensional with a sufficiently high degree of accuracy. Therefore, in processing interferogram for the purpose of calculating the temperature field near the plate it was assumed, that the temperature does not depend on the coordinate z in the direction of propagation of the probing radiation in the interferometer, except for the plate edges, where the end effects manifest themselves.

To calculate the temperature corresponding to the kth interference band, the following relation obtained in [2.143] was used:

$$T_k = \frac{(n_0 - 1)P_\infty T_0 T_\infty z}{(n_0 - 1)P_\infty T_0 z - k\lambda P_0 T_\infty} \,. \tag{2.7.7}$$

In Fig. 2.7.12, along side the corresponding interferograms the graphs of the temperature distribution in the boundary layer near the vertical plate with obstacles are given. A teflon obstacle in this case is not a heat generating element, and it is heated from a steel plate, therefore the surface temperature of the obstacle is somewhat lower than the surface temperature of the plate. In the case of one obstacle, it follows from the interferogram and temperature field graph given in Fig. 2.7.12a that the temperature distribution across the boundary layer (along the coordinate y) is monotonic in the region upstream and downstream of the obstacle. Moreover, the thermal boundary layer thickness in these regions increases approximately twice compared with the nondisturbed region of free-convective flow upstream the obstacle. The temperature distribution upstream the obstacle has a wavy character, which

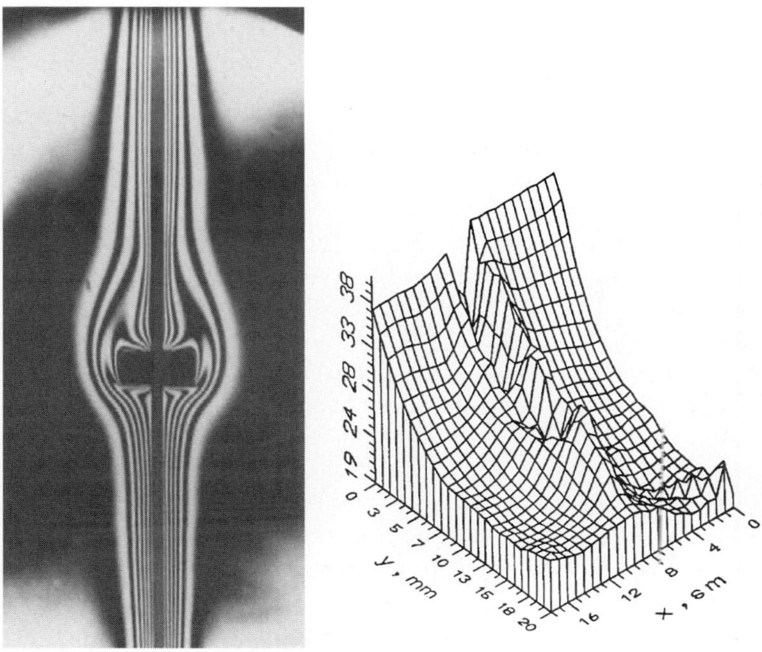

Fig. 2.7.12a. Interferogram and corresponding temperature distribution on a plate with a constant heat flux in the presence of one obstacle

is observed in the interferogram as alternating bands with neighboring numbers in this region. This character of the change in the temperature can be explained by inflow of a hotter gas into the space between the boundary layer on the obstacle surface and the colder region of the ambient gas as a result of retarding effect of the obstacle. Note should be made of the relatively low transverse temperature gradient in the region located just behind the obstacle, indicating the presence of intensive convective heat transfer in it. ***In the case of two or three obstacles on a vertical plate*** (Figs. 2.7.12b, 2.7.12c) the temperature field near the obstacles is similar to the temperature distribution near one obstacle, but here a decrease in the transverse temperature gradient in the boundary layer downstream on the obstacles is observed. The disturbing effect of the obstacles leads to the wavy character of temperature distribution along the boundary layer. The regions of the temperature field between the obstacles are characterized by a smaller transverse temperature gradient. This is due to the presence of convective mixing of the gas layers having different temperatures. It should be noted that in the presence of two and, especially, three obstacles the transverse temperature gradient on the plate increases at some distance after the obstacles in comparison with the gradient on the plate upstream the obstacles.

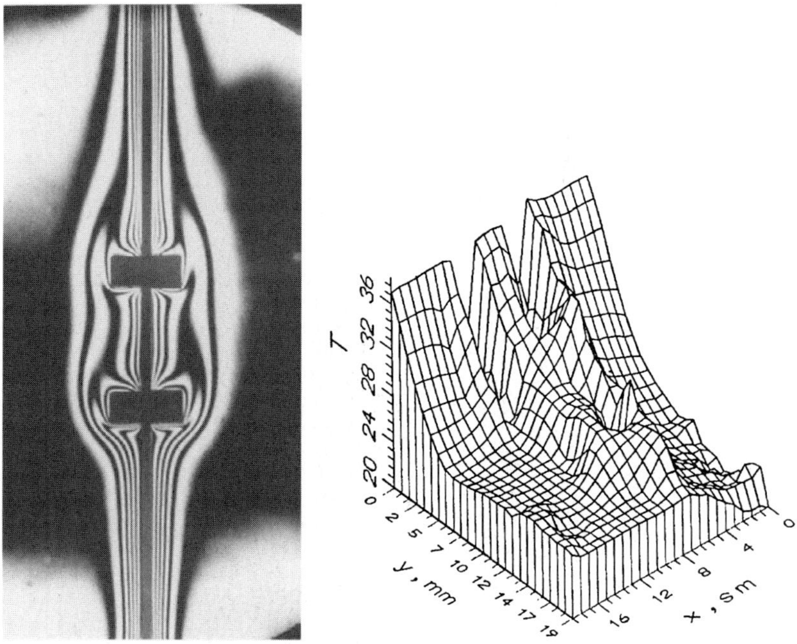

Fig. 2.7.12b. Interferogram and corresponding temperature distribution on a plate with a constant heat flux in the presence of two obstacles

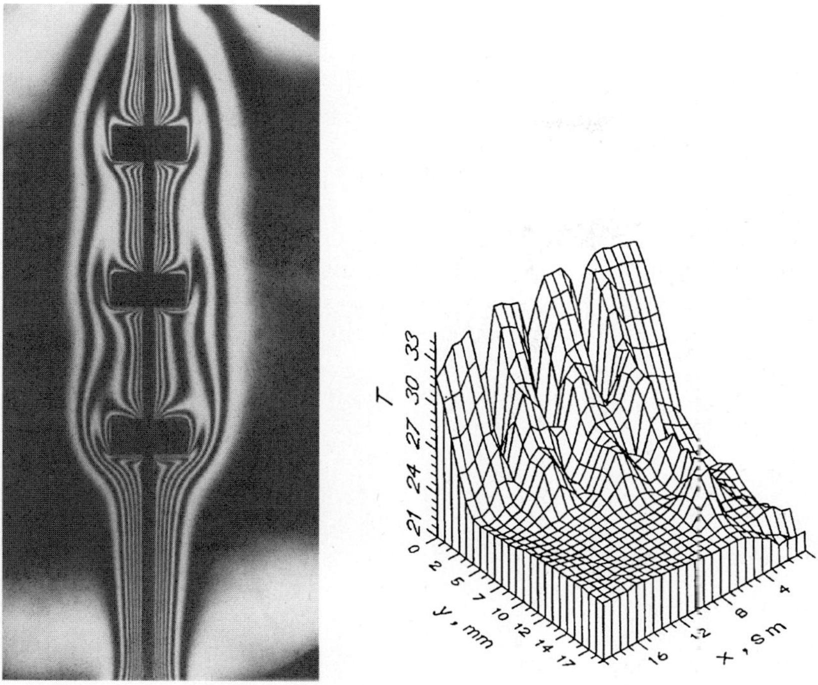

Fig. 2.7.12c. Interferogram and corresponding temperature distribution on a plate with constant heat flux in the presence of three obstacles

The photographs of track visualization of a free-convective flow on a vertical plate with a constant heat flux in the presence of one or several obstacles spaced 20 mm apart are presented in Fig. 2.7.13. As is seen from the figure, the presence of the obstacles on the plate causes separation of the dynamic boundary layer which smoothly envelopes the region of the obstacles. In the separation zone immediately behind the obstacle a near-wall circulating flow consisting of two oppositely rotating vortices is observed.

Obviously, the presence of a rather low transverse temperature gradient in the region between the obstacles registered in interferometric measurements is caused by the existence of these vortical flows ensuring intensive convective heat transfer. It is seen from Fig. 2.7.14, where a magnified image of the region between two obstacles is given, the vortical gas motion occupies the whole space between the steps, though one should expect the occurrence of stagnant zones before the second and third obstacles.

Figure 2.7.15 presents flow patterns obtained as a result of flow visualization on a vertical plate with different spaces between two obstacles, which was carried out for to determine its effect on the vortical flow induced. The distance between the steps in the given photographs was 10 mm and 30 mm. For the distance of 10 mm (the region between the obstacles has the shape of

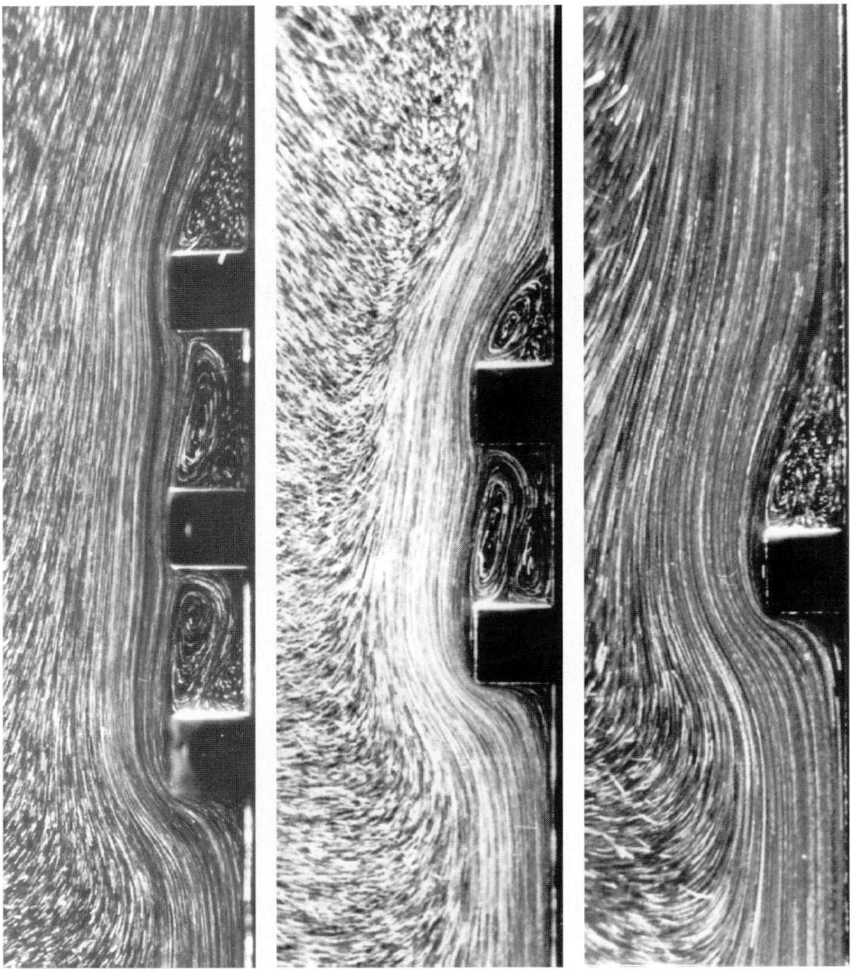

Fig. 2.7.13. Track visualization results for a free-convective flow on a vertical plate with a constant heat flux in the presence of one or several obstacles

the square) the appearing separated flow represents two oppositely rotating circular vortices. When the spacing is increased to 30 mm (which corresponds to the ratio of the spacing width to the obstacle height equal to 1:3) the shape of vortices becomes more extended, and in the region adjacent to the second step a small stagnant zone is formed. A further increase in the spacing width leads to an increase in the size of the stagnant zone and is terminated by the adhering of the dynamic boundary layer to the plate in the region between the obstacles.

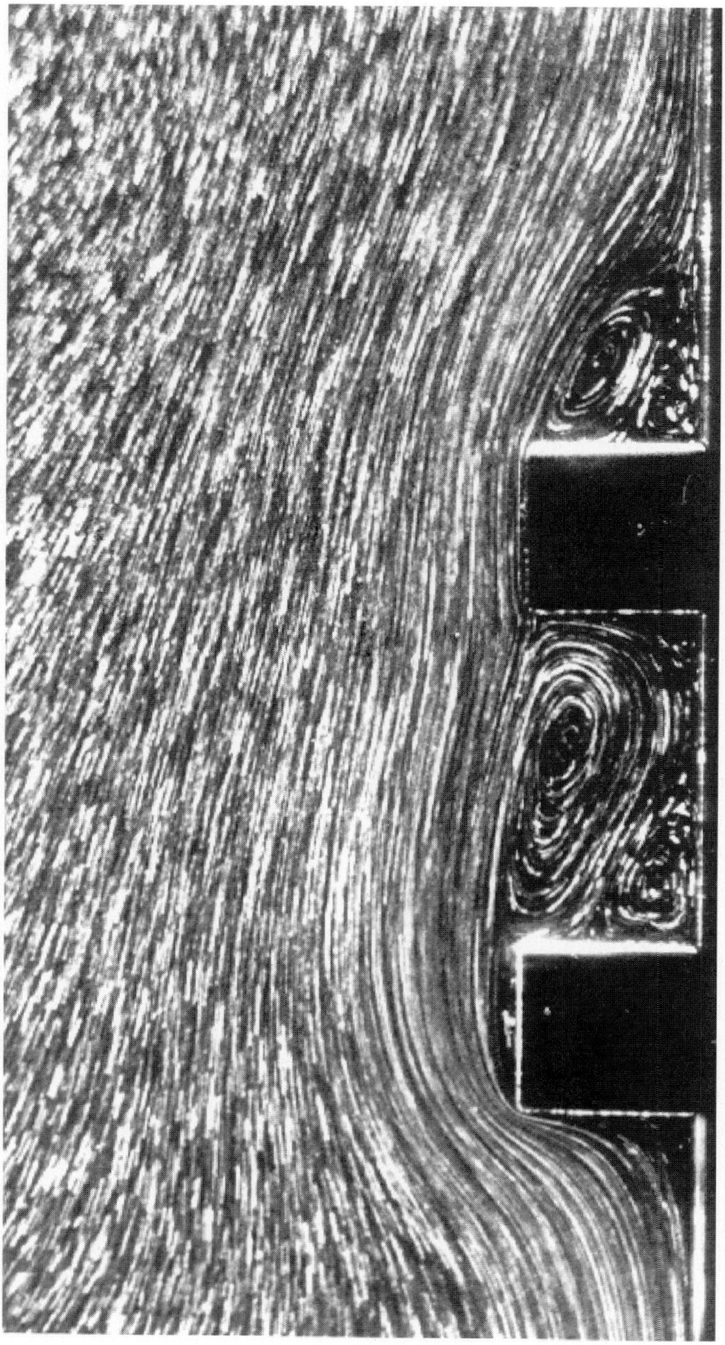

Fig. 2.7.14. A magnified image of a free-convective flow between obstacles

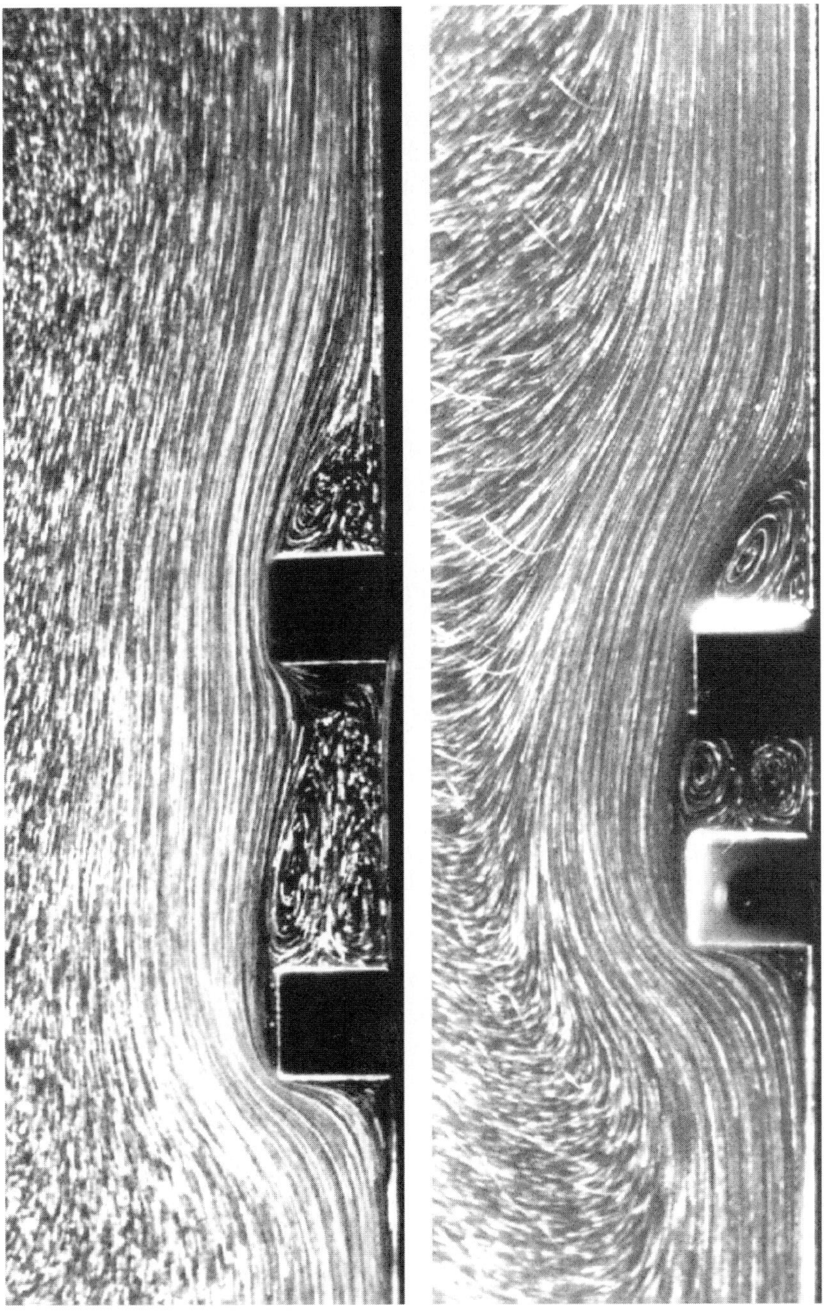

Fig. 2.7.15. Visualization of a free-convective flow on a plate with two steps (the spacing between the obstacles is 10 mm and 30 mm)

2.8 Free Convection Near a Vertical Surface in a Variable Field of Mass Forces

Generally, the buoyancy force, which is the reason of free-convective flows, is encountered in a nonuniform field of mass forces in the presence of a density gradient. Natural-convective flows can be caused by both gravitational and other force fields, in particular, by the Coriolis, centrifugal, electromagnetic forces, etc. Similar situations are frequently encountered in practice and therefore represent an independent interest for investigation. Effects of this kind appear, for example, in electrically conducting liquids placed in an electrical or magnetic field. The motion of electrically conducting liquid in electromagnetic fields is studied in magnetohydrodynamics. In the flows which develop in rotating systems, relative motions of liquid give rise to centrifugal and Coriolis forces, which is characteristic for ocean flows or motions of atmospheric air masses. These two forces interact with the forces of gravitational convection caused by a change in the liquid density. The density gradient can be local or exist in the whole flow region. In both cases, the motion of liquid particles along the curved trajectory leads to the occurrence of additional displacements depending on density distribution. The normal component of the pressure gradient necessary for maintaining the flow is represented as $\partial p/\partial n = -\rho r\Omega^2$, where r and Ω is the local curvature radius of the streamline and the angular rotation velocity, respectively. For example, in the conditions of vertical stratification in the z direction, the value $\partial p/\partial z$ exists due to stratification. If the effect of the forces caused by rotation exceeds the effect of the hydrostatic pressure gradient $-g\rho$, which tends to stabilize the system, a free-convective motion of liquid appears in this system.

The centrifugal and Coriolis forces appearing in rotation must be taken into account in the force and momentum balance equations. A change in the density in the Boussinesq approximation has the form $\Delta\rho = \rho\beta(T - T_r)$, where T_r is the characteristic temperature. During rotation around a vertical axis with the angular velocity Ω in the presence of the axial symmetry of the boundary conditions the transfer equations are written in a cylindrical coordinate system

$$u\frac{\partial u}{\partial r} + w\frac{\partial u}{\partial z} - \frac{v^2}{r} - 2\Omega\nu = -\frac{1}{\rho}\frac{\partial p}{\partial r} + \nu\left(\nabla^2 - \frac{1}{r^2}\right)u\ , \qquad (2.8.1)$$

$$u\frac{\partial v}{\partial r} + w\frac{\partial v}{\partial z} - \frac{vu}{r} + 2\Omega u = \nu\left(\nabla^2 - \frac{1}{r^2}\right)\nu\ , \qquad (2.8.2)$$

$$u\frac{\partial w}{\partial r} + w\frac{\partial w}{\partial z} - g\beta(T - T_r) = -\frac{1}{\rho}\frac{\partial p}{\partial z} + \nu\nabla^2 w\ , \qquad (2.8.3)$$

$$\frac{1}{r}\frac{\partial(ur)}{\partial r} + \frac{\partial w}{\partial z} = 0\ , \qquad (2.8.4)$$

$$u\frac{\partial t}{\partial r} + w\frac{\partial t}{\partial z} = \alpha\nabla^2 T\ , \qquad (2.8.5)$$

where p is the difference of the local values of the static and hydrodynamic pressures at the temperature T_r. The additional physical effects connected with rotation are characterized by a dimensionless parameter representing a gradient of the vertical velocity of convection per unit height L related to the angular velocity $\Omega(U_c/\Omega L)$. Assuming $U_c = \sqrt{gL\beta\Delta T}$, we can transform this parameter as

$$\frac{\sqrt{gL\beta\Delta T}}{\Omega L} = \sqrt{\mathrm{Gr}}\frac{\nu}{\Omega L^2} = 2\sqrt{\mathrm{Gr}}\mathrm{Ek} = \frac{2\sqrt{\mathrm{Gr}}}{\sqrt{\mathrm{Ta}}}\ , \tag{2.8.6}$$

where the Taylor and Eckman numbers characterize the coupling between the forces of internal friction and centrifugal forces:

$$\mathrm{Ta} = \frac{4\Omega^2 L^4}{\nu^2} = \mathrm{Ek}^{-2}, \quad \mathrm{Ek} = \frac{\nu}{2\Omega L^2}\ . \tag{2.8.7}$$

At small values of the Ta numbers ($\nu^2 \gg \Omega^2 L^4$), which corresponds to high viscosity and/or slow rotation, the fluid rotates as a solid body. The large Ta values refer to the flows of the boundary-layer type.

Sometimes in investigation of the effect of centrifugal forces on free convection the Rossby number is used:

$$\mathrm{Ro} = U_{\mathrm{c}}/\Omega L\ , \tag{2.8.8}$$

which characterizes the interaction of inertial and centrifugal forces [2.7].

The development of space engineering required investigation of the problems associated with a variable acceleration of the gravity force. Considering as a model a cooled flat plate passing through in the center of rotation of an artificial satellite, going around its orbit, one can establish that the cooled fluid is rejected along the radius of rotation under the action of centrifugal forces. Here, the variable acceleration is equal to $a = \omega^2 x$, where x is the distance from the rotation center along the plate. The surrounding medium rotates with the same angular velocity as the plate. This statement of the problem differs from the heat transfer of a plate rotating in a motionless air.

In investigation of a vertical plate ***with a constant heat flux*** at $a = \omega^2 x$ the following formula for calculating heat transfer is obtained [2.64]:

$$\mathrm{Nu}_x = 0.616\,(\mathrm{Pr} + 1.143)^{-1/5}\left(\mathrm{Gr}_x^*\,\mathrm{Pr}^2\right)^{1/3}\ . \tag{2.8.9}$$

Heat transfer of a laminar flow on a vertical surface ***of constant temperature*** in a gravitational field $g(x)$ is [2.26]

$$\frac{\mathrm{Nu}_x}{\mathrm{Ra}_x^{1/4}} = \left(\frac{f}{25}g^{1/3}x \Big/ \int\limits_0^x g^{1/3}(x)\,dx\right)^{1/4}, \tag{2.8.10}$$

where

$$f(x,\mathrm{Pr}) = \left(\frac{3}{4} + \frac{5}{7\,\mathrm{Pr}} g^{4/15} \int\limits_0^x g^{1/3}(x)\,dx \Bigg/ \int\limits_0^x g^{3/5}(x)\,dx\right)^{-1}$$

is determined approximately assuming a weak dependence on x.

For a hot and a cold plate rotating near an axis located at the distance x_0 from the leading edge, for $g(x) = \omega^2 x_0 (1 \pm x)$,

$$\frac{\mathrm{Nu}_x}{\mathrm{Ra}_x^{1/4}} = \left\{\frac{4}{45} x \frac{(1 \pm x)^{1/3}}{(1 \pm x)^{4/3} - 1} \Bigg/ \left[1 + \frac{8}{7\,\mathrm{Pr}} (1 \pm x)^{4/15} \frac{(1 \pm x)^{4/3} - 1}{(1 \pm x)^{8/5} - 1}\right]\right\}^{1/4}, \tag{2.8.11}$$

$$\mathrm{Nu}_x = -\left(\frac{\mathrm{Gr}_{x0}}{4}\right)^{1/4} \sum_{n=0}^{\infty} (1 \pm x)^n \frac{\theta_n'(0)}{n!}, \tag{2.8.12}$$

where $\mathrm{Gr}_{x0} = \omega^2 x_0 \beta x^3 \theta_{\mathrm{w}} / \nu^2$, and the values of $\theta_n'(0)$ for different values of the Pr number are given in Table 2.8.1.

Table 2.8.1.

Pr	$n=0$	$n=1$	$n=2$	$n=3$
0.1	–0.23014	–0.05530	0.03255	–0.04291
0.7	–0.49951	–0.12890	0.08493	–0.12748
1	–0.56715	–0.14844	0.09977	–0.15289
10	–1.16933	–0.32736	0.23938	–0.39906

For $g = \omega^2 x$

$$\mathrm{Nu}_x = 0.456 \left(1 + \frac{8}{7\,\mathrm{Pr}}\right)^{-1/4} \mathrm{Ra}_x^{1/4}. \tag{2.8.13}$$

The error of the approximate formula (2.8.13) is 3–5%.

In rotation of a ***triangular heat conductive rib of length l and base*** b_l with the angular velocity ω around the axis $x = x_0$ and $g = \omega^2 x_0 (1 - x/x_0)$, ***when heat is supplied to the base***, the heat transfer is characterized by the additional conjugation criteria:

$$\mathrm{P} = \frac{\lambda_\infty l}{\lambda_\mathrm{b} b_l} \mathrm{Ra}_l^{1/4} \quad \text{and} \quad l/x_0\,,$$

where $\mathrm{Ra}_l = \frac{\omega^2 x_0 \beta l^3}{\nu^2} \theta_{\mathrm{w}}$.

In the investigated range of parameters for $\mathrm{Pr} = 0.72$, the effect of l/x_0 on heat transfer is small. The value of $\mathrm{Nu}_{x0}/\mathrm{Ra}_l^{1/4}$ is a linear function of l/x_0 at the given Pr values [2.26].

2.9 Free-Convective Heat Transfer on a Plane Inclined Surface

In the case of inclined free-convective flows, the transfer equations are characterized by the presence of two buoyancy force (B) components: the tangential B_t and normal B_n. If the value of the buoyancy force is equal to B, then $B_t = B\cos\vartheta$, and $B_n = B\sin\vartheta$. The component B_t leads to the tangential movement of the fluid, the normal B_n, to the transverse one. Figure 2.9.1 shows the coordinate x systems for different cases of an inclined flow. If the surface temperature T_w is higher than the ambient temperature T_∞, the induced flow will be ascending. Then the coordinate x is reckoned from the lower leading edge. When $T_\infty > T_\mathrm{w}$, a descending flow appears, and the coordinate is reckoned from the upper leading edge. In any case the coordinate y is reckoned from the surface. The inclination angle ϑ is considered positive above the surface and negative under it, when $T_\infty > T_\mathrm{w}$, and vice versa, when $T_\mathrm{w} > T_\infty$, which reflects the real physical conditions, and this allows one to use the same system of equations for describing all kinds of the problem statement.

Let us consider various kinds of the problem statement:

(a) ***a heated surface facing upwards***: $T_\mathrm{w} > T_\infty, \vartheta > 0$, and the force B_n is directed from the surface ($B_n > 0$); an ascending motion of fluid with a positive gradient of the pressure $\partial p/\partial y$ appears in the vertical direction;
(b) ***a heated surface facing downwards***: $T_\mathrm{w} > T_\infty, \vartheta < 0$, and the force B_n is directed to the surface ($B_n < 0$); the pressure gradient tends to increase the normal velocity component v;
(c) ***flow under a cooled surface***: $T_\infty > T_\mathrm{w}, \vartheta > 0$, and B_n is again directed from the surface;
(d) ***flow over a cooled surface***: $T_\infty > T_\mathrm{w}, \vartheta < 0$, and B_n tends to increase the velocity component directed to the surface.

The laminar free convection of ***an incompressible fluid*** on a plane surface inclined at an angle ϑ to the vertical in a stationary statement of the problem includes the following equations:

$$u\frac{\partial u}{\partial x} + \nu\frac{\partial u}{\partial y} = g\beta\Delta T\cos\vartheta + \frac{1}{\rho}\frac{\partial p}{\partial x} + \nu\left(\frac{\partial^2 u}{\partial x^2} + \frac{\partial^2 u}{\partial y^2}\right), \tag{2.9.1}$$

$$u\frac{\partial \nu}{\partial x} - \nu\frac{\partial \nu}{\partial y} = g\beta\Delta T\sin\vartheta + \frac{1}{\rho}\frac{\partial p}{\partial y} + \nu\left(\frac{\partial^2 \nu}{\partial x^2} + \frac{\partial^2 \nu}{\partial y^2}\right), \tag{2.9.2}$$

$$\frac{\partial u}{\partial x} + \frac{\partial \nu}{\partial y} = 0, \tag{2.9.3}$$

$$u\frac{\partial T}{\partial x} + \nu\frac{\partial T}{\partial y} = a\left(\frac{\partial^2 T}{\partial x^2} + \frac{\partial^2 T}{\partial y^2}\right) + \frac{1}{\rho c_p}\left(u\frac{\partial p}{\partial x} + \nu\frac{\partial p}{\partial y}\right) + \frac{Q_\nu}{\rho c_p}$$
$$+\frac{\nu}{c_p}\left[\left(\frac{\partial u}{\partial x}\right)^2 + \left(\frac{\partial \nu}{\partial y}\right)^2 + \left(\frac{\partial u}{\partial y} + \frac{\partial \nu}{\partial x}\right)^2\right] \tag{2.9.4}$$

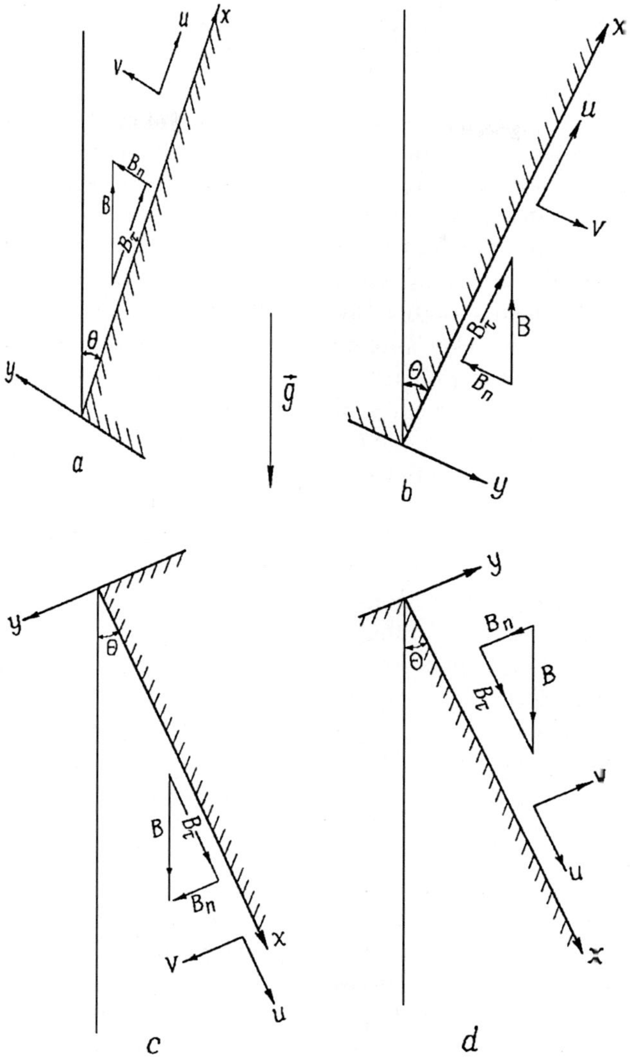

Fig. 2.9.1. Different kinds of a free-convective flow on an inclined surface and the corresponding coordinate systems [2.7]

with the corresponding boundary conditions. The approximate boundary-layer equations for this system can be obtained using the characteristic scale $\Delta = \delta/L$ and neglecting the terms of the first and higher order of smallness in Δ. In the general case

$$u\frac{\partial u}{\partial x} + \nu\frac{\partial u}{\partial y} = g\beta\Delta T\cos\vartheta - \frac{1}{\rho}\frac{\partial p}{\partial x} + \nu\left(\frac{\partial^2 u}{\partial y^2}\right), \tag{2.9.5}$$

$$0 = g\beta\Delta T \sin\vartheta - \frac{1}{\rho}\frac{\partial p}{\partial y} . \tag{2.9.6}$$

If as the Grashof number we use the value $\mathrm{Gr}_{x\vartheta} = \mathrm{Gr}_x \cos\vartheta$ for inclined surfaces, then, provided that $\Delta \mathrm{tg}\vartheta \ll 1$, all self-similar solutions for vertical surfaces can be applied to inclined flows. However, this method has an essential drawback: it does not take into account the special features of the flow above the surface and under it.

In [2.144], by the method of disturbances and using the value $(\mathrm{Gr}\cos\vartheta)^{1/4}$ as the small parameter, an approximate solution was obtained for a plate at small inclination angles. In this method the simple solution for the inclined surface for small inclination angles was used as a zero approximation. The correction of the first order of smallness has a significant effect on the velocity field, but virtually it has no effect on the temperature field. The main difference from the boundary-layer solution is in the presence of asymmetry for the cases of flow above and under the surface.

The numerical solutions ***for air*** [2.145] in the range of the angles ϑ from –30° to 90° showed that for ϑ not exceeding ±30°, the surface inclination has an insignificant effect on the velocity and temperature profiles.

Using the method of local non-self-similarity, the solutions for an inclined ***isothermal surface*** in the range of the angles ϑ from 0 to almost 90° are obtained [2.146]. The calculation results show the validity of the use of the correlation dependence for a vertical surface, where Gr_x is replaced by an equivalent value $\mathrm{Gr}_x \cos\vartheta$. However, this rule is inapplicable for the large inclination angles and small values of the Pr number. For the surface facing upwards, when $B_n > 0$, the flow velocity u is higher, and for $B_n < 0$ (the surface facing downwards), it is correspondingly smaller than for an equivalent vertical surface.

By the method of matched asymptotic expansions an approximate analytical solution was obtained for the case of ***cooling an inclined heated plate*** located at an angle α to the direction of gravity force and at an angle β with respect to the ***wall with the ambient temperature*** [2.147].

Figures 2.9.2 and 2.9.3 represent the interferograms and the corresponding temperature distributions on an inclined plate of ***constant temperature*** (the inclination angle $\vartheta = 45°$), ***with the heated surface facing downwards*** (Fig. 2.9.2) ***and upwards*** (Fig. 2.9.3). As is seen from the figures, the flow on the inclined plate with the heated surface facing upwards becomes unstable along coordinate directed upstreame. Figure 2.9.4 presents the results of track visualization of a flow on an inclined plate of ***constant temperature: (a) the heated surface facing upwards; (b) the heated surface facing downwards***. In the second case the flow has a high degree of stability. Figure 2.9.5 presents a sequence of interferograms of thermal boundary-layer development on an inclined plate ***with a constant heat flux on the wall*** (inclination angle $\vartheta = 45°$). Figures 2.9.6 and 2.9.7 presents the results of track visualization for a convective flow on an inclined plate ***with***

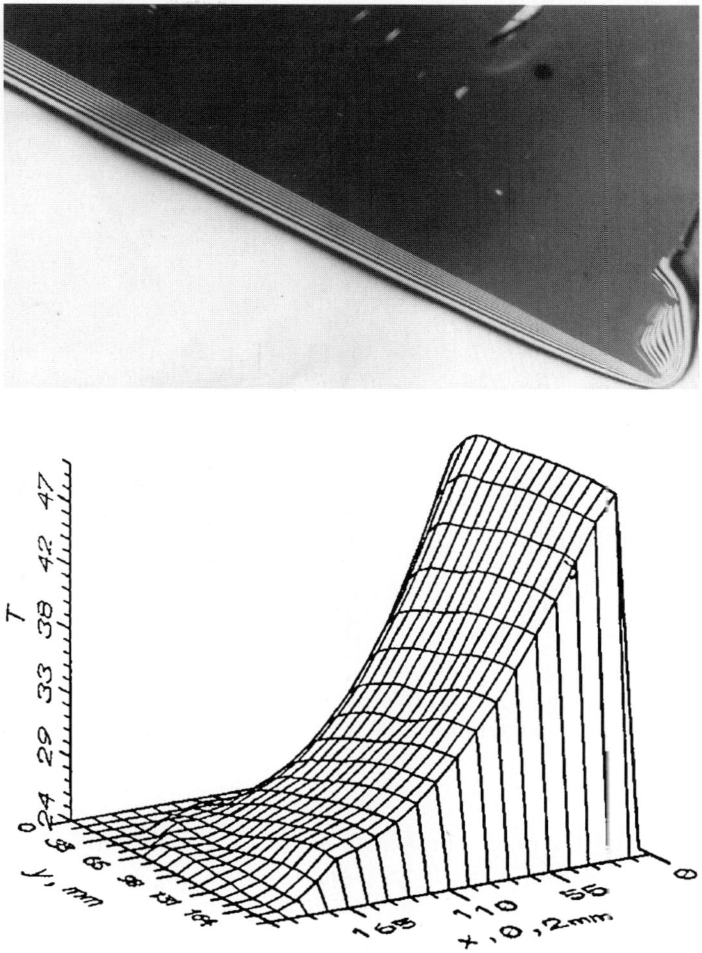

Fig. 2.9.2. The interferogram and corresponding temperature distribution on an inclined plate of constant temperature ($\vartheta = 45°$) with the heated surface facing downwards

a constant heat flux on the surface facing upwards (Fig. 2.9.6) ***and downwards*** (Fig. 2.9.7).

The experimental data for the coefficient of heat transfer from inclined plates at the angles ϑ from 0 to 40°, obtained in [2.148], confirm the validity of the use of correlation dependences for equivalent vertical surfaces with Gr_x, replaced by $\mathrm{Gr}_x \cos\vartheta$. A good agreement of experimental data ***for air*** with calculated formulas was obtained in [2.144] both ***over and under a heated surface*** at inclination angles of up to 45°. The local coefficient of heat transfer from an inclined ***isothermal plate in air*** was measured both for $T_\mathrm{w} > T_\infty$ and $T_\mathrm{w} < T_\infty$ in the range of the angles $-\pi/2 \le \vartheta \le \pi/2$ [2.149].

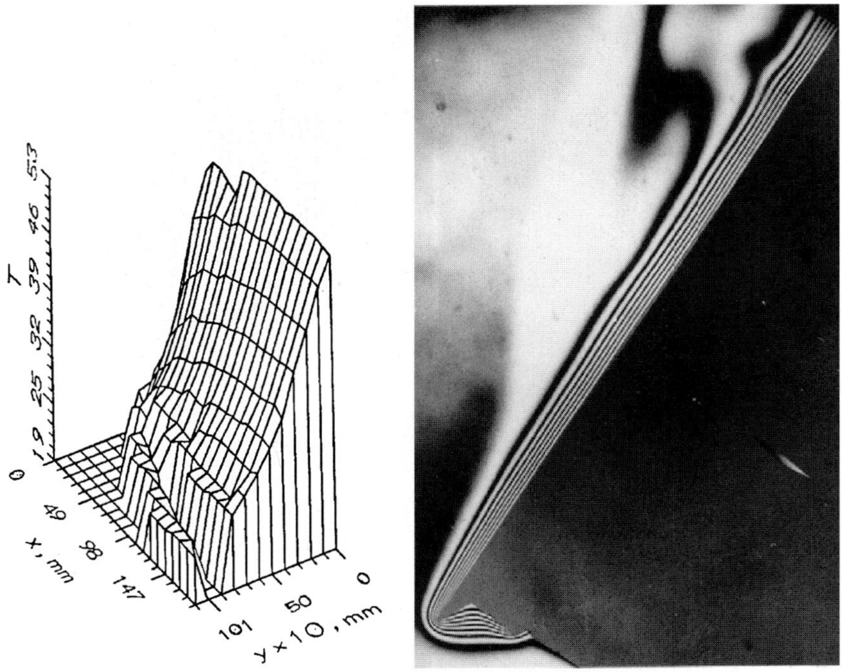

Fig. 2.9.3. The interferogram and corresponding temperature distribution on an inclined plate of constant temperature ($\vartheta = 45°$) with the heated surface facing upwards

It was shown that the range of applicability of the dependences for equivalent vertical surfaces is 0 – 75° for $B_n > 0$ and 0–60° for $B_n < 0$.

The measurements carried out ***in water*** near an inclined surface ***with a constant heat flux density*** [2.150] showed a good agreement with the calculation from the formula for equivalent vertical surfaces up to the angle values of –70°, and in [2.151] this range of applicability was extended to $\vartheta = -85°$ for the case, when the force B_n is directed to the surface. If B_n is directed from the surface, then at some distance along the flow the flow separates, and the length of the part of the plate, where the flow remains laminar, decreases with increase in the inclination angle. Experimental measurement of the parameters of transfer for inclined surfaces ***at a constant heat flux*** for large Prandtl numbers was carried out in [2.152]. In the range of inclination angles from 0 to ±30° the effect of these angles on the velocity field is insignificant for ***surfaces facing both upwards and downwards***. The experimental results are in good agreement with the calculations carried out in [2.146].

The effect of the Prandtl number on the velocity and temperature profiles in natural convection near inclined surfaces begins to manifest itself when the angle of deviation from the vertical attains a certain value, when there appears

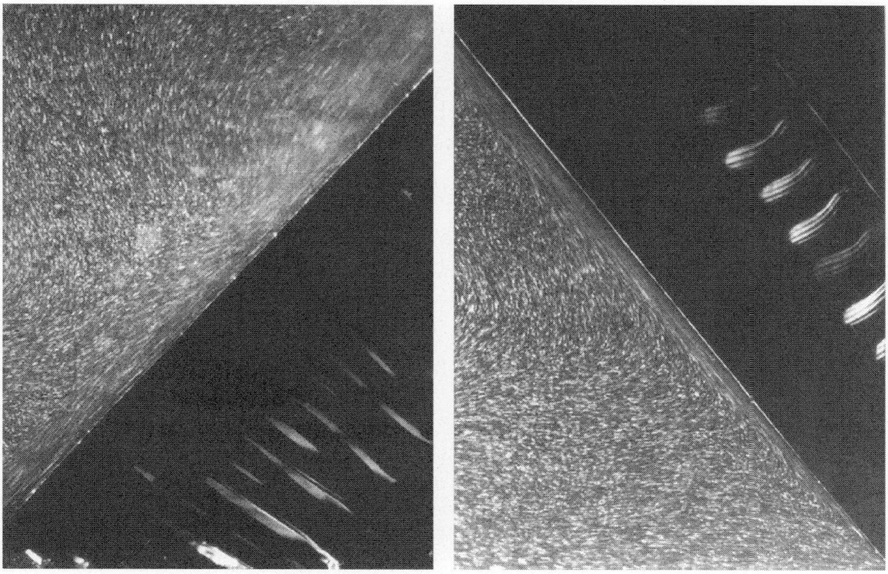

Fig. 2.9.4. The results of track visualization of a free-convective flow on an inclined plate of constant temperature ($\vartheta = 45°$): (**a**) heated side of the plate is facing upwards, (**b**) the heated side of the plate is facing downwards

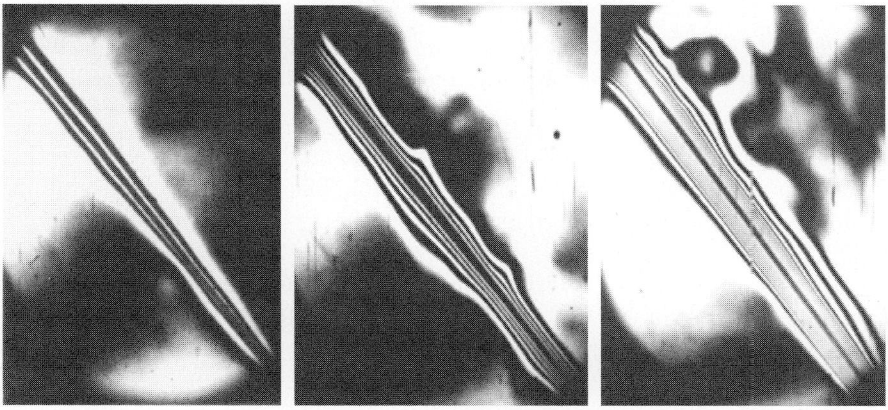

Fig. 2.9.5. The sequence of the interferograms of thermal boundary-layer development on an inclined plate ($\vartheta = 45°$) with a constant heat flux on the wall

a substantial difference in comparison with the results of calculation from the formula for equivalent vertical surfaces. With increase in the Prandtl number, this difference increases. In the limit $\mathrm{Pr} \to \infty$, the effect of the deviation angle on the heat flux and velocity field is determined by the value $g \cos \vartheta$, and then the solutions for equivalent vertical surfaces are applicable [2.7].

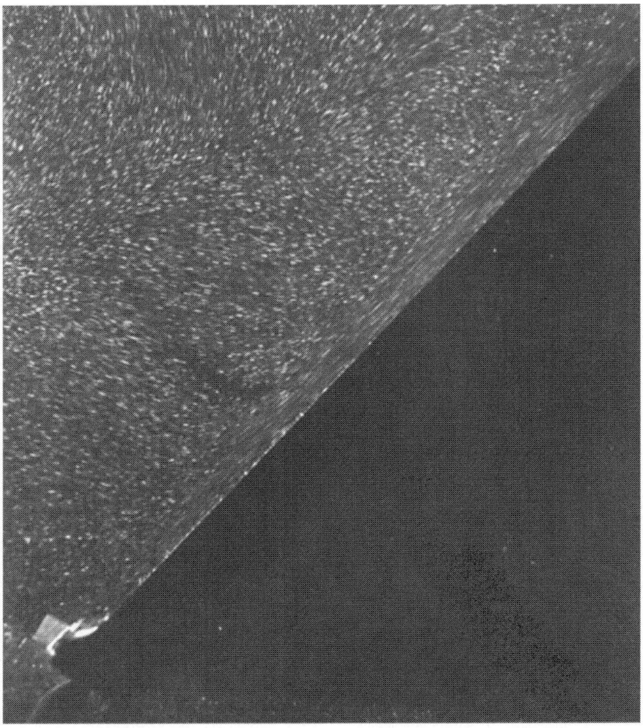

Fig. 2.9.6. Track visualization of a dynamic boundary layer on an inclined plate with the heated surface facing upwards and a constant heat flux on the wall

On a plate with a constant temperature the method of equivalent vertical surfaces is applicable for $-85° < \vartheta < 75°$ [2.26]. The questions of the stability of flows near inclined plates depending on the Pr values were investigated in [2.149, 2.153].

Heat transfer from an ***inclined disk of constant temperature with a laminar regime*** for $\mathrm{Pr} \gg 1$ is calculated from [2.154]

$$\mathrm{Nu}_{\mathrm{av}} = 0.705\,\mathrm{Ra}_{d\vartheta}^{1/4}\,. \tag{2.9.7}$$

Heat transfer from an inclined surface ***with a constant heat flux in a laminar regime*** is determined with an accuracy of 7–8% in the range of the angles $\vartheta = 60 - (-86.5°)$ [2.150]

$$\mathrm{Nu}_x = 0.630\,\mathrm{Pr}^{2/5}\,(\mathrm{Pr}\, + 0.8 - 0.15\,\mathrm{Pr}\sin\vartheta)^{-1/5}\,\mathrm{Gr}_{\vartheta x}^{1/5}\,. \tag{2.9.8}$$

For turbulent convection for the same problem the following generalized dependences are obtained:

$$\begin{aligned} \mathrm{Nu}_x &= 0.889\,\mathrm{Ra}_x^{*0.205}, \qquad \mathrm{Pr} = 0.7 - 10, \quad \vartheta = 0 - (-32°)\,, \\ \mathrm{Nu}_x &= 0.17\,\mathrm{Ra}_x^{*0.25}, \qquad \mathrm{Pr} = 0.7, \quad \vartheta = 0 - 30°\,. \end{aligned} \tag{2.9.9}$$

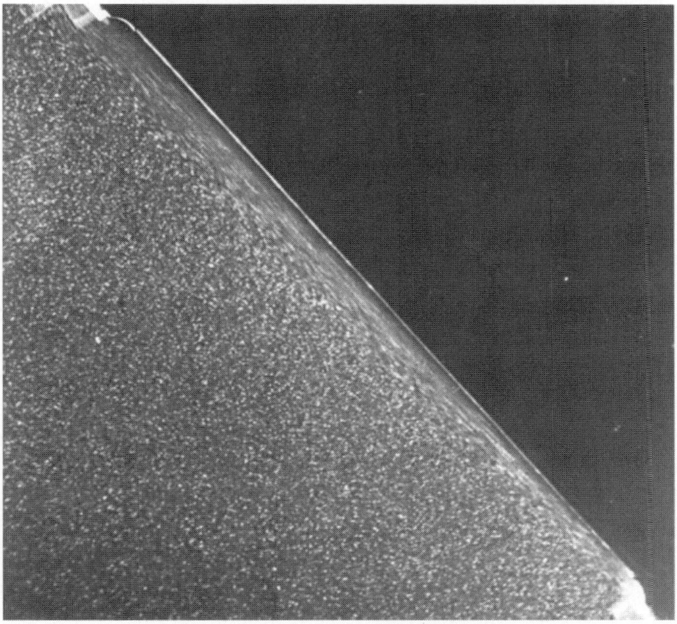

Fig. 2.9.7. Track visualization of a dynamic boundary layer on an inclined plate with the heated surface facing downwards and a constant heat flux on the wall

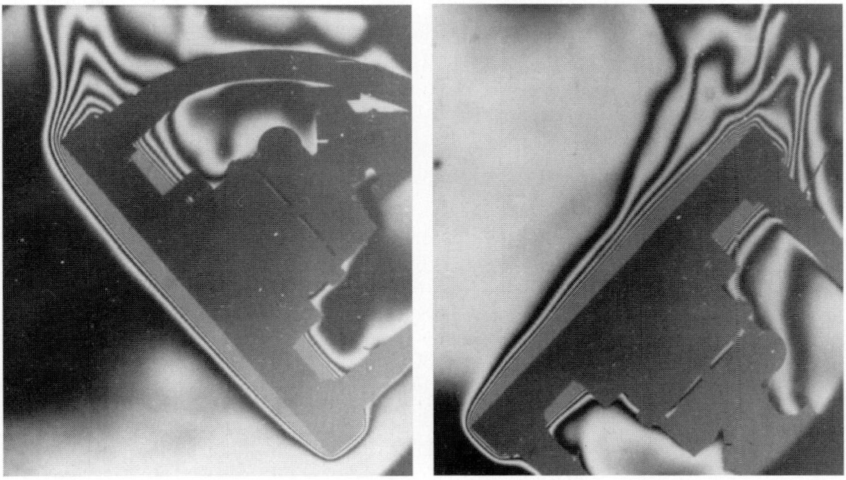

Fig. 2.9.8. Interferograms of a free-convective flow on an inclined round disk of constant temperature

The error of formulas (2.9.9) is 6.5% [2.26].

In [2.155] a new model for a simplified analytical solution of the problem of free convection above an inclined flat ***isothermal plate*** is suggested. The simulation presupposes division of the entire flow region, as it follows from visual patterns, into two parts with different regimes: (1) ***the flow direction coincides with the direction of the buoyancy force component parallel to the plate*** and (2) ***these two directions are opposite***. The obtained analytical solutions are applicable for any inclination angles in the range $0 \leq \vartheta \leq \pi/2$ and are in good agreement with the experimental data presented in Table 2.9.1 ***for a round isothermal inclined plate in glycerin***.

An analytical and numerical investigation of a transient free convection along a ***flat moist inclined plate in air*** was carried out in [2.156]. The calculated data for the average Nusselt and Sherwood numbers were presented. As it follows from these data, the heat and mass transfer rate from the plate decreases with increase in the inclination angle from the vertical, and in the

Table 2.9.1. The Values of the Average Heat Transfer Coefficient for a Round Inclined Isothermal Plate in Glycerin

ϑ[deg]	T_w [°C]	T_∞ [°C]	α[W/m^2 · deg]	Nu_d	Ra_d
		$q = 2.514$ kW/m^2			
0	32.57	15.67	89.993	22.072	$6.206 \cdot 10^6$
1	32.57	15.91	91.264	22.383	$6.206 \cdot 10^6$
2	32.59	16.09	92.176	22.605	$6.224 \cdot 10^6$
3	32.47	16.28	93.917	23.032	$6.134 \cdot 10^6$
4	32.57	16.49	94.605	23.199	$6.202 \cdot 10^6$
5	32.80	16.75	94.744	23.231	$6.379 \cdot 10^6$
6	32.94	16.97	95.162	23.333	$6.486 \cdot 10^6$
8	33.11	17.25	95.868	23.504	$6.612 \cdot 10^6$
10	33.60	17.46	94.191	23.090	$7.017 \cdot 10^6$
12	33.81	17.60	93.781	22.988	$7.199 \cdot 10^6$
		$q = 7.288$ kW/m^2			
0	54.28	20.45	130.246	31.797	$5.478 \cdot 10^7$
1	54.17	20.35	130.337	31.820	$5.418 \cdot 10^7$
2	54.07	20.17	129.975	31.733	$5.359 \cdot 10^7$
3	54.07	19.98	129.257	31.559	$5.339 \cdot 10^7$
4	54.21	19.79	128.021	31.257	$5.378 \cdot 10^7$
5	54.07	19.51	127.498	31.132	$5.289 \cdot 10^7$
6	54.07	19.25	126.550	30.902	$5.260 \cdot 10^7$
8	53.86	18.85	125.870	30.739	$5.129 \cdot 10^7$
10	53.30	18.57	126.893	30.993	$4.874 \cdot 10^7$
12	53.20	18.21	125.954	30.767	$4.799 \cdot 10^7$

course of time it reaches some quasi-steady value, which, in turn, depends substantially on the diffusion rate in the boundary layer.

A theoretical and numerical analysis of momentum, heat and mass transfer in a laminar free-convective boundary layer above an inclined flat plate ***at a constant heat flux density on a wall*** was carried out in [2.157]. ***The plate is designed so that it has two moist zones divided by a dry one.*** The determining criterion affecting heat transfer and the boundary-layer flow is the length of the dry zone.

An analysis of vortical instability of a laminar boundary layer on an inclined ***flat plate heated from below for the surface temperature changing by the law*** $T_w(x) - T_\infty = Ax^n$ carried out on the basis of the linear theory shows that for the given Pr and n values the flow approaches an unstable vortical regime with increase in the inclination angle ϕ with respect to the horizontal [2.158]. The main stable flow is considered two-dimensional, and the amplitude of the disturbance function is assumed dependent on the distance down the flow. The critical values of the Grashof numbers and of the corresponding wave numbers are obtained for Pr = 0.7 and 7 in the range of the inclination angles relative to the horizontal $0° \leq \phi \leq 70°$ and for n from $-1/3$ to 1. It is also shown that the dependence of the disturbance function amplitude on the distance along the flow leads to a stabilization of the main flow.

An experimental measurement of the local coefficient of heat transfer from inclined flat plates ***heated by a constant heat flux*** for the inclination angle changing from 0 to 75° was carried out in [2.159]. The experimental data obtained are presented in Table 2.9.2 and are generalized by the empirical dependence

$$\mathrm{Nu}\,(x) = 0.587\,(\mathrm{Ra}_x^* \cos\vartheta)^{1/5} . \tag{2.9.10}$$

Example 2.9. Determine heat transfer from a wall of size $a \times b = 1 \times 2\,\mathrm{m}^2$, located on the larger dimension at an angle of $\vartheta = 60°$ with respect to the

Table 2.9.2. The Values of the Constant Coefficients of the Correlation Equation $\mathrm{Nu}(x) = C(\mathrm{Ra}_x^* \cos\vartheta)^m$ for Different Inclination Angles ϑ

ϑ	$\log(\mathrm{Ra}_x^* \cos\vartheta)$ Laminar Regime	A Choice of Two Coefficients		A Choice of One Coefficient for $m = 1/5$
		C	m	
0	4.7–12.8	0.585	0.200	0.584
15	5.0–12.0	0.592	0.199	0.586
30	5.7–10.0	0.573	0.201	0.584
45	5.2–9.0	0.615	0.195	0.576
60	4.5–8.0	0.524	0.205	0.568
75	4.2–7.1	0.237	0.259	0.553

vertical, if the temperature of the heated surface is $t_{\mathrm{w}} = 80°$C, and the temperature of the surrounding air is $t_\infty = 20°$C.

The determining temperature is $t_r = (80+20)/2 = 50°$C, the characteristic dimension is $l = b = 2$ m. The thermophysical characteristics are:

$$
\begin{aligned}
\nu &= 17.95 \cdot 10^{-8}\ \mathrm{m}^{-2}/\mathrm{sec},\\
\beta &= 1/323.16\ 1/\mathrm{deg},\\
\lambda &= 2.83 \cdot 10^{-2}\ \mathrm{W/m \cdot deg},\\
\mathrm{Pr} &= 0.698.
\end{aligned}
$$

The dimensionless determining complex Ra_l is

$$
\mathrm{Ra}_l = \frac{9.81 \cdot 8 \cdot 60 \cdot 0.698}{323.168 \cdot (17.95)^2} \cdot 10^{12} = 6.53 \cdot 10^{12}\ ,
$$

$$
\mathrm{Ra}_{l\vartheta} = \mathrm{Ra}_l \cos\vartheta = 3.26 \cdot 10^{10}\ .
$$

The average value of the coefficient of heat transfer from the upper and lower surfaces of the plate can be determined by formula (2.1.7) with Ra_l substituted by $\mathrm{Ra}_{l\vartheta}$

$$
\mathrm{Nu}_{\mathrm{av}}^{1/2} = 0.825 + \frac{0.387 \cdot (3.26 \cdot 10^{10})^{1/6}}{\left[1 + (0.492/0.698)^{9/16}\right]^{8/27}} = 366\ ,
$$

$$
\alpha_{\mathrm{av}\theta} = 5.15\ \mathrm{W/m^2 \cdot deg}\ .
$$

To calculate heat transfer from the upper and lower surfaces we use the formula

$$
\mathrm{Nu} = 0.262\,(\mathrm{Ra}\ \mathrm{Pr})^{0.35}\ ,
$$

and determine all the Nu values by (2.1.7): $\mathrm{Ra}_{l0} = 4 \cdot 10^9$, $\mathrm{Nu}_{0T} = 450$, $\mathrm{Nu}_0 = 189$, $\vartheta = 60°$: $\mathrm{Ra}_{l\vartheta} = 3.3 \cdot 10^{-6}$, $\mathrm{Nu}_{l\vartheta} = 20.4$, $\mathrm{Nu}_{\mathrm{av}\vartheta} = 281.4$, $\alpha_{\mathrm{av}\vartheta} = 3.98$ $\mathrm{W/m^2 \cdot deg}$; $\vartheta = -60°$: $\mathrm{Ra}_{l\vartheta} = 2.2 \cdot 10^9$, $\mathrm{Nu}_{l\vartheta} = 157$, $\mathrm{Nu}_{\mathrm{av}\vartheta} = 418$, $\alpha_{\mathrm{av}\vartheta} = 5.91$.

The average value from both surfaces is $\alpha_{\mathrm{av}\vartheta} = 5.08\ \mathrm{W/m^2 \cdot deg}$.

2.10 Horizontal and Almost Horizontal Surfaces

The transfer equation for free-convective flows near horizontal surfaces are obtained from the general (2.9.1)–(2.9.4) for $\vartheta = \pm\pi/2$. Considering the boundary layer equations for inclined surfaces, we note that the buoyancy force B_t in (2.9.3) in the case of a horizontal orientation of the plate turns to zero, and only the normal component B_n is a source of motion. This component is directed from the plate for the heated surface facing upwards ($\vartheta = \pi \mathrm{p}/2$) and to the plate for the heated surface facing downwards ($\vartheta = -\pi/2$).

The experimental data for flow and heat transfer ***of water*** above flat horizontal rectangular plates of width from 20 to 1500 mm ***heated by a constant heat flux*** for Ra numbers from 10^6 to 10^{15} are presented in [2.160]. Flow visualization was carried out using a dye, the plate surface temperature was measured using liquid crystals. The results of investigation of free convection along wide plates show the presence of four different flow regimes begining from the leading edge along the flow: (I) laminar boundary layer; (II) transient region; (III) turbulent flow; (IV) region of collisions of turbulent vortices. It should be noted that the transient region occupies a significant part of the overall plate surface, the local heat transfer coefficient decreases with increase of the distance from the leading edge, and the plate width exerts a strong enough effect on the heat transfer rate even for very high Ra numbers achieved in the given experiments. The experimental data for the Nusselt number are generalized by the dependences

local heat transfer coefficient

(I): $10^2 < \mathrm{Ra}_x^* < 10^6$

$$\mathrm{Nu}(x) = 0.66\,(\mathrm{Ra}_x^*)^{1/6}\ , \tag{2.10.1}$$

(II): $10^6 < \mathrm{Ra}_x^* < 10^7$

$$\mathrm{Nu}(x) = 0.066\,(\mathrm{Ra}_x^*)^{1/3}\ , \tag{2.10.2}$$

(III): $8 \cdot 10^7 < \mathrm{Ra}_x^* < 8 \cdot 10^{10}$

$$\mathrm{Nu}(x) = 0.7\,(\mathrm{Ra}_x^*)^{1/5}\ , \tag{2.10.3}$$

(IV): $8 \cdot 10^{10} < \mathrm{Ra}_x^* < 10^{14}$

$$\mathrm{Nu}(x) = 0.2\,(\mathrm{Ra}_x^*)^{1/4}\ , \tag{2.10.4}$$

average heat transfer coefficient

(I): $1.6 \cdot 10^3 < \mathrm{Ra}^* < 1.6 \cdot 10^7$

$$\mathrm{Nu} = 1.25\,(\mathrm{Ra}^*)^{1/6}\ , \tag{2.10.5}$$

(II): $1.6 \cdot 10^7 < \mathrm{Ra}^* < 8 \cdot 10^8$

$$\mathrm{Nu} = 0.04\,(\mathrm{Ra}^*)^{1/3} + 9.7\ , \tag{2.10.6}$$

(III): $8 \cdot 10^8 < \mathrm{Ra}^* < 1.3 \cdot 10^{12}$

$$\mathrm{Nu} = (\mathrm{Ra}^*)^{1/5} - 13.5\ , \tag{2.10.7}$$

(IV): $1.3 \cdot 10^{12} < \mathrm{Ra}^* < 1.6 \cdot 10^{15}$

$$\mathrm{Nu} = 0.2\,(\mathrm{Ra}^*)^{1/4} + 37.0\ . \tag{2.10.8}$$

In [2.161] the problem of laminar free convection ***above a heated horizontal surface of an infinitely long sheet with adiabatically insulated sides and uniform temperature distribution in the central region*** and ***above a rectangular plate of finite size for a constant surface temperature and different values of*** γ (a ratio of the area of the heated region to the overall plate area) in the case of high Ra numbers is solved numerically. The calculation results for $\mathrm{Pr} = 0.7$ are given in Table 2.10.1.

Table 2.10.1. The Average Nu Values for Different Ratios of the Area of the Heated Region to the Overall Area of the Plate

γ	0.2	0.3	0.4	0.5	0.6	0.7	0.8	0.9	1.0
$\mathrm{Nu}/\mathrm{Gr}^{1/5}$	0.098	0.137	0.173	0.206	0.237	0.279	0.315	0.361	0.412

The numerical data for the local and average coefficients of free-convective heat transfer of a ***semi-infinite horizontal surface with the heat flux on it changing by the law*** $q_\mathrm{w}(x) = bx^m$, where the coordinate x is reckoned from the leading edge along the plate, for $0.1 \le \mathrm{Pr} \le 100$ and $-0.4 \le m \le 1.0$ are generalized by the dependences [2.162, 2.163]

$$\mathrm{Gr}_x^* = g\beta q_\mathrm{w}(x)\, x^4 / \left(k\nu^2\right) ,$$

$$\mathrm{Nu}(x)\left(\mathrm{Gr}_x^*\right)^{-1/6} = \alpha(1+V) , \tag{2.10.9}$$

$$\mathrm{Nu}\left(\mathrm{Gr}_L^*\right)^{-1/6} = \alpha \frac{6}{m+4}(1+V) , \tag{2.10.10}$$

where L is the plate length,

$$\alpha = \left(\frac{\mathrm{Pr}}{6}\right)^{1/6} \frac{\sqrt{\mathrm{Pr}}}{0.12 + 1.195\sqrt{\mathrm{Pr}}}, \qquad V = \frac{m}{A1} - \frac{m^2}{A2} + \frac{m^3}{A3} ,$$

$$A1 = 9.12 \cdot 10^{-2} \ln(\mathrm{Pr}) + 2.468, \quad A2 = 1.8 \cdot 10^{-1} \ln(\mathrm{Pr}) + 6.17 ,$$

$$A3 = 3.3710^{-1} \ln(\mathrm{Pr}) + 17.37 .$$

An analysis of the vortical instability of a laminar boundary layer on a horizontal ***flat plate heated from below for the surface temperature changing by the law*** $T_\mathrm{w}(x) - T_\infty = Ax^n$, carried out on the basis of the linear theory applied to a model of a non-parallel flow shows that for the given Pr the flow stability with respect to vortical disturbances increases with n.

Moreover, the stability of flows with high Pr numbers is lower, than for fluids with low Pr [2.164]. The main stable flow is considered to be two-dimensional, and the amplitude of the disturbance function is dependent on the distance down the flow. The critical values of the Grashoff numbers and of the corresponding wave numbers are obtained for $0.7 \le \mathrm{Pr} \le 100$ and for

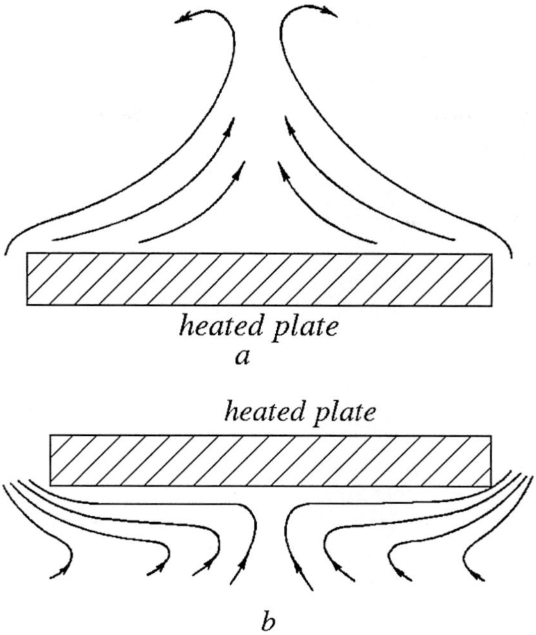

Fig. 2.10.1. Flow structure in free convection on a horizontal surface: (**a**) heated surface facing upwards, (**b**) heated surface facing downwards

n from $-1/3$ to 1. It is also shown that the dependence of the disturbance function amplitude on the distance along the flow leads to stabilization of the main flow. The results of numerical calculations are presented in Table 2.10.2.

The results of an experimental investigation of natural-convective heat transfer ***from a heated step located on a vertical or horizontal surface*** are given in [2.165]. In the presence of an obstacle on the vertical surface a near-wall jet in close vicinity to it appears. When an obstacle is located on a horizontal surface, a free buoyancy plume is formed. The features of flows and characteristics of heat transfer in such systems depend greatly on the energy supply rate and on the step thickness. The longitudinal thermal conductivity of the plate has also a substantial effect. The obtained experimental data for the average Nusselt number for $H/L = 0.4$ and $5{\cdot}10^5 < \mathrm{Ra} < 10^7$ are generalized by the correlation dependences

$$\mathrm{Nu} = 0.900\ \mathrm{Gr}^{0.170} \ \textbf{\textit{for vertical configuration}} \tag{2.10.11}$$

and

$$\mathrm{Nu} = 0.431\ \mathrm{Gr}^{0.217} \ \textbf{\textit{for horizontal configuration}} \tag{2.10.12}$$

An analytical and numerical investigation of a problem of stable natural-convective flow ***along a horizontal and a vertical plate with the surface***

Table 2.10.2. The Critical Values of the Grashoff Number and Wave Number and the Magnitude of the Local Heat Transfer Coefficient $\mathrm{Nu}(x)\,(\mathrm{Gr}_x/5)^{-1/5} = -\theta''(0)$

n	Gr_x^*	α^*	$-\theta'(0)$
	Pr = 0.7		
1	1164.8	0.62066	0.82703
3/4	1075.5	0.63690	0.76234
1/2	989.56	0.65379	0.68840
1/3	933.99	0.66405	0.63186
1/4	907.65	0.67004	0.60063
0	834.52	0.68803	0.48933
–1/4	780.19	0.70939	0.32818
–1/3	771.92	0.72046	0.25067
	Pr = 7		
1	117.70	0.90566	1.39955
3/4	100.99	0.91200	1.29837
1/2	85.252	0.91887	1.18280
1/3	75.264	0.92484	1.09432
1/4	70.411	0.92839	1.04535
0	56.326	0.94275	0.86956
–1/4	42.616	0.96876	0.60739
–1/3	37.951	0.98318	0.47532
	Pr = 100		
1	13.470	1.6635	2.45011
3/4	11.502	1.6641	2.27896
1/2	9.6599	1.6654	2.08333
1/3	8.4967	1.6675	1.93332
1/4	7.9327	1.6691	1.85018
0	6.2972	1.6775	1.55059
–1/4	4.6925	1.6978	1.09815
–1/3	4.1372	1.7111	0.86632

heated by the harmonic law in a porous medium saturated with a liquid was carried out in [2.166, 2.167]. ***In the case of horizontal surfaces*** for small Ra the analytical solution assuming ***either constant ambient temperature depending on Ra numbers or adiabatic insulation*** as boundary conditions at infinity is obtained. For an arbitrary value of the constant ambient temperature the analytical solution is impossible, however, it is possible to obtain a solution for the case of constant ambient temperature at some finite distance from the plate, say $y = d$, then as $d \to \infty$, the solution tends to the obtained analytical distribution for adiabatic conditions at infinity. ***In the case of a vertical surface***, for $\mathrm{Ra} \geq 40$ and small times two recirculation flow regions develop at the point of collision of two boundary layers.

In [2.168] the effect of inertial forces on a flow in a free-convective boundary layer near ***an isothermal horizontal plate*** placed ***in a porous medium saturated with a liquid*** is investigated numerically. It is shown that near the leading edge the inertial effects exert a significant effect on the flow, but further, as the distance from the leading edge increases, the flow obeys the Darcy law.

Using the analogy between the processes of viscous friction and heat or mass transfer, the analytical and numerical data for mixed convection from a vertical and a horizontal plate for cases of ***constant temperature (UWT) and constant heat flux (UHF)*** were obtained in [2.169, 2.170]. In the region of predominance of free convection in the range of the Prandtl numbers $0.7 \le \mathrm{Pr} \le \infty$ and of the parameter ξ which describes the correlation of the buoyancy and inertia forces in the range $0.6 \le \xi \le 1$ the following generalized correlation holds:

$$\frac{C_N/2}{\mathrm{Nu}\ \mathrm{Ra}^{2/n}}\left(\frac{\mathrm{Pr}}{1+\mathrm{Pr}}\right)^{(n-4)/2n} = c\,, \tag{2.10.13}$$

where $c = 1.15$ and $c = 1.20$ for the cases of ***a constant temperature and a constant heat flux on the vertical surface***, respectively; $c = 1.05$ and $c = 1.02$ ***for the same boundary conditions on the horizontal surface***.

The numerical data for different values of Pr and ξ are presented in Table 2.10.3.

It should be noted that horizontal flows are not particular cases of inclined ones, so it is impossible to obtain the boundary-layer equations for horizontal flows from (2.9.5) and (2.9.6), because their construction is based on the use of scales characteristic for vertical flows. The effect of the normal component of the buoyancy force leads to the occurrence of a vertical gradient of the pressure $\partial p_m/\partial y$, which affecting indirectly the fluid through its component $\partial p_m/\partial x$, creates a horizontal flow, usually weaker compared with the vertical one.

The boundary-layer equations for horizontal flows are [2.7]

$$u\frac{\partial u}{\partial x} + \nu\frac{\partial u}{\partial y} = \nu\frac{\partial^2 u}{\partial y^2} - \frac{1}{\rho}\frac{\partial p_m}{\partial x}\,, \tag{2.10.14}$$

$$0 = g\beta(T - T_\infty) - \frac{1}{\rho}\frac{\partial p_m}{\partial y} \quad \text{for } \vartheta = \pi/2\,, \tag{2.10.15}$$

$$0 = -g\beta(T - T_\infty) - \frac{1}{\rho}\frac{\partial p_m}{\partial y} \quad \text{for } \vartheta = -\pi/2\,. \tag{2.10.16}$$

Like for inclined flows, there are two types of flows near a horizontal surface: (1) ***heated surface facing upwards and equivalent cooled surface facing downwards*** and (2) ***cooled surface facing upwards and heated surface facing downwards***. In the first case $\vartheta > 0$, and the force B_n is

Table 2.10.3. The Results of Numerical Calculations of the Value $[(C_N/2)(\Pr/(1+\Pr))^{(n-4)/2n}]\ \mathrm{Nu}\ \mathrm{Ra}^{2/n}$ in the Region of Predominance of Natural Convection

ξ	Pr = 0.7	1	10	100	1000	10000
			Vertical plate, UWT case ($n = 4$)			
0.6	1.122	1.157	1.196	1.213	1.220	1.222
0.7	1.138	1.139	1.155	1.171	1.178	1.180
0.8	1.136	1.134	1.140	1.155	1.162	1.165
0.9	1.137	1.133	1.135	1.150	1.158	1.161
1	1.137	1.133	1.134	1.149	1.157	1.160
			Vertical plate, UHF case ($n = 5$)			
0.6	1.111	1.127	1.189	1.208	1.213	1.214
0.7	1.158	1.168	1.206	1.217	1.220	1.220
0.8	1.183	1.189	1.215	1.221	1.222	1.223
0.9	1.194	1.199	1.218	1.222	1.223	1.224
1	1.197	1.201	1.218	1.222	1.223	1.224
			Horizontal plate, UWT case ($n = 5$)			
0.6	1.022	1.032	1.087	1.100	1.099	1.106
0.7	1.022	1.028	1.062	1.074	1.078	1.079
0.8	1.027	1.031	1.055	1.066	1.070	1.071
0.9	1.030	1.032	1.053	1.063	1.068	1.069
1	1.030	1.033	1.052	1.063	1.067	1.069
			Horizontal plate, UHF case ($n = 6$)			
0.6	0.964	0.967	1.045	1.050	1.052	1.064
0.7	0.990	1.003	1.047	1.047	1.047	1.043
0.8	1.003	1.013	1.049	1.047	1.046	1.046
0.9	1.009	1.018	1.049	1.047	1.045	1.045
1	1.010	1.018	1.049	1.047	1.045	1.045

directed from the surface, in the second one – $\vartheta < 0$, and B_n is directed to the surface.

Consider a horizontal semi-infinite heated surface facing upwards. As the fluid above it is heated, it tends to move up from the surface, which leads to the occurrence of a vertical positive pressure gradient $\partial p_m/\partial y$. In a motionless environment $p_m \approx 0$, so in the upstream of fluid there appears a negative horizontal pressure gradient $\partial p_m/\partial x$, which causes a flow along the surface and formation of a boundary layer increasing from the leading edge.

The existence of such a flow pattern is confirmed by many observations and experimental data [2.171, 2.173].

Figure 2.10.2 presents an interferogram and the restored temperature distribution in a free-convective flow on a horizontal ***plate of constant***

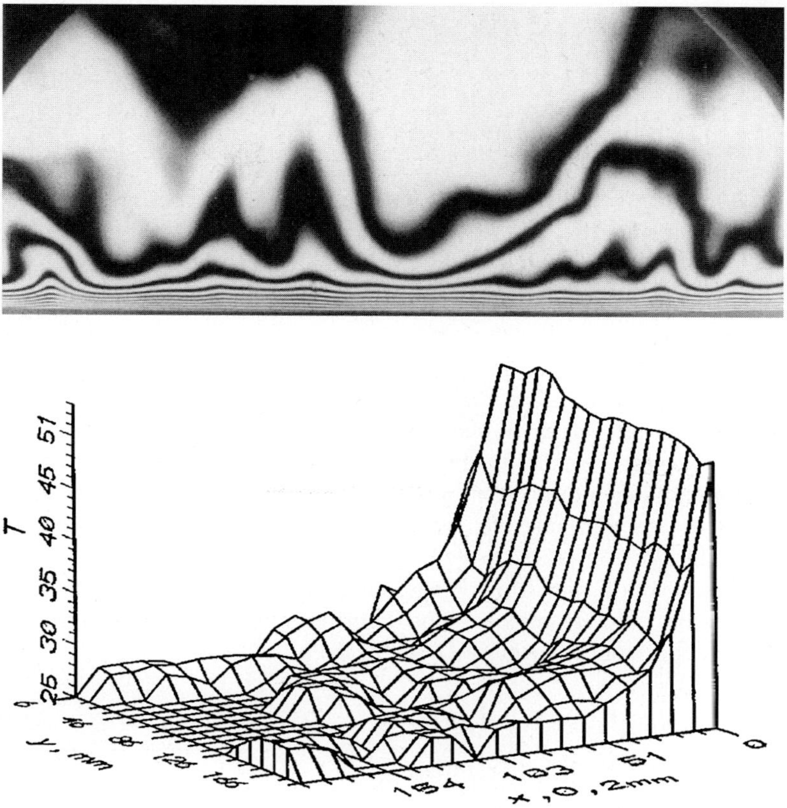

Fig. 2.10.2. Interferogram and corresponding temperature distribution on a horizontal plate of constant temperature with the heated surface facing upwards

temperature with the heated surface facing upwards. As is seen from the presented interferogram and graph of temperature distribution, the flow is strongly unstable.

An analytical investigation of free convection ***for the cases of the heated surface facing upwards and the cooled surface facing downwards*** was carried out in [2.174–2.178]. The criterial dependences for the local Nusselt number are obtained for the surface of constant temperature for different Pr values, including Pr $\to 0$ and Pr $\to \infty$.

If we introduce the self-similar variables [2.18, 2.22]

$$\begin{aligned} &\xi = x/L, \quad \theta(\eta) = (T - T_\infty)/(T_w - T_\infty) \quad \Psi = \zeta^{3/5} f(\eta)\,, \\ &\eta = \mathrm{Gr}_L^{1/5} (y/L) \zeta^{-2/5}, \quad p = (\nu/L)^2 \mathrm{Gr}_L^{4/5} \zeta^{2/5} G(\eta)\,, \end{aligned} \tag{2.10.17}$$

then the boundary-layer equations for a horizontal semi-infinite plate (with the heated surface facing upwards or cold facing downwards) are written as

$$f''' + \frac{3}{5} f'' - \frac{1}{5} f'^2 - \frac{2}{5} (G - \eta G') = 0 \,, \tag{2.10.18}$$

$$G' = \theta \,, \tag{2.10.19}$$

$$\theta'' + \frac{3}{5} \Pr f \theta' = 0 \,. \tag{2.10.20}$$

The boundary conditions corresponding to the constant surface temperature are transformed to

$$\begin{aligned} \eta = 0: &\quad f = f' = 0, \quad \theta = 1 \,, \\ \eta \to \infty: &\quad f = 0, \quad G = \theta = 0 \,. \end{aligned} \tag{2.10.21}$$

The solution of this problem shows a strong instability of the free-convective boundary layer on the surface of a heated semi-infinite horizontal plate; for the large Grashoff numbers the flow is characterized by a cellular structure with longitudinal (extended in the flow direction) vortices. Such a flow structure remains also for small inclination angles of the surface with respect to the horizon [2.22].

In general the application of the self-similarity method leads to a system of ordinary differential equations describing flow and free-convective heat transfer near horizontal surfaces [2.7]:

$$f''' + (n+3) f'' - (2n+1) f'^2 - \frac{4n+2}{5} P - \frac{n-2}{5} \eta P' = 0 \,, \tag{2.10.22}$$

$$P' = \theta \,, \tag{2.10.23}$$

$$\theta'' + \Pr [(n+3) f\theta' - 5n f'\theta] = 0 \,, \tag{2.10.24}$$

$$f(0) = f'(0) = 1 - \theta(0) = f'(\infty) = \theta(\infty) = P(\infty) = 0 \,. \tag{2.10.25}$$

For an isothermal surface $n = 0$. The condition of constant heat flux density on the surface corresponds to the value of $n = 1/3$.

The application of the method of matched asymptotic expansions to investigation of free-convective heat transfer near a semi-infinite heated plate allows one to obtain a series of boundary Dirichlet problems for the Laplace equation in the region of external flow. The finite sizes of the flow region due to the presence of solid borders complicate and transform the solutions of the external problems, which changes boundary conditions on the external edge of the boundary layer.

The approximate solution obtained for a horizontal plate of constant temperature reveals the mechanism of interaction between the boundary layer and external flow and shows that the effect of sharp edges on free-convective heat transfer in this case is determined by another mechanism than for a vertical plate [2.147].

In the case of horizontal rectangular surfaces of finite size, if the force B_n is directed from the surface (Fig. 2.10.1a), for large ratios L_2/L_1 the results of theoretical analysis for the semi-infinite plate can be applied [2.7].

But for the small ratio L_2/L_1 there are large divergences of experimental and calculated data. Comparison of calculations by the boundary layer theory and some experimental results for air in the case of square, rectangular, and round plates shows that the true heat transfer coefficients usually appear to be higher than those calculated: the experimental values of the Nusselt number change as $\mathrm{Ra}^{1/4}$, but not as $\mathrm{Ra}^{1/5}$, as predicted by the boundary-layer theory. In the range of small Ra numbers ($10 < \mathrm{Ra} < 500$) the experimental data can well be described by the dependence $\sim\mathrm{Ra}^{1/6}$. These divergences can be explained by the effect of additional boundary flows, variable physical properties of the fluid in the boundary layer, interaction of flows in the middle of the plate, and by other effects, not taken into account in the theoretical analysis.

For a horizontal surface of finite size with two edges in the absence of lateral flows both edges behave independently like the leading ones, and the flow on the surface develops in the direction to its middle part. Therefore, on each side from the central region there appear two flows similar to the boundary layer flows, which converge in the middle and form an upstream flow. Far from the vertical ascending stream the flow is the same, as in the case of a semi-infinite plate, it does not depend on the flow on the other side. However, the effect of the buoyancy force directed normal to the surface causes a convective instability, which can lead to the flow separation from the surface before it meets the flow moving from the opposite edge. The separation can also happen due to the absence of the pressure gradient [2.18].

On the basis of the numerical calculation carried out for the cases of ***a constant heat flux density and a constant surface temperature*** the following corresponding formulas for the local Nusselt number and for the Prandtl numbers $\mathrm{Pr} \approx 1$ are obtained [2.7]:

$$\mathrm{Nu}_x = 0.501\mathrm{Gr}_x^{1/5}\,\mathrm{Pr}^{1/4}\,, \qquad (2.10.26)$$

$$\mathrm{Nu}_x = 0.394\mathrm{Gr}_x^{1/5}\,\mathrm{Pr}^{1/4}\,. \qquad (2.10.27)$$

When there is a laminar boundary layer on the plate edges, and there are irregular vortices and a convective jet flow near the center in the central part, the rate of heat transfer ***for air*** is determined by the formulas [2.26]

at the edge $\quad \mathrm{Nu}_x = 0.465\ \mathrm{Gr}_x^{1/5}\,, \qquad (2.10.28)$

near the centre $\quad \mathrm{Nu}_x = 0.12\ \mathrm{Gr}_x^{1/3}\,. \qquad (2.10.29)$

For an isothermal surface the average Nusselt number is 5/3 times higher than its local value at $x = L$. In the limiting cases, when $\mathrm{Pr} \gg 1$, Nu changes proportionally to $\mathrm{Pr}^{1/5}$, and when $\mathrm{Pr} \gg 1$ the heat transfer coefficient depends on Pr as $\sim\mathrm{Pr}^{2/3}$ [2.177].

The correlation dependences ***for heated surfaces facing upwards (or cooled surfaces facing downwards)*** are also given in [2.3, 2.179]. There are

$$1 < \mathrm{Ra}_L < 200 \qquad \mathrm{Nu}_L = 0.96\ \mathrm{Ra}_L^{1/6}\ , \tag{2.10.30}$$

$$200 < \mathrm{Ra}_L < 8 \cdot 10^6 \qquad \mathrm{Nu}_L = 0.54\ \mathrm{Ra}_L^{1/4}\ , \tag{2.10.31}$$

$$8 \cdot 10^6 < \mathrm{Ra}_L < 3 \cdot 10^{10} \qquad \mathrm{Nu}_L = 0.14\ \mathrm{Ra}_L^{1/3}\ . \tag{2.10.32}$$

The characteristic scale L in these formulas is the ratio of the surface area to its perimeter. The error of these formulas within the limits of experimental data scatter is within $\pm 5\%$ for laminar and $\pm 7\%$ for turbulent regimes.

For a horizontal surface ***with strips of widths*** l_1 ***and*** l_2 $(l = l_1 + l_2)$ ***with a low and a high temperatures*** $\mathrm{Nu}_{\mathrm{av}}$ increases in the case of laminar convection, when the ratio of the widths of the heated and cooled strips decreases. When a horizontal surface is heated discretely, the heat transfer is more efficient, than in the case of uniform heating [2.26].

The experimental data on the heat transfer ***of a viscous fluid*** in a turbulent flow regime ***near a horizontal surface of size*** $l \times l$ ***and temperature*** T_{w} ***with a heater of size*** $l_b \times l_b$ ***located in the centre*** (its temperature is T_b) can be generalized by the following dependence [2.26]:

$$\mathrm{Nu} = 0.165(\ \mathrm{Ra}_l P_\vartheta)^{1/3}\ , \tag{2.10.33}$$

$$P_\vartheta = \frac{T_b - T_\infty}{T_{\mathrm{w}} - T_\infty}, \quad T_m = \frac{F_b T_b + F_{\mathrm{w}} T_{\mathrm{w}}}{F_b + F_{\mathrm{w}}}, \quad T_{\mathrm{av}} = (T_m + T_\infty)/2\ ,$$

where $\mathrm{Ra}_l = g\beta \mathrm{b} l^3 (T_m - T_\infty)/\nu^2$, F_b and F_{w} are the surface areas of the heater and the wall, respectively.

For flat horizontal plates of different shapes and orientation described above the heat transfer in a laminar and a turbulent regimes is described by the following dimensionless relations:

$$\mathrm{Nu}^T = 0.835 C_1\ \mathrm{Ra}^{1/4}\ , \tag{2.10.34}$$

$$\mathrm{Nu}_l = \frac{1.4}{\ln\left(1 + 1.4/\mathrm{Nu}^T\right)}\ , \tag{2.10.35}$$

$$\mathrm{Nu}_t = C_2\ \mathrm{Ra}^{2/3}\ , \tag{2.10.36}$$

$$\mathrm{Nu} = [(\mathrm{Nu}_l)^m + (\mathrm{Nu}_t)^m]^{1/m}\ . \tag{2.10.37}$$

The values of the coefficients C_1 and C_2 are determined by the expressions [2.77]

$$C_1 = \frac{0.503}{\left[1 + (0.492/\mathrm{Pr})^{9/16}\right]^{4/9}}, \quad C_2 = \frac{0.13\ \mathrm{Pr}^{0.22}}{\left(1 + 0.61\ \mathrm{Pr}^{0.81}\right)^{0.42}}\ .$$

Axisymmetric horizontal flows are formed near an ***extended disk with a local heating or cooling and with the symmetric conditions being satisfied on its surface***. Analyzing this flow, it was assumed that the ascending flow induced by heating is completely-extinct at a rather great distance from the coordinate origin. When $T_{\mathrm{w}} > T_\infty$, the flow is directed inside

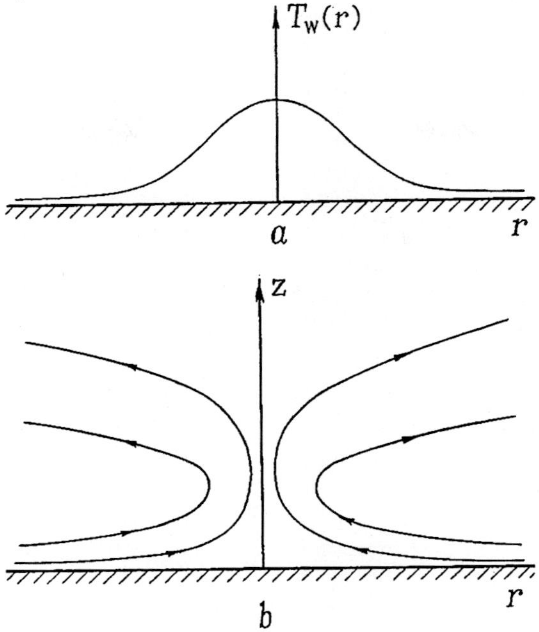

Fig. 2.10.3. Temperature distribution and the free-convective flow structure on a horizontal surface with a local heat source [2.7]

and has the shape of a torch near the symmetry axis. The steamlines of this flow are shown in Fig. 2.10.3b. When $T_\infty > T_w(B_n < 0)$, the flow near the axis is directed to the surface. At some distance from the axis the flow becomes horizontal and directed to the surrounding medium. In the case of the flow directed from the axis, the region of convective flow decreases with increase in the Gr number and its rate increases, which is characteristic for boundary-layer flows. This flow is always stable. In the case of the flow directed to the axis, both the region and rate of the flow increase with the increasing Gr number, and at a certain value of Gr the flow becomes thermally unstable.

In [2.108] the boundary-layer flows ***on a disk with the surface temperature changing according to the power-law*** are considered. The existence of self-similar solutions at $T_w - T_\infty = Nx^n$ is shown.

When $T_w > T_\infty$, physically suitable solutions exist only for $-1/2 \leq n \leq 2$, and when $T_w < T_\infty$, for $-4/3 \leq n \leq -1/2$. Visual observations of a flow ***above a heated disk of finite size*** ($T_w > T_\infty$) showed that the above solutions on an infinitely extended axisymmetric surface are inapplicable for small disks, since in this case unstable flows directed to the disk symmetry axis are observed. The instability of the flows observed is connected with ends effects and thermal stratification. The results of investigation of isothermal horizontal round disks [2.181] showed that on the disk periphery a two-dimensional boundary layer develops, however, closer to the disk center the axisymmetric

Table 2.10.4.

N	$\Delta T/T_r$	$-\theta_n'(0)$
0	0	0.35741
0	0.25	0.36810
0	0.50	0.37767
0	0.75	0.38634
0	1	0.39430
1	0	−0.05006
1	0.25	−0.05115
1	0.50	−0.05217
1	0.75	−0.05312
1	1	−0.05401
2	0	−0.03285
2	0.25	−0.03333
2	0.50	−0.03382
2	0.75	−0.03430
2	1	−0.03476

compression has a substantial effect on the flow, as the result of which the fluid turns upwards and forms the base of an ascending torch. The Nusselt number determined on the edge of the disk base $D = 2a$ is calculated by the formula [2.7]

$$\mathrm{Nu}_D \mathrm{Gr}_D^{-1/5} = -\frac{25}{12} 2^{2/5} C^{-2/3} \frac{\rho_0 \lambda_0}{\rho_r \lambda_r} \left[\theta'(0) + \sum_{n=1}^{\infty} \frac{24\theta_n'(0)}{(5n+3)(5n+8)} \right], \tag{2.10.38}$$

where the values of $\theta'(0)$ are given in Table 2.10.4, and the value of C can be found from the Chapman–Rubezin correlation which defines the dependence of viscosity on temperature:

$$\left(\frac{1}{C}\right) \rho\mu / \rho_r \mu_r = 1 .$$

According to the boundary-layer theory, the results of calculation of heat transfer ***for horizontal round disks*** are very close to the results of calculation for a square plate with the side of the square equal to the diameter of a round plate.

In Fig. 2.10.4 the sequence of interference pictures of the temperature field development ***on a horizontal plate with a constant heat flux on the wall*** is shown. Figure 2.10.5 presents the temperature distributions restored from the lower interferogram of Fig. 2.10.4: in Fig. 2.10.5a ***the heated surface is facing downwards***; in Fig. 2.10.5b ***it is facing upwards***.

In the case of a horizontal surface, the flow is created only by the effect of the normal of pressure gradient. However, even with a small inclination angle

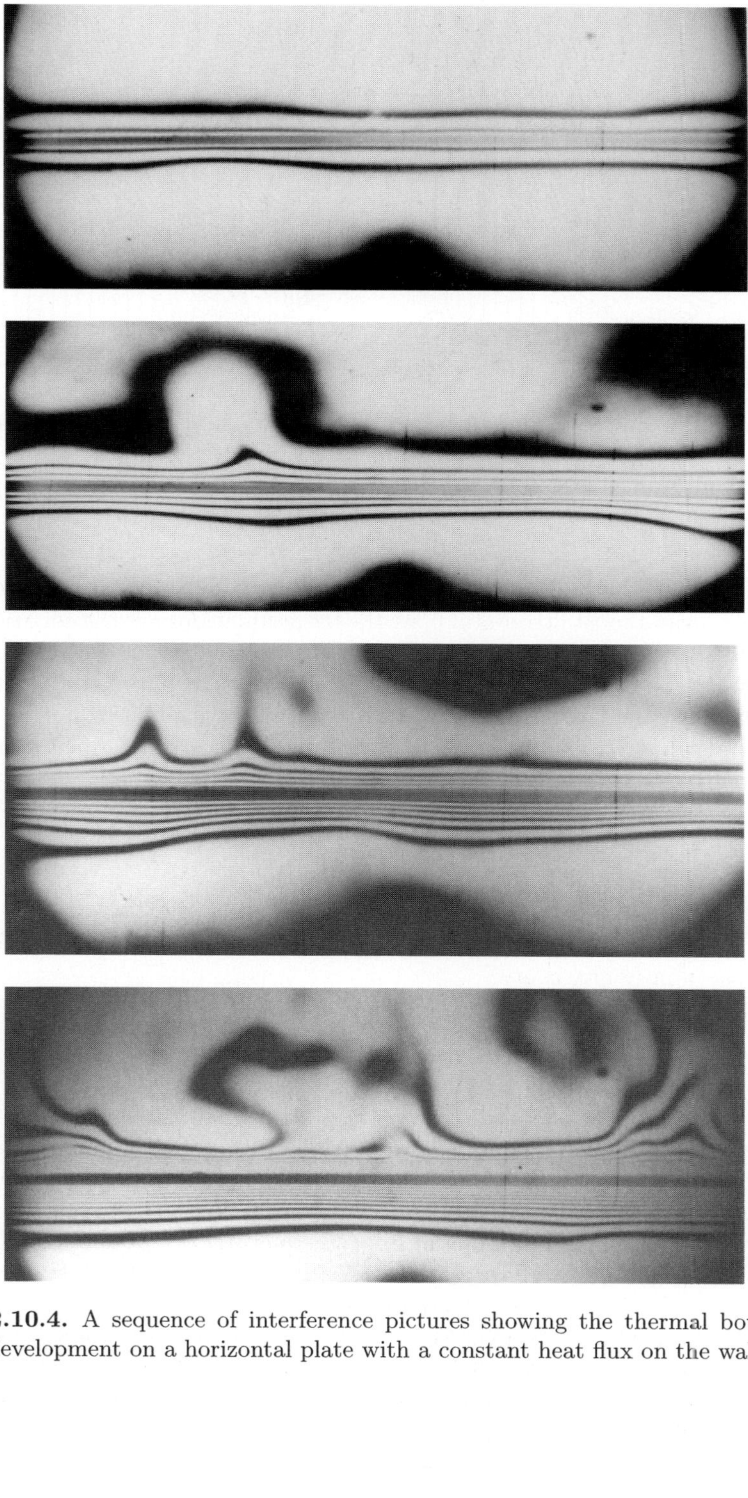

Fig. 2.10.4. A sequence of interference pictures showing the thermal boundary layer development on a horizontal plate with a constant heat flux on the wall

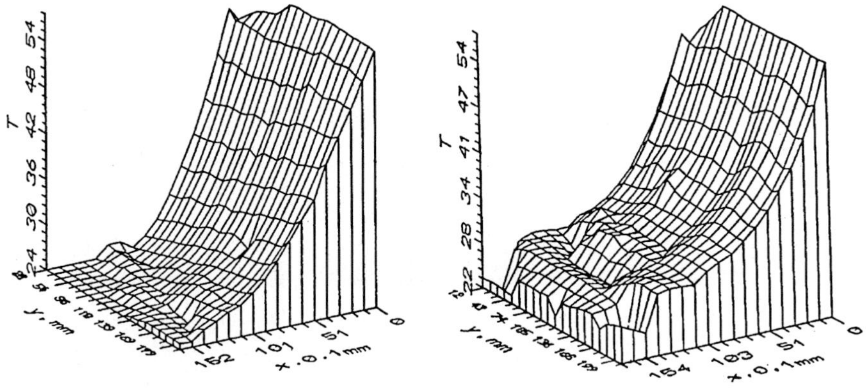

Fig. 2.10.5. Temperature distribution in a boundary layer near a horizontal plate with a constant heat flux restored on processing the interferograms of Fig. 2.10.3: (**a**) thermal layer above the heated surface, (**b**) thermal layer under the heated surface

a component of the buoyancy force B_t parallel to the horizontal is induced. In describing this almost horizontal flow, the basic equations are the equations for an inclined surface, which do not admit a self-similar solution. In this case the solution can be obtained by the method of small perturbations [2.171, 2.182].

It is convenient to characterize this flow by an inclination angle to the horizontal which is expressed as $\xi x = 90° - \vartheta$. The transformed equations then are [2.7]

$$\frac{\partial u}{\partial x} + \frac{\partial \nu}{\partial y} = 0 , \tag{2.10.39}$$

$$u\frac{\partial u}{\partial x} + \nu\frac{\partial u}{\partial y} = g\beta(T - T_\infty)\sin\xi - \frac{1}{\rho}\frac{\partial p}{\partial x} + \nu\left(\frac{\partial^2 u}{\partial x^2}\right) , \tag{2.10.40}$$

$$0 = g\beta(T - T_\infty)\cos\xi - \frac{1}{\rho}\frac{\partial p}{\partial y} , \tag{2.10.41}$$

$$u\frac{\partial \theta}{\partial x} + \nu\frac{\partial \theta}{\partial y} = \alpha\frac{\partial^2 \theta}{\partial y^2} . \tag{2.10.42}$$

The values of the ξ being small, the structure of the boundary layer is similar to that of a boundary layer on a horizontal plate. To take into account the deviations from self-similarity the required functions are expanded in series of the powers of a small parameter

$$\varepsilon\,(x) = (\mathrm{Gr}_{x\,\xi}/5)^{1/5}\,\mathrm{tg}\xi ,$$

where $\mathrm{Gr}_x\xi_x = \mathrm{g}\beta(T_\mathrm{w} - T_\infty)x^3\cos\xi/\nu^2$. The value of ε is determined so that the buoyancy force and pressure gradient along the plate be the values of the same order of magnitude.

The solution of the equations obtained at Pr = 0.7 for the isothermicity condition on the surface showed that even very small inclination angles exert a significant effect on the velocity distribution and a relatively small effect on the temperature distribution [2.171]. The effect of the inclination angle on the heat transfer coefficient cannot be neglected. The isothermal surface with small inclination angle was also investigated by the method of small perturbations in [2.182]. One of the expansions obtained describes the flow in the region of the leading edge, while another at a large distance downstream from the leading edge. In the first region, the prevailing factor causing the motion is a pressure field induced by the normal component of the buoyancy force. In the other region, the prevailing factor is the tangential component of the buoyancy force.

The first experimental data for surfaces with small inclination angles were obtained in [2.171]. The measurements were carried out in a range of inclination angles from 0 to 6°. On the whole, the results obtained are in good agreement with theoretical calculations, however, for $\mathrm{Gr}_x < 10^4$ the measured heat fluxes appeared smaller than the calculated ones. For small Gr_x an appreciable effect is exerted by the phenomena associated with the higher-order terms in the equations of free-convective boundary layer, which can be described by the method of perturbations, using as the small parameter the value $\varepsilon_1 = \mathrm{Gr}_x^{-1/5}$:

$$\varepsilon/\varepsilon_1 = (\mathrm{Gr}_{x\xi}/5^{1/2})^{2/5}\mathrm{tg}\xi \, .$$

For $\varepsilon/\varepsilon_1 \gg 1$ and small Gr_x the corrections of the higher orders of accuracy to the approximated solution obtained by the boundary layer method should be taken into account. At large Gr_x the experimental data appear to be higher than the calculated ones, which can be due to the uncertain position of the effective leading edge in these flows [2.7]. In [2.183] local and average heat transfer coefficients and temperature distributions ***for isothermal heated surfaces in air*** were measured using the Mach–Zehnder interferometer. The coincidence between the results of experimental and theoretical investigations is noted, but due to the occurrence of periodic instabilities random fluctuations of the local Nusselt number of significant amplitude can appear.

Figure 2.10.6 presents an interferogram and the corresponding temperature distribution on a horizontal plate of ***constant temperature with the heated surface facing downwards***. As is seen from the interferogram and graph of temperature distribution, the flow is laminar and stable.

The results of visualization of a free-convective flow on such a plate (Fig. 2.10.1b) ***in water*** [2.184] and ***in air*** [2.185] showed that in this case too there is a rather stable flow similar to the boundary-layer one, though even a small perturbation (plate inclination) has a significant effect on the flow structure in the entire region.

In the case of a heated surface facing downwards or cooled surface facing upwards, the motive force is the normal component of the buoyancy force $B_n < 0$. The pressure gradient is also negative, and under

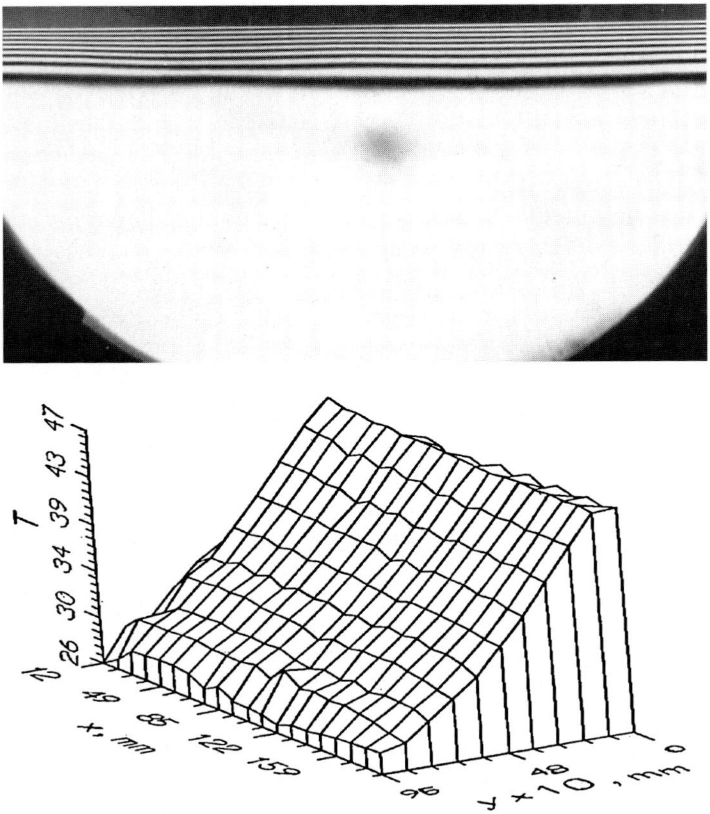

Fig. 2.10.6. The interferogram and corresponding temperature distribution on a horizontal plate of constant temperature with the heated surface facing downwards

the influence of this gradient the fluid flows to the surrounding medium and downstream from the plate edges. In the developed flow the boundary-layer thickness is the greatest in the middle of the plate and decreases to the edges.

A self-similar solution does not exist for this flow. An analytical investigation was carried out by integral methods assuming the boundary-layer thickness being equal to zero at the plate edges [2.186–2.188], which does not correspond to the observed flow pattern. Experiments show that the boundary-layer thickness at the edges is significant [2.187, 2.189, 2.190]. In [2.187] an approximate solution was obtained with account for the finite boundary-layer thickness. The integral method was used to obtain the following dimensionless relations for the local Nusselt number:

$$\mathrm{Nu}_a = -\frac{a}{T_\mathrm{w} - T_\infty}\left(\frac{\partial T}{\partial y}\right)_{y=0} = \frac{2}{y_2}\,\mathrm{Ra}_a^{-1/5}\,, \tag{2.10.43}$$

where a is the halfwidth of the plate, $y_2 = (\delta/a)\mathrm{Ra}_a^{1/5}$.

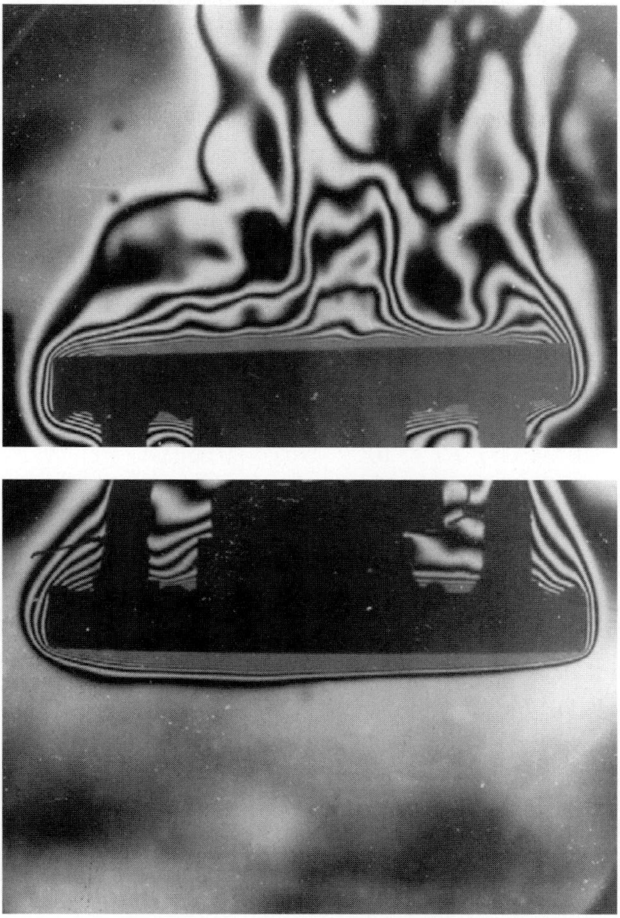

Fig. 2.10.7. Interference pictures of a free-convective flow on a horizontal round disk of constant temperature

For a heated surface facing downwards the local value of the heat transfer coefficient is determined from the formulas

$$\textit{for edges} \quad \mathrm{Nu}_x = 0.45\ \mathrm{Ra}_x^{1/5}\left(2 + \ \mathrm{Pr}^{-1}\right) , \tag{2.10.44}$$

$$\textit{near the centre} \quad \mathrm{Nu}_x = 0.0862\ \mathrm{Ra}_x^{1/3} . \tag{2.10.45}$$

The average Nusselt number is determined by the formula

$$\mathrm{Nu} = 3.88 + 0.0775\,\mathrm{Gr}_l^{1/3} . \tag{2.10.46}$$

The boundary between two zones can be found from the correlation [2.26]

$$\mathrm{Ra} = 2.43 \cdot 10^5 \left(1 + \ \mathrm{Pr}^{-1}\right)^{-3/2} . \tag{2.10.47}$$

The experimental data on the average heat transfer coefficient for a laminar regime are generalized by the dependences

for a constant wall temperature $T_{\text{av}} = (T_{\text{w}} + T\infty)/2$

$$\text{Nu} = 0.68 \, \text{Ra}_l^{1/5} , \tag{2.10.48}$$

for a constant heat flux on the wall

$$\text{Nu} = 0.305 \, \text{Ra}_l^{*1/5} , \tag{2.10.49}$$

for horizontal surfaces of finite length some other empirical formulas are also suggested [2.22, 2.179]:

$$3 \cdot 10^5 < \text{Gr}_L \, Pr < 3 \cdot 10^{10} \quad \text{Nu}_L = 0.27 \, (\text{Gr}_L \, \text{Pr})^{1/4} , \tag{2.10.50}$$

$$10^6 < \text{Gr}_L \, Pr < 10^{11} \quad \text{Nu}_L = 0.58 \, (\text{Gr}_L \, \text{Pr})^{1/5} . \tag{2.10.51}$$

Example 2.10. Determine the ratio between the coefficients of heat transfer to air from a horizontal plate ($a \times b = 1 \times 2$ m) with the heated surface facing upwards and downwards, if the wall temperature is $t_{\text{w}} = 80°$C, and the ambient temperature is $t_\infty = 20°$C.

The determining temperature is $t_{\text{av}} = (20 + 80)/2 = 50°$C. The thermophysical characteristics are:

$$\beta = 1/323.161/\text{deg},$$
$$\nu = 17.95 \cdot 10^{-6} \text{M}^2/\text{sec},$$
$$\lambda = 2.83 \cdot 10^{-2} \text{ W/m} \cdot \text{deg},$$
$$\text{Pr} = 0.698.$$

For a horizontal plate with the heated surface facing downwards, from (2.10.32) we get

$$l = \frac{ab}{2(a+b)} = \frac{1}{3}\text{m} ,$$

$$\text{Ra}_l = \frac{9.81 \cdot 60 \cdot 0.698}{323.16 \cdot 27 \cdot 17.95^2} \cdot 10^{12} = 1.485 \cdot 10^8 ,$$

$$\text{Nu} = 0.15(1.485 \cdot 10^8)^{1/3} = 79.45 ,$$

$$\alpha = \text{Nu}\lambda/l = 79.45 \cdot 3 \cdot 28.3 \cdot 10^{-2} = 6.75 \text{ W/m}^2 \cdot \text{deg} .$$

For a horizontal plate with the heated surface facing downwards, from (2.10.50) we have

$$l = a = 1\text{m} ,$$

$$\text{Ra}_l = \frac{9.81 \cdot 60 \cdot 0.698}{323.16 \cdot 17.95^2} \cdot 10^{12} = 3.97 \cdot 10^9 ,$$

$$\text{Nu} = 0.27 \left(3.97 \cdot 10^8\right)^{1/4} = 67.8 ,$$

$$\alpha = 67.8 \cdot 2.83 \cdot 10^{-2} = 1.92 \text{ W/m}^2 \cdot \text{deg} .$$

Thus, the heat transfer coefficient for a horizontal surface with the heated side facing upwards is 3.5 times higher than that for the surface with the heated side facing downwards [2.26].

2.11 Spatial Flow on a Plane Surface

A spatial free-convective flow near a vertical or inclined flat plate is set up in the presence of a nonuniform mass force.

In the case of ***a plane plate with a nonhorizontal leading edge*** the heat transfer coefficient can be determined by using the calculations of a problem of two-dimensional flow near a vertical flat plate with a horizontal leading edge. However, the distance from the leading edge in the direction of the buoyancy force should be used instead of the vertical distance. The numerical calculations and visual observations show that at small and large dimensionless values of $\overline{z}$ the heat transfer characteristics are determined by solutions for a two-dimensional convection [2.191].

For a three-dimensional flow ***near the flow retardation point on a nose*** the heat transfer rate decreases by 25% for $\mathrm{Pr} = 0.72$ [2.192, 2.193]. When $\mathrm{Pr} \to \infty$, the three-dimensional flow passes into a two-dimension one, because the inertia components in the momentum equations become small [2.154].

The laminar heat transfer of ***a vertical plane plate of arbitrary section*** (Fig. 2.11.1a) ***for constant wall and ambient temperatures*** is determined approximately by the thin layer method from the formula [2.194]

$$\mathrm{Nu} = C_l \ \mathrm{Ra}_l^{1/4} l^{1/4} \frac{\int_0^B (x_B - x_H)^{3/4} \, dz}{\int_0^B (x_B - x_H) \, dz} \, . \tag{2.11.1}$$

A disk of diameter d (Fig. 2.11.1b)

$$\mathrm{Nu} = 1.052 C_l \ \mathrm{Ra}_d^{1/4} \, . \tag{2.11.2}$$

A rhomb of height H with the side A (Fig. 2.11.1c)

$$\mathrm{Nu} = \frac{8}{7} C_l \ \mathrm{Ra}_H^{1/4} \, . \tag{2.11.3}$$

A parallelogram with the sides B and C and inclination angle ϑ (Fig. 2.11.1d)

$$\mathrm{Nu} = C_l \left(1 + \frac{8}{7} C \cos\vartheta\right) \ \mathrm{Ra}_H^{1/4} \quad B \le C \cos\vartheta \, . \tag{2.11.4}$$

On arbitrary change in the wall temperature $\theta_\mathrm{w}(x, z)$ for the average value of heat transfer coefficient the following dependence is suggested [2.194]:

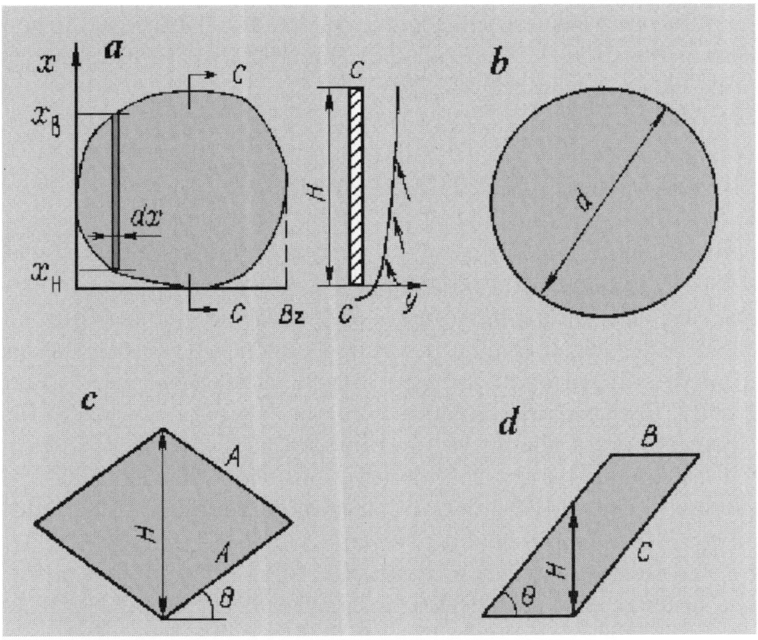

Fig. 2.11.1. Geometrical configurations of a vertical flat surface

$$\mathrm{Nu} = \frac{\frac{3}{4} C_l \, \mathrm{Ra}_l^{1/4} l^{1/4}}{\int_0^B (x_B - x_H)\, dz} \int_0^B \int_{x_H}^{x_B} \left(\frac{g_x}{g} \theta_\mathrm{w}^2\right)^{1/3} x^{-1/4} dx$$
$$: \left(\frac{\theta_\mathrm{w}}{x} \int_{x_H}^{x_B} \left(\frac{g_x}{g} \theta_\mathrm{w}^5\right)^{1/3} dx\right)^{1/4} dz \,. \qquad (2.11.5)$$

The spatial character of a flow on a plate of width b is determined by introducing the additional criterion Gr_b. The average value of the heat transfer rate ***for a narrow vertical plate with a constant temperature in a laminar regime*** (Gr_l, $\mathrm{Gr}_b = 10^5$–10^8) is [2.26]

in water for $\mathrm{Pr} = 2$

$$\frac{\mathrm{Nu}}{\mathrm{Cr}_l^{0.25}} = 0.67 + \frac{330}{\mathrm{Gr}_b^{0.4}} \,, \qquad (2.11.6)$$

in air for $\mathrm{Pr} = 0.7$

$$\frac{\mathrm{Nu}}{\mathrm{Gr}_l^{0.25}} = 0.53 + \frac{150}{\mathrm{Gr}_b^{0.4}} \,. \qquad (2.11.7)$$

The local and average values of the heat transfer rate for ***a vertical disk with the constant heat flux density on the surface*** are [2.195]

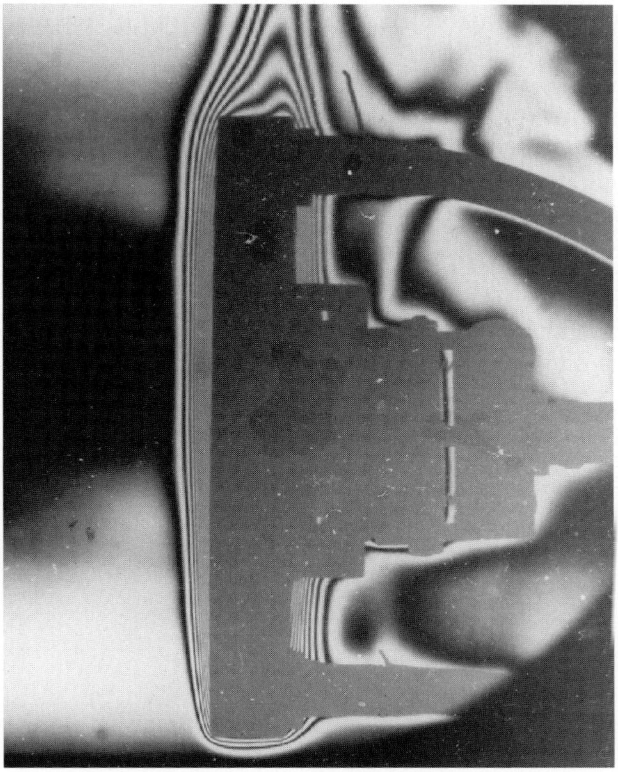

Fig. 2.11.2. Interference pictures of free convection near a round horizontal disk of constant temperature

$$\begin{aligned} \mathrm{Pr} = 0.7, \mathrm{Gr}_d^* = 10^5 - 10^8, \quad T_{\mathrm{av}} = (T_{\mathrm{w}} + T_\infty)/2 \;, \\ \mathrm{Nu}_x = 1.003\ \mathrm{Gr}_x^* \;, \\ \mathrm{Nu} = 1.467\ \mathrm{Gr}_d^{*0.2} \;, \end{aligned} \tag{2.11.8}$$

where x is the vertical coordinate reckoned from the lower edge of the disk.

The account for the spatial flow character is especially important for horizontal plates, since the presence of edges favors the jet flow with a higher heat transfer coefficient [2.196].

For horizontal plates with the heated surface facing upwards in a turbulent flow the heat transfer coefficient on the plate of width 5 cm is 23% higher than the corresponding value for the plate of width 30 cm [2.151]. The difference in the heat transfer coefficients decreases with increase in the inclination angle relative to the horizontal.

On a heated surface facing downwards at a constant heat flux on the wall the heat transfer coefficient ***on the disk*** is higher, than ***on a square and semi-infinite plates*** [2.197].

The spatial effects are absent in heat transfer of inclined plates and disks, when $\mathrm{Pr} \to \infty$ [2.154].

A numerical calculation of the distributions of air temperature and local Nu in natural convection ***near a vertical surface with an adjacent horizontal discretely heated plate*** is confirmed by the experimental data obtained using infrared thermal visualization [2.198].

Example 2.11. Find the value of the heat transfer coefficient of a plate of height $l = 0.5$ m in air for widths $b_1 = 0.1$ m and $b_2 = 0.5$ m, if the temperature of the heated surface is 80°C, and the ambient temperature is 45°C.

The determining temperature in air is $t_{\mathrm{av}} = 80 - 0.38 \cdot 35 = 66.7$°C, the characteristic dimension is $l = 0.5$ m. Then

$$\begin{aligned} \beta &= \frac{1}{339.86}\ \mathrm{deg}^{-1} , \\ \nu &= 19.67 \cdot 10^{-6}\ \mathrm{m}^2/sec , \\ \lambda &= 2.94 \cdot 10^{-2}\ \mathrm{W/m \cdot deg} , \\ \mathrm{Pr} &= 0.694 , \\ \mathrm{Gr}_l &= \frac{9.81 \cdot 0.125 \cdot 35}{339.86 \cdot (19.67)^2} \cdot 10^{12} = 3.28 \cdot 10^8 , \\ \mathrm{Gr}_{b1} &= 2.62 \cdot 10^6, \mathrm{Gr}_{b2} = 3.28 \cdot 10^8 . \end{aligned}$$

In the region of laminar free convection for a plate of finite width we use formula (2.11.7):

$$\mathrm{Nu}_1 = \left(0.53 + \frac{150}{(2.62 \cdot 10^6)^{0.4}}\right) \left(3.28 \cdot 10^8\right)^{1/4} = 126.1$$

$$\mathrm{Nu}_2 = 79.3 .$$

The heat transfer coefficients

$$\alpha_1 = \frac{\mathrm{Nu}_1 \lambda}{l} = 126.1 \cdot \frac{2.94 \cdot 10^{-2}}{0.5} = 7.42\ \mathrm{W/m}^2 \cdot \mathrm{deg} ,$$

$$\alpha_2 = 4.66\ \mathrm{W/m}^2 \cdot \mathrm{deg} .$$

differ by 60%. According to (2.1.7) for these conditions $\alpha = 4.62$ W/m^2·deg. This confirms the validity of the formulae obtained before for wide enough plates.

2.12 Compressibility and Variability of Thermophysical Properties

In considering free-convective flows it is usually assumed that the thermophysical properties of the medium (except for the density) do not change with the

Table 2.12.1.

Thermophysical Parameters	Gas				
	A	B	C	D	E
ρ	p/RT				
λ	$\sim T^{3/4}$	$\sim T^{2/3}$	$\sim 1/\rho$	$T^{3/2}/(T+A_1)$	
μ				$T^{3/2}/(T+A_2)$	
c_p	Const				$a+bT$
Pr	Const				Variable

pressure and temperature, and in dimensionless relations the rate of the heat transfer is calculated from certain characteristic values. The numerical calculations of the rate of heat transfer on a vertical surface with a constant temperature for certain hypothetical gases (Table 2.12.1) yield the characteristic temperature $T = T_{\text{w}} - 0.38(T_{\text{w}} - T_\infty)$ and $\beta = 1/T_\infty$. Moreover, for a laminar regime the scattering of the data on local and average heat transfer values does not exceed 1% [2.25]. Similar calculations ***for mercury*** showed that $T = T_{\text{w}} - 0.3(T_{\text{w}} - T_\infty)$. For a large temperature difference ***in gases and liquids with high viscosity*** the simplex $\text{P} = (T_{\text{w}} - T_\infty)/T_\infty$ appears in the dimensionless equations [2.199].

Heat transfer from ***a vertical plane plate of constant temperature in laminar free convection*** **[2.26]**:

in two-atomic ideal gases $(\rho/\rho_\infty = T_\infty/T, \mu/\mu_\infty = (T/T_\infty)^r, \lambda/\lambda_\infty = (T/T_\infty)^k, c_p/c_{p\infty} = (T/T_\infty)^m, T_{\text{av}} = T_\infty)$

$$\text{Nu}_x = 0.354(T_{\text{w}}/T_\infty)^{3(r-1)/8}\text{Gr}_x^{1/4}\,, \tag{2.12.1}$$

in air $(k = l = 0, m = 0, T_{\text{av}} = T_{\text{w}})$

$$\text{Nu}_x = 0.354\left[1 + 0.38(T_{\text{w}} - T_\infty)/T_\infty\right]^{-1/8}\text{Gr}_x^{1/4}\,, \tag{2.12.2}$$

in an ideal compressible gas $(\rho = \rho RT, \lambda/\lambda_\infty = \mu/\mu_\infty = T/T_\infty, c_p,$ $\text{Pr} = \text{const}, p = p_0\exp(-\xi), \xi = gx/RT)$

$$\text{Nu}_x = -e^{-\xi/4}\left[\theta_0' + \xi\theta_1'(0)\right]\text{Gr}_x^{1/4}\,, \tag{2.12.3}$$

the values of θ_i' are given in Table 2.12.2,

in steam

$$\text{Nu} = \left(\frac{\text{Ra}}{4}\right)^{1/4}(0.4748 + 0.1251\,\text{Pr} - 0.0328\,\text{Pr}^2) \times\left[1 - 0.0176p\left(\frac{T_\infty}{T_{\text{w}}} - 1\right)\right], \tag{2.12.4}$$

where $T_{\text{av}} = T_{\text{w}} - 0.46(T_{\text{w}} - T_\infty), \beta = \beta(T_\infty), p$ – in MPa.

Table 2.12.2.

Pr	0.72	1
θ_0'	−0.4995	−0.5671
T_{w}/T_∞	$-\theta_1'(0)$	
1.2	0.3350	–
1.5	0.1166	0.1274
2	0.0438	0.0458
3	0.0074	0.0050
∞	−0.0290	−0.0358

The heat transfer in laminar free convection ***on a horizontal surface of width b and length l with the heated surface facing upwards*** [2.200]:

$$\frac{\mathrm{Nu}_\infty}{\mathrm{Ar}_{b\infty}^{1/5}} = 2.1992\left(1 + 0.25\frac{b}{l}\right)\frac{\mathrm{Nu}_{x\infty}}{\mathrm{Ar}_{b\infty}^{1/5}}\,, \tag{2.12.5}$$

the values of $\mathrm{Nu}_{x\infty}/\mathrm{Ar}_{b\infty}^{1/5}$ ***for an ideal gas*** ($\rho/\rho_\infty = T_\infty/T, \mu/\mu_\infty = \lambda/\lambda_\infty = (T/T_\infty)^r, c_p = \mathrm{const}, \mathrm{Pr} = 0.72$) are presented in Table 2.12.3, ***for water***, in Table 2.12.4.

The heat transfer by laminar free convection ***on a vertical surface of constant temperature in a medium with variable viscosity***:

linear dependence of viscosity on temperature [2.201]:

$$\mu = \mu_{\mathrm{av}}\left(1 + P_\mu\theta\right), \quad P_\mu = \frac{\theta_{\mathrm{w}}}{\mu_{\mathrm{av}}}\left(\frac{d\mu}{dT}\right)_{\mathrm{av}}, \quad T_{\mathrm{av}} = (T_{\mathrm{w}} + T_\infty)/2\,,$$

$$\mathrm{Nu}_x = -\frac{\theta'(0)}{\sqrt{2}}\mathrm{Gr}_x^{1/4}\,; \tag{2.12.6}$$

Table 2.12.3.

R	1	0.76
T_c/T_∞	$\mathrm{Nu}_{x\infty}/\mathrm{Ar}_{x\infty}^{1/4}$	
1	0.3574	0.3574
1.2	0.3661	0.3604
1.4	0.3740	0.3631
1.6	0.3812	0.3655
1.8	0.3880	0.3677
2	0.3943	0.3698
2.2	0.4002	0.3716
2.4	0.4058	0.3734
2.6	0.4111	0.3750
2.8	0.4161	0.3765
3	0.4209	0.3780

Table 2.12.4.

T_∞,°C	θ_w,°C	$\Pr_\infty$	$\mathrm{Nu}_{x\infty}/\mathrm{Ar}_{x\infty}^{1/5}$
20	10	6.9255	0.6404
20	20	6.9255	0.6557
20	30	6.9255	0.6736
20	40	6.9255	0.6874
20	50	6.9255	0.6982
20	60	6.9255	0.7125
20	70	6.9255	0.7232
20	80	6.9255	0.7323
40	10	4.2607	0.5747
40	20	4.2607	0.5830
40	30	4.2607	0.5902
40	40	4.2607	0.6023
40	50	4.2607	0.6102
40	60	4.2607	0.6167
60	10	2.9340	0.5215
60	20	2.9340	0.5343
60	30	2.9340	0.5388
60	40	2.9340	0.5427
80	10	2.1867	0.4733
80	20	2.1867	0.4832

Table 2.12.5.

	$-\theta'(0)$			
P_μ	Pr = 1	Pr = 10	Pr = 100	Pr = 1000
−1.6	0.6514	1.4076	2.7168	4.9827
−0.8	0.5965	1.2476	2.3600	4.2887
0	0.5671	1.1693	2.1914	3.9654
0.8	0.5469	1.1190	2.0843	3.7602
1.6	0.5315	1.0843	2.0117	3.6178

the values of $\theta'(0)$ are given in Table 2.12.5;

petroleum, $T_\infty > T_w, T_{av} = T_\infty, \mathrm{Gr}_{x\infty} = 10^4 - 8.5 \cdot 10^9$, [2.202]:

$$\mathrm{Nu}_{x\infty} = 0.497\ \mathrm{Ra}_{x\infty}^{1/4}(\nu_\infty/\nu_w)^{0.17}\ . \tag{2.12.7}$$

The heat transfer from a ***vertical surface with the constant heat flux density to petroleum***, $T_{av} = T_\infty, \mathrm{Gr}_{x\infty}^* = 10^5 - 10^{10}$ [2.202]:

$$\mathrm{Nu}_{x\infty} = 0.62\ \mathrm{Ra}_{x\infty}^{*1/5}\ (\nu_\infty/\nu_w)^{0.21}\ . \tag{2.12.8}$$

Table 2.12.6 lists the numerical values of $\mathrm{Nu}_{x\infty}/\mathrm{Gr}_{x\infty}^{*1/5}$ ***for the linear dependence of viscosity on temperature*** [2.203]. The discrepancy in

Table 2.12.6.

Pr	P_μ	P_∞	$\mathrm{Nu}_{x\,\infty}/\mathrm{Gr}_{x\,\infty}^{*1/5}$
50	–0.8	70	1.602
	–0.4	60	1.480
	0	50	1.344
	0.4	40	1.202
	0.8	30	1.048
100	–0.8	140	1.869
	–0.4	120	1.715
	0	100	1.556
	0.4	80	1.391
	0.8	60	1.211
500	–0.8	700	2.611
	–0.4	600	2.393
	0	500	2.170
	0.4	400	1.937
	0.8	300	1.686

the whole range of the parameters between the calculated data and formula (2.12.8) does not exceed 2.5%.

A vertical wall with the temperature increasing along the height in a medium of variable viscosity [2.26]:

$$\theta_\mathrm{w} = \theta_0 x^n, n = 0 - 1, T_\mathrm{av} = T_\infty, \Pr_\infty = 200 - 5000, \mathrm{Ra}_{x\infty} = 10^7 - 4 \cdot 10^{12} ,$$

$$\mathrm{Nu}_{x\infty} = 0.5\,(1 + 0.84\,n)\ \mathrm{Ra}_{x\infty}^{1/4}\,(\Pr_\infty/\Pr_\mathrm{w})^{1/4} . \qquad (2.12.9)$$

It is experimentally shown that in this case the motion regime remains laminar.

For a liquid with linear dependence of the volumetric expansion coefficient β ***on temperature*** $\beta = \beta_\infty(1 + P_\beta\theta)$ the density from the hydrostatics equations is determined as $\rho = \rho_\infty \exp[-\beta_\infty\theta(1 + P_\beta\theta/2)]$. The ratio of the heat transfer coefficients for the variable values of β and ρ and constant values of the volumetric expansion coefficient and the density ***on the vertical plate of constant temperature*** are [2.204]

$$\frac{\mathrm{Nu}_x}{\mathrm{Nu}_{x0}} = \left[\Pr + \frac{20}{21}\frac{1 + \beta_\infty\theta_\mathrm{w}\,(7/12 + 13P_\beta/33)}{1 + \beta_\infty\theta_\mathrm{w}\,(13/28 + P_\beta/6)}\right]^{-1/4} \{(\Pr + 20/21)$$
$$\times\,(1 + 3P_\beta/10)\,[1 - \beta_\infty\theta_\mathrm{w}\,(15/28 + P_\beta/6)]\}^{1/4} . \qquad (2.12.10)$$

The values of β_∞ and $c = P_\beta/\theta_\mathrm{w}$ for some fluids are given in Table 2.12.7. The effect of $\beta(T)$ on heat transfer is more complex, than the change in other thermophysical parameters dependent on temperature. For laminar free convection the correction for the heat transfer coefficient due to the change in β can be within ±15% [2.204].

Table 2.12.7.

Coefficient	Water	Ethanol	Saturated freeon-12	Benzene	Mercury
$\beta_\infty \cdot 10^3, \mathrm{deg}^{-1}$	0.255	1.037	2.787	1.205	0.182
$c \cdot 10^3, \mathrm{deg}^{-1}$	35.50	5.207	2.867	1.395	0.043

The heat transfer from a ***vertical plane surface in a medium with variable viscosity and compressibility*** $\mu = \mu_{\mathrm{av}}(1 + P_\mu \theta), \beta = \beta_{\mathrm{av}}(1 + P_\beta \theta), T_{\mathrm{av}} = (T_{\mathrm{w}} + T_\infty)/2$ is given by [2.205]

$$\mathrm{Nu}_x = -\theta'(0)\,(0.75\ \mathrm{Ra}_x)^{1/4}\ , \tag{2.12.11}$$

the values of $\theta'(0)$ ***for water*** are given in Table 2.12.8.

Table 2.12.8.

T_{av}, °C	Pr	θ_{w}	$-\theta'(0)$
20	7.03	0	0.490
		10	0.498
		20	0.508
30	5.45	0	0.485
		10	0.492
		20	0.499
50	3.59	0	0.474
		10	0.479
		20	0.484

A numerical investigation of the effect of the temperature dependence of viscosity on the characteristics of flow and heat transfer, in particular, on vortical instability of the free-convective boundary layer ***along the heated horizontal surface in a saturated porous medium*** showed that for the case of liquid heating the exponential change of viscosity with temperature leads to an increase in the heat transfer rate and to the flow destabilization; in the case of flow of a heated gas the reverse tendency is observed [2.206].

The results of the numerical analysis of the effect of temperature dependence of viscosity on the free-convective heat transfer ***along a vertical surface in a saturated porous medium*** show that for $\theta_{\mathrm{w}} < 0$ the heat transfer rate is higher than that for the case of constant viscosity; for $\theta_{\mathrm{w}} > 0$ there is a reverse situation [2.207]. Note that θ_{w} is negative for liquids and positive for gases.

The problem of free convection of ***water along an isothermal vertical flat plate*** is solved numerically. The solution take into account the dependence of viscosity and thermal conductivity coefficients on temperature, which leads to a system of nonlinear equations [2.208]. The results of the calculation for the velocity field for different temperatures are confirmed by the experimental data obtained by the laser-Doppler anemometer method. On the basis of the temperature gradient obtained the following computational dependence is suggested for the local Nu number:

$$\mathrm{Nu}(x) = -\left[\frac{1}{4}\mathrm{Gr}\,(x)\right]^{1/4} \left.\frac{d\theta}{d\eta}\right|_{\eta=0} , \tag{2.12.12}$$

where $(d\theta/d\eta)|_{\eta=0}$ is determined from the dependence,

$$\left.\frac{d\theta}{d\eta}\right|_{\eta=0} = 0.5764 + 0.1797 \ln \mathrm{Pr}_\infty + 0.0331 \ln^2 \mathrm{Pr}_\infty .$$

Example 2.12. Determine the local and average coefficients of heat transfer from oil to a vertical wall of height $l = 2$ m. The wall temperature is $t_\mathrm{w} = 20°$C; the oil temperature is $t_\infty = 50°$C.

For highly viscous materials the determining temperature is $t_\mathrm{av} = t_\infty$, the characteristic size on the vertical plate is $l = 2$ m. Then

$$\begin{aligned}
&\beta_\infty = 6.46 \cdot 10^{-4}\,\mathrm{deg}^{-1} ,\\
&\nu_\infty = 1.53 \cdot 10^{-4}\,\mathrm{m}^2/\,\mathrm{sec} ,\\
&\lambda_\infty = 0.13\ \mathrm{W/m} \cdot \mathrm{deg} ,\\
&\mathrm{Pr}_\infty = 2180 ,\\
&\nu_\mathrm{w} = 1.125 \cdot 10^{-3}\mathrm{m}^2/\,\mathrm{sec} ,
\end{aligned}$$

$$\mathrm{Ra}_{x\infty} = \frac{9.81 \cdot 6.46 \cdot 10^{-4} \cdot 8}{1.53^2 \cdot 10^{-8}} \cdot 30 \cdot 2180 = 1.299 \cdot 10^{11} .$$

The local value of heat transfer from the liquid to the wall in the section $x = 2$ m will be determined from (2.12.7):

$$\mathrm{Nu}_{x\,\infty} = 0.497 \cdot (1.299 \cdot 10^{11})^{1/4} \left(\frac{153}{1125}\right)^{0.17} = 213.3 ,$$

$$\alpha_x = \mathrm{Nu}_x \frac{\lambda}{x} = 213.3 \frac{0.13}{2} = 13.9\ \mathrm{W/m}^2 \cdot \mathrm{deg} .$$

The average heat transfer coefficient on the plate of height $l = 2$ m is

$$\alpha_{\mathrm{av}\infty} = 4/3\alpha_{x\infty} = 4/3 \cdot 13.9 = 18.25\ \mathrm{W/m}^2 \cdot \mathrm{deg} .$$

2.13 Energy Dissipation and the Work of Compression

Free convection can be accompanied by a large viscous dissipation in devices with significant velocity gradients or operating at high rotation speeds. The relative value of the viscous dissipation effect is determined by the number Ec. The effects of viscous dissipation and of the work of compression appear in a laminar flow regime and in the majority of cases exert a small effect on heat transfer.

The ratio of heat transfer coefficients ***on a vertical plane surface for a constant wall temperature***, $\mathrm{Ec} \ll 1$ is [2.34]

$$\frac{\mathrm{Nu}_x}{\mathrm{Nu}_{x0}} = 1 \pm 4B\,\mathrm{Ec}_x\,, \qquad \frac{\mathrm{Nu}}{\mathrm{Nu}_0} = 1 \pm \frac{12}{7} B\,\mathrm{Ec}_l\,, \tag{2.13.1}$$

The values of the coefficient B are given in Table 2.13.1. The dissipation reduces heat transfer, when the heat flux is directed from the surface to the fluid ($B < 0$) and vice versa. The effect of viscous dissipation is equal to zero at the leading edge and increases along the surface.

Table 2.13.1.

Pr	10^{-2}	0.72	10^2	10^4
B	–0.0434	–0.1487	–0.2226	–0.2378

The heat transfer ***from a vertical surface at*** $\theta_{\mathrm{w}} = \theta_0 e^{-mx}$ ***with account for viscous dissipation*** is [2.49, 2.209]

$$\mathrm{Nu}_x = \frac{\theta'(0)}{\sqrt{2}} \left(mx\mathrm{Gr}_x\right)^{1/4}\,, \qquad \frac{\mathrm{Nu}}{\mathrm{Gr}_l^{1/4}} = -\theta'(0)\, \frac{2\sqrt{2}}{5} \frac{(ml)^{1/2}\left(e^{\frac{5}{4}ml} - 1\right)}{\left(e^{ml} - 1\right)^{5/4}}\,. \tag{2.13.2}$$

Table 2.13.2 lists the values of $-\theta'(0)$ for different values of Ec_x/mx.

Table 2.13.2.

Pr	$\mathrm{Ec}_x/mx = 0$	$\mathrm{Ec}_x/mx = 0.125$	$\mathrm{Ec}_x/mx = 0.25$	$\mathrm{Ec}_x/mx = 0.375$	$\mathrm{Ec}_x/mx = 0.5$
0.72	0.7411	0.7104	0.6789	0.6467	0.6138
1	0.8235	0.7874	0.7502	0.7120	0.6728
10	1.6172	1.5237	1.4255	1.3224	1.2139
100	2.980	2.786	2.579	2.360	2.126
∞	0.9566 $\mathrm{Pr}^{1/4}$	0.8907 $\mathrm{Pr}^{1/4}$	0.8205 $\mathrm{Pr}^{1/4}$	0.7453 $\mathrm{Pr}^{1/4}$	0.6647 $\mathrm{Pr}^{1/4}$

Figure 2.13.1 presents calculation of heat transfer $\mathrm{Nu}_x/\mathrm{Gr}_{x0}^{1/4}$ ***on a vertical surface*** for $\mathrm{Pr} = 0.7$ ***and*** $\theta_{\mathrm{w}} = \theta_0 e^{-mx}$ [2.210]. A change in the wall

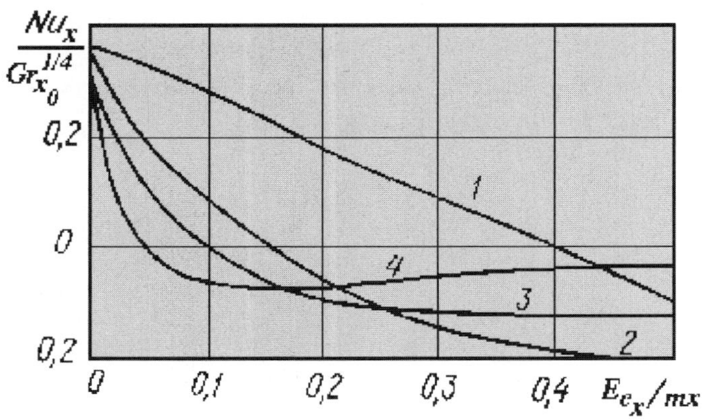

Fig. 2.13.1. Distribution of the local heat flux for Pr = 0.7: (1) $\mathrm{Ec}_x/mx = 0.2$, (2) 0.5, (3) 1, (4) ∞

temperature has a more substantial effect on heat transfer, than viscous dissipation.

The heat transfer ***at a constant heat flux on a vertical surface***, $\mathrm{Ec}_x \ll 1, \mathrm{Pr} \gg 1$ is [2.34]

$$(\mathrm{Ra}_x^*/5)^{0.2} = \mathrm{Nu}_x \left(1.1476 + 0.2268\ \mathrm{Pr}^{-1/2} - 0.0304\ \mathrm{Pr}^{-1}\right) \times (1 + 5r\mathrm{Ec}_x) \ , \tag{2.13.3}$$

where $r = 0.1030 - 0.088\ \mathrm{Pr}^{-1/2} + 0.1025\ \mathrm{Pr}^{-1}$.

In the limiting case ($\mathrm{Pr} \to \infty$)

$$\mathrm{Nu}_x/\mathrm{Nu}_{x0} = \mathrm{Ec}_x^{0.2} \left(e^{\mathrm{Ec}_x} - 1\right)^{-0.2} . \tag{2.13.4}$$

The viscous dissipation introduces an additional component into the calculation of the system stability [2.49].

An investigation of natural convection ***along an isothermal vertical wall in a saturated porous medium*** carried out in [2.211] shows that the effect of viscous dissipation leads to a decrease in the heat transfer rate approximately by 10% irrespective of the flow regime. In all regimes the effect of viscous dissipation increases with the dispersion parameter.

The work of compression, like viscous dissipation, is determined by the Ec number. In the free convection equations the work of pressure forces is proportional to Ec, the dissipation component – to $\mathrm{Ec}/\mathrm{Gr}^{1/2}$. For a laminar regime of motion the heat transfer due to viscous dissipation is much smaller than that due to the work of compression [2.212, 2.213]. The work of compression can lead to both generation and absorption of heat in the boundary layer. It is convenient to characterize this value by the parameter $\mathrm{M} = gl/c_p\theta_0$.

The heat transfer from ***a vertical plate with a constant temperature with account for the work of compression*** is given by [2 213]

$$\frac{\mathrm{Nu}_x}{\mathrm{Nu}_{x0}} = 1 - \frac{56\,(\mathrm{Pr}+40/C_1C_2)}{3\,(\mathrm{Pr}+20/C_1C_2)}\left(1+\frac{15T_\infty}{16C_1\theta_0}\right)\mathrm{Ec}_x\,,$$

$$\frac{\mathrm{Nu}}{\mathrm{Nu}_0} = 1 - \frac{8\,(\mathrm{Pr}+40/C_1C_2)}{\mathrm{Pr}+20/C_1C_2}\left(1+\frac{15T_\infty}{16C_1\theta_0}\right)\mathrm{Ec}_l \qquad (2.13.5)$$

for $P > 1$: $C_1 = 10.5\ \mathrm{Pr}^{-1}(5 - 4\ \mathrm{Pr}^{-1/2} + \mathrm{Pr}^{-1})$, $C_2 = \mathrm{Pr}^{-1/2}$;
for $P < 1$: $C_1 = 10.5\ \mathrm{Pr}^{-1}(5 - 4\ \mathrm{Pr}^{1/2} + \mathrm{Pr}^{1})$, $C_2 = 1$.

A vertical plate with a linearly increasing temperature [2.64]:

$$\mathrm{Nu}_x = -\theta'(0)\mathrm{Gr}_x^{1/4} \qquad (2.13.6)$$

for $\mathrm{M} \gg 1$

$$-\theta'(0) = \left(\frac{1}{4}\,\mathrm{Pr}\,M\right)^{1/4}\left[1+\frac{1}{M}\,\frac{2\,\mathrm{Pr}-1}{10\,\mathrm{Pr}}+O\left(\frac{1}{M^2}\right)\right].$$

The values of $\theta'(0)$ are given in Table 2.13.3.

Table 2.13.3.

	$-\theta'(0)$	
M	Pr = 0.72	Pr = 1
0.0	0.5332	0.5951
0.5	0.6315	0.7009
1	0.7020	0.7757
4	0.9572	1.0252

The heat flux increases with M, because a portion of the thermal energy is spent on the work of compression. The specified effect is especially important in the case of small temperature difference, when convection and heat conduction become comparable. For $\mathrm{Ec} \gg 1$ the heat transfer is calculated from the limiting dependence [2.214]:

$$\mathrm{Nu}_x = \left(\frac{1}{4}\,\mathrm{Ra}_x M\right)^{1/4}. \qquad (2.13.7)$$

The result obtained has a practical importance in meteorology. For example, in an equilibrium atmosphere the temperature decreases by 1° at a distance of 100 m.

2.14 Effect of Volumetric Heat Generation on Free Convection

Some processes in which heat is generated or absorbed may proceed in equipment and in the surrounding medium, for example, heat generation in the fuel elements of nuclear reactors due to the deceleration of nuclear fission fragments and slowing down of neutron fluxes; during a number of chemical reactions, etc. The volumetric heat generation Q_v in dimensionless equations is determined by the number Os.

The ratio of the heat transfer coefficients ***on a vertical surface of constant temperature in the presence of regularly distributed heat sources and without them in the volume*** is [2.215]

$$\frac{\mathrm{Nu}_x}{\mathrm{Nu}_{x0}} = \frac{3.94\ \mathrm{Pr}^{-1/2}\,(0.953+\mathrm{Pr})^{1/4}}{\sum_{n=0}^{\infty}\left(\dfrac{\mathrm{Os}_x}{\mathrm{Pr}\ \mathrm{Gr}_x^{1/2}}\right)^n f_n\,(\mathrm{Pr})}\,, \tag{2.14.1}$$

where $f_0\,(\mathrm{Pr}) = 3.94\ \mathrm{Pr}^{-1/2}\,(0.953+\mathrm{Pr})^{1/4}$, the subsequent components $f_n(\mathrm{Pr})$ are determined from the transcendental equations.

For free convection ***in water with uranium fuel*** ($\mathrm{Pr} = 1.7$) (2.14.1) takes the form

$$\frac{\mathrm{Nu}_x}{\mathrm{Nu}_{x0}} = \left[1 - 3.33\frac{\mathrm{Os}_x}{\mathrm{Pr}\ \mathrm{Gr}_x^{1/2}} + 12.2\left(\frac{\mathrm{Os}_x}{\mathrm{Pr}\ \mathrm{Gr}_x^{1/2}}\right)^2\right]^{-1}. \tag{2.14.2}$$

The heat transfer coefficient increases slightly with increase in the heat source power Q_v.

Combustion is accompanied by a chemical reaction with a volumetric heat generation $Q_v \sim \exp(-E/RT)$. In a ***laminar stabilized two-component flow*** of a ***propane-air mixture*** the ignition temperature decreases with increase in the propane concentration, and the minimum limit is 2.25%. The ignition temperature also decreases with increase in the initial mixture temperature [2.216]. The free convection is the determining mechanism underlying the quenching of the combustion process of volatile explosives [2.217].

2.15 Injection and Suction on a Plane Surface

The condition $\nu = \nu_0$ on a permeable surface for $y = 0$ on nondimensionalizing the boundary-layer equations leads to the complex $\mathrm{Re}_{cr}/\mathrm{Gr}^{1/4}$. In injuction $\nu_0 > 0$ ($\mathrm{Re}_{cr}/\mathrm{Gr}^{1/4} > 0$), in suction $\nu_0 < 0$ ($\mathrm{Re}_{cr}/\mathrm{Gr}^{1/4} < 0$).

The relative heat transfer ***with uniform injection or suction on a vertical surface of constant temperature*** for a laminar flow is

$$\frac{\mathrm{Nu}_x}{\mathrm{Nu}_{x0}} = 1 - 0.912 \frac{\rho_{\mathrm{w}}}{\rho_\infty} \frac{\mathrm{Re}_{crx}}{\mathrm{Gr}_x^{1/4}} \, . \tag{2.15.1}$$

In the limit $\mathrm{Re}_{crx}/\mathrm{Gr}_x^{1/4} \to -\infty$ the boundary layer becomes so thin that its parameters are independent of the transverse coordinate. and the heat transfer is determined by the formula [2.218]

$$\frac{\mathrm{Nu}_x}{\mathrm{Gr}_x^{1/4}} = -\frac{1}{\sqrt{2}} \mathrm{Pr} \frac{\mathrm{Re}_{crx}}{\mathrm{Gr}_x^{1/4}} \, . \tag{2.15.2}$$

The effect of injection and suction on the transient region is determined experimentally. The beginning of the transient region is determined by

$$\mathrm{Gr}_{xt} = 10^7 \left(1 - 0.424 \frac{\mathrm{Re}_{crx}}{\mathrm{Gr}_{xt}^{1/4}}\right) \, . \tag{2.15.3}$$

The beginning of the developed turbulent flow is determined by

$$\mathrm{Gr}_{xcr} = 10^8 \left(1 - 0.282 \frac{\mathrm{Re}_{crx}}{\mathrm{Gr}_{xcr}^{1/4}}\right) \, . \tag{2.15.4}$$

The dimensionless heat transfer by turbulent free convection ***in air on a permeable surface*** is

$$\frac{\mathrm{Nu}_x}{\mathrm{Nu}_{x0}} = \left[1 - 1.682 \left(\frac{\mathrm{Re}_{crx}^2}{\mathrm{Gr}_x}\right)^{1/3}\right]^2 \, . \tag{2.15.5}$$

When heat transfer from a surface is accompanied by mass transfer (evaporation or condensation), the boundary layer is binary, and the buoyancy force is determined by the temperature and concentration differences. The concentration component has a substantial effect on heat transfer, when the molecular masses of the components are different.

Heat transfer ***on a vertical surface in evaporation and condensation of steam from a steam-air mixture*** is described by [2.26]

$$\frac{\mathrm{Nu}_x}{\mathrm{Ra}_x^{1/4}} = 0.508 \left(1 + \frac{\beta_{\mathrm{w}}}{\beta} \frac{c_{1\mathrm{w}} - c_{1\infty}}{\theta_{\mathrm{w}}}\right)^{1/4}$$

$$\times \left[\frac{\mathrm{Pr}}{\eta\,(0.953\eta + \mathrm{Pr})}\right]^{1/4} \left[1 + \mathrm{Le}^{-1} \mathrm{Du}\,(c_{1\mathrm{w}} - c_{1\infty})\right] , \tag{2.15.6}$$

$$\eta = 1 + \frac{\rho_{\mathrm{w}}}{\rho_\infty} \frac{c_{1\mathrm{w}} - c_{1\infty}}{\mathrm{Le}} \left(\mathrm{Du} + \frac{1}{1 - c_{1\mathrm{w}}} + \frac{2}{3}\right) .$$

If the evaporation and condensation heat transfer is accompanied by a phase change ($q_{\mathrm{w}}^* = q_{\mathrm{w}} + r m_{1\mathrm{w}}$), then the Nusselt number is

$$\frac{\mathrm{Nu}_x^*}{\mathrm{Ra}_x^{1/4}} = 0.508\left(1+\frac{\beta_\mathrm{w}}{\beta}\frac{c_{1\mathrm{w}}-c_{1\infty}}{\theta_\mathrm{w}}\right)^{1/4}\left[\frac{\mathrm{Pr}}{\eta(0.953\eta+\mathrm{Pr})}\right]^{1/4} \times\left[1+\frac{c_{1\mathrm{w}}-c_{1\infty}}{\mathrm{Le}}\left(\mathrm{Du}+\frac{k}{1-c_{1\mathrm{w}}}\right)\right] . \quad (2.15.7)$$

The heat and mass transfer ***in a turbulent binary boundary layer, with the thermodiffusion effect and the difference of the component thermal capacities being neglected***, is given by [2.26]

$$\frac{\mathrm{Nu}_x}{\mathrm{Nu}_{x0}} = \left(\frac{\rho_\mathrm{w}}{\rho_\infty}\right)^{1/3}\left(1+\frac{\beta_\mathrm{w}}{\beta}\frac{c_{1\mathrm{w}}-c_{1\infty}}{\theta_\mathrm{w}}\frac{1}{\xi}\right)^{1/3} \times\left[1+\left(\frac{\mathrm{Pr}}{\mathrm{Sc}}\right)^{2/3}\mathrm{Du}\,(c_{1\mathrm{w}}-c_{1\infty})\,\xi^{1/2}\right]\left\{1-\frac{2.14}{2.14+\ \mathrm{Pr}^{2/3}}\left(\frac{\mathrm{Pr}}{\mathrm{Sc}}\right)^{2/3} \times\frac{c_{1\mathrm{w}}-c_{1\infty}}{1-c_{1\mathrm{w}}}\left[1+\mathrm{Du}(1-c_{1\mathrm{w}})\xi^{1/2}\right]\right\}^{-1/3} , \quad (2.15.8)$$

$$\frac{\mathrm{Sh}_x}{\mathrm{Nu}_x} = \left(\frac{\mathrm{Sc}}{\mathrm{Pr}}\right)^{1/3}\xi^{1/2}\,(1-c_{1\mathrm{w}})^{-1}\left[1+\left(\frac{\mathrm{Pr}}{\mathrm{Sc}}\right)^{2/3}\mathrm{Du}\,(c_{1\mathrm{w}}-c_{1\infty})\,\xi^{1/2}\right] . \quad (2.15.9)$$

In the presence of the phase change in a turbulent boundary layer it is given by

$$\frac{\mathrm{Nu}_x^*}{\mathrm{Nu}_{x0}} = \frac{1+\left(\frac{\mathrm{Pr}}{\mathrm{Sc}}\right)^{2/3}(c_{1\mathrm{w}}-c_{1\infty})\left(\mathrm{Du}+\frac{k}{1-c_{1\mathrm{w}}}\right)\xi^{1/2}}{1+(\mathrm{Pr}\,/\mathrm{Sc})^{2/3}\,\mathrm{Du}(c_{1\mathrm{w}}-c_{1\infty})\xi^{1/2}} , \quad (2.15.10)$$

$$\mathrm{Le} < 1\,;$$

$$\left(\frac{\xi}{0.309}\right)^{1/2} = \frac{c_{1\mathrm{w}}-c_{1\infty}}{1-c_{1\mathrm{w}}}\left[1+\mathrm{Du}\,(1-c_{1\mathrm{w}})\right] + \sqrt{\left\{\frac{c_{1\mathrm{w}}-c_{1\infty}}{1-c_{1\mathrm{w}}}\left[1+\mathrm{Du}\,(1-c_{1\mathrm{w}})\right]\right\}^2+3.6\left(\frac{1-c_{1\mathrm{w}}}{1-c_{1\infty}}\mathrm{Le}^{2/3}-1\right)} .$$

The transient zone is determined by (2.15.3) and (2.15.4) for

$$\mathrm{Re}_{crx} = (c_{1\mathrm{w}}-c_{1\infty})\,\mathrm{Sh}_x, \quad \mathrm{Gr}_x^* = \frac{g\beta x^3}{\nu^2}\left[\beta\theta_\mathrm{w}+\beta_\mathrm{w}(c_{1\mathrm{w}}-c_{1\infty})\right] .$$

The local heat transfer ***on a horizontal permeable surface in the boundary zone for a heated surface facing upwards and downwards*** is generalized by the universal dimensionless relation [2.26]

$$\frac{\mathrm{Nu}_x}{\mathrm{Nu}_{x0}} = 1-0.847\frac{\mathrm{Re}_{crx}}{\mathrm{Gr}_x^{1/5}} . \quad (2.15.11)$$

with the corresponding value of Nu_{x0} (2.10.29) or (2.10.44).

The transition boundary between zones and the heat transfer in the central zone depend on the heat transfer surface orientation.

The heated surface facing upwards:

$$\mathrm{Gr}_{x1} = 5.4 \cdot 10^4 \left(1 - 0.344 \frac{\mathrm{Re}_{crx}}{\mathrm{Gr}_x^{1/5}}\right) ,$$

$$\frac{\mathrm{Nu}_x}{\mathrm{Nu}_{x0}} = 1 - 2.52 \frac{\mathrm{Re}_{crx}}{\mathrm{Gr}_x^{1/3}} + 5.63 \left(\frac{\mathrm{Re}_{crx}}{\mathrm{Gr}_x^{1/3}}\right)^2 . \qquad (2.15.12)$$

The heated surface facing downwards:

$$\mathrm{Gr}_{x1} = 2 \cdot 10^5 \left(1 - 0.344 \frac{\mathrm{Re}_{crx}}{\mathrm{Gr}_x^{1/5}}\right) ,$$

$$\frac{\mathrm{Nu}_x}{\mathrm{Nu}_{x0}} = 1 - 6.82 \frac{\mathrm{Re}_{crx}}{\mathrm{Gr}_x^{1/3}} + 14.16 \left(\frac{\mathrm{Re}_{crx}}{\mathrm{Gr}_x^{1/3}}\right)^2 . \qquad (2.15.13)$$

The experimental data for free-convective heat transfer ***from a heated vertical surface in mercury in the presence of gas injection*** are presented in [2.219]. The range of constant heat flux density on the surface is $370 \leq q_\mathrm{w} \leq 16000\ \mathrm{W/m^2}$(the corresponding modified Boussinesq number is $10^5 \leq \mathrm{Bo} \leq 10^9$). The gas injection rate changed in the range $0.9 \leq Q \leq 9.2\,\mathrm{cm^3/sec}$. The data obtained in experiments for low and high heat flux densities point to the existence of three boundary layers different from the thermal one, namely: a single-phase internal layer near the surface, a two-component layer, where there is a majority of bubbles of the injected gas, and an external single-phase layer. An analysis of experimental data for the heat transfer rate without gas injection shows, that transition to turbulence occurs at $\mathrm{Ra} \sim 2 \cdot 10^7$ in contrary to the expected values of $\sim 10^9$.

A numerical investigation of the transient and quasi stationary regimes of total heat and mass transfer of ***an air-helium mixture near a vertical surface*** shows that the dimensionless heat flux density on the wall reaches its minimum at dimensionless times ~ 2.5, and at the dimensionless time exceeding five the flow becomes quasi stationary [2.220]. In the quasi stationary regime the heat flux on the wall decreases with increase in the ratio of temperatures and the ratio of the Gr numbers for flows in the presence and absence of the mass flux on the wall. For the small values of the ratio of temperatures the heat flux decreases with increase of $\mathrm{Gr}/\mathrm{Gr}_0$ (Gr_0 is the Grashoff number in the absence of mass transfer) and eventually changes the heat transfer direction to the opposite, which can, in turn, lead to a decrease in the boundary layer temperature to the values smaller than the liquid temperature in the main flow.

The self-similar solutions for heat and mass transfer by free convection ***from a vertical plate placed in a saturated porous medium in the presence*** of injection are obtained for the cases: 1) ***constant temperature and concentration on the wall*** and 2) ***constant heat and mass fluxes***

on the vertical surface [2.221]. The determining parameters for this problem are the ratio of buoyancy forces caused by temperature and concentration gradients (N) and the Lewis number (Le). Depending on the sign, the concentration gradient can both intensify and decelerate the flow caused by the temperature gradient. The average values of the Nu and Sh numbers are determined for wide ranges of parameters: $-1 < N < 10, 0.1 \leq \mathrm{Le} \leq 100$.

A numerical analysis of natural convection ***near a vertical and horizontal isothermal surfaces in the presence of uniform directed mass transfer on it*** for $\mathrm{Pr} = 0.7$ was carried out in [2.222]. For both cases of the surface arrangement, an increase in the normal component of the surface injection ($\nu_{\mathrm{w}} > 0$) leads to a decrease in the local surface heat transfer rate for both directions (down and up the flow) of the longitudinal mass transfer component ($u_{\mathrm{w}} > 0$ and $u_{\mathrm{w}} < 0$, respectively). An increase in the normal mass transfer component in the case of suction ($\nu_{\mathrm{w}} < 0$) leads to an increase in the local surface heat transfer rate. For a fixed value of the normal component of injection or suction, an increase in the longitudinal mass transfer component directed downwards leads to an increase in the local heat transfer rate on the surface; for the opposite direction the reversed tendency is observed.

References

1. A. Bejan, *Convection Heat Transfer*, Wiley Publ., New York (1984).
2. M.-J. Huang, C.-K. Chen, *Int. J. Heat Mass Transfer*, **33**, 119 (1974).
3. W. H. McAdams, *Heat Transmission*, 3rd edn., McGraw-Hill, New York (1957).
4. D. Sucker, *VDI-Forschungsheft,* No. 585, 1 (1978).
5. D. A. Frank-Kamenetskii, *DAN SSSR*, **17**, 9 (1937).
6. S. S. Kutateladze, *Near-Wall Turbulence*, Nauka Press, Novosibirsk (1973).
7. B. Gebhart, Y. Jaluria, R. L. Manajan, and B. Sammakia, *Buoyancy-Induced Flows and Transport*, Hemisphere Publ. Corp., Washington (1988).
8. S. W. Churchill and H. H. S. Chu, *Int. J. Heat Mass Transfer*, **18**, 1323 (1975).
9. A. Oberbeck, *Ann. Phys. Chem.*, **7**, 271 (1879).
10. L. Lorenz, *Ann. Phys. Chem.*, **13**, 582 (1881).
11. O. A. Saunders, *Proc. Roy. Soc., Ser. A*, **55**, 172 (1939) London.
12. E. Schmidt and W. Beckmann, *Tech. Mech. Thermodyn.*, **1**, 391 (1930).
13. H. Schuh, *Boundary layers of temperature*, in: W. Tollmein (ed.), *Boundary Layers*, Sec. B6, British Ministry of Supply, German Doc. Cent. Ref. 3220T, (1948).
14. S. Ostrach, *An Analysis of Laminar Free Convection Flow and Heat Transfer about a Flat Plate Parallel to the Direction of the Generating Body Force*, NACA TR 1111, 953.
15. H. Schuh, *Gottinger Monographien*, Bd. B, Grenzschichten (1946).
16. E. M. Sparrow and J. L. Gregg, *Details of Exact Low Pr Boundary Layer Solutions for Forced and Free Convection*, NACA Memo 2-27-59E, (1959).
17. E. J. Le Fevre, *Proc. 9th Int. Cong. Appl. Mech.*, Brussels, **4**, 168 (1956).
18. Y. Jaluria, *Natural Convection Heat and Mass Transfer*, Pergamon, Oxford, U. K. (1980).

19. H. B. Squire, in: S. Goldstein (ed.), *Modern Developments in Fluid Dynamics*, Oxford Univ. Press, New York (1938).
20. A. J. Ede, *Advances in Free Convection – Advances in Heat Transfer*, **4**, 1 (1967).
21. S. W. Churchill and R. Usagi, *AIChE J.*, **18**, 1121 (1972).
22. T. Cebeci and P. Bradshaw, *Convective Heat Transfer*, Mir Press, Moscow (1987).
23. A. Mongruel, M. Cloitre, and C. Allain, *Int. J. Heat Mass Transfer*, **39**, 3899 (1996).
24. S. K. Rastogi and D. Poulikakos, *Int. J. Heat Mass Transfer,* **38**, 935 (1995).
25. E. M. Sparrow and G. L. Gregg, *Trans ASME, J. Heat Transfer*, **80**, 879 (1958).
26. O. G. Martynenko and Yu. A. Sokovishin, *Free-Convective Heat Transfer. Handbook*, Nauka i Tekhnika Press, Minsk (1982).
27. E. M. Sparrow and J. L. Gregg, *Trans. ASME, J. Heat Transfer*, **78**, 435 (1956).
28. E. M. Sparrow, *Laminar Free Convection on a Vertical Plate with Prescribed Nonuniform Wall Heat Flux or Prescribed Nonuniform Wall Temperature*, NACA TN 3508, July 1955.
29. H. B. Squire, in: S. Goldstein (ed.), *Modern Developments in Fluid Dynamics*, Vol. 2, Dover, New York (1965), pp. 641–643.
30. K. S. Chang, R. G. Akins, L. Burris, and S. G. Bankoff, *Argonne National Laboratory Rept.* ANLG835, 1964.
31. J. R. Selman, *J. Heat Transfer*, **93**, 465 (1971).
32. S. W. Churchill and H. Ozoe, *Trans. ASME, J. Heat Transfer*, **95**, 540 (1973).
33. T. Fujii and M. Fujii, *Int. J. Heat Mass Transfer*, **19**, 121 (1976).
34. B. Gebhart, *J. Fluid Mech.*, **14**, 225 (1962).
35. T. Fujii and H. Uehara, *Int. J. Heat Mass Transfer*, **13**, 607 (1970).
36. H. K. Kuiken, *J. Eng. Math.*, **2**, 95 (1968).
37. H. K. Kuiken, *Int. J. Heat Mass Transfer*, **11**, 1141 (1968).
38. Z. M. Qureshi and B. Gebhart, *Int. J. Heat Mass Transfer*, **21**, 1467 (1978).
39. N. Sheriff and N. W. Daries, in: *Proc. 6th Int. Heat Transfer Conf.*, vol. 5, Toronto (1978), pp. 131–136.
40. J. P. Dotson, *Heat Transfer from a Vertical Plate by Free Convection*, Purdue University, USA (1954).
41. G. C. Vliet and C. K. Liu, *Trans. ASME, J. Heat Transfer*, **91C**, 517 (1965).
42. K. S. Chang and R. G. Akins, *Int. J. Heat Mass Transfer*, **15**, 513 (1972).
43. D. V. Julian and R. G. Akins, *Ind. Eng. Chem. Fundam.*, **8**, 641 (1969).
44. A. A. Zhukauskas and O. G. Martynenko (eds.), *Advances in Heat Transfer*, Mokslas Press, Vilnius (1989).
45. V. M. Kapinos, A. F. Slitenko and I. V. Volovelski, *Inzh.-Fiz. Zh.*, **26**, 411 (1974).
46. T. Y. Na, *Appl. Sci. Res.*, **33**, 519 (1978).
47. V. V. Slavnov, in: *Collection of Sci.Papers of Perm Polytechnic Inst.*, **1**, 87 (1970).
48. Yu. A. Sokovishin and S. L. Semin, *Tr. LPI*, 81 (1984).
49. B. Gebhart and J. C. Mollendorf, *J. Fluid Mech.*, **38**, 97 (1959).
50. A. A. Hayday, D. A. Bowlus, and R. A. McGraw, *Trans. ASME, J. Heat Transfer*, **89C**, 244 (1967).
51. S. Patankar, *Numerical Solution Methods of Heat Transfer and Fluid Dynamics Problems*, Energatomizdat Press, Moscow (1984).

52. J. A. Schetz and R. Eichorn, *J. Fluid Mech.*, **18**, 167 (1964).
53. M. Zalesak and J. Kaspar, *Stavebnicky Casopis*, **40**, 929 (1982).
54. V. N. Piskunov, Yu. A. Sokovishin, and V. F. Stepanov, *Inzh.-Fiz. Zh.*, **19**, 202 (1970).
55. T. T. Kao, G. A. Domoto, and H. G. Jr. Elrod, *Trans. ASME, J. Heat Transfer*, **99C**, 72 (1977).
56. Yu. A. Sokovishin and A. G. Semjenov, *Inzh.-Fiz. Zh.*, **39**, 724 (1980).
57. A. G. Semjenov, *Inzh.-Fiz. Zh.* (1980); Dep. in VINITI 25 February 1980, No. 687-80.
58. P. P. Smyshljaev, in: *Mathematical Methods of Optimization and Systems Structuring*, Kalinin (1979) pp. 138–143.
59. R. Bradean, D. B. Ingham, P. G. Heggs, and I. Pop, *Int. J. Heat Mass Transfer*, **39**, 2245 (1996).
60. M. Tribus, *Trans. ASME, J. Heat Transfer*, **80C**, 1180 (1958).
61. N. N. Tran, *G. R. Acad. Sci.*, **275A**, 1123 (1972).
62. T. S. Chen, H. C. Tien, and B. F. Armaly, *Int. J. Heat Mass Transfer*, **29**, 1465 (1968).
63. O. G. Martynenko and Yu. A. Sokovishin, *Free-Convective Heat Transfer on a Vertical Surface (II Kind Boundary Conditions)*, Nauka i Tekhnika Press, Minsk (1978).
64. Yu. A. Sokovishin, A. N. Kaljutik, A. G. Semjenov, and T. N. Aitov, *Izv. VUZov, Energetika*, No. 11, 120 (1981).
65. E. M. Sparrow, *Laminar Free Convection on a Vertical Plate with Prescribed Nonuniform Wall Temperature*, NACA Techn. Note, No. 3508 (1955).
66. R. P. Bobko, *J. Aero/Space Sci.*, **29**, 846 (1959).
67. N. N. Kortikov, Yu. A. Sokovishin, and M. V. Shapiro, in: *Collection of Works of ScientificTtechnical Conference*, Kuibyshev (1972), pp. 148–150.
68. Y. Jaluria and B. Gebhart, *J. Fluid Mech.*, **66**, 309 (1974).
69. R. Cheesewright, *Int. J. Heat Mass Transfer*, **10**, 1847 (1967).
70. K. T. Yang, J. L. Novotny, and Y. S. Cheng, *Int. J. Heat Mass Transfer*, **15**, 1097 (1972).
71. R. Eichhorn, *Prog. Heat Mass Transfer*, **2**, 41 (1969).
72. R. H. Hubbel and B. Gebhart, *Proc. 24th Heat Transfer and Fluid Mech. Inst.*, Corvallis, Oregon (USA) (1974).
73. A. R. Tenner and B. Gebhart, *Int. J. Heat Mass Transfer*, **14**, 2051 (1971).
74. L. Prandtl, *Essentials of Fluid Dynamics*, Hafner, NewYork (1952).
75. P. A. Iyer, *Boundary Layer Met.*, **5**, 53 (1973).
76. A. E. Gill and A. Davey, *J. Fluid Mech.*, **35**, 775 (1969).
77. W. M. Rohsenow, J. P. Hartnett, and E. N. Ganic (eds.), *Handbook of Heat Transfers Fundamentals*, 2nd edn., McGraw-Hill, New York (1985).
78. C. C. Chen and R. Eichhorn, *Trans. ASME, J. Heat Transfer*, **98C**, 446 (1976).
79. M. C. Jischke, *Trans. ASME, J. Heat Transfer*, **99C**, 676 (1977).
80. T. Fujii, M. Takeuchi, and I. Morioka, *Proc. 5th Int. Heat Transfer Conf.*, vol. 3, Tokyo (1974), p. 44.
81. J.-M. Piau, *C.r. Acadsci.*, **272A**, 63 (1971).
82. G. D. Raithby and K. G. T. Hollands, *Trans. ASME, J. Heat Transfer*, **100C**, 278 (1978).
83. Y. Jaluria, *Lett. Heat Mass Transfer*, **2**, 151 (1975).
84. S. S. Kutateladze, V. P. Ivakin, A. G. Kirdjashkin, and A. N. Kekalov, *Teplofiz. Vys. Temp.*, No. 3, 545 (1977).

85. J. Tanny and J. Cohen, *Int. J. Heat Mass Transfer*, **41**, 2125 (1998).
86. D. Angirasa and G. P. Peterson, *Int. J. Heat Mass Transfer*, **40**, 4329 (1997).
87. K. Kapoor and Y. Jaluria, *Int. J. Heat Mass Transfer*, **39**, 3977 (1996).
88. A. V. Luikov, A. A. Aleksashenko, V. A. Aleksashenko, *Conjugate Problems of Convective Heat Transfer*, BGU Press, Minsk (1971).
89. L. B. Gdalevich and V. E. Fertman, *Inzh.-Fiz. Zh.*, **33**, 539 (1977).
90. O. G. Martynenko, A. G. Semjenov, and Yu. A. Sokovishin, *Parametrical Methods in Free Convection*, Nauka i Tekhnika Press, Minsk (1984).
91. M. D. Kelleher and K. T. Yung, *Appl. Sci. Res.*, **17**, 249 (1967).
92. M. D. Kelleher and K. T. Yung, *Quart. J. Mech. Appl. Math.*, **25**, 447 (1972).
93. M. Mijamoto, J. Sumikawa, T. Akiyoshi, and T. Nakamura, *Int. J. Heat Mass Transfer*, **23**, 1545 (1980).
94. K. Yamasaki, *Bull. Kochi Techn. Coll.*, No. 17, 25 (1981).
95. A. E. Zinnes, *Trans. ASME, J. Heat Transfer*, **92C**, 528 (1970).
96. A. M. Andreev, M. P. Mikhailov, V. V. Ris, and Yu. A. Sokovishin, *Tr. LPI*, No. 323, 11 (1972).
97. C. G. Advanti and S. P. Sukhatme, *Int. J. Technol.*, **11**, 105 (1973).
98. K. Himasekhar, *Int. J. Heat Mass Transfer*, **30**, 201 (1987).
99. A. M. Andreev, V. N. Piskunov, Yu. A. Sokovishin, and V. S. Yuferev, in: *Thermal gas lens and thermal hydrodynamic optical guides*, Minsk (1974), pp. 168–185.
100. A. M. Andreev, Yu. A. Sokovishin, and V. S. Yuferev, in: *Heat and mass transfer*, vol. 1, part 1, Minsk (1972), pp. 351–354.
101. V. I. Lebedevitch, Yu. A. Sokovishin, and V. S. Yuferev, *Armaturoctroenie*, No. 3, 57 (1973).
102. K. Yamasaki, *Trans. JSME*, **49B**, 181 (1983).
103. G. S. Lock and J. C. Gunn, *Trans. ASME, J. Heat Transfer*, **90C**, 63 (1968).
104. O. G. Martynenko and Yu. A. Sokovishin, *Introduction to the Theory of Free-Convective Heat Transfer*, LGU Press, Leningrad (1982).
105. G. S. H. Lock and R. S. Ko, *Int. J. Heat Mass Transfer*, **16**, 2087 (1973).
106. A. Bejan and R. Anderson, *Trans. ASME, J. Heat Transfer*, **102C**, 630 (1980).
107. E. M. Sparrow and M. Faghri, *Trans. ASME, J. Heat Transfer*, **102C**, 402 (1980).
108. E. M. Sparrow and C. Prakash, *Int. J. Heat Mass Transfer*, **24**, 895 (1981).
109. R. Viskanta and D. Lankford, *Int. J. Heat Mass Transfer*, **24**, 1171 (1981).
110. A. Bejan and R. Anderson, *Int. J. Heat Mass Transfer*, **24**, 1611 (1981).
111. F. J. Higuera, *Int. J. Heat Mass Transfer*, **40**, 3157 (1997).
112. F. J. Higuera and I. Pop, *Int. J. Heat Mass Transfer*, **40**, 123 (1997).
113. C. Trevino, F. Mendez, and F. J. Higuera, *Int. J. Heat Mass Transfer*, **39**, 2231 (1996).
114. J. H. Merkin and I. Pop, *Int. J. Heat Mass Transfer*, **39**, 1527 (1996).
115. M.-I. Char and C.-L. Chang, *Int. J. Heat Mass Transfer*, **40**, 3641 (1997).
116. H. T. Chen and S. M. Chang, *Int. J. Heat Mass Transfer*, **39**, 383 (1996).
117. S. Kimura, A. Okajima, and T. Kiwata, *Int. J. Heat Mass Transfer*, **41**, 3203 (1998).
118. W.-S. Yu and H.-T. Lin, *Int. J. Heat Mass Transfer*, **36**, 1303 (1993).
119. M. Sakakibara, H. Amaya, S. Mori, and A. Tanimoto, *Int. J. Heat Mass Transfer*, **35**, 2289 (1992).
120. M. Vynnycky and S. Kimura, *Int. J. Heat Mass Transfer*, **39**, 1067 (1996).

121. M. Vynnycky and S. Kimura, *Int. J. Heat Mass Transfer*, **37**, 229 (1994).
122. M. Vynnycky and S. Kimura, *Int. J. Heat Mass Transfer*, **38**, 219 (1995).
123. E. M. Sparrow and M. Faghri, *Trans. ASME, J. Heat Transfer,* **102C**, 623 (1980).
124. K. T. Yang, *J. Appl. Mech.*, **31**, 131 (1964).
125. N. E. Hardwick and E. K. Levy, *Trans. ASME, J. Heat Transfer*, **95C**, 289 (1973).
126. E. M. Sparrow, S. V. Patankar, and R. M. Abdel-Wahed, *Trans. ASME, J. Heat Transfer*, **100C**, 184 (1978).
127. Y. Jaluria, *Trans. ASME, J. Heat Transfer*, **104**, 223 (1982).
128. I. P. Warneford and P. E. Fussey, *5th Int. Heat Transfer Conf.*, N.C.1.7, Japan (1974).
129. V. P. Ivakin and A. N. Kekalov, in: *Some Problems of Hydrodynamics and Heat Transfer*, Novosibirsk (1976), pp. 23–28.
130. C. E. Kwak and T. H. Song, *Int. J. Heat Mass Transfer*, **41**, 2517 (1998).
131. H. Y. Wang, F. Penot, and J. B. Sauliner, *Int. J. Heat Mass Transfer*, **40**, 1509 (1997).
132. S. H. Bhavnani and A. E. Bergles, *Int. J. Heat Mass Transfer*, **33**, 965 (1990).
133. C. K. Hsien and R. W. Coldewey, *Trans. ASME, J. Heat Transfer*, **99C**, 439 (1977).
134. S. G. Moulic and L. S. Yao, *Int. J. Heat Mass Transfer*, **35**, 2931 (1992).
135. E. M. Sparrow and L. F. A. Azevedo, *Trans. ASME, J. Heat Transfer*, **107C**, 977 (1985).
136. A. A. Khalatov, V. V. Orlanski, and A. F. Vasiljev, *Prom.Teplotekh.*, **10**, 41 (1988).
137. Y. H. Hung and W. M. Shiau, *Int. J. Heat Mass Transfer*, **32**, 863 (1989).
138. Y. H. Hung and W. M. Shiau, *Int. J. Heat Mass Transfer*, **31**, 1279 (1988).
139. S. A. M. Said and R. J. Krane, *Int. J. Heat Mass Transfer*, **33**, 1121 (1990).
140. T. Misuki and K. Kitamura, *Trans. JSME,* **B 53**, 1072 (1987).
141. T. Misuki and K. Kitamura, *Trans. JSME,* **B 56**, 115 (1990).
142. T. Inagaki and K. Komori, *Trans. JSME,* **B 57**, 3873 (1991).
143. I. A. Vatutin, V. F. Vinokurov, O. G. Martynenko, P. P. Khramtsov, and I. A. Shikh, *Exp. Heat Transfer*, **6**, 69 (1993).
144. W. T. Kierkus, *Int. J. Heat Mass Transfer*, **11**, 241 (1968).
145. J. B. Lee and G. S. H. Lock, *Trans. CSME*, **1**, 197 (1972).
146. M. M. Hasan and R. Eichhorn, *Trans. ASME, J. Heat Transfer*, **101C**, 642 (1979).
147. O. G. Martynenko, A. A. Berezovsky, and Yu. A. Sokovishin, *Asymptotic Methods in the Theory of Free-Convective Heat and Mass Transfer*, Nauka i Tekhnika Press, Minsk (1979).
148. B. R. Rich, *Trans. ASME, J. Heat Transfer*, **75C**, 489 (1953).
149. K. E. Hassan and S. A. Mohamed, *Int. J. Heat Mass Transfer*, **13**, 1873 (1970).
150. D. E. Fussey and I. P. Warneford, *Int. J. Heat Mass Transfer*, **21**, 119 (1978).
151. T. Fujii and H. Imura, *Int. J. Heat Mass Transfer*, **15**, 755 (1972).
152. A. F. Emery, A. Yang, and J. R. Wilson, *ASME Paper,* (1976), 76-HT-46.
153. S. P. Beschastnov, *Free-convective heat transfer of supercritical liquids in large volume*, Abstract of Doctoral Dissertation (in Technical Sciences), Moscow (1974).
154. W. E. Stewart, *Int. J. Heat Mass Transfer*, **21**, 119 (1978).

155. W. M. Lewandowski, *Int. J. Heat Mass Transfer*, **34**, 875 (1991).
156. B. Zeghmati, M. Daguenet, and G. L. Palec, *Int. J. Heat Mass Transfer*, **34**, 899 (1991).
157. M. Mammou, M. Daguenet, and G. L. Palec, *Int. J. Heat Mass Transfer*, **35**, 2277 (1992).
158. H. R. Lee, T. S. Chen, and B. F.Armaly, *Int. J. Heat Mass Transfer*, **35**, 207 (1992).
159. J. A. King and D. D. Reible, *Int. J. Heat Mass Transfer*, **34**, 1901 (1991).
160. K. Kitamura and F. Kimura, *Int. J. Heat Mass Transfer*, **38**, 3149 (1995)
161. F. J. Higuera, *Int. J. Heat Mass Transfer*, **36**, 3565 (1993).
162. B. F. Armaly, T. S. Chen, and N. Ramachandran, *Int. J. Heat Mass Transfer*, **30**, 405 (1987).
163. W. R. Risbeck, T. S.Chen, and B. F.Armaly, *Int. J. Heat Mass Transfer*, **37**, 699 (1994).
164. H. R. Lee, T. S. Chen, and B. F. Armaly, *Int. J. Heat Mass Transfer*, **34**, 305 (1991).
165. B. H. Kang and Y. Jaluria, *Int. J. Heat Mass Transfer*, **33**, 1347 (1990).
166. R. Bradean, D. B. Ingman, P. J. Heggs, and I. Pop, *Int. J. Heat Mass Transfer*, **39**, 615 (1996).
167. R. Bradean, D. B. Ingman, P. J. Heggs, and I. Pop, *Int. J. Heat Mass Transfer*, **40**, 665 (1997).
168. D. A. S. Rees, *Int. J. Heat Mass Transfer*, **39**, 3425 (1996).
169. H.-T. Lin and Y.-H. Chen, *Int. J. Heat Mass Transfer*, **37**, 1683 (1994).
170. H.-T. Lin, *Int. J. Heat Mass Transfer*, **35**, 1325 (1993).
171. L. Pera and B. Gebhart, *Int. J. Heat Mass Transfer*, **16**, 1131 (1973).
172. E. Schmidt, *VD-Forschungsh.*, **3**, 181 (1932).
173. R. Weise, *VDI-Forschungsh.*, **6**, 28 (1935).
174. L. F. Fatt, *J.I.E.M*, **20**, 24 (1976).
175. W. N. Gill, D. W. Zeh, and E. D. Casal, *ZAMP*, **16**, 539 (1965).
176. Z. Rotem, *Proc. 1st Can. Nat. Cong. Appl. Mech.*, vol. 2-b, (1967), p. 309.
177. Z. Rotem, L. Claassen, *J. Fluid Mech.*, **39**, 173 (1969).
178. K. Stewartson, *ZAMP*, **9**, 276 (1958).
179. J. R. Welty, C. E. Wicks, and R. E. Wilson, *Fundamentals of Momentum, Heat, and Mass Transfer*, John Wiley & Sons, Inc., 3rd edn., New York (1984).
180. P. Blanc, B. Gebhart, *Proc. 5th Int. Heat Transfer Conf.*, Paper NC1.5, Tokyo (1974).
181. M. Zakerullah and J. A. D. Ackroyd, *ZAMP*, **30**, 427 (1979).
182. D. R. Jones, *Q.J. Mech. Appl. Math.*, **26**, 77 (1973).
183. W. W. Yousef, J. D. Tarasuk, and W. J. McKeen, *Trans. ASME, J. Heat Transfer*, **104C**, 493 (1982).
184. T. Aihara, Y. Yamada, and S. Endo, *Int. J. Heat Mass Transfer*, **15**, 2535 (1972).
185. S. N. Singh, R. C. Birkebak, and R. M. Jr. Drake, *Prog. Heat Mass Transfer*, **2**, 87 (1969).
186. S. Levy, *J. Appl. Mech.*, **22**, 515 (1955).
187. S. N. Singh and R. C. Birkebak, *ZAMP*, **20**, 454 (1969).
188. C. Wagner, *J. Appl. Mech.*, **23**, 320 (1956).
189. R. C. Birkebak and A. Abdulkadir, *Proc. 4th Int. Heat Transfer Conf.*, Paper NC2.2, Paris (1970).

190. M. Fishenden and O. A. Saunders, *An Introduction to Heat Transfer,* Oxford Univ. Press, London (1950).
191. P. H. Oosthuizen, *Trans. ASME, J. Heat Transfer*, **98C**, 570 (1976).
192. W. H. H. Banks, *J. Eng. Math.*, **8**, 45 (1974).
193. W. H. H. Banks, *J. Eng. Math.*, **6**, 109 (1972).
194. G. D. Raithby and K. G. T. Hollands, *Proc. 6th Int. Heat Transfer Conf.*, vol. 2, Toronto (1978), pp. 187–192.
195. D. Pushpanathan and B. Rao, *Ind. Eng.*, **22**, 1 (1979).
196. H. P. Weiss, *ZAMP*, **28,** 409 (1977).
197. T. Fujii, H. Honda, and I. Morioka, *Int. J. Heat Mass Transfer*, **16**, 611 (1973).
198. Z.-Z. Lin, B.-X. Wang, and J.-Y. Xu, *Int. J. Heat Mass Transfer*, **34**, 2813 (1991).
199. A. A. Gukhman, *Introduction to Similarity Theory,* Vysshaya Shkola Press, Moscow (1973).
200. J. A. D. Ackroyd, *Proc. Roy. Soc., Ser. A*, **352**, 249 (1976) London.
201. V. P. Carey and J. C. Mollendorf, *Proc. 6th Int. Heat Transfer conf.*, vol. 2, Toronto (1978), pp. 211–216.
202. I. Suhara, *Japan Shipbuild. Mar. Engng.*, **5**, 5 (1970).
203. V. P. Carey and J. C. Mollendorf, *Int. J. Heat Mass Transfer*, **23**, 95 (1980).
204. A. Brown, *Trans. ASME, J. Heat Transfer*, **97C**, 133 (1975).
205. J.-M. Piau, *Int. J. Heat Mass Transfer*, **17**, 465 (1974).
206. J.-Y. Jang and J.-S. Leu, *Int. J. Heat Mass Transfer*, **36**, 1287 (1993).
207. F. C. Lai and F. A. Kulacki, *Int. J. Heat Mass Transfer*, **33**, 1028 (1990).
208. D.-Y. Shang, B.-X. Wang, Y. Wang, and Y. Quan, *Int. J. Heat Mass Transfer*, **36**, 3411 (1993).
209. S. Roy, *Ind. J. Phys.*, **46**, 277 (1972).
210. V. N. Piskunov, Yu. A. Sokovishin, and V. F. Stepanov, *Izv. VUZov, Energetika*, No. 11, 116 (1970).
211. P. V. S. Murthy and P. Singh, *Int. J. Heat Mass Transfer*, **40**, 1251 (1997).
212. J. A. D. Ackroyd, *JFM*, **62**, 677 (1974).
213. A. Brown, *Trans. Inst. Chem. Eng.*, **56**, 77 (1978).
214. L. G. Napolitano, *Aerotechn. Missili e Spazio*, **55**, 62 (1976).
215. I. E. Randall and A. Sesonske, *AIChE J.*, **5**, 150 (1959).
216. S. Ono, H. Kowano, H. Niho, and G. Fukuyama, *Bull. JSME*, **19**, 676 (1976).
217. E. A. Shtessel, in: *Heat and Mass Transfer in Burning.* Chernogolovka (1980).
218. P. G. Parikh, R. J. Moffat, W. M. Kays, and D. Bershader, *Int. J. Heat Mass Transfer*, **17**, 1465 (1974).
219. A. T. Tokuhiro and P. S. Lykoudis, *Int. J. Heat Mass Transfer*, **37**, 997 (1994).
220. Z. Dursunkaya and W. M. Worek, *Int. J. Heat Mass Transfer*, **35**, 2060 (1992).
221. F. C. Lai and F. A. Kulacki, *Int. J. Heat Mass Transfer*, **34**, 1189 (1991).
222. T. S. Chen, W. P. Buchanan, and B. F. Armaly, *Int. J. Heat Mass Transfer*, **36**, 479 (1993).

3

Free Convection on Curved Surfaces

Many interesting practical problems on natural convection heat transfer deal with the bodies of an arbitrary and complex shape. In most cases, the surface on which the flow develops is curvilinear and the angle γ between the gravity vector and its projection to a plane tangent to the surface changes as the flow moves from the leading edge along the surface. In some cases the curvature of the surface is small, and then the surface can approximately be considered as plane and thus numerous available data on vertical and inclined plane surfaces can be used.

The special features of heat transfer on a curved surface manifest themselves in a thin boundary layer. The effects caused by the body curvature are especially significant at small and moderate Grashof numbers. In the layer adjacent to the surface, the main mechanism of heat transfer is heat conduction.

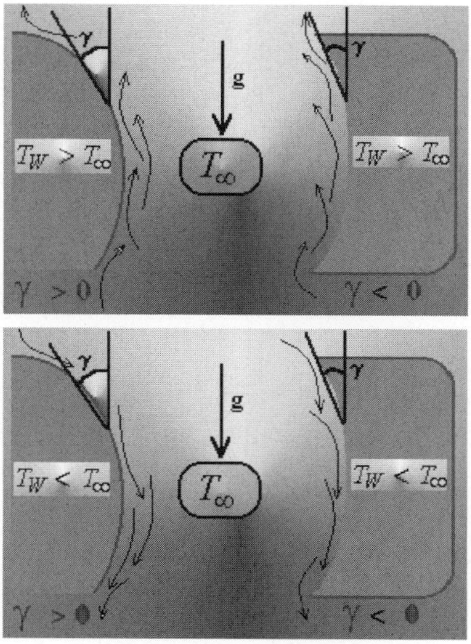

This underlies the Langmuir approximate method of a thermal layer [3.1], according to which the rate of the heat transfer from a curved surface is determined by the formula

$$\mathrm{Nu} = 2/\ln(1 + \delta_T/r) \ . \tag{3.1}$$

The thermal layer method was finally developed in [3.2]. The method in the main gives qualitative results because of the approximate determination of the thermal boundary-layer thickness δ_T. The Nusselt number can be calculated in the approximation of the thin layer method [3.3]:

$$\mathrm{Nu}_x = \mathrm{Nu}_{xp}(r\theta_{\mathrm{w}}g_x/g)^{1/3}\left[1/x\theta_{\mathrm{w}}\int\limits_0^x \theta_{\mathrm{w}}^{4/3}r^4(g_x/g\theta_{\mathrm{w}})^{1/3}dx\right]^{-1/4} , \tag{3.2}$$

$$g_x/g = [1 - (dr/dx)^2]^{1/2} ,$$

where Nu_{xp} is the Nusselt number calculated for a plane surface (Chap. 2).

The process of nonstationary free-convective heat transfer of a two-dimensional (horizontal cylinder) and an axisymmetric (sphere) bodies ***in a linearly stratified medium*** was studied in [3.4].

3.1 Vertical Cylinder

Free-convective flow part a vertical cylinder is encountered in many applied problems (bundles of pipes in nuclear reactors, cylindrical heating elements, etc.), therefore the special features of these flows are of separate interest for investigation.

At large values of D/L the boundary-layer thickness δ is small in comparison with the diameter of the cylinder $D = 2R$, the flow may approximately be considered similar to the flow near a flat plate. The maximum boundary-layer thickness corresponding to the cylinder height L is proportional to $L/(\mathrm{Gr})^{1/4}$, therefore the condition $\delta \ll D$ is similar to the condition $L/(\mathrm{Gr})^{1/4} \ll D$. In this case, for a vertical cylinder the general form of the solution is [3.5]

$$\mathrm{Nu} = \mathrm{Nu_w}\,(1 + \gamma\delta/R) , \tag{3.1.1}$$

$$\gamma = \gamma(\mathrm{Pr}) \ .$$

Since the boundary-layer thickness is a function of the Grashof number, the error of calculations by the formulas for a flat plate also depends on the Grashof number and on the ratio D/L. The criterion for estimating the error can be the expression [3.6]

$$\frac{D}{L} \geq \frac{35}{\mathrm{Gr}_L^{1/4}} \ . \tag{3.1.2}$$

If this expression is satisfied, then the difference between the calculations by the formulas for a flat plate and the experimental data for a cylinder does not exceed 5%, when Pr = 0.72 and Pr = 1,0.

For convective heat transfer of ***a vertical cylinder with heat sources*** q_w ***distributed over the surface*** the parameter of curvature is the quantity $(L/D)\mathrm{Gr}_L^{*-1/5}$ which is proportional to $L^{2/5}$.

The influence of the transverse curvature increases with the distance from the bottom of the cylinder increases. In processing the results of calculations and experiments the following classification is introduced according to the value of the indicated parameters: short cylinders or flat plates, long cylinders and wires [3.7, 3.8].

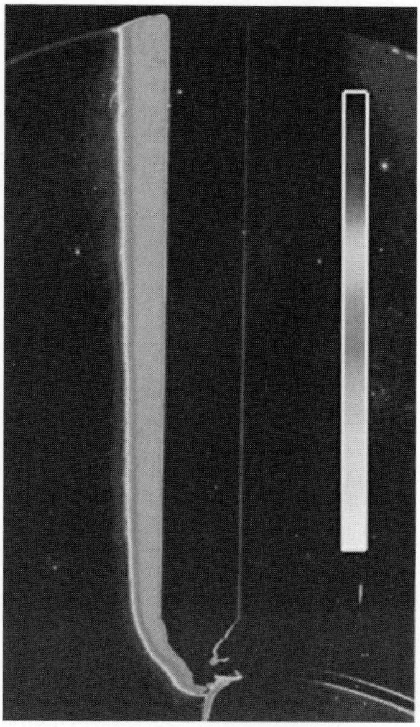

In the case of a variable temperature along the cylinder surface given by the relation

$$T_w - T_\infty = Nx^n \, , \tag{3.1.3}$$

the value of the local Nusselt number can be calculated from the formula [3.9]

$$(\mathrm{Nu}_x)_w = (\mathrm{Nu}_x)_p + 0.435\frac{x}{R} \, , \tag{3.1.4}$$

$$\frac{x}{R} \le 0.7\,(\mathrm{Nu}_x)_p \, .$$

In [3.10] the following dependence is suggested for the power-law distribution of temperature:

$$\frac{\mathrm{Nu}_x}{\mathrm{Gr}_x^{1/4}} = -(4h(x)^5)^{-1/4} \sum_{n=0}^{\infty} \theta'_n(0)\, \xi^n \,, \tag{3.1.5}$$

where

$$\xi = \frac{2L}{cR} x^{\frac{1-m}{4}}, \quad c = \left(\frac{g\beta N L^3}{4\nu^2}\right)^{1/4},$$

$$T_\mathrm{w} = T_\infty + N x^m h(x), \; h(x) = 1 + a_1\xi + a_2\xi^2 + \dots \,,$$

and the coefficients $\theta'_n(0)$ are presented in Table 3.1.1.

Table 3.1.1.

Pr	$4^{1/5}\{\theta'_0(0)\}^{-4/5}$	$\theta'_0(0)$	$\theta'_1(0)$	$\theta'_2(0)$
0.7	2.068533	−0.570081	−0.29424	0.1063
1	1.872838	−0.645490	−0.26064	0.0830
2	1.559112	−0.811743	−0.20879	0.0529
3	1.408018	−0.922050	−0.18467	0.0413
4	1.312465	−1.006712	−0.16969	0.0348
5	1.244133	−1.076295	−0.15910	0.0306
7	1.149541	−1.188117	−0.14461	0.0253
10	1.059004	−1.316415	−0.13096	0.0203

For a variable heat flux on a cylinder surface $q_\mathrm{w} = Mx^m$

$$(\mathrm{Nu}_x)_\mathrm{w} = (\mathrm{Nu}_x)_p + 0.345\frac{x}{R} \,, \tag{3.1.6}$$

where $(\mathrm{Nu}_x)_p$ is the local Nusselt number for a plane vertical surface (Chap. 2).

The use of the integral method in view of the transverse curvature yields the following dimensionless relation for the average Nusselt number [3.11]:

$$\mathrm{Nu} = \frac{4}{3}\left[\frac{7\mathrm{Gr}_L\,\mathrm{Pr}^2}{5\,(20 + 21\,\mathrm{Pr})}\right]^{1/4} + \frac{4\,(272 + 315\,\mathrm{Pr})\,L}{35\,(64 + 63\,\mathrm{Pr})\,D} \,. \tag{3.1.7}$$

In particular, in the case of a linear dependence of the surface temperature on the longitudinal coordinate in the range Pr = 0.7–100

$$\mathrm{Nu} = 1.058\ \mathrm{Gr}^{1/4}\left(\frac{\mathrm{Pr}^2}{4 + 7\,\mathrm{Pr}}\right)^{1/4}, \tag{3.1.8}$$

$$\mathrm{Gr} = g\beta D^4 N/\nu^2 \,,$$

where N is the coefficient of proportionality of the linear dependence of a temperature on a coordinate, but the local heat transfer coefficient does not depend on a longitudinal coordinate.

Application of the Langmuir approximate thermal layer method gives the following dimensionless equation for calculating the average heat transfer coefficient [3.12]:

$$\mathrm{Nu} = 0.6\,(D/L)^{1/4}\,\mathrm{Ra}^{1/4}\exp\left(\frac{2}{\mathrm{Nu}}\right) . \tag{3.1.9}$$

When $D \to \infty$,

$$\mathrm{Nu} = 0.6\ \mathrm{Ra}^{1/4} , \tag{3.1.10}$$

which corresponds to free convection for a flat plate.

The heat transfer due only to the thermal conductivity of the cylinder ***of constant temperature***, length L, and diameter D in a cavity of diameter d can be calculated from the formula [3.13]

$$\mathrm{Nu}_D = 2\ln\left(\frac{2d/D}{1+\sqrt{(d/L)^2+1}}\right) . \tag{3.1.11}$$

The minimum experimental value of the heat transfer coefficient for a vertical cylinder was obtained at $\mathrm{Ra} = 3.8\cdot 10^{-9}$ and was equal to $\mathrm{Nu} = 0.18$ [3.14].

The local value of the heat transfer coefficient of a vertical cylinder ***with a constant wall temperature*** at $\mathrm{Pr} = 0.7$–100, $x\mathrm{Gr}_x^{-1/4}/d < 2$ is [3.7,3.15, 3.16]

$$\mathrm{Nu}_x = \mathrm{Nu}_{xp} + 0.68\frac{x}{D} . \tag{3.1.12}$$

This dependence can also be used ***for a stratified medium for a linearly decreasing temperature of the fluid and constant temperature of the vertical cylinder*** [3.17].

For $\mathrm{Pr} = 0.7$, $x\mathrm{Gr}_x^{-1/4}/D = 0 - 1.5\cdot 10^3$

$$\mathrm{Nu}_x = \mathrm{Nu}_{xp}\left[1 + 0.766\left(\frac{x}{D\mathrm{Nu}_{xp}}\right)^{5/6}\right] . \tag{3.1.13}$$

For $\mathrm{Pr} \to \infty$

$$\mathrm{Nu}_x = \mathrm{Nu}_{xp} + 1.056x/D . \tag{3.1.14}$$

Approximate dependences with an error of 5% for the average heat transfer coefficient on a the vertical cylinder ***of constant temperature*** are [3.8]

$$T_{\mathrm{av}} = (T_{\mathrm{w}} + T_\infty)/2, \quad \mathrm{Pr} = 0.7 - 250 ,$$

$$\mathrm{Nu}_x = C_1\ \mathrm{Ra}_x^{n_1}\,(x/D)^{n_2} , \tag{3.1.15}$$

$$\mathrm{Nu}_{\mathrm{av}} = C_2\ \mathrm{Ra}_L^{n_1}\,(L/D)^{n_2} , \tag{3.1.16}$$

$$\mathrm{Nu}_{\mathrm{av}} = C_2\,(\mathrm{Ra}_D D/L)^{n_2} . \tag{3.1.17}$$

Table 3.1.2.

$\mathrm{Ra}_d d/l$	C_1	C_2	n_1	n_2
$\geq 10^4$	0.428	0.57	0.25	0
0.05–10^4	1.093	1.3	0.16	0.36
10^{-4}–0.05	0.827	0.87	0.05	0.8

The coefficients C_i and n_i for laminar free convection are given in Table 3.1.2.

For heat transfer in a gas ***of a very thin vertical wire*** ($D = 10$–$100\,\mu$m) through which an electric current flows the following dependences are suggested which give an accuracy within ±2% [3.18]:

$$\mathrm{Pr} = O(1)\,, \quad T_{\mathrm{av}} = (T_{\mathrm{w}} + T_\infty)/2, \quad \frac{L/D}{\mathrm{Nu}_{L\mathrm{w}}} = 5 - 10^3\,,$$

$$\frac{\mathrm{Nu}_L}{\mathrm{Nu}_{L\mathrm{w}}} = 1 + 0.762\left(\frac{L/D}{\mathrm{Nu}_{L\mathrm{w}}}\right)^{5/6}, \tag{3.1.18}$$

where $\mathrm{Ra}_D \frac{D}{L} = 10^{-11} - 10^{-4.5}$

$$\frac{2}{\mathrm{Nu}_D} = \ln\left[1 + \frac{4.47}{(\mathrm{Ra}_D D/L)^{0.26}}\right]. \tag{3.1.19}$$

For experimentally determined region of transition from a laminar boundary layer to a turbulent one on a vertical cylinder ***of constant temperature*** [3.19] $\mathrm{Pr} = 0.7$–10^3, $l/d = 15$, $\mathrm{Gr}_x = 1.85 \cdot 10^9 \mathrm{Pr}^{-3/5}$. The generalized dependence of heat transfer ***in a gas for laminar and turbulent free convective flows on a vertical cylinder*** [3.20]

$$d/l = 0.01 - 1, \quad \mathrm{Pr} = O(1), \quad \mathrm{Ra}_d = 10^{-5} - 10^{13}, \quad T_{\mathrm{av}} = (T_{\mathrm{w}} + T_\infty)/2\,,$$

$$\mathrm{Nu}_D = 2/\ln\left[1 + 2/\left\{\left(0.67\,(\mathrm{Ra}_D D/L)^{1/4}\left[1 + (0.599/\mathrm{Pr})^{3/5}\right]^{-\frac{5}{12}}\right)^{15} + \left(0.1\,\mathrm{Ra}_D^{1/3}\right)^{15}\right\}^{1/15}\right]. \tag{3.1.20}$$

Transition from a laminar boundary layer to a turbulent one on a vertical thin wire occurs at a greater value of the Gr_x number than in the case of a flat plate, and this value increases as the heat flux decreases. For example, the value of the critical Ra number ***for a vertical platinum wire*** with $D = 51.4$ µm is $\mathrm{Ra}_{xk} = 5 \cdot 10^{10}$, when $I = 0.2$ A, $\mathrm{Ra}_{xk} = 2.3 \cdot 10^{10}$, when $I = 0.335$ A and $\mathrm{Ra}_{xk} = 1 \cdot 10^{10}$, when $I = 0.375$ A [3.21]. The local heat transfer coefficient ***of the thin vertical wire in air for turbulent free***

convection when $D = 51.4\ \mu\text{m}$, $L/D = (1.38 - 4.31) \cdot 10^4$, $\text{Ra}_x > \text{Ra}_{xk}$, $T_{\text{av}} = (T_{\text{w}} + T_\infty)/2$, $\beta = 1/T_\infty$ is

$$\text{Nu}_x = 0.103\ \text{Ra}_x^{1/3} + 0.671\,(x/D)^{5/6}\ \text{Ra}_x^{1/18}\,. \tag{3.1.21}$$

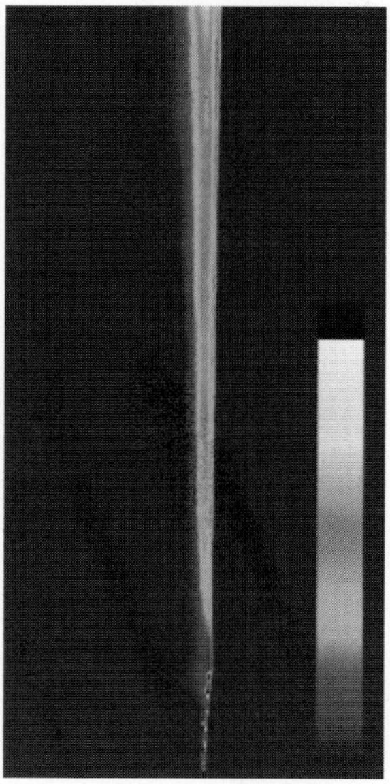

The average heat transfer coefficient ***in the presence of a laminar starting length*** is

$$\begin{aligned}\text{Nu}_{\text{av}} &= 0.103\left(\text{Ra}_L^{1/3} - \text{Ra}_{Lk}^{1/3}\right)\\ &\quad + 0.671\left[(L/D)^{5/6}\ \text{Ra}_L^{1/18} - (L_k/D)^{5/6}\ \text{Ra}_{Lk}^{1/18}\right]\\ &\quad + 0.515\ \text{Ra}_{Lk}^{1/4} + 0.682\,(L_k/D)^{5/6}\ \text{Ra}_{Lk}^{1/24}\,.\end{aligned} \tag{3.1.22}$$

The average heat transfer coefficient for ***a short vertical cylinder with insulated ends*** (the cylinder is mounted by its top or bottom to a horizontal organic glass plate) is [3.22]

$$L/D = 1 - 6, \quad \text{Pr} = 0.7, \quad \text{Ra}_D = 5 \cdot 10^4 - 3 \cdot 10^7, \quad T_{\text{av}} = (T_{\text{w}} + T_\infty)/2\,.$$

$$\frac{\text{Nu}_L}{\text{Nu}_{L0}} = \frac{1}{1 + 0.25D/L} + \frac{A\,(L/D)^{0.25}}{1 + 4L/D}\,, \tag{3.1.23}$$

where Nu_{L0} is the Nusselt number for a vertical infinite cylinder; $A = 0.98$ for a plate attached to the bottom of the cylinder, $A = 0.5$ for a plate attached to the top of the cylinder. The heat transfer coefficient increases, when both ends of the cylinder are insulated.

The heat transfer of a vertical cylinder ***of constant temperature with exponentially decreasing viscosity of a dropping liquid (94%-solution of glycerine)*** is given by [3.23]

$$T_w = 13 - 70°\text{C}, \quad L/D = 30, \quad T_\infty = 15 - 40°\text{C} ,$$

$$\nu = 6.06 \cdot 10^{-2} \exp\left(-0.864\sqrt{T}\right), \qquad \text{Ra}_{Lw} = 5 \cdot 10^4 - 5 \cdot 10^8 ,$$

$$\text{Nu}_{\text{av,w}} = 0.87 \left(\nu_{\text{av}}/\nu_\infty\right)^{0.6} \text{Ra}_{Lw}^{0.25} \cdot , \tag{3.1.24}$$

where ν_{av} is the kinemat viscosity for $T_{\text{av}} = (T_w + T_\infty)/2$, and ν_∞ for T_∞.

The heat transfer of a vertical cylinder ***with a variable wall temperature*** with laminar free convection is described [3.24]

for $\theta_w = \theta_0 x^m$ by

$$\text{Nu}_{\text{av}} = C \left(\text{Ra}_D D/L\right)^m . \tag{3.1.25}$$

The values of the coefficients C and m are given in Table 3.1.3.

Table 3.1.3.

	$n = 2$		$n = -0.5$	
$\text{Ra}_D\ D/L$	C	m	C	m
$\geq 10^4$	0.97	0.25	0.25	0.25
$0.05 - 10^4$	1.75	0.18	1.3	0.18
≤ 0.05	1.16	0.05	1.23	0.05

for $\theta_w = \theta_0 e^{mx}$ by

$$\text{Nu}_{\text{av}} = B \frac{e^m - e^{-km}}{m^{1-k}} \left(\text{Ra}_D \frac{D}{L}\right)^k . \tag{3.1.26}$$

The values of the coefficients B and k are given in Table 3.1.4.

Table 3.1.4.

$\text{Ra}_D D/L$	B	k
$\geq 10^4$	0.97	0.25
$0.05 - 10^4$	1.86	0.16
$10^{-4} - 0.05$	1.074	0.05

When the heat flux decreases exponentially on a vertical cylinder $q_w = q_0 \exp(1 + x/m)$, the approximate dependence of the heat transfer with an error of about 2% is given by the expression

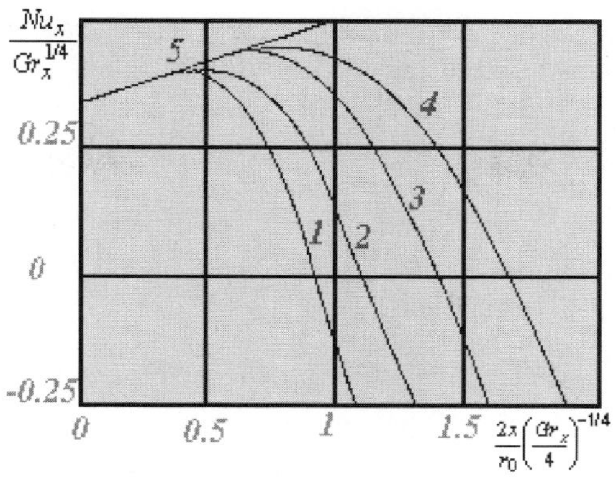

Fig. 3.1.1. The local heat transfer coefficient for a linearly decreasing wall temperature $\theta_w = \theta_0 (1 - x/l)$, Pr = 0.7: (1) $4l^4\ \mathrm{Gr}_l^{-1}/r_0^4 = 1$; (2) 2 (3) 5; (4) 10; (5) ∞

$$\mathrm{Nu}_x = (x/R) \Big/ \ln\left(1 + \frac{x}{R\mathrm{Nu}_{xp}}\right) , \tag{3.1.27}$$

where

$$\mathrm{Nu}_{xp} = \left(\frac{\mathrm{Pr}\ \mathrm{Ra}_x^*}{4 + 4\ \mathrm{Pr}^{1/2} + 10\,\mathrm{Pr}}\right)^{1/5} \left(\frac{x}{m}\right)^{1/5} \left(e^{x/m} - 1\right) .$$

The dependence of the coefficient of heat transfer from a vertical cylinder ***in air with a linearly decreasing wall temperature*** is presented in Fig. 3.1.1 and ***for an exponential change in the wall temperature*** in Fig. 3.1.2.

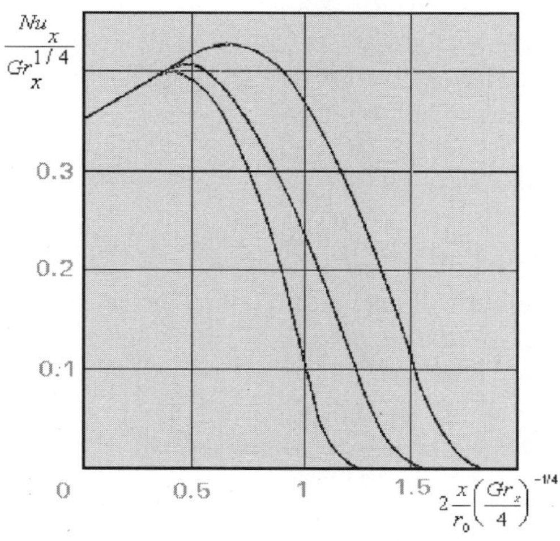

Fig. 3.1.2. Heat transfer from a cylinder in air (Pr = 0.7) for $\theta_w = \theta_0 \exp(-mx)$

The average heat transfer coefficient ***at a constant heat flux on the surface*** of a vertical cylinder is

$$\mathrm{Nu}_{\mathrm{av}} = C_1 \left(\mathrm{Ra}^*_D D/L\right)^{m_1} , \quad \mathrm{Nu}_{\mathrm{av}} = C_2 \left(\mathrm{Ra}^D_D/L\right)^{m_2} . \qquad (3.1.28)$$

The values of the coefficients C_i and m_i are presented in Table 3.1.5.

Table 3.1.5.

$\mathrm{Ra}_D D/L$	$\geq 10^4$	$0.05 - 10^4$	≤ 0.05
C_1	0.55	1.33	0.90
m_1	0.20	0.14	0.048
C_2	0.60	1.37	0.93
m_2	0.25	0.16	0.05

The local heat transfer coefficient on a vertical cylinder ***in the presence of weak mass transfer on the wall and uniform surface temperature*** is [3.25]

$$D/L = 0.5, \quad \mathrm{Pr} = 0.71, \quad \mathrm{Sc} = 0.63, \quad \mathrm{Gr}_x = 3 \cdot 10^5 - 2 \cdot 10^8 ,$$
$$\mathrm{Gr}_{xD} = 6 \cdot 10^4 - 1 \cdot 10^8, \quad \mathrm{Pe}_{xw}/\mathrm{Gr}_x^{1/2} \ll 1 ,$$
$$\mathrm{Nu}_x/\mathrm{Gr}_x^{1/4} = 0.52 \left(1 + \mathrm{Gr}_{xD}/\mathrm{Gr}_x\right)^{1/4} . \qquad (3.1.29)$$

The heat transfer rate increases with the cylinder curvature. Moreover, the value of the number Nu_x increases or decreases depending on the direction of buoyancy force caused by the mass diffusion on the surface. In the case of simultaneous mass and heat diffusion, heat transfer rate for a uniform heat (mass) flux on the surface is higher than at a constant temperature (concentration) on the wall.

The experiments on heat transfer of a vertical cylinder ***in mercury*** with an error of 2% are generalized by the dependence [3.5]

$$D/L = 0.15 - 0.55, \quad \mathrm{Pr} = 0.023, \quad \mathrm{Gr}^*_x = 10^5 - 10^9, \quad T_{\mathrm{av}} = (T_{\mathrm{w}} + T_\infty)/2 ,$$

$$\mathrm{Nu}_x = 0.226 \,(D/L)^{0.032} \,\mathrm{Gr}_x^{*0.183} \,(D/L)^{-0.032} . \qquad (3.1.30)$$

The results of numerical calculations of heat transfer for ***a rotating*** vertical cylinder ***of constant temperature*** are given in [3.26] for various values of Prandtl and Taylor numbers.

The influence of the bottom edge of a vertical cylinder on the intensity of heat transfer ***in air and water*** was analyzed in detail in [3.27].

Example 3.1. A vertical cylinder of diameter $D = 4$ mm and length $L = 0.5$ m is in air at $t_\infty = 80°$C. What will be the power of a heater?

For the characteristic temperature $t_{\text{av}} = (20 + 80)/2 = 50°\text{C}$ we have

$\nu = 17.96 \cdot 10^{-6}$ m^2/sec,
$\lambda = 2.83 \cdot 10^{-2}$ W/m$^2 \cdot$ deg,
$\text{Pr} = 0.698$,

$$\text{Gr}_D\,\text{Pr} = 9.81 \cdot 64 \cdot 10^{-9}/323/(1.795)^2 \cdot 10^{10} \cdot 60 \cdot 0.698 = 253\ ,$$
$$\text{Ra}_D D/L = 253 \cdot 4 \cdot 10^{-3}/0.5 = 2.024\ .$$

According to the classification given, this case can be considered as a long cylinder, and we can use the following relation:

$$\text{Nu}_{\text{av}} = C_2\,(\text{Ra}_D D/L)^{n_2} = 1.3\,(2.024)^{0.36} = 1.652\ .$$
$$\alpha_{\text{av}} = \text{Nu}_{\text{av}}\lambda/D = 1.652 \cdot 2.82 \cdot 10^{-2}/4 \cdot 10^{-3} = 11.69\ \text{W/m}^2 \cdot \text{deg}\ .$$

The power of the heater is

$$Q = \alpha_{\text{av}}\theta_{\text{w}}\pi DL = 11.69 \cdot 60 \cdot 3.14 \cdot 4 \cdot 10^{-3} \cdot 0.5 = 4.37\,\text{kW}.$$

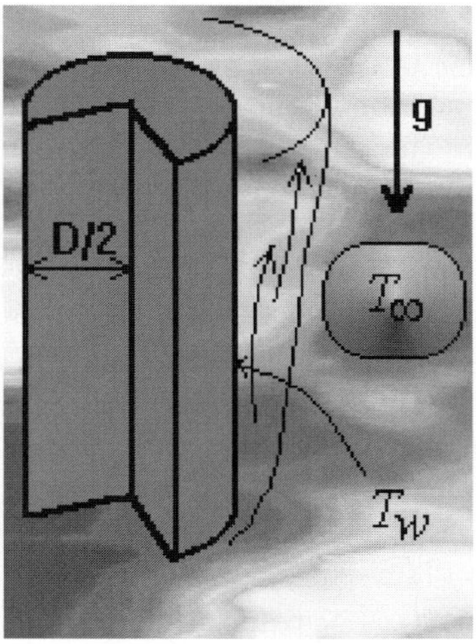

3.2 Horizontal Cylinder

Free-convective flows near a horizontal circular cylinder have been rather adequately investigated in a wide range of transfer parameters due to the abundance of this geometry in nature and engineering. However, determination of local heat transfer characteristics on a horizontal cylinder by numerical methods involves significant computational difficulties when solving complete equations of motion and energy, and in the majority of cases the results obtained are not liable to generalize action. The general trends in the approach

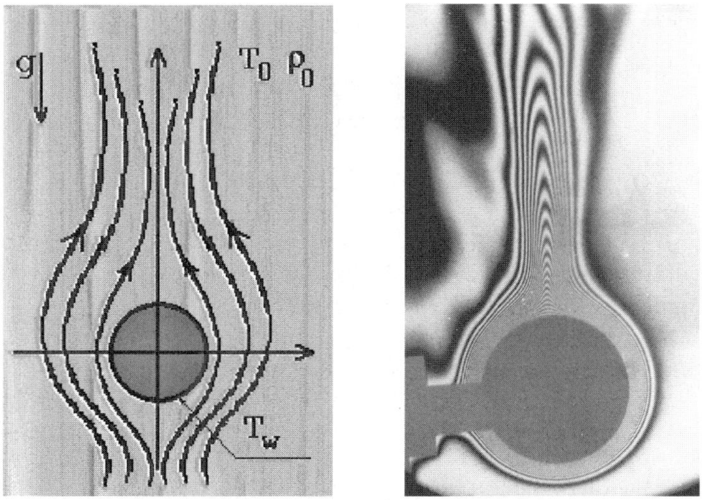

Fig. 3.2.1. Structure of streamlines and interferogram of the temperature field in free-convective flow near a horizontal cylinder of constant temperature

to these problems, noted by many researchers [3.28, 3.29], follow from the limitatedness of the application of the boundary-layer method. Under certain conditions the solution of the problem of flow around a horizontal cylinder in the approximation of a boundary layer is possible. However, at larger angles of flow around the top of the cylinder two enveloping flows separate from the surface, and then they merge and form an ascending plume.

This mechanism should be distinguished from separation of forced convection flow caused by an external pressure field and characterized by retardation and reversal of flow. The change in the direction of flow in the upper part and the formation of a plume above the cylinder causes an increase in the flow velocity and an increase in the heat transfer coefficient in spite of the thickening of the boundary layer. In this region, one must not neglect the influence of curvature, and the boundary-layer approximation becomes unsuitable for correct description of the mechanisms of transfer [3.28]. The boundary-layer curvature is the essential factor determining heat transfer of a horizontal cylinder also for flows at small values of the Grashof number.

Numerical solution for ***a horizontal*** cylinder of radius R in boundary-layer approximation gives the following dimensionless relations between the local Nusselt number and the azimuthal parameter $\xi = x/R$ [3.28]:

for $\mathrm{Pr} = 0.7$:

$$\mathrm{Nu}_R\,(x)\,/\mathrm{Gr}_R^{1/4} = 0.37023 - 0.01609\xi^2 - 0.00009\xi^4\ , \tag{3.2.1}$$

for $\mathrm{Pr} = 1$:

$$\mathrm{Nu}_R\,(x)\,/\mathrm{Gr}_R^{1/4} = 0.42143 - 0.01861\xi^2 - 0.00011\xi^4\ . \tag{3.2.2}$$

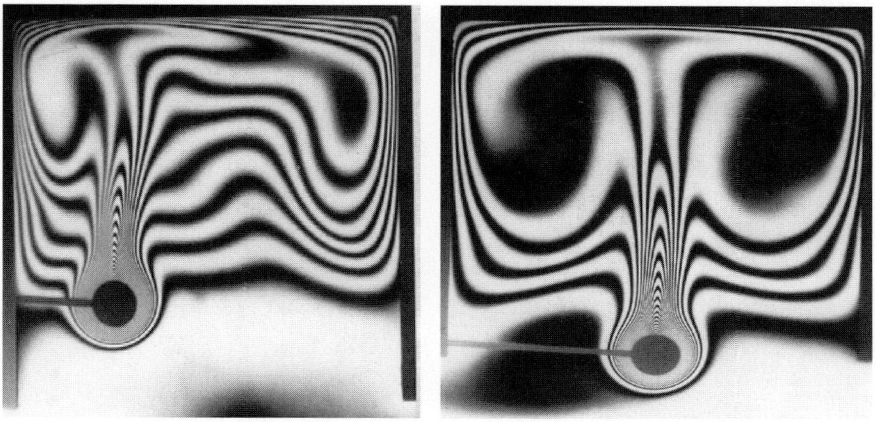

Fig. 3.2.2. Interferograms of free-convective flow near a horizontal cylinder in the space limited by a rectangular cavity

The average Nusselt number Nu_D obtained by the integral method can be calculated from the following formulas [3.29]:

$$\mathrm{Nu}_D = \frac{2}{\pi} \int_0^{\pi/2} \mathrm{Nu}_D \left(\frac{x}{D}\right) d\left(\frac{x}{D}\right) = F\,(\mathrm{Pr})\,(\mathrm{Gr}_D\,\mathrm{Pr})^{1/4} \,, \qquad (3.2.3)$$

where the Nusselt and Grashof numbers are calculated based on the cylinder diameter D, or

$$\mathrm{Nu}_R = 2^{-1/4} F\,(\mathrm{Pr})\,(\mathrm{Gr}_R\,\mathrm{Pr})^{1/4} \,, \qquad (3.2.4)$$

when the Nusselt and Grashof numbers are calculated based on the radius R.

The dimensionless relation taking into account the curvature in the flow direction is

$$\frac{2}{\mathrm{Nu}_D} = \ln\left[1 + \frac{4.065}{F\,(\mathrm{Pr})\,\mathrm{Ra}_D^m}\right] , \qquad (3.2.5)$$

where $m = 1/4 + 1/(10 - 4\,\mathrm{Ra}_D^{1/8})$, $\mathrm{Ra}_D = \mathrm{Gr}_D\mathrm{Pr}$, $\mathrm{Gr}_D = g\beta(T_0 - T_\infty)D^3/\nu^2$ is applicable in the range $10^{-10} \leq \mathrm{Ra}_D \leq 10^7$ [3.28].

As a result of the survey of a large number of experimental investigations a set of formulas is suggested for calculating heat transfer ***from long horizontal cylinders*** in a wide range of Rayleigh numbers [3.30, 3.31]:

$$\begin{aligned}
\mathrm{Nu} &= 0.675\,\mathrm{Ra}_D^{0.058}, & \mathrm{Ra}_D &= 10^{-10} - 10^{-2}\,,\\
\mathrm{Nu} &= 1.02\,\mathrm{Ra}_D^{0.148}, & \mathrm{Ra}_D &= 10^{-2} - 10^{2}\,,\\
\mathrm{Nu} &= 0.85\,\mathrm{Ra}_D^{0.188}, & \mathrm{Ra}_D &= 10^{2} - 10^{4}\,,\\
\mathrm{Nu} &= 0.48\,\mathrm{Ra}_D^{0.25}, & \mathrm{Ra}_D &= 10^{4} - 10^{7}\,,\\
\mathrm{Nu} &= 0.125\,\mathrm{Ra}_D^{0.333}, & \mathrm{Ra}_D &= 10^{7} - 10^{10}\,,
\end{aligned} \qquad (3.2.6)$$

where all the parameters are determined at $T_{\mathrm{av}} = (T_0 + T_\infty)/2$.

The approximation dependences for the local value of heat transfer in a laminar boundary layer in the range of angle $\varphi = 0 - \pi$ and $\mathrm{Gr}_D = 10^5$–10^7 are [3.32]

$$\mathrm{Nu}_\varphi = \left[C_0 + C_1 (\varphi + \pi/2)^2 + C_2 (\varphi + \pi/2)^4\right] \mathrm{Ra}_{D\varphi} , \tag{3.2.7}$$

for Pr = 0.1–1:

$$\begin{aligned} C_0 &= (-0,0234 \ln \ \mathrm{Ra}_D + 0,573) \\ &\quad \times \left[0,114(1 + \ln \mathrm{Pr})^2 + 0,118(1 + \ln \mathrm{Pr}) + 0,831\right] , \\ C_1 &= (0,00447 \ln^2 \ \mathrm{Ra}_D - 0,05605 \ln \ \mathrm{Ra}_D + 0,173) \\ &\quad \times \left[6,382(1 + \ln \mathrm{Pr})^2 - 3,39(1 + \ln \mathrm{Pr}) - 0,754\right] , \\ C_2 &= (0,00048 \ln \ \mathrm{Ra}_D - 0,0067) \\ &\quad \times \left[-0,302(1 + \ln \mathrm{Pr})^2 + 0,433(1 + \ln \mathrm{Pr}) + 0,8\right] , \end{aligned}$$

for Pr = 1–100:

$$\begin{aligned} C_0 &= (-0,0234 \ln \ \mathrm{Ra}_D + 0,573) \\ &\quad \times \left[-0,374(1 + \ln \mathrm{Pr})^2 + 1,94(1 + \ln \mathrm{Pr}) + 0,855\right] , \\ C_1 &= (0,00447 \ln^2 \ \mathrm{Ra}_D - 0,05605 \ln \mathrm{Ra}_D + 0,173) \\ &\quad \times \left[-286,7(1 + \ln \mathrm{Pr})^2 + 1335,5(1 + \ln \mathrm{Pr}) - 1429,5\right] , \\ C_2 &= (0,00048 \ln \ \mathrm{Ra}_D - 0,0067) \\ &\quad \times \left[0,05265(1 + \ln \mathrm{Pr})^2 - 0,5(1 + \ln \mathrm{Pr}) + 1,363\right] . \end{aligned}$$

The asymptotic values of the rate of heat transfer for $\mathrm{Pr} \gg 1$ are

at a constant temperature of the cylinder surface [3.33]

$$\mathrm{Nu}_\varphi = 0.711 \ \mathrm{Ra}_{D\varphi}^{1/4} \sin^{1/3} \varphi \left(\int_0^\varphi \sin^{1/3} \varphi d\varphi \right)^{-1/4} , \tag{3.2.8}$$

at a constant density of the heat flux on the cylinder surface [3.34]

$$\mathrm{Nu}_\varphi = 0.682 \ \mathrm{Ra}_{D\varphi}^{1/5} \sin^{1/4} \varphi \left(\int_0^\varphi \sin^{1/4} \varphi d\varphi \right)^{-1/5} . \tag{3.2.9}$$

The deviation of the calculated data from the experiment, for example, ***in mineral oil*** (Pr = 1000) was ±2% for $\varphi < 150°$.

At a constant temperature of the surface for laminar free convection the general dimensionless relation is applicable [3.35, 3.36]:

$$\mathrm{Nu} = C\mathrm{Ra}_D^m \ \mathrm{Pr}^n . \tag{3.2.10}$$

Table 3.2.1.

Ra_D	Pr	T_{ov}	C	m	n
$3\cdot10^2 - 2\cdot10^7$	0.7–3090	$1/2(T_{\mathrm{w}}+T_\infty)$	0.474	0.25	0.47
$5\cdot10^2 - 7.1\cdot10^4$	0.01–5	$1/2(T_{\mathrm{w}}+T_\infty)$	1.458	0.125	0.1
$7.1\cdot10^4 - 10^8$	0.01–100	$1/2(T_{\mathrm{w}}+T_\infty)$	0.375	0.25	0.125
$3\cdot10^2 - 2\cdot10^7$	0.7–3090	T_{w}–$0.68\theta_{\mathrm{w}}$	0.478	0.25	0.05

The error of calculations was within ±1.5%. The regions of applicability, the values of the coefficients C, m, and n and also the characteristic temperatures are given in Table 3.2.1.

The standard error of calculations for a laminar mode of the flow is 1.5%, when $T_{\mathrm{av}} = (T_0 + T_\infty)/2$.

Using the thin-layer method for $\mathrm{Ra}_D > 10^{-2}$ for a laminar and a turbulent flow ***at a constant temperature of the surface of the cylinder*** the following dependence was obtained [3.37]:

$$\mathrm{Nu}^{3.337} = \left[2/\ln\left(1+\pi/1.294^{3/4}C_l\,\mathrm{Ra}_D^{1/4}\right)\right]^{3.337} + \left(0.72C_t\,\mathrm{Ra}_D^{1/3}\right)^{3.337}, \tag{3.2.11}$$

where

$$C_l = 0.50\left[1+(0.49/\mathrm{Pr})^{9/16}\right]^{-4/9}, \quad C_t = \left[0.14\,\mathrm{Pr}^{0.084}, 0.15\right]^* .$$

The operator $[,]^*$ means that $[A,B]^* = A$ if $A \le B$ and $[A,B]^* = A$ if $A > B$.

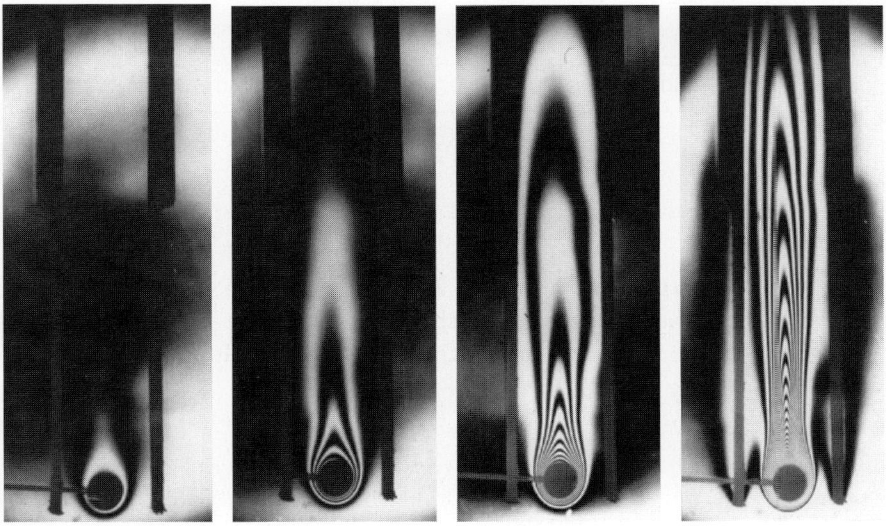

Fig. 3.2.3. Interferograms of free-convective flow near a horizontal cylinder (located in a vertical rectangular channel) with a constant heat flux on the cylinder surface

The solution of this very problem by the thermal conductive layer method gives the dependence [3.38]

$$\mathrm{Nu} = 2/\ln\left(1 + 2/\mathrm{Nu}_D\right) , \qquad (3.2.12)$$

where

$$\mathrm{Nu}_D = \left[\left(0.518\ \mathrm{Ra}_D^{1/4}\left[1 + (0.559/\mathrm{Pr})^{3/5}\right]^{-5/12}\right)^{15} + \left(0.1\ \mathrm{Ra}_D^{1/3}\right)^{15}\right]^{1/15} .$$

In the region of a laminar regime for $\mathrm{Ra}_D > 1$ the approximate dependence is applicable [3.39]:

$$\mathrm{Nu}^{1/2} = 0.36^{1/2} + \left[(\mathrm{Ra}_D/300) \Big/ \left\{1 + (0.5/\mathrm{Pr})^{9/16}\right\}^{16/9}\right]^{1/6} , \qquad (3.2.13)$$

which gives an error of 4.5%.

A similar formula for $\mathrm{Ra}_D > 10^{-6}$ can be used both in a laminar and a turbulent flow regimes [3.40]:

$$\mathrm{Nu}_D = 0.60 + 0.387 \left\{\mathrm{Ra}_D \Big/ \left[1 + (0.559/\mathrm{Pr})^{9/16}\right]^{16/9}\right\}^{1/6} . \qquad (3.2.14)$$

The results of numerical calculations of the local Nusselt number for a horizontal cylinder of ***constant temperature in air*** are given in Table 3.2.2 [3.41].

Table 3.2.2.

Ra	$\vartheta = 0°$	$\vartheta = 30°$	$\vartheta = 60°$	$\vartheta = 90°$	$\vartheta = 120°$	$\vartheta = 150°$	$\vartheta = 180°$	Nu_D
10^3	3.813	3.772	3.640	3.374	2.866	1.975	1.218	3.024
10^4	5.995	5.935	5.750	5.410	4.764	3.308	1.534	4.826
10^5	9.675	9.557	9.278	8.765	7.946	5.891	1.987	7.898

At a constant density of the heat flux on the cylinder surface heat transfer can be calculated from the formulas

for a laminar regime in the ranges $\mathrm{Pr} = 0.01$–100, $\mathrm{Gr}_D = 10^3$–10^8 [3.36]:

$$\mathrm{Nu} = 0.563\ \mathrm{Ra}_D^{0.2}\ \mathrm{Pr}^{0.04} , \qquad (3.2.15)$$

for a laminar and a turbulent regimes in the ranges $\mathrm{Pr} = 0.022$–7640, $\mathrm{Ra}_D > 1, T_{\mathrm{av}} = (T_{\mathrm{w}} + T_\infty)/2$ [3.39]:

$$\mathrm{Nu}^{1/2} = 0.36^{1/2} + \left[(\mathrm{Ra}_D/300) \Big/ \left\{1 + (0.44/\mathrm{Pr})^{9/16}\right\}^{16/9}\right]^{1/6} \qquad (3.2.16)$$

using the thin-layer method, when $\mathrm{Ra}_D > 0.1$ [3.37]:

$$\mathrm{Nu}^{3.337} = \left[2/\ln\left(1 + 1/0.38^{3/4} C_l\ \mathrm{Ra}_D^{1/4}\right)\right]^{3.337} + \left(0.72 C_t\ \mathrm{Ra}_D^{1/3}\right)^{3.337} , \tag{3.2.17}$$

$$\mathrm{Nu}^{3.337} = \left[1.64/\ln\left(1 + 1.64/1.05\ C_l\ \mathrm{Ra}_D^{1/5}\right)\right]^{3.337} + \left(0.72 C_t\ \mathrm{Ra}_D^{1/4}\right)^{3.337} . \tag{3.2.18}$$

In the case of a variable temperature of the cylinder surface:

$$\theta_\mathrm{w} = \theta_0 \left(1 + a_1 x^2 + a_2 x^4\right), \quad \mathrm{Gr}_R = g\beta\theta_0 R^3/\nu^2 ,$$

the laminar free convective heat transfer is calculated from the formulas [3.42]

for $\mathrm{Pr} = 0.7$

$$\mathrm{Nu}(x)/\mathrm{Gr}_R^{1/4}(x) = 0.3702 + (0.7569 a_1 - 0.0161)\, x^2 + \left(0.9285 a_2 + 0.0947 a_1^2 - 0.0289 a_1 - 0.0001\right) x^4 , \tag{3.2.19}$$

for $\mathrm{Pr} = 1$

$$\mathrm{Nu}(x)/\mathrm{Gr}_R^{1/4}(x) = 0.4214 + (0.852 a_1 - 0.0186)\, x^2 + \left(1.0411 a_2 + 0.107 a_1^2 - 0.0331 a_1 - 0.0001\right) x^4 . \tag{3.2.20}$$

In the case of a variable density of the heat flux on the surface [3.43]

$$q_\mathrm{W} = q_0 \left(1 + b_1 x^2 + b_2 x^4\right), \quad \mathrm{Gr}_R = g\beta q_0 R^4/\lambda\nu^2 ,$$

for $\mathrm{Pr} = 0.7$

$$\theta_\mathrm{W}\lambda\mathrm{Gr}_R^{1/5}(x)/(Rq_0) = 2.2142 + (1.0831 b_1 + 0.0471)\, x^2 + \left(0.8829 b_2 - 0.0541 b_1^2 + 0.029 b_1 + 0.0016\right) x^4 , \tag{3.2.21}$$

for $\mathrm{Pr} = 1$

$$\theta_\mathrm{W}\lambda\mathrm{Gr}_R^{1/5}(x)/(Rq_0) = 1.9963 + (0.9874 b_1 + 0.0436)\, x^2 + \left(0.8081 b_2 - 0.0502 b_1^2 + 0.0269 b_1 + 0.0015\right) x^4 . \tag{3.2.22}$$

The average value of heat transfer of ***a horizontal wire*** in an infinite space for small values of the Gr_D number in the range $\mathrm{Ra}_D\mathrm{Pr} \le 2 \cdot 10^{-3}$ [3.44]

$$\mathrm{Nu} = 6 \ln \left(\frac{49.6\,(\mathrm{Pr}+9.4)^{1/2}}{\mathrm{Nu}\ \mathrm{Ra}_D\,\mathrm{Pr}} \right)^{-1} . \tag{3.2.23}$$

It should be noted that formula (3.2.23) is inapplicable in rarefied gases.

For a cylinder of infinite length (***vertically located disk***) the intensity of heat transfer ***in air for laminar free convection*** on the end surface can be determined from the formula [3.45]

$$\mathrm{Nu} = 1.052 C_l\ \mathrm{Ra}_D^{1/4} . \tag{3.2.24}$$

When $l/D > 10$, the end heat losses may not be taken into account.

The intensity of heat transfer from horizontal cylinders ***to air, water, and silicon oil*** is calculated from the correlation formula [3.33]

$$\mathrm{Nu} = 0.478\ \mathrm{Ra}_D^{0.25}\ \mathrm{Pr}^{0.05} . \tag{3.2.25}$$

The parameters of the problem at $T_{\mathrm{av}} = T + 0.32(T_0 - T_\infty)$ lie in the ranges $\mathrm{Pr} = 0.7 - 3090, \mathrm{Ra}_D = 2.5 \cdot 10^2 - 1.8 \cdot 10^7$.

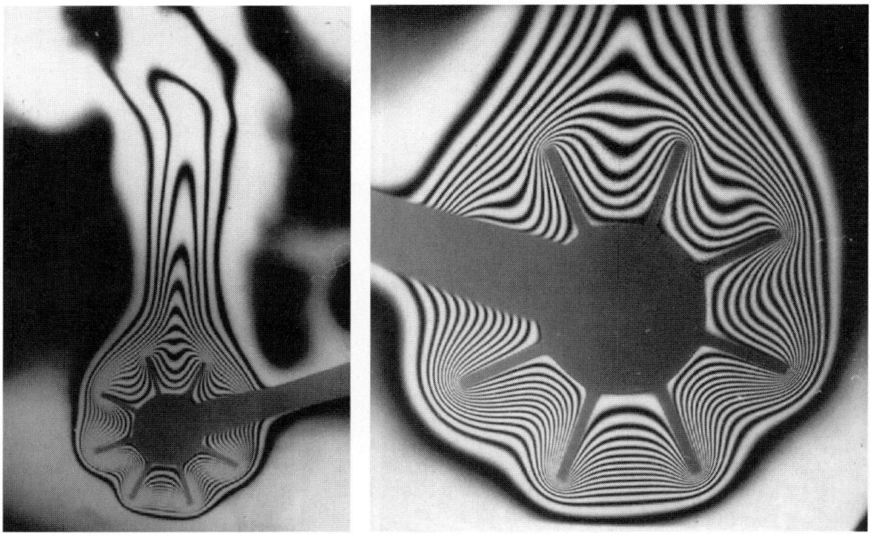

Fig. 3.2.4. Interferograms of free convection on a finned cylinder of constant temperature

The average coefficient of heat transfer of a horizontal cylinder with an ***artificial roughness*** ($\Delta/D = 0.002 - 0.04$), [3.46,3.47] ***at a constant temperature of the surface*** is determined by the generalized experimental dependence.

$$\mathrm{Nu} = \frac{2}{\ln B}\left[1 - \frac{A}{\ln B\mathrm{Ra}_D^{1/4}}\left\{\left(1 + \frac{\ln B\mathrm{Ra}_D^{1/4}}{A}\right)^{1/2} - 1\right\}\right], \tag{3.2.26}$$

$$A = \frac{0.51}{\mathrm{Pr}\left[1 + (8.6/\mathrm{Pr})^{7/16}\right]^{16/7}}, \quad B = 1 + 3.84\frac{\left[1 + (0.46/\mathrm{Pr})^{3/4}\right]^{1/3}}{\mathrm{Ra}_D^{1/4}}.$$

The diameter of the cylinders is $D = 10$–76 mm. An increase in the coefficient of heat transfer of a rough cylinder amounts to 7–10% in comparison with a smooth cylinder.

The Pr and Ra_D numbers change in the ranges $\mathrm{Pr} = 0.01$–10^3, $\mathrm{Ra}_D = 4 \cdot 10^4$–$2 \cdot 10^8$. An increase in the effective surface area by means of fine knurling was 10%, and rough knurling was up to 51%.

The relative heat transfer coefficient ***of a long horizontal isothermal cylinder*** of diameter $D = 21.5$ mm ***with circular and longitudinal grooves*** of width $b = 1$ mm, depth $t = 0.5$ mm, and spacing $B = 3.2$–7.5 mm [3.48] ***in air*** is calculated from the formula

for circular grooves 25–30 in number

$$\frac{\mathrm{Nu}}{\mathrm{Nu}_0} = \frac{b}{B+b}\left[\frac{B}{b} + 0.98\left(\frac{B+b}{b}\right)^{0.3}\right], \tag{3.2.27}$$

for longitudinal grooves 8–16 in number ($\mathrm{Gr}_D = 2 \cdot 10^4 - 10^8$)

$$\frac{\mathrm{Nu}}{\mathrm{Nu}_0} = 1 + \frac{7.38b}{B+b}\mathrm{Gr}_D^{-1/4}, \tag{3.2.28}$$

The increase in the heat transfer coefficient was 1–7%.

Numerical calculation of free-convective heat transfer ***on a finned horizontal cylinder of constant temperature*** was made in [3.49]. The results of calculation of the average Nusselt number for various values of the number of edges, dimensionless lengths of the edges $\gamma = l/D$, and Ra number are given in Table 3.2.3.

Free convection from a horizontal ***isothermal cylinder*** with the temperature T_w in a ***layer of water*** of height H with temperature T_∞. The cylinder is located at the height $H_1 < H$ from a free surface. The jet of heated liquid rises upwards and along the free surface with the temperature T_s and forms a thermal boundary layer. For calculating heat transfer, the experimental dependence of [3.50] is used with an error of 19%:

$$\mathrm{Nu} = 0.525(1 + 16.75\, P_s^{0.295})\, \mathrm{Ra}_D^{0,25}, \tag{3.2.29}$$

where $P_s = \beta(T_\mathrm{w} - T_s)D/(H_1 + D/2)$.

The parameters of the problem change in the ranges $H/D = 3.8 - 15$, $\mathrm{Pr}_\infty = 5 - 7$, $\mathrm{Ra}_D = 9.8 \cdot 10^4 - 1.1 \cdot 10^6$, $P_s = 4.3 \cdot 10^{-5} - 5 \cdot 10^{-4}$.

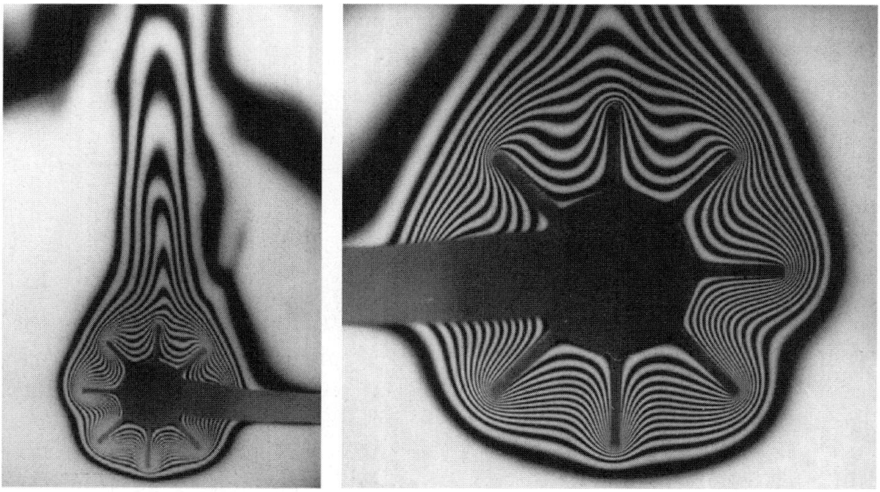

Fig. 3.2.5. Interferograms of free convection on a finned cylinder of constant temperature

Table 3.2.3.

Ra	γ	0	1	3	5	11
10^3	0.25	2.967	2.676	2.303	2.203	2.141
	0.5	2.967	2.433	1.863	1.769	1.721
10^4	0.25	4.788	4.330	3.382	3.032	2.819
	0.5	4.788	4.052	2.716	2.274	2.095
10^5	0.25	8.065	7.509	5.579	4.655	3.668
	0.5	8.065	7.221	5.150	4.388	2.549
10^6	0.25	13.724	13.177	–	–	7.130
	0.5	13.724	13.004	–	–	6.072

The relative value of the local heat transfer coefficient ***on the circumference of a horizontal cylinder in a stratified medium*** with laminar free convection is determined by the dimensionless relation [3.51]

$$\frac{\mathrm{Nu}\,(x)}{\mathrm{Nu}\,(R)} = \frac{(T_\mathrm{w} - T_\infty)^{17/12}}{(T_\mathrm{w} - T_\infty)_0} (F/G)^{1/4} \sin^{1/3} X \,, \tag{3.2.30}$$

$$F = \exp\left[\frac{4}{3}(k+1)\int\limits_0^X \frac{dT_\infty/dX}{(T_\mathrm{w} - T_\infty)} dX\right] ,$$

$$G = \int\limits_0^X \sin^{1/3} X \,(T_\mathrm{w} - T_\infty)^{5/3} F dX \,,$$

$$T_\infty = T_\infty(z), \quad z = R[1 - \cos(X)], \quad X = x/R \,.$$

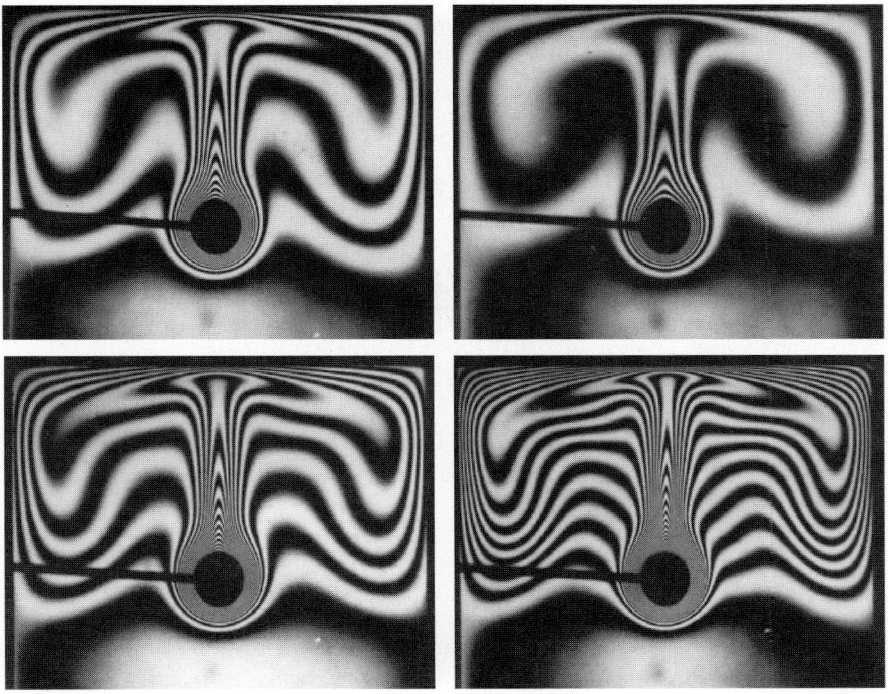

Fig. 3.2.6. Dynamics of temperature field development for natural convection near a horizontal cylinder with a constant heat flux on the surface in the space limited by a rectangular cavity

The deviation of the calculated data from experiment does not exceed 6%. The dependence of the coefficient k on the Pr number is given in Table 3.2.4.

Table 3.2.4.

k	2	2.5	3.5
Pr	10	5	0.7

Distribution of the local Nusselt number on a horizontal cylinder ***of constant temperature located in a rectangular cavity near the upper wall*** is given in [3.52].

The average heat transfer coefficient ***of an isothermal cylinderin air*** for laminar free convection ***with volumetric heat sources*** in a boundary layer depends on the thickness of the latter δ. In the range $\mathrm{Gr}_D = 10^4$–10^9 [3.53]

$$\mathrm{Nu} = 1.5\ \mathrm{Gr}_D^{1/4}/\delta\ . \qquad (3.2.31)$$

The stength the volumetric heat sources is proportional to: 1) the temperature difference $Q_v = \gamma\theta$ and 2) the temperature gradient $Q_v = -\gamma_1 \lambda \partial\theta/\partial y$. The conditions of heat transfer are defined by the criteria $\mathrm{Gr}_D = g\beta\theta_\mathrm{w} D^3/\nu^2$, Pr, and $\mathrm{Os}_D = Q_{\mathrm{v}0} D^2/(\lambda\theta_\mathrm{w})$. The thickness δ for these two cases can be found from the equations

(1)

$$\begin{aligned}
&\frac{0.000096}{\mathrm{Gr}_D^{1/2}}\delta^{10} + \frac{0.000192}{\mathrm{Gr}_D^{1/4}}\delta^9 + 0.00012\delta^8 - 0.014444\frac{\mathrm{Os}_D}{\mathrm{Gr}_D^{3/4}}\delta^7 \\
&- 0.01933\frac{\mathrm{Os}_D}{\mathrm{Gr}_D^{1/2}}\delta^6 - \frac{0.187938}{\mathrm{Gr}_D^{1/4}}\delta^5 + \left(2.219285\frac{\mathrm{Os}_D^2}{\mathrm{Gr}_D} - 0.359154\right)\delta^4 \\
&+ 22.043326\frac{\mathrm{Os}_D}{\mathrm{Gr}_D^{1/2}} + 52.664747 = 0\,,
\end{aligned}$$

(2)

$$\begin{aligned}
&\frac{0.000096}{\mathrm{Gr}_D^{1/2}}\delta^{10} + \frac{0.000192}{\mathrm{Gr}_D^{1/4}}\delta^9 + 0.00012\delta^8 + 0.038512\frac{\mathrm{Os}_D}{\mathrm{Gr}_D^{1/2}}\delta^6 \\
&- \left(\frac{0.187938}{\mathrm{Gr}_D^{1/4}} - 0.05055\frac{\mathrm{Os}_D}{\mathrm{Gr}_D^{1/4}}\right)\delta^5 - 0.359154\delta^4 + 15.781581\frac{\mathrm{Os}_D^2}{\mathrm{Gr}_D^{1/4}}\delta^2 \\
&- 58.782203\frac{\mathrm{Os}_D}{\mathrm{Gr}_D^{1/4}}\delta + 52.664747 = 0\,.
\end{aligned}$$

The influence of viscous dissipation in calculating heat transfer from horizontal cylinders in the ranges of the parameters $\mathrm{Pr} = 0.7 - 4 \cdot 10^4$ and $\mathrm{Ra}_D = 10^{-8} - 10^8$ is taken into account in the correlation formula [3.54]

$$\begin{aligned}
\mathrm{Nu} = {} & 0.4\,\mathrm{Pr}^{0.0432}\,\mathrm{Ra}_D^{0.25} + 0.503\,\mathrm{Pr}^{0.0334}\,\mathrm{Ra}_D^{0.0816} \\
& + 0.958\,\mathrm{Ge}^{0.122}/\mathrm{Pr}^{0.06}\,\mathrm{Ra}_D^{0.0511}\,,
\end{aligned} \tag{3.2.32}$$

where $\mathrm{Ge} = g\beta D/c_p$ is the parameter of viscous dissipation.

The heat transfer rate ***of a horizontal cylinder in the presence of injection or suction of the same gas on the surface*** is given by the following equations [3.36]:

the temperature of the surface is constant

$$\frac{\mathrm{Nu}}{\mathrm{Nu}_0} = 1 - 0.55\frac{\mathrm{Pe}_{D\mathrm{w}}}{\mathrm{Gr}_D^{1/4}} + 0.062\left(\frac{\mathrm{Pe}_{D\mathrm{w}}}{\mathrm{Gr}_D^{1/4}}\right)^2, \tag{3.2.33}$$

$$\frac{\mathrm{Pe}_{D\mathrm{w}}}{\mathrm{Gr}_x^{1/4}} = (-1.72) - 1.35, \quad \mathrm{Gr}_D = 10^5 - 10^8\,,$$

the heat flux density on the surface is constant

$$\frac{\mathrm{Nu}}{\mathrm{Nu}_0} = 1 - 0.28\frac{\mathrm{Pe}_D}{\mathrm{Gr}_D^{1/5}} + 0.037\left(\frac{\mathrm{Pe}_D}{\mathrm{Gr}_D^{1/5}}\right)^2 , \tag{3.2.34}$$

$$\frac{\mathrm{Pe}_D}{\mathrm{Gr}_x^{1/5}} = -0.86 - 0.686, \quad \mathrm{Gr}_D = 10^5 - 10^7 .$$

The results of numerical investigation of the local heat transfer on a horizontal cylinder ***of constant temperature*** are given in Table 3.2.5 [3.55].

Numerical calculations of heat transfer of a horizontal cylinder ***in a limited space*** (in the presence, at some distance, of a flat inclined wall) show that the coefficient of heat transfer for small distances increases due to heat conduction and decreases at large distances when the prevailing mechanism of heat transfer is convection [3.56, 3.57].

Table 3.2.5.

Ra	Pr	$\vartheta = 0°$	30	60	90	120	150	180	Nu_D
10^0	0.7	1.41	1.37	1.25	1.08	0.87	0.68	0.56	1.04
10^1	0.7	1.83	1.79	1.67	1.47	1.21	0.94	0.81	1.40
10^2	0.7	2.71	2.66	2.51	2.23	1.80	1.27	0.97	2.05
10^3	0.7	3.89	3.85	3.72	3.45	2.93	2.01	1.22	3.09
10^4	0.7	6.24	6.19	6.01	5.64	4.82	3.14	1.46	4.94
10^5	0.7	10.15	10.03	9.65	9.02	7.91	5.29	1.72	8.00
10^6	0.7	16.99	16.78	16.18	15.19	13.60	9.38	2.12	13.52
10^7	0.7	29.41	29.02	27.95	26.20	23.46	16.48	2.51	23.32
10^4	0.01	3.63	3.56	3.17	2.51	1.74	1.13	0.93	2.40
10^4	0.1	5.25	5.16	4.89	4.34	3.26	1.84	1.12	3.78
10^4	1.0	6.40	6.33	6.10	5.69	4.91	3.36	1.48	5.06
10^4	5.0	6.89	6.82	6.59	6.19	5.55	4.35	1.74	5.66
10^4	10.0	7.01	6.93	6.69	6.29	5.71	4.67	1.79	5.81

The heat transfer ***of an electrically heated horizontal cylinder*** in the presence of two vertical unheated walls with the temperature equal to the temperature of the environment is calculated from the generalized experimental dependence [3.58]

$$\mathrm{Nu} = 0.06\ \mathrm{Ra}_D^{0.21}\,(H/D)^{0.14}\,(t/D)^{-0.05} , \tag{3.2.35}$$

$$H/D = 5 - 128, \quad t/D = 2 - 20, \quad \mathrm{Pr} = 0.7 - 10 ,$$

where t is the distance between the walls ($t/D > 1$) and H is the height of the walls. The increase in the heat transfer coefficient is appreciable even at large distances between the walls. For example, when $t/D = 20$, heat transfer coefficient grows by 30%.

Example 3.2. To heat a room, the amount of heat $Q = 1\,\text{kW}$ is required from a horizontal pipeline with $D = 25\,\text{mm}$. The temperature of the heater surface is $t_{\text{w}} = 100°\text{C}$, the temperature of air in the room is $t_\infty = 20°\text{C}$. The length of the pipeline is to be calculated.

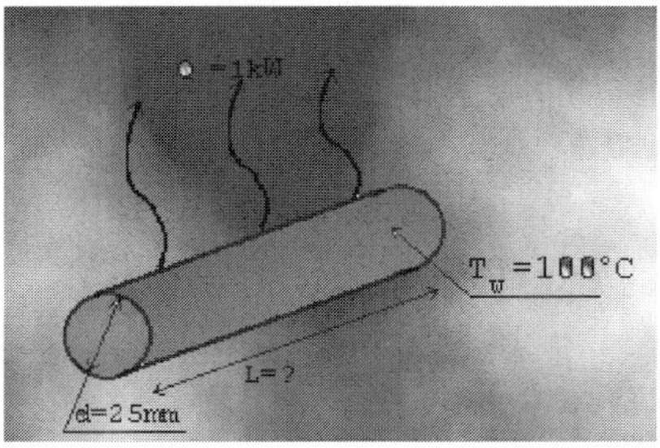

The determining temperature of air is $t_{\text{av}} = (100 + 20)/2 = 60°\text{C}$. For the temperature t_{av} we have

$\beta = 1/333.16\ 1/\text{deg}$,
$\nu = 18.97 \cdot 10^{-6}\ \text{m}^2/\text{sec}$,
$\lambda = 2.9 \cdot 10^{-2}\ \text{W/m} \cdot \text{deg}$,
$\text{Pr} = 0.696$.

The basic determining parameters are

$$\text{Gr}_D = \frac{9.81 \cdot (2.5)^3\, 10^{-6}}{333 \cdot (18.97)^2\, 10^{-12}} 80 = 1.025 \cdot 10^5, \quad \text{Ra}_D = 0.713 \cdot 10^5 \ .$$

The Nu number is given by the formula

$$\text{Nu}^{1/2} = 0.6 + \left[\left(0.713 \cdot 10^5/300\right) \Big/ \left\{ 1 + \left(\frac{0.5}{0.696}\right)^{9/16} \right\}^{16/9} \right]^{1/6}, \quad \text{Nu} = 7.2 \ .$$

The coefficient of heat transfer from a lateral surface is

$$\alpha = \frac{\text{Nu}\lambda}{D} = \frac{7.2 \cdot 2.9 \cdot 10^{-2}}{2.5 \cdot 10^{-2}} = 8.35\ \text{W/m}^2 \cdot \text{deg} \ .$$

The length of the pipeline can be found from the heat balance equation:

$$L = Q/\alpha\theta_{\text{w}}\pi D = 1 \cdot 10^3 / \left(8.38 \cdot 80 \cdot 3.14 \cdot 2.5 \cdot 10^{-2}\right) = 19.1\,\text{m} \ .$$

From formula (3.2.11) $L_2 = 19.15$ m, from (3.2.12) $L_3 = 19.17$ m.

3.3 Inclined Cylinder

Free-convective flows near an inclined cylinder are distinguished by the three-dimensional character of flow caused by the presence of the axial component of velocity (being the consequence of the appearance of the axial component of the buoyancy force). The increase in the thickness of the boundary layer with inclination of the cylinder leads to reduction of the convective heat transfer component. Among the general laws governing the flow, we can note the decrease in the coefficient of heat transfer on increase in the inclination angle φ reckoned from the horizontal [3.28]. The heat transfer of the inclined cylinder near its bottom is determined by the value of the velocity characteristic for the flow along the cylinder. On its upper part the flow approaches the case of the flow past a horizontal cylinder. For small inclination angles this effect is insignificant. For example, when the angle increases from 0 to 45°, Nu decreases by 8%. When the axis of the cylinder approaches the vertical, its heat transfer rate decreases rapidly [3.14]. The boundary-layer equations for a three-dimensional flow can be reduced to a two-dimensional problem using the assumption concerning the dependence of the main flow on the degree of the contour curvature [3.59–3.61], which is confirmed by numerical calculations of the flow past an inclined cylinder with a constant temperature and by flow visualization.

The Grashof number for an inclined cylinder is determined by the formula

$$\mathrm{Gr}_D = g\beta\,(T_0 - T_\infty)\,D^3 \cos\varphi/\nu^2\ . \tag{3.3.1}$$

For the number Pr = 0.7 and axial length of the heated part L we can use the following dimensionless dependence [3.28]:

$$\begin{aligned} L^* > 10, &\quad \mathrm{Nu}_D/\left(\mathrm{Gr}_D \cos\varphi\right)^{1/4} = 0.42\ ,\\ L^* < 1, &\quad \mathrm{Nu}_D/\left(\mathrm{Gr}_D \cos\varphi\right)^{1/4} = 0.55\ ,\\ 1 < L^* < 10, &\quad \mathrm{Nu}_D/\left(\mathrm{Gr}_D \cos\varphi\right)^{1/4} = 0.42\left[1 + \left(1.31/L^{*0.25}\right)^8\right]^{0.125}\ ,\\ &\quad L^* = L/(D\,\mathrm{tg}\varphi)\ . \end{aligned} \tag{3.3.2}$$

For large values of L^* or of the axial coordinate, the flow is almost two-dimensional and is characterized by two components of velocity u and v. The heat transfer coefficient in this case practically coincides that in the case of flow past a horizontal cylinder. At small values of L^* and near the lower part of the cylinder of any length the effective coefficient of heat transfer coincides with the coefficient for flow past a vertical cylinder.

The application of the boundary-layer method to three-dimensional problems at large Prandtle numbers for various bodies gives the generalized formula for the average heat transfer coefficient [3.62]:

$$\mathrm{Nu} = 0.5402\left(|\mathrm{Gr}_L\,\mathrm{Pr}|\right)^{1/4} K\ , \tag{3.3.3}$$

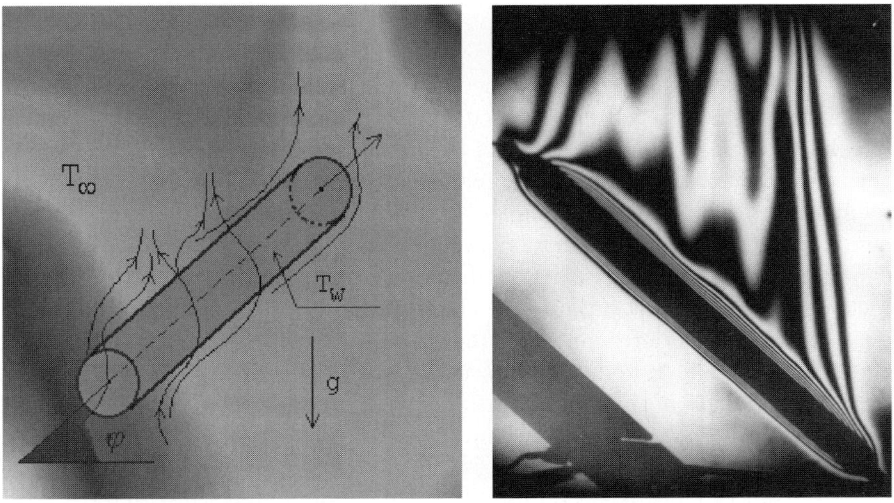

Fig. 3.3.1. Structure of streamlines and interferogram of free-convective flow near an inclined cylinder with a constant heat flux density on the surface

where K is the coefficient dependent on the geometrical configuration of the problem. In particular, ***for an inclined cylinder with the closed ends***

$$K \to 0.9581\,(\cos\varphi)^{1/4}\,, \quad \left(\frac{L}{D}\cos\varphi \to \infty\right)\,,$$

$$K \to \left(\frac{4}{3}\right)^{3/4}\left(\frac{L}{D}\right)^{-1/4}(\sin\varphi)^{1/4}\,, \quad \left(\frac{L}{D}\cos\varphi \to 0\right)\,.$$

The given dependence with an error of 3% describes also experimental data for an inclined wire ***in glycerine*** [3.63].

The experimental data for the average coefficient of heat transfer from a cylinder ***in air***, when the position of the cylinder changes from a horizontal to a vertical one, are generalized by the dimensionless relation [3.64]

$$\mathrm{Nu}_D/\left(\mathrm{Gr}_D\cos\varphi\right)^{1/4} = 0.42\left(1 + 8.673/L^{*2}\right)^{1/8}\,, \tag{3.3.4}$$

$$\mathrm{Gr}_D = 4\cdot10^4 - 9\cdot10^4\,, \quad L/D = 8, 10.67, \quad \text{and } 16\,.$$

When $L^* > 10$, the heat transfer coefficient practically coincides with the coefficient for the flow past a horizontal cylinder, and when $L^* < 1$, it tends to its value for a vertical cylinder.

The local value of the heat transfer coefficient for an inclined cylinder of constant temperature is calculated, with an error of ±4.5%, as [3.65]

$$\mathrm{Nu}(x) = \left[0.545 - 0.387(\cos\varphi)^{1.462}\right]\mathrm{Ra}^{1/4+\cos\varphi^{1,75}/12}(x)\,, \tag{3.3.5}$$

$$\mathrm{Pr} = 0.7, \quad \mathrm{Ra}_k = 1.482\cdot10^8 + 4.5\cdot10^8\mathrm{ctg}\varphi\,,$$

$$L/D = 25; \quad \varphi = 0 - 90°; \quad T_{\mathrm{av}} = \left(T_{\mathrm{W}} + T_\infty\right)/2\,.$$

The average value of the heat transfer coefficient ***for an inclined cylinder of constant temperature in air*** is [3.64, 3.65]

$$\mathrm{Nu} = 0.42\left[1 + \left(1.31/L^{*0.25}\right)^8\right]^{1/8} \mathrm{Gr}_D^{0.25}\,, \tag{3.3.6}$$
$$\mathrm{Pr} = 0.7, \quad \mathrm{Gr}_D = 10^4 - 10^5\,,$$
$$L/D = 8 - 16, \quad T_{\mathrm{av}} = (T_{\mathrm{W}} + T_\infty)/2,$$
$$L^* = L\mathrm{tg}\varphi/D = 1 - 10\,,$$

and also

$$\mathrm{Nu} = 0.6 - 0.488(\cos\varphi)^{1.03}\,\mathrm{Ra}_L^{1/4+(\cos\varphi)^{1,75}/12}\,, \tag{3.3.7}$$
$$\mathrm{Pr} = 0.7, \quad \mathrm{Ra}_L = 2\cdot 10^6 - 1.2\cdot 10^9\,,$$
$$L/D = 25, \quad T_{\mathrm{av}} = (T_{\mathrm{W}} + T_\infty)/2.$$

It is shown experimentally that transition from a laminar to a turbulent regime occurs at $\mathrm{Ra}_{Lk} = (1-2)\cdot 10^9$ [3.60, 3.61].

The experimental data on heat transfer rate ***from isothermal inclined cylinders in air***, with an error of no more than 8%, are generalized as follows [3.66]:

for a laminar regime

$$\mathrm{Pr} = 0.7, \quad \mathrm{Ra}_L = 2\cdot 10^6 - 1.2\cdot 10^9\,,$$
$$L/D = 25, \quad T_{\mathrm{av}} = (T_{\mathrm{w}} + T_\infty)/2\,,$$

the average Nu number

$$\mathrm{Nu} = \left[2.9 - 2.23\,(\sin\varphi)^{0.8}\right](\mathrm{Gr}_D)^{-1/12}\,(\mathrm{Gr}_L\,\mathrm{Pr})^{1/4+\frac{1}{12}(\sin\varphi)^{1.2}}\,, \tag{3.3.8}$$

the local Nu number

$$\mathrm{Nu}(x) = \left[2.3 - 1.72(\sin\varphi)^{0.8}\right](\mathrm{Gr}_D)^{-1/12}\,(\mathrm{Gr}_x\,\mathrm{Pr})^{1/4+\frac{1}{12}(\sin\varphi)^{1.2}}\,, \tag{3.3.9}$$

for a turbulent regime

$$\mathrm{Gr}_D = 1.08\cdot 10^4 - 6.9\cdot 10^5, \quad \varphi = 0 - 90^\circ\,,$$

the average Nu number

$$\mathrm{Nu} = \left[0.47 + 0.11\,(\sin\varphi)^{0,8}\right](\mathrm{Gr}_D)^{-1/12}\,(\mathrm{Gr}_L\,\mathrm{Pr})^{1/3}\,, \tag{3.3.10}$$

the local Nu number

$$\mathrm{Nu}\,(x) = \left[0.42 + 0.16\,(\sin\varphi)^{0,8}\right](\mathrm{Gr}_D)^{-1/12}\,(\mathrm{Gr}_x\,\mathrm{Pr})^{1/3}\,. \tag{3.3.11}$$

The average value of the heat transfer coefficient ***for an inclined wire with electric heating*** is [3.67]

$$\mathrm{Pr} = 0.7, \quad T_{\mathrm{av}} = (T_{\mathrm{w}} + T_\infty)/2, \quad \mathrm{Ra}_D = 4 \cdot 10^{-4} - 10^{-3}, \quad (3.3.12)$$
$$L/D = 5.4 \cdot 10^3, \quad \varphi = 0 - 84.6^\circ,$$
$$\mathrm{Nu} = 1/[1 - 0.175 \ln \mathrm{Ra}_D] .$$

The correlation formula for calculating laminar free convection ***for an inclined long cylinder in air*** for $\varphi < \mathrm{arctg}\,(L/D)$ is obtained on the basis of the experimental data of [3.68]:

$$\mathrm{Nu} = 0.515\,(\mathrm{Pr}\ \mathrm{Gr}_D)^{0.25}, \qquad (3.3.13)$$
$$\varphi = 0 - 75^\circ, \quad \mathrm{Pr}\ \mathrm{Gr}_D = 4.17 \cdot 10^5 - 2.75 \cdot 10^7 .$$

Example 3.3. How will the length of pipelines change in the previous example if the pipe is set at an angle of 45° to the horizon?

The heat transfer coefficient of the cylinder changes insignificantly on deviation from the horizontal up to 45°, therefore in the first approximation we may assume that the length of the pipeline is about 19.1 m. Then

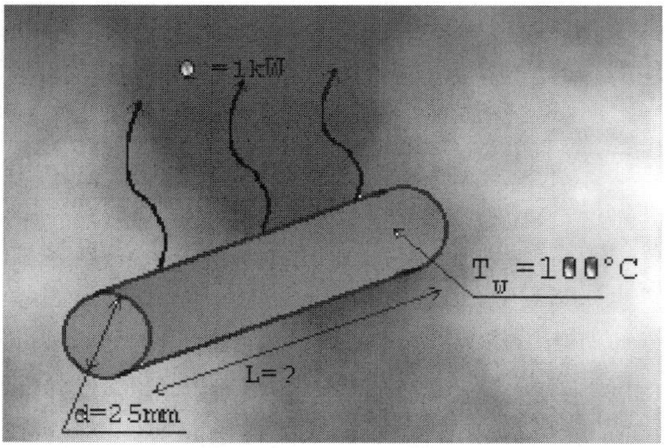

$$L^* = \frac{L}{D}\mathrm{tg}\varphi = \frac{19.1}{2.5 \cdot 10^{-2}} = 764 \gg 10 .$$

The influence of the buoyancy force component directed parallel to the axis of the cylinder is negligibly small, therefore

$$\mathrm{Nu} \cong 0.42\ \mathrm{Gr}_{D\varphi}^{0.25} = 0.42 \left(\frac{9.81 \cdot (2.5)^3 \cdot 10^{-6}}{333 \cdot (18.97)^3 \cdot 10^{-12}} 80 \cdot 0.707 \right)^{1/4} = 6.93 .$$

The coefficient of heat transfer from the lateral surface is

$$\alpha = \lambda \mathrm{Nu}/D = 6.93 \cdot 2.9 \cdot 10^{-2}/2.5 \cdot 10^{-2} = 8.04\ \mathrm{W/m^2} \cdot \mathrm{deg} .$$

The length of the pipeline is $1 \cdot 10^3 (8.04 \cdot 80 \cdot 3.14 \cdot 2.5 \cdot 10^{-2})^{-1} = 19.83\,\mathrm{m}$, that is by 3.8% larger than the pipeline placed horizontally. The second approximation is not necessary, since in the example considered $L^* \gg 10$.

3.4 Cone

The boundary-layer theory approximation for investigation of axisymmetric cone-shaped surfaces is applicable, when the thickness of the boundary layer d is small in comparison with the local radius of curvature. The cone is one of the axisymmetric bodies admitting self-similar solutions.

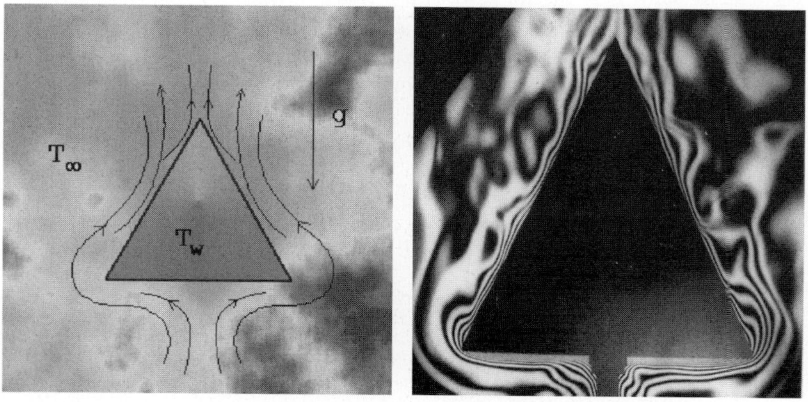

Fig. 3.4.1. Structure of streamlines and interferogram of the temperature field for natural convection near a vertical cone of constant temperature

In the case of ***power-law distribution of the temperature of the cone surface*** (the apex angle is $2\phi \leq \pi$)$T_\mathrm{w} - T_\infty = Nx^m$, the local value of the Nusselt number can be determined from the relation [3.10]

$$\frac{\mathrm{Nu}_x}{\mathrm{Gr}_x^{1/4}} = -h^{-5/4} \sum_{n=0}^{\infty} \theta'_n(0)\, \xi^n \,, \tag{3.4.1}$$

where

$$\xi = \frac{2}{c_1 \tan\phi} x^{-\frac{m+3}{4}}, \quad c_1 = \left(\frac{g\beta N L^3 \cos\phi}{\nu^2}\right)^{1/4}$$

and L is the length of the cone genarator.

For a cone of constant temperature ($m = 0$, $h = 1$), the values of the coefficients $\theta'_0(0)$ are presented in Table 3.4.1.

On the basis of formula (3.4.1) ***for an isothermal vertical cone*** at $\mathrm{Pr} = 0.7$ the dependence of the average Nusselt number on the parameter of curvature ε was obtained [3.69]:

$$\frac{\mathrm{Nu}}{\mathrm{Gr}_L^{1/4}} = 0.51554 + 0.412\varepsilon + 0.022\varepsilon^2 \,, \tag{3.4.2}$$

where $\varepsilon = \frac{2}{\mathrm{Gr}_L^{1/4} \mathrm{tg}\,\phi}$.

Table 3.4.1.

Pr	$\theta_0'0$	$\theta_1'0$	$\theta_2'0$
0.7	−0.451095	−0.20605	−0.0028
1	−0.510400	−0.20547	−0.0010

Numerical solutions of the boundary-layer equations ***for an isothermal surface*** ($\boldsymbol{m = 0}$) ***and for a linear law of change in the surface temperature*** ($\boldsymbol{m = 1}$) at $\Pr = 0.7$ [3.70] result in the following formulas for calculating the local heat transfer coefficient:

$$m = 0: \qquad \mathrm{Nu}\,(x) = 0.4511\ \mathrm{Gr}_x^{1/4}\ , \tag{3.4.3}$$

$$m = 1: \qquad \mathrm{Nu}\,(x) = 0.56699\ \mathrm{Gr}_x^{1/4} \tag{3.4.4}$$

and the average heat transfer coefficient:

$$m = 0: \qquad \mathrm{Nu} = 0.51554\ \mathrm{Gr}_L^{1/4}\ , \tag{3.4.5}$$

$$m = 1: \qquad \mathrm{Nu} = 0.37799\ \mathrm{Gr}_L^{1/4}\ , \tag{3.4.6}$$

For the power-law distribution of temperature on the cone surface $T_\mathrm{w} - T_\infty = (T_0 - T_\infty)x^m$, the local heat transfer coefficient is calculated from the formula [3.71]

$$\frac{\mathrm{Nu}\,(x)}{(\mathrm{Gr}_x\ \Pr^2)^{1/2}} = -\theta'(0) \tag{3.4.7}$$

and the average heat transfer coefficient from

$$\frac{\mathrm{Nu}}{(\mathrm{Gr}_L\ \Pr^2)^{1/2}} = \frac{8}{5m+7}[-\theta'(0)]\ . \tag{3.4.8}$$

The values of $\theta'\ (0)$ for different m and Pr are given in Table 3.4.2

The local and average values of the heat transfer coefficient ***for a vertical cone of height H for the power-law change in the wall temperature*** $\theta_\mathrm{w} = \theta_0 x^m$ in a laminar regime is calculated, with an error of 9%, from the formulas [3.69, 3.70, 3.72]

$$\mathrm{Nu}\,(x)\,/\mathrm{Gr}_x^{1/4} = 2/a_0\ , \tag{3.4.9}$$

$$\frac{\mathrm{Nu}}{\mathrm{Gr}_L^{1/4}} = \frac{8}{(5m+7)}\{\{(5m+7)[1+2a_0\varepsilon/(4-m)]^3\ \Pr^2\}/A\}^{1/4}\ , \tag{3.4.10}$$

where

$$A = \{63\,(m+3)\,[1+2.4a_0\varepsilon/\,(4-m)] + (5m+7)\,[1+2a_0\varepsilon/\,(4-m)]\,\Pr\}\ ,$$

$$a_0 = \left[1008\frac{m+3+(5m+7)\Pr}{(5m+7)^2\ \Pr^2}\right]^{1/4}\ .$$

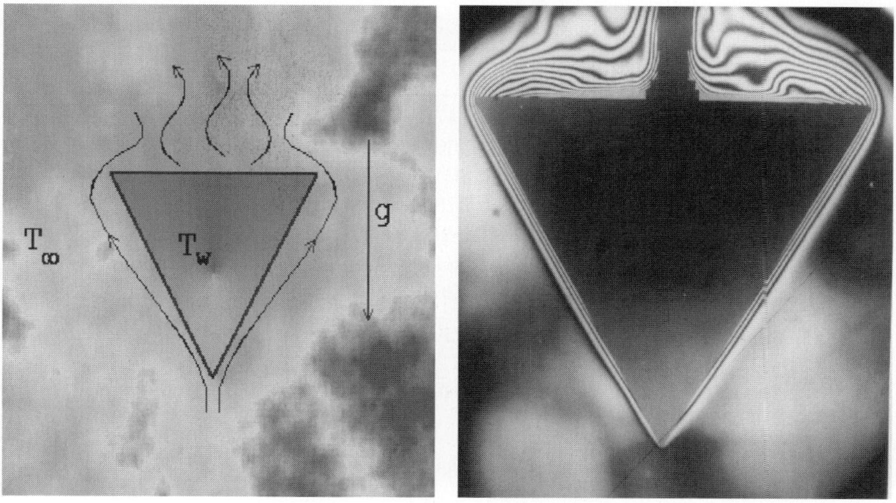

Fig. 3.4.2. Structure of streamlines and interferogram of free-convective flow near a vertical cone of constant temperature

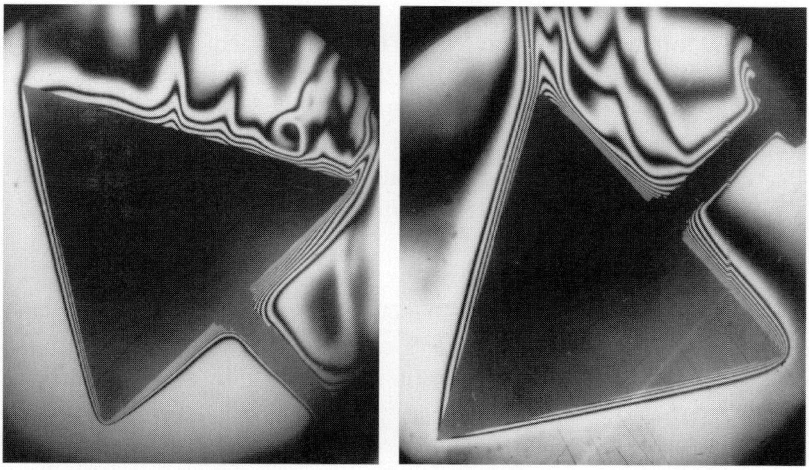

Fig. 3.4.3. Interferograms of free-convective flow near an inclined cone of constant temperature

The case of $m = 0.2$ corresponds to the constant value of the heat flux density on the surface [3.72].

The empirical dependences for calculating heat transfer ***in air for a cone with a small apex angle and a rounding at the end*** (error is 3.5%) are [3.73]

$$\mathrm{Nu} = 0.63\,(1 + 0.72\varepsilon)\,\mathrm{Gr}_L^{1/4}\,, \qquad (3.4.11)$$
$$2\phi = 3.5 - 11.5^\circ, \quad \mathrm{Gr}_L = 10^7 - 10^9$$

Table 3.4.2.

Pr	0	0.001	0.003	0.01	0.03	0.1	0.7	1
$m = 0$	0.7916	0.7803	0.7669	0.7475	0.7185	0.6683	0.5392	0.5104
−0.5	0.5920	0.5863	0.5780	0.5667	0.5493	0.5182	0.4304	0.4094
0.2	0.8567	0.8423	0.8280	0.8058	0.7726	0.7159	0.5730	0.5148
1	1.063	1.037	1.020	0.9886	0.9415	0.8663	0.6777	0.6389
2	1.245	1.210	1.189	1.149	1.089	0.9919	0.7693	0.7240
4	1.491	1.444	1.416	1.364	1.287	1.165	0.8945	0.8406
8	1.792	1.733	1.696	1.629	1.533	1.381	1.053	0.9889

and ***for a cone with a large apex angle with a sharp end*** are [3.74]

$$\mathrm{Nu}\,(x) = 0.49\left[(x\mathrm{tg}\,\phi/L)^{-3/4}\,\mathrm{Gr}_x\right]^{1/4}, \tag{3.4.12}$$

$$\mathrm{Nu} = 0.6\left[(\mathrm{tg}\,\phi)^{-3/4}\,\mathrm{Gr}_L\right]^{1/4},$$

$$2\phi = 30^\circ, 60^\circ, 90^\circ, \quad \mathrm{Gr}_L = 10^4 - 10^9, \quad \mathrm{Pr} = 0.7\,. \tag{3.4.13}$$

The local values of the heat transfer coefficient $\mathrm{Nu}_{x'}/\mathrm{Gr}_{x'}^{1/4}$ ***for a cone truncated at the distance x_0 along the cone generator for a constant temperature of the surface*** are given in Table 3.4.3. The coordinate $x' = x - x_0$ is reckoned from the base of the truncated cone [3.75–3.77].

The change in the temperature along the height of the cone $\theta_\mathrm{w}\lambda\mathrm{Gr}_{x\phi}^{*1/5}/(q_\mathrm{w}x')$ ***at a constant heat flux on the surface*** is given in Table 3.4.4

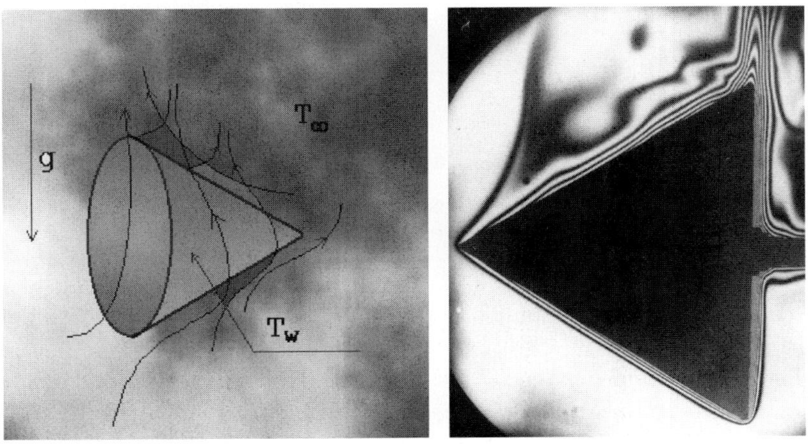

Fig. 3.4.4. Structure and interferogram of free-convective flow near a horizontal cone of constant temperature

Table 3.4.3.

Pr	0.01	0.7	1	10	100
$x'/x_0 = 0$	0.05742	0.35320	0.40110	0.82690	1.54930
0.25	0.06135	0.37202	0.42185	0.86336	1.61215
0.75	0.06536	0.39204	0.44404	0.90293	1.68088
1.75	0.06895	0.41366	0.46830	0.95023	1.76690
3.75	0.07148	0.42893	0.48525	0.98229	1.82496
7.75	0.07307	0.43958	0.49749	1.00866	1.87572
15.75	0.07397	0.44462	0.50312	1.01835	1.89189
31.75	0.07444	0.44822	0.50726	1.02848	1.91286
63.75	0.07468	0.44916	0.50827	1.02889	1.92150
127.75	0.07481	0.45052	0.50987	1.03369	1.92189
∞	0.07493	0.45101	0.51039	1.03397	1.92197

Table 3.4.4.

Pr	0.01	1	10	100
$x'/x_0 = 0$	3.7952	1.8729	1.0589	0.6425
0.25	3.6970	1.8282	1.0363	0.6294
0.75	3.5954	1.7841	0.0142	0.6168
1.75	3.4876	1.7325	0.9855	0.5995
3.75	3.4016	1.6930	0.9642	0.5870
7.75	3.3454	1.6649	0.9473	0.5764
15.75	3.3128	1.6498	0.9393	0.5718
31.75	3.2950	1.6406	0.9333	0.5678
63.75	3.2864	1.6368	0.9317	0.5671
127.75	3.2813	1.6339	0.9295	0.5654
∞	3.2781	1.6329	0.9336	0.5738

The relative rate of heat transfer from ***a vertical cone of constant temperature in the presence of injection or suction of the same gas on the surface*** is given by [3.78]

$$\frac{\mathrm{Nu}\,(x)}{\mathrm{Nu}_0} = 1 - 0.775 \left[\frac{\nu_{\mathrm{w}} x}{2\nu c}\right]^{1/4} ,$$

$$c = \left[\frac{g\beta \cos\phi\,(T_{\mathrm{w}} - T_\infty)}{\nu^2}\right]^{1/4} , \tag{3.4.14}$$

where ν_{w} is the velocity of injection or suction.

The heat transfer of a horizontally located cone with a small apex angle and with the base of diameter D ***in a laminar flow regime*** is described by the relation [3.79]

$$\mathrm{Nu}_D = 0.35 + 0.25\,\mathrm{Ra}_D^{0.125} + 0.45\,\mathrm{Ra}_D^{0.25} , \tag{3.4.15}$$

$$\mathrm{Pr} = 0.7, \quad 2\phi = 3.5 - 11.5^\circ, \quad \mathrm{Gr}_D = 10^4 - 10^9 .$$

3.5 Vertical Needle

The great attention which is paid to free-convective heat transfer on thin wires and needles is due to the widespread application of thermoanemometric methods of measurements in flows. The geometric needle represents a body of revolution $r = r\ (x)$. The basic parameter used also for calculating heat transfer of a thin cylinder is the complex $D\mathrm{Ra}_D/L$ [3.80–3.82].

The heat transfer by free convection in a laminar regime of flow ***past a vertical needle*** is approximated, with an error of 1%, by the generalized dependence [3.83]

$$\mathrm{Nu} = \frac{8}{\pi}/\ln\left[1 + 2.51/C_l\left(\frac{D}{L}\,\mathrm{Ra}_D\right)^{1/4}\right], \qquad (3.5.1)$$

where

$$C_l = \frac{2}{3}/\left[1 + (0.49/\mathrm{Pr})^{9/16}\right]^{4/9}, \quad D/L \le 0.05\,,$$

D is the diameter, L is the length of the needle.

For thin needles, the heat transfer coefficient is higher than for cylinders, because the diameter of the needle is smaller than D and a correction for a thick layer is larger.

The numerical solution of the equations ***for an isothermal needle*** gives the following dependence for the average Nusselt number [3.81]:

$$\mathrm{Nu} = -\frac{8}{3}(C)^{1/2}\,\theta'(C)\,\mathrm{Gr}^{1/4}\,, \qquad (3.5.2)$$

where the Nusselt and Grashof numbers are calculated from the characteristic length L, C is a constant determined from the equation of the surface of a concrete needle.

3.6 Cylinder of Arbitrary Cross Section and Prism

A horizontal cylinder of elliptic section is a generalizing geometry for a vertical and a horizontal plates and horizontal circular cylinder. Depending on the orientation of the large axis $2b = B$ ($2c = C$ is the small axis of the ellipse), there are horizontally (oblate ellipse) and vertically (polate ellipse) located figures.

The average coefficient of heat transfer ***from an isothermal elliptic cylinder*** can be calculated from the approximated dependence [3.37]

$$\mathrm{Nu}_B = 2^{1/4}\left(f_2^{3/4} f_1\right) C\,(\mathrm{Pr})\ \mathrm{Ra}_B^{1/4}\,, \qquad (3.6.1)$$

where

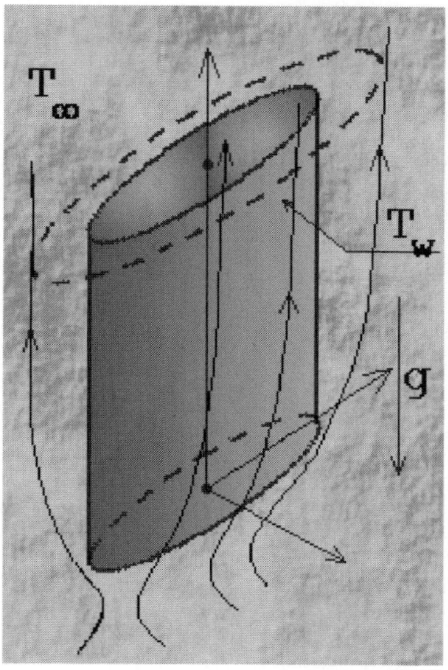

$$C(\mathrm{Pr}) = \frac{2}{3\left[1 + (0.492/\mathrm{Pr})^{9/16}\right]^{4/9}} .$$

From this formula, when $A/B \to 1$ (elliptic cylinder becomes circular), for an isothermal horizontal cylinder with the diameter B we get

$$\mathrm{Nu}_B = 0.77C(\mathrm{Pr})\ \mathrm{Ra}_B^{1/4} . \quad (3.6.2)$$

When $A/B \to 0$, the elliptic cylinder degenerates into a flat plate of height B for which

$$\mathrm{Nu}_B = C(\mathrm{Pr})\ \mathrm{Ra}_B^{1/4} . \quad (3.6.3)$$

The correlation dependence ***for a thin vertically located cylinder*** approximates, within the limits of 5%, the results of numerical and experimental data for the average value of the heat transfer coefficient at arbitrary values of the Pr number and $T_{\mathrm{av}} = (T_{\mathrm{w}} + T_\infty)/2$ in the ranges of the parameters: $\mathrm{Ra}_B = 10^{-2} - 10^{12}$; $C/B = 0 - 1$ [3.37]

$$\mathrm{Nu}^m = \left\{\frac{2\pi k/f_1}{\ln\left[1 + 2^{3/4}\pi k/\left(f_2^{3/4} C_l\ \mathrm{Ra}_B^{1/4}\right)\right]}\right\}^m + \left(0.72\ C_t\ \mathrm{Ra}_B^{1/3}\right)^m , \quad (3.6.4)$$

where

$$m = 3.5\sqrt{C/B + 1}, \quad k = [0.91; (C/B)^{0.041}]_{\max} .$$

Definitions of C_t and C_l are given in the Sect. 3.1, formula (3.2.11). The values of the functions f_1 and f_2 are given in Table 3.6.1.

Table 3.6.1.

C/B	f_1	f_2
0	2	2
0.1	2.032	2.019
0.2	2.101	2.059
0.4	2.301	2.169
0.6	2.553	2.300
0.8	2.836	2.441
0.9	2.978	2.514
1	3.142	2.587

Horizontally located modified (rounded at the edges) triangular and square prisms. The experimental data for the average value of the heat transfer coefficient are generalized by the relation [3.84]

$$\begin{aligned} \mathrm{Pr} = 4 - 6, \quad \mathrm{Ra}_D &= 10^6 - 10^9 \,, \\ T_{\mathrm{av}} &= T_{\mathrm{W}} - (T_{\mathrm{W}} - T_\infty)/4 \,, \\ \mathrm{Nu} &= 0.55 \, \mathrm{Ra}_D^{0.25} \,, \end{aligned} \tag{3.6.5}$$

where the coefficient β is determined from the temperature $(T_{\mathrm{w}} + T_\infty)/2$.

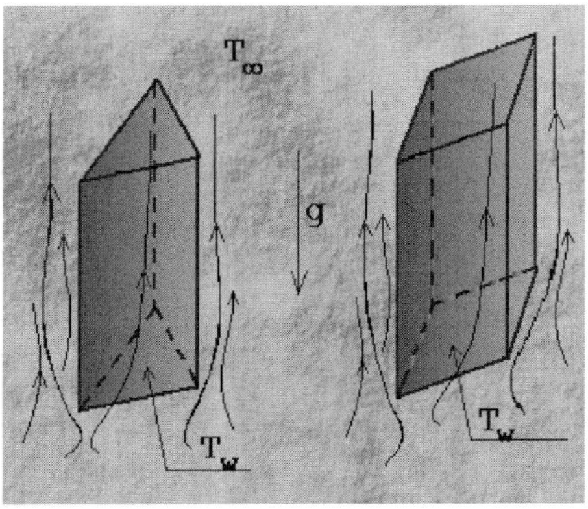

The local values of the heat transfer coefficient $\mathrm{Nu}_x/\mathrm{Gr}_B^{1/4}$ ***for a horizontal elliptic cylinder with a constant temperature of the wall*** for $\mathrm{Pr} = 1$ are presented in Table 3.6.2 [3.85]. When a cylinder is placed horizontally, the value of the dimensionless heat transfer rate in the lower part of the ellipse increases and then, on reaching the maximum, begins to decrease.

The results of numerical calculations of heat transfer on ***a horizontal cylinder of elliptic section with the large axis located vertically*** are given in [3.86] for $\mathrm{Pr} = 0.7$ and $10 < Ra < 1000$.

Table 3.6.2.

C/B	0.1	0.25	0.5	0.75
$\varphi = 0$	0.2369	0.2079	0.3542	0.3920
0.2	0.2382	0.2994	0.3555	0.3925
0.4	0.2421	0.3039	0.3593	0.3940
0.6	0.2490	0.3118	0.3657	0.3961
0.8	0.2599	0.3240	0.3747	0.3986
1.0	0.2767	0.3418	0.3861	0.4004
1.2	0.3031	0.3673	0.3984	0.4006
1.4	0.3497	0.4008	0.4081	0.3975
1.6	0.4142	0.4244	0.4088	0.3897
1.8	0.3731	0.4070	0.3958	0.3766
2.0	0.3206	0.3670	0.3713	0.3585
2.2	0.2748	0.3241	0.3407	0.3364
2.4	0.2361	0.2840	0.3081	0.3112
2.6	0.2030	0.2476	0.2752	0.2838
2.8	0.1735	0.2136	0.2418	0.2535
3.0	0.1444	0.1791	0.2056	0.2186
π	0.1206	0.1504	0.1746	0.1873

The values of the average Nusselt number for heat transfer ***on a finned horizontal cylinder with a constant heat flux density with transverse arrangement of the fins*** can be calculated from the formula [3.87]

$$5 \cdot 10^4 \leq \mathrm{GrPr}\, b/D \leq 10^5$$

$$\mathrm{Nu} = 0.28\,(\mathrm{Gr}_{D,eff}\,\mathrm{Pr}\,b/D)^{1/3}\ , \tag{3.6.6}$$

where $\mathrm{Gr}_{D,eff}$ is the Grashof number calculated from the effective diameter $D_{eff} = D + f$; f is the diameter of a fin, and b is the distance between the nearby fins.

The average value of the heat transfer coefficient for laminar flow of fluid ***past a horizontally located torr*** of inner diameter d and outer diameter D is

for a constant wall temperature [3.88]

$$\mathrm{Nu}/\mathrm{Nu}_0 = 1 + 0.337\, d/D\ , \tag{3.6.7}$$

for a constant heat flux [3.89]

$$\mathrm{Nu}/\mathrm{Nu}_0 = 1 + 0.645\, d/D\ , \tag{3.6.8}$$

where Nu_0 is the value for a horizontal cylinder of diameter d.

Table 3.6.3.

C/B	0.1	0.25	0.5	0.75
$\varphi = 0$	0.4212	0.5150	0.5953	0.8359
0.2	0.4204	0.4828	0.5826	0.7682
0.4	0.4182	0.4733	0.5519	0.6617
0.6	0.4145	0.4596	0.5159	0.5788
0.8	0.4093	0.4436	0.4819	0.5187
1.0	0.4025	0.4271	0.4522	0.4745
1.2	0.3942	0.4108	0.4270	0.4409
1.4	0.3843	0.3952	0.4058	0.4109
1.6	0.3727	0.3804	0.3878	0.3943
1.8	0.3594	0.3661	0.3724	0.3779
2.0	0.3443	0.3519	0.3589	0.3646
2.2	0.3270	0.3373	0.3465	0.3538
2.4	0.3073	0.3212	0.3342	0.3447
2.6	0.2847	0.3022	0.3204	0.3363
2.8	0.2581	0.2784	0.3019	0.3266
3.0	0.2252	0.2463	0.2731	0.3084
π	0.1963	0.2145	0.2407	0.2785

The local values of the heat transfer coefficient $\mathrm{Nu}_x/\mathrm{Gr}_B^{1/4}$ ***for an elliptic cylinder located vertically with a constant wall temperature*** for $\mathrm{Pr} = 1$ are presented in Table 3.6.3 [3.85], from which we can see that the parameter $\mathrm{Nu}_x/\mathrm{Gr}_B^{1/4}$ monotonically decreases, and the thickness of the boundary layer increases.

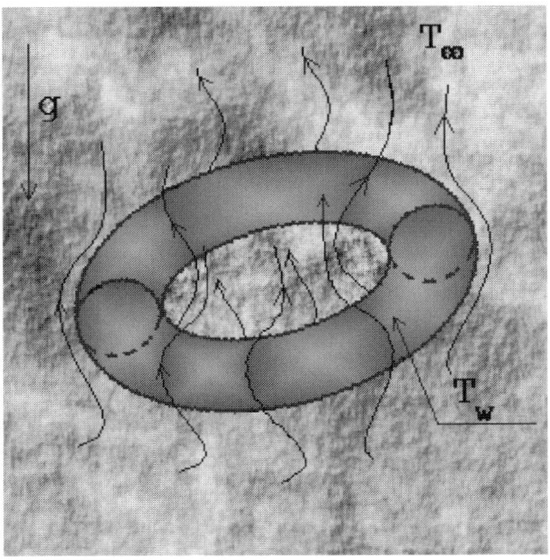

In the case of a constant heat flux,the dimensionless wall temperature $\theta_\mathrm{w} = (\mathrm{Nu}_x/\mathrm{Gr}_B^{1/4})$ for horizontal orientation first decreases, whereas for vertical arrangement it evenly increases. The average value of the heat transfer coefficient for a vertical cylinder is higher than for a horizontal one (Table 3.6.4). It coincides with the experimental data ***for air*** for laminar flow past an elliptic cylinder [3.90].

Table 3.6.4.

Horizontal orientation of the large axis				
C/B	0.1	0.25	0.5	0.75
$\mathrm{Nu}/\mathrm{Gr}_B^{1/4}$	0.2373	0.2938	0.3440	0.3496
Vertical orientation of the large axis				
C/B	1	0.75	0.5	0.25
$\mathrm{Nu}/\mathrm{Gr}_B^{1/4}$	0.3551	0.3799	0.4075	0.4925

The generalized dependences ***for a vertically located elliptic cylinder with a constant heat flux on the surface*** are [3.37]

$$T_{\mathrm{av}} = (T_{\mathrm{w}} + T_{\infty})/2, \quad C/B = 0-1, \quad \mathrm{Ra}_B^* > 0.1\,,$$

$$\mathrm{Nu}^{7.82} = \left[\frac{1.49}{\ln\left(1 + 1.49/1.3C_l\,\mathrm{Ra}_B^{*0,2}\right)}\right]^{7.82} + \left(C_t\,\mathrm{Ra}_B^{*1/4}\right)^{7\,82}. \quad (3.6.9)$$

3.7 Sphere and Spheroid

Spherical geometry is frequently encountered in various technological units in chemical and electronic industry, nuclear power engineering, and other areas. Just as in the case of a cylinder, there are no self-similar solutions for a sphere. However, application of the method of small perturbations enables us to obtain approximate analytical dependences to calculate heat exchange of such surfaces. In particular, it was found that the local heat transfer coefficient for a sphere is higher than for a cylinder due to the influence of the shape of the body which is responsible for divergence in space of streamlines on the lower part of the sphere. On the upper part of the sphere, the boundary-layer thickness increases more rapidly than in the case of a cylinder, because of the reduction in the boundary surface area and increase in the flow rate in the flow direction. Consequently, the local heat transfer coefficient for a sphere decreases more rapidly and eventually becomes smaller than in the case of a cylinder. As the Pr number increases, the Nusselt number becomes smaller due to reduction of the boundary-layer thickness [3.29].

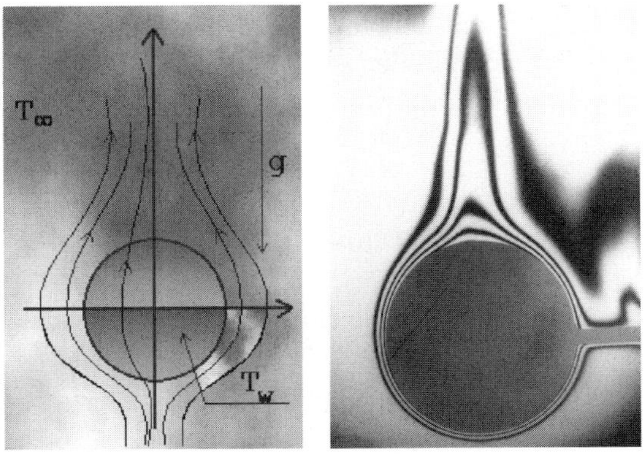

Fig. 3.7.1. Structure of streamlines and interferogram of free-convective flow near a sphere of constant temperature

The average heat transfer coefficient ***for a sphere of constant temperature*** is [3.91]

$$\mathrm{Gr}_D = 0 - 1, \quad \mathrm{Pr} = O(1)\,,$$
$$\mathrm{Nu} = 2 + \mathrm{Gr}_D + \left(0.139 - 0.4519\,\mathrm{Pr} + 1.1902\,\mathrm{Pr}^2\right)\mathrm{Gr}_D^2\,. \qquad (3.7.1)$$

On the frontal part of the sphere in a laminar flow regime at $\mathrm{Pr} = 0.7$ the distribution of heat transfer is quadratic [3.92]:

$$\mathrm{Nu}_D\,(x)\,/\mathrm{Gr}_D^{1/4} = 0.4576 - 0.03402\left(\frac{2x}{D}\right)^2\,. \qquad (3.7.2)$$

The generalized dependence for the average heat transfer coefficient in the case of ***laminar free convection in air*** ($\mathrm{Pr} = 0.7$) for the values $\mathrm{Gr}_D = 10^{-2}$–10^6 is obtained with an error of 0.5% [3.93]:

$$\mathrm{Nu} = 2 + 0.223\,\mathrm{Gr}_D^{\left[1/4 + 1/\left(4 + 6.9\,\mathrm{Gr}_D^{0.075}\right)\right]}\,. \qquad (3.7.3)$$

For any values of Pr and $\mathrm{Ra}_D = 1$–10^5 the average heat transfer coefficient is calculated with an error of 10% from the formula [3.92]

$$\mathrm{Nu} = 2 + 0.56\,[\mathrm{Pr}\,/\,(0.846 + \mathrm{Pr})]^{1/4}\;\mathrm{Ra}_D^{1/4}\,. \qquad (3.7.4)$$

The generalized empirical dependence for calculating heat transfer of a sphere ***in air and water*** is [3.94]

$$\mathrm{Nu} = \left(2^{0.816} + 0.152\,\mathrm{Ra}_D^{0.227}\right)^{1/0.816}\,, \qquad (3.7.5)$$

for $\mathrm{Pr} = 0.7$–6, $\mathrm{Ra}_D = 10^{-6}$–10^4, $T_{\mathrm{av}} = (T_{\mathrm{w}} + T_\infty)/2$.

As a result of numerical investigation of laminar natural convection on ***an isothermal sphere***, the following dependences are suggested to calculate heat transfer rate ***in air and water*** [3.95]:

$$\mathrm{Nu} = 2 + 0.282\,(\mathrm{Gr}_D\,\mathrm{Pr})^{1/4}\,, \quad \mathrm{Gr}_D\;\mathrm{Pr} \le 10^2\,, \tag{3.7.6}$$

$$\mathrm{Nu} = 2 + 0.5\,(\mathrm{Gr}_D\,\mathrm{Pr})^{1/4}\,, \quad 0^2 \le \mathrm{Gr}_D\;\mathrm{Pr} \le 10^6\,. \tag{3.7.7}$$

The average Nusselt number ***for an isothermal sphere*** can be calculated from the formula [3.96]

$$\mathrm{Nu}_D = \frac{0.589\;\mathrm{Ra}_D^{1/4}}{\left[1 + (0.5/\,\mathrm{Pr})^{9/16}\right]^{4/9}}\,, \tag{3.7.8}$$

which asymptotically passes into

$$\mathrm{Nu}_D = 0.714\;\mathrm{Ra}_D^{1/4}\;\mathrm{Pr}^{1/4}, \quad \text{for Pr} \to 0\,, \tag{3.7.9}$$

$$\mathrm{Nu}_D = 0.6\;\mathrm{Ra}_D^{1/4}, \quad \text{for Pr} \to \infty\,. \tag{3.7.10}$$

For a fluid with variable viscosity at $\mathrm{Pr} = 7\text{–}2.4\cdot 10^4$, $\mathrm{Gr}_D = 5.7\cdot 10^{-3}\text{–}10^7$, $T_{\mathrm{av}} = T_\infty$ [3.97]

$$\mathrm{Nu} = 2 + 0.44\;\mathrm{Ra}_D^{0.25}(\mu_\infty/\mu_{\mathrm{w}})^{0.25}\,. \tag{3.7.11}$$

An ***ellipsoid of revolution with the large axis $2b = B$ and small axis $2c = C$ for a constant temperature of the surface***. The generalized dependence for the average heat transfer coefficient ***of vertically and horizontally located spheroids*** is [3.98]

$$\mathrm{Nu}^n = \left[\mathrm{Nu}_T + \frac{2^{1/4} f_2^{3/4}}{f_1} C_l \left(\frac{A}{B}\;\mathrm{Ra}_B\right)^{1/4}\right]^n + \left[C_t f_3 \left(\frac{A}{B}\;\mathrm{Ra}_B\right)^{1/3}\right]^n. \tag{3.7.12}$$

The values of f_1,f_2,f_3, n, and A are presented in Tables 3.7.1 and 3.7.2.

For vertical position (prolate spheroid) for $n = 6$, $A = B$

Table 3.7.1.

C/B	f_1	f_2	f_3
1.0	2	1.683	0.73
0.9	1.935	1.653	0
0.8	1.873	1.624	0.73
0.7	1.814	1.596	0
0.6	1.759	1.570	0.72
0.5	1.709	1.545	0.71
0.4	1.665	1.524	0.71
0.3	1.627	1.506	0
0.2	1.598	1.491	0.71
0.1	1.578	1.481	0.71
0.05	1.573	1.479	0.71
0.01	1.571	1.478	0.71

For horizontal position (oblate spheroid) for $A = C$

Table 3.7.2.

C/B	f_1	f_2	f_3	n
1.0	2	1.683	0.73	6
0.9	1.868	1.601	0	–
0.8	1.739	1.518	0.79	5
0.7	1.615	1.435	0	–
0.6	1.494	1.352	0.87	5
0.5	1.380	1.269	0.93	4
0.4	1.274	1.187	0.99	3
0.3	1.177	1.108	0	–
0.2	1.094	1.033	1.20	2.5
0.1	1.030	0.969	1.44	2.5
0.05	1.009	0.945	1.73	2.5

For an prolate ellipsoid

$$\mathrm{Nu} = -\frac{4\sqrt{1-(C/B)^2}}{f_1} \frac{1}{(C/B)\ln th\,(\eta_i/2)} . \tag{3.7.13}$$

For an oblate ellipsoid

$$\mathrm{Nu} = \frac{4\sqrt{1-(C/B)^2}}{f_1} \frac{1}{\pi/2 - \mathrm{arctg}\,(sh\eta_i)} , \tag{3.7.14}$$

where $\eta_i = \frac{1}{2}\ln\frac{1+C/B}{1-C/B}$.

The error in determining the intensity of heat transfer from (3.7.14) in comparison with experiments in air for an ellipsoid of revolution with various eccentricities and for a sphere for any values of the Pr number is 5% without any restrictions on the Gr_B numbers. For $C = B$ the formula can be used for calculating heat exchange of a sphere of constant temperature.

An isothermal hemisphere immersed in

dehydrated glycerine [3.99]:

$$6.04 \cdot 10^3 < \mathrm{Ra}_D < 1.63 \cdot 10^6,\ \mathrm{Pr} = 120 ,$$

$$\mathrm{Nu}_D = 0.510\ \mathrm{Ra}_D^{1/4} , \tag{3.7.15}$$

technical glycerine:

$$1.2 \cdot 10^4 < \mathrm{Ra}_D < 2.0 \cdot 10^6, \mathrm{Pr} = 700 ,$$

$$\mathrm{Nu}_D = 0.501\ \mathrm{Ra}_D^{1/4} , \tag{3.7.16}$$

distilled water:

$$4.4 \cdot 10^4 < \mathrm{Ra}_D < 4.1 \cdot 10^6,\ \mathrm{Pr} = 6\ ,$$

$$\mathrm{Nu}_D = 0.444\ \mathrm{Ra}_D^{1/4}\ , \qquad (3.7.17)$$

A sphere with cosine arrangement of heat sources on the surface for $\theta_{\mathrm{w}} = \theta_0 \cos\varphi$ [3.100]:

$$\mathrm{Nu} = 0.13\,[\,\mathrm{Ra}_D \exp\left(0.59\,(\varphi - 0.15\pi)\right)]^{1/3}\ , \qquad (3.7.18)$$

$$\varphi = \left(-\frac{\pi}{2}\right) - \frac{\pi}{2}, \quad \mathrm{Pr} = 0.7 - 10, \quad \mathrm{Ra}_D = 10^8 - 10^{14}\ .$$

The local heat transfer coefficient for ***a sphere of constant temperature in a stratified environment*** is given by [3.51]

$$\frac{\mathrm{Nu}(x)}{\mathrm{Nu}_0(x)} = \frac{(T_{\mathrm{w}} - T_\infty)^{17/12}}{(T_{\mathrm{w}} - T_\infty)_0} \sin^{2/3}\bar{x} \left(\frac{F}{G}\right)^{1/4} \qquad (3.7.19)$$

$$T_\infty = T_\infty(z), \quad z = R_0\,(1 - \cos\bar{x})\ , \quad \bar{x} = x/R_0\ ,$$

$$F = \exp\left[\frac{4}{3}\,(k+1)\int_0^{\bar{x}} \frac{dT_\infty/d\bar{x}}{(T_{\mathrm{w}} - T_\infty)}\,d\bar{x}\right]\ ,$$

$$G = \int_0^{\bar{x}} \sin^{5/3}\bar{x}\,(T_{\mathrm{w}} - T_\infty)^{5/3}\,F d\bar{x}\ .$$

The values of the coefficient k are given in Table 3.7.3. The deviation of the calculated results for the average heat transfer coefficient from the experimental values does not exceed 8%.

Table 3.7.3.

Pr	0.7	5	6	10
k	3.5	2.5	2.5	2

Laminar free convection ***of a binary mixture (mixture of air with steam and the vapor of bensol) near a heated sphere. On the surface of sphere, the temperature and the concentration are held constant.*** In the case of a small concentration, when the buoyancy forces coincide, the heat transfer rate on the surface increases with $\mathrm{Gr}_D/\mathrm{Gr}$. For a large concentration, the heat transfer rate decreases with increase in $\mathrm{Gr}_D/\mathrm{Gr}$ up to a certain value [3.101].

The average heat transfer coefficient ***for injection or suction of the same gas on the surface*** for Pr = 0.7 and 10 is approximated by the dependences [3.102]

for a laminar mode of flow:

$$\frac{\mathrm{Pe}_D}{\mathrm{Gr}_D^{1/4}} = 0 - 0.4 \,, \tag{3.7.20}$$

$$\frac{\mathrm{Nu}}{\mathrm{Nu}_0} = 1 - 1.18\frac{\mathrm{Pe}_D}{\mathrm{Gr}_D^{1/4}} \,,$$

for a turbulent mode of flow:

$$\frac{\mathrm{Pe}_D}{\mathrm{Gr}_D^{2/5}} = 0 - 0.2 \,, \tag{3.7.21}$$

$$\frac{\mathrm{Nu}}{\mathrm{Nu}_0} = 1 - 1.03\frac{\mathrm{Pe}_D}{\mathrm{Gr}_D^{2/5}} \,,$$

the average heat transfer coefficient of a ***heated rotating sphere*** [3.103]:

$$\mathrm{Nu}^3 = \mathrm{Nu}_F^3 + \mathrm{Nu}_N^3 \,, \tag{3.7.22}$$

$$\mathrm{Nu}_N = 2 + 0.392\ \mathrm{Gr}^{0.31} \quad (\mathrm{Gr} > 3 \cdot 10^5) \,,$$

$$\mathrm{Nu}_F = 2 + 0.175\ \mathrm{Re}^{0.583} \quad (\mathrm{Gr} = 0,\ \mathrm{Re} < 33320, \mathrm{Pr} = 0.71) \,.$$

Example 3.7. A tank for storing gas and petroleum is a sphere of diameter $D =$ 1 m and is located in the surrounding medium at $t_\infty = 10°\mathrm{C}$. The temperature on the surface is kept at $t_\mathrm{w} = 8°\mathrm{C}$. It is required to determine heat fluxes from the environment to the tank due to free convection.

The determining air temperature is $t_\mathrm{av} = (10 + 8)/2$. For the temperature t_av we have

$\beta = 1/282.16$ 1/deg,
$\nu = 14.16 \cdot 10^{-6}$ m^2/sec,
$\lambda = 2.51 \cdot 10^{-2}$ W/m · deg,
Pr $= 0.707$.

We calculate the determining criteria:

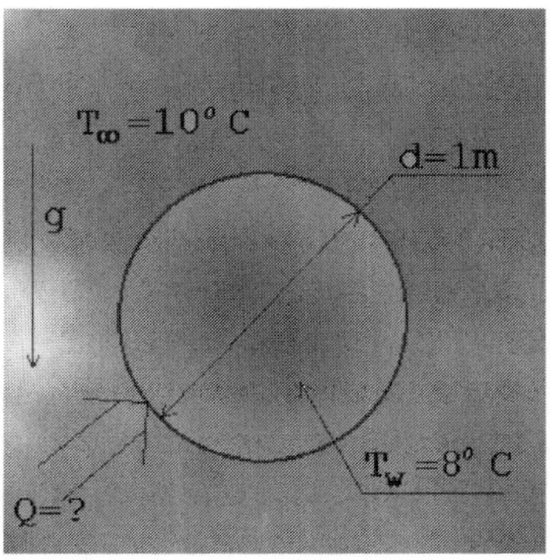

$$\mathrm{Gr}_D = \frac{9.81 \cdot 2}{282 \cdot (14.16)^2} \cdot 10^{12} = 3.84 \cdot 10^8 \,,$$

$$\mathrm{Ra}_D = 3.48 \cdot 10^8 \cdot 0.707 = 2.46 \cdot 10^8 \,.$$

To calculate the average value of the heat transfer coefficient, we use formula (3.7.12):

$$\mathrm{Nu}^6 = \left[\frac{2^{1/4} \cdot 1.683^{3/4}}{2} \frac{0.667}{\left[1 + (0.49/\,\mathrm{Pr})^{9/16}\right]^{4/9}} \left(2.46 \cdot 10^8\right)^{1/4} \right]^6 + \left[0.14\,\mathrm{Pr}^{\,0.081} \left(2.46 \cdot 10^8\right)^{1/3}\right]^6 ,$$

$$\mathrm{Nu} = 70.35 \,.$$

The average value of the heat transfer coefficient and heat flux on the lateral surface are

$$\alpha = 70.35 \cdot 2.51 \cdot 10^{-2}/1 = 1.77\,\mathrm{W/m^2} \cdot \mathrm{deg} \,,$$

$$Q = \alpha \theta_{\mathrm{w}} F_\sigma = 1.77 \cdot 2\pi = 11\,\mathrm{W} \,.$$

3.8 Curved Surface of Complex Geometry

The methods of calculation of free-convective heat transfer on bodies of any shape are based on various approaches, many of which lead to significant errors and require additional calculations. Self-similar solutions are applicable only in some special cases of the body geometry [3.104, 3.105]. Mangler's transformation [3.106] reduces the problem to a self-similar solution. Using integral methods with an assumption of local self-similarity [3.84] gives the underestimated heat transfer results and requires certain updating depending on the body shape.

Using the analogy between the processes of heat transfer from a surface of complex geometry and heat transfer in film condensation in [3.3], a general expression for a local Nusselt number for two-dimensional ($i = 0$) and axisymmetric ($i = 1$) bodies is suggested:

$$\mathrm{Nu}\,(x) = C\,(\mathrm{Pr})\ \mathrm{Ra}_x^{1/4} \left\{ (1/x) \int\limits_0^x \left[r^{4i} \left(g_\parallel/g\right) \right]^{1/3} dx \right\}^{-1/4} \Bigg/ \ (1/x) \int\limits_0^x r^i dx \,, \tag{3.8.1}$$

where $g_\parallel$ is a component of acceleration along a curved surface,

$$C\,(\mathrm{Pr}) = \frac{0.503}{\left[1 + (0.492/\,\mathrm{Pr})^{9/16}\right]^{4/9}} .$$

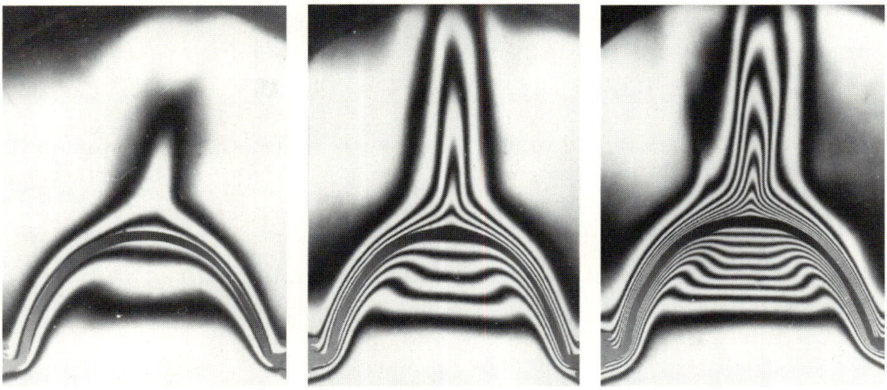

Fig. 3.8.1. Interferograms of free-convective heat transfer dynamics on a horizontal surface bent as an arc of a circle

Let us consider a parametric method for calculating heat transfer on a curved surface. Choose the coordinate x along the surface from the front critical point and the coordinate y along the normal to the surface. For axisymmetric bodies r represents the distance from surface element to the symmetry axis. The component $g_{||}(x)$ of local acceleration is directed to the side of an increasing x. Introduce the parametric function

$$\Lambda(\overline{x}) = 2\frac{\xi}{\overline{u_1}}\frac{d\overline{u_1}}{d\xi} = 2\frac{\overline{g_{||}}\xi}{\overline{r^2 u_1^3}}, \tag{3.8.2}$$

where

$$\overline{u_1} = \left(2\int_0^{\overline{x}} \overline{g_{||}} d\overline{x}\right)^{1/2}, \quad \xi = \int_0^{\overline{x}} \overline{r^2 u_1} d\overline{x}.$$

The value $\Lambda(\overline{x})$ is constant for self-similar problems: for an inclined plate, $\overline{g_{||}} = \sin\varphi$, $\Lambda = 2/3$; for a vertical circular cone $\Lambda = 2/7$; for the frontal critical point of flat and axisymmetric bodies $\Lambda = 1/2$ and $\Lambda = 1$; the body contour [3.104–3.106]

$$\frac{d\overline{r}}{d\overline{x}} = \left(1 - \overline{g_{||}}^2\right)^{1/2}, \quad \overline{g_{||}} = \left(\frac{m+1}{m}\overline{x}\right)^{\frac{(m-3)}{(m+1)}}, \quad \Lambda = \frac{m-1}{m}.$$

The local heat flux density is

$$\frac{\mathrm{Nu}}{\mathrm{Gr}_x^{1/4}} = \frac{\overline{ru_1}}{(2\xi)^{1/2}}\left[\theta_0'(\Lambda, 0) + \xi\frac{d\Lambda}{d\xi}\theta_1'(\Lambda, 0) + \xi^2\frac{d^2\Lambda}{d\xi^2}\theta_2'(\Lambda, 0) + \cdots\right]. \tag{3.8.3}$$

The values of $\theta_i'(\Lambda, 0)$ for $\mathrm{Pr} = 0.72$ are given in Table 3.8.1.

In the case of $\Lambda = \mathrm{const}$ free-convective movement is self-similar, and only the first term in (3.8.3) remains.

Table 3.8.1.

Λ	$-\theta_0'(\Lambda,0)$	$-\theta_1'(\Lambda,0)$	$-\theta_2'(\Lambda,0)$
1.4	0.3965	0.00628	0.00367
1.2	0.3863	0.00621	0.00400
1.0	0.3741	0.00592	0.00436
0.99	0.3734	0.00590	0.00438
0.98	0.3727	0.00587	0.00440
0.97	0.3720	0.00584	0.00442
0.95	0.3706	0.00579	0.00446
0.93	0.3692	0.00573	0.00452
0.90	0.3670	0.00563	0.00457
0.85	0.3631	0.00544	0.00467
0.80	0.3589	0.00520	0.00479
0.75	0.3545	0.00490	0.00490
0.70	0.3498	0.00455	0.00502
2/3	0.3465	0.00427	0.00511
0.65	0.3447	0.00412	0.00515
0.60	0.3393	0.00359	0.00528
0.57	0.3358	0.00321	0.00536
0.55	0.3333	0.00294	0.00542
0.54	0.3321	0.00279	0.00545
0.53	0.3308	0.00263	0.00548
0.52	0.3295	0.00247	0.00551
0.51	0.3282	0.00231	0.00554
0.50	0.3268	0.00213	0.00556
0.40	0.3118	–0.00159	0.00589
0.30	0.2928	–0.00400	0.00629
2/7	0.2896	–0.00476	0.00635

Numerical calculations by formula (3.8.3) are in good agreement with the data on heat transfer for laminar convection near a vertical round cone, a flat plate rotating around an axis which passes through its front edge, a long horizontal cylinder of circular and elliptic sections, a sphere, and an ellipsoid of revolution [3.33].

The values of $\theta_i'(\Lambda, 0)$ for Pr =100 are presented in Table 3.8.2.

When $\mathrm{Pr} \to \infty$, free-convective flow is self-similar one relative to the body shape. The local Nusselt number can be calculated from the formula

$$\mathrm{Nu}\,(x) \,/\, (\mathrm{Gr}_x\,\mathrm{Pr})^{1/4} = 0.5028\overline{r g_{\|}} \Bigg/ \left[\int_0^x \left(\overline{r}^4 \overline{g_{\|}}\right)^{1/3} d\bar{x} \right]^{1/4} . \tag{3.8.4}$$

The local and average values of heat transfer ***in liquid metals*** ($\mathrm{Pr} \ll 1$) near ***heated bodies with the convexity facing downwards*** [3.107] is calculated from the formulas:

Table 3.8.2.

Λ	$-\theta_0'(\Lambda,0)$	$-\theta_1'(\Lambda,0)$	$-\theta_2'(\Lambda,0)$
1.4	1.7994	0.06102	0.02382
1.2	1.7343	0.06796	0.02662
1.0	1.6599	0.07719	0.03034
0.99	1.6559	0.07773	0.03056
0.98	1.6518	0.07828	0.03079
0.97	1.6477	0.07885	0.03101
0.95	1.6395	0.08000	0.03148
0.93	1.6311	0.08121	0.03197
0.90	1.6182	0.08309	0.03273
0.85	1.5960	0.08648	0.03410
0.80	1.5727	0.09024	0.03561
0.75	1.5483	0.09441	0.03730
0.70	1.5225	0.09910	0.03920
2/3	1.5046	0.10256	0.04059
0.65	1.4953	0.10440	0.04134
0.60	1.4664	0.11046	0.04379
0.57	1.4482	0.11453	0.04543
0.55	1.4356	0.11745	0.04661
0.54	1.4292	0.11898	0.04723
0.53	1.4227	0.12057	0.04787
0.52	1.4161	0.12220	0.04853
0.51	1.4093	0.12389	0.04922
0.50	1.4025	0.12564	0.04992
0.40	1.3278	0.14722	0.05865
0.30	1.2371	0.18085	0.07224
2/7	1.2223	0.18730	0.07484

for the given distribution of surface temperature

$$\mathrm{Nu}_{R_0}(x)\,/\,\left(\mathrm{Gr}_{R_0}\,\mathrm{Pr}^2\right)^{1/4} = -[g_0'(0) + g_1'(0)\overline{x^2} + g_2'(0)\overline{x^4}]\,, \tag{3.8.5}$$

$$\mathrm{Nu} = -\frac{g_0'(0) + g_1'(0)\,\overline{x_m^2}/3 + g_2'(0)\,\overline{x_m^4}/5}{1 + a_1\overline{x_m^2}/3 + a_2\overline{x_m^4}/5}\left(\mathrm{Gr}_{R_0}\,\mathrm{Pr}^2\right)^{1/4}\,, \tag{3.8.6}$$

for the given distribution of a heat flux on the surface

$$\mathrm{Nu}_{R_0}(x)\,/\,\left(\mathrm{Gr}^*_{R_0}\,\mathrm{Pr}^2\right)^{1/5} = [g_0(0) + g_1(0)\overline{x^2} + g_2(0)\overline{x^4}]\,, \tag{3.8.7}$$

$$\mathrm{Nu} = \left[\frac{g_0(0) + g_1(0)\,\overline{x_m^2}/3 + g_2(0)\,\overline{x_m^4}/5}{1 + a_1\overline{x_m^2}/3 + a_2\overline{x_m^4}/5}\right]^{-1}\left(\mathrm{Gr}^*_{R_0}\,\mathrm{Pr}^2\right)^{1/5}\,, \tag{3.8.8}$$

where R_0 is the radius of the curvature at the critical point; $\overline{x_m}$ is the maximum value of the curved coordinate.

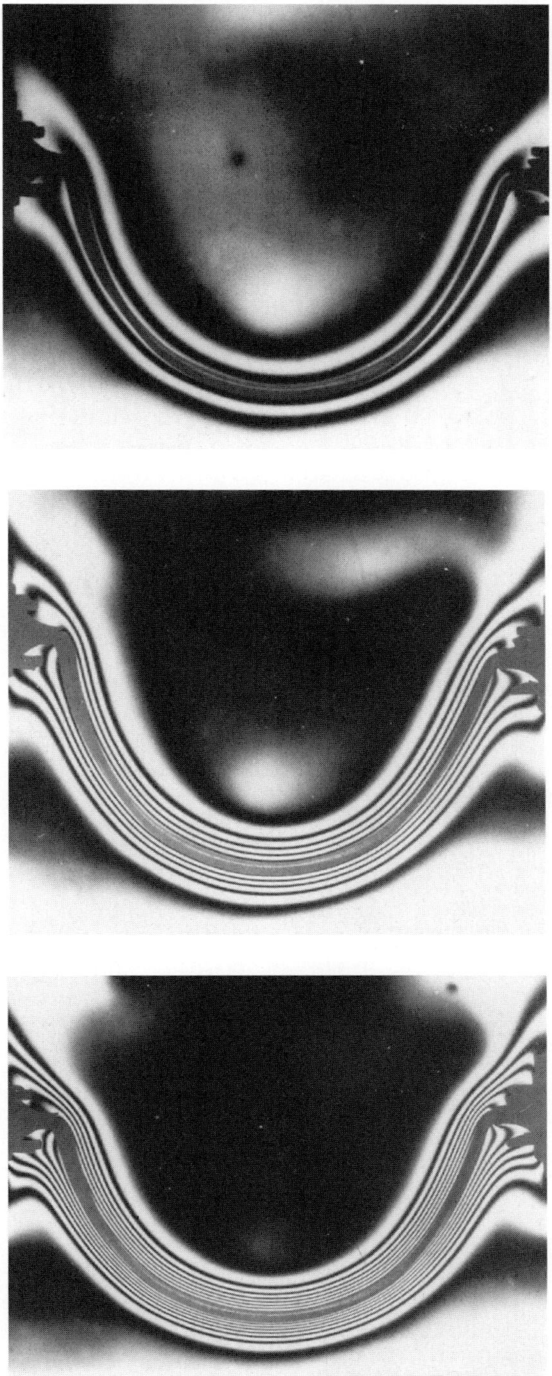

Fig. 3.8.2. Interferograms of the free-convective heat transfer dynamics on a vertical surface bent as an arc of a circle

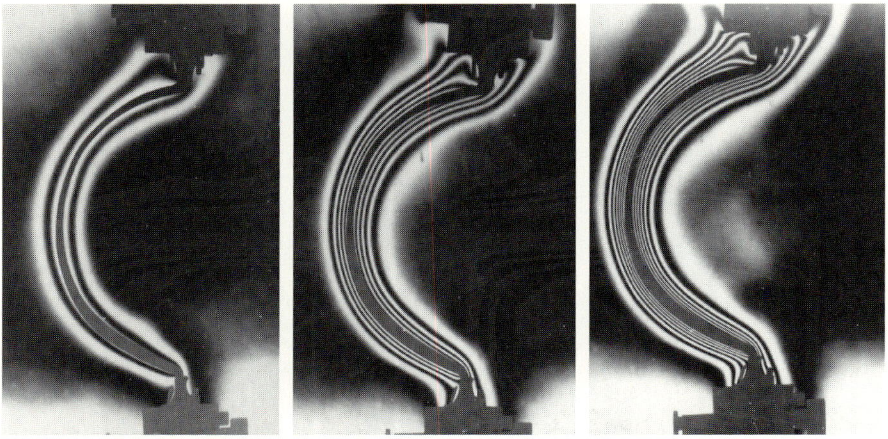

Fig. 3.8.3. Interferograms of the free-convective heat transfer dynamics on a vertical surface bent as an arc of a circle

For two-dimensional bodies:

$$g_0 = G_1, \quad g_1 = a_1 G_1 + s_1 G_3, \quad g_2 = a_2 G_4 + a_1^2 G_5 + a_1 s_1 G_6 + s_1^2 G_7 + s_2 G_8 .$$

The values of $G_j(0)$ and $G'_j(0)$ are given in Table 3.8.3.

Table 3.8.3.

Pr	0	0.004	0.006	0.008	0.01
			$G'_j(0)$		
$j = 1$	1.476	1.512	1.520	1.527	1.533
2	0.640	0.643	0.650	0.656	0.661
3	−0.142	−0.146	−0.148	−0.150	−0.151
4	0.488	0.489	0.496	0.502	0.506
5	−0.0206	−0.0229	−0.0235	−0.0240	−0.0245
6	−0.0717	−0.0775	−0.0793	−0.0808	−0.0821
7	0.0415	0.0450	0.0463	0.0473	0.0482
8	−0.0853	−0.0898	−0.0918	−0.0935	−0.0950
			$G_j(0)$		
$j = 1$	−0.616	−0.596	−0.592	−0.589	−0.586
2	−1.480	−1.404	−1.387	−1.373	−1.361
3	−0.137	−0.135	−0.135	−0.134	−0.134
4	−1.978	−1.850	−1.820	−1.797	−1.777
5	−0.153	−0.155	−0.155	−0.154	−0.153
6	−0.251	−0.253	−0.252	−0.252	−0.251
7	0.031	0.0321	0.0323	0.0324	0.0326
8	−0.108	−0.109	−0.110	−0.110	−0.110

Table 3.8.4.

Pr	0	0.004	0.006	0.008	0.01
			$G'_j(0)$		
$j = 1$	1.211	1.243	1.251	1.257	1.263
2	0.660	0.662	0.669	0.674	0.679
3	−0.150	−0.153	−0.155	−0.156	−0.158
4	0.0257	0.0256	0.0258	0.0259	0.0260
5	0.528	0.525	0.532	0.537	0.542
6	−0.0317	−0.0333	−0.0340	−0.0345	−0.0350
7	−0.0866	−0.0900	−0.0916	−0.0929	−0.0949
8	0.0560	0.0578	0.0588	0.0598	0.0606
9	−0.104	−0.106	−0.108	−0.109	−0.111
10	0.0395	0.0388	0.0390	0.0392	0.0393
11	0.0035	0.0036	0.0037	0.0038	0.0039
12	0.0074	0.0073	0.0074	0.0074	0.0075
			$G_j(0)$		
$j = 1$	−0.790	−0.762	−0.756	−0.751	−0.747
2	−1.501	−1.432	−1.416	−1.402	−1.391
3	−0.181	−0.176	−0.175	−0.174	−0.173
4	0.0310	0.0295	0.0291	0.0288	0.0286
5	−1.913	−1.809	−1.783	−1.762	−1.744
6	−0.173	−0.171	−0.170	−0.169	−0.168
7	−0.292	−0.288	−0.287	−0.285	−0.284
8	0.0517	0.0511	0.0509	0.0509	0.0508
9	−0.156	−0.154	−0.153	−0.153	−0.153
10	0.0671	0.0634	0.0625	0.0618	0.0612
11	0.0164	0.0169	0.0169	0.0169	0.0169
12	0.0127	0.0119	0.0117	0.0116	0.0114

For axisymmetric bodies:

$$g_0 = G_1, \quad g_1 = a_1 G_2 + s_1 G_3 + G_4 \,,$$
$$g_2 = a_2 G_5 + a_1^2 G_6 + a_1 s_1 G_7 + s_1^2 G_8 + s_2 G_9 + s_1 G_{10} + a_1 G_{11} + G_{12} \,.$$

The values of $G_j(0)$ and $G'_j(0)$ are given in Table 3.8.4.

The stabilizing influence of the heated bodies with the convexity facing downwards displaces the beginning of the transition to a turbulent flow regime to larger values of Gr numbers.

A vertical surface of length $L = \pi D/4$ is curved as an arc of a circle. Heat transfer ***from the internal part of the surface of constant temperature*** is given by [3.108]

$$\mathrm{Nu}_{D\varphi} = 0.744\,(1 - 0.418\varphi)\ \mathrm{Ra}_D^{1/4} \,, \tag{3.8.9}$$

$$\mathrm{Nu} = 0.522\ \mathrm{Ra}_D^{1/4}\ , \tag{3.8.10}$$

for $\varphi = \pi/12 - \pi/2$, $\mathrm{Pr} = 0.7$, $\mathrm{Gr}_D = 10^6 - 10^{10}$.

The average heat transfer coefficient ***of the internal surface of a horizontal bath in air with the heating surface facing upwards*** is [3.109]

$$\mathrm{Nu}\,(x) = 0.2\ \mathrm{Ra}_x^{1/3}\ , \tag{3.8.11}$$

for $T_\mathrm{w} = 600-800°\mathrm{C}$, $\mathrm{Ra}_x = 5\cdot10^7 - 1.5\cdot10^8$, $\mathrm{Pr} = 0.68$, $T_\mathrm{av} = (T_\mathrm{w} + T_\infty)/2$.

A piece of a wire located in a vertical plane bent as a semiring with the convex part facing downwards [3.110]. The calculations carried out at the values of the parameters: $\mathrm{Pr} = 0.7$, $R/D = 3 - 1000, R = 0.5, \mathrm{Gr}_D^{0.25} = 3 - 100$ (R is the ring radius, D is the wire diameter), show that Nu_D decreases as R/D and R increase.

A horizontally located wire spiral of diameter d, coil diameter D and step h: $d = 0.2 - 2.5\,\mathrm{mm}$, $D = 1.89 - 15.1\,\mathrm{mm}$, $h/d = 1.5 - 8$. The free-convective component of heat transfer ***in air*** for $T_\infty = 20°\mathrm{C}$ and $T_\mathrm{w} < 1200°\mathrm{C}$ is calculated from the formulas ***for a horizontal cylinder*** of characteristic size $d_r = d + d(\pi D - d)/h$ [3.111].

The heat transfer of ***a vertical helically curved pipe with a constant heat flux*** can be calculated from [3.112]

$$4.8\cdot10^{12} \le \mathrm{Ra}_L \le 4\cdot10^{14},\quad D/d = 13.923\ ,$$

$$\mathrm{Nu}_L = 5.11\ \mathrm{Ra}_L^{0.214} N^{0.296}\ , \tag{3.8.12}$$

$$2.8\cdot10^{12} \le \mathrm{Ra}_L \le 1\cdot10^{14},\quad D/d = 9.914\ ,$$

$$\mathrm{Nu}_L = 2.64\ \mathrm{Ra}_L^{0.228} N^{0.324}\ , \tag{3.8.13}$$

$$3\cdot10^{12} \le \mathrm{Ra}_L \le 2\cdot10^{14},\quad D/d = 19.957\ ,$$

$$\mathrm{Nu}_L = 0.258\ \mathrm{Ra}_L^{0.247} N^{1.037}\ , \tag{3.8.14}$$

$$5\cdot10^{11} \le \mathrm{Ra}_L \le 3\cdot10^{13},\quad D/d = 9.914\ ,$$

$$\mathrm{Nu}_L = 0.002\ \mathrm{Ra}_L^{0.431} N^{0.375}\ , \tag{3.8.15}$$

where d is the pipe diameter; D is the spiral coil diameter; L is the spiral length; N is the number of spiral coils.

The heat transfer ***of an electric motor for horizontally and vertically oriented cases*** of length L and diameter D for the parameters $L/D = 1.1-2$, $\mathrm{Pr} = 0.7$, $\mathrm{Ra}_D = 10^5 - 10^9$ is given by [3.113]

without an output shaft:

$$\mathrm{Nu} = 0.783\,\mathrm{Ra}_D^{0.226}\ , \tag{3.8.16}$$

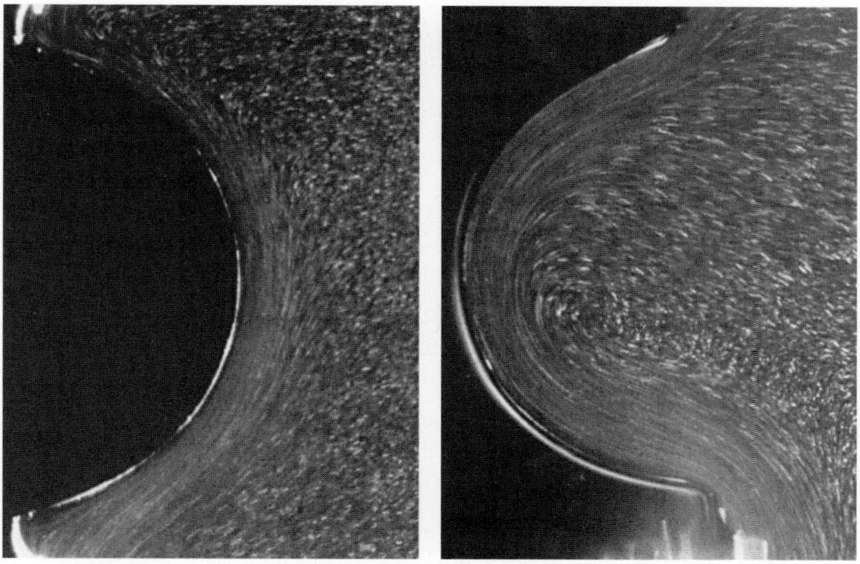

Fig. 3.8.4. Track visualization of free-convective flow near a vertical surface bent as an arc of a circle with a constant heat flux on a wall

in the presence of free end of a shaft:

$$\mathrm{Nu} = 0.94\,\mathrm{Ra}_D^{0.226}\,. \tag{3.8.17}$$

The heat transfer is influenced by the nonuniformity of the motor surface heating, ambient disturbance by rotation of the shaft, vibration, roughness, etc.

The average coefficient of heat transfer ***of the flat surface of an isothermal radiator shaped as a disk on a horizontal pipe*** is [3.62]

$$\mathrm{Nu} = 0.702\,|\,\mathrm{Ra}|^{1/4}\;, \tag{3.8.18}$$

where the Ra number is calculate from the diameter of the radiator.

The average heat transfer coefficient ***of the flat surface of an isothermal square radiator on a horizontal pipe*** is [3.62]

$$\mathrm{Nu} = 0.67\,|\,\mathrm{Ra}|^{1/4}\;, \tag{3.8.19}$$

where the Ra number is calculated on the side of the radiator.

The results of numerical calculation of the Nusselt number ***on a vertical plate of a sinusoidal shape*** ($y = \alpha \sin 2\pi x$), immersed ***in a non-Newtonian fluid*** are given in [3.114].

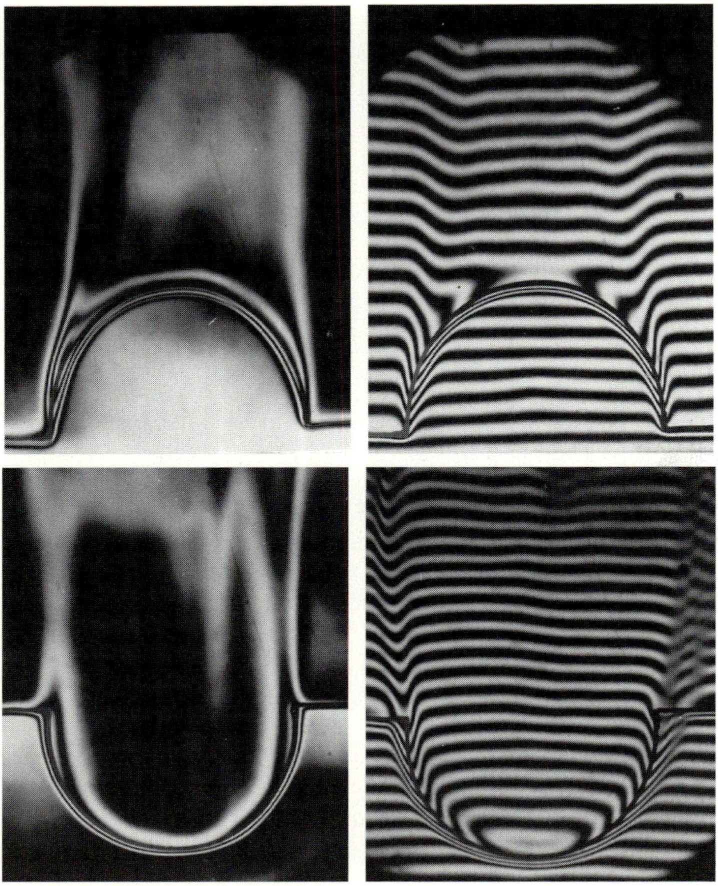

Fig. 3.8.5. Interferograms and shadow pictures of the free-convective heat transfer dynamics on a horizontal wire bent as an arc of a circle

Experimental results for ***a steady-state natural convection from uniformly heated helically coiled tubes in air*** [3.115] show that the average heat transfer coefficient decreases with increase of the number of coils up to the middle of the spiral and then a transition to turbulence begins. The correlation describing the natural convection regime of horizontally located coils for all heat fluxes is

$$\mathrm{Nu} = a\mathrm{Ra}_d^b\,, \tag{3.8.20}$$

where d is the tube diameter, a and b are constants given in Table 3.8.5. Using the distance along the horizontal axis to the corresponding coil x as the characteristic length, we may obtain the general correlation for all coils and any values of heat fluxes:

$$\mathrm{Nu}_x = 0.9125\ \mathrm{Ra}_x^{0.301}\,. \tag{3.8.21}$$

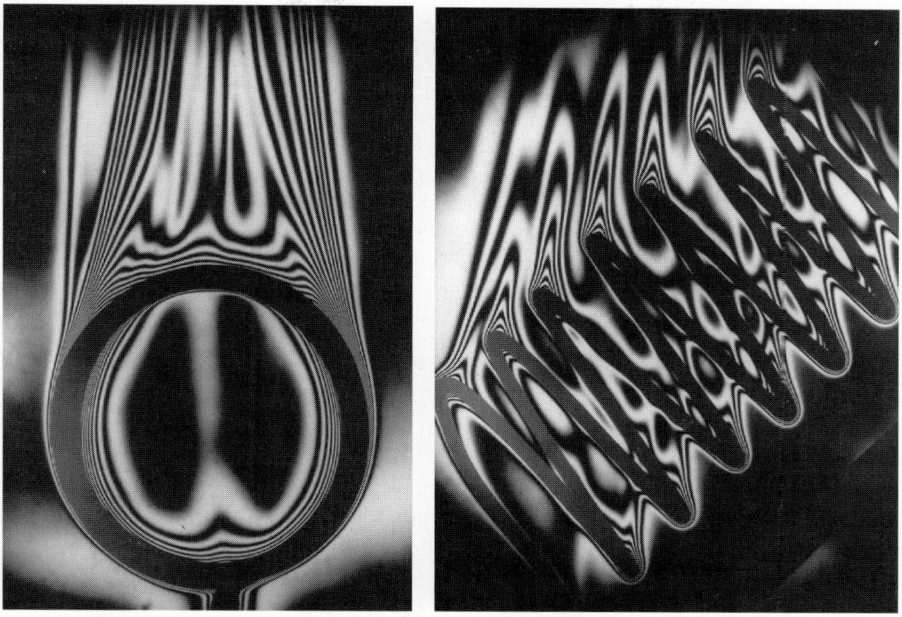

Fig. 3.8.6. Interferograms of free-convective flow near a spiral heated by an electric current

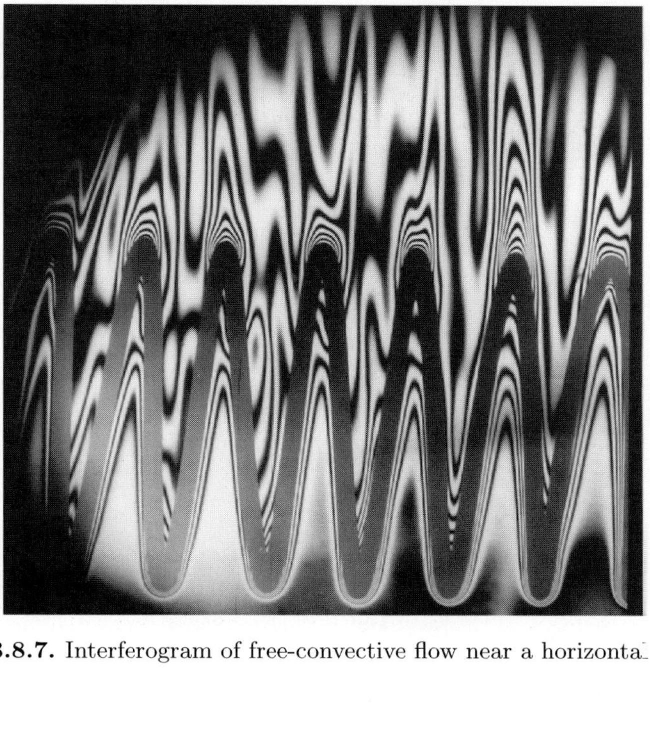

Fig. 3.8.7. Interferogram of free-convective flow near a horizontal spiral

Table 3.8.5. Values a, b and Ra for Various Heat Fluxes

a	b	Ra
$1.198 \cdot 10^{9}$	-2.678	1119 – 1277
$1.832 \cdot 10^{12}$	-3.632	1296 – 1486
$1.253 \cdot 10^{18}$	-5.431	1437 – 1554
$1.198 \cdot 10^{45}$	-14.061	1531 – 1567

References

1. I. Langmuir, *Phys. Rev.*, **32**, 401 (1912).
2. H. Sentfleben, *Z. Aangew. Phys.,* **3**, 361 (1951).
3. G. D. Raithby and K. G. T. Hollands, *Adv. Heat Transfer,* **11**, 265 (1975).
4. R. K. Tripathi and G. Nath, *Trans. ASME, Ser. C, J. Heat Transfer*, **115**, 130 (1993).
5. O. G. Martynenko and Yu. A. Sokovishin, *Free-Convective Heat Transfer on a Vertical Surface (II Kind Boundary Conditions)*, Nauka i Tekhnika Press, Minsk (1978).
6. E. M. Sparrow and J. L. Gregg, *Trans. ASME, Ser. C, J. Heat Transfer,* **78**, 1823 (1956).
7. O. G. Martynenko, Yu. A. Sokovishin, and M. V. Shapiro, *Inzh.-Fiz. Zh.*, **33**, 311 (1977).
8. H. R. Nagendra, M. A. Tirunarayanan, and A. Ramachandran, *ASME Paper*, No. 69-HT–G, 1 (1969).
9. T. Fujii and H. Uehara, *Int. J. Heat Mass Transfer*, **13**, 607 (1970).
10. H. K. Kuiken, *Int. J. Heat Mass Transfer*, **11**, 1141 (1968).
11. E. J. Le Fevre and A. J. Ede, *Proc. 9th Int. Cong. Appl. Mech.*, Vol. 4, Brussels (1956), p. 175.
12. W. Elenbass, *J. Appl. Physics*, **19**, 1148 (1948).
13. G. A. Ohman, *Acta acad. aboen.*, *Ser. B*, **30**, 1 (1970).
14. V. T. Morgan, *Adv. Heat Transfer,* **11**, 199 (1975).
15. T. Fujii and S. Koyama, *Trans. JSME,* **44,** 118 (1978).
16. L. J. Crane, *J. Eng. Math.*, **10**, 115 (1976).
17. C. C. Chen and R. Eichorn, *Trans. ASME, Ser. C, J. Heat Transfer*, **98**, 446 (1976).
18. V. A. Zaitsev and Yu. A. Sokovishin, in: *Heat and Mass Transfer–VI*, vol. 1, part 3, Minsk (1980), pp. 82–86.
19. T. Fujii and M. Fujii, *Repts. Res. Inst. Ind. Sci. Kyushu Univ.,* No. 69, 15 (1979).
20. T. H. Kuehn, *Trans. ASME, Ser. C, J. Heat Transfer*, **100**, 374 (1978).
21. T. Fujii, S. Koyama, and M. Fujii, *Repts. Res. Inst. Ind. Sci. Kyushu Univ.*, No. 71, (1980), pp. 37–43.
22. P. H. Oosthuizen, *Trans. CSME*, **5**, 231 (1978–1979).
23. V. S. Antonova, S. I. Isataev, and V. P. Kashkarov, *Fizika (Alma-Ata)*, No. 5, 61 (1971).
24. H. R. Nagendra, M. A. Tirunarayanan, and A. Ramachandran, *Nucl. Eng. Des.*, **16**, 153 (1970).

25. F. A. Bottemanne, *Appl. Sci. Res.*, **25**, 372 (1972).
26. M. Prud'Homme, T. Hung Nguen, and P. G. Mao, *Int. J. Heat Mass Transfer*, **36**, 2837 (1993).
27. K. Velusamy and V. K. Garg, *Int. J. Heat Mass Transfer*, **36**, 1853 (1993).
28. B. Gebhart, Y. Jaluria, R. L. Manajan, and B. Sammakia, *Buoyancy-Induced Flows and Transport*, Hemisphere Publ. Corp., Washington (1988).
29. Y. Jaluria, *Natural Convection Heat and Mass Transfer*, Pergamon, Oxford, U.K., (1980).
30. V. T. Morgan, *Adv. Heat Transfer*, **11**, 199 (1975).
31. D. C. Collis and M. J. Williams, *Aerodynamics Note* 140, *Aeronautical Research Laboratories,* Melbourne (1954).
32. V. G. Malinin, *Collection of Papers of Institute of Forest Engineering*, No. 93, 138 (1977) Moscow.
33. F. N. Lin, B. T. Chao, *Trans. ASME, Ser. C, J. Heat Transfer*, **96**, 435 (1974); addendum, **98**, 344 (1976).
34. C. B. Kim, T. J. Pontikes, and D. E. Wollersheim, *Trans. ASME, Ser. C, J. Heat Transfer*, **97**, 129 (1975).
35. R. M. Fand, E. W. Morris, and M. Lum, *Int. J. Heat Mass Transfer*, **20**, 1173 (1977).
36. P. M. Brdlik, V. S. Kuptsova, and V. G. Malinin, *Collection of Papers of Institute of Forest Engineering*, No. 102, 58 (1977) Moscow.
37. G. D. Raithby and K. G. T. Hollands, *Trans. ASME, Ser. C, J. Heat Transfer*, **98**, 72 (1976).
38. T. H. Kuehn and R. J. Goldshtein, *Int. J. Heat Mass Transfer*, **19**, 1127 (1976).
39. S. W. Churchill and H.-J. Theten, *Chem.-Ing. Techn.*, **47**, 453 (1975).
40. W. Churchill and H. H. S. Chu, *Int. J. Heat Mass Transfer*, **18**, 1323 (1975).
41. T. Saitoh and T. Sajiki, *Int. J. Heat Mass Transfer*, **36**, 1251 (1993).
42. J. C. Y. Koh and J. F. Price, *Trans. ASME, Ser. C, J. Heat Transfer*, **87**, 237 (1965).
43. J. C. Y. Koh, *Int. J. Heat Mass Transfer*, **7**, 811 (1964).
44. S. Nakai and T. Okazaki, *Int. J. Heat Mass Transfer*, **18**, 387 (1975).
45. V. I. Subbotin, L. N. Sorokin, D. M. Ovechkin, and A. P. Kudrjavtsev, *Metal Boilingunder the Conditions of Free-Convective Heat Transfer*, Nauka Press, Moscow (1969).
46. Z. S. Prasolov, *Inzh.-Fiz. Zh.*, **4**, 3 (1961).
47. N. Heya, M. Takeuchi, and T. Fujii, *Trans. JSME*, **44**, 2789 (1978)
48. T. Tsubouchi and H. Masuda, *Repts. Res. Inst. Sci. Tohoku Univer. High Speed Mech.*, vol. 18B, (1966/1967), pp. 211–242.
49. G. N. Facas and H. L. Brown, *Int. J. Heat Mass Transfer*, **39**, 2077 (1996).
50. M. A. Yaghoubi and F. P. Incopera, *Proc. 6th Int. Heat Transfer Conf.*, Toronto, vol. 2, (1976), pp. 269–274.
51. C. C. Chen and R. Eichorn, *Trans. ASME, Ser. C, J. Heat Transfer*, **101**, 566 (1979).
52. H. Koizumi and I. Hosokawa, *Int. J. Heat Mass Transfer*, **39**, 1081 (1996).
53. O. A. Arnas, *Appl. Sci. Res.*, **27**, 81 (1972)
54. R. M. Fand and J. Brucker, *Int. J. Heat Mass Transfer*, **26**, 709 (1983).
55. T. H. Kuehn and R. G. Goldstein, *Int. J. Heat Mass Transfer*, **23**, 971 (1980).
56. A. Sesterhenn and U. Muller, *Arch. Mech.*, **28**, 903 (1976).
57. H. Jungbluth and U. Muller, *Arch. Mech.*, **32**, 859 (1978).

58. G. F. Marsters, *Can. J. Chem. Eng.*, **53**, 144 (1975).
59. K. Kato, *Mem. Ehime Univ.*, **4**, sec. 3, 1 (1960).
60. J.-L. Peube and D. Blay, *Int. J. Heat Mass Transfer*, **21**, 1125 (1978).
61. J.-L. Peube and D. Blay, *C. r. Acad. Sci.*, **287B**, 223 (1978).
62. W. E. Stewart, *Int. J. Heat Mass Transfer*, **14**, 1013 (1971).
63. R. C. L. Boswort, *Heat Transfer Processes*, GITL Press, Moscow (1957).
64. P. H. Oosthuizen, *Trans. ASME, Ser. C, J. Heat Transfer*, **98**, 672 (1976).
65. M. Al-Arabi and J. K. Salman, *Int. J. Heat Mass Transfer*, **23**, 45 (1980).
66. M. Al-Arabi and M. Khamis, *Int. J. Heat Mass Transfer*, **25**, 3 (1982).
67. T. Fujii, M. Fujii, and S. Koyama, *Repts. Res. Inst. Ind. Sci. Kyushu Univ.*, No. 70, 31 (1979).
68. J. Chand and D. Vir, *Lett. Heat Mass Transfer*, **7**, 213 (1980).
69. M. Alamgir, *Trans. ASME, Ser. C, J. Heat Transfer*, **101**, 174 (1979).
70. R. G. Hering and R. J. Grosh, *Int. J. Heat Mass Transfer*, **5**, 1059 (1962).
71. R. G. Hering, *Int. J. Heat Mass Transfer*, **8**, 1333 (1965).
72. F. N. Lin, *Lett. Heat Mass Transfer*, **3**, 49 (1976).
73. P. H. Oosthuizen and E. Donaldson, *Trans. ASME, Ser. C, J. Heat Transfer*, **94**, 330 (1972).
74. H. Koyama, S. Osawa, and R. Izumi, *Trans. JSME*, **43**, 1867 (1977).
75. T. Y. Na and J. P. Chiou, *Wu S*, **12**, 83 (1979).
76. T. Y. Na and J. P. Chiou, *Wu S*, **13**, 73 (1980).
77. T. Y. Na and J. P. Chiou, *Appl. Sci. Res.*, **35**, 409 (1979).
78. R. Manvi, *Ind. J. Technol.*, **9**, 207 (1971).
79. P. H. Oosthuizen, *Trans. ASME, Ser. C, J. Heat Transfer*, **95**, 409 (1973).
80. T. Cebeci and T. Y. Na, *Phys. Fluids*, **12**, 463 (1969); erratum, **13**, 536 (1970); comments, **15**, 211 (1972).
81. J. P. Narain and M .S. Uberoi, *Phys. Fluids*, **15**, 928 (1972).
82. T. Convindarajulu, *Ind. Chem. Eng.*, **14**, 41 (1972).
83. G. D. Raithby and K. G. T. Hollands, *Trans. ASME, Ser. C, J. Heat Transfer*, **98**, 522 (1976).
84. H. Nakamura and Y. Asako, *Bull. JSME,* **21**, 471 (1978).
85. J. H. Merkin, *Trans. ASME, Ser. C, J. Heat Transfer*, **99**, 453 (1977).
86. H. M. Badr and K. Shamsher, *Int. J. Heat Mass Transfer*, **36**, 3593 (1993).
87. E. Hahne and D. Zhu, *Int. J. Heat Mass Transfer*, **37**, Suppl. 1, 59 (1994).
88. G. F. Shilin, E. F. Kharchenko, G. A. Kalachevskaya, N. A. Blanutsa, in: *Heat Transfer Problems and Determination of Heat Transfer Characteristics*, Tomsk (1971), pp. 103–105.
89. G. F. Shilin and E. F. Kharchenko, in: *Heat and Mass Transfer in Bodies and Systems under Different Boundary Conditions*, No. 4, Omsk (1973), pp. 102−105.
90. K. Kato and H. Ito, *Trans. JSME*, **23**, Rep. 1, 918 (1957).
91. M. A. Hossain and B. Gebhart, in: *Heat Transfer*, vol. 4, (1970), NC 1.6, Paris–Versailles.
92. C. J. Cremers and D. L. Finley, in: *Heat Transfer*, vol. 4, (1970), NC 1.5, Paris–Versailles.
93. T. Fujii, M. Fujii, and T. Honda, *Trans. JSME, Ser. B*, **45**, 993 (1979).
94. A. Campo, *Wu S*, **13**, 93 (1980).
95. H. Gia and G. Gogos, *Int. J. Heat Mass Transfer*, **39**, 1603 (1996).
96. K. Jafarpur and M. M. Yovanovich, *Int. J. Heat Mass Transfer*, **35**, 2195 (1992).

97. A. Yamanaka, T. Yuki, and N. Mitsuishi, *J. Chem. Eng., (Japan)*, **9**, 445 (1976).
98. G. D. Raithby, A. Pollard, K. G. T. Hollands, and M. M. Yovanovich, *Trans. ASME, Ser. C, J. Heat Transfer*, **98**, 452 (1976).
99. W. M. Lewandowski, P. Kubski, J. M. Khubeiz, H. Bieszk, T. Wilczewski, and S. Szymanski, *Int. J. Heat Mass Transfer*, **40**, 101 (1997).
100. E. V. Kudrjavtsev, *Izv. Akad. Nauk SSSR, Otd. Tekh. Nauk*, No. 1, 53 (1948).
101. M. K. Kosvintseva, *Konvektivnye Techenija*, No. 1, 95 (1979), Perm.
102. M. K. Kosvintseva, *Gidrodinamika*, No. 7, 47 (1974), Perm.
103. S. M. Tieng and A. C. Yan, *Int. J. Heat Mass Transfer*, **36**, 599 (1993).
104. N. N. Kortikov, Yu. A. Sokovishin, and V. N. Piskunov, in: *Energy Transfer Problems in Inhomogeneous Media*, Minsk (1975), pp. 172–182.
105. V. N. Nikitin and A. M. Pylajev, in: *Heat and Mass Transfer Investigation*, No. 4, (1979), pp. 100–111.
106. M. G. Alishaev, *Vestnik MGU, Matematika Mekhanika*, No. 3, 57 (1963).
107. G. M. Harpole and I. Catton, *Trans. ASME, Ser. C, J. Heat Transfer*, **98**, 208 (1976).
108. K. Kishinami, H. Saito, and I. Tokura, *Trans. JSME, Ser. B*, **45**, 1613 (1979).
109. V. D. Afonin, V. P. Brusov, *Collection of Papers of Scientific-Reseach Institute of Sanitary Engineering*, No. 40, 95 (1973), Moscow.
110. P. H. Oosthuizen, *NHT*, **2**, 477 (1979).
111. Z. Nhojnacki and J. Hauser, *Elektrowärme Inst., Ser. B*, **36**, 94 (1978).
112. M. E. Ali, *Int. J. Heat Mass Transfer*, **37**, 665 (1994).
113. M. N. Uljanitskii, *Elektrichestvo*, No. 5, 71 (1965).
114. E. Kim, *Int. J. Heat Mass Transfer*, **40**, 3069 (1997).
115. M. E. Ali, *Int. J. Heat Mass Transfer*, **41**, 2175 (1998).

4

Natural Convection in Enclosures

The initial stage of heating a motionless fluid in a cavity is characterized by an increase in the wall temperature and by a distribution of a temperature wave in the mass of the fluid. After some characteristic time the temperature wave reaches the center, and then the difference between the wall and center temperatures begins to tend to a constant value. When the temperature difference becomes approximately constant, the natural-convective process turns to a quasi-steady regime of flow and heat transfer.

In the general picture of temperature distribution in the middle and bottom parts of the cavity two regions can be distinguished: a core characterized by a horizontally uniform temperature and a boundary layer near the wall, where the main change in the temperature occurs. In most cases, these two areas are coupled, and this factor strongly complicates the task of studying the problem of free-convective flows inside closed cavities.

An additional criterion for calculation of heat transfer inside a cavity is the Fourier number.

The process of free-convective heat transfer in a cavity is accompanied by thermal stratification in the flow core, with the temperature of the upper liquid layers exceeding the average values. The stratification can be significant in the case of nonuniform heating of the cavity, heat generation in the liquid or heat and mass transfer on the free surface.

In calculating free-convective heat transfer in the case of a long process of heat supply it becomes necessary to take into account the influence of a change in a pressure during the process of heating.

4.1 Spherical and Cylindrical Cavities

A spherical cavity ***fully filled with a fluid*** with coordinate-uniform and time-independent boundary conditions. For $\mathrm{Ra}_D < 10^3$ the heat transfer in the cavity occurs due to thermal conductivity, and $\mathrm{Nu}_D \to 2$ [4.1]. Generalization of experimental data ***for distilated water at a constant rate of***

increase in the temperature difference between the environment and the sphere centre yields the dimensional relation

$$\mathrm{Nu} = 0.429\ \mathrm{Ra}_D^{0.3}\ , \tag{4.1.1}$$

$$\mathrm{Pr} = 2 - 10, \quad \mathrm{Ra}_D = 6 \cdot 10^5 - 9 \cdot 10^7, \quad T_{\mathrm{av}} = 0.75\ T_\infty + 0.25\ T_{\mathrm{w}}\ .$$

The experiments with ***water and a petrolsolvent*** with ***an equidistant change in the wall and fluid temperature*** (heating and cooling) are approximated as [4.2]

$$\mathrm{Nu} = 0.46\ \mathrm{Ra}_D^{0.3}\ , \tag{4.1.2}$$

$$\mathrm{Pr} = 1 - 20, \quad \mathrm{Ra}_D = 10^6 - 10^9, \quad T_{\mathrm{av}} = 0.5\,(T_\infty + T_{\mathrm{w}})\ .$$

The time of the heating of the fluid in the case of laminar free convection is

$$\tau_l = 0.362 \frac{d^2}{a}\ \mathrm{Ra}_D^{-0.25}\ .$$

Visual observations show that a thin boundary layer is formed near the wall with a large temperature gradient. The temperature changes also inside the core along the vertical. The temperature fields in the horizontal sections of the core are practically uniform. When $\mathrm{Ra}_D > 10^8$, a transient flow regime is observed. Generalization of the experimental data ***for water*** gives [4.1]

$$\mathrm{Nu} = 0.098\ \mathrm{Ra}_D^{0.345}\ , \tag{4.1.3}$$

$$\mathrm{Pr} = 2 - 10, \quad \mathrm{Ra}_D = 10^8 - 10^{11}, \quad T_{\mathrm{av}} = 0.5(T_\infty + T_{\mathrm{W}})\ .$$

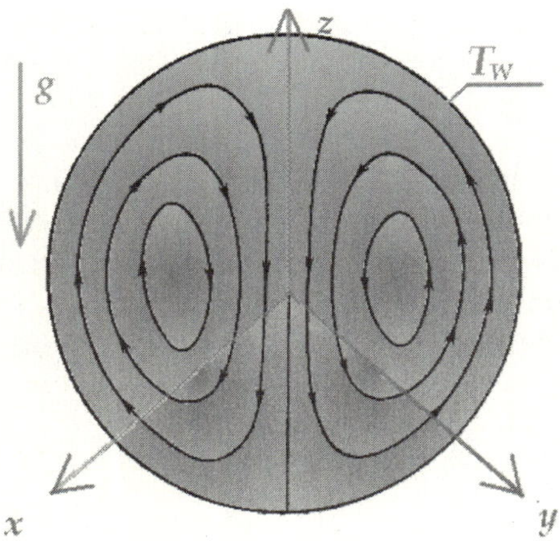

In the case of cooling of ***deuterium, hydrogen, and nitrogen*** [4.3]

$$\mathrm{Nu} = 0.104\ \mathrm{Ra}_D^{0.345}\ , \tag{4.1.4}$$

$$\mathrm{Pr} = 1.2 - 2.5, \quad \mathrm{Ra}_D = 7 \cdot 10^8 - 6 \cdot 10^{11}, \quad T_{\mathrm{av}} = T_\infty\ .$$

For the numbers Pr = 1–10 the critical value Ra_{Dk} of the transition to the turbulent regime changes within the range $2.2 \cdot 10^9 - 1.8 \cdot 10^{10}$ [4.2]. The experimental dependence of heat transfer ***in the turbulent regime*** is:

$$\mathrm{Nu} = 0.012\ \mathrm{Ra}_D^{0.345}\,, \qquad (4.1.5)$$
$$\mathrm{Pr} = 1 - 10, \quad \mathrm{Ra}_D > 10^{10}, \quad T_{\mathrm{av}} = 0.5\,(T_\infty + T_{\mathrm{w}})\ .$$

The time of liquid heating ***in the turbulent regime*** is

$$\tau_l = 14\frac{d^2}{a}\ \mathrm{Ra}_D^{-0.4}\ .$$

In the regime of cooling in the bottom region of the cavity a stagnant zone appears in which cold liquid is collected and which increases in the vertical direction. In the heating regime, the stagnant zone is formed in the upper part of the cavity. The presence of the stagnant zones reduces the heat transfer and can lead to wall overheating.

Heat transfer of a spherical cavity on ***instantaneous supply of a heat flux to the surface*** is described by [4.4]

$$\mathrm{Nu} = 1.044\ \mathrm{Ra}_D^{*0.1933} \qquad (4.1.6)$$
$$\mathrm{Pr} = 1 - 5, \quad \mathrm{Ra}_D = 10^7 - 10^{11}, \quad T_{\mathrm{av}} = 0.5\,(T_\infty + T_{\mathrm{w}})\ .$$

Heat transfer of a ***nonuniformly heated sphere***: the spherical cavity ***filled with water*** is firmly clamped between a refrigerator and a heater. For different positions of the refrigerator and heater, the average heat transfer coefficient is determined by the dependence [4.5]

$$\mathrm{Nu} = C_\vartheta \mathrm{Ra}_D^{0.25}\,, \qquad (4.1.7)$$
$$\mathrm{Pr} = 2 - 10, \quad \mathrm{Ra}_D = 4 \cdot 10^5 - 4 \cdot 10^7, \quad T_{\mathrm{av}} = 0.5\,(T_\infty + T_{\mathrm{w}})\ ,$$

where $C_\vartheta = 0.3, 0.275$ and 0.225 for $\vartheta = 0, \pi/4$, and $\pi/2$, respectively (the angle ϑ is counted from below). The dependence of Nu on Ra_D becomes weaker when $\vartheta > \pi/2$ and passes to $\mathrm{Nu} = 2$ for $\vartheta = \pi$.

When a spherical cavity ***filled with a gas*** is emptied, the average heat transfer coefficient is determined from the formula [4.6]

$$\mathrm{Nu} = 0.095\ \mathrm{Ra}_D^{1/3}\,, \qquad (4.1.8)$$
$$\mathrm{Pr} = 0.72, \quad \mathrm{Ra}_D = 1 \cdot 10^9 - 2 \cdot 10^{11},$$
$$T_{\mathrm{av}} = 0.5\,(T_\infty + T_{\mathrm{w}})\,, \quad p = (4 - 30) \cdot 10^5\ \mathrm{Pa}\ .$$

The stationary value of the surface-averaged Nusselt criterion in a spherical cavity ***with uniformly distributed heat sources and isothermal walls*** is [4.7]

$$\mathrm{Nu} = 10 + 1.5 \cdot 10^{-11}\,\mathrm{Gr}_{vD}^2 \quad \text{for } \mathrm{Gr}_{vD} < 10^6, \tag{4.1.9}$$

$$\mathrm{Nu} = 2.51\,\mathrm{Gr}_{vD}^{0.13} \quad \text{for } \mathrm{Gr}_{vD} = 10^6 - 10^{10}\,, \tag{4.1.10}$$

$$\mathrm{Pr} = 0.7, \quad T_{\mathrm{av}} = \frac{1}{V}\int\limits_V T dV, \quad \mathrm{Nu} = \frac{q_{\mathrm{w}}}{T_{\mathrm{av}} - T_{\mathrm{w}}}\frac{D}{\lambda}\,.$$

When $\mathrm{Gr}_{vD} > 10^{10}$, the boundary layer becomes unstable and is turbulized in the upper part of the cavity.

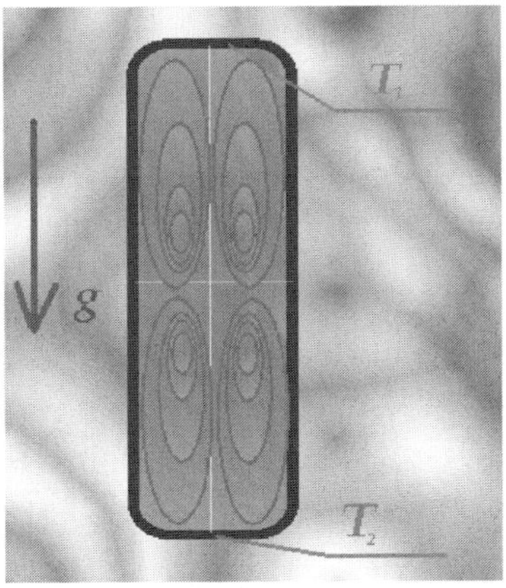

A laminar natural convection ***in a vertical cylinder of height H*** was mainly investigated by numerical methods of solving the full Navier-Stokes and energy equations [4.8,4.9]. For $\mathrm{Ra}_H \leq 6 \cdot 10^9$ ***heat supply to the lateral surface***, the heat transfer rate from the wall quickly decreases, and then fluctuates about some average value

$$\mathrm{Nu}/\mathrm{Ra}_H^{1/4} \cong 0.54\,. \tag{4.1.11}$$

The fluctuations of heat transfer correspond to the fluctuations of the velocity and temperature fields connected with the appearance and disappearance of secondary vortieces near the cavity axis.

Experimental and theoretical investigations ***in a vertical cylinder and semicylinder filled with fluids on instantaneous increase in the temperature*** yield [4.10,4.11]

$$\mathrm{Nu} = 0.524\,\mathrm{Ra}_D^{1/4}\,, \tag{4.1.12}$$

$$H/D = 0.25 - 2, \quad \mathrm{Pr} = 5 - 80000, \quad \mathrm{Ra}_D = 8 \cdot 10^5 - 4 \cdot 10^{10}\,,$$

where the Ra_D number is determined from the temperature difference between the wall and the geometrical center of the cavity.

A numerical investigation of a vertical cylindrical tank with $H/D = 1$, $\mathrm{Pr} = 1$, and $q_{\mathrm{w}} = \mathrm{const}$ shows that in the beginning of the process of natural convection development its velocity increases, reaches a maximum, and then decreases to a certain stationary value. The convective flow brings the hot liquid to the upper layer and noticeably distorts the temperature field which becomes two-dimensional or three-dimensional due to this effect. When a quasi-stationary regime is attained in the upper part of the cavity into which the hot liquid enters, the wall temperature can be higher compared with the heat transfer only due to the thermal conductivity. The time of the attainment of the quasi-stationary regime at $\mathrm{Ra}_H^* \geq 10^5$ is determined from the equation

$$Fo_H = 40/\mathrm{Ra}_H^{*0.4}$$

and decreases with an increase in Ra_H^*. The approximation of numerical calculations yields [4.9]

$$\mathrm{Nu} = 0.55\ \mathrm{Ra}_H^{*0.2}\,, \qquad (4.1.13)$$
$$\mathrm{Pr} = 1, \quad \mathrm{Ra}_H^* = 10^5 - 10^{10}\,.$$

When For $\mathrm{Ra}_H^* \leq 10^5$, the dependence $\mathrm{Nu} = f(\mathrm{Ra}_H^*)$ deviates from (4.1.13) and tends to the value of Nu = const, corresponding to the quasi-stationary regime of the thermal conductivity of a motionless liquid. In an intermediate regime, the temperature stratification increases with an increase of Ra_H^*, reaches a maximum at $\mathrm{Ra}_H^* = 10^5$, and decreases in the boundary layer regime.

As the cylinder height increases the stratification depends less on H/D [4.3, 4.9, 4.10].

Dependence (4.1.5) is used to calculate heat transfer rate in a vertical cylinder in the transient regime of natural convection.

The data of experimental and numerical investigation of laminar free convection in a vertical cylinder ***with adiabatic horizontal and isothermal vertical walls with uniformly distributed internal heat sources*** for the average Nusselt number are generalized by the dependence [4.12]

$$\mathrm{Nu} = 0.576\ \mathrm{Ra}^{0.2024}(H/D)^{-0.186}\,, \qquad (4.1.14)$$
$$3 \cdot 10^{10} < \mathrm{Ra} < 10^{13}\,.$$

For $10^{13} < \mathrm{Ra} < 10^{14}$ a transition to a turbulent regime is observed, with the flow turbulization rate strongly depending on the boundary conditions.

The results of numerical simulation of laminar convection ***in a vertical cylinder heated from the side and cooled from above at the same value of the heat flux density on the surface*** are generalized by the dependences [4.13]

$$\mathrm{Nu} = 0.55\,A^{-1.25}\ \mathrm{Ra}^{0.5}\,, \qquad (4.1.15)$$
$$10^2 \leq \mathrm{Ra} \leq 10^6, \quad 1/2 \leq A \leq 2\,,$$
$$\mathrm{Nu} = 0.54\,A^{-0.74}\ \mathrm{Ra}^{0.2}\,, \qquad (4.1.16)$$
$$10^4 \leq \mathrm{Ra} \leq 10^6, \quad 1/5 \leq A \leq 1/3\,,$$

where $A = R/H$. In the transient regime, when $A \leq 1/3$, two vortices are observed; in a stationary flow there is one circulation cell in the entire range of investigated parameters. The average heat transfer value in the range $1/5 \leq A \leq 1/2$ does not depend on the Prandtl number.

A vertical cylinder ***filled with water heated from below and insulated from the side***. The average heat transfer value with an error of 2% ***with turbulent free convection*** is generalized by the dependence [4.14]

$$\mathrm{Nu} = 0.0556\ \mathrm{Ra}_H^{1/3}\ , \tag{4.1.17}$$

$$H/D = 0.22 - 1.75, \quad \mathrm{Pr} = 5.7 - 6.65\ , \mathrm{Ra}_H = 1 \cdot 10^8 - 2.26 \cdot 10^{11}\ .$$

Under mixed boundary conditions ($q =$ const on the lateral surface and on the base and $q = 0$ on the free surface) in a cavity filled up to the height h the effect on heat transfer is limited to a narrow near-wall area on the bottom of the cavity [4.15]. The flow directed vertically exerts the main effect on the liquid stratification. For $H/D = 0.5 - 2, \mathrm{Pr} = 0.5 - 2, \mathrm{Gr}_D = 4 \cdot 10^8 - 1.6 \cdot 10^9$ the number Pr has a greater effect on the heat transfer rate, than a change in the Gr_D number. The convection rate decreases due to the vertical heat flux, and its effect on heat transfer in the cavity decreases substentially.

A long cylinder with the temperatures T_1 and T_2 on the end-face surfaces. An inclined cylinder filled with **He** ***or*** **Ar**. The empirical dependence for two geometrical configurations [4.16] (the angle ϑ is counted from the vertical; $\vartheta = 0 - 90°$; $p = 5 - 20$ MPa; $H/D = 14.75, 16.7$) is

$$\mathrm{Ra}_H = 10^8 - 10^{12}\ ,$$

$$\mathrm{Nu}_D = 10^{q_1}\ \mathrm{Ra}_H^{q_2} (H/D)^{q_3}\ , \tag{4.1.18}$$

where

$$q_1 = -6.6 + 0.174\vartheta - 15.4 \cdot 10^{-3}\vartheta^2 + 6.11 \cdot 10^{-6}\vartheta^3\ ,$$

$$q_2 = 0.77 - 0.105\vartheta + 9.44 \cdot 10^{-5}\vartheta^2\ ,$$

$$q_3 = -0.35 - 0.03\vartheta + 5.26 \cdot 10^{-6}\vartheta^2 - 3.38 \cdot 10^{-6}\vartheta^3\ .$$

The numerical calculations of natural convection ***in a vertical asymmetrically heated cylinder*** showed, that the degree of the asymmetry appearing due to the non-uniform heating of the flow depends on the ratio between the amplitudes of the maximal azimuthal and maximal vertical temperature differences, and at a constant value of this ratio the flow asymmetry increases with the Rayleigh number [4.17]. The calculations were carried out for $\mathrm{Pr} = 0.01$ and 1 and for $\mathrm{Ra} = 250$, 2500 and 15000.

A vertical cylinder ***with uniformly distributed heat sources with the density Q_v. The end-face surfaces are maintained at the constant temperature $T_1 = T_2$***. The numerical calculations and experiments where performed for the cases of ***insulated lateral walls***. The data on heat transfer are generalized by the dependence [4.18]

$$\mathrm{Nu} = C\mathrm{Ra}_{vH}^{n} \ . \tag{4.1.19}$$

The lateral surface temperature is equal to the temperature of the upper and lower walls;

$H/D = 0.5$:

for the upper surface	for $\mathrm{Ra}_{vH} > 5.5 \cdot 10^6$,	$C = 0.808$	$n = 0.176$;
	for $\mathrm{Ra}_{vH} < 5.5 \cdot 10^6$,	$C = 0.523$	$n = 0.203$;
for the lower surface		$C = 1.204$	$n = 0.081$;
for the lateral surface		$C = 0.311$	$n = 0.217$.

The lateral wall is insulated;

$H/D = 0.5$:			
for the upper surface		$C = 0.414$,	$n = 0.225$;
for the lower surface		$C = 1.406$,	$n = 0.09$;
$H/D = 1$:			
for the upper surface	for $\mathrm{Ra}_{vH} > 5.5 \cdot 10^6$,	$C = 2.102$,	$n = 0.131$;
	for $\mathrm{Ra}_{vH} < 5.5 \cdot 10^6$,	$C = 0.167$,	$n = 0.284$;
for the lower surface		$C = 1.466$,	$n = 0.073$;
$H/D = 2.5$:			
for the upper surface	for $\mathrm{Ra}_{vH} > 5.5 \cdot 10^6$,	$C = 3.099$,	$n = 0.129$;
	for $\mathrm{Ra}_{vH} < 5.5 \cdot 10^6$,	$C = 0.109$,	$n = 0.47$;
for the lower surface		$C = 0.278$,	$n = 0.185$.

A vertical cylinder ($R = H$) *with heat sources* of diameter $2c = R/3$ with the given temperature. ***The source is located coaxially with the cylinder at its bottom. The wall temperature increases linearly with height***, the flow in the entire region is laminar. The fluid stratification S exerts its influence on heat transfer. The heat flux increases with Gr_c and for $\mathrm{Gr}_c > 10^5$ $Q \sim \mathrm{Gr}_c^{1/2}$. When $\mathrm{Gr}_c = 10^5$ at $S = 1, Q$ tends to an asymptotic value [4.19].

A vertical cylinder with ***a volumetric heat generation*** at small Rayleigh numbers in a wide range of the radius-to-height ratio $A = R/H$. For $0.2 < A < 1$ a single vortex of toroidal shape is formed in the cavity, which occupies almost the entire volume at $A = 1$ and which gradually moves to the upper region of the cylinder as A decreases to 0.2. Two vortices are seen in the cavity when $1 < A < 2$ and three when at $A = 3$ [4.20].

The processes of heat transfer ***in a completely filled horizontal cylinder*** are considered two-dimensional at $L/D \geq 2.5$ [4.21, 4.22]. The developed free convection regime is characterized by the presence of a boundary layer near the walls and a central core with almost isothermal horizontal sections and a stratified temperature profile in the vertical direction. The increase in the thickness of the thermal boundary layer along the walls from the lower pole to the upper one leads to a nonuniform distribution of the heat transfer coefficient in this direction (heat transfer rate in the lower region is higher, than in the upper one). The average value of the heat transfer coefficient after the initial decrease is stabilized for a while, and then slowly decreases.

The transient period is characterized by a fast temperature increase. Its duration is determined by the number Ra_{d0} calculated for the initial time [4.22]:

$$Fo = 0 - Fo_1, \quad Fo_1 = 0.4\,\mathrm{Ra}_{d0}^{0.25}\,,$$
$$\mathrm{Pr} = 0.7 - 25, \quad f(\mathrm{Pr}) = 1 - 0.21\exp(-0.247\ \mathrm{Pr})\,.$$
$$\mathrm{Nu} = \frac{0.279 f(\mathrm{Pr})\,\mathrm{Ra}_{d0}^{0.15}}{Fo\left[Fo^{-0.575} - 0.486 f(\mathrm{Pr})\,\mathrm{Ra}_{d0}^{0.25}\right]}\,, \tag{4.1.20}$$
$$\mathrm{Nu} = \frac{0.486 f(\mathrm{Pr})\,Fo_1^{0.575}}{Fo_1\left[1 - 0.309 f(\mathrm{Pr})\,Fo_1^{0.575}\,\mathrm{Ra}_{d0}\right]}\,. \tag{4.1.21}$$

Quasi-stationary regime

$$\mathrm{Pr} = 0.7 - 2.5, \quad \mathrm{Ra}_d = 10^5 - 10^{10}\,,$$
$$\mathrm{Nu} = 0.59\,\mathrm{Ra}_d^{0.25}\,. \tag{4.1.22}$$

The generalized dependence for heat transfer ***with laminar and turbulent free convection inside the horizontal cylinder*** is [4.23]

$$\mathrm{Nu} = -2/\ln\left(1 - \frac{2}{\mathrm{Nu}_d}\right)\,, \tag{4.1.23}$$

where

$$\mathrm{Nu}_d = \left\{\left(\left[\left(\frac{2}{1-e^{-0.25}}\right)^{5/3} + \left(0.587G\,\mathrm{Ra}_d^{1/4}\right)^{5/3}\right]^{3/5}\right)^{15} + \left(0.1\,\mathrm{Ra}_d^{1/3}\right)^{15}\right\}^{1/15}\,,$$
$$G = \left[\left(1 + 0.6\,\mathrm{Pr}^{-0.7}\right)^{-5} + \left(0.4 + 2.6\,\mathrm{Pr}^{0.7}\right)^{-5}\right]^{-1/5}\,.$$

Numerical calculations for a transient turbulent convection in a horizontal cylindrical container which were confirmed by the experimental data of [4.24] reveal the presence of three periods: a regime of heat conduction, a developing flow, and a steady quasistationary flow. The container is filled with ***water***, ***a partially-continuous constant heat flux*** is applied to the outer wall. The number $\mathrm{Ra} = 4.01 \cdot 10^{12}$. The flow structure changes during the transition from two-vortical to a rather unstable four-vortical accompanied by an asymmetric motion near the container bottom, and then again to the two-vortical phase. The results of calculations and experiments show a significant effect of the thermal conductivity of the wall on the characteristics of flow and heat transfer.

In the problems of conjugate heat transfer on the surface, additional thermal resistances of the wall $(\Sigma\delta_i/\lambda_i)$ and environment $(1/\alpha_\mathrm{w})$ should be taken into account. The dimensional relation for the quasistationary heat transfer regime in this case is [4.25]

$$\mathrm{Nu} = f(\mathrm{Bi}_R)\mathrm{Nu}_\infty\,, \tag{4.1.24}$$

$$f(\mathrm{Bi}_R) = 1 - 0.39\ \mathrm{Bi}_R^{-0.51}\,, \quad \mathrm{Bi}_R = \frac{R}{\lambda(1/\alpha_\mathrm{w} + \Sigma\delta_i/\lambda_i)}, \quad \mathrm{Pr} = 7\,,$$

where Nu_∞ refers to the assignment of a constant temperature on the surface.

For small values of Bi_R the heat transfer is determined by the value $k_\mathrm{w} = (1/\alpha_\mathrm{w}+\Sigma\delta_i/\lambda_i)$. When $\mathrm{Bi} > 10$, the coefficients k_w and α_w are commensurable, and their effect on heat transfer is approximately the same. When $\mathrm{Bi} \to \infty$, the process of heating a fluid in the cylinder is fully determined by the inner conditions of the problem. By the dimensional analysis method for a conjugate problem it is shown that the additional parameter is Bi/k, where $k = [(d+2\delta)^2 - d^2]\rho_\mathrm{w}c_\mathrm{w}/(2d^2\rho_l c_l)$ [4.26]. In practical problems the parameter k changes in the range 0.03–0.2.

The dimensional relations for different values of the parameters are

$$\mathrm{Nu}_R = 0.476\ \mathrm{Ra}_R^{0.334}\,, \tag{4.1.25}$$

$$R = 0.05 - 0.25\ \mathrm{m},\ \mathrm{Bi}_R = 10^5,\ \mathrm{Ra}_R = 10^7 - 10^{10}\,.$$

$$\mathrm{Nu}_R = 0.44\ \mathrm{Ra}_R^{0.25}\,, \tag{4.1.26}$$

$$R = 0.025 - 0.032\ \mathrm{m},\ \mathrm{Bi}_R = 160,\ \mathrm{Ra}_R = 5.5\cdot 10^5 - 5.5\cdot 10^9\,.$$

$$\mathrm{Nu}_R = 0.14\ \mathrm{Ra}_R^{0.338}\,, \tag{4.1.27}$$

$$R = 0.025 - 0.032\ \mathrm{m},\ \mathrm{Bi}_R = 143 \quad (\mathrm{Bi}_R/k = 284)\,,$$

$$\mathrm{Ra}_R = 5\cdot 10^5 - 8\cdot 10^6\,.$$

Quasistationary heat transfer inside ***a horizontal cylinder with a constant increase of the wall temperature*** ($d\theta_\mathrm{w}/d\tau = \mathrm{const}$). In this case $\theta_\mathrm{w} - \theta_l \cong \mathrm{const}$ during the whole heat transfer process. Experimental investigations [4.27] and numerical calculations [4.28], with an error of 5%, show that for $\mathrm{Pr} > 0.7$ and $\mathrm{Ra}_D < 10^7$

$$\mathrm{Nu} = 4\sqrt{2}\left[\frac{2\cdot 10^{-4}\ \mathrm{Ra}_D}{\sqrt{(1+2\cdot 10^{-4}\ \mathrm{Ra}_D)} - 1}\right]^{1/2}\,. \tag{4.1.28}$$

Heat transfer ***in a long horizontal pipe with different temperatures T_1 and T_2 at the end-face surfaces*** with allowance for heat conduction around the tube perimeter of thickness δ_w is given by [4.29]

$$\mathrm{Nu}_R\frac{l}{R} = 1 + \frac{2+7P_\mathrm{w}}{1+P_\mathrm{w}}\frac{\left(\mathrm{Ra}_R\frac{R}{l}\right)^2}{46080} + O\left[\left(\mathrm{Ra}_R\frac{R}{l}\right)^4\right]\,, \tag{4.1.29}$$

$$P_\mathrm{w} = \frac{\lambda R}{\lambda_\mathrm{w}\delta}, \quad \mathrm{Ra}_R\frac{R}{l} \ll 1, \quad \frac{R}{l} \ll 1, \quad \frac{\delta_\mathrm{w}}{R} \ll 1\,.$$

Experimental investigations show that for $\mathrm{Bi}_R/k = 200-500$ the criterion Nu is independent of Ra_R ($\mathrm{Ra}_R > 10^6$) and is determined only by the complex

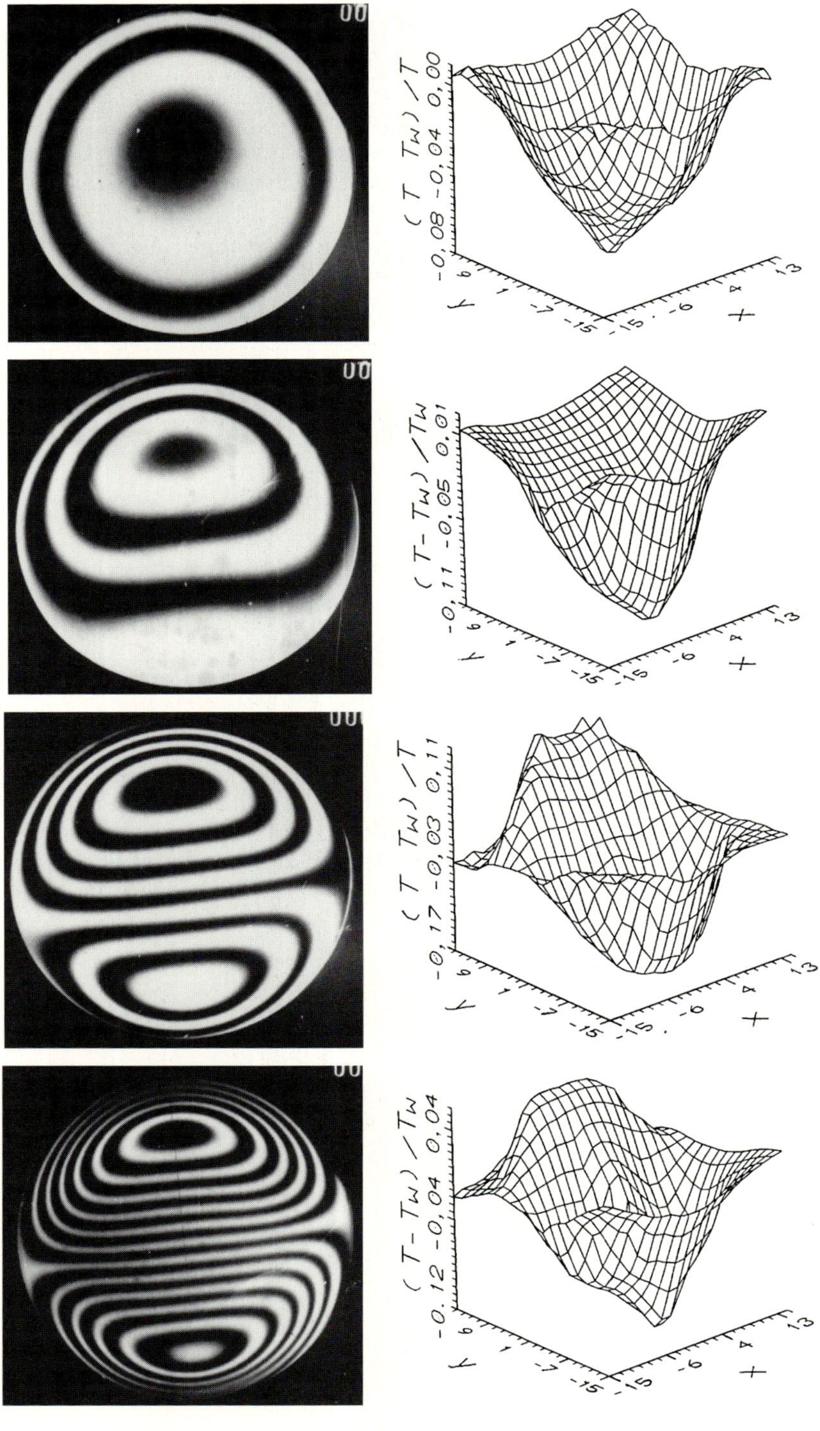

Fig. 4.1.1. Dynamics of temperature field development in a horizontal tube of circular section with a constant heat flux on the wall

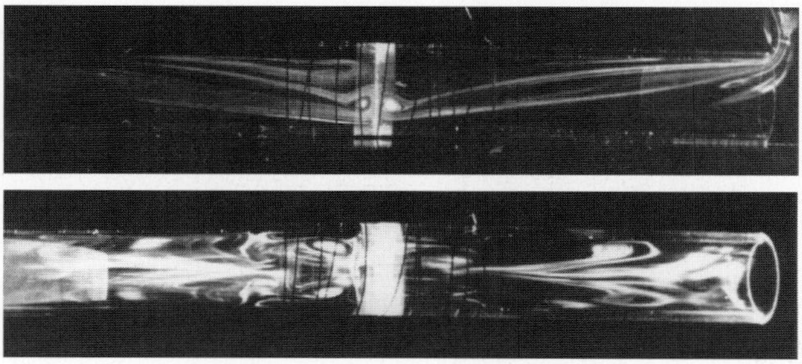

Fig. 4.1.3. Smoke visualization of a free-convective flow in a horizontal tube with a constant heat flux on the wall in the vertical and horizontal central sections

Bi_R/k [4.24]. Heat transfer inside the horizontal cylinder ***with a constant heat flux on the surface*** is [4.27]

$$\mathrm{Nu} = 1.215\ \mathrm{Ra}_D^{0.21}\,, \qquad (4.1.30)$$
$$\mathrm{Pr} > 0.7, \quad \mathrm{Ra}_D = 10^3 - 3 \cdot 10^7\,.$$

The experimental investigations carried out ***on liquid oxygen and nitrogen in a horizontal toroidal cavity filled to a different degree with these fluids*** show that heat transfer in it is accompanied by stratification of the liquid and gas [4.30]. The value of the temperature difference in the liquid decreases with increase in the distance from its free surface. To calculate heat transfer for different degrees of filling one can use formula (4.1.8) for a sphere. The distribution of the heat flux in the system is independent of the value of the supplied power and is determined by the degree of the filling of the volume by the liquid. The amount of thermal energy spent to evaporate the liquid and heat the walls increases with decrease of the degree of filling.

The cylindrical cavities with hemispheres on end-face surfaces, used for transportation and storage of fuel, were investigated to determine the temperature field in the liquid volume and to find efficient methods of elimination or substential reduction of the temperature distribution nonuniformity [4.22,4.30,4.31]. ***Under the conditions of heat supply to the lateral surface, corresponding to the part of the cylindrical cavity filled with a fluid to the height h*** in a certain period of time τ a certain difference between the temperature in the liquid volume and that of the free surface is established. This difference can be calculated by the formula [4.8]

$$\frac{T_\mathrm{w} - T}{q_\mathrm{w} h}\lambda = \frac{h}{R}\mathrm{Zh}\left\{\left[I\,\mathrm{Pr}\left\{\left[1 - \left(1 + 0.0924\frac{h}{R}\right)\frac{\mathrm{Zh}}{\mathrm{Pr}^{2/3}}\right]\right.\right.\right.$$
$$\left.\left.\left.\times\left(\frac{\mathrm{Gr}_h^*}{1 + 0.443\ \mathrm{Pr}^{2/3}}\right)^{2/7}\right\}^{-7}\right\}^{-1}, \qquad (4.1.31)$$

$$h/R = 1 - 3, \quad \mathrm{Pr} = 1.3 - 32, \quad \mathrm{Ra}_h^* = 10^8 - 10^{15}, \quad I \cong 0.35 - 0.65\,,$$

where I is the integral of the dimensionless temperature difference distribution along the thermal boundary layer thickness:

$$I = \int_0^1 \frac{1}{\delta_\theta} \frac{T_z - T}{T_\mathrm{w} - T} dz$$

and $\mathrm{Zh} = \nu\tau/h^2$ is the Zhukovskii criterion.

With supply of the heat flux Q_H to the spherical part of the base and Q_w to the lateral surface the distribution of the excess temperature along the height is

$$\frac{T_\mathrm{w} - T}{Q_\mathrm{w} H}\lambda = 0.048\ \mathrm{Ra}_H^* \left\{ \left[\mathrm{Gr}_H^* \frac{Q_H}{Q_\mathrm{w}} \left(\frac{H^2\ \mathrm{Pr}^3}{D^2 \mathrm{Gr}_H^*} \right) \right]^{5/6} \left(\frac{D}{H} \right)^2 \right\}^{2/7}, \quad (4.1.32)$$

$$D/H = 1 - 3, \quad H = h + R, \quad Q_H/Q_\mathrm{w} = 0.01 - 0.5\ ,$$

$$\left[\mathrm{Gr}_H^* \frac{Q_H}{Q_\mathrm{w}} \left(\frac{H^2\ \mathrm{Pr}^3}{D^2 \mathrm{Gr}_H^*} \right) \right]^{5/6} \left(\frac{D}{H} \right)^2 = 10^{15} - 2 \cdot 10^{19}\ .$$

To eliminate the temperature field nonuniformity in the volume partitions on the internal wall of the enclosure are used [4.31]. In moving ***along the wall with barriers*** the thermal boundary layer is destroyed on striking a barrier and then develops again downstream. The average value of the heat transfer coefficient for n partitions is

$$\alpha_n/\alpha = (n + 1)^{-0.2}\ , \quad (4.1.33)$$

where α_n and α are the heat transfer coefficients with partitions and without them, respectively. Dependence (4.1.33) is confirmed by experimental data for ***liquid nitrogen and water in vertical enclosures*** for $\mathrm{Gr}_D = 10^{11} - 6.3 \cdot 10^{14}, n = 0 - 6, \mathrm{Pr} = 2 - 7.5$. The effect of partitions on the rate of liquid stratification becomes essential, when their width exceeds the local thickness of the thermal boundary layer.

The height of the un-stratified region in a vertical cylinder ***filled with a viscous liquid at a constant temperature of the lateral surface*** is [4.10, 4.11]

$$z_{T \to \infty} = D/(D + 4H)\ .$$

To calculate heat transfer inside the region, formula (4.1.12) can be used.

The results of numerical calculations for the average heat transfer coefficient ***in a vertical cylindrical cavity heated from below are obtained for different cases of thermal boundary conditions on lateral surfaces (from adiabatically insulated to fully conducting walls)*** in the ranges of the parameters $1 \le A \le 5$ and $10^{-2} \le \mathrm{Pr} \le 10^5$ ($A = H/2R, H, R$ are the height and radius of the cylinder). The numerical data for $\mathrm{Pr} = 1$ are well approximated by the correlation obtained in [4.32], applied to the case of a vertical cylinder,

$$\mathrm{Nu} = 1 + 0.323A^{9/4}\left(\frac{A-0.1}{0.9}\right)^{-1/2} + 0.42\left[\left(\frac{\mathrm{Ra}}{A^3}\right)^{1/4} - 5000^{1/4}\right] . \quad (4.1.34)$$

In most cases of technical application of this kind of a geometrical configuration for $\mathrm{Pr} \geq 0.5$, the effect of the Prandtle number on the heat transfer characteristics is negligibly small. It is significant only when $\mathrm{Pr} < 0.1$. For small values $\mathrm{Pr} \leq 5 \cdot 10^{-2}$ the oscillatory flow regimes appear, whereas for $\mathrm{Pr} > 0.1$ the presence of the oscillatory character of the flow has not received experimental confirmation in the investigated ranges of parameters.

The results of numerical calculations for the free convection developing ***in a vertical porous cylinder due to internal heat generation*** show that for large values of the Ra number the flow structure is characterized by the presence of a uniform ascending flow in the central part and a thin descending boundary layer on the cooled surface [4.33].

A numerical investigation of natural convection ***in a horizontal circular porous cylinder with nonuniform heating of lateral walls*** leading to a linear temperature distribution in the vertical direction, ***with the end-face surfaces being insulated*** was carried out in [4.34]. When $L > 0.86$, a three-dimensional flow appears. For cylinders of small length ($L < 0.86$) two-dimensional convection develops, with two variants of solution consisting of two or three vortices existing for supercritical Rayleigh numbers. It is shown that any of these structures and also their compositions are stable. The flow is defined unambiguously in the case of external thermal effect.

4.2 Rectangular Cavities and Interlayers

Natural convection in a rectangular cavity of height H and width B under different boundary conditions on the surfaces is characterized in addition by the ratio H/B and by relationships between temperatures and heat fluxes. Heat transfer in a laminar regime is effectively investigated by using numerical methods [4.21, 4.35, 4.36].

The vertical surfaces are maintained at constant temperatures $T_0(y = 0)$ *and* $T_B(y = B)$ ***and two horizontal surfaces insulated are***. In the laminar regime of free convection in the cavity ***filled with gas*** the maximal temperature in the centre $T_{\max}$ changes by the law

$$\theta_{\max} = \frac{T_{\max} - T_B}{T_0 - T_B} = 0.2 + 0.6\frac{x}{H} ,$$

where the coordinates x and y correspond to the vertical and horizontal directions, respectively. The local value of heat transfer on the hot surface is determined by the dependence [4.37]

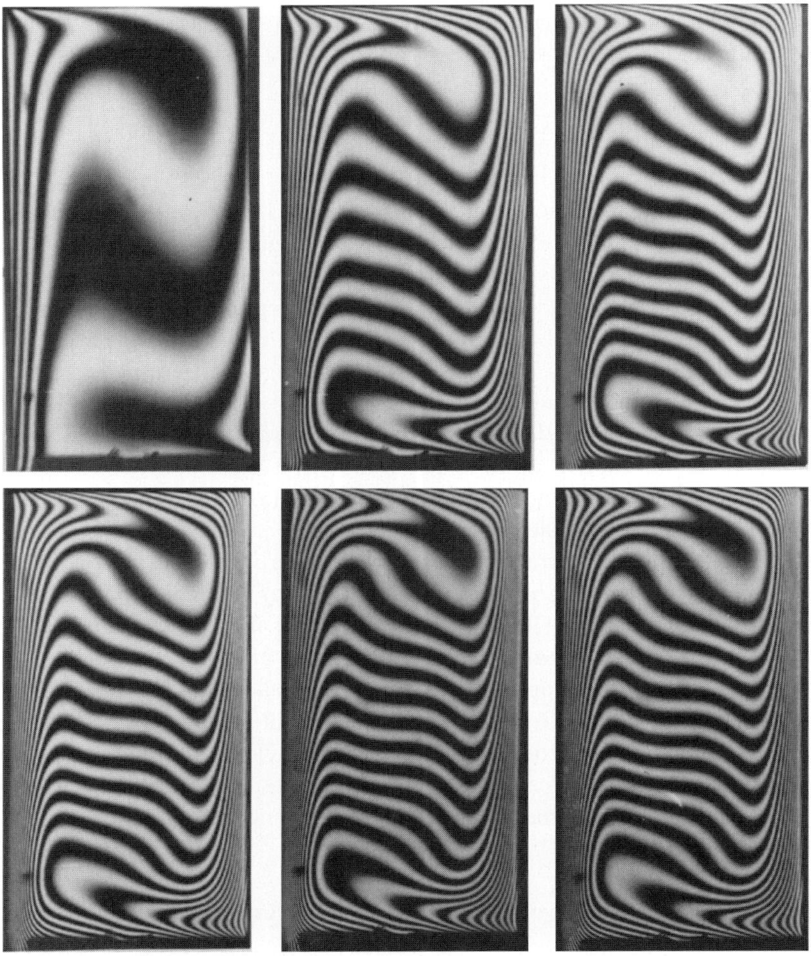

Fig. 4.2.1. Dynamics of the temperature field development in a horizontal channel of rectangular section at a constant difference of the temperatures between the opposite vertical walls

$$\mathrm{Nu}_{x,y=0} = 0.256 \left(\frac{H}{B}\right)^{-0.1} \mathrm{Ra}_B^{0.3} \left(0.8 - 0.6\frac{x}{H}\right) \left(\frac{x}{H}\right)^{-0.1} , \qquad (4.2.1)$$

$$\mathrm{Pr} = 0.71, \quad \mathrm{Ra}_B = 1.94 \cdot 10^3 - 2.3 \cdot 10^5, \quad H/B = 1.25 - 20 .$$

Due to the symmetry of the problem on the cold surface

$$\mathrm{Nu}_0(x) = \mathrm{Nu}_B(H - x) .$$

For $H/B \ll 1$ the average heat transfer coefficient depends slightly on Pr [4.38, 4.39]. The generalized dependence for $H/B \ll 1$ and $\mathrm{Ra}_H < 10^9$ is [4.39]

$$\mathrm{Nu}\frac{B}{H} = 1 + \left\{ \left[\frac{(\mathrm{Ra}_H H/B)^2}{362880} \right]^{-0.368} + \left(0.623\ \mathrm{Ra}_H^{1/5} \frac{B}{H} \right)^{-0.368} \right\}^{-1/0.368} . \tag{4.2.2}$$

The approximation of the calculated data with an error of ±7% yields [4.38]

$$\mathrm{Nu} = 0.28\ \mathrm{Pr}^{0.024} \left(\frac{H}{B} \right)^{1.75} \mathrm{Ra}_H^{1/2} , \tag{4.2.3}$$

$$H/B = 0.03 - 0.2, \quad \mathrm{Pr} = 1 - 10^3, \quad \mathrm{Ra}_H = 10^2 - 10^5 .$$

$$\mathrm{Nu} = 0.851\ \mathrm{Pr}^{0.024} \left(\frac{H}{B} \right)^{1.02} \mathrm{Ra}_H^{1/4} , \tag{4.2.4}$$

$$H/B = 0.03 - 0.2, \quad \mathrm{Pr} = 1 - 10^3, \quad \mathrm{Ra}_H = 10^5 - 10^6 .$$

$$\mathrm{Nu} = 0.223\ \mathrm{Pr}^{0.024} \left(\frac{H}{B} \right)^{0.19} \mathrm{Ra}_H^{1/4} , \tag{4.2.5}$$

$$H/B = 0.2 - 1, \quad \mathrm{Pr} = 1 - 10^3, \quad \mathrm{Ra}_H = 10^3 - 10^6 .$$

For $H/B > 1$ the heat conduction regime occurs when $\mathrm{Ra}_B B/H < 100$ with a linear temperature distribution between the heated walls and Nu→ 1 [4.40, 4.41]. The average heat transfer coefficient can be determined from the formulas [4.40–4.42]

$$\mathrm{Nu} = 1 + \mathrm{Ra}_B^2 \cdot 10^{-6} \left[\frac{0.8}{(H/B)^2 + 21(H/B)^{-3}} + \frac{10}{(H/B)^2 + 2^{10}(H/B)^{-1}} \right] , \tag{4.2.6}$$

$$H/B = 1 - 100, \quad \mathrm{Pr} = 0.1 - 10, \quad \mathrm{Ra}_B < 5 \cdot 10^3 .$$

$$\mathrm{Nu} = 0.33\ \mathrm{Pr}^{0.27} \left(\frac{H}{B} \right)^{0.06} \mathrm{Ra}_H^{1/4} , \tag{4.2.7}$$

$$H/B = 5 - 47.5, \quad \mathrm{Pr} = 3 - 4 \cdot 10^4, \quad \mathrm{Ra}_H = 10^7 - 4 \cdot 10^9 .$$

$$\mathrm{Nu} = 0.084\ \mathrm{Pr}^{0.051}\ \mathrm{Ra}_H^{1/3} , \tag{4.2.8}$$

$$H/B = 5 - 47.5, \quad \mathrm{Pr} = 1 - 200, \quad \mathrm{Ra}_H = 4 \cdot 10^9 - 4 \cdot 10^{12} .$$

$$\mathrm{Nu} = 0.039\ \mathrm{Ra}_H^{1/3} , \tag{4.2.9}$$

$$H/B = 5 - 47.5, \quad \mathrm{Pr} = 1 - 4, \quad \mathrm{Ra}_H > 2 \cdot 10^{12} .$$

A rectangular cavity with the aspect ratio $H/B = 20$ and $L/B = 6$ ***filled with water or silicon oil at different constant temperatures of the opposite lateral surfaces*** was investigated experimentally for high *Ra* numbers in the range $1.58 \cdot 10^8 - 1.48 \cdot 10^{10}$. The flow structure was visualized by a laser-Doppler velocimeter. The empirical dependences for the average heat transfer coefficient were obtained for two flow regimes [4.43]

for laminar

$$\mathrm{Nu} = 0.224\ \mathrm{Ra}^{0.257}\ \mathrm{Pr}^{0.056}, \quad \mathrm{Ra} < 3 \cdot 10^9\ , \tag{4.2.10}$$

for turbulent

$$\mathrm{Nu} = 0.059\ \mathrm{Ra}^{0.315}\ \mathrm{Pr}^{0.056}, \quad \mathrm{Ra} \geq 3 \cdot 10^9\ . \tag{4.2.11}$$

The generalized dependence ***for the heat conduction regime, laminar and turbulent free convection*** is [4.44]

$$\mathrm{Nu} = \left[1, C_l\left(\mathrm{Ra}_{B\varphi} B/H\right)^{1/4},\ 0.29 C_t\left(\mathrm{Ra}_{B\varphi}\right)^{1/3}\right]_{\max}\ . \tag{4.2.12}$$

The use of formula (4.2.12) is recommended for deviation from the vertical by an angle of $\varphi = \pm 20°$. For large Pr values formula (4.2.12) can be used to calculate heat transfer for the given heat flux on the surface.

A long horizontal cavity with differently heated end surfaces filled with liquid for $0 < A < 0.25, \mathrm{Ra} < A^{-12}, \mathrm{Pr} > 1$. The longitudinal temperature gradient and the boundary layer thickness for this geometry are determined by the value of $\mathrm{Ra}A^2$. The main heat transfer mechanisms are vertical heat conduction and longitudinal convection. A theoretical analysis presupposes the presence of three heat transfer regimes: thermal conductive, transient, and convective. The boundaries between the regimes are determined by the values $\mathrm{Ra}A^2 = 10^2$ and $\mathrm{Ra}A^2 = 10^4$, respectively [4.45].

An experimental investigation of free convection of ***air*** in a large cavity (height 0.94 m, width 1.04 m, depth 0.7 m) ***with two opposite heated and two insulated vertical walls*** ($\Delta T = T_h - T_\mathrm{w} = 20°, \mathrm{Ra} = 1.69 \cdot 10^9$) was carried out. The vertical distribution of temperature in the cavity center due to the central symmetry of the problem allows one to achieve a regime with five linearly stratified zones [4.46]. The study of ascending and descending thermal boundary layers shows the presence of symmetric temperature profiles connected with the phenomenon of relaminarization after which follows the appearance of slightly turbulent thermal fields. An investigation of velocity profiles clearly shows a difference in the behavior of ascending and descending dynamic boundary layers with earlier development of turbulence along the hot wall compared with the cold one.

Local heat transfer in an interlayer filled with ***a liquid metal with the same heat fluxes on the vertical surfaces*** is described by [4.47]

$$H/B = 2 - 19, \quad \mathrm{Pr} = 0.023, \quad T_\mathrm{av} = 0.7\ T_\mathrm{w} + 0.3\ T_\infty$$

$$\mathrm{Nu}(x)\frac{B}{H} = 0.298\left[\mathrm{Gr}_x^*\left(\frac{B}{H}\right)^5\right]^{0.141}, \quad \mathrm{Gr}_x^*\left(\frac{B}{H}\right)^5 = 10^{-2} - 10^3\ , \tag{4.2.13}$$

$$\mathrm{Nu}(x)\frac{B}{H} = 0.268\left[\mathrm{Gr}_x^*\left(\frac{B}{H}\right)^5\right]^{0.165}, \quad \mathrm{Gr}_x^*\left(\frac{B}{H}\right)^5 = 10^3 - 10^9\ . \tag{4.2.14}$$

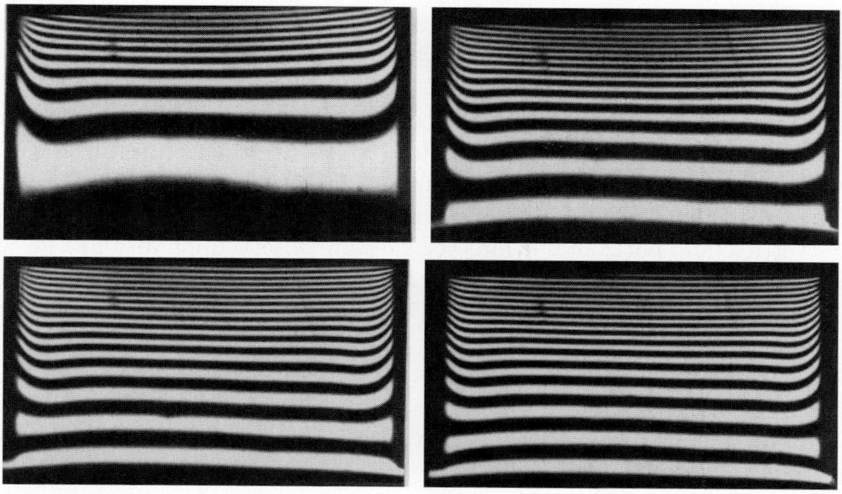

Fig. 4.2.5. Stable air stratification in a horizontal rectangular channel heated from above

In the case where ***a heat flux is given on the one vertical surface, and the temperature on the other***, the formulas for calculating heat transfer have the form [4.48]

$$\mathrm{Nu} = 0.0714\ \mathrm{Ra}_B^{*0.262}\,, \tag{4.2.15}$$

$$H/B = 1.5 - 6,\quad \mathrm{Pr} = 0.018 - 30.6\,,\quad \mathrm{Ra}_B^* = 8.12 \cdot 10^4 - 1.01 \cdot 10^{12}\,.$$

$$\mathrm{Nu} = 0.042\ \mathrm{Pr}^{0.012} \left(\frac{B}{H}\right)^{0.3} \mathrm{Ra}_B^{0.25}\,, \tag{4.2.16}$$

$$H/B = 1 - 40,\quad \mathrm{Pr} = 1 - 2 \cdot 10^4,\quad \mathrm{Ra}_B = 10^4 - 10^7\,.$$

$$\mathrm{Nu} = 0.046\ \mathrm{Ra}_B^{1/3}\,, \tag{4.2.17}$$

$$H/B = 1 - 40,\quad \mathrm{Pr} = 1 - 20,\quad \mathrm{Ra}_B = 10^6 - 10^9$$

Nonuniform heating from above, $\theta_H = \theta_0 \sin \pi y$. The average heat transfer value is [4.49, 4.50]

$$\mathrm{Nu} = 0.26\ \mathrm{Ra}_H^{0.19}\,, \tag{4.2.18}$$

$$H/B = 1,\quad \mathrm{Pr} < 10^5,\quad \mathrm{Ra}_H = 5 \cdot 10^4 - 10^{10},\quad \mathrm{Ra}_H < 10^{6.7}\ \mathrm{Pr}^{2.2}\,.$$

$$\mathrm{Nu} = 0.2 (H/B)^{0.3}\ \mathrm{Ra}_H^{0.16}\,, \tag{4.2.19}$$

$$H/B = 0.3 - 1,\quad \mathrm{Pr} = 1,\quad \mathrm{Ra}_H = 5 \cdot 10^4 - 10^7\,.$$

The temperature field in a rectangular cavity of square section is determined by the number Ra_H for $\mathrm{Ra}_H < 10^{6.7}\ \mathrm{Pr}^{2.2}$. The effect of the parameter H/B on thermoconvective processes appears to be appreciable in the range from 0 to 1. In vertical liquid layers with $H/B > 1$ the total heat flux and its convective component between the vertical walls virtually cease to depend on the geometrical shape of the region.

A numerical investigation of free convection ***in a container of square section with a variable distribution of the temperature of the lateral walls*** showed a marked effect of the flow structure (Pr = 1) on the heat transfer inside the cavity [4.51]. Two pairs of stable circulation flows with vertical and horizontal symmetry axes are observed for very low Ra and symmetric boundary conditions on the lateral surfaces. An increase in Ra and the presence of a nonuniform wall temperature distribution leads to a marked disturbance of the flow symmetry and to a change in the character and rate of heat transfer.

A rectangular cavity ($H \times B \times L = 1.04$ m $\times$ 0.94 m $\times$ 0.7 m) ***filled with air at different constant temperatures of the vertical walls and insulated horizontal surfaces***. Experimental investigation of temperature distribution along the horizontal surfaces, a spectral analysis of the time dependence of the temperature in the cavity, and the flow visualization by using laser tomography in the range $\mathrm{Ra} = (1.69 \pm 0.01) \cdot 10^9$ show the presence of instability and complex interaction between internal gravitational waves, thermal disturbances along the lower surface and the Tollmien–Schlihting waves in the vertical boundary layer near the heated wall [4.52].

Heat transfer in a cubic cavity with the upper boundary of temperature T_T moving with a constant velocity U_0, with the lower one having the temperature T_B and the other walls being insulated is determined by a ratio of two parameters: $\mathrm{Re} = U_0 H/\nu$ and $\mathrm{Ri} = \mathrm{Gr}\mathrm{Re}^{-2}$ [4.53]. The distribution of the average Nusselt number for the upper boundary of the cavity depending on the Re and Ri numbers is presented in Table 4.2.1.

Table 4.2.1.

Re	Ri = 0.001	Ri = 1.0	Ri = 10.0
10^2	1.82	1.33	1.08
$4 \cdot 10^2$	3.99	1.50	1.17
10^3	7.03	1.80	1.37

The average value of heat transfer rate in a rectangular cavity of square section ***with heating from the side*** ($\theta_{0H} = 0$, $\theta_H = \theta_0$, $\theta_{0B} = \theta_B = \theta_0(1 - x/H)$) is [4.49]

$$\mathrm{Nu} = 0.2\, \mathrm{Ra}_H^{0.29}\,, \qquad (4.2.20)$$

$$H/B = 1, \quad \mathrm{Pr} < 10^5, \quad \mathrm{Ra}_H = 5 \cdot 10^4 - 10^{10}, \quad \mathrm{Ra}_H < 10^{5.25}\, \mathrm{Pr}^{2.1}\,.$$

The account of ***the temperature dependence of the liquid viscosity*** for large Prandtl numbers with natural convection in a ***narrow vertical***

rectangular cavity with the aspect ratio equal to 15 gives a modified correlation for the average Nu number [4.54] obtained on the basis of the dimensiomless equation for the case of constant viscosity [4.55]:

$$\mathrm{Nu} = 0.1631\ \mathrm{Ra}^{0.2537}/\delta\left(\nu_n/\nu_m\right) , \qquad (4.2.21)$$

where $\delta\left(\nu_n/\nu_m\right) = \max\left\{1, 1.210 - 0.4776\nu_n/\nu_m\right\}, \nu_n/\nu_m$ is the ratio of viscosities at the heated wall temperature and the average cavity temperature. As an investigated liquid, **70, 80, *and* 90% *aqueous solution of glycerine*** with different viscosities was used.

The relative contribution of heat transfer due to the heat conduction in the solid wall compared with heat release by natural convection in the cavity ***filled with liquid*** was calculated numerically in [4.56]. Investigation of conjugated heat transfer was carried out in the ranges of the Ra numbers from 10^2 to 10^7, the ratios of thermal conductivities of the solid body and liquid R_k from 0.5 to 50, the dimensionless thickness of the solid wall $W = w/B$ (w and B are the wall thickness and the width of the cavity including the wall) from 0.05 to 9.25, and the cavity aspect ratios $A = H/B$ from 0.1 to 10. In the case of low Ra and high R_k and W for small and large values of A the heat transfer process is determined by the thermal conductivity in the solid wall. For high Ra and low R_k values for moderate values of A the interaction between the heat conduction in the wall and the convection in the liquid has a substantial effect on the heat transfer rate.

A vertical cavity with ***one wall representing a plate of finite thickness*** δ (the coefficient of the transverse thermal conductivity in it λ_δ considerably exceeds the coefficient of the longitudinal one λ). ***The vertical walls are maintained at constant temperatures***. The average heat transfer value in such a cavity depends also on the conjugation parameter $P' = \lambda_\delta B/(\lambda\delta)$ [4.57]:

$$\mathrm{Nu} = C\mathrm{Gr}_B^n(H/B)^{-0.12} , \qquad (4.2.22)$$
$$H/B = 1 - 6, \quad \mathrm{Pr} = 0.72, \quad \mathrm{Gr}_B = 5\cdot 10^3 - 5\cdot 10^6, \quad P' = 7 - \infty .$$

The values of the coefficients C and n are given in Table 4.2.2.

The coefficient of heat transfer between the vertical walls in the cavity decreases in the presence of vertical and horizontal partitions The greatest effect is achieved by mounting installing a partition at the top near the cold wall or at the bottom near the hot one [4.58].

Table 4.2.2.

P'	C	N
∞	0.225	0.25
14	0.276	0.196
7	0.334	0.174

The results of numerical investigation of the averaged fields of temperature, velocity, and average Nusselt number for a turbulent natural convection in a ***two-dimensional rectangular cavity with partitions and without them*** show a strong decrease in the Nu number along the nonadiabatic walls with a stepwise increase in the beginning of transition to turbulence [4.59]. The optimum variant of geometrical arrangement and heights of the partition are calculated for different boundary conditions (including ***heating of lateral surfaces and heating from below***) for transient and high Rayleigh numbers ($\mathrm{Ra} = 10^{10} - 10^{12}$).

Convective heat transfer ***for air*** in rectangular cavities ***without a partition and with a vertical partition with different constant temperatures on the opposite walls*** was calculated numerically [4.60]. The aspect ratio changed from 0.1 to 16, and the Rayleigh number from $3.5 \cdot 10^3$ to $3.5 \cdot 10^7$ (for cavities without partitions) and from $1 \cdot 10^5$ to $1.6 \cdot 10^8$ (for cavities with partitions), with the partition thickness and thermal conductivity being varied. The thermal boundary conditions on the end walls corresponded to a ***linear temperature profile or adiabatic conditions***. Comparison of the effect of a partition on the heat transfer rate for an "ideal" (infinitely thin isothermal partition) and real (partition with finite thickness and thermal conductivity) cases shows that in the investigated range of parameters the difference does not exceed ~12% for adiabatic boundary conditions and ~3% for a linear temperature profile.

A numerical calculation of the dependence of the average heat transfer coefficient for a vertical rectangular channel with two partitions on the opposite sides on the Ra number, the aspect ratio $A = H/L$, and the location of the partitions L_1 and L_2 showed that the average Nu number for the partially obstructed channel is lower compared with a smooth channel. ***The lateral walls are maintained at a constant temperature exceeding that of the surrounding medium***. The Rayleigh number changed in the range $10^2 < \mathrm{Ra} < 5 \cdot 10^4$; $A = 0.2$ and 0.3; the ratios of the partition height and thickness to the channel height H remained constant and equal to $h = 0.0667$ and $d = 0.01334$; two cases of the relative heights of the partitions were considered: (1) $L_1 = 0.25, L_2 = 0.75$ and (2) $L_1 = 0.5, L_2 = 0.5$. The maximal decrease in the average Nu due to partitions in the investigated range was observed at $\mathrm{Ra} = 5 \cdot 10^2$ and was approximately equal to 31%. The minimal decrease in heat transfer approximately equal to 3% was noted at $\mathrm{Ra} = 10^2$ [4.61].

The flow and heat transfer ***in a partially partitioned chamber of square section*** were investigated numerically in [4.62]. Three cases of the partition height and location were considered. For small Ra ($10^5 - 10^6$), the weak flow is observed in a stratified area with a tendency for flow separation behind the partition. As Ra increases ($\mathrm{Ra} > 10^9$), the stratification is enhanced, and the flow separates upstream the partition. Analysis of numerical data shows that the thermal stratification between the partition and the cold wall and the ratio of the sections $A_p = (H - h)/B$ have the strongest effect on the total heat transfer (H, B are the chamber height and width, h is the partition height). The results of numerical calculations for the average Nusselt number can be generalized by the dependence

$$\mathrm{Nu} = 0.091\,\mathrm{Ra}^{0.318} A_p^{0.462}\,, \qquad (4.2.23)$$
$$1/3 < A_p < 1, \quad 10^4 \le \mathrm{Ra} \le 3.55 \cdot 10^5, \quad \mathrm{Pr} = 0.71\,.$$

The position of the partition has a little effect on the average coefficient of heat transfer from the cavity. The values of the average Nusselt number for different locations of a partition are shown in Table 4.2.3.

The presence of through and nonthrough air filtration ($H/B = 4 - 16; \mathrm{Re} > 2 \cdot 10^3$) leads to a linear change in the wall temperature and local heat flux [4.63].

Table 4.2.3.

Position	Ra	Nu				
		$A_p = 0$	$A_p = 1/3$	$A_p = 1/2$	$A_p = 2/3$	$A_p = 1$
1	10^4		1.14	1.24	1.47	
	10^5		1.92	2.62	3.02	
	$3.55 \cdot 10^5$		3.11	4.02	4.56	
2	10^4	1.05	1.05	1.14	1.39	1.75
	10^5	1.65	1.92	2.73	3.07	3.36
	$3.55 \cdot 10^5$	2.25	3.24	4.23	4.66	4.90
3	10^4		1.15	1.21	1.41	
	10^5		1.91	2.50	2.90	
	$3.55 \cdot 10^5$		3.38	4.19	4.52	

In a cavity ***formed by horizontal plates with the temperatures*** T_0 ***and*** T_H ***and insulated vertical walls*** free convection is absent, if the temperature of the upper plate is higher, than that of the lower one. ***With heating from below***, free convection appears, when $\mathrm{Ra}_H > \mathrm{Ra}_k \cong 1700$, and it is characterized by a cellular structure. As the Ra_H number increases, the ordered cellular flow is destructed and passes into a number of intermediate discrete states ($\mathrm{Ra}_H \approx 3 \cdot 10^4 - 10^7$) until it becomes fully turbulent [4.35].

The average value of the heat transfer coefficient is [4.64]

$$\mathrm{Nu} = 1\,, \qquad (4.2.24)$$
$$\mathrm{Ra}_H < 1740\,,$$
$$\mathrm{Nu} = 1 + 1.44\left(1 - \frac{1740}{\mathrm{Ra}_H}\right)\,, \qquad (4.2.25)$$
$$\mathrm{Pr} = 0.7 - 11760, \quad \mathrm{Ra}_H = 1740 - 4 \cdot 10^3\,.$$
$$\mathrm{Nu} = 0.24\,\mathrm{Ra}_H^{0.25}\,, \qquad (4.2.26)$$

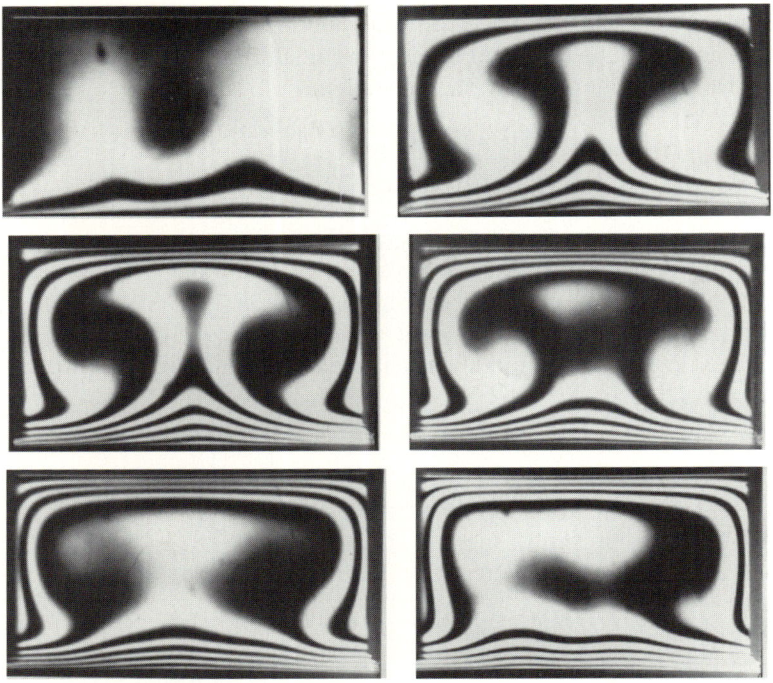

Fig. 4.2.6. Dynamics of the temperature field development in a horizontal channel of rectangular section with heating from below

$$\begin{gathered} \mathrm{Pr} = 0.7 - 11760, \quad \mathrm{Ra}_H = 4 \cdot 10^3 - 3 \cdot 10^6 \,, \\ \mathrm{Nu} = 0.115\ \mathrm{Ra}_H^{0.25} \,, \\ \mathrm{Pr} = 0.7 - 6, \quad \mathrm{Ra}_H = 3 \cdot 10^6 - 10^{10} \,. \end{gathered} \tag{4.2.27}$$

The following generalized dependences were obtained

for air for $\mathrm{Ra}_H < 10^8$ [4.65]:

$$\mathrm{Nu} = 1 + 1.44 \left(1 - \frac{1708}{\mathrm{Ra}_H}\right) + \left[\left(\frac{\mathrm{Ra}_H}{5830}\right)^{1/3} - 1\right] , \tag{4.2.28}$$

for water for $\mathrm{Ra}_H < 5 \cdot 10^9$:

$$\begin{aligned} \mathrm{Nu} = 1 + 1.44 \left(1 - \frac{1708}{\mathrm{Ra}_H}\right) + \left[\left(\frac{\mathrm{Ra}_H}{5830}\right)^{1/3} - 1\right] \\ + 2 \left(\frac{\mathrm{Ra}_H^{1/3}}{140}\right)^{1 - \ln\left(\frac{\mathrm{Ra}_H^{1/3}}{140}\right)} . \end{aligned} \tag{4.2.29}$$

A cubic or rectangular cavity with the ***lower hot and upper cold surfaces being maintained at constant temperatures***. In the case of a rectangular cavity ($25 \times 14.6 \times 12.3$ mm) for Ra = 11000 and Pr = 130 the calculated data for the lines of equal velocities W and U in the central section, where the velocity component $V = 0$, given in Fig. 4.2.7, show a very good qualitative and quantitative agreement with the experimental data of [4.66]. ***The adiabatic conditions on four lateral surfaces*** of the cubic cavity for small numbers Ra = 3500 – 10000 (Pr = 0.71) cause the occurrence of four different steady convective flows, three of which are simple ring vortices with horizontal axes of rotation and the vorticity axes either parallel to two opposite vertical walls (vortices S_1 and S_3), or oriented along the diagonal of the lower and upper boundaries (S_2 and S_4). The fourth vortical ring represents an approximately toroidal structure with a descending flow connected with the effect of four vertical ribs and general ascending flow along the vertical cavity axis [4.67]. For simple vortical rings the average Nusselt number increases with Ra, and changes from 0.7 to 0.4 in the range of the investigated parameters. A similar tendency is also observed for the toroidal ring, but in this case the heat transfer rate is approximately 65% lower. The distributions of the local Nusselt number along the lower and upper boundaries of the cavity have the maxima which corresponds to the critical values of Ra_{cr}. For simple vortices (S_1, S_2, and S_3) $\mathrm{Ra}_{cr} = 3500$, for a toroidal ring (S_4) $\mathrm{Ra}_{cr} = 7700$.

Numerical data for convection ***in a rectangular cavity heated from below with symmetrically cooled lateral walls*** show that in the range of Ra numbers from 10^3 to 10^7 for two Prandtl numbers (Pr = 0.7 and Pr = 7.0) at different length-to-height ratios (from 1 to 9) the flow structure represents one circulating cell in all the investigated cases, except for some variants of the assignment of a constant temperature on the lower surface in a narrow cavity, when small secondary cells near to vertical walls are observed. For a cavity of square section, different boundary conditions prescribed on the lower surface (constant temperature or constant heat flux) exert a slight effect on the flow structure and heat transfer character. For the narrow cavity, on the contrary, in the case of a constant heat flux the streamlines and isotherms are more uniformly distributed in the cavity volume, than in the case of a constant temperature even for low Ra values [4.68].

A square cavity whose two walls are at the temperature T_H and two other at T_B. The heat transfer average coefficient ***for air*** is [4.69]

$$\mathrm{Nu} = 2.917\ \mathrm{Ra}_H^{0.0275}\,, \qquad (4.2.30)$$

$$\mathrm{Ra}_H < 2.07 \cdot 10^4\,,$$

$$\mathrm{Nu} = 0.525\ \mathrm{Ra}_H^{0,2}\,, \qquad (4.2.31)$$

$$\mathrm{Ra}_H = 2.07 \cdot 10^4 - 2.6 \cdot 10^6\,.$$

A steady two-dimensional free-convective flow in the narrow rectangular cavity ***with regularly distributed internal heat sources. The horizontal walls are maintained at constant temperature, the vertical ones are***

a b

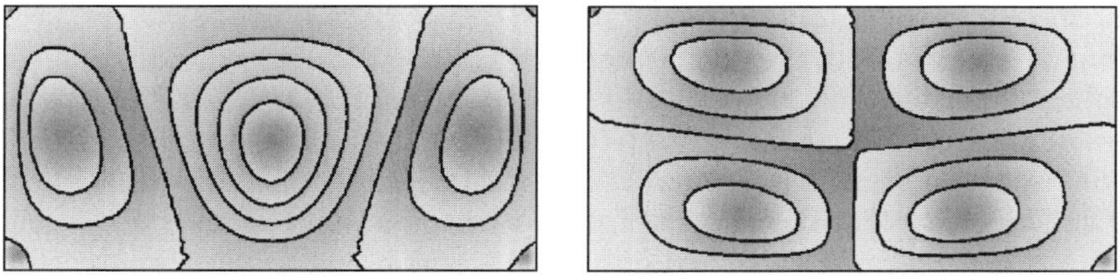

Fig. 4.2.7. The lines of constant velocity components W (**a**) and U (**b**) in the mid section of a rectangular cavity [4.66]

adiabaticlly insulated. The cavity aspect ratio is equal to 5. The Rayleigh number changes from 0 to 4000. The influence of nonlinear effects on the flow structure and temperature field inside the cavity near both end walls leads to the occurrence of new local vortices in addition to the main two-cellular flow in the central region of the cavity [4.70].

Free convection ***between horizontal surfaces in the presence of volumetric heat sources in the liquid. The horizontal surfaces are maintained at the temperature*** $T_{\mathbf{w}}$, ***the lateral surfaces are insulated***. The average value of heat transfer rate ***for a uniform distribution of the heat sources*** in the cavity at $\mathrm{Pr} = 1 - 80, \mathrm{Gr} = 10^5 - 10^8, T_{\mathrm{av}} = \frac{1}{F}\int_F T dF$ is [4.71]

on the upper wall

$$\mathrm{Nu} = 0.78\ \mathrm{Ra}_{vH}^{0.2}\ , \tag{4.2.32}$$

on the lower bottom wall

$$\mathrm{Nu} = 2.14\ \mathrm{Ra}_{vH}^{0.1}\ . \tag{4.2.33}$$

The lower wall is insulated, the heat generated in the volume is removed through the upper surface, which leads to an increase in the temperature difference. The process is described by the dimensionless heat transfer equation with an error of 7% [4.72]

$$\mathrm{Os}_H = 0.268\ \mathrm{Ra}_{vH}^{0.237}\ , \tag{4.2.34}$$
$$\mathrm{Pr} = 0.7, \quad \mathrm{Ra}_{vH} = 10^6 - 10^{11}\ .$$

Two horizontal walls with different temperatures. The heat flux Q is removed from both lower (Q_0) ***and upper*** (Q_H) ***walls***. The ratio of the heat fluxes is determined from the equation [4.73–4.76]

$$\frac{Q_0}{Q} = -1.38\frac{|\mathrm{Ra}_H|}{\mathrm{Ra}_H}\left(\frac{|\mathrm{Ra}_H|}{\mathrm{Ra}_{vH}}\right)^{0.719} + 1.916\ \mathrm{Ra}_{vH}^{-0.0785}\ , \tag{4.2.35}$$
$$\mathrm{Pr} > 0.7, \quad \mathrm{Ra}_{vH} = 10^4 - 10^{11}\ .$$

A cubic cavity ***filled with air*** is investigated at the inclination angles 0, 45° and 90°. ***Two opposite lateral surfaces are maintained at constant temperatures T_h and T_w and two other ones have a linear temperature distribution from T_h to T_w***. The Rayleigh number is $4 \cdot 10^4$. Detailed experimental and numerical data for the average Nusselt number are given in [4.77].

The numerical data ***for the turbulent natural convection*** of ***air*** in a rectangular cavity ***for different inclination angles*** (from 0 to 180°) are generalized by the dependences of the average Nusselt number on the aspect ratio $A = H/L$, the ratio of the thermal conductivities of the massive wall and air $\lambda_r = \lambda_w/\lambda_f$ and the wall thickness W [4.78]:

$$\mathrm{Nu}_{\varphi=90} = a\,\mathrm{Ra}^b A^c k_r^d W^e \,, \tag{4.2.36}$$

$$\mathrm{Nu}_{\varphi} = \mathrm{Nu}_{\varphi=90} (\sin\varphi)^f \,, \tag{4.2.37}$$

where $a = 0.1307, b = 0.1, c = -1/9, d = 0.75, e = -0.75, f = -0.01$ for $0 \le \varphi \le 90°$; $f = 6.0$ for $90° \le \varphi \le 180°$. ***One of the lateral surfaces is heated by a constant heat flux, the opposite massive wall (with finite thickness and thermal conductivity) is maintained at a constant temperature***. The range of the Rayleigh numbers is $10^8 \le \mathrm{Ra} \le 10^{12}$.

An experimental investigation of heat transfer of an ***air layer in a rectangular inclined cavity heated from above*** is carried out for different inclination angles φ (from 0 to 180°) and the values of $H/B = 20$ and 80. In the range of φ from 120° to 180° the measurements were carried out for different Ra numbers (from 10^2 to $2 \cdot 10^6$). The average heat transfer coefficient is approximated by the following relations [4.79]:

$$\mathrm{Nu} = \left[1 + \left(0.21\,\mathrm{Ra}^{0.136}\right)^{11}\right]^{1/11} \,, \quad H/B = 20, \varphi = 180° \,. \tag{4.2.38}$$

$$\mathrm{Nu} = \left[1 + \left(0.0566\,\mathrm{Ra}^{0.332}\right)^{4.76}\right]^{1/4.76} \,, \quad H/B = 20, \quad \varphi = 120° \,. \tag{4.2.39}$$

$$\mathrm{Nu}\,(\varphi) = \mathrm{Nu}\,(180°) + \frac{180 - \varphi}{60}\left[\mathrm{Nu}\,(120°) - \mathrm{Nu}\,(180°)\right] \,, \quad H/B = 20, 120° \le \varphi \le 180° \,. \tag{4.2.40}$$

A rectangular cavity $H \times B \times S$ located at an angle of φ relative to the horizon. The average value of the heat transfer coefficient of the ***air layer heated from below***, with the large ratio H/B being large, is [4.80]

$$\mathrm{Nu} = 1 + 1.44\left(1 - \frac{1708}{\mathrm{Ra}_H}\right) + \left[\left(\frac{\mathrm{Ra}_H}{5830}\right)^{1/3} - 1\right] \times \left(1 - \frac{1708(\sin 1.8\varphi)^{1.6}}{\mathrm{Ra}_{H\varphi}}\right) \,, \tag{4.2.41}$$

$$B/H > 10, \quad \varphi = 0 - 60°, \quad \mathrm{Pr} = 0.7, \quad \mathrm{Ra}_H < 10^5 \,.$$

For $\varphi < 72°$ the heat conduction regime is preserved when $\mathrm{Ra}_{H\varphi} < 1708$, whereas for $\varphi > 72°$ this boundary is the function of the inclination angle and B/H. The heat transfer for $\varphi > 45°$ is determined from the dimensionless equation [4.81]

$$\mathrm{Nu} = 0.118\left[\mathrm{Ra}_H \cos^2(\varphi - 45)\right]^{0.29}, \tag{4.2.42}$$
$$B/H = 9 - 30, \quad S/H = 2.3 - 8, \quad \varphi = 45–90°,$$
$$\mathrm{Pr} = 0.7, \quad \mathrm{Gr}_H = 4 \cdot 10^3 - 3.1 \cdot 10^5 .$$

The average value of the intensity of heat transfers calculated on the basis of the interferometer investigation of the temperature distribution for $\mathrm{Gr}_H > 10^5$ is [4.82]

$$\mathrm{Nu} = C\mathrm{Gr}_H^{1/3} \tag{4.2.43}$$
$$B/H = 6 - 27, \quad \mathrm{Pr} = 0.7, \quad \varphi = 0 - 90°, \quad \mathrm{Gr}_H = 10^5 - 4 \cdot 10^6 .$$

The values of the coefficient C are given in Table 4.2.4.

Table 4.2.4.

$\varphi°$	0	10	20	30–50	60	70	80	90
C	0.08	0.079	0.075	0.074	0.072	0.069	0.068	0.064

To calculate the average heat transfer coefficient ***in inclined rectangular cavities*** the following dependences are recommended [4.83]:

$$\mathrm{Nu}_{H\varphi} = \mathrm{Nu}_{H\varphi=0}(\cos\varphi)^{0.29}, \quad \varphi = 0 - 75°, \tag{4.2.44}$$
$$\mathrm{Nu}_{H\varphi} = \mathrm{Nu}_{H\varphi=90}(\sin\varphi)^{0.25}, \quad \varphi = 75° - 90°, \tag{4.2.45}$$
$$\mathrm{Nu}_{H\varphi} = 1 + (\mathrm{Nu}_{H\varphi=90} - 1)\sin\varphi, \quad \varphi = 90° - 180°. \tag{4.2.46}$$

The detailed results of the numerical analysis of steady natural convection in a two-dimensional square channel ***with different constant temperatures on the opposite walls*** show the effect of the temperature gradient and channel inclination angle on the flow regime [4.84]. If the temperature of the upper walls is higher than that of the lower ones, local recirculation flows appear in the channel. In the presence of the inclination of the channel a single global cell is formed.

Two vertical walls are maintained at the temperatures T_0 and T_B, and two other are insulated ($\varphi = 0$). The numerical calculations are performed for different cavity orientations, $\varphi = (-\pi) - \pi$ and $\mathrm{Gr}_B = 5 \cdot 10^4, \mathrm{Pr} = 0.71$. The thermophysical characteristics satisfy the Sutherland equation. As the angle φ decreases to $-\pi/2$, the heat transfer decreases essentially, and for the case of the cavity heated from above

($\varphi = -\pi/2$) the gas passes to the state of hydrostatic equilibrium. The heat transfer regime is determined by heat conduction. The effect of variable thermophysical properties is approximated by the power-law formulas [4.85]:

$$\mathrm{Nu} = C_T\,\mathrm{Ra}_B^{0.272}\left(\frac{T_0}{T_B}\right)^{0.17}\left(\frac{C}{T_B}\right)^{0.013}\left(\frac{c_p}{c_v}\right)^{-0.057}, \tag{4.2.47}$$

$$H/B = 0.2 - 2,\quad \mathrm{Gr}_H = (0.63 - 5)10^4,\quad \mathrm{Pr} = 0.71\,,$$
$$\varphi = 0,\quad \pi/2,\quad T_{\mathrm{av}} = (T_0 + T_B)/2,\quad C = 75 - 673^\circ\mathrm{C}\,,$$
$$T_0/T_B = 1 - 4,\quad C/T_B = 0.25 - 3,\quad c_p/c_v = 1 - 4\,.$$

In a cavity of ***rectangular section two horizontal and one vertical walls are insulated, the other vertical surface is maintained at the temperature*** T_0. On the insulated vertical surface at the distance of s from the upper wall a ***heating element*** of width b and temperature T_b is located. The main mechanism of heat transfer for $\mathrm{Ra}_H < 10^3$ is heat conduction even in the presence of certain circulation in the cavity [4.86]. The approximation of the calculated and experimental data for the average heat transfer ***from the heating element in air*** yields [4.87]

$$\mathrm{Nu} = C\mathrm{Gr}_H^n\,, \tag{4.2.48}$$

$$H/B = 1,\quad s/H = 0.5,\quad \mathrm{Pr} = 0.7,\quad \mathrm{Gr}_H = (5 - 9)10^4\,,$$
$$T_{\mathrm{av}} = (T_0 + T_B)/2\,.$$

The values of the coefficient C are presented in Table 4.2.5.

Table 4.2.5.

b/H	C	n
0.125	0.045	0.33
0.25	0.069	0.32
0.5	0.097	0.31
1	0.119	0.3

The influence of the parameter H/B on heat transfer is insignificant at $b/H = 0.25$ and $s/H = 0.125 - 0.875$. The heat transfer also increases a little with increase in the height of the heating element for $b/H > 0.25$. The maximum circulation rate in the cavity for the given heat transfer was observed at $H/B = 0.5 - 2$, $s/H = 0.4$, and $b/H = 0.25$.

Dependence (4.2.48) can be used for calculating any location of the heating element on one of the vertical insulated walls at $\mathrm{Gr}_H = 10^5 - 10^7$ with an error of 10% [4.87].

Stationary heat and mass transfer in a square cavity ***filled with air and exposed to combined influence of the temperature and concentration***

gradients in the horizontal direction. The Rayleigh number calculated from the temperature difference is $\mathrm{Ra}_t = 107$, the Rayleigh number calculated from the concentration difference is $\mathrm{Ra}_s = 10^5 - 5 \cdot 10^7$. The generalized empirical dependence for the average Nusselt number for cocurrent and opposite flows is [4.88]

$$\mathrm{Nu} = 0.22\,(\mathrm{Ra}_t(1+N))^{0.27}\ , \tag{4.2.49}$$

where $N = (\beta_s \Delta C^*)/(\beta_t \Delta T^*)$; $\Delta C^* = C_l^* - C_r^*$; $\Delta T^* = T_l^* - T_r^*$; T_l^*, T_r^* are the temperature of the left and right walls of the cavity; C_l^*, C_r^* are the concentrations of the solution near the left and right walls of the cavity. The dependence of the average Nusselt and Sherwood numbers (along the vertical axis) depending on N is shown in Table 4.2.6 ($\mathrm{Le} = \alpha/D = 1, \mathrm{Pr} = \nu/\alpha = 0.7, \mathrm{Ra}_t = 10^7$, where α, ν, and D are the coefficients of thermal diffusivity, viscosity and diffusion).

The convective motion ***in the solution in the presence of a vertical density gradient and lateral heating*** in a rectangular area has a cellular character with a step density change between cells. Inside the cells there is a rather fast equilibration of density due to convective mixing [4.89].

Table 4.2.6.

Ra_s	10^5	10^6	$2 \cdot 10^6$	$5 \cdot 10^6$	$8 \cdot 10^6$	$9 \cdot 10^6$	$1.5 \cdot 10^7$	$5 \cdot 10^7$
N	−0.01	−0.1	−0.2	−0.5	−0.8	−0.9	a−1.5	−5.0
Nu	16.4	16.0	15.5	13.6	10.6	8.8	13.6	23.7
Sh	−16.4	−16.0	−15.5	−13.6	−10.6	−8.8	−13.6	−23.7

A binary gas mixture is exposed to ***opposite horizontal temperature and concentration gradients in a rectangular cavity***. The interaction of the gradients leads to the occurrence of an oscillatory natural-convective flow [4.90]. In certain range of the relationships between the gradients ($N = 0.9 - 1.122$) the cellular structure of secondary flows is observed. The parameters ranges of the problem change in the following ranges: $\mathrm{Ra}_T = 10^5, \mathrm{Pr} = 1$, Le $= 2, N = 0.0 - 2.0 (A = 2)$, where $N = [\beta_c(C_h - C_l)]/[\beta_T(T_h - T_c)]$; $A = H/B$; β_T, and β_c are the coefficients of temperature and concentration volumetric expansion; C_h and C_l are the high and low concentrations; T_h and T_c are the temperatures of the hot and cold walls.

The results of numerical simulation of a two-dimensional convection ***in liquid metals*** in a square cavity ***with simultaneous acting horizontal temperature and concentration gradients*** [4.91] reveal the presence of two regions with different heat and mass transfer regimes ***in the double melt*** (**Pb** – **Sn**). ***The opposite side walls are maintained at different constant temperatures*** T_h ***and*** T_c ***and different of concentrations of the light (saturated*** **Sn*****) and heavy (saturated*** **Pb*****) melts***. High Sc numbers (Sc = 150, Le = 7500) and low values of Pr = 0.03, corresponding to large concentration gradients and small temperature gradients, are characteristic for the given problem ($N = [\beta_c(C_h - C_l)/[\beta_T(C_h - C_c)] \cong 10$). The number Ra = 100 − 5000. The flow structure is characterized by formation of a boundary layer near the vertical walls with the predominance of mass transfer and a transient zone at the center of the cavity, where the thermal and concentration mechanisms of free convection are of the same order of magnitude due to which the rate of mass transfer here is much lower, than near the walls. With time, the heavy liquid saturated with **Pb** is concentrated at the bottom, and the light one saturated with **Sn** moves upwards, and after the lapse of the time $t = t'\nu/L \cong 3.15$ (t' is the time, L is 1/2 of the cavity height or length) the concentration stratification develops in the volume.

A numerical investigation of the simultaneous effect of ***the temperature and concentration gradients in a vertical cavity of square section*** on the flow and heat transfer of ***a humid air with a low steam concentration*** (Pr = 0.7, Le = 0.8) shows a strong dependence of the temperature field structure and characteristics of heat and mass transfer on the parameter which characterizes the ratio between the buoyancy forces caused by a change in the temperature and concentration ($N = \beta_c \Delta C_{\max}/\beta_T \Delta T_{\max}$) [4.92]. ***The vertical walls are maintained at constant temperatures and different concentrations***. The finite thickness, thermal conductivity, and the permeability of the horizontal dividing walls are taken into account in the calculation. The cases of the same and opposite directions of the temperature and concentration gradients are considered. The aspect ratio $L/B = 20$ (L and B are the length of the cavity side and the thickness of the dividing walls). The number Ra = 10^5.

The method of dimensional analysis and the approximation of the numerical data give a correlation for the average rate of heat and moisture transfer by laminar free convection in a rectangular cavity applicable in the case of ***isothermal vertical and adiabatic horizontal walls for nonsaturated humid air at constant relative humidity*** [4.93]:

$$\mathrm{Ra} = 53.5 \frac{1 + \mathrm{Pr}}{\mathrm{Pr}} \mathrm{Nu} \left(\mathrm{Nu} + A^{-1}\right)^3 , \tag{4.2.50}$$

where $A = H/B$. The Ra number changes in the range $10^3 - 10^6$

The free-convective flow of ***two nonmixing fluids with different viscosities*** in a rectangular cavity under the action of a ***horizontal temperature gradient*** represents a cellular structure consisting of three or four vortices (the latter case is typical of small values of the Vi number which characterizes the ratio of viscosities), the occurrence of which is caused by two mechanisms. First, there is an ascending flow of the fluid along the hot wall and a descending one along the cold walls; second, the interaction of two fluid layers at their interface leading to the occurrence of two oppositely rotating vortices with the reversal of rotation on passing the interface. The effect of the viscosity discontinuity at the interface of the layers is insignificant for Vi $\ll 1$; when Vi increases (Vi > 1), the viscosity jump begins to play a substantial role, especially in the vicinity of the interface, and for Vi > 10 the interface behaves as a rigid surface [4.94].

Investigation of linear stability of a fully developed natural-convective flow of ***air*** in a high vertical rectangular cavity ***at different constant temperatures on two opposite vertical walls (the other four walls are adiabatic)*** under the conditions different from the Boussinesq approximation revealed two different mechanisms of instability [4.95]. The instability is of fluctuational character, and, depending on the magnitude of the temperature gradient, along with the well-known shear mechanism a new type of instability appears which is directly connected with the effect of the nonlinear change in density depending on temperature.

The experimental data for the average Nusselt number with free convection of ***air in an inclined cubic cavity*** were obtained in [4.96] for the Rayleigh numbers Ra $= 10^4, 10^5, 10^6, 10^7$, and 10^8 and three values of the inclination angle of isothermal surfaces: $\varphi = 0, 45$, and $90°$. ***Two lateral opposite sides are maintained at different constant temperatures T_h and T_c, and two other sides have the temperatures which change linearly within these limits***. The results of the experimental investigation are given in Table 4.2.7 and are generalized by the following dependences

for $\varphi = 90°$

$$\mathrm{Nu} = 0.08461\ \mathrm{Ra}^{0.3125}, \quad 10^4 \le \mathrm{Ra} \le 10^8\ , \tag{4.2.51}$$

for $\varphi = 45°$

$$\mathrm{Nu} = 0.1492\ \mathrm{Ra}^{0.2955}, \quad 10^6 \le \mathrm{Ra} \le 10^8\ , \tag{4.2.52}$$

$$\mathrm{Nu} = 0.07309\ \mathrm{Ra}^{0.3358}, \quad 10^4 \le \mathrm{Ra} \le 10^5\ , \tag{4.2.53}$$

for $\varphi = 0°$

$$\mathrm{Nu} = 0.1194\ \mathrm{Ra}^{0.3021}, \quad 10^5 \le \mathrm{Ra} \le 10^8\ . \tag{4.2.54}$$

A numerical investigation of free-convective heat and mass transfer ***in a rectangular cavity with combined effect of the horizontal temperature and concentration gradients for the cases of oppositely*** [4.97] ***and cocurrently*** [4.98] ***acting corresponding buoyancy forces*** points to the presence of four types of the steady flow for both cases of the changing R_p. The ratio of the buoyancy forces $R_p = (\beta_c \Delta C)/(\beta_t \Delta t)$ changed from 0.5 to 30 due to the change in the thermal Rayleigh number R_t, whereas the concentrational Rayleigh number was equal to $R_s = 1.38 \cdot 10^8$ for the gradients acting in the same direction and $R_s = 6 \cdot 10^7$ for the opposite direction, with Le $= 100$, Pr $= 7$, and $H/L = 2(H, L$ are the height and length of the cavity). At moderate values of the ratio of the buoyancy forces R_p in both cases of the action of the gradient a multilayered flow structure was noticed in the region of the internal core.

Table 4.2.7.

Ra	Nu		
	$\varphi = 0°$	$\varphi = 45°$	$\varphi = 90°$
10^4	1.246 ± 0.013	1.614 ± 0.015	1.520 ± 0.015
$4 \cdot 10^4$	2.018 ± 0.017	2.650 ± 0.027	2.337 ± 0.020
10^5	3.509 ± 0.035	3.492 ± 0.034	3.097 ± 0.028
$3 \cdot 10^5$	3.916 ± 0.042	–	–
10^6	7.883 ± 0.091	8.837 ± 0.101	6.383 ± 0.070
10^7	15.38 ± 0.19	17.50 ± 0.21	12.98 ± 0.16
10^8	3.22 ± 0.43	34.52 ± 0.42	26.79 ± 0.34

For the oppositely acting buoyancy forces caused by the temperature and concentration gradients, with increase in the value of R_p the average Nu number monotonically decreases to the values corresponding to conductive heat transfer, and the Sh number attains the minimum values for moderate R_p.

The results of numerical calculations for the average of Nusselt and Sherwood numbers in a stationary regime are presented in Tables 4.2.8 and 4.2.9.

The steady natural convection ***in a stably stratified salt-water solution, caused by lateral heating of a rectangular cavity*** with the aspect ratio 3.0 was experimentally investigated in [4.99]. Depending on the ratio between the buoyancy forces N, which represents the value of the ratio between the solution stratification to the buoyancy force due to the temperature difference, four flow regimes are distinctly observed: a unicellular flow for $N < 10$, simultaneously formed flow layers for $10 \leq N < 40$, successively formed flow

Table 4.2.8. The Values of the Average Nusselt and Sherwood Numbers for Cocurrently Acting Temperature and Concentration Gradients

R_p	Sh	Nu Thermoconcentrational	Nu Thermal Convection
0.5	166.84	36.41	36.21
1.0	149.79	30.36	30.16
4.0	120.55	20.44	20.98
20.0	116.13	12.34	13.80
41.0	113.19	8.417	11.44
60.0	111.43	5.857	10.36
100.0	109.78	3.334	9.057
200.0	109.21	2.486	7.541
500.0	108.87	2.196	5.907
1000.0	108.72	2.134	4.897
3000.0	108.61	2.108	3.624
10000.0	108.59	2.104	2.610

Table 4.2.9. The Values of the Average Nusselt and Sherwood Numbers for Oppositely Acting Temperature and Concentration Gradients

R_p	Sh	Nu Thermoconcentrational	Nu Thermal Convection
0.5	133.09	28.79	29.10
1.0	104.94	23.47	24.26
3.0	68.996	14.43	18.22
6.0	70.449	9.510	15.21
9.0	78.180	6.937	13.69
12.5	81.778	4.951	12.57
20.0	83.450	1.965	11.11
30.0	84.125	1.713	9.994

layers when $40 \le N < 55$, and a stagnant flow for $N \ge 55$. Formation and growth of the layered flow structure with the corresponding distributions of temperatures and concentrations in each layer were visually observed.

4.3 Cylindrical Interlayers

The character of heat transfer in a cylindrical interlayer is determined by the Ra number, the ratio between the length of the cooled part to the layer thickness, the ratio between the diameters of the cylinders, possible eccentricity, and the surface curvature. In the simplest case, on the basis of the similarity theory, the dimensionless relation for calculating heat transfer can be presented as [4.100]

$$\mathrm{Nu} = f\left(\mathrm{Gr}, \mathrm{Pr}, d_1/d_2\right) . \tag{4.3.1}$$

Free convection between vertical cylinders with constant temperatures T_{w1} (inner cylinder) and T_{w2} (outer cylinder), when the width of the annulus considerably exceeds the boundary layer thickness. Depending on the surface curvature of the inner cylinder the following dimensionless formulas were obtained [4.101]:

$$\mathrm{Nu} = 0.48\,\mathrm{Ra}_l \left[6830 \left(\frac{l}{d_2}\right)^4 \frac{d_1}{l} + \mathrm{Ra}_l^{3/4}\right]^{-1} , \tag{4.3.2}$$

$$\frac{l}{d_1}\,\mathrm{Ra}_l^{-1/4} < 0.1 ,$$

$$\mathrm{Nu} = 1.19\,\mathrm{Ra}_l \left[16900 \left(\frac{l}{d_2}\right)^4 \frac{d_1}{l} + \mathrm{Ra}_l^{0.84} \left(\frac{d_1}{l}\right)^{0.36}\right]^{-1} , \tag{4.3.3}$$

$$\frac{l}{d_1}\,\mathrm{Ra}_l^{-1/4} = 0.1 - 0.738 ,$$

$$\mathrm{Nu} = 0.845\,\mathrm{Ra}_l \left[12000 \left(\frac{l}{d_2}\right)^4 \frac{d_1}{l} + \mathrm{Ra}_l^{0.95} \left(\frac{d_1}{l}\right)^{0.8}\right]^{-1}, \qquad (4.3.4)$$

$$\frac{l}{d_1}\,\mathrm{Ra}_l^{-1/4} > 0.738 \,.$$

A cavity of height H formed by ***vertically located coaxial cylinders with constant temperatures*** ($T_{w1} > T_{w2}$) ***and insulated end surfaces***. The numerical calculations of the average heat transfer coefficient are approximated by the power-law dependences with an error of 5% [4.102]:

$$\mathrm{Ra}_{r_{av}} < 2 \cdot 10^5, \quad \mathrm{Pr} = 0.5 - 10^4, \quad H/r_{av} = 1 - 33,$$
$$d_2/d_1 = 1 - 33, \quad r_{av} = (r_1 + r_2)/2 :$$

$$\mathrm{Nu} = 0.595\,\mathrm{Ra}_{r_{av}}^{0.101}\,\mathrm{Pr}^{0.024} \left(\frac{H}{r_{av}}\right)^{-0.052} \left(\frac{r_2}{r_1}\right)^{0.505}, \qquad (4.3.5)$$

$$\mathrm{Ra}_{r_{av}} r_{av}/H < 400 \,,$$

$$\mathrm{Nu} = 0.202\,\mathrm{Ra}_{r_{av}}^{0.294}\,\mathrm{Pr}^{0.097} \left(\frac{H}{r_{av}}\right)^{-0.246} \left(\frac{r_2}{r_1}\right)^{0.423}, \qquad (4.3.6)$$

$$\mathrm{Ra}_{r_{av}} r_{av}/H = 400 - 3 \cdot 10^3 \,,$$

$$\mathrm{Nu} = 0.286\,\mathrm{Ra}_{r_{av}}\,\mathrm{Pr}^{0.006} \left(\frac{H}{r_{av}}\right)^{-0.238} \left(\frac{r_2}{r_1}\right)^{0.442}. \qquad (4.3.7)$$

$$\mathrm{Ra}_{r_{av}} r_{av}/H \ge 3 \cdot 10^3 \,.$$

Formulas (4.3.5)–(4.3.7) give the most exact results for $\mathrm{Pr} = O(1)$ and $r_2/r_1 = 1 - 4$.

The outer cylinder and the bottom basis of the vertical cylindrical innerlayer are insulated. The temperature of the inner cylinder is T_{w1}. The average value of heat transfer from the inner cylinder to **He** ***on compression*** ($dp/d\tau = (6.9 - 35) \cdot 10^5$ Pa/s, $l = 0.354$ m, $d_1 = 0.248$ m, $d_2 = 0.311$ m) is given by [4.103]

$$\mathrm{Nu} = 0.171\,\mathrm{Ra}_{d_{ef}}^{0.303}\,, \qquad (4.3.8)$$

$$d_{ef} = \frac{V}{\sum F}, \quad \mathrm{Ra}_{d_{ef}} = 1.2 \cdot 10^3 - 3.5 \cdot 10^5 \,,$$

$$\Delta T = (Q_{conv} + Q_p) \left(\frac{\lambda \sum F}{2l}\right)^{-1} .$$

A vertical cavity ***with insulated coaxial cylindrical surfaces and end surfaces with constant temperatures*** T_0 ***and*** T_H. The numerical and experimental data for heat transfer of ***carbonic acid and water at a high pressure*** are approximated by the dependence ($p = 0.35 - 3.1$ Mpa, $l = 0.61 - 1.22$ m, $T_0 - T_H = 20 - 60°$C) [4.104]

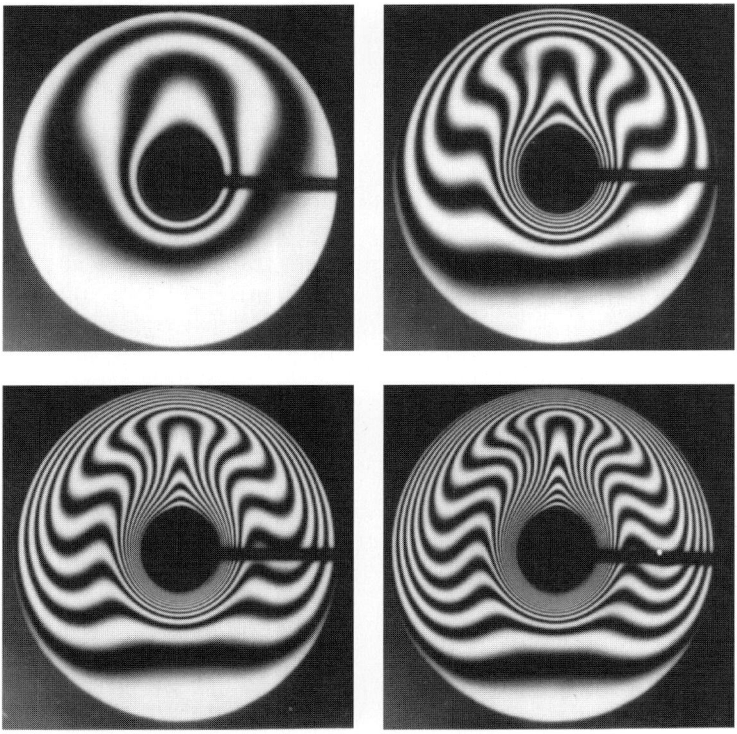

Fig. 4.3.1. Interferograms of nonstationary convection in a horizontal cylindrical annulus at a constant heat flux on the inner cylinder surface

$$\mathrm{Nu} = 5 \cdot 10^{-4} \left(\mathrm{Gr}_{d_{\mathrm{av}}}\ \mathrm{Pr}^2\right)^{0.79} (1 - d_1/d_2)^{0.41} \ , \tag{4.3.9}$$
$$d_1/d_2 = 0.914 - 0.997, \quad \mathrm{Pr} = 1.5 - 3.5 \ ;$$
$$d_{\mathrm{av}} = (d_1 + d_2)/2, \quad \mathrm{Gr}_{d_{\mathrm{av}}}\ \mathrm{Pr}^2 = 2 \cdot 10^4 - 5 \cdot 10^{10} \ .$$

A narrow annulus in the presence of a temperature difference between inner and outer cylinders and insulated horizontal walls. The average value of heat transfer is [4.105]

$$\mathrm{Nu} = \frac{l}{d_1 \ln\left(\frac{d_2}{d_1}\right)} \left\{ 1 + 5.52 \cdot 10^{-6} \frac{1}{d_1} \frac{(\mathrm{Ra}_l l)^2}{(d_2 - d_1) \ln\left(\frac{d_2}{d_1}\right)} + O\left[\left(\frac{l}{d_2 - d_1}\right)^3\right] \right\} . \tag{4.3.10}$$

The heat transfer between ***horizontal cylinders of constant temperature*** for $\mathrm{Pr} = 0.005 - 0.1$ is described by [4.106]

$$\lambda_{\mathrm{ef}}/\lambda = 1 \ , \tag{4.3.11}$$

$$r_2/r_1 = 1.1, \quad \mathrm{Gr}_{r_1} < 10^6 .$$

$$\lambda_{\mathrm{ef}}/\lambda = 1 + 5 \cdot 10^{-8}\,\mathrm{Ra}_{r_1}^{2.014} , \tag{4.3.12}$$

$$r_2/r_1 = 2, \quad \mathrm{Gr}_{r_1} < 5 \cdot 10^3, \quad \mathrm{Ra}_{r_1} < 50 .$$

$$\lambda_{\mathrm{ef}}/\lambda = 1 + 1 \cdot 10^{-4}\,\mathrm{Ra}_{r_1}^{1.935} , \tag{4.3.13}$$

$$r_2/r_1 = 4, \quad \mathrm{Gr}_{r_1} < 5 \cdot 10^3, \quad \mathrm{Ra}_{r_1} < 50 .$$

In the case of rarefaction, the effect of free convection is observed for $\mathrm{Gr}_\delta < 10^{-3}$ [4.107]. The experimental data on heat transfer in **Ar** ***and air*** are generalized as

$$\lg \frac{\lambda_{\mathrm{ef}}}{\lambda} = 0.0128 + 0.00945 \lg \mathrm{Gr}_\delta + 0.00155 (\lg \mathrm{Gr}_\delta)^2 + 0.000279 (\lg \mathrm{Gr}_\delta)^3 , \tag{4.3.14}$$

$$d_1/d_2 = 1.9 - 9.4, \quad T_{\mathrm{w}1}/T_{\mathrm{w}2} = 0.6 - 0.85, \quad \mathrm{Gr}_\delta = 10^{-1} - 10^{-4} .$$

The generalizing dependence for calculating heat transfer ***between cylinders of constant temperature in the presence of eccentricity*** is [4.23]

$$\mathrm{Nu}''_{d_1} = \left[\left(\mathrm{Nu}'_{d_1}\right)^{15} + \mathrm{Nu}_T^{15} \right]^{1/15} \tag{4.3.15}$$

where

$$\mathrm{Nu}'_{d_1} = 2/\ln \frac{1 + 2/\mathrm{Nu}_{d_1}}{1 - 2/\mathrm{Nu}_{d_{12}}} ,$$

$$\mathrm{Nu}_T = \frac{2}{\mathrm{arch}\left[\left(d_2^2 + d_1^2 - 4e^2\right)/2d_1 d_2\right]} .$$

The values of Nu_{d_1} are determined by the formula

$$\mathrm{Nu}_{d_1} = \left[\left(0.518\,\mathrm{Ra}_{d_1}^{1/4} \left[1 + (0.559/\mathrm{Pr})^{3/5} \right]^{-5/12} \right)^{15} + \left(0.1\,\mathrm{Ra}_{d_1}^{1/3} \right)^{15} \right]^{1/15} \tag{4.3.16}$$

and the values of $\mathrm{Nu}_{d_{12}}$ by formula (4.1.23). The fluid temperature T_∞ is determined from the thermal balance equation

$$\left(T_\infty - T_{\mathrm{w}2}\right) / \left(T_{\mathrm{w}1} - T_\infty\right) = \mathrm{Nu}'_{d_1} / \mathrm{Nu}_{d_{12}} \tag{4.3.17}$$

by the method of successive iterations. Three-four approximations are sufficient for practical calculations.

The heat transfer between ***coaxial cylinders in air*** for $\mathrm{Ra}_{d_1} = 5 \cdot 10^3$ with the maximal error of 1.7% is generalized by the dependence [4.108]

$$\mathrm{Nu}_{d_1} = \left[\left(\mathrm{Nu}''_{d_1}\right)^{15} + \mathrm{Nu}_T^{15} \right]^{1/15} , \tag{4.3.18}$$

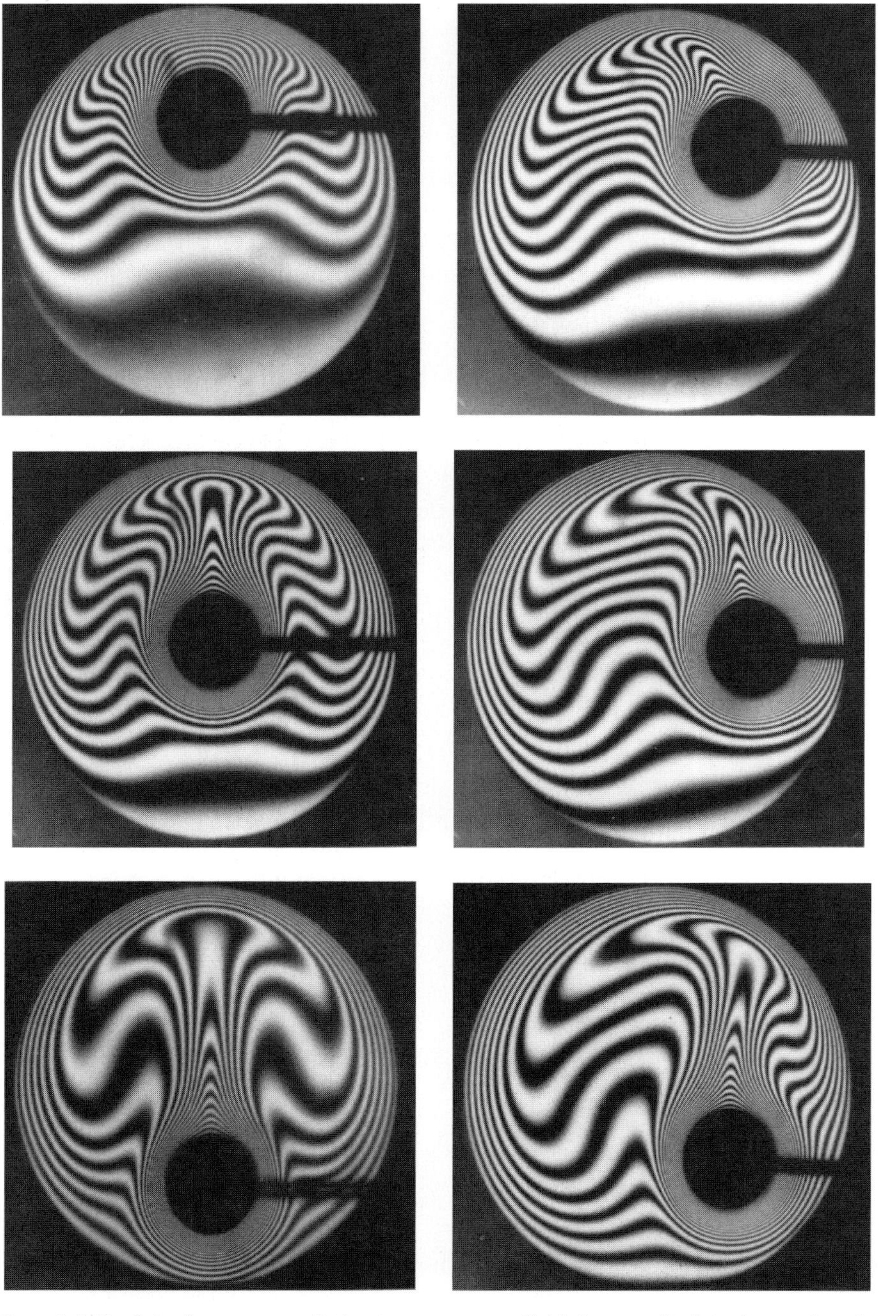

Fig. 4.3.2. Interferograms of the temperature field in a cylindrical annulus for different values and directions of eccentricity

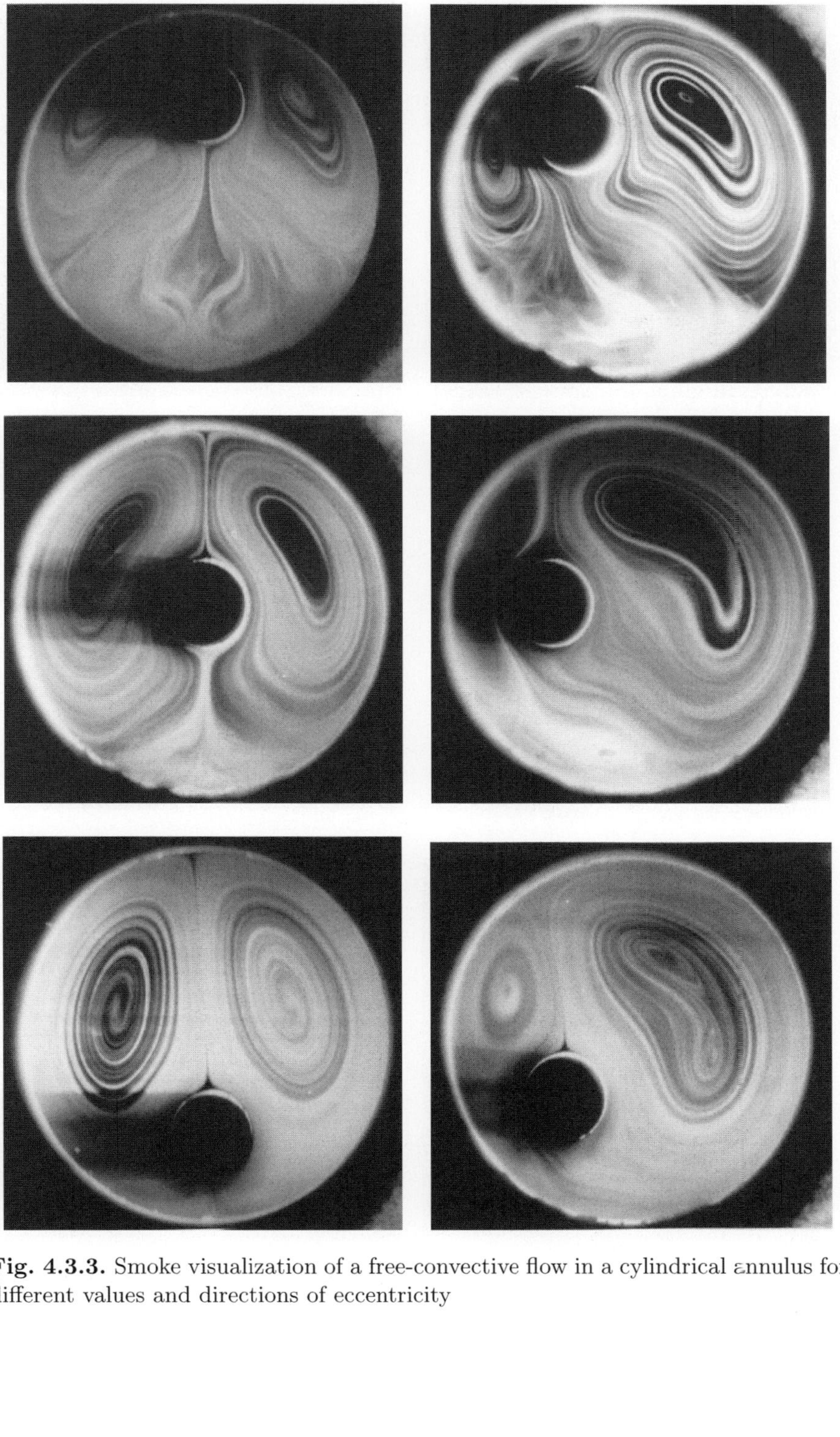

Fig. 4.3.3. Smoke visualization of a free-convective flow in a cylindrical annulus for different values and directions of eccentricity

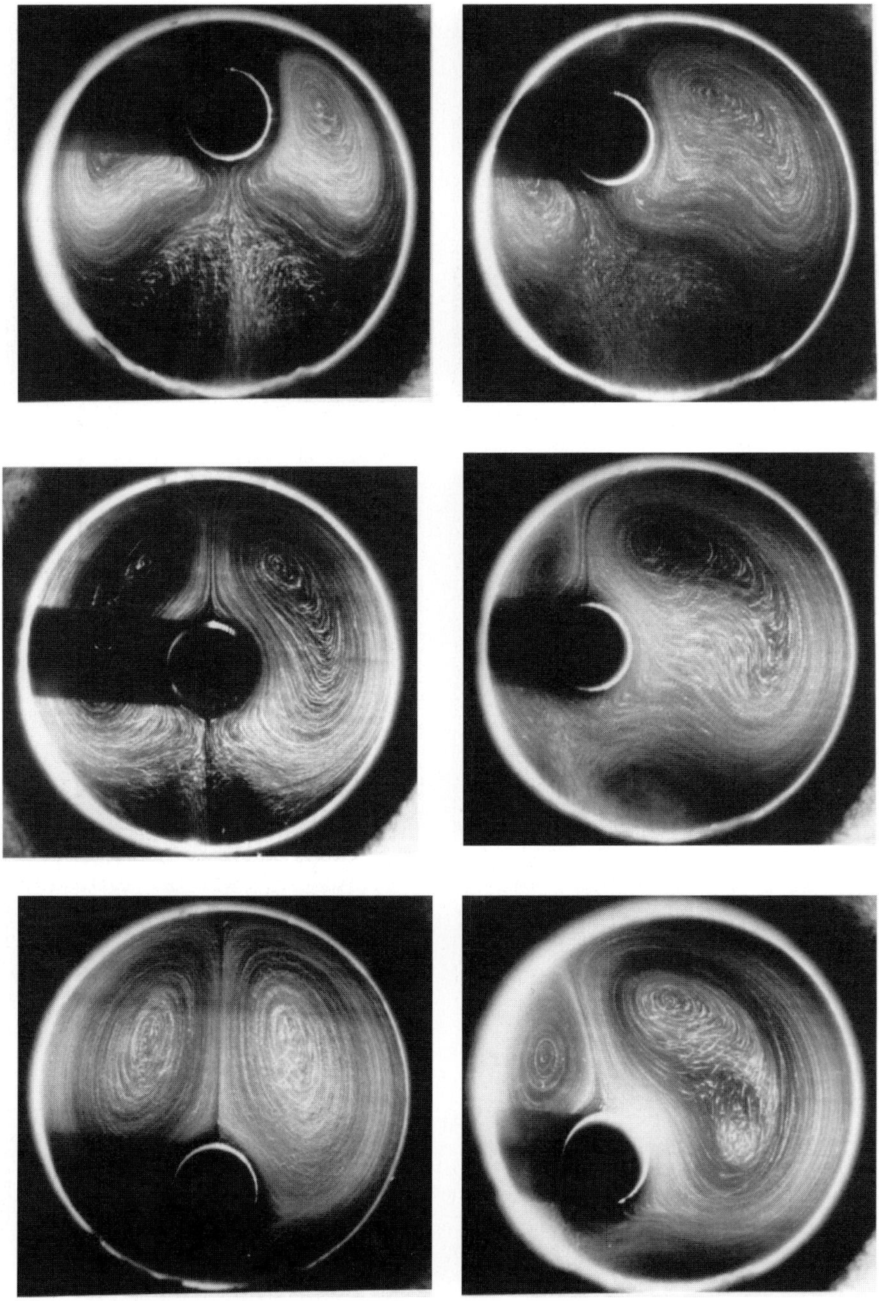

Fig. 4.3.4. Track visualization of a free-convective flow in a horizontal cylindrical annulus for different eccentricity degrees and orientations

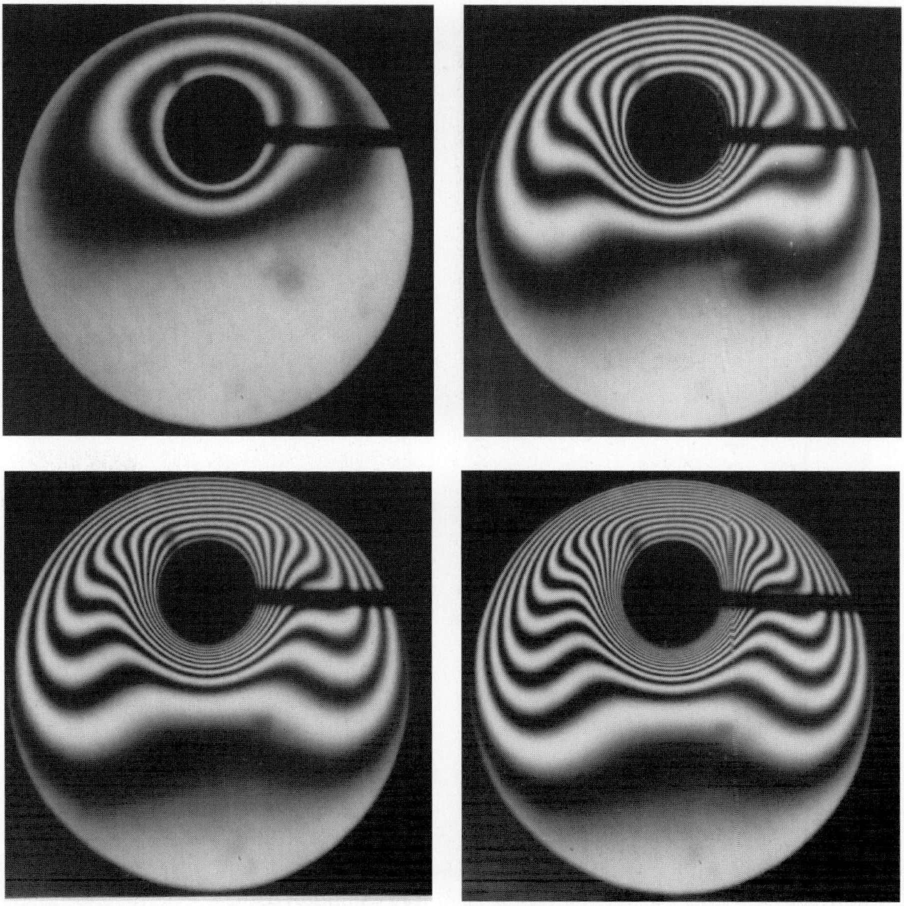

Fig. 4.3.5. Dynamics of temperature field development in a horizontal cylindrical annulus with a constant heat flux on the inner cylinder surface (eccentricity angle and value are $\varphi = 0°$ and $\delta = D/4$)

where

$$\mathrm{Nu}''_{d_1} = \left[1/\mathrm{Nu}'_{d_1} + 1/\mathrm{Nu}'_{d_2}\right]^{-1} ,$$

$$\mathrm{Nu}_T = 2/\ln\left(d_2/d_1\right) ,$$

$$\mathrm{Nu}'_{d_1} = 2/\ln\left\{1 + 2\left[\left(0.5\,\mathrm{Ra}_{d_1}^{1/4}\right)^{15} + \left(0.12\,\mathrm{Ra}_{d_1}^{1/3}\right)^{15}\right]^{-1/15}\right\} ,$$

$$\mathrm{Nu}'_{d_2} = -2/\ln\left\{1 - 2\left[\left(\mathrm{Ra}_{d_2}^{1/4}\right)^{15} + \left(0.12\,\mathrm{Ra}_{d_2}^{1/3}\right)^{15}\right]^{-1/15}\right\} ,$$

$$\left(T_\infty - T_{\mathrm{w2}}\right) / \left(T_{\mathrm{w1}} - T_{\mathrm{w2}}\right) = \mathrm{Nu}'_{d_1} / \left(\mathrm{Nu}'_{d_{12}} + \mathrm{Nu}'_{d_1}\right) .$$

Fig. 4.3.6. Dynamics of temperature field development in a horizontal cylindrical annulus with a constant heat flux on the inner cylinder surface (eccentricity angle and value are $\varphi = 45°$ and $\delta = D/4$)

The approximate formula for calculating heat transfer, with the error of 8.6%, laminar free convection between ***coaxial horizontal cylinders with a constant temperature*** not requiring iterative calculations of the temperature of the investigated body, is [4.109]

$$\mathrm{Nu}_{d_1} = \left[(\mathrm{Nu}_{d_1})^{15} + \mathrm{Nu}_T^{15}\right]^{1/15} , \tag{4.3.19}$$

$$\mathrm{Nu}_T = 2/\ln(d_2/d_1) ,$$

$$\mathrm{Nu}_{d_1} = 2/\ln\left[\left\{1 + 3.86\,[1 + (0.559/\mathrm{Pr})^{3/5}]^{5/12}\,\mathrm{Ra}_{d_1}^{-1/4}\right\}/\left(1 - 3.41\,\mathrm{Ra}_{d_1}^{-1/4}\right)\right] ,$$

$$T_{\mathrm{av}} = (T_{\mathrm{w}1} + T_{\mathrm{w}2})/2, \quad d_2/d_1 = 1 - \infty, \quad \mathrm{Pr} = 0.001 - 1000 .$$

For the given heat flux on a horizontal helical coil in a glass cavity filled with ***air and* Ar** ($p = 4 \cdot 10^3 - 1.01 \cdot 10^5$ Pa) free convection was observed when $\mathrm{Ra}^*_{d_1} < 10^{-4}$. The average value of the coefficient of heat transfer from the helical coil is [4.110]

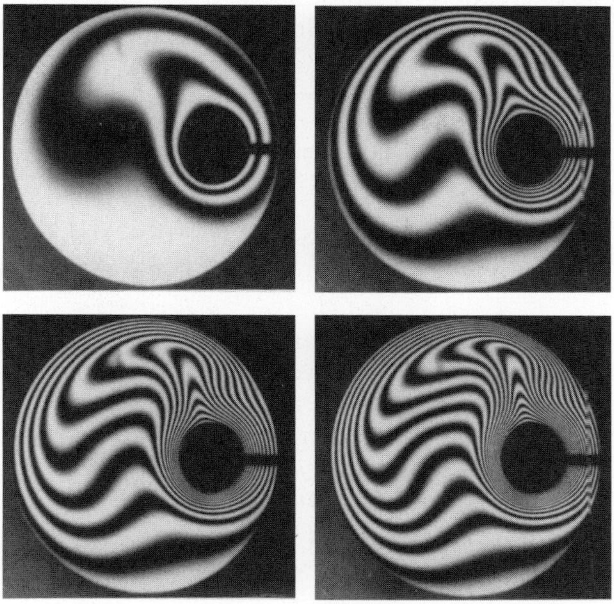

Fig. 4.3.7. Dynamics of temperature field development in a horizontal cylindrical annulus with a constant heat flux on the inner cylinder surface (eccentricity angle and value are $\varphi = 90^\circ$ and $\delta = D/4$)

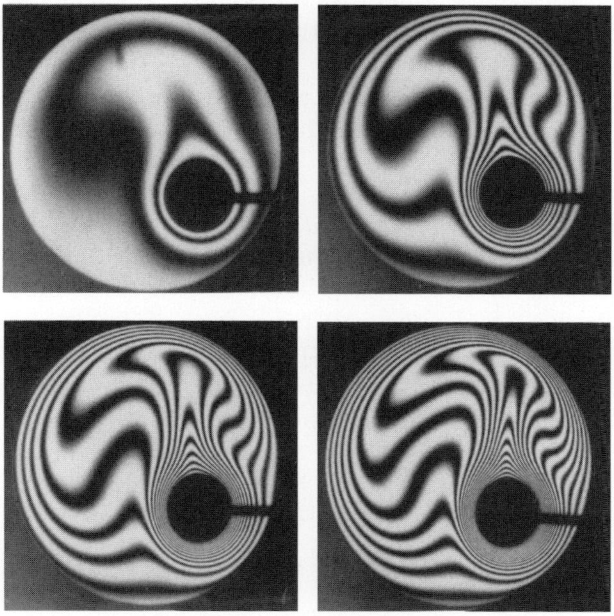

Fig. 4.3.8. Dynamics of temperature field development in a horizontal cylindrical annulus with a constant heat flux on the inner cylinder surface (eccentricity angle and value are $\varphi = 135^\circ$ and $\delta = D/4$)

$$\frac{1}{Nu} = \frac{1}{2} \ln \frac{9.5}{\mathrm{Ra}_{d_1}^{*1/4}} , \tag{4.3.20}$$

$$\mathrm{Ra}_{d_1}^{*} = 10^{-4} - 10^{-1}, \quad T_{\mathrm{w2}}/T_{\mathrm{w1}} = 0.333 - 0.707, \quad \mathrm{Pr} = 0.6 - 0.8 .$$

The average value of the heat transfer coefficient of $\mathbf{SF}_6$ ***in a layer of high-voltage cables in a turbulent regime*** [4.111]

$$\mathrm{Nu} = (0.502 - 0.177\delta/d_1)\, \mathrm{Ra}_{\delta}^{0.225+0.0254\delta/d_1} , \tag{4.3.21}$$

$$\delta/d_1 = 0.95 - 2.18, \quad \mathrm{Pr} = 0.6 - 1, \quad \mathrm{Ra}_{\delta} = 10^8 - 10^{11} .$$

For small values of Gr_{δ}, with allowance for the thermal conductivity of the inner cylinder one can estimate the effect of the conjugation parameter on the heat transfer between the cylinders [4.112].

The average value of the heat transfer coefficient ***in water*** between the ***eccentrically located horizontal cylinders in the case of a constant heat flux on the surface of the inner cylinder*** is [4.113].

$$\frac{\lambda_{\mathrm{ef}}}{\lambda} = 1.34\, \mathrm{Ra}_{\delta}^{0.11} \left(\frac{h_1 \delta}{h_2 d_2} \right)^{0.3} , \tag{4.3.22}$$

$$d_1(e) = 50(155) - 129(75)\,\mathrm{mm},\ d_2 = 299\,\mathrm{mm},\ \mathrm{Pr} = 2 - 10,\ \mathrm{Ra}_{\delta} = 10^6 - 10^9,$$

where h_1 and h_2 are the distances from the axis of the inner cylinder to the upper and lower water levels, respectively. The local value of the temperature is determined by the same criteria as heat transfer. The temperature distribution is mainly determined by the location of the internal cylinder and slightly depends on the heat flux value.

The largest thermal resistance of air between the cylinders with a horizontal axis in the temperature range 20–160°C is observed, when the axis of the inner cylinder is displaced by $e = 0.12(d_2 - d_1)$ in the vertical direction relative to the outer cylinder [4.114].

A helical partition or six axial partitions along the whole length and height in the air space between coaxial cylinders [4.115]. For $\mathrm{Gr}_{d1} = 5 \cdot 10^5$ the average value of the heat transfer coefficient decreases by 20% due to heat conductive partitions. With increase in the Gr_{d1} number, the presence of obstacles does not practically affect the heat transfer rate, however, it leads to a uniform distribution of the local heat transfer coefficient on the circumference of the inner cylinder.

The heat transfer ***in air*** between cylinders ***with a constant temperature*** at $d_2/d_1 = 1.5$, $\mathrm{Pr} = 0.71$, $\mathrm{Gr}_{\delta} = 5 \cdot 10^3 - 2 \cdot 10^4$ (angle φ is counted from the lower base of the cylinders) [4.116]

without partitions

$$\mathrm{Nu} = 0.16\, \mathrm{Gr}_{\delta}^{0.3} , \tag{4.3.23}$$

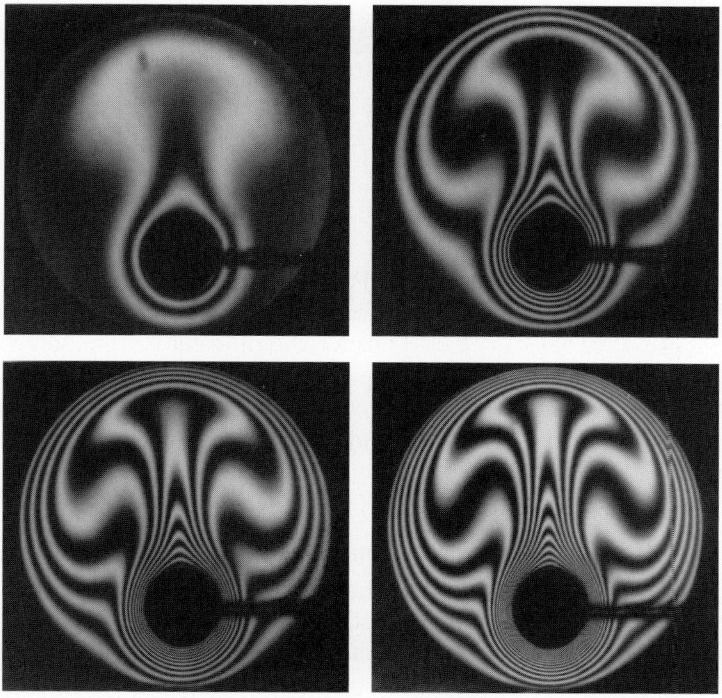

Fig. 4.3.9. Dynamics of temperature field development in a horizontal cylindrical annulus with a constant heat flux on the inner cylinder surface (eccentricity angle and value are $\varphi = 180°$ and $\delta = D/4$)

two horizontal partitions in the axial plane ($\varphi = 90°$)

$$\mathrm{Nu} = 0.144\,\mathrm{Gr}_\delta^{0.3}\,, \tag{4.3.24}$$

two vertical partitions at the slot height ($\varphi = 0, 180°$)

$$\mathrm{Nu} = 0.173\,\mathrm{Gr}_\delta^{0.3}\,. \tag{4.3.25}$$

In the presence of partitions the maximal value of the local heat transfer coefficient of the inner cylinder is observed at $\varphi = 30°$. Its minimal value ***in the case of horizontal location of the partitions*** is observed at $\varphi = 80°$; ***in the case of the vertical location of the partitions***, at $\varphi = 165°$.

The experimental and numerical investigation of natural turbulent convection ***in a horizontal annular slot*** with the ratio of radii $R_1/R_2 = 0.5$ for $\mathrm{Pr} = 0.71$ in different flow regimes including oscillatory and turbulent ones was carried out in [4.117]. The flows with the Rayleigh numbers up to $6 \cdot 10^5$ are implemented experimentally. The calculation results for the average Nusselt number are given in Table 4.3.1.

Table 4.3.1. The Values of the Average Nusselt Number for Different Flow Regimes

Ra_L	$4.0 \cdot 10^3$	$2.0 \cdot 10^4$	$6.0 \cdot 10^4$	$1.0 \cdot 10^5$	$1.1 \cdot 10^5$	$1.5 \cdot 10^5$	$3.0 \cdot 10^5$	$6.0 \cdot 10^5$
Nu	1.477	2.509	3.015	3.522	3.607	3.975	5.012	6.027

4.4 Spherical Interlayers

An investigation of the effect of the Prandtl number on the linear stability of axisymmetric disturbances of stationary natural convection between two concentric spherical casings carried out for the range of Pr numbers from 0 to 100, for a relative width of the slot equal to 0.1; 0.075 and 0.05 showed, that for every value of the slot width there exists a critical value of the stability parameter R_{cr} depending on the Prandtl number [4.118]:

$$R_{cr} = C(\varepsilon)/\sqrt{\mathrm{Pr}}\ ,$$

where $C(\varepsilon)$ is the proportionality coefficient depending on the slot width. The experimentally observed changes in the main flow are caused by the process non-stationarity. For each value the slot width there exists a transient value of the Prandtl number, so that the flow at higher Pr values turns to stationary regime, while at its low values the bifurcation flows remain periodic in time.

The detailed data of numerical and experimental investigation of the structure of flow and heat transfer in a concentric spherical layer with heated internal surface for $\mathrm{Gr} < 10^6$ and $\mathrm{Pr} = 0.7$ show that at low Gr numbers the crescent-shaped vortices are observed [4.119]. In the range $3.7 \cdot 10^3 < \mathrm{Gr} <$

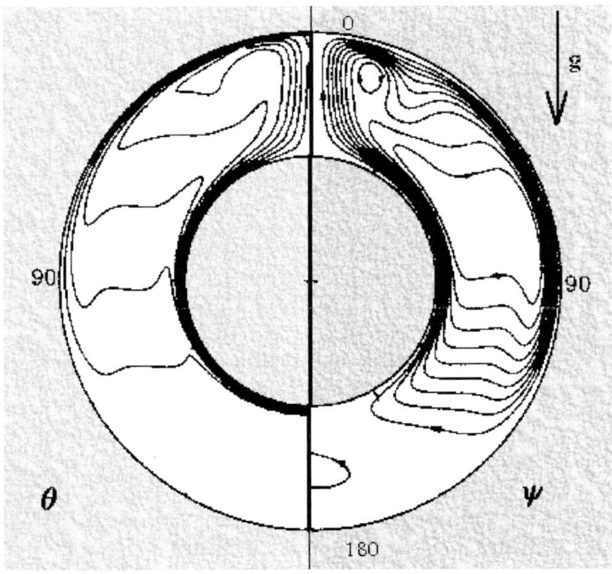

$2.65 \cdot 10^4$ the flow character becomes multicellular, changing from a steady two-cellular to an oscillating five-cellular one depending on the values of the Gr number and the parameter $\gamma = L/R_i$, where L is the slot width, R_i is the inner sphere radius. For $\gamma = 0.2$ and 0.25 the flow becomes noncellular again, and at $\gamma = 0.25$ hysteresis is observed. The numerical and experimental data show a substantial effect of the flow structure on the behavior of the local and global heat transfer coefficients.

An analysis of the conditions of motion and heat transfer with natural convection in a volume between eccentrically located spheres ***of constant temperatures*** allows one to find a dependence for the average value of the heat transfer coefficient as [4.100, 4.120]

$$\mathrm{Nu} = f\left(\mathrm{Ra}_{d1}, \mathrm{Pr}; \frac{d_1}{d_2}, \frac{2e}{d_2 - d_1}\right) . \tag{4.4.1}$$

Processing of experimental data for heat transfer between concentric spheres of ***constant temperature for gases*** leads to the power-law dependence [4.100]

$$\frac{\lambda_{\mathrm{ef}}}{\lambda} = 0.316\left(\frac{d_1}{d_2}\right)^{0.133} \mathrm{Ra}_{d1}^{0.25}\,\mathrm{Ra}_{d2}^{0.039} , \tag{4.4.2}$$

$$d_1/d_2 = 0.118 - 0.605, \quad \mathrm{Ra}_{d1} = \mathrm{Ra}_{dT} - 10^{10}, \quad T_{\mathrm{av}} = T_{\mathrm{w}2} .$$

The lower applicability limit of formula (4.4.2) is determined by the transition from the convective character of heat transfer through the layer to the heat conduction:

$$\mathrm{Ra}_{dT} = \left\{6.33\left[\left(1 - \frac{d_1}{d_2}\right)\left(\frac{d_1}{d_2}\right)^{0.133}\right]^{-1}\right\}^{3.46} . \tag{4.4.3}$$

A numerical solution of the problem of stationary natural-convective flow in an annular channel ***with isothermal walls*** formed by concentric spheres with the ratio of diameters equal to 2, carried out in the range of the Rayleigh numbers $10^2 < \mathrm{Ra} < 10^6$ ***for air, water, and liquid metals*** ($\mathrm{Pr} = 0.02$, 0.7 and 6) [4.121], shows the presence of the multicellular flow structure for high Ra numbers for water and air and for very low Ra numbers for liquid metals.

The numerical calculations of heat transfer between coaxial spheres of ***constant temperature*** yields

for an incompressible medium [4.122]

$$\frac{\lambda_{\mathrm{ef}}}{\lambda} = 0.45\left(\frac{d_1}{d_2}\right)^{0.031} \mathrm{Ra}_{d1}^{0.226} , \tag{4.4.4}$$

$$d_1/d_2 = 0.5 - 0.83, \quad \mathrm{Pr} = 0.2 - 5, \quad \mathrm{Ra}_{d1} = 8 \cdot 10^3 - 8 \cdot 10^7 .$$

compressible air with allowance for the compression work and energy dissipation [4.123, 4.124]

$$\frac{\lambda_{\mathrm{ef}}}{\lambda} = 0.143\ \mathrm{Ra}_\delta^{0.273}\,(1 + e/\mathrm{Ec})^{1/2}\ , \tag{4.4.5}$$

$$d_1/d_2 = 0.33 - 0.83,\quad \mathrm{Pr} = 0.71,\quad \mathrm{Ra}_\delta = 0.21 \cdot 10^4 - 10^5\ .$$

The average coefficient of heat transfer between spheres of ***constant temperature with a vertical eccentricity*** e [4.120]

$$\frac{\lambda_{\mathrm{ef}}}{\lambda} = C\,\mathrm{Ra}_\delta^m\ , \tag{4.4.6}$$

$$\delta = r_2 - R_1,\quad \mathrm{R}_1 = (r_1 + r_2)/\left[1 - bc + \sqrt{(1-b^2)(1-c^2)}\right],$$

$$b = (e + r_1)/r_2\ ,$$

$$c = (e - r_1)/r_2,\quad T_{\mathrm{av}} = \frac{\left(R_{\mathrm{av}}^3 - R_1^3\right)T_{\mathrm{w1}} + \left(r_2^3 - R_{\mathrm{av}}^3\right)T_{\mathrm{w2}}}{r_2^3 - R_1^3}\ ,$$

$$R_{\mathrm{av}} = (R_1 + r_2)/2\ ,$$

$$\delta/r_1 = 0.09 - 1.81,\quad e/\delta = (-0.75) - 0.75,\quad \mathrm{Pr} = 0.7 - 4148\ .$$

The values of the coefficients C and m are given in Table 4.4.1.

Table 4.4.1.

Pr	Ra_δ	C	m
0.7	$1.4 \cdot 10^4 - 2.5 \cdot 10^6$	0.117	0.276
4.7–12.1	$2.4 \cdot 10^4 - 5.4 \cdot 10^8$	0.033	0.328
148–336	$2.4 \cdot 10^4 - 9.7 \cdot 10^7$	0.031	0.353
1954–4148	$1.3 \cdot 10^3 - 5.6 \cdot 10^6$	0.056	0.330

A simplified formula for $\mathrm{Ra}_\delta \delta/r_1 = 1.2 \cdot 10^2 - 1.1 \cdot 10^9$ is [4.120]

$$\frac{\lambda_{\mathrm{ef}}}{\lambda} = 0.228\,(\mathrm{Ra}_\delta \delta/r_1)^{0.226}\ . \tag{4.4.7}$$

4.5 Cavities of Complex Geometry

In cavities of complex geometry, a substantial effect on process of heat transfer is exerted by the ratios between the sizes which enter into the dimensionless relations as the parameters P_{ik}.

Parallelepiped $\boldsymbol{H \times B \times \delta}$ ***with one surface having the temperature*** T_{w1} ***and the remaining surfaces*** $T_{\mathrm{w2}}(T_{\mathrm{w1}} > T_{\mathrm{w2}})$. The average value of the heat transfer coefficient in the cavity is [4.125]

Fig. 4.4.1. Dynamics of the change in the temperature field in a horizontal cylindrical annulus on acceleration and deceleration of rotation of the outer cylinder

$$\frac{\lambda_{\text{ef}}}{\lambda} = 0.24N[6.25 - 5.25\,(1+\delta/l)^{-1.67}]\,\text{Ra}_l^{0.25}\,, \tag{4.5.1}$$

$$l = \sqrt{HB}, \quad \text{Pr} = 0.7, \quad \delta/l = 0.2 - 0.8\,,$$

$$\text{Ra}_l = 2.5\cdot 10^5 - 1\cdot 10^8\,, \quad T_{\text{av}} = (T_{\text{w1}} + T_{\text{w2}})/2\,,$$

where $N = 1$ for the vertical orientation of the heated surface; $N = 1.3$ for its horizontal orientation.

In the parallelepiped with two heated vertical parallel surfaces ($\text{Pr} = 0.7, \text{Ra}_H = 3\cdot 10^8 - 3\cdot 10^{10}$) the presence of two characteristic heat transfer regions can be assumed: first, where the heat conduction processes prevail ($\text{Ra}_H < 10^9$), and second, where the heat transfer is carried out mainly by convection ($\text{Ra}_H > 3\cdot 10^9$) [4.126].

A uniformly heated horizontal cylinder placed in a large rectangular cavity with isothermal walls filled with air for different intensities of heating the cylinder. The heat transfer rate from the cavity reaches maximum near the upper wall. For all heat flux values investigated the velocity of air is maximal at the distance equal to about 9 cylinder diameters along the vertical axis. The correlations for the average heat transfer coefficient can be generalized by the empirical dependences [4.127]

$$\text{Nu}_L = 1.81\,\text{Ra}_L^{*\,0.207}\,, \tag{4.5.2}$$

$$\text{Nu}_b = 0.604\,\text{Ra}_b^{0.2083}\,, \tag{4.5.3}$$

where $\text{Ra}_L^* = \text{Ra}_L\,(L/r)\,, r$ is the cylinder radius, $b = \pi r, L = 48b$.

A cavity of triangular section with the vertex facing upwards. The height of the isosceles triangle is H, the width of the base is B, the inclination angle of the lateral surface is θ_1, the apex angle is θ_2, the side size $L = H/\sin\theta_1$.

The parameters of the investigated region are given in Table 4.5.1 [4.128].

Table 4.5.1.

B, m	H, m	B/H	θ_1°	θ_2°
0.1078	0.0933	0.865	60	60
0.1524	0.0762	0.5	45	90
0.1867	0.0539	0.289	30	120

The coordinate x is directed from the base along the lateral heated surface. The temperature of the hot lateral surface is $T_H = 30 - 80°\text{C}$, of the cold one is $T_0 = 0°\text{C}$, the base is insulated. The local and average values of the heat transfer coefficient of the hot surface ***in air*** are

$$\text{Nu} = C_2\text{Gr}^n_{L,\theta_2/2}, \quad T_{\text{av}} = (T_H + T_0)/2\,, \tag{4.5.4}$$

$$\text{Nu}(x) = C_1\text{Gr}^n_{x,\theta_2/2}, \quad T_{\text{av}} = T_H - 0.38\,(T_H - T_\infty)\,, \tag{4.5.5}$$

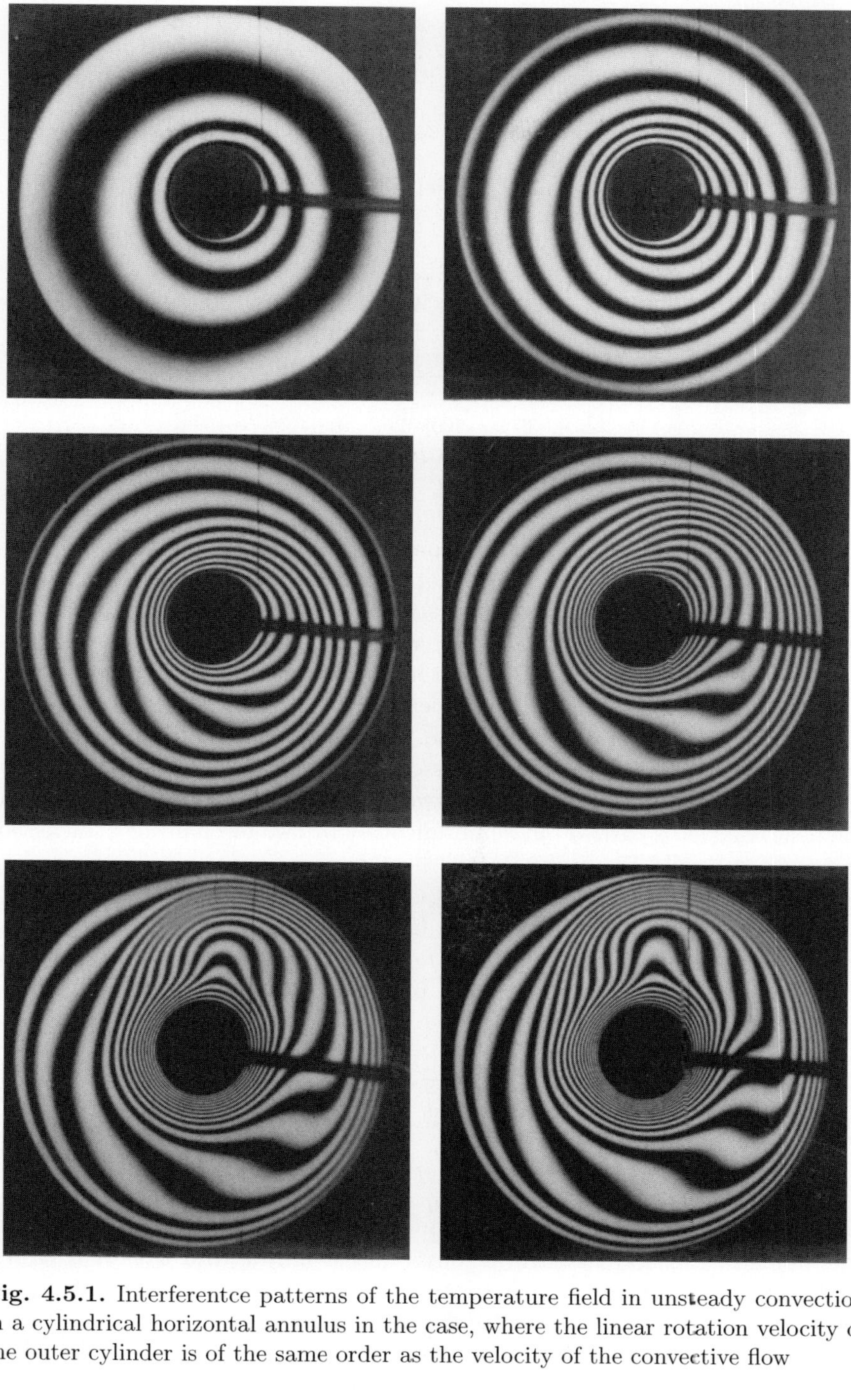

Fig. 4.5.1. Interferentce patterns of the temperature field in unsteady convection in a cylindrical horizontal annulus in the case, where the linear rotation velocity of the outer cylinder is of the same order as the velocity of the convective flow

$$\mathrm{Pr} = 0.7, \quad \mathrm{Gr}_{L,\theta_2/2} = 2.9 \cdot 10^6 - 9 \cdot 10^6 ,$$
$$(T_\infty - T_0)\,(T_H - T_0) = \frac{1}{2}[1 + (x/L)^2] ,$$
$$\mathrm{Gr}_{x,\theta_2/2} = g\beta_\infty\,(\cos\theta_2/2)\,(T_H - T_\infty)\,x^3/\nu^2 ,$$

where the coefficients C_i and n are determined by the value of the angle at the apex θ_2 (Table 4.5.2).

Table 4.5.2.

θ_2°	C_1	C_2	n	C_3
60	0.372	0.0921	0.31	0.15
90	0.339	0.0502	0.34	0.19
120	0.158	0.0284	0.37	0.22

The value of Nu_x depends slightly on the value of the angle at the apex, the local heat transfer coefficient for the hot wall is higher, than for the isothermal plate, which is caused by the liquid stratification in the middle part of the cavity and by the inclination of the walls. For the cold wall the value of Nu_x is in agreement with the data on heat transfer of an isothermal inclined surface in the infinite space. The number $\mathrm{Gr}_{x,\theta_2/2}$ first increases with x/L, and when $x/L > 0.76$ it starts to decrease. This effect is caused by the complex character of interaction of the processes of momentum and heat transfer in the region, where the flow becomes two-dimensional. At $x = (0.76 - 0.9)L$, the heat transfer is carried out simultaneously by two mechanisms: convection and radiation, and for $x > 0.9L$ the heat transfer is mainly determined by heat conduction.

A cavity of triangular section filled with ***air with the lateral walls maintained at the temperature*** $T_H, (T_H > T_0)$ ***and the base at the temperature*** T_0. The average value of the heat transfer coefficient on the cold wall is [4.129]

$$\mathrm{Nu} = 4.5 , \tag{4.5.6}$$
$$\mathrm{Gr}_H < 8 \cdot 10^6, \quad T_H > T_0 .$$
$$\mathrm{Nu} = C_3\,\mathrm{Gr}_H^{0.3} , \tag{4.5.7}$$
$$\mathrm{Gr}_H = 7 \cdot 10^4 - 5 \cdot 10^5, \quad T_H < T_0, \quad T_{\mathrm{av}} = (T_0 + T_H)/2 .$$

The numerical results for natural convection of ***water and air in a cavity of triangular section with a cooled inclined upper surface*** (Fig. 4.5.2) for the values Ra from 10^2 to 10^5 for the aspect ratio in the range 0.1–1.0 show that at low Ra the flow has a simple circulating character and at high Ra the structure of the free-convective movement inside the cavity becomes multicellular [4.130].

In the case, where the ratio of the maximal height to the cavity length is very low (which corresponds to small angles between the upper and lower surfaces), the three-dimensional problem can be reduced to the local two-dimensional problem of a flow between two parallel planes on each part of the length, and thus to determine the critical value of the Rayleigh number, above which the flow inside the cavity loses stability [4.131].

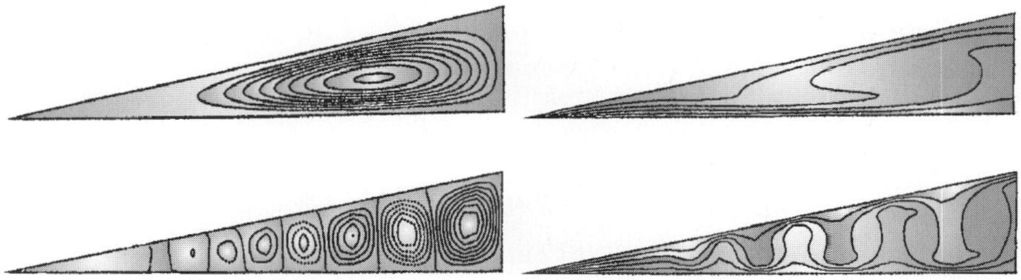

Fig. 4.5.2. Isotherms and streamlines of a free-convective flow in a cavity of triangular section [4.130]

A semispherical cavity $(\boldsymbol{H/B} > \mathbf{10})$ ***with the given temperatures on the vertical surfaces and insulated bases. On the upper part of the cold surface there is a cylindrical obstacle*** with $B_{\max} = B + d/2$. In the presence of the obstacle, the thermal resistance in air (Pr = 0.71) increases with the value of the thermal resistance of the rectangular cavity [4.132].

A theoretical analysis shows [4.133] that the structure of the free-convective flow ***in a toroidal channel under the conditions of periodic change of the wall temperature***, changes from the full absence of motion to periodic

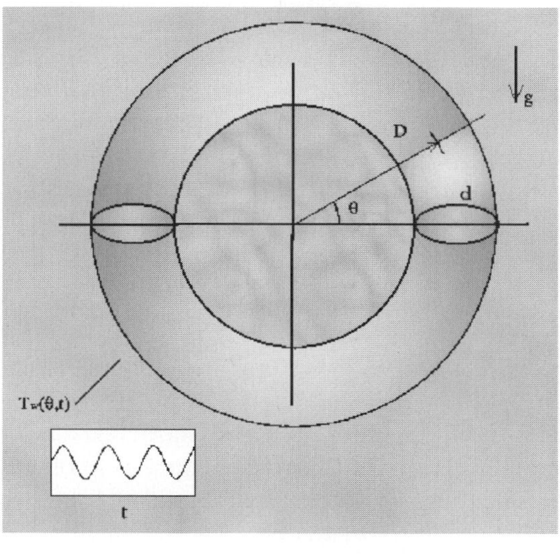

pulsations and further to a chaotic flow with decrease in the Rayleigh number. The heating and cooling of the channel walls are symmetric with respect to the axes parallel to the gravity force vector. On the average, the temperature of the upper half of the contour exceeds the temperature of the lower half. The temperature field inside the contour is steadily stratified.

A vertical cylinder of diameter d and maximal height H with a spherical segment of height h and a polar angle θ in the base. The temperature of the upper flat disk T_H is constant, the lateral walls are insulated, and there is an electrical heater in the base. The temperature of the lower surface is maximal in the centre and is equal to T_0. There are heat sources of density Q_v regularly distributed in the cavity. The average value of the coefficient of heat transfer to the upper flat disk [4.134]

$$\mathrm{Nu} = 4\, Q_v H / \left[\lambda \pi d^2 \left(T_0 - T_H\right)\right] = C\, \mathrm{Ra}_{vH}^n \,, \qquad (4.5.8)$$

$$H = 0.12 - 0.73\,\mathrm{m}, \quad h = 0.12\,\mathrm{m}, \quad d = 0.914\,\mathrm{m}, \quad \theta = 30 - 60^\circ \,.$$

The values of C and n are given in Table 4.5.3.

Table 4.5.3.

H/d	$\mathrm{Ra}_{vH} \cdot 10^{10}$	Pr	C	n
0.108	$0.912 \cdot 10^{-2}$–1.268	2.09–5.94	2.278	0.127 ± 0.06
0.148 – 0.158	$3.72 \cdot 10^{-2}$– 35.4	2.7–7.01	2.53	0.25 ± 0.05
0.403	3.5–$1.35 \cdot 10^{-3}$	2.27–6.82	1.225	0.66 ± 0.08
0.686 – 0.703	55.4–$1.55 \cdot 10^{-4}$	1.9–5.76	0.328	0.27 ± 0.06

The heat transfer is independent of the shape of the base with increase in H/d, it tends to the value of Nu for horizontal disks. When $H/d < 0.15$, the main factor determining heat transfer is the segment geometry.

A rectangular cavity B × H with a rectangular heater b × h protruding inside located in the base centre. On the surface of the obstacle and on the upper base a constant temperature is maintained. The lateral sides and the part of the bottom without the obstacle are insulated. Processing of the results of calculations of the average value of the heat transfer coefficient leads to the dependence [4.135]

$$\mathrm{Nu} = 0.115\, \mathrm{Gr}_d^{0.33} \,, \qquad (4.5.9)$$

$$B/H = 1/2 - 2, \quad h/H = 1/6 - 5/6, \quad b/B = 1/6 - 2/3 \,,$$
$$d = H - hb/B, \quad \mathrm{Pr} = 0.71, \quad \mathrm{Gr}_d = 10^4 - 10^6 \,.$$

The values of the power index 1/3 are characteristic for the regimes of developed laminar convection. In a cavity of complex geometry it is connected with the earlier appearance of secondary flows and complication of the flow structure.

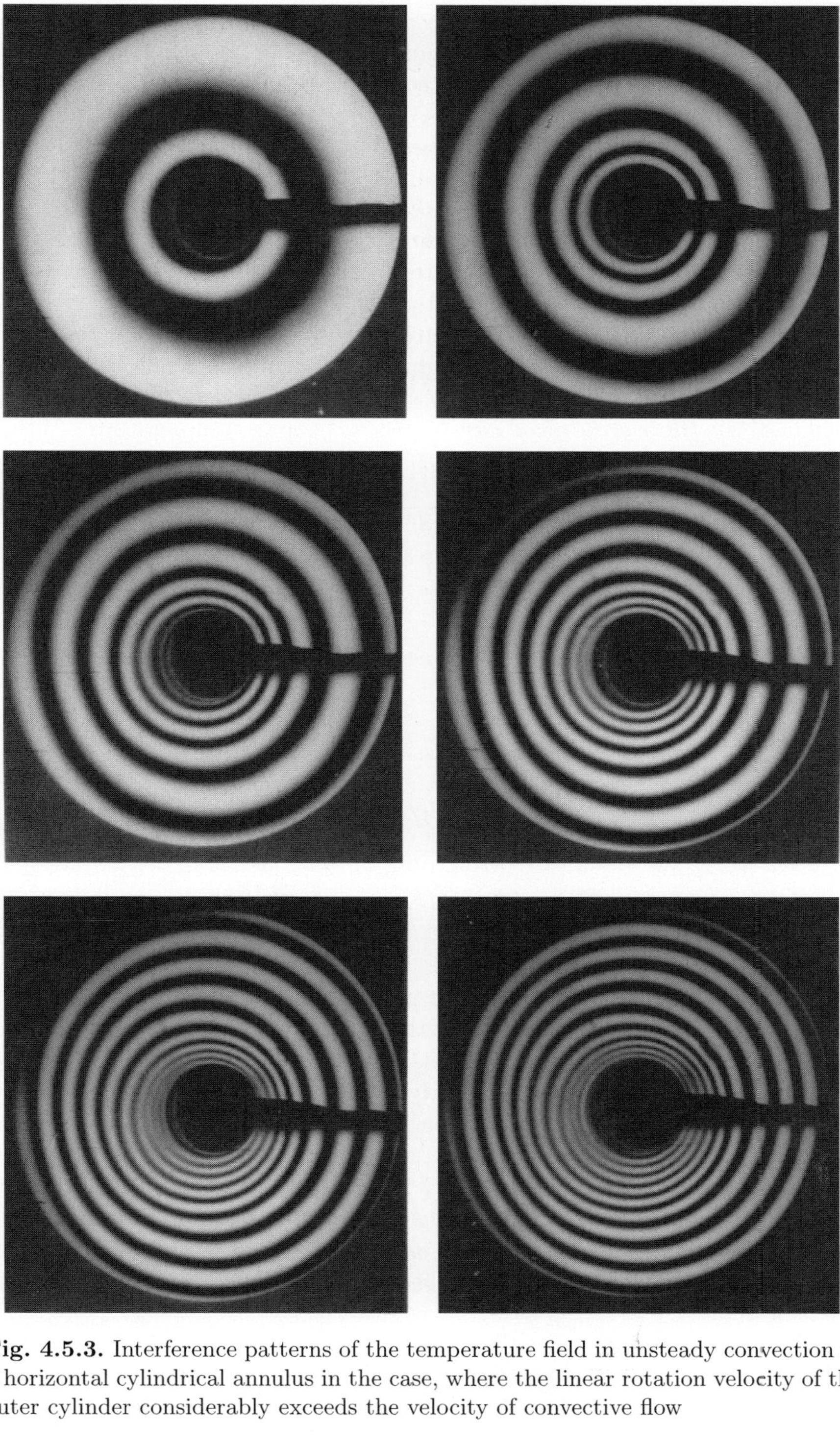

Fig. 4.5.3. Interference patterns of the temperature field in unsteady convection in a horizontal cylindrical annulus in the case, where the linear rotation velocity of the outer cylinder considerably exceeds the velocity of convective flow

The calculated dependences for the average Nusselt number ***for different wall materials*** in the case of natural-convective immersion cooling of a ***single and several heat sources located in parallel coupled cavities filled with the test fluid freon*** (Pr = 12.4) were obtained in [4.136]. The cases of high (***alund ceramics***) and low (***bakelite***) thermal conductivity of the substrate are considered. The Rayleigh number changed from $6.8{\cdot}10^8$ to $5.5 \cdot 10^{10}$. For the low thermal conductivity (***bakelite***) the flow and temperature field are asymmetric, whereas for high (***alund ceramics***) it is symmetric.

An investigation of the flow structure ***in a region with two obstacles*** ($h_1/H = 1/3, h_2/H = 2/3$) shows that the zone of intensive convective motion is located above the end surface of the lower heater.

A vertical heated cylinder of height h and diameter d is located in the centre of an insulated base of a cavity. The average value of the heat transfer coefficient ***in the cylindrical cavity*** $H \times D$ of lower temperature is [4.137]

$$\mathrm{Nu} = 0.128\,\mathrm{Ra}_h^{0.25+0.00751\,\lg \mathrm{Ra}_h} \left(\frac{D-d}{d}\right)^{0.485} \left(\frac{H}{h}\right)^{0.67} \left(\frac{D}{2H}\right)^{0.8} , \quad (4.5.10)$$

$$\mathrm{H}/h = 1/2 - 2, \quad D/d = 1.43 - 6, \quad \mathrm{Pr} = 0.7 ,$$
$$\mathrm{Ra}_h = 3 \cdot 10^4 - 7 \cdot 10^5, \quad T_{\mathrm{av}} = (T_{\mathrm{w1}} + T_{\mathrm{w2}})/2 ,$$

and ***in the cavity having the shape of a parallelepiped*** $H \times B \times L$ it is [4.137, 4.138]

$$\mathrm{Nu} = 0.39\,\mathrm{Ra}_h^{0.25+0.00751\,\lg \mathrm{Ra}_h} \left(\frac{B-d}{2h}\right)^{0.3} \left(\frac{H}{h}\right)^{-0.1} , \quad (4.5.11)$$

$$H/h = 1.24 - 4.45, \quad (B-d)/h = 1.14 - 5.66 ,$$
$$\mathrm{Ra}_h = 4 \cdot 10^4 - 7 \cdot 10^5, \quad \mathrm{Pr} = 0.7 .$$

$$\mathrm{Nu} = 0.288(H/L)^{-0.1}\,\mathrm{Ra}_h^{0.3} , \quad (4.5.12)$$

$$H/h = 1.3 - 2.5, \quad \mathrm{Ra}_h = 5 \cdot 10^5 - 10^7, \quad \mathrm{Pr} = 0.7 .$$

While the heat-generating element moves along the diagonal or the shortest distance between the casing and the heater, the heat transfer does not change up to some limiting eccentricity value e_{lim}. Until $e < e_{\mathrm{lim}}$, the structure of the convective flow in the cavity is determined by two free developing boundary layers. The heat transfer in this case is calculated from formulae (4.5.11). The limiting eccentricity depends on the dimensions of the casing and heater and lies in a range $e_{\mathrm{lim}} = 0.9 - 1.0$.

In the centre of a square cavity of height H with constant temperature there is a square heater with the side of length h. The values of the full coefficient of heat transfer ***between coaxial tubes*** at $H/h = 5$ are given in Table 4.5.4 [4.139].

The total heat flux from the heater surface increases with the increase of Ra_h and decrease of Pr. Starting from Pr $\sim$ 6.8 its effect on heat transfer become insignificant. The similar character of the local and average heat

Table 4.5.4.

Pr	0.7	1	6.8	1000
Ra_h		Nu_h		
1700	2.21	2.20	2.20	2.19
2000	2.26	2.26	2.24	2.23
9800	3.67	3.50	3.41	3.40
12000	3.83	3.62	3.53	3.53
20000	–	4.16	3.95	3.95

transfer is observed ***in the case of rarefaction of a compressible gas in the cavity*** [4.140].

Throttles and transformers of low power of height H with the heat transfer surface area F of maximum (B) and minimum (L) sizes mounted at three mutually orthogonal locations in the space. The lower side of the investigated samples was located parallel to the horizontal plane in the centre of a cube of height 0.7 m made of a steel tube of diameter 5 mm. The average value of the heat transfer coefficient with an error of 7–9% for three types of samples ($k = 1 + H/B + H/L, \mathrm{Pr} = 0.7, T_{\mathrm{av}} = T_\infty$) is given by [4.141]

shielded transformers and throttles

$$\mathrm{Nu} = \frac{\mathrm{Ra}_H^{0.3265}}{2.24 \cdot 10^{0.0017k} (F \cdot 10^3)^{0.198}} , \qquad (4.5.13)$$
$$\mathrm{Ra}_H = 5 \cdot 10^4 - 3 \cdot 10^6 ,$$

pivot-shaped transformers and throttles

$$\mathrm{Nu} = \frac{\mathrm{Ra}_H^{0.3753}}{2.59 \cdot 10^{0.01378k} (F \cdot 10^3)^{0.3936}} , \qquad (4.5.14)$$
$$\mathrm{Ra}_H = 8 \cdot 10^4 - 6.3 \cdot 10^6 ,$$

toroidal transformers and throttles

$$\mathrm{Nu} = \frac{\mathrm{Ra}_H^{0.3143}}{1.68 \cdot 10^{-0.0066k} (F \cdot 10^3)^{0.3}} , \qquad (4.5.15)$$
$$\mathrm{Ra}_H = 4 \cdot 10^3 - 2 \cdot 10^6 .$$

Depending on the orientation of the samples in space, the horizontal and vertical dimensions can change 2–3 times. The average value of the heat transfer coefficient changes within the limits 3–10%.

The average value of heat transfer ***for little radioelements randomly located on an assembly panel*** is [4.142]

$$\mathrm{Nu} = 1.45\, \mathrm{Ra}_l^{0.23} , \qquad (4.5.16)$$

where $l = H$ ***for vertical lamps, relays, and transformers***; $1/l = (1/H + 1/d)$ ***for horizontal resistors***.

A heat genegating horizontal cylinder of size d × l inside a horizontal cylindrical casing of size D × L with the eccentricity created by displacement of the longitudinal axis of the inner cylinder in the vertical plane. ***The temperatures of the inner cylinder*** (T_{w1}) ***and casing*** (T_{w2})($T_{w1} = T_{w2}$) ***are constant***. The ranges of the experimentally investigated relative characteristic sizes and parameters are [4.143]: $d/D = 0.06-0.5, L/d = 2-16.6, \ l/d = 1-1.5, \mathrm{Ra}_d = 5\cdot10^3 - 1.25\cdot10^6, \mathrm{Pr} = 0.7$. Depending on the eccentricity, three heat transfer regions are established [4.144]:

a region of weak eccentricity effect (displacement of the inner cylinder $\pm e \le 0.5(D-d) - 6$)

$$\mathrm{Nu} = 0.834\left(\frac{d}{D}\right)^{-0.182}\left(\frac{l}{d}\right)^{0.318}\left(\frac{L}{d}\right)^{-0.147}\mathrm{Ra}_d^{0.206}\,, \tag{4.5.17}$$

a region of positive eccentricity close to the limiting one $e = [0.5(D-d) - 6] - [0.5(D-d) - 1]$:

$$\mathrm{Nu} = 1.527\left(\frac{d}{D}\right)^{-0.122}\left(\frac{l}{d}\right)^{0.446}\left(\frac{L}{d}\right)^{-0.222} \times\left(\frac{D-d-2e}{2d}\right)^{-0.185}\mathrm{Ra}_d^{0.133}\,, \tag{4.5.18}$$

a region of negative eccentricity $(-e) = [0.5(D-d)-6]-[0.5(D-d)-1]$

$$\mathrm{Nu} = 0.597\left(\frac{d}{D}\right)^{-0.131}\left(\frac{l}{d}\right)^{0.613}\left(\frac{L}{d}\right)^{-0.154} \times\left(\frac{D-d-2e}{2d}\right)^{-0.074}\mathrm{Ra}_d^{0.236}\,. \tag{4.5.19}$$

Inside a sphere of diameter d_2 ***there is a coaxially located vertical cylinder of height H and diameter*** d_1 ***with spherical end surfaces. The surface temperatures*** T_{w1} ***for the sphere and*** T_{w2} ***for the cylinder are maintained constant***. The ratio between convection and diffusion heat fluxes is determined by the dependence [4.145]

$$q_{con}/q_{dif} = 0.228\,(\mathrm{Ra}_l l/r_1)^{0.226}\,, \tag{4.5.20}$$

$$T_{\mathrm{av}} = \frac{2\int_0^{\pi}\left[T_{\mathrm{w1}}\left(r_{\mathrm{av}}^3 - R_1^3\right) - T_{\mathrm{w2}}\left(r_2^3 - r_{\mathrm{av}}^3\right)\right]d\varphi}{4r_2^3 + 2r_1^3 + 3r_1^2 H}\,,$$

$$l = r_2 - \frac{r_1}{2}\left(\frac{6H}{r_1} - 4\right)^{1/3}, \quad r_{\mathrm{av}} = \frac{r_1 + r_2}{2}\,.$$

$$d_1/d_2 = 0.458 - 0.92, \quad H/d_1 = 1.5 - 1.036, \quad d_2 = 0.2496\,\mathrm{m}\,,$$

where $R_1(\varphi)$ is a radial coordinate of the inner cylinder.

A closed layer of height H and width B with louvers of width b, spaced h apart, and with the angle of inclination to the horizon ϑ. The vertical layer divides two cavities of different temperatures T_0 and T_B. The louvers are heated by an outer electrical heater to the temperature T_w. The average value of the heat transfer coefficient at $T_{av} = [(T_0 + T_B)/2 + T_w]/2$. $\Pr = 0.7$ is [4.146]:

on the surface of glasses in a window with the louvers between the glasses

$$\mathrm{Nu} = 0.183\,\mathrm{Ra}_H^{1/3}\,, \quad \mathrm{Ra}_H = 3.9 \cdot 10^7 - 5.5 \cdot 10^8\,, \tag{4.5.21}$$

on the louvers surface

$$\mathrm{Nu} = (0.29 + 0.32 b \cos\vartheta/h)\,\mathrm{Ra}_b^{0.29}\,, \quad \mathrm{Ra}_b = 1.5 \cdot 10^5 - 6 \cdot 10^5\,. \tag{4.5.22}$$

The calculation parameters change in the range

$$b\cos\vartheta/h = 1.15;\quad 0.7;\quad 0.43 \quad \text{for} \quad \vartheta = 0\,,$$
$$b\cos\vartheta/h = 0.632 \quad \text{for} \quad \vartheta = 15^\circ\,,$$
$$b\cos\vartheta/h = 0.47 \quad \text{for} \quad \vartheta = 35^\circ\,.$$

A cavity with a section as an equilateral trapezium of height H with the length of the lower base b, the upper one B $(b < B)$, and the inclination angle of the lateral wall relative to the vertical ϑ. The trapezoidal base is located at an angle of φ with respect to the horizon. The lower base has the temperature T_0, the upper one T_H, the lateral walls are insulated. The average value of the heat transfer coefficient is [4.147]

$$\mathrm{Nu} = C_\varphi \mathrm{Ra}_H^{0.375}\,, \tag{4.5.23}$$
$$H = 7.68\,\mathrm{sm}\,,\quad b = 2.54\,\mathrm{sm}\,,\quad \vartheta = 30^\circ,\quad p = 0.67 \cdot 10^3 - 10^6\,\text{Па}\,,$$
$$T_0 = 35 - 66^\circ\mathrm{C},\quad T_H = 28^\circ\mathrm{C}, \Pr = 0.7,\quad \mathrm{Ra}_H = 10^4 - 10^7\,,$$

where $C_\varphi = 0.074$ for $\varphi = 0^\circ; C_\varphi = 0.076$ for $\varphi = 15 - 30^\circ; C_\varphi = 0.073$ for $\varphi = 45 - 75^\circ; C_\varphi = 0.07$ for $\varphi = 90^\circ$.

The numerical and experimental investigations of free-convective flow of a ***gas mixture in a trapezoidal cavity with simultaneous effect of the temperature and concentration gradients*** show the presence of unstable thermal stratification [4.148]. For different gas injection rates (gas is a mixture of equal amounts of **Ar** and $\mathbf{N_2}$) both regimes with the prevailing temperature free convection (oscillatory turbulent flow) and with the prevailing solutal convection (laminar flow) were observed. The temperature of the cavity bottom is $T_h = 57 \pm 3^\circ$C, the temperature of the lateral surfaces is

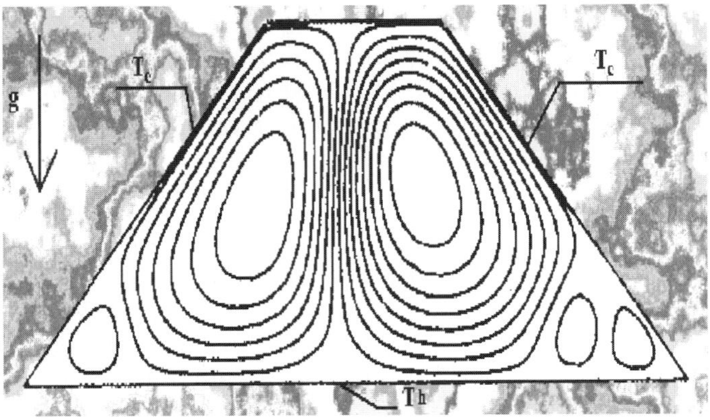

$T_w = 13 \pm 2°C$. The temperature and concentration Grashof numbers are: $Gr_t = 2.6 \cdot 10^8$, $Gr_c = 6.4 \cdot 10^8$. The Lewis number is Le = 1.16.

A vertical cavity with a rectangular base ($B \times L = 376 \times 678$ mm) of height $H = 450, 630$ mm ***is bounded by cylindrical surfaces*** of radius $R = 450$ mm and central angle 60 and 90°. The upper horizontal surface is free. The cavity is filled with a **$ZnSO_4 - H_2O$ *solution*** with the ***heat sources*** regularly distributed inside. The average value of the coefficient of heat transfer from the solution to the surface [4.149]

for the bottom base is

$$\mathrm{Nu} = 5.34\,\mathrm{Ra}_{vH}^{0.068} , \qquad \mathrm{Ra}_{vH} = 2 \cdot 10^{10} - 4 \cdot 10^{11} , \tag{4.5.24}$$

for the curved lateral surface is

$$\mathrm{Nu} = 0.7\,\mathrm{Ra}_{vH} , \qquad \mathrm{Ra}_{vH} = 2 \cdot 10^{11} - 6 \cdot 10^{12} . \tag{4.5.25}$$

A numerical and experimental investigation of natural convection of ***liquid gallium in a narrow rectangular cavity placed in a rectangular cavity with different constant temperatures on lateral surfaces*** testifies to the significant contribution of the mechanism of heat conduction by solid walls to the process of convective heat transfer of liquid metals. The empirical dependence for the average Nusselt number based on the three-dimensional numerical model is [4.152]

$$\mathrm{Nu} = 0.882 + 9.312 \cdot 10^{-7}\,\mathrm{Gr}^{0.909} . \tag{4.5.26}$$

The experimental data for the flow structure for $\mathrm{Ra} = 15.8 \cdot 10^6$ were obtained by the method of radioscopic visualization.

The experimental data for the heat transfer coefficient with heating of the lateral surface by a constant heat flux in an open horizontal annular cavity with the ratio of the radii equal to 0.375, the ratio of the length to the outer cylinder radius – 0.35 and the ratio of the wall thickness to the outer cylinder radius – 0.025 in the range of the modified numbers Ra = $g\beta q R_0^4/(\lambda a \nu) = 1.3 \cdot 10^9 - 5.1 \cdot 10^9$ are generalized by the dependences [4.151]

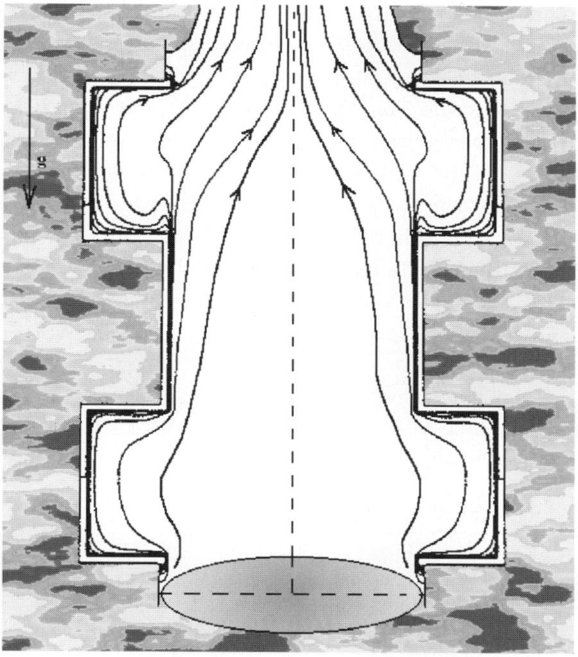

for the outer cylinder

$$\mathrm{Nu} = 0.0136\,\mathrm{Ra}^{0.376}\;; \tag{4.5.27}$$

in the region of the top angle between the lateral surface of the outer cylinder and the annular surface

$$\mathrm{Nu} = 0.0133\,\mathrm{Ra}^{0.379}\;, \tag{4.5.28}$$

for the inner cylinder

$$\mathrm{Nu} = 0.0113\,\mathrm{Ra}^{0.381}\;, \tag{4.5.29}$$

in the region of the bottom angle between the lateral surface of the inner cylinder and the annular surface

$$\mathrm{Nu} = 0.0119\,\mathrm{Ra}^{0.378}\;, \tag{4.5.30}$$

in the cylindrical cavity

$$\mathrm{Nu} = 0.0131\,\mathrm{Ra}^{0.378}\,. \tag{4.5.31}$$

The experimental investigation of the effect exerted by the size of the inlet and outlet slots on the characteristics of free-convective flow and heat transfer between two parts of a ***partially ventilated step cavity with a heater near the entrance aperture*** shows that an increase in the outlet aperture sizes leads to a decrease of the average cavity temperature and the temperature of the circulating ascending and descending flows. The presence of ventilation increases the mass flow rate of air, however, the heat losses of the upper and lower parts and of the whole cavity decrease with increase in the outlet aperture height, but only up to the value of 0.02 m [4.152]. With a further increase in the size, there is a tendency towards an increase in the total thermal losses. The picture is reversed for the heat transfer characteristics in the circulating flow inside the cavity.

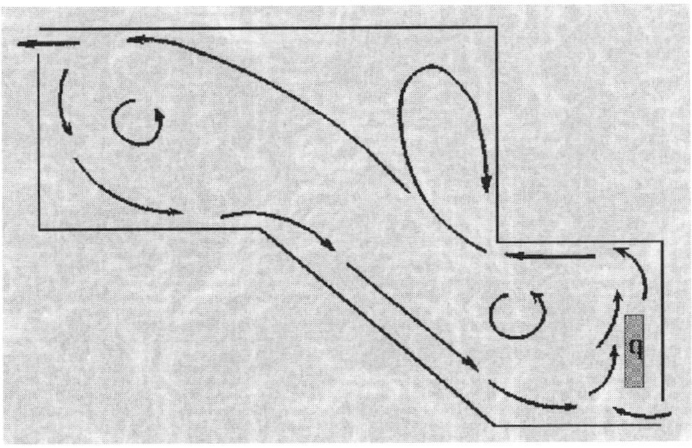

References

1. M. Y. Chow and R. G. Akins, *Trans. ASME, Ser. C, J. Heat Transfer*, **97**, 54 (1975).
2. N. N. Haziev, *Investigation of Convective Heat Transfer in Spherical Enclosures*, Abstract of Doctoral Dissertation (in Technical Sciences), Perm (1969).
3. D. E. Daney, *Int. J. Heat Mass Transfer*, **19**, 431 (1976).
4. Y. A. Kirichenko, V. N. Shchelgunov, and P. S. Chernjakov, *Inzh.-Fiz. Zh.*, **16**, 977 (1969).
5. G. F. Shaidurov, *Zh. Tekh. Fiz.*, **28**, 855 (1958).
6. L. G. Chirva, A. V. Polshin, and V. S. Kostenko, *Izv. VUZov, Aviats. Tekh.*, No. 3, 165 (1978).

7. R. J. Kee, C. S. Landram, and G. C. Miles, *Trans. ASME, Ser. C, J. Heat Transfer*, **98**, 55 (1976).
8. I. A. Clark, *Adv. Heat Transfer*, **5**, 325 (1968).
9. V. I. Polezhaev, *Mekh. Zhidkosti Gaza*, No. 4, 109 (1970).
10. J. Hiddink, *Agr. Res. Repts.*, No. 839, 1 (1975).
11. J. Hiddink, J. Schenk, and S. Bruin, *Appl. Sci. Res.*, **32**, 217 (1976)
12. M. Holzbecher and A. Steiff, *Int. J. Heat Mass Transfer*, **38**, 2893 (1995).
13. A. Lemembre and J.-P. Petit, *Int. J. Heat Mass Transfer*, **41**, 2437 (1998).
14. R. J. Goldstein, *Int. J. Heat Mass Transfer*, **23**, 738 (1980).
15. G. N. Tovarnykh, *Collection of Papers of Moscow State Technical University*, No. 293, 25 (1979).
16. J. F. Stelzen and G. P. Scheidler, *Proc. 6th Int. Heat Transfer Conf.*, vol. 2, Toronto (1978), pp. 263–267.
17. J. P. Pulicani, S. Krukowski, J. I. D. Alexander, J. Quazzani, and F. Rosenberger, *Int. J. Heat Mass Transfer*, **35**, 2119 (1992).
18. N. K. Lambha, S. A. Korpela, and F. A. Kulacki, *Proc. 6th Int. Heat Transfer Conf.*, vol. 2, Toronto (1978), pp. 311–316.
19. K. E. Torrance, *J. Fluid Mech.*, **95**, 477 (1979).
20. K. F. Wu, H. Combeau, and J. P. Brancher, *Int. J. Heat Mass Transfer*, **40**, 1535 (1997).
21. S. Ostrach, *Adv. Heat Transfer*, **8**, 161 (1972).
22. G. B. Petrazhitskii (ed.), *Investigation of Heat Transfer Processes in Design elements*, vol. 1, MVTU Press, Moscow (1973).
23. T. H. Kuehn and R. J. Goldstein, *Int. J. Heat Mass Transfer*, **19**, 1127 (1976).
24. J. L. Xia, B. L. Smith, G. Yadigaroglu, U. Gatner, and B. Sigg, *Int. J. Heat Mass Transfer*, **41**, 3635 (1998).
25. O. G. Martynenko and Yu. A. Sokovishin, *Free-Convective Heat Transfer. Handbook*, Nauka i Tekhnika Press, Minsk (1982).
26. W. Hauf and U. Grigull, *WuS*, **9**, 21 (1976).
27. F. K. Deaver and E. R. G. Eckert, in: *Heat Transfer*, vol. 4, NC 1.1, Paris–Versailles (1970).
28. M. Takeuchi and K. Cheng, *WuS*, **9**, 215 (1976).
29. A. Bejan and C. L. Tien, *Int. J. Heat Mass Transfer*, **21**, 701 (1978)
30. Y. A. Kirichenko and Zh. A. Suprunova, *Experimental Investigation of Convective Heat Transfer in Torroidal Enclosure*, Preprint No. 4–77 of Physical and Technical Institute of Low-Temperature Problems, Kharkov (1977).
31. V. A. Pukhov, A. I. Borisenko, F. M. Pozvonkov, et al., in: *Heat and Mass Transfer*, vol. 1, part 2, Minsk (1972), pp. 395–401.
32. S. Schneider and J. Straub, *Int. J. Heat Mass Transfer*, **35**, 545 (1992).
33. Y. F. Rao and B. X. Wang, *Int. J. Heat Mass Transfer*, **34**, 247 (1991).
34. L. Storesletten and M. Tveitereid, *Int. J. Heat Mass Transfer*, **34**, 1959 (1991).
35. G. B. Petrazhitskii and V. I. Polezhaev, in: *Investigation of Heat and Mass Transfer Processes in Design Elements*, vol. 3, Moscow (1976), pp. 27–66.
36. B. Roux, J. C. Grondin, P. Bontoux, and B. Gilly, *NHT*, **1**, 331 (1978).
37. I. P. Jones, *A Numerical Study of Natural Convection in an Airfilled Cavity: Comparison with Experiment*, *AERE*, Harwell (1978), R. 9346, pp. 1–22.
38. N. Seki, S. Fukusako, and H. Inaba, *Bull. Fac. Engng Hockaido Univ.*, No. 87, 75 (1978).
39. A. Bejan and C. L. Tien, *Trans. ASME, Ser. C, J. Heat Transfer*, **100**, 641 (1978).

40. A. Bejan, *Int. J. Heat Mass Transfer*, **23**, 723 (1980).
41. M. Shimizu, T. Kokugau, F. Watanabe, S. Sugiyama, *J. Chem. Eng., (Japan)*, **5**, 203 (1972).
42. N. Seki, S. Fukusako, and H. Inaba, *Bull. JSME*, **21**, 246 (1978).
43. S.-S. Hsieh and S.-S. Yang, *Int. J. Heat Mass Transfer*, **40**, 1467 (1997).
44. G. D. Raithby, K. G. T. Holland, and T. E. Unny, *Trans. ASME, Ser. C, J. Heat Transfer*, **99**, 287 (1977).
45. B. Boehrer, *Int. J. Heat Mass Transfer*, **40**, 4105 (1997).
46. S. Mergui and F. Penot, *Int. J. Heat Mass Transfer*, **40**, 3427 (1977).
47. R. G. Colwell and J. R. Welty, *Trans. ASME, Ser. C, J. Heat Transfer*, **96**, 448 (1974).
48. A. F. Lillie and H. B. Nottage, *Chem. Eng. Prog. Symp. Ser.*, **64**, 181 (1968).
49. B. M. Berkovskii and V. K. Polevikov, *Proc. 5th Int. Heat Transfer Conf.*, vol. 3, Tokyo (1974), pp. 85–89.
50. A. V. Luikov and B. M. Berkovskii, *Convection and Heat Waves*, Energya Press, Moscow (1974).
51. C. C. Jahnke, V. Subramanyan, and D. T. Valentine, *Int. J. Heat Mass Transfer,* **41**, 2307 (1998).
52. S. Mergui and F. Penot, *Int. J. Heat Mass Transfer*, **39**, 563 (1996).
53. R. Iwatsu and J. M. Hyun, *Int. J. Heat Mass Transfer*, **38**, 3319 (1995).
54. Y. Y. Jin and C. F. Chen, *Int. J. Heat Mass Transfer*, **39**, 2663 (1996).
55. Y. Y. Jin and C. F. Chen, *Trans. ASME, Ser. C, J. Heat Transfer*, **118**, (1996).
56. Z.-G. Du and E. Bilgen, *Int. J. Heat Mass Transfer*, **35**, 1969 (1992).
57. L. B. Gdalevich, E. F. Nogotov, and V. E. Fertman, *Int. J. Heat Mass Transfer*, **22**, 1601 (1979).
58. H. E. Janikowski, J. Ward, and S. D. Probert, *Proc. 6th Int. Heat Transfer Conf.*, vol. 2, Toronto (1978), pp. 257–262.
59. K. Hanjalic, S. Kenjeres, and F. Durst, *Int. J. Heat Mass Transfer*, **39**, 1407 (1996).
60. T. G. Karayiannis, M. Ciofalo, and G. Barbaro, *Int. J. Heat Mass Transfer*, **35**, 1645 (1992).
61. P. Viswatmula and M. R. Amin, *Int. J. Heat Mass Transfer*, **38**, 2039 (1995).
62. S. Acharya and R. Jetli, *Int. J. Heat Mass Transfer*, **33**, 931 (1990).
63. E. M. Litinskii and E. A. Sidorov, *Inzh.-Fiz. Zh.*, **30**, 270 (1976).
64. A. S. Lyalikov and N. A. Tsvetkova, *Inzh.-Fiz. Zh.*, Dep. in VINITI 18 April 1978, No. 1562–78.
65. K. G. T. Hollands, G. D. Raithby, and L. Konicek, *Int. J. Heat Mass Transfer*, **18**, 879 (1975).
66. M. P. Arroyo and J. M. Saviron, *J. Fluid Mech.*, **245**, 319 (1992).
67. J. Pallares, I. Cuesta, F. X. Grau, and F. Giraet, *Int. J. Heat Mass Transfer*, **39**, 3233 (1996).
68. M. M. Ganzarolli and L. F. Milanez, *Int. J. Heat Mass Transfer*, **38**, 1063 (1995).
69. A. S. Nenishev, A. I. Groshev, and G. F. Shilin, in: *Heat and Mass Transfer in Bodies and Systems under Different Boundary Conditions*, vol. 4, Omsk (1973), pp. 80–91.
70. P. G. Daniels and O. K. Jones, *Int. J. Heat Mass Transfer*, **41**, 3979 (1998).
71. M. Jahn and H. H. Reineke, *Proc. 5th Int. Heat Transfer Conf.*, vol. 3, Tokyo (1974), pp. 74–78.

72. F. B. Cheung, *Int. J. Heat Mass Transfer*, **20**, 499 (1977).
73. P. Boon-Long, T. W. Lester, and R. E. Faw, *Int. J. Heat Mass Transfer*, **22**, 437 (1979).
74. R. F. Bergholz, M. M. Chen, and F. B. Cheung, *Int. J. Heat Mass Transfer*, **22**, 763 (1979).
75. F. B. Cheung, *Trans. ASME, Ser. C, J. Heat Transfer*, **100**, 763 (1978).
76. R. Shramm and H. H. Reineke, *Proc. 6th Int. Heat Transfer Conf.*, vol. 2, Toronto (1978), pp. 299–304.
77. W. H. Leong, K. G. T. Hollands, and A. P. Brunger, *Int. J. Heat Mass Transfer*, **41**, 3817 (1998).
78. R. B. Yedder and E. Bilgen, *Int. J. Heat Mass Transfer*, **38**, 1879 (1995).
79. S. M. Elsherbiny, *Int. J. Heat Mass Transfer*, **39**, 3925 (1996).
80. K. G. T. Hollands, T. E. Unny, G. D. Raithby, and L. Konicek, *Trans. ASME, J. Heat Transfer*, **98C**, 189 (1976).
81. K. R. Randall, J. W. Mitchell, and M. M. El-Wakil, *Trans. ASME, Ser. C, J. Heat Transfer*, **101**, 120 (1979).
82. W. M. M. Schinkel and C. J. Hoogendoorn, *Proc. 6th Int. Heat Transfer Conf.*, vol. 2, Toronto (1978), pp. 287–292.
83. I. Catton, *Proc. 6th Int. Heat Transfer Conf.*, vol. 6, Toronto (1978), pp. 13–31.
84. E. Ramos and A. Castrejen, *Int. J. Heat Mass Transfer*, **33**, 917 (1990).
85. G. B. Petrazhitskii and V. I. Polezhaev, in: *Investigation of Heat and Mass Transfer Processes*, vol. 4, Moscow (1979), pp. 40–99.
86. H. H. Chu, S. W. Churchill, and C. V. S. Patterson, *Trans. ASME, Ser. C, J. Heat Transfer*, **98**, 194 (1976).
87. B. L. Turner and R. D. Flack, *Trans. ASME, Ser. C, J. Heat Transfer*, **102**, 236 (1980).
88. C. Beghein, F. Haghighat, and F. Allard, *Int. J. Heat Mass Transfer*, **35**, 833 (1992).
89. M. D. Oreshina, L. I. Savel'eva, N. N. Khaziev, and G. N. Shemyakina, 123 (1971). *Prikl. Mekh. Tekh. Fiz.*,
90. T. Nishimura, M. Wakamatsu, and A. M. Morega, *Int. J. Heat Mass Transfer*, **41**, 1601 (1998).
91. T. L. Bergman and M. T. Hyun, *Int. J. Heat Mass Transfer*, **39**, 2883 (1996).
92. V. A. F. Costa, *Int. J. Heat Mass Transfer*, **40**, 4061 (1997).
93. G. D. McBain, *Int. J. Heat Mass Transfer*, **40**, 3005 (1997).
94. V. C. Regnier, P. M. Parmentier, G. Lebon, and J. K. Platten, *Int. J. Heat Mass Transfer*, **38**, 2539 (1995).
95. S. A. Suslov and S. Paolucci, *Int. J. Heat Mass Transfer*, **38**, 2143 (1995).
96. W. H. Leong, K. G. T. Hollands, and A. P. Brunger, *Int. J. Heat Mass Transfer*, **42,** 1979 (1999).
97. J. W. Lee and J. M. Hyun, *Int. J. Heat Mass Transfer*, **33**, 1619 (1990).
98. J. W. Lee and J. M. Hyun, *Int. J. Heat Mass Transfer*, **33**, 1605 (1990).
99. J. W. Lee, M. T. Hyun, and Y. S. Kang, *Int. J. Heat Mass Transfer*, **33**, 869 (1990).
100. A. S. Lyalikov, *Inzh.-Fiz. Zh.*, **29**, 897 (1976).
101. H. R. Nagendra, M. A. Tirunarayanan, and A. Ramachandran, *Chem. Eng. Sci.*, **25**, 605 (1970).
102. R. W. de Vahl Davis G. Thomas, in: *Heat Transfer*, vol. 4, (1970), NC 2.4, Paris–Versailles.

103. R. P. Bobko, *AIAA Paper*, No. 1536, 1 (1980).
104. J. G. Collier, B. E. Boyce, A. S. de Forge Dedman, and R. Khanna, in: *Heat Transfer,* vol. 4, (1970), NC 2.6, Paris–Versailles.
105. G. P. Merker and L. G. Leal, *Int. J. Heat Mass Transfer*, **23**, 677 (1980).
106. J. R. Custer and E. L. Shaughnessy, *Trans. ASME, Ser. C, J. Heat Transfer*, **99**, 675 (1977).
107. Yu. A. Koshmarov and A. E. Ivanov, *Izv. Akad. Nauk SSSR, Energetika Transport*, No. 5, 164 (1972).
108. T. H. Kuehn and R. J. Goldstein, *Trans. ASME, Ser. C, J. Heat Transfer*, **100**, 635 (1978).
109. T. H. Kuehn and R. J. Goldstein, *Trans. ASME, Ser. C, J. Heat Transfer*, **102**, 768 (1980).
110. V. I. Bolshakov and L. S. Kokarev, in: *Vopr. Teplofiz. Yadern. Reaktorov*, vol. 6, Moscow (1977), pp. 26–28.
111. L. I. Strokovskii and V. P. Borisoglebskii, *Izv. Akad. Nauk SSSR, Energetika Transport*, No. 5, 68 (1975).
112. Z. Rotem, *Int. J. Heat Mass Transfer*, **15**, 1679 (1972).
113. E. Van de Sande and B. J. Hamer, *Int. J. Heat Mass Transfer*, **22**, 361 (1979).
114. S. D. Probert, D. Sadhu, and D. Syed, *Appl. Energy*, **1**, 145 (1975).
115. J. Lis, *Proc. 3rd Heat Transfer conf.*, vol. 2, (1966), pp. 196–204, Chicago.
116. M. J. Shilton and S. D. Probert, *Appl. Energy*, **4**, 21 (1978).
117. K. Fukuda, Y. Miki, and S. Hasegawa, *Int. J. Heat Mass Transfer*, **33**, 629 (1990).
118. R. W. Douglass, K. G. Tebeest, S. A. Trogdon, and D. R. Gardner, *Int. J. Heat Mass Transfer*, **33**, 2533 (1990).
119. C. Thamire and N. T. Wright, *Int. J. Heat Mass Transfer*, **41**, 4121 (1998).
120. N. Weber, R. E. Rowe, E. H. Bishop, and J. A. Scanlan, *Trans. ASME, Ser. C, J. Heat Transfer*, **95**, 47 (1973).
121. V. K. Garg, *Int. J. Heat Mass Transfer*, **35**, 1935 (1992).
122. S. V. Solov'ev and A. S. Lyalikov, *Inzh.-Fiz. Zh.*, **36**, 807 (1979).
123. G. B. Petrazhitskii and N. M. Stankevich, *Prikl. Mekh. Tekh. Fiz.*, No. 5, 100 (1976).
124. G. B. Petrazhitskii and N. M. Stankevich, in: *Heat and Mass Transfer–V*, vol. 1, part 2, Minsk (1976), pp. 300–304.
125. Yu. G. Volodin and G. N. Dulnev, *Inzh.-Fiz. Zh.*, **9**, 603 (1965).
126. Yu. I. Kozlov, E. Yu. Merkel, and N. D. Krishtal, in: *Heat and Mass Transfer in Bodies and Systems under Different Boundary Conditions*, vol. 4, Omsk (1973), pp. 122–126.
127. N. K. Ghaddar, *Int. J. Heat Mass Transfer*, **35**, 2327 (1992).
128. R. D. Flack, T. T. Konopnicki, and J. H. Rooke, *Trans. ASME, Ser. C, J. Heat Transfer*, **101**, 648 (1979).
129. R. D. Flack, *Trans. ASME, Ser. C, J. Heat Transfer*, **102**, 770 (1980).
130. H. Salmun, *Int. J. Heat Mass Transfer*, **38**, 351 (1995).
131. H. Salmun, *Int. J. Heat Mass Transfer*, **38**, 363 (1995).
132. A. Cabelli, in: *2nd Austral. Conf. on Heat and Mass Transfer*, Sydney (1977), pp. 257–264.
133. Y.-Z. Wang and H. H. Bau, *Int. J. Heat Mass Transfer*, **35**, 111 (1992).
134. J. H. Min and F. A. Kulacki, *Proc. 6th Int. Heat Transfer Conf.*, vol. 5, Toronto (1978), pp. 155–160.

135. G. B. Petrazhitskii, F. V. Klyushnikov, and E. V. Bekneva, *Proc. 5th Int. Heat Transfer Conf.*, vol. 3, Tokyo (1974), pp. 110–113.
136. V. Behnia, A. A. Dehghan, H. Mishima, and W. Nakayama, *Int. J. Heat Mass Transfer*, **41**, 797 (1998).
137. Yu. A. Korolenko and I. V. Esina, *Inzh.-Fiz. Zh.* Dep. in VINITI 9 Jull 1979, No. 2486–79.
138. Yu. A. Korolenko and I. V. Esina, *Collection of Papers of Chelyabinsk Polytechnic Institute*, No. 122, (1973), pp. 151–156.
139. K. Rashid and S. Kuwabara, *J. Phys. Soc.* (*Japan*), **45**, 2014 (1978).
140. K. Rashid and S. Kuwabara, *J. Phys. Soc.* (*Japan*), **45**, 2021 (1978).
141. G. F. Shilin, A. I. Groshev, B. A. Kalachevskii, V. F. Lyarskii, and E. F. Kharchenko, in: *Heat and Mass Transfer in Bodies and Systems under Different Boundary Conditions*, vol. 4, Omsk (1973), pp. 26–38.
142. A. D. Kraus, *Cooling of Electronic Equipment*, Energya Press, Leningrad (1971).
143. A. S. Lyalikov, P. A. Lyukshin, and L. G. Kalugina, in: *Heat Heat Transfer Problems and Determination of Heat Transfer Characteristics*, Tomsk (1971), pp. 21–25.
144. S. V. Solov'ev and A. B. Devyatkin, in: *Thermophysics and Optimization of Heat Processes*, vol. 4, Kuibyshev (1978), pp. 106–108.
145. C. T. McCoy, R. E. Powe, E. H. Bishop, N. Weber, and J. A. Scanlan, *Proc. 5th Int. Heat Transfer Conf.*, vol. 3, Tokyo (1974), pp. 105–109.
146. E. S. Gulkarov and E. T. Artykpaev, *Inzh.-Fiz. Zh.* Dep. in VINITI, 14 Jull 1976, No. 2901–76.
147. L. Lyican, L. C. Witte, and Y. Bayazitoglu, *Trans. ASME, Ser. C, J. Heat Transfer*, **102**, 648 (1980).
148. J. T. Eyden, T. H. Meer, and K. Hanjalic, *Int. J. Heat Mass Transfer*, **41**, 1885 (1998).
149. J. D. Gabor, L. Ir. Baker, J. C. Cassulo, D. J. Erskine, and J. G. Warner, *Trans. ASME, Ser. C, J. Heat Transfer*, **102**, 519 (1980).
150. R. Dereball and J. N. Koster, *Int. J. Heat Mass Transfer*, **40**, 1169 (1997).
151. C. P. Desai and K. Vafai, *Int. J. Heat Mass Transfer*, **39**, 2053 (1995).
152. S. Ergin-Ozkan, M. R. Mokhtarzadeh-Dehghan, and A. J. Reynolds, *Int. J. Heat Mass Transfer*, **38**, 2159 (1995).

5

Free Convection in Tubes and Channels, on Ribbed Surfaces and in Tube Bundles

The flow structure and the character and rate of heat transfer with natural convection in a limited volume depend substantially on the dimensions of the area. For example, in a tube, starting from some distance from the entrance, the fluid in the whole cross section is exhibited to the slowing effect of viscosity forces, and the fluid temperature changes both in the cross section and along the channel. A boundary layer gradually develops near the walls and its thickness increases. Over the hydrodynamically stable part of the tube the boundary layer fills up the whole cross section. When the boundary layers of a fluid with constant liquid physical properties join up, constant dimensionless velocity distribution typical of the given flow regime is established. Heat exchange in the cooling of surfaces is intensified by means of ribs. The heat exchange of such structures can be calculated by modifying the calculated data for smooth channels.

To calculate heat transfer in tubes of noncircular section, a method is used which is based on the calculation of heat transfer in some equivalent tube of circular section.

The fluid flow in tube bundles and cells is of complex character. The flow past the first row of tubes is similar to the flow past a single cylinder; for the subsequent rows the flow depends substantially on the arrangement of the cylinders and on the distance between them due to the interaction of the streams flowing past them. The geometry of the channel or of the enclosure containing the bundle has also a strong influence on the flow structure and heat transfer.

5.1 Rectangular Tube

Vertical plates of height H with the wall temperatures T_{w} ***are located at the distance B from each other. Depending on the sign of the temperature difference*** $\Delta T = T_{\mathrm{w}} - T_{\infty}$ (T_{∞} is the ambient temperature), there is either an ascending or descending flow. The average heat transfer coefficient of a laminar regime with an error of up to 2% is determined by the formula [5.1]

$$\mathrm{Nu} = \frac{1}{24}\frac{B}{H}\mathrm{Ra}_B\left[1-\exp\left(-35\frac{H}{B}\mathrm{Ra}_B\right)\right]^{3/4} , \tag{5.1.1}$$

$$\mathrm{Pr} > 0.7, \quad T_{\mathrm{av}} = T_{\mathrm{w}}, \quad \frac{B}{H}\mathrm{Ra}_B = 0.2 - 10^5$$

or by the more precise formula [5.2]

$$\mathrm{Nu} = \frac{1}{24}\frac{B}{H}\mathrm{Ra}_B\left[1-\exp\left(-32.7\frac{H}{B}\mathrm{Ra}_B\right)\right]^{3/4} . \tag{5.1.2}$$

The optimum distance between the plates and heat transfer are described as [5.3]

for the maximal heat flux from the plate

$$\mathrm{Nu}_{\mathrm{max}} = 0.42\left(\frac{B}{H}\mathrm{Ra}_B\right)^{1/4}_{\mathrm{max}} , \tag{5.1.3}$$

$$\frac{B}{H}\mathrm{Ra}_B \cong 50 ,$$

for the minimal temperature difference between the wall and flow

$$\mathrm{Nu} \cong 3.35 ,$$

$$\frac{B}{H}\mathrm{Ra}_B = 400\text{–}800 . \tag{5.1.4}$$

The minimal plate width providing the minimal temperature difference is 1.5–2 times larger than for the case of the maximal heat flux from the plate.

The average heat transfer calculated by the method of a thin layer is [5.4]

$$\mathrm{Nu} = C_l\left(\frac{B}{H}\mathrm{Ra}_B\right)^{1/4}\sum_{n=1}^{\infty}(-1)^{n+1}\frac{3}{(4n-1)\,(n-1)!}\left(\frac{H}{B}\frac{C_1}{\mathrm{Ra}_B}\right)^{n-1} ,$$

$$\mathrm{Nu} = \frac{\Gamma(7/4)}{C_1^{3/4}}C_l\frac{B}{H}\mathrm{Ra}_B , \quad \text{for } \frac{B}{H}\mathrm{Ra}_B \to 0 , \tag{5.1.5}$$

where $C_1 = [24C_l\Gamma(7/4)]^{4/3}$ ***for equal surface temperatures***; $C_1 = [12C_l\Gamma(7/4)]^{4/3}$ ***in the case of one of the plates being insulated***.

At small values of $\mathrm{Ra}_B B/H$, ***the heat transfer is sensitive to the presence of horizontal limiting surfaces at the channel entrance and exit effects***. When $\mathrm{Ra}_B B/H < 4$, ***the heat transfer in the channel closed from above*** decreases by 15–20%, in the channel closed from below and from above – by 30% [5.5]. When $\mathrm{Ra}_B B/H > 20$, this effect can be neglected, and the heat transfer can be calculated from formulas (5.1.1)–(5.1.5).

In the channel vertical discontinuous walls are installed which are short vertical plates of height h having the temperature T_∞.

The temperature ***in the vertical channel*** increases by a factor of two at $\mathrm{Ra}_{dr} d_r/H = 10^3 – 10^6$ ($d_r = 4b, 2b$ is the distance between discontinuous vertical walls) [5.6].

For ***staggered and in-line arrangement of discontinuous plates in a vertical channel*** the character of heat transfer depends little on the arrangement and number of plates along a vertical row [5.7]. For the given temperature difference between heated surfaces and outer medium and the same areal of the heat transfer surface the heat transfer coefficient increases by 80–90% in comparison with the case of free convection in the channels between continuous plates. For the given values of the heat transfer area and heat loading the decrease in the temperature difference is 35–40%. For the given loading and temperature difference the height of the vertical channel with the plates decreases by 50%.

The average heat transfer in a vertical channel with the wall temperatures T_{w1} and T_{w2} [5.8]

$$\mathrm{Nu} = \frac{70}{17}\left(\theta_3 + \theta_2 \frac{1 - e^{-\omega}}{\omega}\right) , \tag{5.1.6}$$

$$\frac{B}{H}\mathrm{Ra}_B = \frac{120}{\omega^2 \,\mathrm{Pr}} \frac{\frac{14}{17}\left(\frac{14}{17} + \omega\,\mathrm{Pr}\right)}{\theta_1 - \frac{14}{17}\theta_2\left(1 - e^{-\omega}\right)/\omega} ,$$

$$\theta_0 = \frac{T_\infty}{T_{\mathrm{w1}}}, \quad \theta_1 = \frac{\frac{3}{10} + \frac{7}{10}\xi - \theta_0}{\left(1 + \xi\right)/2 - \theta_2} ,$$

$$\theta_2 = \frac{\frac{9}{35} + \frac{26}{35}\xi - \theta_0}{\left(1 + \xi\right)/2 - \theta_0} ,$$

$$\xi = \frac{T_{\mathrm{w2}}}{T_{\mathrm{w1}}} , \quad \omega = \frac{140}{17}\frac{6\overline{\varepsilon}(1 - \overline{\varepsilon})aH}{\omega_{oc}B^2} ,$$

$$\overline{\varepsilon} = \frac{\varepsilon}{B} = \frac{1}{2} + \cos\left(\frac{1}{3}\right)\left(1 + \arccos\frac{1 - \xi}{1 - \xi + N}\right) ,$$

$$N = \frac{70}{17}\left[\left(\frac{9}{35} + \frac{26}{35}\xi\right) - \theta_0\right]\exp\left(-\omega\overline{x}\right) ,$$

where $\overline{x} = x/B; \omega_{oc}$ is the velocity on the neutral plane; ε is the distance between the neutral plane and axis x.

In a semi-infinite channel the entrance temperature T_∞ is lower than the constant wall temperatures T_{w1} and T_{w2}, the entry velocity is ω_o, the channel width and height are B and H. The local and average heat transfer coefficients for Pr = 0.7 are generalized by the formulas [5.9]

$$\frac{\mathrm{Re}_B}{\mathrm{Gr}_B} = 0.0003 - 0.03 , \quad \theta_{\mathrm{w1}} = \frac{T_{\mathrm{w1}} - T_\infty}{T_{\mathrm{av}} - T_\infty} = 0.15 - 1.1 ,$$

$$\mathrm{Nu}(x)\frac{\mathrm{Pr}}{\mathrm{Gr}_B} = a_1 \exp\left(a_2\theta_{\mathrm{w1}}\right)\overline{x}^{a_3 + a_4\theta_{\mathrm{w1}}} , \tag{5.1.7}$$

$$\mathrm{Nu}_B = a_5 + a_6\theta_{\mathrm{w1}} , \tag{5.1.8}$$

where

$$H/B = 10^{a_7}\theta_{\rm w1}^{-0.4081+0.3211\exp(-480.6{\rm Re}_B/{\rm Gr}_B)} ,$$
$$T_{\rm av} = [T_{\rm w1}, T_{\rm w2}]_{\max}, \quad \overline{x} = x/B ,$$
$$a_i = a_{i1} + a_{i2}\log({\rm Re}_B/{\rm Gr}_B) + a_{i3}\log({\rm Re}_B/{\rm Gr}_B)^2 + a_{i4}\log({\rm Re}_B/{\rm Gr}_B)^3 ,$$

and the values of the coefficients a_{ij} are given in Table 5.1.1.

Table 5.1.1.

j	1	2	3	4
$i = 1$	−2.096	−2.796	−1.095	−0.1386
$i = 2$	8.961	8.807	2.948	0.3126
$i = 3$	−1.292	−3.273	−1.223	−0.1513
$i = 4$	1.768	1.956	0.6997	0.0814
$i = 5$	−8.605	−9.372	−2.972	−0.5229
$i = 6$	1.547	0.5623	1.213	0
$i = 7$	1.35	2.834	0.2505	0

The numerical calculations of laminar free convection ***in a rectangular channel for asymmetric heating*** at $\Pr = 0.7$–3 show that the edge effects on heat transfer are insignificant [5.5,5.10]; to calculate heat transfer rate, the formulas obtained for vertical layers are recommended.

Laminar convection of ***mercury in a vertical channel for two cases of the heat flux on a wall:*** (1) ***one wall is insulated and is a constant heat flux on another one;*** (2) ***constant values of heat fluxes on both walls***. The local and average heat transfer coefficients are determined by the dependences [5.11] ($H/B = 2$–$19; \Pr = 0.023, T_{\rm av} = 0.7T_{\rm w} + 0.3T_\infty$)

for ${\rm Gr}_x^*\,(B/H)^5 = 10^{-2}$–$10^3$

$$\mathrm{Nu}(x)\frac{B}{H} = 0.247\left[{\rm Gr}_x^*\left(\frac{B}{H}\right)^5\right]^{0.141} , \tag{5.1.9}$$

for ${\rm Gr}_x^*\,(B/H)^5 = 10^3 - 10^9$

$$\mathrm{Nu}(x)\frac{B}{H} = 0.194\left[{\rm Gr}_x^*\left(\frac{B}{H}\right)^5\right]^{0.18} , \tag{5.1.10}$$

for ${\rm Gr}_x^*\,(B/H) = 10^{-2} - 10^3$

$$\mathrm{Nu}\frac{B}{H} = 0.438\left[{\rm Gr}_x^*\left(\frac{B}{H}\right)^5\right]^{0.141} , \tag{5.1.11}$$

for $\mathrm{Gr}_x^* (B/H) = 10^3 - 10^9$

$$\mathrm{Nu}\frac{B}{H} = 0.227 + 0.269 \left[\mathrm{Gr}_x^* \left(\frac{B}{H}\right)\right]^{0.18} . \qquad (5.1.12)$$

A narrow channel ($H/B < 20$) is more effcient for heat transfer than the wide one. The upper limit for Gr_x^* is taken with a reserve, and dependences (10)–(13) are also applicable in the transient region.

The average value of the coefficient of heat transfer ***between vertical copper plates in a package of trunks heated by an electric current*** is [5.12]

$$\mathrm{Nu} = 0.75\mathrm{Ra}_H^{0.25} \left\{1 - \exp\left[-\left(0.24 + 0.174\ \mathrm{Ra}_H^{0.56} \left(\frac{B}{H}\right)^{2.5}\right)\right]\right\} , \qquad (5.1.13)$$

$$H/B = 0 - 83, \quad \mathrm{Pr} = 0.7, \quad \mathrm{Ra}_H = 10^5 - 10^9, \quad T_{\mathrm{av}} = T_\infty .$$

The local value of the coefficient of heat transfer ***on a vertical plate with a constant heat flux*** $q_{\mathrm{w}1}$ ***in the presence of a parallel plate with a constant heat flux*** $q_{\mathrm{w}2}$ is [5.13]

$$\mathrm{Nu}(x) = \frac{(B/x)\mathrm{Ra}_{B1}^{\bullet}}{[(B/x)\mathrm{Ra}_{B1}^{\bullet}]^{1/2}} \times \frac{1 - \exp\left\{-k\left(1 + \frac{q_{\mathrm{w}2}}{q_{\mathrm{w}1}}\right)^{3/4} \left(\frac{(B/x)\mathrm{Ra}_{B1}^{\bullet}}{\left[(B/x)\mathrm{Ra}_{B1}^{\bullet}\right]^{1/2}}\right)^{n}\right\}}{\left[24\left(1 + \frac{q_{\mathrm{w}2}}{q_{\mathrm{w}1}}\right)\right]^{1/2}} , \qquad (5.1.14)$$

where k =2.84, $n = -3/5$ for $\mathrm{Pr} = 0.7$; $n = -2/3$ for $\mathrm{Pr} = 10$.

In the conjugate problem which includes convection ***in a vertical channel and heat conduction through the wall of thickness*** 2δ ***with uniformly distributed internal heat sources of density*** Q_v due to heating by an electric current, the temperature distribution in walls and the maximal temperature are calculated from the equations [5.14]

$$\mathrm{Gr}_B = \mathrm{Gr}_{\nu B} P_\lambda \left[\frac{H\delta/B^2}{2P_\lambda r_1} - \left(\frac{1}{3} + \frac{r_2}{2r_1}\right)\left(\frac{H}{B}\right)^2 + \left(1 - \frac{r_2}{r_1}\right)\frac{xH}{B^2} - \frac{1}{2}\left(\frac{x}{B}\right)^2 + \frac{r_3 - r(x/B)}{r_1}\frac{H}{B}\right], \qquad (5.1.15)$$

$$\mathrm{Gr}_{B,\max} = \mathrm{Gr}_{\nu B} P_\lambda \left[\frac{H\delta/B^2}{2P_\lambda r_1} + \left(\frac{1}{6} + \frac{r_2}{2r_1}\right)\left(\frac{H}{B}\right)^2 + \frac{r_3 - r(H/B)}{r_1}\frac{H}{B}\right] , \qquad (5.1.16)$$

where

$$P_\lambda = \lambda_\infty/\lambda_\mathrm{w}, \quad \gamma_i^2 = \frac{(2i-1)^2\pi^2}{4\overline{u_0}\,\mathrm{Pr}}, \quad \beta_i^2 = \gamma_i^2\,\mathrm{Pr}\,,$$

$$r_1 = \sum_{i=1}^{\infty} \frac{1-\exp\left(-\gamma_i^2 H/B\right)}{\gamma_i^2}, \quad r_2 = \sum_{i=1}^{\infty} \frac{\exp\left(-\gamma_i^2 H/B\right)}{\gamma_i^2},$$

$$r_3 = \sum_{i=1}^{\infty}\left[\frac{1}{\gamma_i^4} - \frac{1-\exp\left(-\gamma_i^2 H/B\right)}{\gamma_i^6}\right], \quad r(x) = \sum_{i=1}^{\infty} \frac{1-\exp\left(-\gamma_i^2 x\right)}{\gamma_i^4}\,.$$

The dimensionless velocity $\overline{u_0} = u_0 B/\gamma$ is determined from the equation

$$\sum_{i=1}^{\infty} \frac{1}{(2i-1)^2}\left\{\exp\left(-\beta_i^2 H/B\right) + \frac{\gamma_i^2}{\overline{u_0}} \int_0^{H/B} \left(\int_0^{\xi} \mathrm{Gr}_B(\tau)\exp\left[-\gamma_i^2(\xi-\tau)\right]d\tau\right)\right.$$
$$\left.\times\, \exp\left[\beta_i^2(H/B-\xi)\right]d\xi\right\} = \pi^2/8\,. \qquad (5.1.17)$$

For ***copper trunks*** and $\delta/H > 0.02$, the difference between the maximum and average temperatures does not exceed 5% ($\mathrm{Gr}_{v\delta}P_\lambda = 10^{-2}$). The optimum ***conditions of cooling of aluminum walls*** correspond to $B/H = 0.03$–0.04.

The data of numerical and experimental investigations of free-convective heat transfer of ***air*** flow ***in a rectangular vertical channel without obstacles and in the presence of a single obstacle*** are presented in [5.15]. The boundary conditions consisted in the maintenance of either constant temperature or constant heat flux on a vertical wall. The number Ra changed from 10^2 to 10^4. In conditions of constant temperature for the channel with an obstacle for $L_1 = 0.5$ (the distance from the channel entrance to the obstacle) the calculations show that as Ra increases, the heat transfer rate also increases. The study of the effect of the value $A_p = b/L(L, b$ are the channel height and thickness) on the average Nusselt number shows that for $\mathrm{Ra} > 10^3$ its effect is insignificant, but it becomes significant for $\mathrm{Ra} < 10^3$. In the case of constant wall temperatures the presence of an obstacle leads to a decrease in the average Nusselt number by 5% for the value of Ra = 104 and by 40% for Ra = 10. Displacement of the obstacle from the inlet to outlet causes a decrease in the average heat transfer rate.

The numerical analysis of the main features of natural convection ***in open cavity on a vertical wall*** showed the presence of linear dependence between the periodic Nusselt number fluctuations and the central vortex fluctuations and also the region of its localization inside the cavity [5.16]. In the Relay number range $3.5\cdot10^5 \le \mathrm{Ra} \le 5.5\cdot10^5$ for different types of flow there correspond certain types of fluctuations: asymmetric harmonic, and deformed W-shaped. The fluctuation frequency of the Nusselt number (or of the central vortex inside the cavity) increases linearly with increase of Ra. For $\mathrm{Ra} \geq 9 \cdot 10^5$, the central vortex fluctuation frequency increases substantially, and its behavior becomes chaotic, which, together with flow separation, leads to flow turbulization.

Experimental investigation of natural convection of ***an air layer in vertical channels with asymmetrically located heated ribs*** (channel A corresponds to the flow direction opposite to buoyancy forces, channel B – to the cocurrent one) is carried out in [5.17]. Based on smoke visualization of the flow structure and interferograms of temperature fields, it is shown that the flow is turbulent in both channels for $1.29 \cdot 10^7 < \mathrm{Ra_w} < 9.43 \cdot 10^9$, where $\mathrm{Ra_w}$ is the Rayleigh number calculated by the channel width W. The generalized dependence for the average heat transfer coefficient is

$$\mathrm{Nu} = 0.12\,(\mathrm{Ra_w})^{0.335}\,, \tag{5.1.18}$$

$$1.43 \cdot 10^5 < \mathrm{Ra_w} < 1.4 \cdot 10^6$$

$$\mathrm{Nu} = 0.58\,(\mathrm{Ra_w})^{0.201}\,, \tag{5.1.19}$$

for channels A and B, respectively.

The data on heat transfer and the photographs obtained on flow visualization point to two main factors affecting the rib temperature: the geometry of the ribs and flow stratification. The character of power-law dependence (5.1.32) shows that in the specified range of $\mathrm{Ra_w}$ the flow in the channel A is turbulent and in the channel B it is laminar:

$$\mathrm{NU} = 0.33\mathrm{Gr}^{0.261}(G/W)^{0.175}\,\mathrm{Pr}^{0.345}\,, \tag{5.1.20}$$

$$H/W = 5.0, \quad B/W = 1.33, \quad 0.33 \leq G/W \leq 1.0\,,$$

$$10^4 \leq \mathrm{Gr} \leq 10^7, \quad 0.7 \leq \mathrm{Pr} \leq 20\,,$$

where B is the length of the inlet and outlet ventilating windows; G is the width of the inlet and outlet ventilating slits; H is the cavity height; W is the cavity width; $\mathrm{Gr} = g\beta(T_\mathrm{w} - T_0)W^3/\nu^2$.

Horizontal plates in a package of bars heated electrically for $H/B = 0$–83, $\mathrm{Pr} = 0.7$, $\mathrm{Ra}_H = 10^5$–10^9 [5.12]

$$\mathrm{Nu} = 0.75\mathrm{Ra}_H^{0.25}\left\{1 - \exp\left[-\left(0.24 + 0.55\,\mathrm{Ra}_H^{0.3}\left(\frac{B}{H}\right)^{2.5}\right)\right]\right\}. \tag{5.1.21}$$

Comparison of formulas (5.1.13) and (5.1.21) reveals a larger efficiency of the heat transfer process in the package of bars in the case of vertical arrangement of the channels in comparison with a horizontal one.

The average heat transfer coefficient ***for shaped horizontal plates in a package of a resistor with a step H (distance between plates) in a infinite space*** for $B/H = 1.5$–7, $L/H = 4$–7, $H = 11$–$22\,\mathrm{mm}$, $\mathrm{Pr} = 0.7$, $\mathrm{Ra}_{2H} = 10$–10^6, $T_\mathrm{av} = T_\infty$ is [5.18]

$$\mathrm{Nu} = C\,\mathrm{Ra}_{2H}^{0.25}\,, \tag{5.1.22}$$

where $C = 0.54$, 0.59, and 0.64 for $H = 11$, 16, and 19, respectively.

In the presence of a casing, the heat transfer increases on the average by 20% due to the draft (mixed convection).

Theoretical analysis of turbulent thermal convection in a layer of fluid between two ***horizontal isothermal surfaces of different temperatures (the temperature of the lower surface exceeds the temperature of the upper one)*** is carried out in an assumption that the layer of fluid can be divided into parts with different mechanisms of heat and momentum transfer [5.19]. The results of calculation show the presence of four main layers for the profile of the averaged temperature gradient: a layer with heat conduction; a transient layer in which the diffusion and convective components of the mechanism of heat and momentum transfer have the same order, and their ratio is determined by the transient Prandtl number ($\mathrm{Pr}_0 \sim 0.1$) [5.20]; a convective layer and a zone of destruction of the free-convective flow. From the data obtained for the averaged temperature gradient it follows that the convective or self-similar layer with an inclination, equal to –4/3 ($d\theta/dz = 0.6z^{-4/3}$), appears approximately at the distance $z \sim 15$, where θ and z are the dimensionless temperature and distance in the corresponding scales.

Free convection ***between tanks of temperatures $T_{\infty 1}$ and $T_{\infty 2}$ through a horizontal channel consisting of two parallel insulated plates***. When $\mathrm{Ra}_H < 72(H/B)^2$, the average value of heat transfer is determined from the equation [5.21]

$$\mathrm{Nu}\frac{B}{H} = 1 + \frac{1}{362880}\left(\mathrm{Ra}_H\frac{H}{B}\right)^2 . \tag{5.1.23}$$

Figure 5.1.1 shows the behavior of Nu for $\mathrm{Ra_H}H/B = 10^2$ and 10^6 and $H/B = 0$–0.2.

Experimental investigations of heat transfer ***in a closed horizontal rectangular channel of square section*** ($L/H = 1.25$–$7.5, \mathrm{Ra}_L = 10^6$–$10^{10}$) show that the coefficients C and n of the generalizing dependence $\mathrm{Nu} = C\,\mathrm{Ra}_L^n$ depend substantially on the channel geometry ($C = 2.12 \cdot 10^{-3}$–$5.25 \cdot 10^{-9}$, $n = 0.5$–1.34) [5.22].

Numerical data for the heat transfer of ***Newtonian and non-Newtonian fluids*** in a ***horizontal rectangular channel*** with the aspect ratio 2:1 are presented in [5.23]. The heat transfer enhancement in the region of the entrance is mainly caused by perturbations of the axial velocity, which are due to the temperature dependence of viscosity. The influence of buoyancy-induced secondary flows is much weaker in the case of the Separan solution than for water owing to the relatively high viscosity of the fluid in the central zone of the rectangular channel.

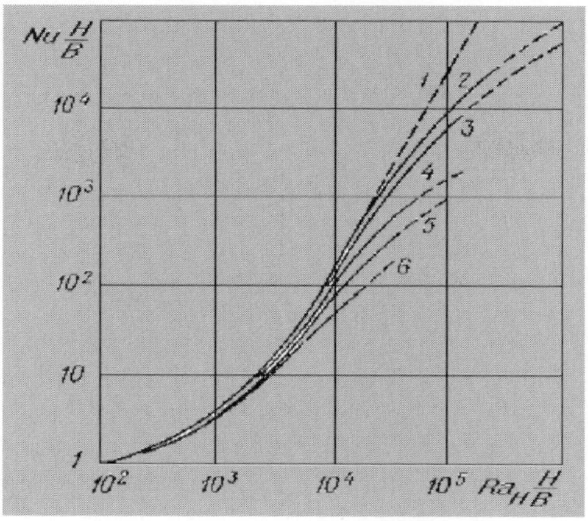

Fig. 5.1.1. Variation of Nu during heat transfer between vessels of different temperatures: (1) $H/B = 0$; (2) 0.01; (3) 0.02; (4) 0.05; (5) 0.1; (6) 0.2

The average coefficient of heat transfer ***in a vertical tube of equivalent diameter d_r and wall temperature*** $T_{\rm w}$ for $\mathrm{Ra}_{d_r}(d_r/H) = 10^{-1}$–$10^5$, $\mathrm{Pr} > 0.7$, $T_{\rm av} = T_{\rm w}$ with an error of 10% is determined by the formula [5.1]

$$\mathrm{Nu} = \frac{1}{8\Psi}\frac{d_r}{H}\mathrm{Ra}_{d_r}\left\{1 - \exp\left[-\Psi\left(\frac{8H}{d_r\mathrm{Ra}_{d_r}}\right)^{3/4}\right]\right\}. \qquad (5.1.24)$$

The values of the coefficient Ψ ***for channels of different cross sections*** are given in Table 5.1.2.

Table 5.1.2.

Channel Section	Ψ
Square, $B \times B$	14.225
Rectangle, $B \times 2B$	15.55
$B \times 5B$	19.05
Equilateral triangle (side $2B$)	13.33

The average heat transfer coefficient for a regime of maximum heat transfer ***from a surface for an optimum hydraulic diameter*** $\mathrm{Ra}_{d_r}(d_r/H) = 11.52\,\psi^{4/3}$ is

$$\mathrm{Nu}_{\max} = 0.42\left(\frac{d_r}{H}\mathrm{Ra}_{d_r}\right)^{1/4}. \qquad (5.1.25)$$

A partially opened vertical channel of height H and width B, with the lateral walls of length L and one limiting vertical wall H×B with

the thickness of the walls δ and temperatures of the channel walls and surrounding medium T_w and T_∞ in the ranges: $(\Delta/H)\mathrm{Ra}_B = 10 - 6\cdot10^3$, $H/L = 1.82 - 6.96$, $T_{av} = T_w$, $\delta/B = 0.047 - 0.455$ is described by the following equations [5.24]:

vertical surface $H \times B$

$$\mathrm{Nu}_B = \frac{B}{70H}\mathrm{Ra}_B\left\{1 - \exp\left[-\left(\frac{57H}{B}\mathrm{Ra}_B\right)^{2/3}\right]\right\} \times\left\{1 - \exp\left[-10\left(\frac{57H}{4B}\mathrm{Ra}_B\right)^{1/4}\right]\right\}, \tag{5.1.26}$$

vertical lateral surfaces $L \times H$

$$\mathrm{Nu}_B = \frac{B}{24H}\mathrm{Ra}_B\left[\left\{1 - \exp\left[-\left(\frac{32.7H}{B}\mathrm{Ra}_B\right)^{3/4}\right]\right\} \times\left(1 - \frac{H\,\mathrm{Ra}_B}{L}\right)^{1/4}\right], \tag{5.1.27}$$

lower end-face surface of width δ

$$\mathrm{Nu}_\delta = 1.6\left(\frac{\delta}{B+\delta}\right)^{0.75}\left(\frac{B}{H}\mathrm{Ra}_B\right)^{0.4}\left\{12 + \frac{1}{7}\frac{B}{H}\mathrm{Ra}_B\left[1 + 1.8\left(\frac{H}{L}\right)^{0.88}\right] \times\left\{1 - \exp\left[-\left\{\frac{250H\,\mathrm{Ra}_B}{B}\left[1 + 1.8\left(\frac{H}{L}\right)^{0.88}\right]\right\}^{0.6}\right]\right\}\right\}^{-0.4}, \tag{5.1.28}$$

vertical end-face surface of width $\delta \times H$

$$\mathrm{Nu}_\delta = \frac{1}{30}\frac{\delta}{H}\mathrm{Ra}_B\left(1 + \frac{\delta}{B}\right)^{-1/4}\left(1 + \frac{H}{3L}\right) \times\left\{1 - \exp\left[-\left\{\frac{17H\,\mathrm{Ra}_B}{B}\left[1 + \left(\frac{H}{3L}\right)\right]\right\}^{3/4}\right]\right\}, \tag{5.1.29}$$

upper end-face surface of width δ

$$\mathrm{Nu}_\delta = 0.94\left(\frac{\delta}{H}\mathrm{Ra}_\delta\right)^{0.1}\left\{1 + 0.018\frac{L}{H}\mathrm{Gr}_B\left(\frac{\delta}{B}\right)^{2.5}\left(\frac{H}{B\,\mathrm{Ra}_B}\right)^{0.4} \times\left[1 - \exp\left(-50\left(\frac{L}{H}\right)^{0.25}\left(\frac{H}{B\,\mathrm{Ra}_B}\right)^{0.5}\right)\right]\right\}. \tag{5.1.30}$$

The total heat flux from the channel surface is determined as an algebraic sum of individual components, the heat transfer ***from the inner channel surface*** can be found from the equation

$$\mathrm{Nu}_B = \frac{\mathrm{Nu}_1 \frac{B}{2L} + \mathrm{Nu}_2}{\frac{B}{2L} + 1} . \tag{5.1.31}$$

The data of numerical calculations for the average coefficient of heat transfer ***in a rectangular cavity open from above and heated from below under various boundary conditions*** show that in the regime of heat conduction ($\mathrm{Gr}_B < 10^3$) the heat transfer decreases with increase in the height of insulated walls [5.25]. Figure 5.1.2 present the data of numerical calculations on the average value of the coefficient of heat transfer from the lower horizontal surface under various boundary conditions. This effect is strongest for constant heat flux on the surface. In the region of developed convective heat transfer ($\mathrm{Gr}_B > 10^5$) the height of confining walls exerts a little effect on heat transfer of the horizontal surface, which is confirmed by numerical calculations for air [5.26, 5.27].

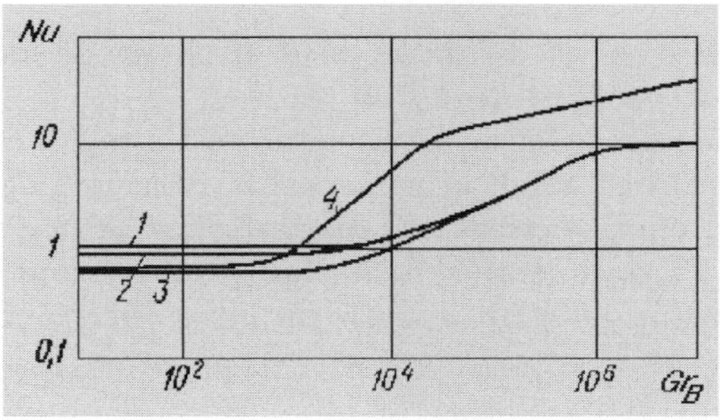

Fig. 5.1.2. Heat transfer from the horizontal surface of an open channel with insulated vertical walls: T_w = const: (1) $H/B = 0.5$; (2) 0.69; (3) 1; q_w = const: (4) $H/B = 0.69$

An parallelepiped open from above (H × B × L = 0.5 × 0.75 × 0.5 m) with insulated vertical walls, filled with water up to the height $H' = 0.45$ m, is heated from below. The average value of the coefficient of heat transfer from a free water surface to air in the ranges: $\mathrm{Pr} = 3 - 8$, $\mathrm{Ra}_B = 3 \cdot 10^8 - 4 \cdot 10^9$ is [5.28]

$$\mathrm{Nu} = 0.156\, \mathrm{Ra}_B^{1/3} . \tag{5.1.32}$$

A technique of numerical calculation of free convection ***in a vertical channel with non-parallel walls*** has been developed. The streamlines and temperature distribution were calculated for $\mathrm{Gr}_B = 10^2$ and $\mathrm{Pr} = 0.2$ [5.29].

The numerical investigation of free convection ***in a vertically-ventilated cavity with one of the walls maintained at a constant temperature*** T_w shows a strong nonuniformity of the local heat transfer coefficient distribution along the isothermal wall. For the average Nusselt number the following dependence holds [5.30]:

$$\mathrm{Nu} = \left[\frac{144}{(\mathrm{Ra}\, W/H)^2} + \frac{2.8}{(\mathrm{Ra}\, W/H)^{1/2}} \right]^{-1/2} , \tag{5.1.33}$$

where W and H are the height and width of the cavity.

The numerical investigation of the effect exerted by the parameter b/L on the characteristics of flow and heat transfer ***in a vertical rectangular channel*** cooled by an ***air*** free-convective flow (b, L are the channel width and height) for the Rayleigh number from 10^2 to 10^5 showed that the optimum spatial location of the channel walls (from the viewpoint of the heat transfer efficiency) is connected with the Grashof number by the generalized dependence

$$(b/L)_{\mathrm{opt}} = 4.57\ \mathrm{Gr}^{-0.192} . \tag{5.1.34}$$

The results were obtained ***for the case of channel walls heated symmetrically by a constant heat flux*** [5.31].

The correlation dependences for the average Nusselt and Sherwood numbers were obtained as a result of numerical investigation of laminar natural convection ***in an open vertical rectangular channel with uniform temperature (UWT) or uniform heat flux (UWC) and also with uniform concentration (UHF) or uniform mass flux (UMF)*** as boundary conditions [5.32]:

for the cases (UHF/UMF)

$$\mathrm{Nu} = \frac{0.37874\eta^{0.62}}{(1 + 0.39538\eta^{0.756})^{1/1.8}} , \tag{5.1.35}$$

where

$$\eta = \mathrm{Gr}(1 + N/\mathrm{Le}), \quad N = \frac{\beta_c m_{\mathrm{W}}/D}{\beta_{\mathrm{T}} q_{\mathrm{W}}/k} ,$$

m_{w} and q_{w} are the mass and heat fluxes at the interface, D and k are the coefficients of diffusion and thermal conductivity.

$$\mathrm{Sh} = \frac{0.4573\eta^{0.4}}{(1 + 0.41345\eta^{0.378})^{1/1.8}} , \tag{5.1.36}$$

where $\eta = \mathrm{GrLe}^2(1{+}N/\mathrm{Le})$,

for the cases (UWT/UWC)

$$\mathrm{Nu} = \frac{C_1\,\eta^{0.982}}{\left(1 + C_2\eta^{1.2896}\right)^{1/1.8}}\,, \tag{5.1.37}$$

where

$$\eta = \mathrm{Gr}(1 + N), \quad N = \frac{\beta_c \Delta C}{\beta_T \Delta T}\,.$$

$$\mathrm{Sh} = \frac{C_1\eta^{0.995}}{\left(1 + C_2\eta^{1.3104}\right)^{1/1.8}}\,, \tag{5.1.38}$$

where $\eta = \mathrm{GrLe}(1 + N)$, and the values of the coefficients C_1 and C_2 are given in Table 5.1.3.

Table 5.1.3. The Values of Constant Coefficients in the Correlation Equations for the Cases UWT/UWC

		Sc = 0.2		Sc = 0.6		Sc = 1.3	
		C_1	C_2	C_1	C_2	C_1	C_2
Q	$N = 0$	1.0161	66.531	1.0161	66.531	1.0161	66.531
	$N = 1$	1.4631	105.86	1.0679	70.715	0.81496	50.337
	$N = 2$	1.6307	120.66	1.1119	74.949	0.77142	47.211
Nu	(N=0, 1, 2)	0.025779	0.0050237	0.023626	0.0050237	0.021860	0.0050237
Sh	$N = 0$	0.014727	0.0046032	0.017766	0.0046032	0.018456	0.0046032
	$N = 1$	0.020674	0.0046032	0.020846	0.0046032	0.020845	0.0046032
	$N = 2$	0.023584	0.0046032	0.022665	0.0046032	0.022361	0.0046032

Natural-convective heat and mass transfer between vertical plane parallel surfaces with unheated inlet and exit parts is studied analytically and numerically. ***The cases of constant temperature/concentration (UWT/UWC), and also of constant heat/mass fluxes (UHF/UMF) in the mid part of surface of the plates*** are considered. The results for the dimensionless liquid flow rate Q, average Nusselt (Nu) and Sherwood (Sh) numbers are obtained at different values of the ratio between the buoyancy forces N, Gr and Sc numbers and also for different ratios of the lengths of heated and unheated portions of the plate E_1 and E_2 (E_1 is the ratio of the unheated part length to the whole plate length; E_2 is the ratio of the heated part length to the whole plate length). The numerical data for Q, Nu and Sh in two limiting cases: when the upper unheated part is longer than the lower one (upper limit) and when the lower unheated part is longer than the upper one (lower limit) are generalized by the dependences

UWT/UWC:

$$\frac{Q}{(1+N)(E_1+E_2)} = \frac{0.333}{[1+0.111X'^{-0.868}]^{1/1.4}} \quad \text{for the upper limit}\,, \tag{5.1.39}$$

$$\frac{Q}{(1+N)(E_1+E_2)} = \frac{0.333}{[1+0.857X'^{0.11}]^{1/2}} \quad \text{for the lower limit}\,, \tag{5.1.40}$$

where $X' = L/(1+N)$,

UWT/UWC:

$$\mathrm{Nu} = \frac{0.33X'}{[1+0.118X'^{1.829}]^{1/2.4}} \quad \text{for the upper limit}\,, \tag{5.1.41}$$

$$\mathrm{Nu} = \frac{0.33X'}{[1+0.484X'^{1.067}]^{1/1.4}} \quad \text{for the lower limit}\,, \tag{5.1.42}$$

where $X' = \mathrm{Gr}_L\,\mathrm{Pr}(1+N)(1+E_1/E_2)$,

UHF/UMF:

$$\mathrm{Sh} = \frac{0.3X'}{[1+1.043X'^{1.791}]^{1/2.35}} \quad \text{for the upper limit}\,, \tag{5.1.43}$$

$$\mathrm{Sh} = \frac{0.3X'}{[1+1.730X'^{0.724}]^{1/0.95}} \quad \text{for the lower limit}\,, \tag{5.1.44}$$

where $X' = \mathrm{Gr}_L Cs^2\,(1+N/\mathrm{Le})\,(1+2E_1/E_2)$.

A method of analysis has been developed in [5.33] to predict induced flow and turbulent heat transfer ***in an asymmetrically heated, vertical plane-parallel channel***. All of the data for different aspect ratios of the channel are correlated by

$$\mathrm{Nu} = 2.1\,[\mathrm{GrPr}\,(W/H)]^{1/5}\,, \tag{5.1.45}$$

where W and H are the channel width and height.

Free convection ***of air in a vertical channel*** is studied experimentally in [5.34] on a laboratory model of height $H = 2.6$ m and rectangular cross-section $b \times s$, where $b = 1.2$ m and the channel width s can be varied. Based on the results of investigation, two empirical formulas were obtained for the dependence of Nu and Re on Ra and on the geometric parameter s/H:

$$\mathrm{Nu} = 0.9282\,\mathrm{Ra}^{0.2035}\,(s/H)^{0.8972}\,, \tag{5.1.46}$$

$$\mathrm{Re} = 0.5014\,\mathrm{Ra}^{0.3148}\,(s/H)^{0.418}\,. \tag{5.1.47}$$

The average heat transfer coefficient for ***parallel vertical disks of diameter D located with at the distance B*** in the ranges: $\mathrm{Pr} = 0.7$, $\mathrm{Ra}_B(B/D) = 3 - 6 \cdot 10^3$, is determined by the dependences [5.35]

$$\mathrm{Nu} = \frac{1}{6\pi}\frac{B}{D}\mathrm{Ra}_B\left\{1 - \exp\left[-\left(25.3\frac{D}{B}\mathrm{Ra}_B\right)^{3/4}\right]\right\}, \quad T_{\mathrm{av}} = T_{\mathrm{w}}\,, \quad (5.1.48)$$

$$\mathrm{Nu} = 0.66\left(\frac{B}{D}\mathrm{Ra}_B\right)^{0.29}, \quad T_{\mathrm{av}} = (T_{\mathrm{w}} + T_\infty)/2\,. \quad (5.1.49)$$

Heat transfer ***in a channel with alternating straight and curvilinear portions*** was studied in [5.36] at Dean numbers from 300 to 750. The channel aspect ratio was 40, the radii ratio was 0.979, and the ratio of the shear layer thickness to the channel inner radius was 0.011.

5.2 Cylindrical Channel

Heat transfer ***in a vertical cylindrical tube of diameter d at a constant wall temperature*** T_{w} for $\mathrm{Pr} > 0.7$, $\mathrm{Ra}_d(d/H) = 10^{-1} - 10^5, T_{\mathrm{av}} = T_{\mathrm{w}}$ within 10% is given by [5.1]

$$\mathrm{Nu} = \frac{1}{128}\frac{d}{H}\mathrm{Ra}_d\left\{1 - \exp\left[-8\left(20\frac{H}{d}\mathrm{Ra}_d\right)^{3/4}\right]\right\}. \quad (5.2.1)$$

The length of the inlet part for a laminar flow regime is $H_0 = 7d\mathrm{Ra}_d/800$ [5.37].

The optimum tube diameter and heat transfer from the lateral surface are determined from formula (5.1.3) for $\psi = 16$.

A vertical pipe with unheated initial part of length l_{C}***(***$T_{\mathrm{w}} = T_l$***, where*** T_l ***is the fluid temperature) and part of length l, where*** $T_{\mathrm{w}} > T_l$. The average heat transfer coefficient for $\mathrm{Pr} > 0.7$, for $T_{\mathrm{av}} = T_\infty$ is determined by the dependences [5.38]

for $l_0/l \le 15$, $\mathrm{Ra}_r r/l < 10^2$

$$\mathrm{Nu} = \frac{r\,\mathrm{Ra}_r}{16\,(l + l_0)}\,, \quad (5.2.2)$$

for $l_0/l = 0$, $\mathrm{Ra}_r r/l = 10^2 - 10^5$

$$\mathrm{Nu} = 0.63\,(\mathrm{Ra}_r r/l)^{0.25}\,, \quad (5.2.3)$$

for $l_0/l = 2$, $\mathrm{Ra}_r r/l = 10^2 - 10^5$

$$\mathrm{Nu} = 0.48\,(\mathrm{Ra}_r r/l)^{0.25}\,, \quad (5.2.4)$$

For $\mathrm{Pr} < 0.7$ the heat transfer depends substantially on the value of Pr. The optimum tube diameter for the maximum heat transfer from the surface depending on l_0/l can be determined from Fig. 5.2.1. There exist such a value

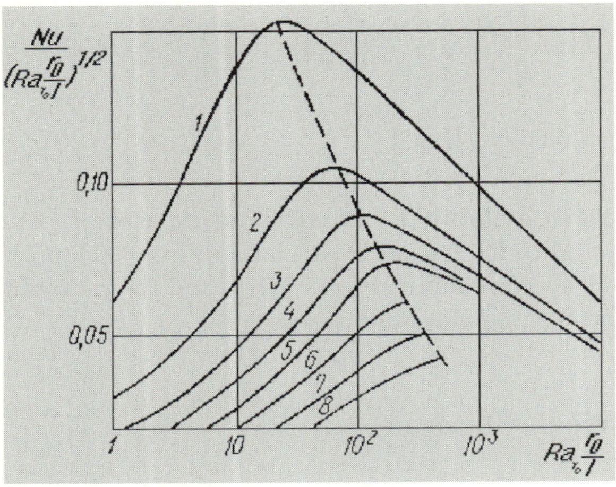

Fig. 5.2.1. Change in the average heat transfer coefficient for a channel of circular section with an unheated initial part: (1) $l_0/l = 0$; (2) 1; (3) 2; (4) 4; (5) 6; (6) 8; (7) 10; (8) 15

of l_0 for which a laminar regime is impossible. It is observed for $l_0/l > 18$, when $\mathrm{Ra}_r\, r/l < 20$ and for $l_0/l = 2$, when $\mathrm{Ra}_r\, r/l > 10^3$.

In designing and constructing ***thermosyphons***, free convection ***in closed vertical channels*** is investigated in which a region of enhanced heat transfer is created, due to convection, with the effective thermal conductivity coefficient λ_{ef}. The heat transfer ***between lateral walls of the thermosyphon and the fluid*** is determined by the dependences [5.39, 5.40]

for $H/d = 3 - 30$, $\mathrm{Pr} = 2.5 - 5.45$, $\mathrm{Ra}_H = 10^8 - 5 \cdot 10^9$

$$\mathrm{Nu} = 3\ \mathrm{Ra}_H^{1/4}\,(d/H)^{0.7}\ , \tag{5.2.5}$$

for $H/r = 9.2–54.2$, $\mathrm{Pr} = 0.01–0.05$, $\mathrm{Ra}_r = 5 \cdot 10^4 - 10^6$

for $\boldsymbol{T}_{\mathrm{w}}$ **= const**

$$\mathrm{Nu} = 0.165\left(\mathrm{Ra}_r\ \mathrm{Pr}^{0.75}\right)^{0.35}(r/H)\ , \tag{5.2.6}$$

for $\boldsymbol{q}_{\mathrm{w}}$ **= const**

$$\mathrm{Nu} = 0.26\left(\mathrm{Ra}_r\ \mathrm{Pr}^{0.75}\right)^{0.35}(r/H)\ . \tag{5.2.7}$$

The presence of partitions in the thermosyphon with ***a liquid metal*** increases its efficiency 10 times at $H/r = 28.5$ [5.39].

Rotating thermosyphons are characterized by two flow regimes. The first one is an intensive helical flow in the flow core along the entire height of the channel with a distinct boundary layer near the walls. Characteristic for this regime is the occurrence of a closed region of rotation in the zone of zero heat supply and closed rotating cells. The temperature changes continuously along the channel length. This mode is characteristic for small flow velocities at $n \sim 400$–$700\,\mathrm{rpm}$ and $H/d = 4$–30. The heat transfer mechanism in this case is the same as in the absence of rotation, therefore it is possible to use formula (5.2.6) for calculations.

Transition to the second flow regime occurs on increase in the channel rotation frequency, and the larger the value of H/d, the larger values of rotation frequency correspond to the transition. ***The heat transfer between the walls and the cooling fluid*** for $\mathrm{Ra}_H < 10^{10}$ proves lower than for the previous regime and is determined from the equation [5.40]

$$\mathrm{Nu} = 0.47\,\mathrm{Ra}_H^{0.29}\,(d/H)^{0.7}\ . \tag{5.2.8}$$

The results of experimental investigation the effect exerted by the inclination angle on the characteristics of flow and heat transfer ***for a closed tubular*** thermosyphon filled ***with water*** (at low Rayleigh numbers) are presented in [5.41]. The numerical model of calculation is based on consideration of interaction between three flow parts: the main part in each tube, the flow in the end parts and over the intermediate part which connects the heated part with the cold one. The secondary flows and angular displacement of both ends are taken into account by modifying the main axisymmetric flow characteristic for a vertical position of the thermosyphon. The transient part, where the buoyancy forces are the sources of flow, is characterized by complex entanglement of streamlines.

Numerical data on laminar natural convection caused by heating of the left-hand side channel in a plane open U-type ***thermosyphon with conducting walls*** are presented in [5.42, 5.43]. The relationship for non-dimensional average heat flux is applicable to the Ra numbers from Ra = 4630 to Ra = $8.3 \cdot 10^5$, which correspond to a laminar regime:

$$\Phi = 0.302\,\mathrm{Ra}^{0.253}\ , \tag{5.2.9}$$

where $\Phi = bq/\lambda\,(T_\mathrm{w} - T_\infty)$ and b is the inner channel width.

When $\mathrm{Ra}_H = 10^9$, the value of Nu is decreased by 20%. In the cooling channels of modern high-rotating gas turbines the flow regime characterized by formula (5.2.8) prevails [5.44].

In a vertical closed coaxial channel filled with water the model of thermal conductivity of [5.45] is applicable to calculate heat transfer rate from the surface for $R_0/R_i = 4.2 - 18$, $H/R_0 = 80$.

Natural convection ***in heated vertical concentric channels*** was studied in [5.46]. ***The outer cylinder is thermally insulated and a constant heat flux was maintained on the inner one***. Under these conditions, the mean temperature of the fluid increases linearly till heat diffusion from the heated surface into the fluid results in a developed temperature distribution in the cross section. Imposing this steady increase in the temperature on the total distribution of temperature results in steady temperature stratification over the cross section that determines the character of fluid motion. At low temperature differences, the flow is steady and parallel, and heat is transferred by conduction between the fluid layers. As the difference becomes larger, the flow becomes unstable and circulation eddies appear in it, which transfer heat by convection.

A vertical concentric cylindrical channel ***with a uniformly heated inner rod and adiabatic casing*** of length $L = 3.68$ m. The radii ratio is $r = R_0/R_i = 1.5$, 2.0 and 2.4. The experimental data for the Nu number average over length in the range $2.6 \cdot 10^3 < \mathrm{Gr} < 8.5 \cdot 10^4$ are generalized by the dependence [5.47]

$$\mathrm{Nu} = \frac{Q}{R_i^2 U_r} A \left(\frac{L}{R_i Gr} \right)^{0.874} , \tag{5.2.10}$$

where Q is the volumetric air flow rate; $U_r = \nu \mathrm{Gr}/R_i$; $A = 32.1$ in the range $2 \cdot 10^{-3} < L/(R_i \mathrm{Gr}) < 3 \cdot 10^{-2}$ for $r = 2.4$. In general the value of the coefficient A depends on r.

Consideration of simultaneous effect of the mechanisms of heat conduction and convection ***in a cylindrical annular cavity*** filled with ***liquid Na*** with cooling from below and heating from above carried out numerically and confirmed by experimental data shows that the temperature profile inside the cavity can be determined by the solving the linear equation of heat conduction, where the thermal convection is considered as a process generating heat in the volume investigated [5.48].

The effect of the elevation of a cooled surface on the characteristics of total heat transfer by natural convection ***from vertical heated rods inside a vertical cylinder*** was investigated experimentally and numerically [5.49]

The effect of thermal and hydrodynamic instability on natural convection ***in a narrow concentric cylindrical channel*** results in a complex multicellular flow structure. As it follows from the numerical calculations for $\mathrm{Pr} = 0.4$ [5.50], the number of cells increases with increasing Gr number. A further increase in Gr again turns the flow character in the vertical part of the channel back to almost monocellular structure.

The results of theoretical and experimental investigation of natural-convective flows in tubes for large Rayleigh numbers ***with allowance for the effect of wall thermal conductivity and temperature dependence of the characteristics of the fluid and tube walls*** show that the parameters of natural-convective heat transfer ***in a vertical tube*** are close to similar parameters for a vertical plate [5.51]. The experimental data for heat transfer are generalized by the dependence

$$\mathrm{Nu} = 0.61 \left(\mathrm{Ra}^*\right)^{1/5} , \tag{5.2.11}$$

$$\mathrm{Ra}^* = \frac{g \beta q_{\mathrm{w}} R_i^5}{l \nu a \lambda} ,$$

where ν, α, and λ are the kinematical viscosity, thermodiffusion coefficient, and the thermal conductivity coefficient, respectively.

The effect of the wall thermal conductivity in mixed convection on a steady laminar ***water*** flow ***in a vertical tube*** was investigated numerically in [5.52]. The fixed values of the numbers were selected for the flow: $\mathrm{Pr} = 7$, $\mathrm{Re} = 50$, and $\mathrm{Gr} = 10000$. It was found that in the majority of considered flows there is a reverse flow causing substantial changes in the velocity and temperature distributions.

The countercurrent fluid motion ***in a horizontal tube of length L, radius r and thickness δ caused by the difference of end temperatures T_1 and T_2*** induces heat transfer along the channel length. The asymptotic value of the average Nu number in view of wall thermal conductivity for $r/L \ll 1, \delta/r \ll 1, \mathrm{Ra}_r \ll 1$ is calculated from the formula [5.53]

$$\mathrm{Nu}\frac{L}{r} = 1 + \frac{7+2P'}{1+P'}\frac{(r\ \mathrm{Ra}_r/L)^2}{46080} + O\left[(r\ \mathrm{Ra}_r/L)^4\right] , \qquad (5.2.12)$$

When $P' = \lambda_\mathrm{w}\delta/(\lambda r) = 0$, the wall is adiabatic; when $P' \to \infty$, the wall temperature is constant. The temperature distribution in the channel is

$$\frac{T-T_1}{T_2-T_1} = \frac{r}{L}\frac{z}{L} + \frac{1}{192}\frac{r}{L}\sin\theta \times \left[\left(\frac{r'}{r}\right)^5 - 3\left(\frac{r'}{r}\right)^3 + 2\frac{2+P'}{1+P'}\frac{r'}{r}\right]\mathrm{Ra}_r , \qquad (5.2.13)$$

where r' is a variable channel radius.

Free convection ***in the space between coaxial horizontal cylinders with different end-face temperatures T_1 and T_2***. The asymptotic values of the average heat transfer coefficient are determined from the graphs of Fig. 5.2.2, where the upper curve corresponds to insulated walls, the lower one – to constant temperatures on both walls, and the middle ones to intermediate cases of boundary conditions.

The numerical and experimental investigation of three-dimensional natural convection in ***a porous medium saturated with a fluid between concentric inclined cylinders*** shows that there is a critical value of the aspect ratio A for which the inclination angle ceases to influence heat transfer [5.54]. ***The temperature or heat flux of the inner cylinder is constant, the outer cylinder is isothermal.*** The parameter ranges are: $\mathrm{Ra} \le 10^3; 1 \le R_0 \le 10; 1 \le A \le 15; 0° \le \varphi \le 90°$ ($R_0 = r_0/r_i$ is the cylinders radii ratio; $A = l/(r_0 - r_i)$; φ is the angle of the inclination to the horizontal).

The linear stability of natural convection ***in vertical annuli with a moving inner cylinder motion*** was studied in [5.55]. The results show that the stability characteristics change substantially due to the inner cylinder motion. Irrespective of the direction of motion of the cylinder, at small Pandtl numbers and small radii ratio, the shear instability is predominant and

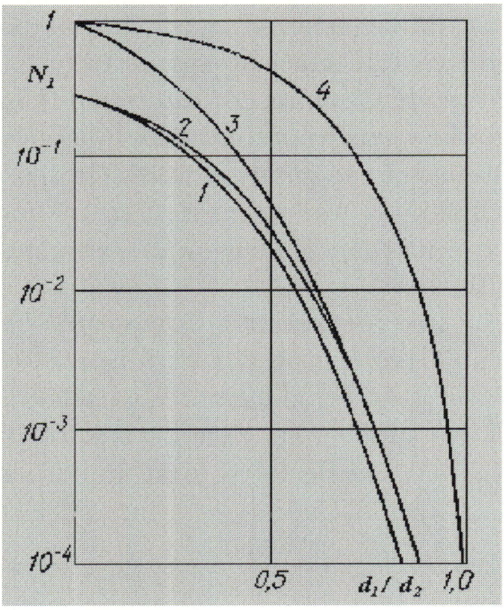

Fig. 5.2.2. The value of the parameter $N_1 = 23040\,(\mathrm{Nu}_{r1}L/r_1 - 1)/(r_1/L)^2$ between coaxial cylinders: (1) $P_{11} = \infty$, $P_{12} = \infty$; (2) ∞, 0; (3) 0, ∞; (4) 0, 0

the most unstable are asymmetric disturbances. For large Prandtl numbers, the main mechanism of instability is the switching from the buoyant mode generated by the boundary layer near the outer cylinder to the buoyant mode induced by the boundary layer near the inner cylinder on increase in the cavity curvature with the inner cylinder motion.

5.3 Finned Surfaces

Horizontal V-shaped corrugated surface (Fig. 5.3.1a). The average value of the heat transfer coefficient ***in air*** for $\mathrm{Ra}_H = 2.5 \cdot 10^4 - 2 \cdot 10^7$ and $T_{\mathrm{av}} = (T_{\mathrm{w}} + T_\infty)/2$ for $\varphi = 30 - 160°, H = 10 - 150\,\mathrm{mm}$, $\mathrm{Pr} = 0.7$ with accuracy of $\pm 8.5\%$ is given by [5.56]

for $\mathrm{Ra}_H = 2.5 \cdot 10^4 - (1.58 - 1.4 \sin\varphi/2) \cdot 10^6$

$$\mathrm{Nu} = \left[\frac{0.46}{\sin(\varphi/2)} - 0.32\right] \mathrm{Ra}_H^{0.148\sin(\varphi/2)+0.187}\,, \tag{5.3.1}$$

for $\mathrm{Ra}_H = (1.58 - 1.4 \sin\varphi/2) \cdot 10^6 - 2 \cdot 10^7$

$$\mathrm{Nu} = [0.09 + 0.054/\,(\sin\varphi/2)]\,\mathrm{Ra}_H^{1/3}\,. \tag{5.3.2}$$

When $\varphi = 160°$, the effect of the surface curvature on heat transfer is insignificant [5.57]. Comparing the data ***for V-shaped corrugated plate***

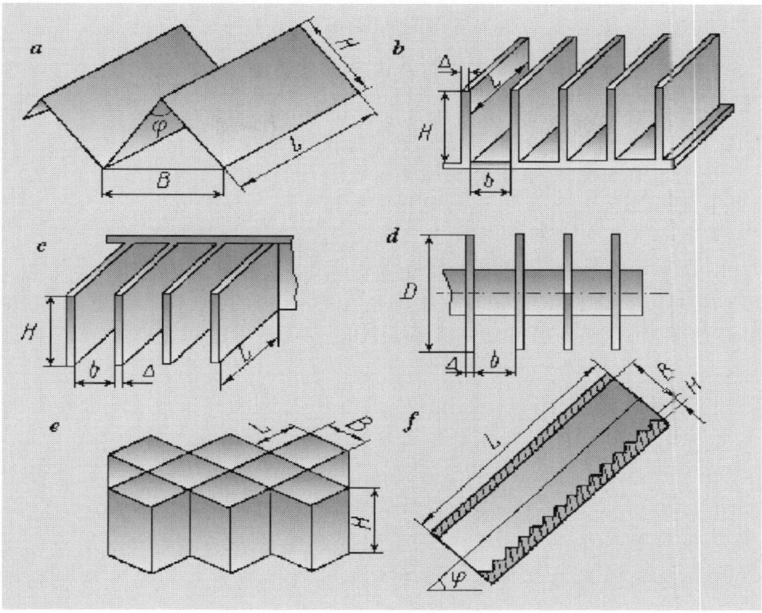

Fig. 5.3.1. Types of finned surfaces

and horizontal plate with vertical edges shows that for $\varphi = 30°$ and for equal heat transfer surfaces the heat flux from the corrugated plate is higher compared with the finned horizontal one. For $\varphi = 60°$ and $H > 30\,\mathrm{mm}$ the finned surface becomes more effective from the point of view of heat transfer rate.

Rectangular fins of constant temperature on a horizontal surface (Fig. 5.3.1b). The experimental data on heat transfer of ***aluminum fins*** with the geometric parameters given in Table 5.3.1 are approximated for ***air*** by the following empirical dependences [5.58–5.61]:

for $(nB/H)\,\mathrm{Ra}_L = 10^6 - 2.5 \cdot 10^7$

$$\mathrm{Nu} = 3.34 \cdot 10^{-3} \left(\frac{nB}{H}\mathrm{Ra}_L\right)^{0.57} \left(\frac{H}{L}\right)^{0.656} \left(\frac{B}{L}\right)^{0.412}, \qquad (5.3.3)$$

for $(nB/H)\,\mathrm{Ra}_L = 2.5 \cdot 10^7 - 1.5 \cdot 10^8$

$$\mathrm{Nu} = 1.241 \cdot 10^{-3} \left(\frac{nB}{H}\mathrm{Ra}_L\right)^{0.745} \left(\frac{H}{L}\right)^{0.656} \left(\frac{B}{L}\right)^{0.412}, \qquad (5.3.4)$$

for $\mathrm{Ra}_B = 2 \cdot 10^2 - 10^6$

$$\mathrm{Nu} = 6.7 \cdot 10^{-4}\mathrm{Ra}_B \left\{1 - \exp\left(\frac{0.746 \cdot 10^4}{\mathrm{Ra}_B}\right)\right\}^{1.7}. \qquad (5.3.5)$$

Table 5.3.1.

N	L, cm	H, cm	B, cm	n	T_{av}	Source
1	12.7, 25.4	0.6 – 4.8	0.635, 0.795	14 – 33	T_{w}	[5.58]
2	25.4	2.39 – 4.93	0.406 – 5.08	5, 7	$(T_{\text{w}} + T_\infty)/2$	[5.59]
3	20.3	0.635, 1.27, 1.9	0.484 – 6.65	1 – 9	T_{w}	[5.60]
4	3.125 – 25	2.5	0.159 – 1.25	–	T_{w}	[5.62]

The optimum distance between the fins is determined from the empirical correlation for $BH = 3.62\,\text{cm}^2$, $(d_r/H)\text{Ra}_{d_r} = 10 - 2.4 \cdot 10^3$

$$\text{Nu} = \frac{8d_r}{H\psi}\text{Ra}_{d_r}\left\{1 - \exp\left[-\left(\frac{8H}{d_r\,\text{Ra}_{d_r}}\right)^{3/4}\right]\right\}, \tag{5.3.6}$$

where

$$d_r = \frac{4HB}{2H + B},$$

$$\psi = 24\left[1 - 0.483\exp\left(-0.17\frac{B}{H}\right)\right]\left\{\left(1 + \frac{B}{2H}\right)^3\left\{1 + \left[1 - \exp\left(-0.83\frac{B}{H}\right)\right]\right.\right.$$
$$\left.\left.\times\left[9.14\left(\frac{B}{H}\right)^{1/2}\exp\left(-0.3\frac{B}{H}\right) - 0.61\right]\right\}^3\right\}^{-1}.$$

The calculated data on heat transfer of ***short vertical fins*** are given in Fig. 5.3.2. For the given Gr and L/H the heat transfer increases with decrease in B/H until flow instability appears [5.62].

The coefficient of heat transfer ***between vertical fins on a horizontal base*** is proportional to the distance between the fins B and slightly depends on the height H. The fin length L and the number of fins n practically do not influence the average heat transfer coefficient.

Laminar natural convection ***near a two-dimensional horizontal downward-facing plate with rectangular grooves*** was studied experimentally and numerically in [5.63]. A Mach–Zehnder interferometer was used in the experiment and the local Nusselt numbers on each groove surface (outer, upper, left and right side surfaces) were measured quantitatively from interferograms. In some cases (grooves of some aspect ratios with a low Rayleigh number), the increase of the total heat transfer rate from grooved surface may not be sufficiently large despite the increased surface area. As revealed by the numerical analysis for the given conditions, secondary recirculation flows are usually observed in the groove. The protrusion prevents the main flow from flowing into the groove and cause recirculations. As they happen, the heat transfer rate at the upper surfaces of the groove is significantly smaller than that at the outer surface. The results are summarized by the following correlation:

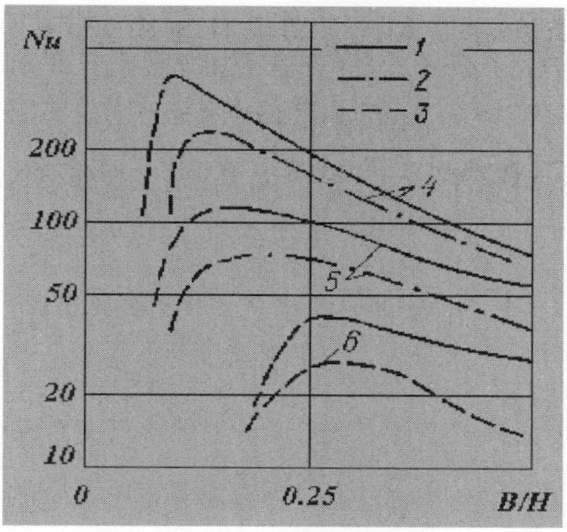

Fig. 5.3.2. Dependence of Nu on B/H: (1) $B/H = 8$; (2) 4; (3) unstable flow region; (4) $\mathrm{Gr}_H = 10^7$; (5) 10^6; (6) 10^5

$$\mathrm{Nu} = 0.27\left(1 + 16.8\frac{(H/W)^{1.45}(W/L)^{1.21}(W_p/W)^{0.33}}{1 + H/W}\right)\mathrm{Ra}_L^{1/4}\,, \quad (5.3.7)$$

where H is the depth of the groove, W is the length of a fin, W_p is the width of the groove, L is the total horizontal length of the finning region.

Vertical surface of the base finned by a system of parallel vertical plates (see Fig. 5.3.1c). Under different thermal conditions, the heat transfer on each surface can be calculated from formulas (5.1.32, 5.1.39, 5.1.40, 5.2.1–5.2.4) [5.24].

The experimental data are approximated by the average value of heat transfer of a ***vertical surface*** with ***aluminum fins in air*** as [5.64, 5.65]

$$I = 2.54,\ 6.35,\ 10\,\mathrm{cm},\ H = 25.4\,\mathrm{cm},\ B = 0.6 - 2.54\,\mathrm{cm}\,,$$
$$\delta = 0.1\,\mathrm{cm},\ T_{\mathrm{av}} = T_{\mathrm{w}} + 0.62(T_\infty - T_{\mathrm{w}})\,,$$

for $(B/H)\,\mathrm{Ra}_B = 1 - 10^4$

$$\mathrm{Nu} = 0.112\left(\frac{B}{H}\mathrm{Ra}_B\right)^{0.534}\left[1 - \exp\left(-\frac{129H}{B\mathrm{Ra}_B}\right)\right]^{0.284}\,, \quad (5.3.8)$$

$$I = 1 - 3\,\mathrm{cm},\quad H = 40; 60\,\mathrm{cm},\quad B = 0.6 - 3.3\,\mathrm{cm}\,,$$
$$\delta = 0.3\,\mathrm{cm},\quad T_{\mathrm{av}} = (T_{\mathrm{w}} + T_\infty)/2\,,$$

for $(B/H)\,\mathrm{Ra}_B = 5 - 2\cdot 10^4$

$$\mathrm{Nu} = 0.65\left(\frac{B}{H}\mathrm{Ra}_B\right)^{0.24} - 0.66\,, \quad (5.3.9)$$

The optimum distance between the fins with an error of 1.5% is determined from the equation $B = 3.15(\mathrm{Ra}_L L/H)^{-0.264}L$ [5.66] or by the thermal boundary layer thickness $B = 0.65\delta_T$ [5.67]. For the fin of minimal mass and volume [5.68]

$$L = \left\{\lambda_c B\delta/\left[\lambda_\infty\left(\mathrm{Ra}_B B/H\right)^{1/4}\right]\right\}^{1/2}.$$

The effect of ***partitions with small thermal conductivity coefficient*** (made of ***plexiglass***) ***attached to the heated vertical plate*** on ***air*** free convection is analyzed experimentally and numerically [5.69]. The effect of such parameters as the Rayleigh number changing from 10^5 to 10^7 and the ratio S/H (where H and S are the fin height and the distance between the fins) is investigated. It is shown that the presence of partitions decreases the rate of free-convective heat transfer from the surface approximately by 75% compared with a smooth plate. For high Ra numbers in the space between the fins there appears a complex cellular flow structure. The calculated results for the average Nusselt number for the plate with partitions are presented in Table 5.3.2.

Table 5.3.2. The Values of the Average Nusselt Number on a Finned Plate

Fin Height (cm)	Number of Fins	S/H	Nu Ra = 10^5	Nu Ra = 10^6	Nu Ra = 10^7
1.0	3	7.00	7.982	13.657	24.415
	5	4.66	7.393	12.096	20.693
	7	3.50	6.931	10.948	18.129
	9	2.80	6.576	10.134	16.380
	11	2.33	6.289	9.513	15.008
3.0	3	2.33	6.118	9.890	20.118
	5	1.55	5.490	7.948	15.142
	7	1.16	5.233	6.826	13.165
	9	0.93	5.081	6.288	11.446
	11	0.77	4.156	5.067	9.869
5.0	3	1.40	5.160	8.883	18.255
	5	0.93	5.097	7.021	14.793
	7	0.70	4.964	5.875	11.806
	9	0.56	4.889	5.399	9.509
	11	0.46	4.827	5.200	7.840
Smooth plate			13.538	20.424	32.918

The experiments on free-convective and ***radiative heat transfer from dense packing pin-fins with a vertical isothermal base plate*** were carried out in [5.70]. The spacing between the vertical pins is used as the characteristic size. The experimental data for the average heat transfer coefficient are generalized by the dependence

$$\left(\frac{\pi d}{2S_V 2}\right)\mathrm{Nu}_{p,s} = \frac{1}{20}\left(\phi \mathrm{Ra}_s^*\right)^{3/4}\left[1-\exp\left(\frac{-120}{\phi\ \mathrm{Ra}_s^*}\right)\right]^{1/2} + \frac{1}{200}\left(\phi\ \mathrm{Ra}_s^*\right)^{1/4} , \quad (5.3.10)$$

where $\phi\ \mathrm{Ra}_s^* = \mathrm{Pr}_b\left[g\beta(T_p - T_\infty)S_h^3/\nu_b^2\right] S_h/H$, T_p is the pin temperature, S_h is the horizontal space between the pins, S_V is the vertical spacing between the pins, $H = (N-1)S_V + d, N = (S_h S_V)^{-1}$ is the pin array density, d is the pin diameter, ν_b is the kinematic viscosity of the fluid at the base plate temperature.

A vertical cylinder of height H, diameter d with vertical flat fins of length l and thickness δ arranged regularly on an outer surface. The average value of heat transfer rate of ***a finned cylinder made of aluminum in air*** is [5.71]

$$H = 158.75\,\mathrm{mm},\quad d = 50.8\,\mathrm{mm},\quad l = 19.05\,\mathrm{mm}\,,$$
$$\delta = 3.08\,\mathrm{mm},\quad n = 8,\quad \mathrm{Gr}_H = 10^7 - 4\cdot 10^7\,,$$
$$\mathrm{Nu} = 0.58\ \mathrm{Ra}_H^{0.25}\,. \quad (5.3.11)$$

A vertical cylinder of height H, diameter d with disk fins of diameter D and thickness δ. The fin spacing is b. The average value of heat transfer of aluminum fins ***in air*** with an error of ±12% is given by [5.72]

$$(D-d) < 200\,\mathrm{mm}, (D-d)/2b = 0 - 8.89, \delta = 8\,\mathrm{mm}\,,$$
$$b = 10\,\mathrm{mm},\ \theta_\mathrm{w} = 40 - 100°\mathrm{C}\,,$$
$$\mathrm{Nu} = 0.7\mathrm{Gr}_H^{0.25}\left[1 + (D-d)/2b\right]^{-0.7}\,. \quad (5.3.12)$$

The numerical calculation of the dependence of heat transfer rate and flow structure in a ***vertical annular cylindrical channel with longitudinal fins on the inner cylinder*** on the number of fins, fin angular thickness δ and the ratio of fin width to fin spacing f_r shows that for fixed Ra, k, f_r, δ and A there exist the optimum number of fins appropriate for the maximum value of Nu ($1 \le k \le 5, k = R_0/R_i, 2° < \delta < 6°, 0.25 \le f_r \le 0.75, A = H/L, H, L$ is the channel height and fin spacing). The Nu number increases with increase in f_r and k and decreases with increase in δ and A. The dependence of the maximum $\mathrm{Nu_{max}}$ and the appropriate optimum ϑ (the angle between the directions of two neighbouring fins) on Ra, A and k for $\delta = 6°$ and $f_r = 0.5$ can be generalized by the following dependences [5.73]:

$$\mathrm{Nu_{max}} = C\ \mathrm{Ra}^l k^m\,,$$
$$\theta_\mathrm{opt} = D\ \mathrm{Ra}^q\ k^r\,. \quad (5.3.13)$$

The values of the exponents for various A are given in Table 5.3.3.

The greatest heat transfer rate among the three values of A corresponds to $A = 1$.

Table 5.3.3.

A	C	l	m	D	q	r
1	0.109	0.35	0.89	55.6	−0.115	0.535
2	0.121	0.333	0.882	57.7	−0.115	0.563
5	0.123	0.316	0.865	59.9	−0.115	0.61

The experimental data for the average Nusselt number in the case of natural convection ***in smooth and finned vertical channels at a constant wall temperature (UWT) and constant heat flux density (UHF)*** are correlated by the following dependence [5.74]:

for UWT boundary condition

$$\begin{aligned} \mathrm{Nu} &= 0.41\ \mathrm{Gr}_b^{0.25} \quad (\text{smooth duct}) , \\ \mathrm{Nu} &= 0.365\ \mathrm{Gr}_b^{0.25} \quad (\text{finned duct}) , \end{aligned} \tag{5.3.14}$$

for UHF boundary condition

$$\begin{aligned} \mathrm{Nu} &= 0.715\,(\mathrm{Gr}_b^{*})^{0.2} \quad (\text{smooth duct}) , \\ \mathrm{Nu} &= 0.64\,(\mathrm{Gr}_b^{*})^{0.2} \quad (\text{finned duct}) . \end{aligned} \tag{5.3.15}$$

The average heat transfer value for ***porcelain finned cover caps in air*** is [5.75]

$$H/d = 6,\ D/d = 1.5,\ F_p/F_{np} = 2.4 - 2.6,\ T_{\mathrm{av}} = (T_{\mathrm{w}} + T_\infty)/2 ,$$
$$d_{\mathrm{w}} = dF_p/F_{np},\ \mathrm{Ra}_{dw} = 10^8 - 10^{10} ,$$
$$\mathrm{Nu} = 0.135\ \mathrm{Ra}_{dw}^{1/3} . \tag{5.3.16}$$

Heat transfer of ***porcelain finned cover caps for 110 kV and above with a finning coefficient 2.4 – 2.6*** for $\theta_{\mathrm{w}} = 40 - 60°\mathrm{C}$ does not depend on the determining size.

The average value of heat transfer for laminar free convection of ***a horizontal cylinder*** is

with a single disk fin (Fig. 5.3.1*d*)

$$\mathrm{Nu} = 0.5402\,C_1\ \mathrm{Ra}_d^{1/4} , \tag{5.3.17}$$

with a square vertical disk with the side B

$$\mathrm{Nu} = 0.5402\,C_2\ \mathrm{Ra}_d^{1/4} , \tag{5.3.18}$$

Table 5.3.4.

$\xi; d/B$	C_1	C_2
0	1.3	7.24
0.25	1.39	1.31
0.5	1.54	1.4
0.75	1.81	1.56
1	$1.27\ (1-\xi)^{-1/4}$	1.94

In Table 5.3.4 the values are listed for the coefficients $C_1(\xi)$ and $C_2(d/B)$ for $Pr \gg 1$ [5.76].

A horizontal cylinder with round ($\xi = D/d$) ***thin thermoconductive disks*** (Fig. 5.3.1d). The average value of heat transfer with rather low fins $B\mathrm{Ra}_B/D = 3 - 6 \cdot 10^3, T_{\mathrm{av}} = T_{\mathrm{w}}$, is given by [5.35]

for $\xi = 0 - 0.6$

$$\mathrm{Nu} = \frac{B}{12\pi D}\mathrm{Ra}_B \left\{ 2 - \exp\left[-\left(\frac{CD}{B\ \mathrm{Ra}_B} \right)^{3/4} \right] \right.$$
$$\left. \times - \exp\left[-\beta_1 \left(\frac{CD}{B\ \mathrm{Ra}_B} \right)^{3/4} \right] \right\}, \tag{5.3.19}$$

where

$$\beta_1 = 0.17\xi + \exp(-4.8\xi)\ ,$$
$$C = \left\{ \left[23.7 - 1.1\left(1 + 152\xi^2\right)^{1/2} \right] \left[1 + 0.17\xi + \exp\left(-4.8\xi\right)\right]^{4/3} \right\}^{-1},$$

for $\xi = 0.6 - 1$

$$\mathrm{Nu} = a\left(\frac{B}{d}\ \mathrm{Ra}_B \right)^m \left\{ 1 - \exp\left[-\left(\frac{C_1 d}{B\ \mathrm{Ra}_B} \right)^{C_2} \right] \right\}^{C_3}, \tag{5.3.20}$$

where $a = -0.15 + 0.3\xi + 0.032\xi^6, C_1 = -180 + 480\xi - 1.4\xi^{-8}, C_2 = 0.04 + 0.9\xi, C_3 = 1.3(1-\xi) + 0.0017\xi^{-12}, m = 0.25 + C_2C_3$.

A horizontal cylinder with round steel and aluminum disks. The average value of the surface heat transfer with an error of $\pm 10\%$, $l = 500\,\mathrm{mm}$, $d = 30 - 56\,\mathrm{mm}$, $D = 50 - 116\,\mathrm{mm}$, $\delta = 0.9 - 1.5\,\mathrm{mm}$, $B = 3.3 - 25\,\mathrm{mm}$, $h_0 = 5.6\,\mathrm{mm}$, $d = \pi(D^2 - d^2)/4D$, $\mathrm{Ra}_d = 10^3 - 10^6$, $\mathrm{Pr} = 0.7$, $T_{\mathrm{av}} = (T_\infty + T_{\mathrm{w}})/2$, $T_{\mathrm{w}} = (-18) - 60°\mathrm{C}$, $T_\infty = 15 - 30°\mathrm{C}$ is calculated from [5.77]

$$\mathrm{Nu} = 0519\left(1.0093 + 0.15d/D\right)\left[1 - 2h_0/\left(D - d\right)\right]\mathrm{Ra}_d\ . \tag{5.3.21}$$

For the optimum distance between the fins $B = (6 - 10)d$ the optimum fin height is $(D - d)/2 \geq 56\,\mathrm{mm}$.

The experimental data with an error of $\pm 10\%$ on the average value of the heat transfer of ***horizontal cylinder with copper vertical fins in air*** $d = 35\,\text{mm}$, $B = 6.35 - 19.05\,\text{mm}$, $\Delta T = 11 - 55$ K, $T_{\text{av}} = (T_\infty + T_{\text{w}})/2$; $\text{Ra}_B(B+d)/(d+d_{\text{av}}) = 2.5 \cdot 10^4 - 4 \cdot 10^5$; is calculated from [5.78]:

round disk fins of diameter $D = 60.2 - 85.7\,\text{mm}$, $d_{\text{av}} = (D+d)/2$

$$\text{Nu}_B = 0.11 \left[\text{Ra}_B \left(B+d\right) / \left(d+d_{\text{av}}\right)\right]^{0.38} , \tag{5.3.22}$$

fins of square section with the side $A = 65.5\,\text{mm}$, $d_{\text{av}} = (\sqrt{4/\pi}A + d)/2$

$$\text{Nu}_B = 0.12 \left[\text{Ra}_B \left(B+d\right) / \left(d+d_{\text{av}}\right)\right]^{0.38} . \tag{5.3.23}$$

On the fins of square section the maximum heat transfer coefficient value is at the fin edge, and the minimum – at its base [5.79]. For small fin spacing the heat transfer rate remains constant as the heat transfer surface area decreases. For large B there is some decrease in the heat transfer rate, but considerably less than the decrease in the surface area.

A horizontal tube with helically located fins of spacing B and inclination angle in respect to the vertical φ. For $\varphi \leq 34°$ dependences (5.3.15) can be used for calculating heat transfer [5.80, 5.81].

The average heat transfer value for ***helically located steel fins on a horizontal cylinder of temperature*** T_0 is given by [5.82]

$$d = 0.038 - 0.057,\ (D-d)/2\,B = 1.2 - 2,\ d/B = 1.9 - 4.75,\ \delta = 0.1\,\text{mm} ,$$
$$T_0 = 100, 200°\text{C}, \text{Pr} = 0.7,\ \text{Gr}_B = (1 - 5.8) \cdot 10^4,\ T_{\text{av}} = (T_\infty + T_{\text{w}})/2 ,$$
$$\text{Nu} = 4.54 \left(\text{Gr}_B \cdot 10^{-5}\right)^{0.147} \left[\left(D-d\right)/2B\right]^{-1.409} \left(d/B\right)^{-0.049} . \tag{5.3.24}$$

On a cylinder of diameter d and length l there is a helical finning of aluminum of thickness δ***, spacing B and outer diameter D. The tube with fins is located at an angle*** ϑ ***with respect to the horizontal.*** The average heat transfer value ***in air*** [5.83]:

$$D = 28\,\text{mm},\quad l = 300\,\text{mm},\quad D = 56\,\text{mm},\quad B = 3\,\text{mm},\quad \delta = 0.75\,\text{mm} ,$$
$$T_{\text{av}} = (T_\infty + T_r)/2, \text{Ra}_d = (4.4 - 12.2) \cdot 10^4 ,$$
$$\text{Nu} = C\ \text{Ra}_d^n . \tag{5.3.25}$$

The values of the coefficients C and n are given in Table 5.3.5. ***For a finned tube located vertically***, the tube height can be chosen as the characteristic size. Dependence (5.3.25) for $\text{Ra}_l = 4.6 \cdot 10^7 - 2.2 \cdot 10^8$ can be presented as

$$\text{Nu} = 0.062\ \text{Ra}_l^{0.27} . \tag{5.3.26}$$

Table 5.3.5.

ϑ^o	C	n
0	0.097	0.25
30	0.0245	0.358
45	0.01	0.434
60	0.003	0.532
90	0.03	0.293

A horizontal cylinder of diameter d with n longitudinal flat fins of height h and thickness δ***, located regularly on the external surface.*** The average heat transfer value for the finned cylinder of ***aluminum in air*** is given by [5.71]

$$d = 50.8\,\text{mm}, l = 19.05\,\text{mm}, \delta = 3.08\,\text{mm}, n = 8, \text{Gr}_d = 1.5 \cdot 10^5 - 1.2 \cdot 10^6 \,,$$

$$\text{Nu} = 0.39\ \text{Ra}_d^{0.25}\,. \tag{5.3.27}$$

In tubular vaporizer a bundle of four horizontal tubes is located in line in continuous square vertical fins. The steel tube diameter is d. The finsare made of brass, steel, and aluminum are of width B, spacing s and height h. The average value of the vaporizer heat transfer coefficient ***in humid air*** is given by [5.84]

$$d = 16\text{ mm}, B = 55-85\text{ mm}, s = 8-22\text{ mm}, \delta = 0.5\text{ mm}, h = 19.5-34.5\text{ mm}\,,$$

for $\text{Gr}_d = 2 \cdot 10^3 - 1.8 \cdot 10^4$

$$\text{Nu} = 0.01\,\text{Gr}_d^{0.6}\,, \tag{5.3.28}$$

for $\text{Gr}_d = 1.8 \cdot 10^4 - 10^5$

$$\text{Nu} = 0.115\ \text{Gr}_d^{0.36}\,, \tag{5.3.29}$$

where $d_e = 4s(B - d)/(4h + s)$.

In calculating d_e, it is necessary to take into account the hoarfrost thickness on the vaporizer surface, because it changes the fin spacing, "increasing" the fin height, the tube outer diameter and affects other geometric sizes. Therefore, the optimum dimensions should be chosen not for a "dry" vaporizer, but for the vaporizer covered with hoarfrost.

An electromagnetic engine of star-shaped type with open base casing. The thermophysical model can be presented as follows. On a ***vertical cylinder of diameter d there are fins of width h and thickness*** δ***. The area of cross section of the space between the fins is*** f_n***, and the perimeter is*** p_n***. The whole system is closed by a casing of diameter D, height H, thickness*** δ' with rectangular windows of height L and width b on a lateral surface in the space between the fins. The average heat transfer coefficient values with an error of 10% is given by [5.72]

$$D = 96\,\text{mm}, \quad H = 112\,\text{mm}, \quad d = 70\,\text{mm}, \quad h = 10\,\text{mm}, \quad \delta = 8\ \text{mm}\ ,$$
$$\delta' = 3\,\text{mm}, b = 3 - 14\,\text{mm}, d_e b\,\text{Gr}\ L/f_p = 0 - 200\ ,$$
$$\text{Nu} = 0.5\frac{L}{d\text{e}}\left(\frac{d_\text{e}b}{f_p}\text{Gr}_L\right)^{1/3}\ , \tag{5.3.30}$$

where $d_e = 4f_p/p_p$ is the heat transfer surface equivalent diameter.

Removal of the casing from an electric engine leads to a significant increase in heat transfer due to the ventilating ability of windows or slots.

The radiator heat transfer surface represents a combination of cylindrical, flat and prismatic elements oriented arbitrary relative to each other in space. Usually the scheme of thermal calculations for such developed surfaces is based on a simplifying approach in determining heat transfer from the formulas for simple single bodies. However, for small values of Ra characteristic for the operational regimes of cooling sets of semi-conductor devices the radiator heat transfer rate is 1.5–2 times less than for a single smooth plate.

The average value of heat transfer of ***prismatic pins of width b, thickness δ, height H of a plate-pin radiator with cross fin spacing s_1 and longitudinal one s_2 in air*** is given as [5.85]

$$B = 2 - 5\,\text{mm}, \delta = 2\,\text{mm}, H = 15 - 55\,\text{mm}, s_1 = 1.5 - 4\ \text{mm}\ ,$$
$$s_2 = 2 - 4\,\text{mm}, d_r = 2\delta b/(b + \delta), \text{Ra}_{dr} = 1 - 10^2\ ,$$
$$\text{Nu} = 0.46\ \text{Gr}_{dr}^{0.25}\ . \tag{5.3.31}$$

Bent prismatic fin of height h, width Δ, thickness δ, a longitudinal slot σ_1 and transverse slot σ_2 on the radiator.

The fins are located over the perimeter of the square base [5.86]:

$$h/(\Delta + \delta) = 1.43 - 1.92, h = 13 - 36.5\ \text{mm}, \Delta = 3.9\ \text{mm}, \delta = 2 - 6\ \text{mm},$$
$$\sigma_1 = 1 - 9\ \text{mm}, \sigma_2 = 1.25 - 10.5\ \text{mm},$$
$$\text{Pr} = 0.7, \text{Ra}_h = 3 \cdot 10^3 - 2 \cdot 10^5\ ,$$
$$\text{Nu} = 0.08\left[h/\left(\Delta + \delta\right)\right]^{0.3}\ \text{Ra}_h^{0.25}\ . \tag{5.3.32}$$

Step arrangement of fins of total number $n = 12 - 32$ [5.86]:

$$h/(\Delta + \delta) = 1.3 - 1.67, h = 12 - 40\,\text{mm}, \Delta = 3 - 12\ \text{mm}, \delta = 1 - 5\ \text{mm},]$$
$$\sigma_1 = 1 - 9\ \text{mm}, \sigma_2 = 2 - 16.5\ \text{mm}, \text{Pr} = 0.7,\ \text{Ra}_h = 4 \cdot 10^3 - 3 \cdot 10^5\ ,$$
$$\text{Nu} = 0.19\left[h/\left(\Delta + \delta\right)\right]^{0.42}\ \text{Ra}_h^{0.29}\ . \tag{5.3.33}$$

The experimental investigations of natural convection of ***five radiators with straight longitudinal fins, loop-wire finning and bent prismatic fins*** showed that the heat transfer can be determined from the equation [5.87]

$$\mathrm{Nu} = A\,\mathrm{Ra}_l \left\{1 - \exp\left[-\frac{B}{\mathrm{Ra}_l^{0.75}}\right]\right\} , \tag{5.3.34}$$

where A and B are the coefficients reflecting an effect of geometrical parameters.

Loop-wire finning of radiators. The wire diameter is d, transverse and longitudinal finning steps are s_1 ***and*** s_2***, and a height is h.*** From formula (5.3.34) for $l = 0.5\pi d$, $d = 0.68\,\mathrm{mm}$, $s_1 = 7.3\,\mathrm{mm}$, $s_2 = 2.5\,\mathrm{mm}$, $h = 70\,\mathrm{mm}$, $h_1 = 15.65\,\mathrm{mm}$, the coefficients $A = 8.5$ and $B = 3.8$.

For loop-wire radiators with bilateral finning the average heat transfer rate with accuracy of ±10% is given by [5.88]

$$d = 0.45 - 1.2\ \mathrm{mm}, s_1 = 2 - 15\ \mathrm{mm}, s_2 = 1 - 6\ \mathrm{mm}, h = 40 - 150\ \mathrm{mm},$$
$$h_1 = 14 - 18\ \mathrm{mm}, \mathrm{Ra}_d = 0.2 - 10, \mathrm{Pr} = 0.7, T_{\mathrm{av}} = (T_\infty + T_{\mathrm{w}})/2\ ,$$
$$\mathrm{Nu} = [0.02 + 0.01\,(s_1/d) + 0.023\,(s_2/d) - 0.00042\,(h/d)]\,\mathrm{Ra}_d^{1/4(h/h_1)^{0.47}} . \tag{5.3.35}$$

For the same temperature difference θ_{w}, the ambient temperature affects slightly the dispersed power. To obtain the best thermal characteristics, the following fin spacing are recommended: for $h = 40$–$70\,\mathrm{mm}$ $s_1 = 7$–$8\,\mathrm{mm}$, for $h = 100$–$150\,\mathrm{mm}$ $s_1 = 10\,\mathrm{mm}$ and $s_2 = 2$–$2.5\,\mathrm{mm}$. When the wire diameter increases, the dispersed power increases insignificantly, but to provide the rigidity of wire loops it is necessary to take $d = 0.6$–$0.7\,\mathrm{mm}$ for copper and about 0.8 mm for aluminum. The coloring by dim enamel increases the dispersed power by 20–30 %.

On a horizontal plate there are the bases of chips of height h, width b and distance between them l. For $h = 0$–$25\,\mathrm{mm}$, $h/b = 0.4$–2.6 and $l = 0$–∞ there are the following recommendations allowing us to use the chip bases of certain geometric sizes in view of thermal regime features [5.89]. The base free-convective heat transfer is most effective for $h/b = 1.3$–1.6. The increase in the distance l between cases located on one plate to more than 10 mm does not lead to an appreciable thermal effect.

The installation of microcircuit cases on the plate does not affect the intensity of heat transfer convective component. The empirical formula for the coefficient of heat transfer from the surface of microcircuits in this case is $Q = 0$–1.2 Wt, $F = (0.18$–$6) \cdot 10^{-3}\ \mathrm{m}^2$ [5.90]: ($[\alpha] = \mathrm{W/m^2{\cdot}deg}$)

$$\alpha = F^{-0.415} - 4\ . \tag{5.3.36}$$

A vertical flat surface with jalousie fins of height 6.85 mm and spacing used in standard water-air heat exchangers. The empirical dependence for calculating the average heat transfer value with an error of 16% is as follows [5.91]:

$$l = 0.048 - 0.1\ \mathrm{m}, \theta_{\mathrm{w}} = 5 - 105°\mathrm{C}, \mathrm{Pr} = 0.7\ ,$$
$$\alpha = 0.09\theta_{\mathrm{w}}^{0.25} l^{-0.85}\ . \tag{5.3.37}$$

The experimental investigation of laminar free convection ***in closed flat layer with lateral heat supply or cooling in the presence of transverse finning on one of the vertical walls*** shows the presence of spatial closed flow structures [5.92]. The temperature measurements revealed the decrease in heat transfer in such a layer.

5.4 Tube Bundles

Geometrically the system is determined by a tube diameter d, longitudinal s_1 and transverse s_2 spacings and by the characteristic size of the outer casing D. From the viewpoint of free-convective heat transfer horizontal, arrangement of tubes in a bundle is most efficient. The conditions of the flow around the first line are nearly the same as for a flow near a single cylinder. All next tubes are located in the vortical zones of the lower-lying tubes, with a stagnant zone between the tubes along the height with a relatively weak circulation of the fluid. The generalizing dependence of heat transfer rate of the tube bundle in a quasi-isothermal flow is

$$\mathrm{Nu} = f\left(\mathrm{Gr}_d, \mathrm{Pr}, s_1/d, s_2/d, D/d, z\right) , \tag{5.4.1}$$

where z is the number of the series in the direction of the flow.

Meshy heaters located horizontally in the lower part of the cavity. Heat transfer intensity in a ***highly-viscous fluid*** is calculated from the formulas for ***a horizontal surface with the hot surface facing upward*** [5.93].

Vertical arrangement of one series of cylinders. Interferometric investigations demonstrate a complex hydrodynamical picture of flow around

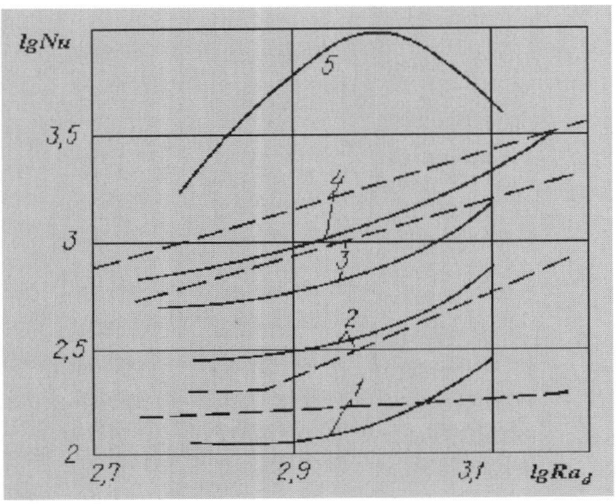

Fig. 5.4.1. The value of Nu for three (*solid line*) and nine (*dashed line*) cylinders in a series: (1) $s_1/d = 2$; (2) 4; (3) 6; (4) 10; (5) 20

cylinders [5.94]. Figure 5.4.1 presents the dependence of heat transfer on the spacing s_1 and the number of cylinders z_1. The Nu number increases with increase in the distance between the cylinders. With the distance between the cylinders being small, the surface temperature increases with the increase in z_1 and decreases with increase along the cylinder series, if the distance between the cylinders is large enough [5.95]. For $z_1 = 2$–4 and $\theta_{\rm w} = 5 - 20°$C experiments show that from the viewpoint of heat transfer the most optimum distance between the tubes is $s_1 = 5d$ [5.96].

The average heat transfer value for a ***vertical line of horizontal tubes in air*** with an error of ±5% is [5.97]

$$s_2/d = 1.5 - 4, \quad z = 5, \quad \mathrm{Pr} = 0.7, \quad \mathrm{Gr}_d = (1-6)\cdot 10^3 \,,$$
$$T_{\rm av} = (T_{\rm w} + T_\infty)/2 \,,$$
$$\mathrm{Nu} = (0.072 + 0.08 s_1/d)\,\mathrm{Gr}_d^{0.38-0.02 s_1/d} \,. \tag{5.4.2}$$

One row of horizontally located tubes. The average heat transfer rate in ***air*** is [5.98]

$$s_2/d = 1 - 6, \mathrm{Pr} = 0.7, \mathrm{Gr}_d = 3.2\cdot 10^3 - 2.3\cdot 10^5, T_{\rm av} = T_\infty \,,$$
$$\mathrm{Nu} = C\mathrm{Gr}_d^m \,, \tag{5.4.3}$$

for $s_2/d < 1.36$	$C = 2.93(s_2/d) - 3.16,$	$m = 0.17$,
for $s_2/d = 1.36 - 1.82$	$C = 0.82,$	$m = 0.17$,
for $s_2/d > 1.82$	$C = 0.47,$	$m = 0.25$.

The effect of cylinders located near each other is not manifested when $s_2/d > 1.82$. This spacing is larger than in bundles with forced convection because of the large thermal boundary layer thickness in a free-convective flow.

Horizontal line of five tubes ($z_2 = 5$) with spacing s_2 in a rectangular enclosure $H \times B$ filled with water up to the height of H_1. ***The tubes are located at the height H_2 from the free surface of the fluid with temperature T_s and at the distance $h_2 = H - H_2 - 2d$ from the base. The lateral walls and base of the enclosure are insulated***. Heat transfer from the free surface of the ***stratified fluid*** to the ambient medium with an error of 33% is given by [5.99]

$$s_2 = 50.8\ \mathrm{mm}, d = 25.4\ \mathrm{mm}, H_1/d = 3 - 15, b/d = 11 - 12 \,,$$
$$\mathrm{Ra}_{H2} = 5\cdot 10^5 - 5\cdot 10^9, \theta_{\rm w} = T_{\rm w} - T_s, T_{\rm av} = (T_{\rm w} + T_s)/2 \,,$$
$$\mathrm{Nu} = 0.52\ \mathrm{Ra}_{H_2}^{0.25} \,. \tag{5.4.4}$$

The average heat transfer coefficient for a vertical row of horizontal tubes is [5.100, 5.101]

$$z_1 \times z_2 = 5 \times 5, l = 48\,\mathrm{cm}, d = 1.96\,\mathrm{cm}, \mathrm{Pr} = 0.7, s_1/d = s_2/d = 1.25 - 2 \,,$$
$$z_{1k} = 1 - 5, z_{2n} = 1 - 5, \mathrm{Gr}_d = 6\cdot 10^3 - 1.4\cdot 10^5 \,,$$

where z_{1k} is the number of the horizontal row, z_{2n} is the number of the vertical row:

in-line bundle:
for $s_1 = s_2 = 1.25$

$$\mathrm{Nu}_{dkn} = \left[0.2e^{-0.5(z_{1k}-1)} - 0.01(z_{2n}-1)\right] \mathrm{Gr}_d^{1/3} , \tag{5.4.5}$$

for $s_1 = s_2 = 1.5 - 2$

$$\begin{aligned} \mathrm{Nu}_{d1n} &= 0.82\,(s/d - 0.5)\, e^{-0.00617(z_{2n}-1)} \mathrm{Gr}_d^{0.32-0.08s/d} , \\ \mathrm{Nu}_{d2n} &= 0.345 e^{1.7(s/d-1.5)-0.115(z_{2n}-1)} \mathrm{Gr}_d^{0.44-0.12s/d} , \\ \mathrm{Nu}_{d,k\geq 3,n} &= 0.156 e^{4(s/d-1.5)-0.1375(z_{2n}-1)} \mathrm{Gr}_d^{0.6-0.2s/d} , \end{aligned} \tag{5.4.6}$$

staggered bundle, $s_1/d = s_2/\sqrt{3}d = 0.635 - 1$:

$$\begin{aligned} \mathrm{Nu}_{d1n} &= \left[0.19e^{3.75S_{d2}} + \frac{1.47(z_{2n}-1)}{z_{2n}^{1.25}}\right] \mathrm{Gr}_d^{0.31-0.55\lg\xi_1} , \\ \mathrm{Nu}_{d2n} &= \{0.0466 + [0.507 + 0.55S_{d1}]\, S_{d2} \\ &\quad + 0.1284\,(z_{2n}-2)\,/z_{2n}^{1.564}\} \,\mathrm{Gr}_d^{0.39-0.55\lg\xi_1} , \\ \mathrm{Nu}_{d3n} &= \{0.0245\xi_1^{5.1} + [0.0805 - 0.0362\lg\xi_1]\, S_{d1} \\ &\quad + 0.0044\,(4 - z_{1k})\, e^{5S_{d2}}\} \,\mathrm{Gr}_d^{0.45-0.762l_1} , \\ \mathrm{Nu}_{d,k=4,5,n} &= \{0.0114\xi_1^{0.6} + [0.053 - 2.36S_{d3}\lg\xi_1]\, S_{d1} \\ &\quad + 0.0016\,(5 - z_{1k})\, \xi_1^{5.813-5.1\xi_1}\} \mathrm{Gr}_d^{0.505-0.99\lg\xi_1} , \end{aligned} \tag{5.4.7}$$

where

$$\begin{aligned} S_{d1} &= (s_1/d - 0.722)\,, & S_{d2} &= (s_1/d - 0.635)\ , \\ S_{d3} &= s_1/d - 0.77, & \xi_1 &= s_1/\,(0.635d)\ . \end{aligned}$$

The average value of the heat transfer coefficient for the bundle is determined as an arithmetic mean:

$$\alpha = \sum_{n,k=1}^{5} \alpha_{dkn}/25\ . \tag{5.4.8}$$

In [5.100] the following dependences are recommended for the average heat transfer coefficient of ***bundles of horizontal tubes***:

$$\mathrm{Pr} = 0.7, \quad \mathrm{Gr}_d = 3.2\cdot 10^3 - 2.3\cdot 10^5, \quad T_{\mathrm{av}} = T_\infty\ ,$$

in-line bundle, $s_1/d > 2, s_2/d \geq 2$

$$\mathrm{Nu} = C_z\left(s_1 s_2/d^2\right)^{0.34} \mathrm{Gr}_d^{0.25} , \tag{5.4.9}$$

for $z_1 = 2 - 6$ $C_z = 0.182 - 0.012(z_1 - 2)$, for $z_1 > 6$ $C_1 = 0.136$,

staggered bundle, $s_1/d > 2, s_2/d > 2.5$

$$\mathrm{Nu} = C_z \left(s_2/d^2\right)^{0.37} \mathrm{Gr}_d^{0.25}, \tag{5.4.10}$$

for $z_1 = 2 - 5$ $C_z = 0.241 - 0.012(z_1 - 2)$, for $z_1 > 5 C_1 = 0.205$.

In [5.102] a mathematical model and a method of calculation of motion of a fluid ***in a bundle with vertical arrangement of tubes*** are suggested. During generation of heat in the tubes, the fluid moves with constant velocity and temperature parallel to the rods. At the outlet the fluid is cooled and moves downwards. The results of calculations coincide well with experiments ***on water*** ($l = 1.232\,\mathrm{m}$, $d = 1.59\,\mathrm{cm}$, $s/d = 1.2 - 2$, $z_1 \times z_2 = 42 \times 42$).

An investigation of free convection of ***bundles in a casing*** was carried out in an application to problems of electrical engineering [5.103], radio engineering [5.100, 5.101], and power engineering [5.104, 5.105].

Inside a horizontal cylinder of diameter d_2 ***there are*** z ***arbitrary located horizontal cylinders of diameter*** d_1 ***with the temperature*** $T_{\mathrm{w}1}$. The average value of heat transfer from the inner cylinder is calculated from the equation [5.103]:

$$\mathrm{Nu}'_{d1} = 2/\ln \frac{\left(1 + 2/\mathrm{Nu}_{d1}\right)^{1/z}}{1 - 2/\mathrm{Nu}_{d2}}, \tag{5.4.11}$$

where Nu_{d1} is given by (3.30) and Nu_{d2} by (4.22), the temperature T_∞ is found from the heat balance equation $(T_\infty - T_{\mathrm{w}2})(T_{\mathrm{w}1} - T_\infty) = z\mathrm{Nu}_{d1}/\mathrm{Nu}_{d2}$. The heat transfer coefficient of the inner cylinder decreases with increase in z. For example, for three cylinders it makes 2/3 of heat transfer of one cylinder in a casing. The volume-average fluid temperature increases with increase in z. The calculation gives low accuracy for $z > 4$ because of the change in the casing temperature.

In three-core oil-filled cable the coefficient of heat transfer from the surface of the core is calculated from the equation

$$\mathrm{Nu} = C\,\mathrm{Ra}_d^{0.25}, \tag{5.4.12}$$

where the coefficient $C = 0.36 - 0.46$ changes depending on the arrangement of the cores in the cable [5.106]. ***Gas-filled three-core cable with cores located over a circle at an angle of 120°.*** The heat flux from the length unit of the cable is [5.107]

$$q_l = k_0 p^{0.6} d_1^{3/4} (T_{\mathrm{w}1} - T_{\mathrm{w}2})^{5/4} \left(\ln \frac{d_2}{2.4 d_1} + 2.2\right)^{-1} \left[1 + 3^{4/5} \left(\frac{d_1}{d_2}\right)^{3/5}\right]^{-5/4},$$

where q_l, W/m; p, Pa; d_i, m; for $\mathbf{SF}_6$ ($p = 0.981 - 6.97{\cdot}10^5$ Pa), $k_0 = 0.247$; for $\mathbf{N}_2$ ($p = 0.334 - 6.97{\cdot}10^5$ Pa), $k_0 = 0.151$.

Inside a horizontal rectangular, hexagonal or cylindrical casing there is a hexagonal packing with different number of rows: 7-rod (central rod and one rows), 19-rod (central rod and two rows), 331-rod (central rod and ten rows). Air, carbonic-acid gas and ***water*** were used as heat carries at a pressure of $0.981 - 6.97 \cdot 10^5$ Pa. The rods of diameter $d_1 = 6, 8, 10\,\mathrm{mm}$ were heated by electric current. The processing of experiment results on gas using the temperature $\theta_{\max} = T_{\max} - T_{w2}$ and characteristic size l equal to the casing perimeter is carried out as

$$\mathrm{Nu}\frac{T_g}{T_{w2}} = f\left(\mathrm{Ra}_l\right) , \tag{5.4.13}$$

$$\frac{s}{d_1} = 1.38, 1.73, 2.3, \quad T_g = \frac{T_{\max} + F_{w2}T_{wc2}/\left(nF_{w1}\right)}{1 + F_{w2}/\left(nF_{w1}\right)} ,$$

$$T_{av} = 0.5\left[T_{\max} + 2F_{w2}T_{w2}/\left(nF_{w1}\right) + T_{w2}\right]/\left[1 + F_{w2}/\left(nF_{w1}\right)\right]$$

as shown in Fig. 5.4.2 [5.108]. ***For water*** ($\mathrm{Ra}_l = 3 \cdot 10^5 - 3 \cdot 10^{10}$) [5.104]

$$\mathrm{Nu} = 0.5\ \mathrm{Ra}_{l\text{Б}}^{0.22} . \tag{5.4.14}$$

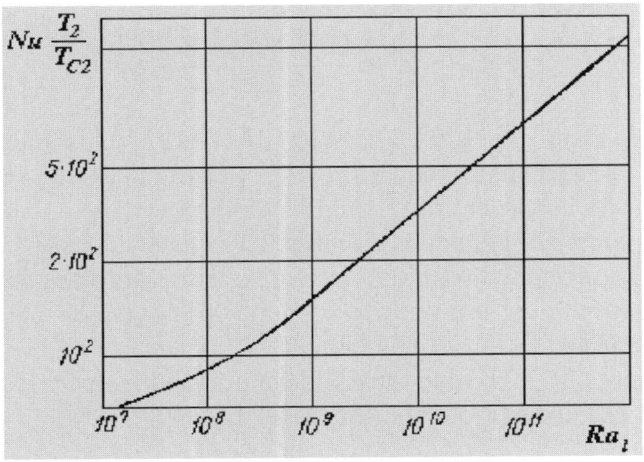

Fig. 5.4.2. Heat transfer of a horizontal bundle of rods in a casing

For $\mathrm{Ra}_l < 6 \cdot 10^6$ the heat is exchanged between the rods and the casing by heat conduction. The dependences suggested allow one to calculate the maximum temperature and the heat carries parameters using the known heat generation in a bundle of horizontal rods with a relative spacing $s/d_1 \geq 1.38$.

In-line and staggered arrays of tube bundles [see (5.4.6), (5.4.7)] in a casing of diameter D. The casing temperature is taken as the determining temperature ($T_{av} = T_{base}$). The average heat transfer for an in-line array of tubes, $\mathrm{Gr}_d = 10^4 - 10^5$ is described by [5.100]

for $D = 350\,\text{mm}$, $s/d = 1.25 - 2$

$$\text{Nu}_{dkn} = [0.088\,(5/z_{1k})^{0.761} - 0.176 S_1 S_4 + 0.416 S_1 S_2 S_4]\text{Gr}_d^{0.25}\,, \qquad (5.4.15)$$

for $D = 610,\ 800\,\text{mm}$, $s/d = 1.25 - 2$

$$\begin{aligned}
\text{Nu}_{d1n} &= \{(0.186 + 0.152\,\text{lg}5/z_{1k}) + [0.32 - 0.0254\ S_{d4}(z_{1K} - 3) \\
&\quad \times (z_{1k} - 4)(z_{1k} - 5)]S_{d2}S_{d3}S_{d4}\}\text{Gr}_d^{0.25}\,, \\
\text{Nu}_{d2n} &= \{[0.145 + 0.48 S_{d2}S_{d3}S_{d4}](5/z_{1k})^{0.45} + 0.55 S_{d1}\} \\
&\quad \times \text{Gr}_d^{0.25}\,, \qquad (5.4.16) \\
\text{Nu}_{d,k\geq 3,n} &= \{0.167 + 0.352\ S_{d2}S_{d3}S_{d4} - (0.044 - 0.044 S_{d1}) \\
&\quad \times (5/z_{1k})^{0.5}\}\text{Gr}_d^{0.25}\,,
\end{aligned}$$

where $S_{d1} = s/d - 1.25$, $S_{d2} = s/d - 1.5$, $S_{d3} = s/d - 1.75$, $S_{d4} = s/d - 2$.

Comparison with experiments shows that the accuracy of formulas (5.4.6), (5.4.7), (5.4.15), and (5.4.16) is within ±15%. ***For a staggered array of 25 tubes in a casing*** the generalizing dependences have a more complex form and a larger error [5.101].

Vertical row of seven tubes located at the corners of an equilateral triangle with a side s. The local and average heat transfer coefficients are [5.105]

$$s/d = 1.1 - \infty,\ \text{Pr} = 0.023,\ \text{Gr}^* = 10^5 - 10^{10},\ T_{\text{av}} = 0.5(T_{\text{w}} + T_\infty),\ \varphi = 0\,,$$

$$\begin{aligned}
\text{Nu}_x &= C_1 \text{Gr}_x^{*m}\,, \qquad (5.4.17) \\
\text{Nu} &= C_2 \text{Gr}_l^{*m}\,.
\end{aligned}$$

The values of the coefficients C_i and m depending on spacing are presented in Table 5.4.1.

Table 5.4.1.

s/d	C_1	C_2	m
∞	0.217	0.287	0.189
1.5	0.305	0.462	0.165
1.3	0.439	0.747	0.147
11	0.898	1.95	0.103

When the spacing decreases, the Nu_x value increases for small Gr_x numbers, and for the range $10^7 - 10^8$ the dependence on spacing is insignificant. For example, for $\text{Gr}_x = 10^5$ the Nu_x number for $s/d = 1.1$ is higher by 54% than for a single cylinder, whereas for $\text{Gr}_x = 10^{10}$, on the contrary, it is 75% lower.

For $s/d = 1.1$ the temperature distribution on the ***cylinder surface*** is substantially nonuniform. With increase in the angle φ, the heat transfer rate decreases ($\varphi = 0 - \pi/6$):

$$\mathrm{Nu}_x = 0.954\,(1 - 0.408\varphi)\,\mathrm{Gr}_x^{*0.0981} \ . \tag{5.4.18}$$

The correction factor $(1 - 0.408\varphi)$ can also be applied for other spacings.

The process of laminar free-convective cooling of a ***staggered bundle of horizontal cylinders*** located inside a fixed volume is investigated numerically, experimentally, and analytically to find the optimum distance between the cylinders corresponding to the maximum heat transfer between the cylinders and ambient medium [5.109]. The optimum distance and maximum free-convective heat removed from the cylinders are generalized by the relations

$$\frac{S_{\mathrm{opt}}}{D} = 2.72\left(\frac{H}{D}\right)^{1/3}\mathrm{Ra}_D^{-1/4} + 0.263 \ ,$$

$$q_{\max} = 0.448\left[\left(\frac{H}{D}\right)^{1/3}\mathrm{Ra}_D^{-1/4}\right]^{-1.6} \ , \tag{5.4.19}$$

where D is the cylinder diameter, H is the bundle height, S is the distance between the cylinders, and q is the relative volumetric density of heat removed from the bundle of cylinders. The patterns of streamlines and isotherms are given in Fig. 5.4.3.

The influence of natural convection on the mass transfer coefficients ***for regular packings and low flow velocities of electrolyte*** has investigated in [5.110]. Correlations for the average mass transfer coefficient were suggested for various geometric configurations of the system:

$$\mathrm{Sh}_L = 0.247\,(\mathrm{Sc}\,\mathrm{Gr}_L)^{0.3} \ , \tag{5.4.20}$$

where $L = DZ/(D\cos\vartheta + Z\sin\vartheta)$, ϑ is the inclination angle with respect to the vertical, D and Z are the width and length of plates, $\mathrm{Gr}_L = g\alpha c_0 L^3/\nu^2$, α is the coefficient of packing, c_0 is the electrolyte concentration in bulk).

The process of heat transfer in the case of steady-state free convection ***for horizontal isothermal cylinders in vertical arrays of two to eight***, at low Ra numbers, was studied experimentally in [5.111]. Heat transfer from the lower cylinder remains the same as that for a single cylinder. However, for other cylinders there may occur both reduction and increase in the heat transfer intensity depending on their mutual position in the array. The results show that there is an optimum spacing between the cylinders which corresponding to the greatest total intensity of heat transfer. A correlation is suggested to calculate the average Nu number for $500 \le \mathrm{Ra} \le 700$, $3.5 \le S/D \le 27.5$, $2 \le n \le 8$

$$\mathrm{Nu}_n = \mathrm{Ra}^{1/4}[0.823 + \exp(-1.5(S/D)^{0.05n})] \ , \tag{5.4.21}$$

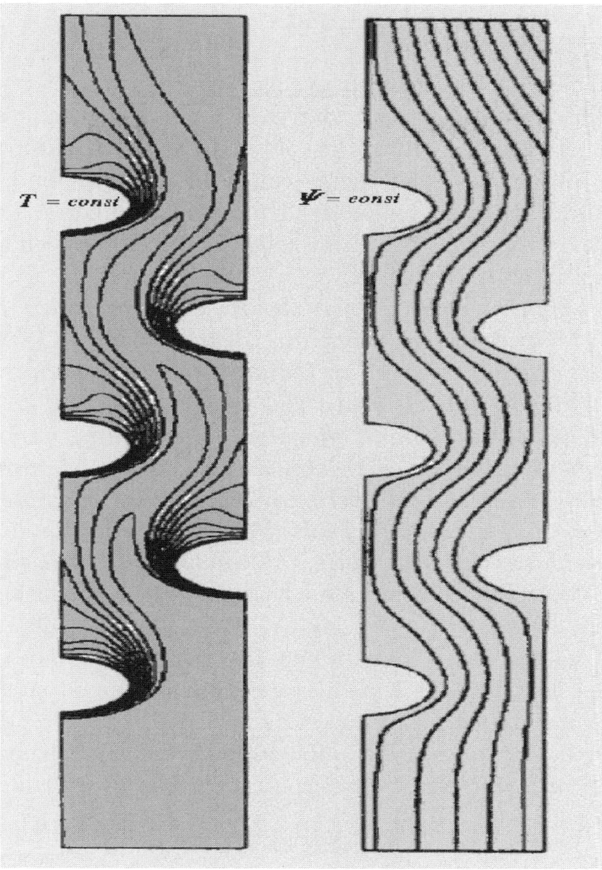

Fig. 5.4.3. Structure of streamlines and isotherms for free-convective cooling of a staggered bundle of horizontal cylinders located inside a fixed volume [5.109]

where S is the spacing between the cylinders, D is the cylinder diameter, n is the number of cylinders in an array.

The results of an experimental and numerical study of natural convection ***around a horizontal tube placed between two isothermal walls of different temperatures*** are presented in [5.112]. The measurement were carried out in the range of Ra numbers: for slot $\mathrm{Ra}_s = 2 \cdot 10^7 - 1 \cdot 10^9$, for tube $\mathrm{Ra}_r = 6 \cdot 10^2 - 9{\cdot}10^2$. The results of investigation show that the hot air near the heated wall entails the layers of hot air from the horizontal tube. Due to the reverse pressure difference and the tendency of the system to equilibrium the layers of the same temperature and density merge. However, this effect depends also on the intensity of heat transfer.

5.5 Panels with Cellular Grids

The main problem in designing solar collectors is the increase of the degree of trapping incident solar energy and the decrease in heat losses from the heat carrier heated by the sun to the cold ambient space. Natural convection in horizontal collectors of solar energy can be effectively suppressed if a thin cellular grid of low thermal conductivity material with correctly chosen sizes and small ratio of the cell width to its height is placed between the absorbing surface and glass coating.

Parallel surfaces heated from below. The distance between the surfaces is H and the inclination to the horizon is φ. Between them there are grids of regular (square, hexagonal or rectangular) shape with sides B and L and thickness δ (see Fig. 5.3.1 e). The cellular grid is usually made of polyethylene film, varnished cardboard or rubberized paper of thickness $\delta = 0.03 - 1.5$ mm [113–118]. Three limiting cases of boundary conditions are possible on the lateral surfaces of the cells: (1) ideal heat conducting lateral walls; (2) ideal insulated lateral walls; (3) negligibly thin lateral walls. The boundary conditions can be characterized by the conjugation parameters $P'_{\text{con1}} = \lambda_{\text{w}}\delta/(\lambda B)$ and $P'_{\text{con2}} = \lambda_{\text{w}} B/(\lambda\delta)$. The first one defines a heat flux in the wall of the cell in the longitudinal direction, the second – in the transverse one. The heat transfer and the value of Ra_x depend on P'_{coni}, but for $P'_{\text{con1}} \ll 1$ (in experiments 0.01 – 0.06) and $P'_{\text{con2}} \gg 1$ (10^4–10^5) this effect can be neglected. For convective motion ***in cells*** the generalizing dependence for calculating heat transfer at quasi-constant values of P'_{coni} is

$$\text{Nu} = f\left(\text{Ra}_H, \text{Pr}, \varphi, B/H, L/H\right) . \qquad (5.5.1)$$

Table 5.5.1 lists the data on the critical number Ra_{Hw} ***in square cells of polyethylene with $\delta = 0.1$ mm*** [5.113]. On origination on convective flow, $\text{Nu} = \text{Nu}_0 + A(\text{Ra} - \text{Ra}_{\text{w}})$ [5.113, 5.114]. The experimental data on heat transfer on development of flow ***in hexagonal and square cells*** are given in Table 5.5.2. Heat transfer ***in a horizontal layer with square cells heated from below***, $\text{Ra} > \text{Ra}_{Hw}$, $\text{Pr} = 0.7$ is [5.113]

$$\text{Nu} = 1 + 0.585\,\text{Ra}_H^{1/3}\left\{1 - \exp\left[-1.19\left(\text{Ra}_H^{1/3}/\text{Ra}_{H_w}^{1/3} - 1\right)\right]\right\} . \qquad (5.5.2)$$

For $\text{Ra}_H > 10^6$ the cells do not exert their effect on heat transfer in the horizontal layer.

Heat transfer ***in inclined square cells of polyethylene and a hexagonal cell of polycarbonate*** with an error of ±7.5% is given by [5.115]

$$H/B \geq 4,\ \varphi = 30 - 90°,\ \text{Ra}_H(B/H)^4 < 6 \cdot 10^3,\ \text{Pr} = 0.7 ,$$

$$\text{Nu} = 1 + 0.89\cos\left(\varphi - \pi/3\right)\left[\text{Ra}_H B^4/\left(2420 H^4\right)\right]^{2.88 - 1.64\sin\varphi} . \qquad (5.5.3)$$

Table 5.5.1.

H, mm	B, mm	H/B	$\mathrm{Ra}_{H\mathrm{w}}\cdot 10^{-4}$	$P'_{\mathrm{con1}}\cdot 10^{-2}$	$P'_{\mathrm{con2}}\cdot 10^{-4}$
6.35	6.35	1	0.58	3	1.33
12.8	6.35	2.01	4.08	3	1.33
12.75	4.25	3	19.29	4.5	0.89
12.75	3.22	3.96	52.9	5.9	0.667
7.65	6.35	1.2	0.809	3	1.33
4.9	6.35	0.771	0.361	3	1.33

Table 5.5.2.

H, mm	B, mm	H/B	Sell Shape	Nu_H	Ra_H
3.17	31.7	1	Hexagonal	$\mathrm{Ra}_H\cdot 1.086\cdot 10^{-4}$	$(1.13–1.4)\cdot 10^4$
22	11	2	–	$\mathrm{Ra}_H\cdot 2.06\cdot 10^{-5}$	$(4.2–4.7)\cdot 10^4$
50.1	24.4	1	Square	$\mathrm{Ra}_H^{1.05}\cdot 6.91\cdot 10^{-4}$	$(8.5–25.5)\cdot 10^4$

If we admit a difference from experiments of 20%, then formula (5.5.3) is applicable also for $H/B = 3$. In the range $\varphi = 0–30°$, linear interpolation between the data given by (5.5.2) and (5.5.3) is possible. ***In hexagonal cells for*** $\varphi = 0$ heat transfer is less intense than in square ones, and for $\varphi > 0$ vice versa. The greatest differences (20%) are observed at high values of Ra_H.

Inclined rectangular cells filled with air with the walls of polyethylene film of thickness $\delta = 0.038$ and 0.15 mm. The average value of heat transfer between the walls with an error of 18% is given by [5.116]

$$H/B = 3 - 10,\ \varphi < 75°, \mathrm{Ra}_H = \mathrm{Ra}_{H\mathrm{w}} - 100\ \mathrm{Ra}_{H\mathrm{w}}\ ,$$

$$\mathrm{Nu} = 1 + 0.131\ \mathrm{Ra}_{H\varphi}^{1/3}[1 - \exp\{-(0.18 + 1.2\cdot 10^{-3}\varphi) \times [(\mathrm{Ra}_H/\mathrm{Ra}_{H\mathrm{w}})^{0.513} - 1]\}]\ , \tag{5.5.4}$$

$$\mathrm{Ra}_{H\mathrm{w}0} = \left(\frac{H}{B}\right)^2\left[768 + 1536\frac{H}{B}P'_{con2} + 144\varepsilon_\mathrm{w}\mathrm{Ki}_H\frac{H}{B}\frac{1-s}{1-s\,(1-\varepsilon_\mathrm{w})}\right],$$

$$s = 1.0102 - 1.4388\frac{B}{H} - 9.4653\left(\frac{B}{H}\right)^2 + 31.44\left(\frac{B}{H}\right)^3 - 27.515\left(\frac{B}{H}\right)^4,$$

$$\mathrm{Ra}_{H\mathrm{w}} = \mathrm{Ra}_{H\mathrm{w}0}/\cos\varphi\ .$$

Inclined rectangular cells filled with silicon oil with walls of cardboard with a polyurethane covering of thickness $\delta = 0.4$ mm. The average coefficient of heat transfer with an error of ±15%, $H/B = 4 - 8$, $L/B = 1 - 24$, is given by [5.117]:

for $\varphi = 0 - 60°$, $\mathrm{Ra}_{H\varphi} < \mathrm{Ra}_{cr}$

$$\mathrm{Nu} = 1 + 1.15\,[1 - \mathrm{Ra}_{Hc1}/\mathrm{Ra}_{H\varphi}] + 1.25\,[1 - \mathrm{Ra}_{Hc1}/\mathrm{Ra}_{H\varphi}]^2\ , \tag{5.5.5}$$

where

$$\mathrm{Ra}_{Hci} = \left(a_i^2 + b_i^2\right)^3 / a_i^2,\ a_i^2 = a_0^2 + b_i^2/2, b_i^2 = (i\pi + 0.85)^2 \ ,$$

$$a_0^2 = \left[\frac{8.5}{1 + 0.25\,(H/B)} + 15\left(\frac{B}{L}\right)^2\right]\left(\frac{H}{B}\right)^2 ,$$

for $\varphi = 90°$

$$\mathrm{Nu} = 1 + \frac{1.1 \cdot 10^{-4}}{1 + 5\,(B/L)^4}\left(\frac{B}{H}\right)^{4.65} \mathrm{Ra}_H^{1.3} \ . \tag{5.5.6}$$

The space between the parallel plates, filled with air, is divided into cells of cardboard of thickness $\delta = 0.18\,\mathrm{mm}$ ***with the angle of partition deviation angle from the normal*** ϑ***. The plates are inclined at an angle*** φ ***to the horizon.*** The average coefficient of heat transfer value with an error of ±7% is [5.118]

$$B/H = 0.5 - 36, L/H = 4 - 16, \mathrm{Ra}_H < 7 \cdot 10^4, \vartheta = 45 - 90° \ ,$$

$$\text{at}\varphi = 90° : \qquad \mathrm{Nu} = C_1 C_2\ \mathrm{Ra}_H^{0.28} \ , \tag{5.5.7}$$

$$\text{at}\varphi \neq 90° : \qquad \mathrm{Nu} = C_3\ \mathrm{Ra}_H^{n} \ . \tag{5.5.8}$$

The values of the coefficients C_i and n are given in Table 5.5.3.

Table 5.5.3.

B/H	0.5	1	2	4
N_1	0.124	0.161	0.166	0.15
$B/H = 0.5 - 4, C_2 = 1 - 0.07\sin(\varphi - \pi/4)$ $B/H = 9 - 36, C_2 = 1 - 0.21\sin(\varphi - \pi/4)$				
$\varphi°$	$\vartheta°$	B/H	C_3	n
60	90	0.5	0.083	0.32
60	60	0.5	0.113	0.28
60	60	1	0.077	0.343
60	60	2	0.137	0.295
45	45	1	0.07	0.34
45	45	1	0.067	0.33

The deviation of the cell partition from the vertical to the plates leads to an appreciable decrease in the heat transfer coefficient. For all the cells of adiabatic partitions the Nu number is 5–35% higher depending on the geometry and the Ra_H number.

In solar collectors V-shaped corrugated cells are also applied. The ***air layer is bounded by a flat plate and a corrugated surface***

(see Fig. 5.3.1f). The expansion angle of corrugation is 60°, the height is H. The average distance between the plates is B. The lower surface is heated. It is shown experimentally that the heat flux is the same ***for the corrugated surface being located both from above and from below***. The average value of heat transfer between the surfaces ***in a cavity at an angle*** φ ***to the horizon*** is [119]

$$B/H = 1 - 4, \varphi = 0 - 60°, \mathrm{Ra}_B = 10 - 4 \cdot 10^6, \mathrm{Pr} = 0.7, T_{\mathrm{av}} = (T_0 + T_B)/2 ,$$

$$\begin{aligned}\mathrm{Nu} = &\left(1 - 0.3025\frac{H}{B} + 0.06825\frac{H^2}{B^2}\right)\frac{B}{H}\ln\frac{2B+H}{2B-H} \\ &+ \frac{2406}{\mathrm{Ra}_{Bc}}\left(1 - 0.195\frac{H}{B} + 5.97\frac{H^2}{B^2} - 4.16\frac{H^3}{B^3}\right) \\ &\times\left(1 - \frac{\mathrm{Ra}_{Bc}}{\mathrm{Ra}_{B\varphi}}\right)\left(1 - \frac{\mathrm{Ra}_{Bc}\,(\sin 1.8\varphi)^{1.6}}{\mathrm{Ra}_{B\varphi}}\right) \\ &+ \left(2.23 - 0.0123\varphi + 0.34 \cdot 10^{-3}\varphi^2\right)\left[\left(\frac{\mathrm{Ra}_{B\varphi}}{\mathrm{Ra}_{Bt}}\right)^{1/3} - 1\right] , \quad (5.5.9)\end{aligned}$$

$$\mathrm{Ra}_{Bc} = 1708\left(1 + 0.036\frac{H}{B} + 2.69\frac{H^2}{B^2} - 1.7\frac{H^3}{B^3}\right) ,$$

The difference between formula (5.5.9) and experiments is 8%. The error increases in a range $\mathrm{Ra}_B = 10^3 - 10^4$ for $B/H = 1$ and reaches 30%.

A trapezoidal solar collector of height 11.5 cm***, base width 7*** cm ***and length 132*** cm ***under different boundary conditions on copper surfaces***. The dividing glass surface is set inside parallel to the base, and the channel itself is located at an angle of φ. The most optimal is the installation of the glass plate at a height of 5–7 cm [5.120].

The three-dimensional problem of laminar natural convection in a honeycomb filled with ***air*** for two cell geometries (***square and triangular cross sections*** with the aspect ratio $L/D = 5$ (L and D are the cell length and width) is solved numerically [5.121]. The effect of rotation of the cell cavity relative to its longitudinal axis and of the inclination of this axis with respect to the horizontal is investigated. The dependence of the average Nu number on Ra is characterized by three ranges with different heat transfer regimes. Typical for $\mathrm{Ra} > 10^6$ is the boundary layer regime, which gradually, when the boundary layer extends, fills the whole cavity volume as the Ra number decreases. For small $\mathrm{Ra} \leq 10^4$, the heat transfer mechanism is fully determined by thermal conductivity, for which the velocity field affects insignificantly the temperature field.

The heat transfer due to natural convection and thermal conductivity ***in an inclined rectangular cavity with a honeycomb adjoining a heated wall*** is investigated numerically in [5.122]. The honeycomb is heated by a ***constant heat flux***, the height to length ratio is $2.5 \leq A \leq 10$, the range of Rayleigh numbers is $10^4 \leq \mathrm{Ra} \leq 10^8$, the range of inclination angles is $5 \leq \varphi \leq 90°$, the ratio of the partition thickness to the cavity height is $e = 0.08$, $\mathrm{Pr} = 0.72$, the ratio of the thermal conductivity coefficients is $10^{-3} \leq \lambda_r \leq 10^5$ ($\lambda_r = \lambda_w/\lambda_f$, where λ_w, λ_f are the thermal conductivity coefficients for wall and fluid). The average heat transfer coefficient increases with increase in the partition height and cavity aspect ratio, especially for low values of this ratio, and decreases with increase in the wall thickness and the ratio of partition thickness to its height (which changes from 0.25 to 2.5), which is also appreciable at small values of the indicated parameters. The effect of the inclination angle on heat transfer in the cavity in the investigated range is insignificant.

References

1. W. Elenbaas, *Physica*, **9**, 865 (1942).
2. T. Aihara, *Trans. JSME*, **29**, 903 (1963).
3. E. K. Levy, P. A. Eichen, W. R. Cintam, and R. R. Shaw, *Trans. ASME, Ser. C, J. Heat Transfer*, **97**, 474 (1975).
4. G. D. Raithby and K. G. T. Hollands, *Adv.Heat Transfer*, **11**, 265 (1975).
5. E. M. Sparrow and P. A. Bahrami, *Trans. ASME, Ser. C, J. Heat Transfer*, **102**, 221 (1980).
6. E. M. Sparrow and C. Prakash, *Trans. ASME, Ser. C, J. Heat Transfer*, **102**, 215 (1980).
7. E. M. Sparrow and C. Prakash, *NHT*, **3**, 89 (1980).
8. P. E. Azaryan, *Izv. VUZov, Energetika*, No. 3, 136 (1978).
9. H. Akbari and T. R. Borgers, *Solar Energy*, **22**, 165 (1979).
10. H. R. Nagendra and M. A. Tirunarayanan, *Nucl. Eng. Des.*, **15**, 17 (1971).
11. R. G. Colwell and J. R. Welty, *Trans. ASME, Ser. C, J. Heat Transfer*, **96**, 448 (1974).
12. S. P. Kharlamova and E. M. Bondarenko, *Elektrovozostroenye* (Novocherkassk), **15**, 92 (1974).
13. O. Miyatake and T. Fujii, *Repts. Res. Inst. Ind. Sci. Kyushu Univ.*, No. 62, 87 (1975).
14. A. N. Nikiforov and V. G. Ushakov, in: *Heat and Gas Supply and Heat and Mass Transfer*, Rostov-na-Donu (1977), pp. 93–95.
15. S. A. M. Said and R. J. Krane, *Int. J. Heat Mass Transfer*, **33**, 1121 (1990).
16. K. Vafai and J. Ettefagh, *Int. J. Heat Mass Transfer*, **33**, 2329 (1990).
17. T.-Y. Lin and S.-S. Hsieh, *Int. J. Heat Mass Transfer*, **33**, 2295 (1990).
18. B. G. Dorofeev, A. I. Bondarenko, V. I. Kolomiitsev, and V. A. Uskov, *Elektrovozostroenye* (Novocherkassk), **15**, 177 (1974).
19. M. K.Chung, H. C. Yun and R. J. Adrian, *Int. J. Heat Mass Transfer*, **35**, 43 (1992).
20. R. H. Kraichann, *Phys. Fluids*, **5** 1374 (1962).
21. A. Bejan and C. L. Tien, *Trans. ASME, Ser. C, J. Heat Transfer*, **100**, 725 (1978).
22. I. Kozhukharov, *Nauchn. Tr. Inst. Mashinostr. Mekhaniz. i Elektrofik. Selsk. Stop.*, **17**, *Ser.* 9, 45 (1975), Ruse.

23. P. Y. Chang, F. C. Chou, and C. W. Tung, *Int. J. Heat Mass Transfer*, **41** 3841 (1998).
24. T. Aihara, *Repts. Res. Inst. Sci. Tohoku Univ., Ser. B,* part 1, **21** (1969/70), pp. 105–134; part 2, **21** (1969/70), pp. 135-162; part 3, **21** (1969/70), pp. 163–187; part 4, **23** (1971), pp. 125–156.
25. H. R. Jacobs and W. E. Jr. Masson, *Proc. 1976 Heat Transfer and Fluid Mech. Inst.*, Stanford (1976), pp. 33–46.
26. H. R. Jacobs, W. E. Jr. Masson, and E. T. Nikida, *Proc.* 5th *Ind. Heat Transfer Conf.*, vol. 3, Tokyo (1974), pp. 89–93.
27. R. Piva and P. Orlandy, in: *Numerical Solutions in Hydrodynamics*, Moscow (1977), pp. 127–134.
28. K. B. Katsaros, W. T. Lin, J. A. Businger, and J. E. Tillman, *J. Fluid Mech.*, **83**, 311 (1977).
29. M. Kageyama and R. Izumi, *Trans. JSME, Ser. B*, **45**, 682 (1979).
30. D. M. Sefcik, B. W. Webb, and H. S. Heaton, *Int. J. Heat Mass Transfer*, **34**, 3037 (1991).
31. B. Morrone, A. Campo, and O. Manca, *Int. J. Heat Mass Transfer*, **40**, 993 (1997).
32. K.-T. Lee, *Int. J. Heat Mass Transfer*, **42**, 4523 (1999).
33. A. G. Fedorov and R. Viskanta, *Int. J. Heat Mass Transfer*, **40**, 3849 (1997).
34. A. L. Pica, G. Rodono, and R. Volpes, *Int. J. Heat Mass Transfer*, **36**, 611 (1993).
35. T. Tsubouchi, H. Masuda, *Repts. Res. Inst. Sci. Tahoku Univ. High Speed Mech., Ser. B*, Rep. 1, **20**, (1968/69), pp. 57–82; Rep. 2, **23**, (1971), pp. 21–59; Rep. 3, **25**, (1973), pp. 143–173.
36. C. R. Hedlund and P. M. Ligrani, *Int. J. Heat Mass Transfer*, **41**, 563 (1998).
37. M. Kageyama and R. Izumi, *Bull. JSME*, **13**, 382 (1970).
38. J. R. Dyer, *Int. J. Heat Mass Transfer*, **21**, 1341 (1978).
39. M. Kaptie and S. Musse, in: *Heat and Mass Transfer*, vol. 1, Moscow (1968), pp. 686–695.
40. G. N. Tkachenko, *Gidromekhanika*, No. 9, 137 (1976).
41. G. S. H.Lock and J. D. Kirchner, *Int. J. Heat Mass Transfer*, **35**, 165 (1992).
42. J. Fleming and M. R. Amin, *Int. J. Heat Mass Transfer*, **39**, 51 (1996).
43. J. Fleming and M. R. Amin, *Int. J. Heat Mass Transfer*, **39**, 49 (1996).
44. L. M. Zysina-Molozhen, L. V. Zysin, and M. P. Polyak, *Heat Transfer in Turbines*, Mashinostroenie Press, Leningrad (1974).
45. H. R. Nagendra, N. A. Tirunarayanan, and A. Ramachandran, *Nucl. Eng. Des.*, **15**, 1 (1971).
46. B. B. Rogers and L. S. Yao, *Int. J. Heat Mass Transfer*, **36**, 35 (1993).
47. A. K. Mohanty and M. R. Dubey, *Int. J. Heat Mass Transfer*, **39**, 2087 (1996).
48. A. W. Stevens, *Int. J. Heat Mass Transfer*, **39**, 3749 (1996).
49. J. F. Lafortune and H. Meneley, *Int. J. Heat Mass Transfer*, **33**, 431 (1990).
50. J. S. Yoo, *Int. J. Heat Mass Transfer*, **42**, 709 (1999).
51. W. M. Yan and T. F. Lin, *Int. J. Heat Mass Transfer*, **34**, 291 (1991).
52. P. J. Heggs and D. B. Ingham, D. J. Keen, *Int. J. Heat Mass Transfer*, **33**, 517 (1990).
53. A. Bejan and C. L. Tien, *Int. J. Heat Mass Transfer*, **21**, 701 (1978).
54. B.-X. Wang and X. Zhang, *Int. J. Heat Mass Transfer*, **33**, 827 (1990).
55. G.-C. Chen, G.-H. Chou, and C.-K. Hsien, *Int. J. Heat Mass Transfer*, **39**, 193 (1996).

56. M. Al-Arabi and M. M. El-Reface, *Int. J. Heat Mass Transfer*, **21**, 357 (1978).
57. R. M. Miller and B. Gebhart, *Int. J. Heat Mass Transfer*, **21**, 1229 (1978).
58. F. Harahap and H. N. Jr. McManus, *Trans. ASME, Ser. C, J. Heat Transfer*, **89**, 32 (1967).
59. C. D. Jones and L. F. Smith, *Trans. ASME, Ser. C, J. Heat Transfer*, **92**, 6 (1970).
60. J. R. Welling and C. B. Wooldridge, *Trans. ASME, Ser. C, J. Heat Transfer*, **87**, 439 (1965).
61. D. N. Van de Pol and J. K. Tierney, *Trans. ASME, Ser. C, J. Heat Transfer*, **95**, 542 (1973).
62. N. K. Sane and S. P. Sukhatme, *Proc. 5*th *Int. Heat Transfer Conf.*, vol. 3, Tokyo (1974), pp. 114–118.
63. C. E. Kwak and T. H. Song, *Int. J. Heat Mass Transfer*, **43**, 825 (2000).
64. J. B. Chaddok, *ASHRAE J.*, **7**, 53 (1970).
65. V. A. Sotchenko, in: *Investigation of Heat and Mass Transfer Processes*, Kiev (1979), pp. 53–58.
66. V. A. Semenyuk, *Vopr. RE., Ser. TRTO*, No. 3, 44 (1973).
67. V. A. Sotchenko, *Sanitarnaya Tekhnika*, **16**, 51 (1976).
68. A. Bar-Cohen, *Trans. ASME, Ser. C, J. Heat Transfer*, **101**, 564 (1979).
69. I. Sezai and A. A. Mohamad, *Int. J. Heat Mass Transfer*, **42**, 2041 (1999).
70. T. Aihara, S. Maruyama, and S. Kobayakawa, *Int. J. Heat Mass Transfer*, **33**, 1223 (1990).
71. R. D. Flack, *Int. J. Mech. Eng. Educ.*, **8**, 89 (1980).
72. E. I. Udler, *Investigation of Free-Convective Heat Transfer of Electromagnetic Backward-Translatory Engines*, Abstract of Doctoral Dissertation (in Technical Sciences), Tomsk (1972).
73. R. Kumar, *Int. J. Heat Mass Transfer*, **40**, 3323 (1997).
74. S. Acharya and A. Mehrotra, *Int. J. Heat Mass Transfer*, **36**, 236 (1993).
75. A. F. Kalichikhin, *Vestnik Elektropromyshlennosti*, No. 7, 31 (1962).
76. W. E. Steward, *Int. J. Heat Mass Transfer*, **14**, 1013 (1971).
77. M. K. El-Ridy, I. G. Chumak, and L. G. Kalinin, *Kholod. Tekh.*, No. 5, 30 (1975).
78. J. G. Knudsen and R. B. Pan, *Chem. Eng. Prog. Symp. Ser.*, **61**, 44 (1965).
79. E. M. Sparrow and P. A. Bahrami, *Int. J. Heat Mass Transfer*, **23**, 1555 (1980).
80. A. A. Nicol and G. B. Babiy, *Can. J. Chem. Eng.*, **45**, 382 (1967).
81. J. Zelazny and G. Kulesza, *Zesz. Polit. Lodzkiej*, No. 306, 83 (1978).
82. S. Nevenkin, *Nauch. Tr. Toplotekh. Vysh. Tekh. Uchebny Zaved.*, vol. 2, (1974) pp. 79–83.
83. M. A. Toporkova, V. B. Kuntysh, and A. E. Piir, *Izv. VUZov, Energetika*, Dep. in NIIEinformenergomash 7 Sept. 1978, No. 29–9P.
84. N. P. Shcheglov, *Holod. Tekh.*, No. 7, 32 (1978).
85. V. I. Efremov, N. K. Lenskaya, and L. P. Proshkina, *Vopr. RE, Ser. TRTO*, No. 2, 42 (1978).
86. V. M. Legkii, D. N. Glushkova, and Yu. K. Tupitsin, *Vopr. RE, Ser. TRTO*, No. 2 (1978).
87. V. M. Legkii and Yu. K. Tupitsyn, in: *Heat Transfer and Hydrodynamics*, Kiev (1977), pp. 189–194.
88. A. P. Ornatskii, B. V. Latenko, Yu. S. Popel, *Teplofizika Teplotekhnika*, No. 23, 53 (1973).

89. G. P. Polyachek, T. V. Leont'ev, and V. A. Shutov, *Collection of Papers of Moscow Institute of Electronic Engineering*, No. 26, 194 (1976).
90. V. B. Gidalevich, V. F. Davydov, and V. N. Meshkov, *Vopr. RE, Ser. TRTO*, No. 3, 20 (1979).
91. B. P. Basylev, V. I. Efimov, V. B. Kalyakin, and N. V. Kachilina, *Vopr. RE, Ser. TRTO*, No. 1, 88 (1980).
92. N. V. Mukhina, *Prikl. Mekh. Tekh. Fiz.*, No. 4, 42 (1979).
93. C. B. Lamm, *J. Inst. Petrol.*, **57**, 278 (1971).
94. J. Lieberman and B. Gebhart, *Int. J. Heat Mass Transfer*, **12**, 1385 (1969).
95. G. F. Marsters, *Int. J. Heat Mass Transfer*, **15**, 921 (1972).
96. E. S. Kurylev and V. I. Machulin, in: *Cooling Devices and Equipment*, Leningrad (1976), pp. 88–94.
97. Yu. A. Zagromov and Yu. A. Korolenko, *Izv. Tomskogo Politekh. Inst.*, **137**, 52 (1965).
98. Yu. A. Korolenko, *Izv. Tomskogo Politekh. Inst.*, **110**, 26 (1962).
99. F. P. Inkropera and M. A. Yaghoubi, *Trans. ASME, Ser. C, J. Heat Transfer*, **101**, 743 (1979).
100. P. P. Shilonosov, A. S. Lyalikov, and G. I. Yuzefovich, in: *Hydrodynamics of Twisted Flows and Dynamics of Shock*, Kemerovo (1970), pp. 93–97.
101. P. P. Shilonosov, A. S. Lyalikov, and G. I. Yuzefovich, in: *Hydrodynamics of Twisted Flows and Dynamics of Shock*, Kemerovo (1970), pp. 98–102.
102. L. P. Davis and J. J. Perona, *Int. J. Heat Mass Transfer*, **16**, 1425 (1973).
103. T. H. Kuehn and R. J. Goldshtein, *Int. J. Heat Mass Transfer*, **19**, 1127 (1976).
104. M. A. Gotovskii, Yu. A. Sokovishin, and V. N. Fromzel, *Tr. CKTI*, No. 142, 27 (1977).
105. J. C. Dutton and J. R. Welty, *Trans. ASME, Ser. C, J. Heat Transfer*, **97**, 372 (1975).
106. R. S. Abdulhadi and J. C. Chato, *Trans. ASME, Ser. C, J. Heat Transfer*, **100**, 37 (1978).
107. K. Itaka, T. Arali, and T. Hara, *IEEE Trans., Power Appar. System*, **97**, 1579 (1975).
108. P. A. Andreev, V. M. Fromzel, T. A. Pervitskaya, and M. A. Gotovskii, in: *Heat Transfer and Hydrodynamics of Single-Phase Flow in Rod Bundles*, Leningrad (1979), pp. 121–129.
109. A. Bejan, A. J. Fowler, and G. Stanescu, *Int. J. Heat Mass Transfer*, **38**, 2047 (1995).
110. M. D. C. Rey, P. Glasserman, and U. Bohm, *Int. J. Heat Mass Transfer*, **41**, 1693 (1998).
111. M. S. Sadeghipour and M. Asheghi, *Int. J. Heat Mass Transfer*, **37**, 103 (1994).
112. W. T. Konka, *Int. J. Heat Mass Transfer*, **43**, 447 (2000).
113. K. G. T. Hollands, *Trans. ASME, Ser. C, J. Heat Transfer*, **95**, 439 (1973).
114. A. C. Kent and J. H. Bowyer, *IEC Fund.*, **11**, 319 (1972).
115. R. F. D. Cane, K. G. T. Hollands, and G. D. Raithby, *Trans. ASME, Ser. C, J. Heat Transfer*, **97**, 86 (1977).
116. D. R. Smart, K. G. T. Hollands, and G. D. Raithby, *Trans. ASME, Ser. C, J. Heat Transfer*, **102**, 75 (1980).
117. D. K. Edwards, J. N. Arnold, and P. S. Wu, *Trans. ASME, Ser. C, J. Heat Transfer*, **101**, 741 (1979).
118. B. A. Meyer, J. W. Mitchel, and M. M. El-Wakill, *Trans. ASME, J. Heat Transfer*, **101C**, 655 (1979).

119. S. M. El-Sherbiny, K. G. T. Hollands, and G. D. Raithby, *Trans. ASME, Ser. C, J. Heat Transfer*, **100**, 410 (1978).
120. R. B. Bannero and H. Nabavi, *AIAA Paper*, No. 896, 1 (1978).
121. G. S. H. Lock and L. Zhao, *Int. J. Heat Mass Transfer*, **35**, 155 (1992).
122. E. K. Lakhal, E. Bilgen, and P. Vasseur, *Int. J. Heat Mass Transfer*, **38**, 1397 (1995).

6

Nonstationary Processes of Free Convection

6.1 Main Dependences for Calculation of Unsteady Free Convection

Occurrence, development, and transient processes in free convection are connected with nonstationarity. On occurrence of a temperature difference between a body and the ambient medium, it is possible to distinguish three heat transfer mechanisms. At the initial moment, the thermal conductivity process is dominating, then there is a transient zone with minimal heat transfer coefficient, and after quite a long time interval a stationary free convection developes. When the unambiguity conditions change, the transient phenomena are possible.

The nonstationary processes are the processes of cooling and heating of a closed cavity at constant values of temperature and flow characteristics on the surface. Under the conditions of an oscillating flow the free-convective heat transfer is also nonstationary, and its rate depends on oscillation parameters. In analyzing of thermal interaction of the oscillating flow with the surface, high-frequency and low-frequency oscillations are distinguished.

Self-similar solutions of the plane unsteady free-convective boundary layer are possible, when the surface temperature is proportional to $(x + a\tau)^{-1}$ and $(x+a)/(1-b\tau)^2$ [6.1]. In the case of large values of x, when the temperature and the heat flux depend only on time (regime of "parallel" flow), the problem can be solved if the temperature is proportional to $(1 + a\tau)^{p}$ or $\exp(p\tau)$ [6.2, 6.3].

For instantaneous change of the wall temperature it is assumed that the process at each point consists of three stages [6.4]. At the first stage the heat is transferred by heat conduction, the velocity and temperature distributions do not depend on x, and the convective heat transfer did not begin as yet. At the third stage, the heat is transferred by convection, the distributions of velocities and temperatures do not depend on time. The second intermediate stage is transient, here the leading edge affects the process. The boundary between the three stages is determined by the lines of the characteristics.

The thermal conductivity stage is accomplished at the time

$$\mathrm{Zh}_x = 1.8(1.5 + \mathrm{Pr})^{1/2}\mathrm{Gr}_x^{-1/2}\,, \tag{6.1.1}$$

and the stationary regime is achieved after the time interval

$$\mathrm{Zh}_x = 5.25(0.952 + \mathrm{Pr})^{1/2}\mathrm{Gr}_x^{-1/2}\,. \tag{6.1.2}$$

The heat transfer coefficient corresponding to the heat conduction is

for constant wall temperature

$$\frac{\mathrm{Nu}_x}{\mathrm{Gr}_x^{1/4}} = \frac{1}{\sqrt{\pi}}\left(\frac{\mathrm{Pr}^2}{\mathrm{Zh}_x^2\mathrm{Gr}_x}\right)^{1/4}\,, \tag{6.1.3}$$

for constant heat flux

$$\frac{\mathrm{Nu}_x}{\mathrm{Gr}_x^{*1/5}} = \frac{\sqrt{\pi}}{2}\left(\frac{\mathrm{Pr}^{5/2}}{\mathrm{Zh}_x^{5/2}\mathrm{Gr}_x^*}\right)^{1/5}\,. \tag{6.1.4}$$

For power-law and exponentially increasing wall temperature [6.3]:

$$\begin{aligned}
\theta_\mathrm{w} = \theta_0 e^{m\tau}:\quad & \frac{\mathrm{Nu}_x}{\mathrm{Gr}_x^{1/4}} = \left(\frac{\mathrm{Pr}^2}{\mathrm{Zh}_x^2\mathrm{Gr}_x}\right)^{1/4}(m\tau)^{1/2} \\
& \times\left\{\mathrm{erf}\sqrt{m\tau} + \frac{1}{\sqrt{\pi m\tau}}\exp\left(-m^2\tau^2\right)\right\}\,, \\
q_\mathrm{w} = q_0\tau^n:\quad & \frac{\mathrm{Nu}_x}{\mathrm{Gr}_x^{*1/5}} = \frac{\Gamma(n+3/2)}{\Gamma(n+1)}\left(\frac{\mathrm{Pr}^{5/2}}{\mathrm{Zh}_x^{5/2}\mathrm{Gr}_x^*}\right)^{1/5}\,, \\
q_\mathrm{w} = q_0 e^{m\tau}:\quad & \frac{\mathrm{Nu}_x}{\mathrm{Gr}_x^{*1/5}} = \left(\frac{\mathrm{Pr}^{5/2}}{\mathrm{Zh}_x^{5/2}\mathrm{Gr}_x^*}\right)^{1/5}\frac{\sqrt{m\tau}}{\mathrm{erf}\,(m\tau)}\,.
\end{aligned} \tag{6.1.5}$$

The transient free-convective flow occurring ***in a boundary layer near a vertical plane surface due to a sudden surface temperature change*** was investigated numerically in [6.5]. The flow characteristics are determined in the main by two parameters: the temperatures before and after the temperature change T_1 and T_2, the ambient temperature T_∞, and by the Prandtl number Pr. The solution obtained is applicable for small times. For the case $\mathrm{Pr} = 1$ the results are presented in a wide range of the parameter R.

The analytical and numerical investigations of the development of flow and temperature field ***in a boundary free-convective layer on an infinite vertical wall induced by a sudden temperature change*** shows the existence of two mechanisms of arrival to a quasi-stationary regime depending on the ratio between the Rayleigh and Prandtl numbers: oscillatory

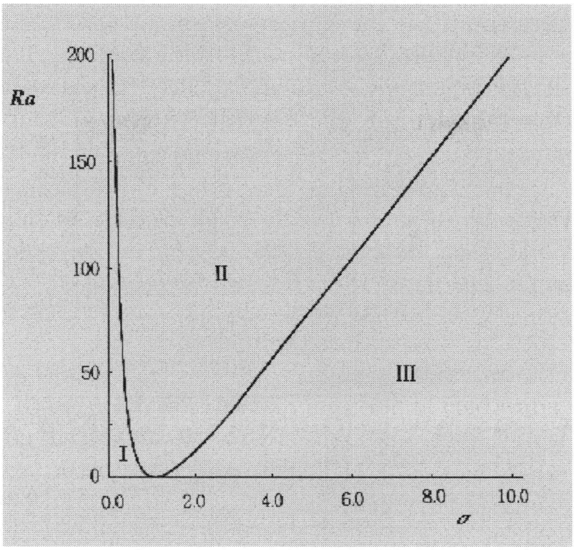

Fig. 6.1.1. The ranges of the parameters of flow in a boundary free-convective layer on an infinite vertical wall corresponding to different regimes of the nonstationary process [6.6]

and laminar [6.6]. Figure 6.1.1 clearly depicts the areas of change in the flow parameters corresponding to different regimes of the nonstationary process. The boundary line on the diagram Ra – Pr is determined by the equation $\mathrm{Ra} = (\mathrm{Pr}-1)^2\pi^4/4\,\mathrm{Pr}$. The areas I and III correspond to the nonoscillatory transition mechanism. In regime I (low values of Pr) the heat conduction mechanism prevails, in regime III this is convective mechanism. Region II (average values of Pr) corresponds to the flow when the magnitudes of horizontal heat conduction and of vertically directed convective flow are of the same order, which leads to the conditions of oscillatory flow. As is seen from the figure, as Ra increases, the area of Pr values, for which the oscillatory process is possible, is extended. The flow characteristics and the temperature field in a nonstationary free-convective boundary layer appear to be strongly dependent on Pr, unlike the quasi-steady regime, whose main parameters practically do not depend on it.

Instantaneous temperature change and ***variable thermophysical characteristics of flow*** [6.7]:

$$\rho T = \rho_\infty T_\infty, \quad \mu T_\infty / (\mu_\infty T) = 1 + \varepsilon\theta, \quad \mathrm{Pr} = 1, \quad \varepsilon \ll 1, \quad T_{\mathrm{av}} = T_\infty \,,$$

$$\frac{\mathrm{Nu}_x}{\mathrm{Gr}_x^{1/4}} = \frac{1}{\left(\mathrm{Zh}_x^2 \mathrm{Gr}_x\right)^{1/4}} \left(0.564 - 0.384\varepsilon + 0.345\varepsilon^2\right) . \qquad (6.1.6)$$

For a semi-infinite plate with variable wall temperature or heat flux it is possible to estimate the time needed for the influence of the leading edge to reach the assigned point on the surface and for the heat conduction regime to cease its existence. The influence of the leading edge is distributed together with the establishment of the velocity field for the regime of a parallel flow. The distance to which this effect is distribute for the time τ is equal to $\int_0^\tau u d\tau$; its maximum value in the case of instantaneous change in the heat flux is $x_{\max} = 0.0579 g\beta q(\tau^5/\lambda\rho c)^{1/2}$. The local values of heat transfer at the point $x = x_{\max}$ ***for the case of an instantaneous change in the temperature and for the power-law change in the heat flux*** are compared with the stationary value for $\tau \to \infty$ in Table 6.1.1.

Table 6.1.1. $\theta = \theta_{\mathrm{w}}$ for $\tau \geq 0$

Pr	$(\mathrm{Nu}_x/\mathrm{Gr}_x^{1/4})_{\tau\to\infty}$	$(\mathrm{Nu}_x/\mathrm{Gr}_x^{1/4})_{x=x_{\max}}$
0.01	0.0574	0.0438
1	0.401	0.315
10	0.827	0.679
100	1.549	1.368

$q_{\mathrm{w}} = q_0\tau^n$ for $\tau \geq 0$

Pr	$(\mathrm{Nu}_x/\mathrm{Gr}_x^{*1/5})_{\tau\to\infty}$	$(\mathrm{Nu}_x/\mathrm{Gr}_x^{*1/5})_{x=x_{\max}}$			
		$n = 0$	$n = 0.5$	$n = 1$	$n = 2$
0.01	0.263	0.238	0.263	0.277	0.294
1	0.534	0.501	0.553	0.584	0.620
10	0.944	0.948	1.045	1.104	1.172
100	1.556	1.641	1.811	1.914	2.032

The temperature burst decreases with increase of Pr; this burst is higher for the ***stepwise temperature change*** than for the ***stepwise increase in the heat flux***. The time of termination of the stage of pure heat conduction for air is $\tau = 2.97(x/g\beta\theta_{\mathrm{w}})^{1/2}$. According to formula (6.1.4) the value of the factor for τ is equal to 2.7 [6.3].

The heat transfer coefficient in the regime of heat conduction is the function of time; it decreases quickly from the initial value equal to infinity to the stationary value. By the time corresponding to the termination of this regime, the coefficient α_x decreases to a some what smaller value in comparison with the stationary regime. The local heat transfer coefficient proved to pass through the minimum shortly before the establishment of the stationary state.

The generalizing dependences of heat transfer for laminar free convection are

instantaneous change in the temperature, Pr = 0.733 [6.8]:

$$\frac{\mathrm{Nu}_x}{\mathrm{Gr}_x^{1/4}} = \frac{1}{\sqrt{\pi}}\left(\frac{\mathrm{Pr}}{\mathrm{Zh}_x^2 \mathrm{Ra}_x}\right)^{1/4} \times \left\{1 + \left(\frac{0.395\sqrt{\pi}\mathrm{Zh}_x^{1/2}\mathrm{Ra}_x^{1/4}}{\mathrm{Pr}^{1/2}\left[1+\left(2.3\,\mathrm{Pr}/\left(\mathrm{Zh}_x \mathrm{Ra}_x^{1/2}\right)\right)^8\right]}\right)^{25}\right\}^{1/25} \quad (6.1.7)$$

instantaneous change in the heat flux, $\mathrm{Pr} > 0.01$ [6.9]:

$$\mathrm{Nu}_x^6 = \left(\frac{\pi}{4}\frac{\mathrm{Pr}}{\mathrm{Zh}_x}\right)^3 + \left(\frac{\mathrm{Ra}_x/10}{\left[1+(0.437/\mathrm{Pr})^{9/16}\right]^{16/9}}\right)^{3/2} . \quad (6.1.8)$$

It was shown by numerical calculations that ***in the heat conduction regime*** [(6.1.4)–(6.1.5)] $T_{\mathrm{av}} = T_\infty + 0.78(T_{\mathrm{w}} - T_\infty)$, $(T_{\mathrm{w}} - T_\infty)/T_\infty = (-0.55) - 3ИT_{\mathrm{av}} = T_\infty + 0.73(T_{\mathrm{w}} - T_\infty)$, $(T_{\mathrm{w}} - T_\infty)/T_\infty = (-0.3) - 3$ for stationary free convection with an error in the heat transfer coefficient not higher than 0.1%.

When $T_{\mathrm{av}} = (T_{\mathrm{w}} + T_\infty)/2$, the calculations by formulas (6.1.7) and (6.1.8) are applicable ***for real media*** with an error of ±0.5% [6.10].

The numerical analysis of the mutual effect of heat conduction and thermodiffusion in the process of transient and steady free convection of a ***binary mixture of air and*** **He** ***near a vertical plate*** carried out in [6.11] showed that in the transient regime this interaction is insignificant. In the stationary regime a decrease in the total dimensionless heat flux on the wall is possible with increase in the dimensionless temperature due to the strong influence of thermal diffusion.

Vertical thin plate ($\mathrm{Bi}_\delta \ll 1$) ***previously heated up to a temperature of 100–120°C by a short-term supply of an electrical current is cooled in an infinite surrounding medium.*** The experimental data with an error of ±15% are generalized by the dependence [6.12]

$$\mathrm{Pr}_0 = 0.7 - 200, \mathrm{Fo}_\delta = 5 - 5.9\cdot 10^4, \mathrm{Gr}_{l0} = 1.8\cdot 10^7 - 10^9 ,$$
$$\lambda_\infty/\lambda_{\mathrm{w}} = 2.1\cdot 10^4 - 3.8\cdot 10^{-2}, T_{\mathrm{av}} = (T_{\mathrm{w}} + T_\infty)/2 ,$$
$$\frac{\mathrm{Nu}}{\mathrm{Nu}_0} = 0.07 + 104\left(\mathrm{Gr}_{l0}\,\mathrm{Pr}_0^{0.5}\right)^{-0.3}\left(\mathrm{Fo}_\delta \frac{\lambda_\infty}{\lambda_{\mathrm{w}}}\frac{\delta}{l}\right)^{-0.15} . \quad (6.1.9)$$

Horizontal heated surface facing upwards on instantaneous heat flux supply q_0. The change in heat transfer from the surface ***in the initial period of development of natural convection*** is given by [6.13]

$$\text{at} \quad \mathrm{Pr} = 1 \quad \frac{q}{q_0} = \frac{P_{\mathrm{con3}}^{1/2}}{1+P_{\mathrm{con3}}^{1/2}}\left[1 + 0.00137\frac{P_{\mathrm{con3}}^{1/2}}{1+P_{\mathrm{con3}}^{1/2}}\mathrm{Gr}_x^*\mathrm{Zh}_x^2\right] . \quad (6.1.10)$$

An analytical and numerical analysis of nonstationary natural convection ***above an inclined moist flat plate in an ambient air*** was carried out in [6.14] on the basis of division into two regions: a boundary layer and a capillary-porous plate. The systems of equations for both regions are coupled by the heat and mass balance on the wall, from which the local and average Nusselt and Sherwood numbers are determined.

The numerical solution of the problem of nonstationary natural convection ***near heat generating vertical cylinders with different radii and heat capacity coefficients*** [6.15] shows that the degree of the influence of the leading edge on the flow structure increases with decrease in the cylinder radius and heat capacity and also with increase in the heat flux on its surface. The thickness of the nonstationary boundary layer exceeds its stationary value, and the value of the nonstationary average heat transfer coefficient attains the minimum which is 53% of the stationary value for the maximum modified Grashof number. The results of numerical calculation of the average heat transfer coefficient for cylinders of different radii ($R_0 = r_0/(\lambda\nu^2/g\beta q)^{1/4}$, where r_0 is the cylinder radius) are given in Table 6.1.2.

Table 6.1.2. The Values of the Average Heat Transfer Coefficient for Cylinders of Different Radii

$\mathrm{Nu}/\mathrm{Gr}_L$	0.745	0.440	0.320
R_0	0.5	1.5	4.0

The flow and heat transfer by transient natural convection ***in a vertical cylinder filled with a fluid-saturated porous medium*** was analyzed numerically in [6.16]. ***The cylinder open from both ends was heated by a constant wall heat flux***. This study is based on the Forshheimer-extended Darcy flow model.

The heat flux is supplied to a vertical cylinder with $\partial T_\mathrm{w}/\partial\tau = 0 - 15\,\mathrm{deg/sec}$. The effect of the nonstationarity is determined with the aid of the parameter k_{T1}. The experimental data on heat transfer are generalized by the dependences [6.17] $l/d = 153$, $\mathrm{Pr} = 0.7$, $T_\mathrm{av} = (T_\mathrm{w} + T_\infty)/2$;

for $\mathrm{Ra}_x = 4\cdot 10^7 - 4\cdot 10^9$, $k_{T1} = 5\cdot 10^{-3} - 2\cdot 10^{-2}$

$$\frac{\mathrm{Nu}_x}{\mathrm{Nu}_{x0}} = 1 + 1.8\cdot 10^4 k_{T1}^{2.5}\left(\frac{4\cdot 10^9}{\mathrm{Ra}_x}\right)^{0.54}, \tag{6.1.11}$$

for $\mathrm{Ra}_x = 4\cdot 10^9 - 3\cdot 10^{10}$, $k_{T1} = 3\cdot 10^{-3} - 10^{-2}$

$$\frac{\mathrm{Nu}_x}{\mathrm{Nu}_{x0}} = 1 + 1.8\cdot 10^{-4} k_{T1}^{2.5}. \tag{6.1.12}$$

Heat transfer ***at a cooling vertical cylinder*** for the main range of time is determined by the appreciable change of the value $\partial(T_\mathrm{w} + T_\infty)/\partial\tau$ [6.17]:

for $\mathrm{Ra}_x = 4\cdot 10^7 - 2\cdot 10^9$, $k_{T2} = (-10^{-3}) - (-4\cdot 10^{-3})$

$$\frac{\mathrm{Nu}_x}{\mathrm{Nu}_{x0}} = 10^9 |k_{T2}|^{4.15} \left(\frac{2\cdot 10^9}{\mathrm{Ra}_x}\right)^{0.59}, \tag{6.1.13}$$

for $\mathrm{Ra}_x = 2\cdot 10^9 - 3\cdot 10^{10}$, $k_{T2} < -4\cdot 10^{-3}$

$$\frac{\mathrm{Nu}_x}{\mathrm{Nu}_{x0}} = 10^9 |k_{T2}|^{4.157}. \tag{6.1.14}$$

Heat transfer ***for an instantaneous change in the temperature of a horizontal cylinder*** $\varphi < 5\pi/6$ is given by [6.18]

$$\frac{\mathrm{Nu}_{\varphi r_0}}{\mathrm{Gr}_{r_0}^{1/4}} \left(\frac{\mathrm{Zh}_{r_1}^2 \mathrm{Gr}_{r_0}}{1+\mathrm{Zh}_{r_0}^2 \mathrm{Gr}_{r_0}}\right)^{1/4} = -\frac{\theta_1'(0)}{2}\left[1 + \left(\frac{\theta_1'(0)}{\theta_0'(0)}\cos\varphi - \frac{1}{4}\right)\mathrm{Zh}_{r_0}^2 \mathrm{Gr}_{r_0} + \left(\frac{5}{32} - \frac{\theta_1'(0)}{4\theta_0'(0)}\cos\varphi + \frac{\theta_{20}'(0)\cos^2\varphi + \theta_{21}'(0)}{\theta_0'(0)}\right)\mathrm{Zh}_{r_0}^4 \mathrm{Gr}_{r_0}^2\right], \tag{6.1.15}$$

$$\mathrm{Pr} = 1, \quad \theta_0'(0) = -1.12833, \quad \theta_1'(0) = -0.04676,$$
$$\theta_{20}'(0) = 0.00433, \quad \theta_{21}'(0) = -0.00328.$$

For developed free convection ($\mathrm{Zh}_{r_0} \to \infty$) dependence (6.1.15) gives the results on heat transfer overestimated by 10%. Table 6.1.3 lists the results of numerical calculations of heat transfer $\mathrm{Nu}_{\varphi r_0}/\mathrm{Gr}_{r_0}^{1/4}$ for $\mathrm{Pr} = 0.7$ and $\varphi = 0 - \pi/3$ [6.19].

The numerical data which characterize the development of nonstationary convection near the stagnation point for ***a heated three-dimensional body placed in a fluid*** are presented in [6.20].

The heat transfer ***on an external surface of a curvilinear body for a power-law change in the wall temperature*** $\theta_\mathrm{w} = \theta_0(\tau/\tau_0)^m$ is given by [6.21]

$$\frac{\mathrm{Nu}_x}{\mathrm{Gr}_x^{1/4}} = \left(\frac{4}{\mathrm{Gr}_x^{1/2}\mathrm{Fo}_x}\right)^{1/2}\left[A + \frac{\mathrm{Zh}_x}{\mathrm{Gr}_x^{1/2}}\left(k + \frac{j}{r}\sin\varphi\right) \times \left[\left(\frac{\mathrm{Gr}_x}{\mathrm{Zh}_x \mathrm{Pr}}\right)^{1/2} + 16B\left(\mathrm{Gr}_x^{1/2}\mathrm{Zh}_x\right)\left(\frac{\tau}{\tau_0}\right)^{m+1/2}\cos\varphi\right] + O\left(\frac{\mathrm{Zh}_x^2}{\mathrm{Gr}_x}\right)\right], \tag{6.1.16}$$

where $k(x)$ and $r(x)$ are the curvature and curvature radius of the body; $j = 0$ for a flat curvilinear body; $j = 1$ for axisymmetrical spatial flow. The values of the coefficients A and B are given in Table. 6.1.4.

Table 6.1.3.

φ°	0	10	20	30	40	50	60
				$(\mathrm{Zh}_{r0}^2 \mathrm{Gr}_{r0})^{1/4}$			
0.1	4.7204	4.7204	4.7204	4.7204	4.7204	4.7204	4.7204
0.2	2.3604	2.3604	2.3604	2.3603	2.3603	2.3603	2.3603
0.3	1.5741	1.5741	1.5741	1.5740	1.5740	1.5740	1.5740
0.4	1.1816	1.1816	1.1815	1.1814	1.1813	1.1811	1.1809
0.5	0.9471	0.9470	0.9469	0.9467	0.9464	0.9460	0.9456
0.6	0.7919	0.7918	0.7916	0.7912	0.7907	0.7901	0.7894
0.7	0.6826	0.6824	0.6821	0.6815	0.6807	0.6797	0.6786
0.8	0.6023	0.6021	0.6016	0.6007	0.5996	0.5982	0.5965
0.9	0.5418	0.5415	0.5408	0.5397	0.5381	0.5362	0.5339
1.0	0.4955	0.4852	0.4944	0.4929	0.4908	0.4883	0.4853
1.2	0.4328	0.4324	0.4311	0.4290	0.4261	0.4224	0.4179
1.4	0.3970	0.3965	0.3950	0.3924	0.3889	0.3843	0.3786
1.6	0.3784	0.3779	0.3762	0.3736	0.3698	0.3648	0.3558
1.8	0.3704	0.3699	0.3683	0.3656	0.3619	0.3571	0.3512
2.0	0.3681	0.3676	0.3680	0.3635	0.3599	0.3555	0.3503
2.5	0.3692	0.3687	0.3672	0.3648	0.3616	0.3580	0.3546
3.0	0.3701	0.3696	0.3681	0.3657	0.3625	0.3589	0.3556
∞	0.3702	0.3697	0.3682	0.3658	0.3524	0.3579	0.3525

Table 6.1.4.

m	A	B
0	1.12838	0.00298
1	2.25676	0.00091
2	3.00901	0.00047
3	3.61081	0.00040
4	4.12664	0.00030
5	4.58516	0.00021
6	5.00199	0.00016
7	5.38676	0.00013
≥ 8	$2\,\mathrm{m}^{1/2} + 0.25\,\mathrm{m}^{-1/2} + 0.0022\,\mathrm{m}^{-3/2}$	0.00010

Heat transfer from a surface of the thin horizontal wire ($d = 30$ and $127\,\mu\mathrm{m}$) with an instantaneously prescribed specific heat per unit length q_l is [6.22]

$$\frac{1}{\mathrm{Nu}_{r_0}} = \frac{16}{\pi^2}\left(\frac{\rho_\infty c_\infty}{\rho_c c_c}\right)^2 \int\limits_0^\infty \frac{1-\exp\left(-\mathrm{Fo}_{r_0}\mu^2\right)}{\mu^3 \Delta(\mu)} d\mu\,, \qquad (6.1.17)$$

$$\Delta\mu = \left[\mu J_0(\mu) - 2\frac{\rho_\infty c_\infty}{\rho_c c_c} J_1(\mu)\right]^2 + \left[\mu Y_0(\mu) - 2\frac{\rho_\infty c_\infty}{\rho_c c_c} Y_1(\mu)\right]^2 .$$

The duration of the nonstationary thermal regime is determined from the equation

$$\mathrm{Fo}_{r0,\infty} - 80.2\mathrm{Ra}_{D0}^{-2/3} = 201\left(\left|\mathrm{Ra}_{Dc}^{-2/3} - \mathrm{Ra}_{D0}^{-2/3}\right|\right) , \qquad (6.1.18)$$

where $\mathrm{Ra}_{D0} \cong 1100$ is the critical value of Ra_D at which the heat conduction regime is ended; $\mathrm{Ra}_{Dc} \cong 12700$ is the limiting number Ra_D at which the heat transfer coefficient reaches its stationary value; D is the penetration depth of the temperature field around a wire.

For air and water the time for the establishment of free convection stationary regime is equal to $\tau \cong 2.271 r_0^2/a$.

Table 6.1.5 presents the data on heat transfer ***in a braking point*** ($\varphi = 0$) ***for the instantaneous sphere temperature change*** for $\mathrm{Pr} = 0.7$ [6.23].

Table 6.1.5.

$(\mathrm{Zh}_{r0}^2\mathrm{Gr}_{r0})^{1/4}$	$\mathrm{Nu}_{r0}/\mathrm{Gr}_{r0}^{1/4}$	$(\mathrm{Zh}_{r0}^2\mathrm{Gr}_{r0})^{1/4}$	$\mathrm{Nu}_{r0}/\mathrm{Gr}_{r0}^{1/4}$
0.1	4.7204	0.9	0.5589
0.2	2.3605	1.0	0.5186
0.3	1.5747	1.2	0.4708
0.4	1.1831	1.4	0.4520
0.5	0.9500	1.6	0.4490
0.6	0.7971	1.8	0.4519
0.7	0.6907	2.0	0.4544
0.8	0.6144	∞	0.4575

Transient natural convection ***in a spherical enclosure consisting of a liquid core and a porous shell fully saturated with the same liquid*** is investigated numerically in [6.24]. The numerical results exhibit remarkable effects along the porous shell-fluid interface; however, the overall heat flux is only sensitive to the ratio of the thermal conductivities of the solid matrix and the liquid.

Buoyancy driven thermal convection due to the presence ***of a sphere of constant heat flux in an unbounded fluid-saturated porous medium*** was study analytically in [6.25]. Of special importance is the finding that for $\mathrm{Ra} > 3$, there appears a second cell below the sphere, so that the flow pattern is multicellular and in this case, the isothermal lines that are formed a little farther away from the sphere do not exhibit closed loops (Fig. 6.1.2). The local Nusselt number is defined by

$$\mathrm{Nu} = 1 - 0.07025\,\mathrm{Ra}^2 + O\left(\mathrm{Ra}^3\right) . \qquad (6.1.19)$$

Numerical results for a transient free convection flow ***from a sphere embedded in a porous medium*** in the range $0.01 \le \mathrm{Ra} \le 200$ were obtained

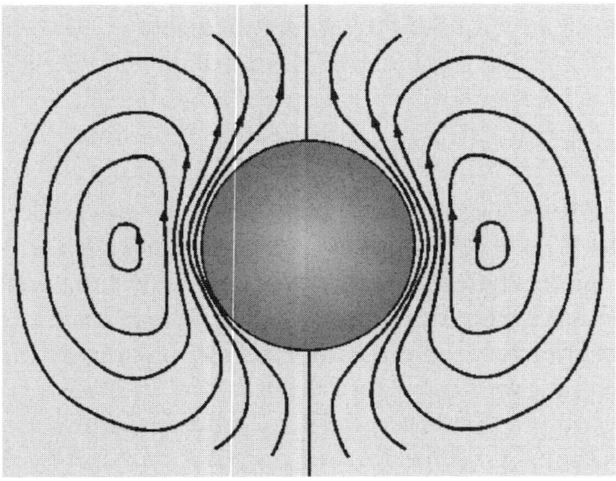

Fig. 6.1.2. Flow pattern of free-convection flow around a heated sphere

in [6.26]. ***On sudden jump either the temperature or the heat flux of the sphere remained constant values over the surface***. For a given value of Ra, as time increases, the temperature vertically above the sphere along the line $\theta = 0$ starts to oscillate. The oscillation starts at a later time in the constant heat flux case than that in the constant temperature case since the convection in this case is mach weaker.

Spherical layer in the absence and in the presence of vertical eccentricity ($R_0/R_i = 2$; $\varepsilon = \pm(0.625)$ ***with inner and outer spheres heated and cooled by step temperature change*** for $\mathrm{Pr} = 0.7$ and $10^3 < \mathrm{Ra} < 5 \cdot 10^5$. In the initial stage the heat transfer and the flow structure depend strongly on Ra and the channel eccentricity. The data for the average Nusselt number are generalized by the dependence

$$\mathrm{Nu} = C\mathrm{Ra}^m , \tag{6.1.20}$$

where the empirical constants C and m are determined from Table 6.1.6 [6.27].

Table 6.1.6.

ε	C	m	Ra
0.625	0.291	0.211	$10^3 < \mathrm{Ra} < 10^5$
0.000	0.210	0.240	$10^3 < \mathrm{Ra} < 5{\cdot}10^5$
–0.625	0.267	0.209	$10^3 < \mathrm{Ra} < 10^5$

Temperature of ***a vertical plate*** changes in time insignificantly, near an average value. The ratio of instantaneous local and average heat flux values $\mathrm{Pr} = 0.72$ is [6.28]

$$\frac{q_x}{q_{x0}} = 1 + 1.53 \frac{1}{\theta_w} \frac{d\theta_w}{d\text{Zh}_x} \text{Gr}_x^{-1/2} ,$$
$$\frac{q}{q_0} = 1 + 0.918 \frac{1}{\theta_w} \frac{d\theta_w}{d\text{Zh}_l} \text{Gr}_l^{-1/2} . \tag{6.1.21}$$

The ratio of average values of heat fluxes ***for a harmonic surface temperature change*** $\theta_w = \theta_0\{1 + \varepsilon \exp(i\omega\tau)\}$ is [6.28]

$$\frac{q}{q_0} = 1 + \frac{5}{4} \left[1 + \frac{108}{125} \left(\frac{348\,\text{Pr} + 1079}{633\,\text{Pr} + 1851} \right)^2 \left(\text{Pr} + \frac{20}{21} \right) \text{Re}_{\omega l}^2 \text{Gr}_l^{-1} \right]^{1/2} \times \varepsilon \cos \omega\tau. \tag{6.1.22}$$

For a large amplitude, high frequency, and large value of Gr_x the time-varying temperature in the inner region adjoining the plate surface causes a reverse heat flux [6.29]. The ratio $\text{Nu}_x/\text{Gr}_x^{1/4}$ based on the average heat flux gives a universal curve for all the values of amplitude and frequency with a limit of 0.354 (Pr = 0.7).

Under actual conditions, the heat capacity of a heated body in liquid exerts its influence on the temperature and heat flux. The problem becomes conjugated, and the step change becomes unattainable. If heat capacity of the body is small, i.e., the step heat supply corresponds closely to the step change in the heat flux, then the surface temperature changes in the main during the initial stage of thermal conductivity. If the body heat capacity is large, then the temperature changes so slowly that the first and the second stages are insignificant. Then the process can be described for any time by substituting the instantaneous surface temperature values to the ordinary stationary solution (quasi-steady process). The conjugation parameter for the given temperature is $P'_{\text{con3}} = \rho_w c_w \delta_w \text{Gr}_l^{1/4} / \rho_\infty c_\infty l$. The range of *the* values of P'_{con3} covering the transient stage is relatively insignificant, and the initial temperature increase is proportional to P'_{con3}. For linear heat supply the quasi-steady condition is reached at smaller values of P'_{con3}.

In a closed vertical flat layer one of the walls is formed by a semi-infinite massif, and the temperature of the other wall changes by step and is maintained constant. The experiments are processed by means of the average heat transfer value [6.30]:

$$H/B = 7.5 - 66, \text{Pr} = 0.722 - 150,\ \text{Fo}_T = 0.052 - 2 ,$$

$$\text{Gr}_B = 1.4 \cdot 10^{-1} - 7 \cdot 10^6 ,$$

$$\text{Nu} = 5.13 \cdot 10^{-4} \left(\text{Ra}_B\, \text{Pr}^{0.5} \right)^{0.5} (a/a_T)^{0.6} (H/B)^{0.36} . \tag{6.1.23}$$

For 70% of experimental points the deviation from the generalized curve makes no more than 20%. The heat conduction regime terminates at $\text{Fo}_T =$ 0.052; almost complete termination of convection occurs at $\text{Fo}_T = 2$.

A vertical layer with the heat flux assigned on one of the surfaces. Let us introduce the concept of nonstationary heat transfer coefficient

$$\alpha' = \left(mc_p \frac{dT_\infty}{d\tau} + Q \right) \Big/ \theta_{\mathrm{w}} F \tag{6.1.24}$$

and the number $\mathrm{Nu}' = \alpha' H/\lambda$. The analysis of experimental data establishes the dependence [6.31]

$$H/B = 2 - 8, \mathrm{Pr} = 2 - 7, \mathrm{Ra}_H = 5 \cdot 10^9 - 7 \cdot 10^{10} \ ,$$

$$\mathrm{Nu}' = 0.075 \mathrm{Ra}_H^{1/3} \left(H/B \right)^{1/4} \ . \tag{6.1.25}$$

The exponent at H/B is positive (unlike the stationary case) and variable. Its maximal value for $H/B = 4$ is equal to 1/3 and it decreases for other values of the ratio H/B. The values of the heat transfer coefficient obtained from formula (6.1.24) are larger than those obtained for the stationary process.

The results of numerical analysis of nonstationary convection ***in a square cavity with differently heated lateral walls*** carried out in [6.32] point to the presence of waves in a vertical thermal boundary layer, propagating from the walls to stratified layers formed on horizontal surfaces. For the Rayleigh number $\mathrm{Ra} = 5 \cdot 10^9$ a shifting region is formed in the stratified layer. It seems that here, there is a transition to turbulence. The overall heat transfer and approximation to the stationary state depend greatly on the presence of internal gravitational waves.

Experimental data for nonstationary free-convective flow and heat transfer ***in a square cavity for instantaneous heating and cooling of the opposite lateral sides*** are given in [6.33]. The ***water-glycerine mixture*** was used as an investigated liquid (originally isothermal) with the Prandtl number changing from 13 to 234. The oscillatory regime of transition from a monotonous flow to a stationary state was investigated for a variable difference of edge temperatures in the specified range of Pr.

Natural convection of ***a dusty fluid in an infinite rectangular channel with differently heated vertical walls and adiabatic horizontal walls*** was studied in [6.34]. It is seen that the heat transfer rate decreases with an increase in the mass concentration of dust particles, but it increases with an increase in the Rayleigh number.

A rectangular container of square section with insulated vertical walls. On horizontal walls the initial temperature T_0 changes instantaneously to the temperature $T_1 < T_0$. The experimental data for the time of transition to cooling due to thermal conductivity are [6.35]

$$H/B = 20, 40 \text{ mm}, \mathrm{Pr} = 5\text{--}5 \cdot 10^3, \mathrm{Ra}_H = 10^4\text{--}10^7 \ ,$$

$$\mathrm{Bi}_H = 400\text{--}650, T_{\mathrm{av}} = T_0, \theta = T_0 - T_1 \ ,$$

$$\mathrm{Fo}_H = 0.148 \, \mathrm{Ra}_H^{-0.28} \, \mathrm{Pr}^{-0.2}, \tag{6.1.26}$$

for the time of complete cooling of the container

$$\mathrm{Fo}_H = 1.22\mathrm{Gr}_H^{-0.66} \,. \tag{6.1.27}$$

The comparison of the time of cooling of the container in the presence of natural convection with the time of cooling only due to heat conduction

$$\mathrm{Fo}_{HT}/\mathrm{Fo}_H = 0.98\mathrm{Gr}_H^{0.06} \tag{6.1.28}$$

shows that ***for transformator oil*** for $\mathrm{Ra}_H = 3.7 \cdot 10^6$. The cooling time decreases due to convection approximately twice.

Transient thermosolutal opposing convection of ***a liquid-water mixture in a square cavity subject to horizontal temperature and concentration gradients*** was numerically investigated in [6.36]. The results for Le = 100 clearly show the double-diffusive nature of the convection. In the initial stage the flow is dominated by the interface velocities at the vertical side walls driven by the concentration gradients there. Later, the flow is governed by the thermal buoyancy. At a much later time, the solutal buoyancy set in inducing new recirculation cells along the side walls. These cells gradually grow and squeeze the thermally driven cell in the core region. A multilayer flow structure is finally formed. The following empirical correlations are proposed:

$$\begin{aligned} \mathrm{Nu} &= 0.112\left[0.993l\ln^2(\mathrm{Le}) - 6.829\ln(\mathrm{Le}) + 13.606\right]\mathrm{Gr}_t^{0.3}\,|N|^{-0.8}\,, \\ \mathrm{Sh} &= 0.011\left[0.0001112\mathrm{Le}^3 - 0.0197\mathrm{Le}^2 + 1.1493\mathrm{Le}\right. \\ &\quad \left.+12.9\right]\mathrm{Gr}_t^{0.285}\,|N|^{0.8}\,, \end{aligned} \tag{6.1.29}$$

where N is the buoyancy ratio.

The numerical investigation of three-dimensional nonstationary convection ***in a cubic container for instantaneous heating and cooling of two opposite vertical walls*** for $\mathrm{Ra} = 10^6$ carried out in [6.37] shows that in the course of time two opposite directed boundary layers and a rather motionless internal core are formed inside the cavity. The heat transfer rate in the cavity depends substantially on the presence of internal gravitational wave motion in it.

Rectangular container with insulated vertical walls and horizontal walls at the temperatures T_0 ***and*** T_H. Approximation of numerical data on the induction time and Nusselt number under the conditions of thermal and solutal convection for an inert liquid yields [6.38]

$$\mathrm{Pr} = Lu = 1, \mathrm{Ra}_H = 10^3 - 5\cdot 10^5, \mathrm{Ra}_{DH} = (-5\cdot 10^5) - 5\cdot 10^5\,,$$

$$\mathrm{Nu}/\mathrm{Nu}_0 = 1 + 0.065\left(\mathrm{Ra}_H + \mathrm{Ra}_{DH}\right)^{4/3} / \left(1.2\cdot 10^4 + \mathrm{Ra}_H + \mathrm{Ra}_{DH}\right), \tag{6.1.30}$$

where Nu_0 is the Nusselt number in the absence of convection.

The nonstationary character of heat transfer ***in a tank*** is provided by injection of ***hot air or nitrogen***. The average value of heat transfer on the surface of a vertical cylindrical tank with wall thickness $\delta = 1.5$ mm is [6.39]

$$H/B = 4, p = 0.5 - 0.6\,\text{МПа}, k_{T1} = 10^{-4} - 1\,,$$
$$\text{Ra}_H = 10^6 - 10^{12}, T_{\text{av}} = (T_{\text{w}} + T_\infty)/2\,,$$
$$\frac{\text{Nu}}{\text{Nu}_0} = 2.29 \cdot 10^4 \text{Ra}_H^{-0.196}\, k_{T1}^{0.867}\,. \qquad (6.1.31)$$

A vertical cylinder of diameter d and height H with a spherical segment of height h and polar angle θ in its base. In a container with regularly distributed heat sources, for the data from Table 6.1.7 the end of the heat conduction stage and the attainment of stationary regime on heating and cooling are determined by the time period [6.40]

$$H/d = 0.108 - 0.861, \text{Ra}_{\nu H} = 6.33 \cdot 10^7 - 1.45 \cdot 10^3\,,$$
$$\text{Fo}_H = A\text{Ra}_{\nu H}^n\,. \qquad (6.1.32)$$

The values of the coefficients A and n are given in Table 6.1.7. For comparison, the table also lists the experimental data on liquid heating ***in a horizontal layer***.

Table 6.1.7.

z/H	Heat Conduction Stage A	n	Stationary Regime A	n
		Container heating		
0	0.951	-0.144 ± 0.015	20	-0.223 ± 0.028
0.5	2.141	-0.191 ± 0.023	48.28	-0.226 ± 0.051
		Container cooling		
0	1.811	-0.193 ± 0.01	20.19	-0.224 ± 0.028
0.5	2.729	-0.217 ± 0.024	47.74	-0.267 ± 0.052
		Heating of the horizontal layer		
0	2.587	-0.268	11.577	-0.219

The numerical analysis of the effect of impermeable end walls on the nonstationary natural convection of ***air in a horizontal annular slot*** point to the decrease in the flow velocity near the ends due to viscous resistance [6.41]. The results obtained for the radii ratio equal to 2.6 show that the temperature distribution inside the core remains constant when the ratio of the channel length to the external radius exceeds 4. The characteristic features of flow and temperature fields ***on instantaneous heating of the inner cylinder*** are also investigated.

Some boundary-layer characteristics obtained by the scale analysis method and the results of numerical calculations of nonstationary laminar natural-convective ***flow from a heated horizontal cylinder under different boundary conditions on a surface*** are given to estimate the development of the region of upstream and surface heat transfer [6.42].

A horizontal cylindrical layer between a platinum wire and a thermally stabilized glass tube. For Fo = 0 the temperature of the ***platinum wire*** increases to some fixed value t_{w1}, which corresponds to the maximum value of Nu. As the liquid is heated in the cavity, there appear weak convective flows intensifying heat transfer. At the $\mathrm{Fo}_1 = 0.117\ \mathrm{Pr}^{1/6}/(0.014\ \mathrm{Ra}_\delta^{2/3} - 1)$, the curve $\mathrm{Nu} = f(\mathrm{Fo})$ reaches its minimum ($\mathrm{Ra}_\delta = 10^2 - 10^6$, $\mathrm{Pr} = 5 - 5 \cdot 10^3$). Developing, the convective motion in time $\mathrm{Fo}_2 = 0.065\ \mathrm{Pr}^{1/4}/(0.119\ \mathrm{Ra}_\delta^{1/3} - 1)$ reaches the quasi-steady condition [6.43].

Nonstationary free convection ***together with mass transfer on a vertical surface on instantaneous change in the temperature and concentration on a wall*** [6.44]. The temperature profiles pass through the maximum value before they attain a stationary state.

From the distribution of the average heat transfer coefficient for Pr = 1 it is possible to determine three regions with the minimum value in the intermediate one (Fig. 6.1.3). When $\mathrm{Gr}_{Dl}/\mathrm{Gr}_l = 0$, the heat transfer does not depend on the Sc number.

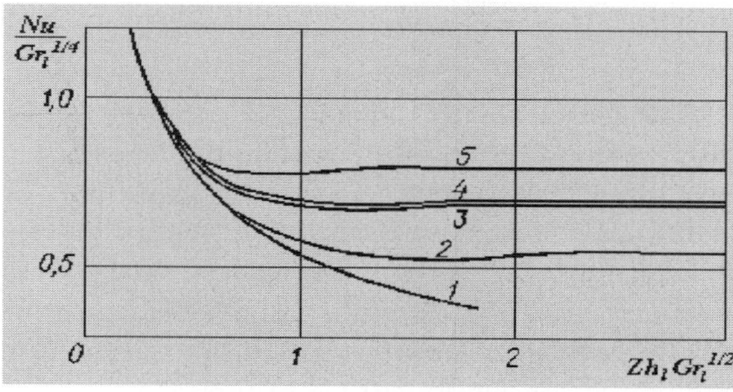

Fig. 6.1.3. Average value of heat transfer from a surface in the presence of mass transfer, Pr = 1: (1) heat conduction; (2) Sc = 0.2, $\mathrm{Gr}_{Dl}/\mathrm{Gr}_l = 0$; (3) 0.2, 2; (4) 0.7, 2; (5) 7, 2

The effect of ***suction and injection on the surface*** is considered for a parallel flow in the limiting cases based on the nonstationarity parameter [6.45–6.48].

Nonstationary convection of ***air*** (Pr = 0.71) in an inclined rectangular cavity ***with different constant temperatures on opposite walls*** for the ratio of length to height equal to 4; 3 and 1. The inclination angle γ changes in the range $0° \leq \gamma \leq 90°$. In the stationary regime, $10^3 \leq \mathrm{Ra} \leq 2 \cdot 10^4$. Investigation of the effect of an inclination angle on the flow and heat transfer characteristics in the transient phase for various Ra numbers showed that for small angles the flow structure consists of four, three, two cells or one cell depending on the value of γ (Fig. 6.1.4). For large inclination angles the cellular structure is destroyed by a strong longitudinal flow [6.49]. For Ra $\geq$ 2000 the hysteresis phenomenon is observed.

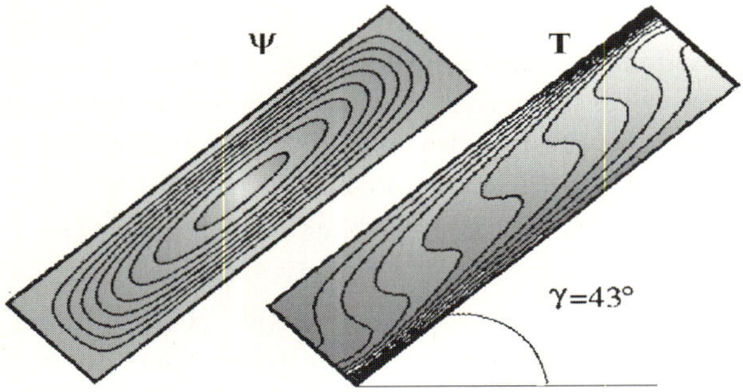

Fig. 6.1.4. Streamlines and isotherms for free convection in a rectangular inclined cavity [6.49]

Experimental investigation of natural convection of ***water suspension of small precipitating particles in a rectangular cell with homogeneous initial temperature and concentration. The suspension is heated from some region on the cell bottom and is cooled on lateral walls***. As the particles settle, some layers are formed, and their number decreases to the formation of one layer due to interaction of temperature and concentrational gradients. The coefficient of heat transfer on the heated wall oscillates cyclically as the layers disappear. The period of the disappearance of the layers decreases decrease in the cell sizes and with increase in the particle sizes [6.50]. The experimental investigation of the dynamics of the velocity and temperature fields for two-dimensional transient natural convection ***in an inclined rectangular cavity*** using multichannel electronic interferometrics and digital processing of the results of track visualization. The nonstationary boundary conditions consist ***of instantaneous switching of heating and cooling of two opposite walls***. The dynamics of flow development to a steady regime is investigated for the values $\mathrm{Pr} = 6.38$, $\mathrm{Ra} = 1.5 \cdot 10^5$ and the aspect ratio equal to one. The cavity inclination angles were equal to $\pi/4$, $\pi/2$, and $3\pi/4$. The flow structure in inclined cavities consists of convective boundary layers near hot and cold walls and a potential flow inside a cavity [6.51].

The experimentally obtained characteristics of the growth of thermal boundary layers for convection of a ***water-glycerine mixture*** with the Prandtl number in the range from 13 to 234 ***in a cavity of square section for instantaneous heating and cooling of opposite lateral walls*** given in [6.52] for the parameter ranges appropriate for transition from a monotonous flow regime to stationary oscillations according to the classification scheme of Patterson and Imberg [6.53] confirm the validity of this classification.

The results of streamline calculation for laminar free convection ***in a cavity of square section with constant temperature on one of the vertical walls and harmonic temperature change on another*** for $\mathrm{Pr} = 7$

and Ra $= 1.4 \cdot 10^5$ show that in the top corner of the cavity near the wall with variable temperature there periodically arises and then disappears a cell of weak secondary flow rotating in the direction opposite to the main stream. The effect of the wall temperature oscillations on the flow inside the cavity is insignificant, the time-averaged heat transfer depends slightly on the nonstationary boundary conditions (see Table 6.1.8) [6.54].

Table 6.1.8.

a	p	Nu
0.4	0.01	5.41
0.2	0.01	5.35
0.8	0.01	5.58
0.4	0.05	5.36
0.4	0.02	5.43

The increase of a or p (dimensionless amplitude or wall temperature oscillations period) leads to the increase of the cell size and intensity of the secondary flow.

A container of ***square section*** filled with ***water on instantaneous switching of heating and cooling of opposite lateral vertical walls. The horizontal walls are insulated***. The results of shadow visualization for Pr $= 6.9 - 7.0$ show that at the initial stage of the process two moving wave packages are observed. In the central region the thermal stratification develops, and near the vertical walls the boundary layer flow structure is observed [6.55]. The range of the Rayleigh numbers is $6.9 \cdot 10^8 \leq \mathrm{Ra} \leq 7.3 \cdot 10^8$. A further development of the flow structure to the steady regime is determined by complex nonlinear interaction between waves in the boundary layer penetrating into the laminar flow in the top part of the container.

The process of destabilization and mixing of the ***two-layer concentration-stratified salt solution under the effect of lateral heating and cooling in three different containers*** was experimentally investigated in [6.56]. The ranges of the parameters are: $\mathrm{Ra}_T = 2 \cdot 10^7 – 7 \cdot 10^8$, $N_i = 0.441 – 0.882$, where Ra_T is the thermal Rayleigh number; N_i is the initial value of the buoyancy ratio ($N_i = (\beta_c \Delta C)/(\beta_T \Delta T)$; β_c and β_T are the coefficients of solutal and thermal expansion, respectively). At the initial stage, before the layer is mixed, the Kelvin-Helmholtz vortices are observed on the interface. Then they completely disappear as the interface inclines. To estimate the dimensionless time of layer mixing, the following correlation dependence is suggested:

$$\mathrm{Fo}_m = (-0.172 \ln \mathrm{Ra}_T + 5.9 N_i + 1.156) \cdot 10^{-3} \,. \tag{6.1.33}$$

The experimental and calculated data for unsteady ***thermal and solutal convection*** of the $\mathbf{H_2O}$**-NaCl** ***solution in a horizontal concentric slot with initially isothermal internal and external cylinders for heating of the inner cylinder*** are given in [6.57]. The initially isothermal solution is exposed during heating to the effect of linear steady stratification. The results of the investigation specify a significant decrease in the heat transfer rate compared with natural convection in the absence of stabilizing concentrational gradient. Apart from that, the experimental data ***for thermal and solutal convection from a vertical cylinder located in a square container*** are obtained.

The time-dependent double-diffusive convection of $\mathbf{H_2O}$**-NaCl** concerning the ***stably stratified salt fluid layer with lateral heating in an inclined cavity*** is investigated in [6.58]. At an early time there are two types of the onset of instability. When $\vartheta < 75°$, the cells rotate strikingly in one sense of rotation, but weakly in the other, while for $\vartheta \geq 75°$ the rotation of comparable vortices in a different sense are generated. The multilayered structure is formed with a sharp solutal difference and a wavy distribution of temperature at all inclined angles. The distribution of local Nusselt number is periodic due to the layered structure.

Numerical investigation of a nonstationary two-dimensional convection in a container of triangular section with insulated bottom base of length L and isothermal inclined lateral surfaces ($T_h > T_w$) was carried out in [6.59]. The aspect ratio for the base is $H/L = 0.5$. The streamlines and isotherms for the stable flow regime are given in Fig. 6.1.5. The shaped lines correspond

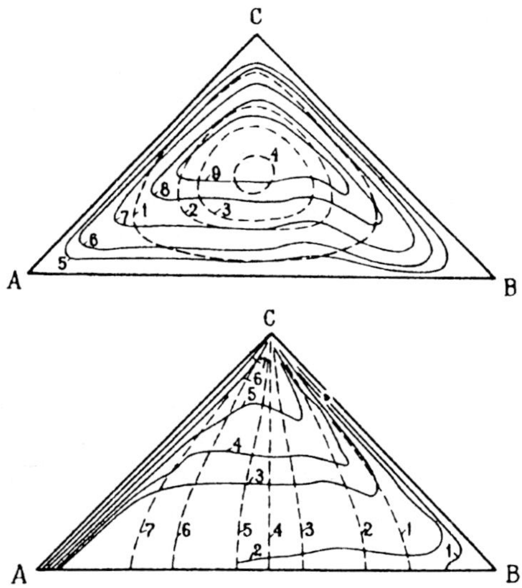

Fig. 6.1.5. Streamlines and isotherms for a steady flow regime in nonstationary two-dimensional convection in an enclosure of triangular section [6.59]

to the value Gr $= 10^3$, continuous to Gr $= 10^8$. As is seen from the figures, for small Grashof numbers the heat transfer in the container is determined by thermal conductivity. With increase in Gr the intensity of convective processes increases, which leads to substantial distortion of the streamlines and isotherms in the container. On the lateral surfaces the clearly expressed gradient zones such as thermal boundary layers are formed. In the central part, the isotherms turn from an almost vertical position (Gr $= 10^3$) to a horizontal one (Gr $= 10^8$), appropriate for the conditions of liquid stratification. When the geometrical parameter increases in the range $0.25 \leq H/L \leq 2$ for the fixed Grashof number, the heat transfer rate in the container increases to some maximal value appropriate for $H/L \cong 1$.

6.2 Free Convection in Oscillating Flows

The vibrational effect created by the motion of a heated surface in an ambient medium or by a supply of disturbances in the form of acoustic or other periodic oscillations in the medium itself can substantially change the rate and characteristics of heat transfer. The generalizing criterial dependence of heat transfer with oscillations is [6.60–6.62]

$$\mathrm{Nu} = f\left(\mathrm{Gr}, \mathrm{Pr}, \mathrm{Re}_\omega, \mathrm{Re}_{\Delta u}, J, P_k\right) . \qquad (6.2.1)$$

In the presence of oscillations, the motion of liquid relative to the surface of heating (Re_ω), the oscillation of the surface of heating ($\mathrm{Re}_{\Delta u}$), and the vibrational acceleration (J) exert their effect on the heat transfer rate. The frequency of the modulating oscillations and the phase difference due to the system inertia can be taken into account by the Re_ω number.

A heated thin vertical plate of constant temperature under transverse harmonic oscillations with the velocity $v_0 = A\omega \sin\omega\tau$. For the local time-averaged Nu number for small value of $(A\omega^2/g)^2 \mathrm{Gr}_x^{-1/2}$ we get [6.61]

$$\frac{\mathrm{Nu}_x}{\mathrm{Nu}_{x0}} = 1 - \left(\frac{A\omega^2}{g}\right)^2 \mathrm{Gr}_x^{-1/2} B\left(\bar{x}, \frac{\mathrm{Re}_{\omega l}}{\mathrm{Gr}_l^{1/2}}\right) . \qquad (6.2.2)$$

The stationary component Nu_x under the conditions of oscillations is smaller than the number Nu_{x0} for a stationary flow. The decrease in the heat transfer coefficient can be explained by the presence of phase difference between velocity and temperature oscillations near the plate. In most cases this decrease is insignificant and, as the calculations show, does not exceed 2%.

Heat transfer ***of a vertical flat plate with transverse oscillations*** [6.63]:

$$\mathrm{Pr} = 0.7, \mathrm{Re}_{\Delta ul} = (1.1 - 6.1)\cdot 10^3, \mathrm{Gr}_l = (1.7 - 6)\cdot 10^6 ,$$
$$\mathrm{Nu}/\mathrm{Nu}_0 = \left(1 + 7.09\, \mathrm{Re}_{\Delta Ul}^2/\mathrm{Gr}_l\right)^{1/4} . \qquad (6.2.3)$$

The linear stability of mechanical quasi-equilibrium flow ***of a prolonged inclined plane fluid layer, in the presence of a constant temperature gradient, subject to a static gravity field and high frequency vibration*** was investigated theoretically in [6.64]. The layer was oriented arbitrarily with respect to the vertical direction. The boundaries of the layer are assumed to be rigid and highly conducting. Each of two vectors – the temperature gradient and the axis of vibration – can have one of the four orientations: vertical, longitudinal, horizontal, and transverse. Thus, a total of sixteen situations are studied. The boundaries of stability and critical disturbance characteristics are determined for all the cases considered.

Heat transfer ***of a horizontal disk for transverse oscillations*** [6.65]:

$$d = 170.5\,\mathrm{mm}, A = 2.71 - 20\,\mathrm{mm}, l = 0.9d\ ,$$
$$f = 40 - 120\,\mathrm{Hz}\ , \mathrm{Ra}_l = 10^5 - 2\cdot 10^7\ ,$$
$$\mathrm{Re}_{\omega l} = fl^2/\nu, \mathrm{Re}_{\omega l}(A/l)^{1/2} = 1.5\cdot 10^3 - 5\cdot 10^5, T_{\mathrm{av}} = (T_{\mathrm{w}} + T_\infty)/2\ ,$$
$$\frac{\mathrm{Nu}}{\mathrm{Gr}_l^{1/4}} = 0.086\left(\mathrm{Re}_{\omega l}\right)^{1/4}(A/l)^{1/8}\ . \tag{6.2.4}$$

For the frequency $f = 50$ Hz and amplitude $A = 3$ mm the heat transfer depending on the sizes and temperature difference increases 3–4 times. The longitudinal oscillations do not cause an increase in heat transfer [6.66].

Heat transfer ***of a disk of diameter d for normal oscillations in a liquid and in air*** [6.67]:

$$d = 70, 130, 190\,\mathrm{mm}, f < 50\ Hz, A < 3\,\mathrm{mm}, \theta_{\mathrm{w}} = 20 - 80^\circ\mathrm{C}\ ,$$
$$T_{\mathrm{av}} = (T_{\mathrm{w}} + T_\infty)/2, \mathrm{Re}_{\omega d} = \omega A d/\nu\ ,$$

in an unlimited space

$$\mathrm{Nu} = C_1\left(\mathrm{Re}_{\omega d}^{0.5}/\mathrm{Gr}_d^{0.313}\right)^{n_1}\ , \tag{6.2.5}$$

between two parallel oscillating disk screens with distance between the disks b $= 5–40\,\mathrm{mm}$

$$\mathrm{Nu} = C_2\left(\mathrm{Re}_{\omega d}^{0.5}\,(b/d)^{0.2}/\mathrm{Gr}_d^{0.26}\right)^{n_{21}}\ . \tag{6.2.6}$$

The values of C_i and n_i are given in Table 6.2.1.

The space orientation of a vibrating disk does not affect heat transfer. It is only necessary to keep its normal orientation relative to the oscillation direction.

Longitudinal oscillations of a plate in a direction of external forces with the velocity $u_{\mathrm{w}} = \varepsilon\Delta u_0 \cos\omega\tau$. The local heat transfer coefficient for large frequencies and small amplitudes of oscillations is [6.61]

Table 6.2.1.

$\mathrm{Re}_{\omega d}^{0.5}/\mathrm{Gr}_d^{0.313}$	C_1	n_1
<8	1	0
8–20	0.354	0.5
20–80	0.143	0.8
$\mathrm{Re}_{\omega d}^{0.5}(b/d)^{0.2}/\mathrm{Gr}_d^{0.26}$	C_2	n_2
< 20	1	0
20–28	0.394	0.31
28–80	0.204	0.5

$$\frac{\mathrm{Nu}_x}{\mathrm{Nu}_{x0}} = 1 + \frac{1}{4}\frac{1+2\,\mathrm{Pr}^{1/2}}{\left(1+\ \mathrm{Pr}^{1/2}\right)^2}\frac{\varepsilon\Delta u_0}{\omega x}\cos\left(\omega\tau - \frac{\pi}{2}\right) - \frac{5}{32}\frac{\mathrm{Pr}}{(1+\mathrm{Pr})\left(1+\ \mathrm{Pr}^{1/2}\right)^2}\left(\frac{\varepsilon\Delta u_0}{\omega x}\right)^2 . \tag{6.2.7}$$

For low-amplitude plate oscillations the time-average coefficient differs little from the quasi-stationary one. For large oscillation amplitudes the increase in the heat transfer coefficient can be more substantial.

The effect of longitudinal harmonic oscillations on heat transfer ***of a vertical plate for the given heat flux***. The graph of the dependence of $\mathrm{Nu}_x/\mathrm{Re}_{\omega x}^{1/2}$ on $\mathrm{Gr}_x/\mathrm{Re}_{\omega x}^2$ (oscillation frequency 10–20 kHz, amplitude A = 6.5 mm, surface temperature $T_\mathrm{w} = 40–95°\mathrm{C}$) with an error of 5% coincides with the corresponding data for mixed convection (Fig. 7.2.1). In the considered range of parameters, the maximum increase in the heat transfer coefficientis 33% [6.68].

Oscillations along a heated round vertical cylinder of diameter 124 mm and height 203 mm. For small oscillation amplitudes the decrease in the heat transfer coefficient agrees with the theoretical formula (6.2.7) [6.61].

A vertical plate performs harmonic oscillations $A = A_0\sin\omega\tau$ relative to the axis located at the height $(l-b)$ and distance h near the plate. The plate represents a lateral surface of a vertical rectangular container $l \times 2h \times z$. The vessel performs oscillations with the aid of curve-pin mechanism located at the base. The local heat transfer value in a viscous liquid with an error of 8% is [6.69]

$$l \times 2h \times z = 1.5 \times 1 \times 0.5\mathrm{m}, \mathrm{Re}_\omega = \omega A_0 x/\nu = (0.3-23)\cdot 10^3 ,$$
$$\mathrm{Pr} = 132 - 6000, \mathrm{Ra}_x = 10^6 - 10^{12}, T_\mathrm{av} = T_\infty ,$$
$$\mathrm{Nu}_x = 0.57C\mathrm{Ra}_x^{0.25}\left(\mathrm{Pr}\,/\,\mathrm{Pr}_c\right)^{0.25} , \tag{6.2.8}$$

for $x/b < 1$, $\mathrm{Re}_\omega \mathrm{Pr}^{0.5} < 1.7 \cdot 10^5$

$$C = (x/b)^{-1.7}\, \mathrm{Re}_\omega\, \mathrm{Pr}^{0.5} ,$$

for $x/b > 1$, $\mathrm{Re}_\omega \mathrm{Pr}^{0.5} > 1.7 \cdot 10^5$

$$C = (x/b)^{-0.3}\, \mathrm{Re}_\omega^{0.33}\, \mathrm{Pr}^{0.15} .$$

The oscillations have a stabilizing effect on the boundary layer. Along the whole height the boundary layer remains laminar for $\mathrm{Ra}_x \le 10^{12}$ obtained in experiments. The effect of the degree of volume filling with a liquid on the local heat transfer was not noted in the range 0.33–1.00, except for end effects that did not exceed 10% of the size of the working surface.

The dimensionless heat transfer value ***for a horizontal Ni-Cr alloy wire with oscillations in horizontal and vertical planes:***

$$d = 0.643, 0.80, 2.06\,\mathrm{mm}, 2A_0 = 1.4 - 5.9\,\mathrm{mm}, f = 39 - 122\mathrm{Hz} ,$$
$$\Delta u = 2A_0 f, \theta_{\mathrm{w.}} \le 145°\mathrm{C}, T_{\mathrm{av}} = (T_{\mathrm{w}} + T_\infty)/2, \mathrm{Pr} = 1 - 10 ,$$
$$\frac{\mathrm{Nu}}{\mathrm{Nu}_0} = 0.75 + 0.0022 \frac{\mathrm{Re}_{d\Delta U}^{2.05} \varepsilon_1^{0.33}}{\mathrm{Pr}^{1.54} \mathrm{Gr}_d^{0.41}} . \tag{6.2.9}$$

Note that the amplitude of the oscillations considerably exceeds the wire diameter. Therefore, the heat transfer problem can be considered for periodic wire oscillations. No substantial difference in heat transfer coefficients for cylinder oscillations in horizontal and vertical planes was noticed.

Heat transfer ***for oscillations of a sublimating horizontal cylinder in air***,

$$d = 0.7 - 1.9\,\mathrm{mm}, \quad f = 20 - 118\,\mathrm{Hz} ,$$
$$\mathrm{Re}_{\Delta u} = (d + A_0) f^2 A_0/\nu, \quad \mathrm{Re}_{\Delta u}\, \mathrm{Pr}^{1.33} = 10 - 500 ,$$
$$\frac{\mathrm{Nu}}{\mathrm{Nu}_0} = 0.038 \mathrm{Re}_{\Delta U}^{0.85}\, \mathrm{Pr}^{1.13} . \tag{6.2.10}$$

The effect of sphere vibration on heat transfer in air,

$$d = 1.905, 2.54\,\mathrm{cm}, A = 4 - 25.5\,\mathrm{cm}, f = 25 - 155\,\mathrm{Hz} ,$$
$$\mathrm{Re}_{\Delta u} = \sqrt{2}\pi A f d/\nu, \mathrm{Re}_{\Delta u} > 220, \mathrm{Gr}_d = 4 \cdot 10^3 - 6 \cdot 10^4 ,$$
$$\frac{\mathrm{Nu}}{\mathrm{Nu}_0} = 0.83 \left[(A/d)^{0.1}\, \mathrm{Re}_{\Delta\omega}^{0.5} / \mathrm{Gr}_d^{0.25} \right]^{1.28} . \tag{6.2.11}$$

For $\mathrm{Re}_{\Delta u} < 220$ the oscillations do not affect the heat transfer coefficient. The maximum increase in the heat transfer coefficient (7 times) was observed for $(A/d)^{0.1} \mathrm{Re}_{\Delta\omega}^{0.5} / \mathrm{Gr}_d^{0.25} = 5$.

Figure 6.2.1 presents the results of stroboscopic framing of the flame glow and the interferograms for the cases of the combustible mixture flow velocity disturbance on exit from a nozzle with frequencies in the range $f = 50$–200 Hz.

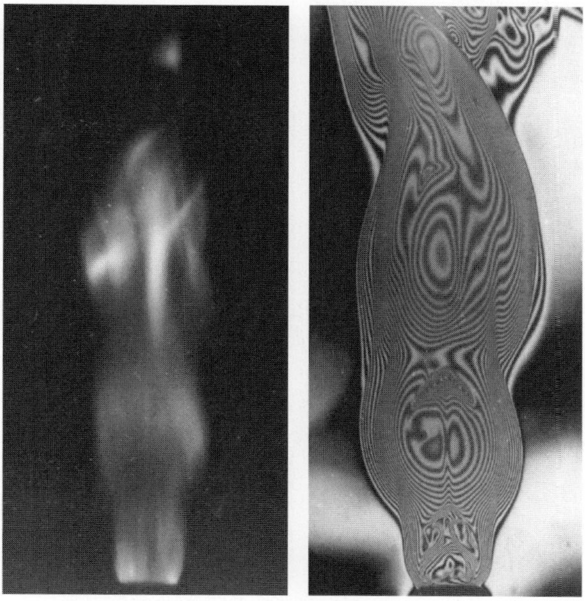

Fig. 6.2.1a. Results of stroboscopic framing of the flame light and interferogram for the case of flow velocity disturbance for a combustible mixture on exit from a nozzle with the frequency $f = 50\,\mathrm{Hz}$

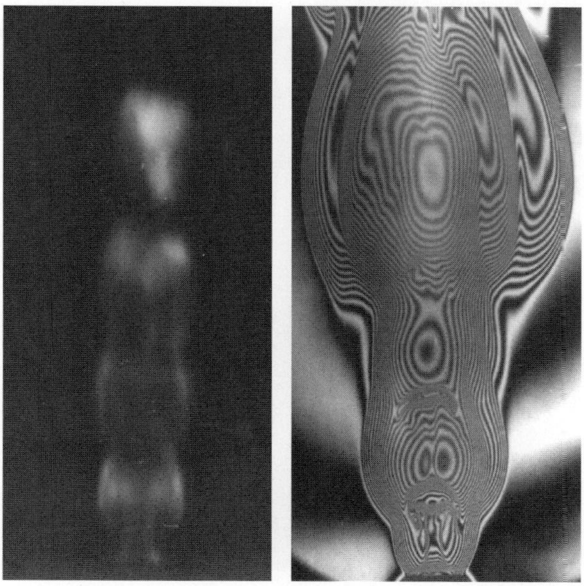

Fig. 6.2.1b. Results of stroboscopic framing of the flame light and interferogram for the case of flow velocity disturbance for a combustible mixture on exit from a nozzle with the frequency $f = 75\,\mathrm{Hz}$

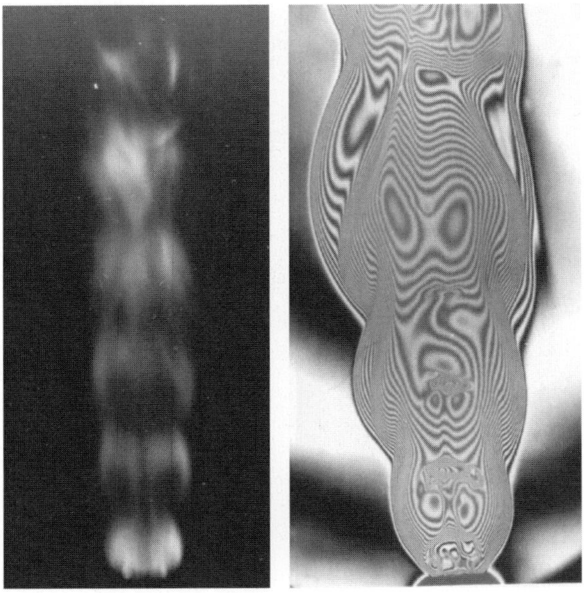

Fig. 6.2.1c. Results of stroboscopic framing of the flame light and interferogram for the case of flow velocity disturbance for a combustible mixture on exit from a nozzle with the frequency $f = 100\,\mathrm{Hz}$

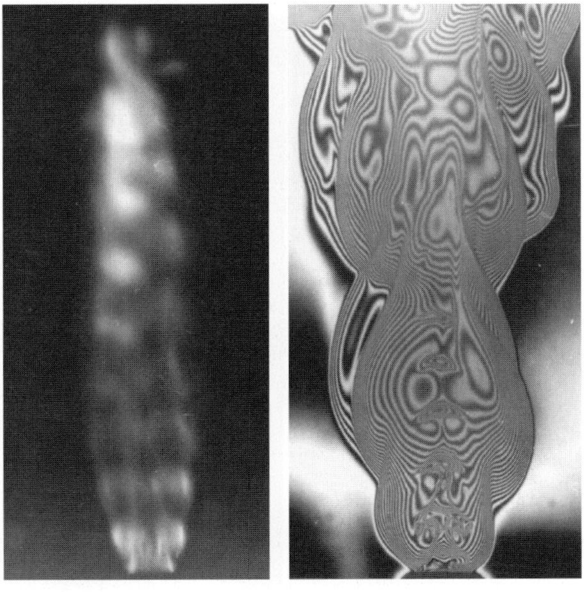

Fig. 6.2.1d. Results of stroboscopic framing of the flame light and interferogram for the case of flow velocity disturbance for a combustible mixture on exit from a nozzle with the frequency $f = 150\,\mathrm{Hz}$

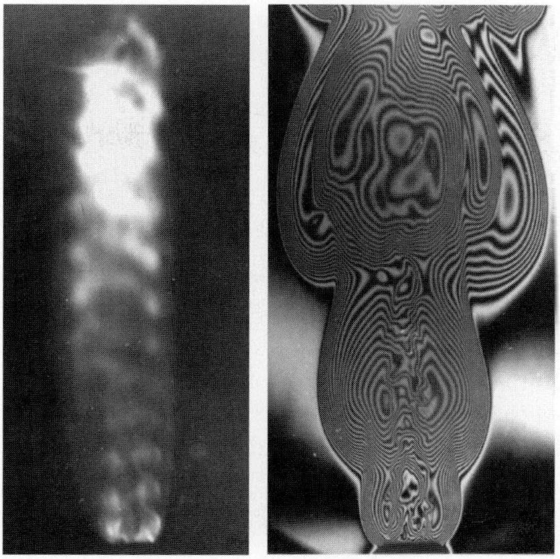

Fig. 6.2.1e. Results of stroboscopic framing of the flame light and interferogram for the case of flow velocity disturbance for a combustible mixture on exit from a nozzle with the frequency $f = 200\,\mathrm{Hz}$

As is seen from the given photographs, two types of vortices can be distinguished in the flame. First, frequency-independent large-scale vortices covering the torch peripheral area and occurring due to the loss of flame convective stability. Second, a row of consecutive torr-shaped vortices induced by a pulse of velocity of mixture supply.

Heat transfer by free convection ***between vertical plates in the presence of vibration in a vertical direction***

$$H/B = 9.4 - 42.7, f = 0 - 400\,\mathrm{Hz}, g_t = 0 - 100\ g\ ,$$
$$\mathrm{Ra}_B = 1.25 \cdot 10^5 - 8.3 \cdot 10^7\ ,$$
$$\mathrm{Nu} = 1.117 \left(\frac{\mathrm{Ra}_B}{10^4}\right)^{0.03106 z \ln(1+g_t)+0.304}\ , \qquad (6.2.12)$$

where

$$g_t = A\omega^2/g,\ z = \frac{f/f_n}{\left\{\left[1-(f/f_n)^2\right]+(1.4654 f/f_n)^2\right\}^{1/2}}$$

and f_n is the resonance oscillation frequency corresponding to the first (main) resonance frequency.

Formula (6.2.12) was obtained in experiments with water, but it can be also recommended for other liquids. Depending on the oscillation frequency, the boundary layer on a plate can be both laminar and turbulent. In a frequency range near the first resonance harmonic the turbulent boundary layer is observed.

Heat transfer ***of a horizontal platinum wire in a vertical cylindrical tank with water***, ***when the whole system (heated surface and liquid) is exposed to vibration***

$$d = 0.81\,\text{mm},\ H \times D = 457 \times 140\,\text{mm},\ f = 10 - 3000\,\text{Hz}\ ,$$
$$\text{Ra}_d = 5 - 2 \cdot 10^3,\ \text{Re}_{\Delta u}^{1/4}\,\text{Pr}^{0.6}/\left(\text{Ra}_d^{0.26}\text{Re}_\omega^{0.4}\right) = 0.004 - 6\ ,$$
$$J^{0.46}/\left(\text{Ra}_d^{0.21}\text{Re}_\omega^{0.1}\right) = 0 - 30\ ,$$
$$\frac{\text{Nu}}{\text{Nu}_0} = 1 + 2.64\frac{\text{Re}_{\Delta U}^{1/4}\,\text{Pr}^{0.6}}{\text{Ra}_d^{0.26}\text{Re}_\omega^{0.4}} + 4 \cdot 10^{-3}\frac{J^{0.46}}{\text{Ra}_d^{0.21}\text{Re}_\omega^{0.1}}\ . \quad (6.2.13)$$

Figure 6.2.2 presents the diagrams of the dependence of the distance between vortices on the frequency of disturbing pulses (a) and of the dependence of the vortex motion velocity V on the disturbance frequency $f(b)$.

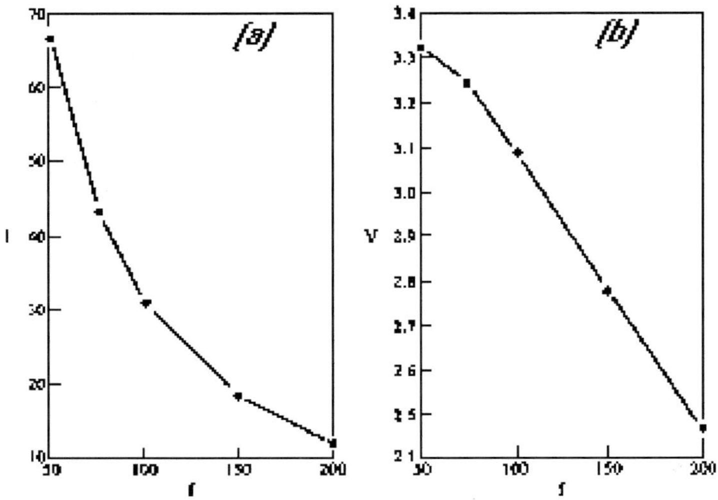

Fig. 6.2.2. The graph of the dependence of the distance between vortices on the frequency of disturbing pulses (**a**) and the graph of the dependence of the vortex motion velocity V on the disturbance frequency f (**b**)

A steel sphere of diameter d is placed in a vessel of diameter D filled with water up to the level $D/2$. The average heat transfer value for the sphere under the conditions of vibration of the whole system with acceleration g_t [6.70]:

$$d = 100\,\text{mm},\ D = 300\,\text{mm},\ g_t = 0 - 8g\ ,$$
$$\text{Ra}_d = 0.8 \cdot 10^7 - 2 \cdot 10^8,\ f = 30 - 80\,\text{Hz}\ ,$$
$$\text{Nu} = 36.5\text{Ra}_d^{0.067}\ . \quad (6.2.14)$$

The resonance frequency of oscillations of the liquid level was at $f_r = 100\,\text{Hz}$. For vibration with this frequency boiling of the liquid due to vibration was observed.

The thermal convection ***in a two-dimensional square cavity*** caused by total action of gravitation and ***vertical vibration*** was investigated numerically in [6.71]. The frequency of vibrations changed from 1 to 10^4 for the Rayleigh numbers equal to 0, 10^4, and 10^6. Depending on the individual characteristics, the behaviour of the thermal convection can refer to one of five types: (I) quasi-static convection; (II) vibroconvection; (III) resonance vibroconvection; (IV) intermediate convection; and (V) high-frequency vibroconvection. For high Rayleigh numbers (10^6) the thermal gravitational convection prevails, and the vibrational motion does not cause strong heat transfer intensification. For low Ra ($\sim 10^4$) everywhere, except for the region of quasi-static convection, vibrational thermal convection dominates, therefore, vibration intensifies heat transfer substantially. The region of quasi-static convection is determined as $\omega = \mathrm{Pr}^{1/2}\mathrm{Ra}^{3/4}(2G)^{-1/4}$, where $G = (\beta b\Omega(T_h - T_c)L^2)/2\nu^2$ is the vibration Grashof number; Ω, ω is the angular and dimensionless vibration frequency; b is the vibration amplitude; L is the cavity length; and T_h and T_c are the temperatures of hot and cold walls.

The results of numerical calculations of non-stationary natural-convective flow of viscous ***thermally stratified liquid in a container of square section with a torsion-oscillating top cover*** show that for the large oscillation frequencies the flow is concentrated in a thin layer near the oscillating surface, whereas for small frequencies the whole cavity volume is under an effect of oscillatory cover movement. ***Between horizontal surfaces the constant temperature difference is maintained, two vertical walls are insulated***, the value of the oscillatory Reynolds number Re = 400, 1000 and 2000, where $\mathrm{Re} = U_0 L/\nu$, $0.1 \le \omega' \le 10$, where $\omega' = \omega L/U_0$; ω and U_0 are the frequency and the maximal speed of the oscillations of the cover, L is the container height (width); $0 < \mathrm{Gr} < 10^6$, $\mathrm{Pr} = 0.7$. The vertical natural-convective flow caused by steady stratification weakens as the parameter $\mathrm{Gr}/\mathrm{Re}^2$ increases, and the vertical heat transfer inside the container is most effective for $\omega' \gg 1$. For certain values of ω' the resonance leading to the flow intensification and the associated heat transfer increase inside the cavity is observed [6.72].

Analytical and numerical investigation of stability of a ***flat horizontal layer of a binary mixture in the presence of the Soret effect*** under the conditions of transverse high-frequency vibration was carried out in [6.73]; it showed that irrespective of the properties of the binary mixture the effect of the transverse vibration always leads to layer stabilization.

Numerical calculation of the parameters of nonstationary free convection ***in a square cavity under the action of temperature and concentration gradients of the same amplitudes and opposite directions*** was carried out in [6.74]. The boundary conditions on lateral walls were selected so that the ratio $N = \mathrm{Gr}_s/\mathrm{Gr}_T = -1$, where Gr_s and Gr_T are the solutal and temperature Grashof numbers. In such a system the stationary convective flow remains steady up to some critical value of Gr_{c1} which depends on the Le number changing from 2 to 45. Beyond the value of Gr_{c1} the oscillatory flow begins, and the beginning of transition to it depends on the value of the Le number. For Gr_T slightly exceeding Gr_{c1}, depending on the value of Le, there can be either an axisymmetric oscillatory flow (for $\mathrm{Le} \geq 17$) or an asymmetric flow (for $\mathrm{Le} \leq 17$). For large values of Gr_T (for fixed values of Gr_s and Le) both flow regimes can take place simultaneously. The calculation shows that Gr_{c1} reaches its minimum value equal to $4.75 \cdot 10^4$ for $\mathrm{Le} \cong 7$.

The effect of a sinusoidal gravity modulation on natural convection of fluids with constant property, ***in vertical slots of aspect ratio of 20 and 15*** was studied numerically in [6.75]. It was found that a low frequency g-modulation had a strong effect on the fluid flow but little effect on the heat transfer rate. The flow of air usually enters a periodic state after two cycles of g-modulation, much faster than that of large Prandtl number fluids. For high Pr number fluids, a flow pattern, consisting of horizontal secondary cells, preceded the onset of the commonly observed vertical secondary cells. At low modulation frequencies, stability of convection is enhanced, as demonstrated by the increase in the critical Gr number.

Heat and momentum transport is investigated theoretically and numerically; ***a rectangular enclosure filled with clear fluid or with fully saturated porous medium***, is considered under the conditions time-periodic horizontal heating [6.76]. Numerical simulations, of various configurations, indicate that the natural convection activity within the enclosure peaks at several discrete frequencies, with the climax attained at a heating frequency referred to as resonance frequency. It is shown that the convection intensity within the enclosure increases linearly with the heating amplitude for a wide range of Rayleigh numbers, Prandtl numbers, enclosure aspect ratios, and Darcy numbers for the porous medium case. Moreover, the flow response to pulsating heat is continuously enhanced as the system becomes more permeable. A general correlation is obtained for predicting the resonance amplitude Nusselt number as a function of the heating amplitude A, Rayleigh number Ra, and Darcy number Da

$$\mathrm{Nu} = [0.055 - 0.0037/(\mathrm{Ra}^{0.05}\mathrm{Da})^{0.317}]A\mathrm{Ra}^{0.18} \,. \qquad (6.2.15)$$

A rectangular cavity heated from the side and ***filled with liquid undergo vertical oscillations according to the harmonic law***. The modulation of the gravity field exerts a significant influence on conditions of the occurrence of convective instability and on the structure of secondary convective flows appearing on loss of stability [6.77].

Heat transfer ***of a cylindrical thermosyphon fully filled with molten sodium oscillating under the conditions of reciprocating movement*** with an error of (–8)–23% is described as [6.78]

$$d = 23\,\text{mm},\ L = 180\,\text{mm},\ l = 20, 50, 90\,\text{mm},\ A = 0 - 20\,\text{mm}\ ,$$
$$f = 0 - 25\,\text{Hz},\ \text{Pr} = 0.006,\ \text{Ra}_l = 20 - 1.1 \cdot 10^3\ ,$$
$$\text{Nu} = 0.86\,(\text{Ra}_l\,\text{Pr})^{0.25}\,(l/l_1)^{0.25}\,\text{Jg}^{0.25}\ , \qquad (6.2.16)$$

where l and l_1 are the lengths of the heated and cooled parts, $\text{Jg} = \omega^2 A/g$.

When the thermosyphon is partially filled, the heat transfer rate on the cooled and heated parts appears to be different. The heat transfer on the heated part is determined by (6.2.16), and on the cooled part it is only a function of the filling degree. When the cavity is filled up to the beginning of the cooled part, the thermal resistance of the volume above the heated part is negligibly small and it can be ignored in heat transfer calculations for $\text{Jg} \geq 0.15$.

A vertical cylindrical container filled with ***liquid*** undergoes vibration in a wide range of frequencies. The number $\text{Ra} = 0$, 10^4, and 10^5. Investigation of frequency dependence of the heat transfer characteristics, especially in the range of resonance frequencies, showed that the thermovibrational convection (the vibration number $\text{Gr}_v = 10^5; 10^6$) considerably increases the heat transfer rate in the vertical cylinder (by 7–15% for $\text{Gr}_v = 10^5$ and by 50–65% for $\text{Gr}_v = 10^6$) [6.79].

The flow and heat transfer characteristics in ***a transversely oscillating cylinder*** were investigated in [6.80] for $0 \leq \text{Re} \leq 300$, $0 \leq S_c \leq 0.3$, $0 \leq A/D \leq 0.7$ ($S_c = f_c D/U_0$ is the dimensionless frequency of oscillations). The Prandtl number was equal to 0.71 or 7.0. The generalizing formula for heat transfer rate as a function of the principal parameters is

$$\text{Nu} = 0.68\text{Re}^{0.485}\,\text{Pr}^{0.4}[1 + 0.0944(A/D) + 0.118(A/D)^2]\ , \qquad (6.2.17)$$

where A is the amplitude of cylinder oscillations.

The characteristics of a vibrational flow ***with a sinusoidal change in the wall temperature in a round tube*** were theoretically investigated in [6.81]. The results obtained can be classified as three generalized formulas for three different regimes depending on the dimensionless oscillation frequency β and the parameter $\lambda = \pi L_S/2L_\text{w}$:

for $\beta \ll 1$ and $\beta\sqrt{\lambda} \ll 1$

$$\text{Nu} = 6\ , \qquad (6.2.18)$$

for $\beta \gg 1$ and $\lambda \ll 1$

$$\text{Nu} = \sqrt{2}\beta\ , \qquad (6.2.19)$$

for $\lambda \gg 1$ and $\beta\sqrt{\lambda} \gg 1$

$$\text{Nu} = 1.52\beta\sqrt{\lambda}\ , \qquad (6.2.20)$$

where $\beta = R\sqrt{(\omega/\nu)\,\text{Pr}}$, R is the inner radius of the tube, $\lambda = \pi L_\text{S}/2L_\text{w}$ is the swept length ratio, L_S is the swept length, L_w is the period of sinusoidal wall temperature distribution.

Experimental data on convective heat transfer ***from a cylinder in an intense acoustic field which induces a strong oscillating flow*** was carried out in [6.82]. Only in the case of low amplitudes it was assumed that the amplitude of the displacement of the particle in the oscillating flow is small in comparison with the cylinder diameter. Two different flow regimes were identified. The first regime is a laminar flow with the corresponding dependence of the Nusselt number on the vibrational Reynolds number (R_s). The second regime is less understood. It is an unstable flow in which vortical motion dominates and intensifies heat transfer rates. The generalized relations for both regimes for the case of air are applicable also for frequently used liquids.

$$\mathrm{Nu} = 0.20 R_s^{0.75} , \tag{6.2.21}$$

where $R_s = \varepsilon^2 \Lambda^2$, Λ and ε is the frequency and amplitude of oscillations; $\varepsilon = a/d$, $\Lambda = d/\delta$, a is the amplitude of the displacement of fluid oscillation, d is the cylinder diameter, $\delta = (\nu/\omega)^{1/2}$ is the Stokes layer thickness, ν is the kinematical viscosity, and ω is the angular frequency.

Acoustic oscillations of air near a horizontal cylinder. For a fully developed turbulent flow there are empirical dimensionless dependences for $d = 22.2\,\mathrm{mm}$, $A_0 = 0–4.05\,\mathrm{mm}$, $f = 54–225\,\mathrm{Hz}$, $\theta_\mathrm{w} \leq 85°\mathrm{C}$, and $\mathrm{Ra}_d > 3{\cdot}10^4$:

vertical oscillations

$$\mathrm{Nu}/\mathrm{Nu}_0 = 3.32\,(Af)\,(\theta_\mathrm{w}/d)^{-0.05} , \tag{6.2.22}$$

transverse horizontal oscillations

$$\mathrm{Nu}/\mathrm{Nu}_0 = 2.67 d^{1/4} \theta_\mathrm{w}^{1/12} (Af)^{2/3} . \tag{6.2.23}$$

The heat transfer rate of ***a horizontal cylinder with transverse acoustic oscillations*** for $d = 19$ mm, $f = 1496$ Hz, $l/d = 6.04$, $\theta_\mathrm{w} < 150°\mathrm{C}$, $M_{\Delta u} = \Delta u/a$ is described as

$$\mathrm{Nu} = 14.2 \left(0.98 \mathrm{Ra}_d M_{\Delta U}^2\right)^{1/2} . \tag{6.2.24}$$

Figure 6.2.3 presents the results of radial scanning of a flame by a chromel-alumel thermocouple at the distance from the nozzle exit of nozzle equal to $3D$. As is seen from the graph, in the case of the absence of disturbances there is a clearly expressed decrease in the average temperature in the torch center. For the frequency of disturbing pulses $f = 50\,\mathrm{Hz}$ the decrease becomes much weaker, and for higher frequencies it practically fully disappears. Such temperature distribution is caused by intense convective heat transfer between the torch center and its peripheral areas and by the displacement of the zone of reaction to the torch axis as a result of vortical motion. Obviously, in these conditions the efficiency of propan-buthan combustion and, hence, the thermal efficiency of the combustion process appears much more comparable with combustion in the absence of disturbance.

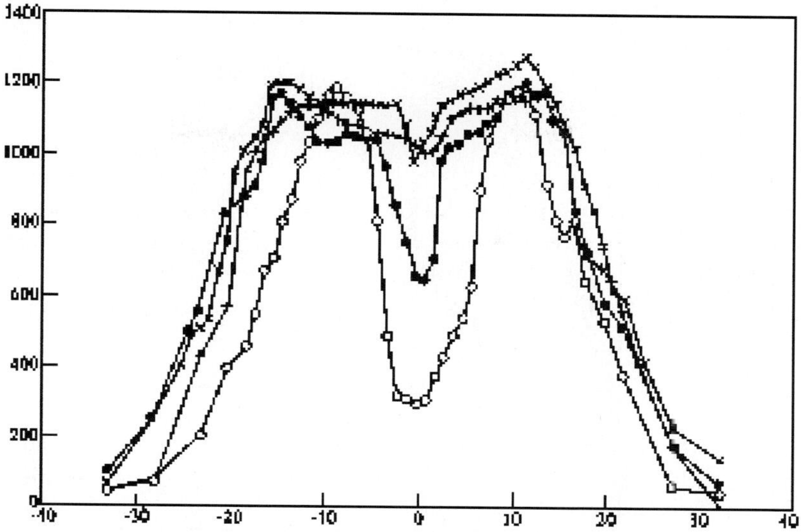

Fig. 6.2.3. Results of radial scanning of a flame by a chromel-alumel thermocouple at a distance from the nozzle exit equal to $3D$

Laminar natural convection ***near a bottom frontal point under periodic pressure oscillations with a wavelength much greater than the diameter of the horizontal cylinder***. The velocity distribution on the external border of a boundary layer is determined by the law of a potential flow. Dimensionless heat transfer from the cylinder surface is given by [6.83]

$$\mathrm{Nu}/\mathrm{Nu}_0 = \left(1 \pm 4\Delta U^2/gd\right)^{1/4} . \tag{6.2.25}$$

The upper sign corresponds to the case of horizontal oscillations, the lower, to vertical. Formula (6.2.25) generalizes the data of experiments (6.2.22)–(6.2.24) with an error of 15%.

A horizontal cylinder in a transverse flow by a high-frequency, low-amplitude sound wave. The average heat transfer on the cylinder surface Nu and in the vicinity of the bottom frontal point Nu_{d0} for $\mathrm{Pr} \leq 1$, $\mathrm{Re}_v = \Delta u^2/\omega\nu$:

$$\mathrm{Nu}^2 = 1.6\mathrm{Re}_v \,\mathrm{Pr} + 4\sqrt{\mathrm{Re}_v^2 \,\mathrm{Pr}^2 + \mathrm{Ra}_d/200} , \tag{6.2.26}$$

$$\mathrm{Nu}_{d0}^2 = 4\mathrm{Re}_v \,\mathrm{Pr} + 4\sqrt{\mathrm{Re}_v^2 \,\mathrm{Pr}^2 + \mathrm{Ra}_d/60} . \tag{6.2.27}$$

For the small oscillation intensity the Nu values tend to the appropriate values for steady free convection, and for the large one the heat transfer is mainly due to an oscillating flow.

In Fig. 6.2.4 there is a longitudinal distribution of the average temperature in the torch center.

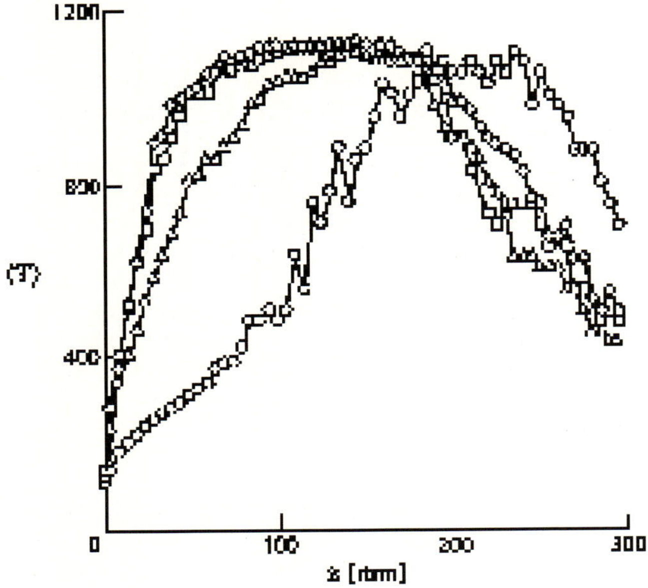

Fig. 6.2.4. Longitudinal distribution of the average temperature in the torch center

Heat transfer of a ***horizontal cylinder in a high-frequency flow of small amplitude*** for $\mathrm{Pr} < 1$ is given by

$$\mathrm{Nu} = 2.18\,(\mathrm{Re}_v\,\mathrm{Pr})^{0.25}\,\mathrm{Gr}_d^{0.278}\,. \tag{6.2.28}$$

For a horizontal cylinder in the presence of the sonic field along the axis in standing acoustic waves the heat transfer coefficient increases more than twice.

In [6.60] for the effect of external parameters (vibration, sound oscillations, electrical field in a conducting liquid, assisting flow, etc.) the approximate method of calculation of heat transfer is suggested. According to this method the free convection equations are supplemented with the criterion $\mathrm{Re}_s^2 = \mathrm{Gr} + \sum \mathrm{Re}_i^2$ instead of the number Gr. For a plate and a horizontal cylinder the error of calculations by the suggested technique does not exceed 20%.

Time-averaged vibrational flows induced by oscillations ***of a heated (or cooled) solid sphere immersed in a liquid*** were investigated in [6.84]. ***The cases of an infinite surrounding liquid and the liquid placed into a spherical rigid envelope are considered. The gravity is absent.*** The interaction of thermovibrational flows and vibrational Schlichting flows was studied. It was found that the vibrational flow depends substantially on the layer thickness and can be rather complicated in the case of small thicknesses. At a fixed layer thickness, the creeping flow structure is defined by only one parameter, which is the ratio of the vibrational Grashof number and Schlichting parameter.

References

1. O. G. Martynenko and Yu. A. Sokovishin, *Free-Convective Heat Transfer on a Vertical Surface (II Kind Boundary Conditions)*, Nauka i Tekhnika Press, Minsk (1978).
2. K. Lal, *Rev.Roun. Sci. Techn.- Mech. Appl.*, **13**, 823 (1968).
3. K. Mizukami, *Int. J. Heat Mass Transfer*, **20**, 981 (1977).
4. R. Siegel, *Trans. ASME, J. Heat Transfer*, **80**, 345 (1958).
5. S. D. Harris, L. Elliot, D. B. Ingham, and I. Pop, *Int. J. Heat Mass Transfer*, **41**, 357 (1998).
6. J. S. Park and J. M. Hyun, *Int. J. Heat Mass Transfer*, **41**, 4393 (1998).
7. I. Pop and V. M. Soundalgekar, *Rev. Roun. Phys.*, **20**, 843 (1975).
8. S. W. Churchill and R. Usagi, *IEC Fund.*, **13**, 39 (1974).
9. S. W. Churchill, *Lett. Heat Mass Transfer*, **2**, 311 (1975).
10. N. Miyamoto, *Int. J. Heat Mass Transfer*, **20**, 1258 (1977).
11. Z. Dursunkaya and W. M. Worek, *Int. J. Heat Mass Transfer*, **35**, 2060 (1992).
12. E. E. Prokhach, *Inzh.-Fiz. Zh.*, **31**, 857 (1976).
13. G. S. Ambrok, in: *Investigation of Unsteady Heat and Mass Transfer*, Minsk (1966), pp. 82–91.
14. B. Zeghmati and M. Daguenet, *Int. J. Heat Mass Transfer*, **34**, 899 (1991).
15. K. Velusamy and V. K. Garg, *Int. J. Heat Mass Transfer*, **35**, 1293 (1992).
16. K. Slimi. S. B. Nasrallah, and J.-P. Fohr, *Int. J. Heat Mass Transfer*, **41**, 1113 (1998).
17. E. K. Kalinin, G. A. Dreitser, A. S. Neverov, in: *Hydraulic*, Moscow (1976), pp. 101–111.
18. T. L. Elliott, *Quart. J. Mech. Appl. Math.*, **23**, 153 (1970).
19. M. Katagiri and I. Pop, *Wu S*, **12**, 73 (1979).
20. A. Slaouti, H. S. Takhar, and G. Nath, *Int. J. Heat Mass Transfer*, **41**, 3397 (1998).
21. M. Arunachalam and N. R. Rajappa, *Acta Mech.*, **35**, 231 (1980).
22. J. R. Parsans and J. C. Mulligan, *Trans. ASME, Ser. C, J. Heat Transfer*, **100**, 423 (1978).
23. N. Katagiri and I. Pop, in: *Numerical Methods in Laminar and Turbulent Flow*, Swansea (1978), pp. 821–829.
24. H. D. Nguyen, S. Paik, and I. Pop, *Int. J. Heat Mass Transfer*, **40**, 379 (1997).
25. R. Ganapathy, *Int. J. Heat Mass Transfer*, **40**, 1551 (1997).
26. B. Yan, I. Pop, and D. B. Ingham, *Int. J. Heat Mass Transfer*, **40**, 893 (1997).
27. C. P. Chiu and W. R. Chen, *Int. J. Heat Mass Transfer*, **39**, 1439 (1996).
28. A. J. Ede, *Adv. Heat Transfer*, **4**, 1 (1967).
29. J. W. Yang, C. Scaccia, and J. Gugman, *Trans. ASME, Ser. C, J. Heat Transfer*, **96**, 9 (1974).
30. I. V. Il'inskii, E. E. Prokhach, and V. P. Pershin, *Teplofizika Teplotekhnika*, No. 2, 136 (1973).
31. A. Bar-Cohen and A. Herman, *Lett. Heat Mass Transfer*, **4**, 111 (1977).
32. S. W. Armfield and J. C. Patterson, *Int. J. Heat Mass Transfer*, **34**, 929 (1991).
33. C. G. Jeevaraj and J. C. Patterson, *Int. J. Heat Mass Transfer*, **35**, 1573 (1992).
34. D. C. Dalal, N. Datta, and S. K. Mukherjea, *Int. J. Heat Mass Transfer*, **41**, 547 (1998).
35. E. P. Kostogorov, E. A. Shtessel, and A. G. Merzhanov, *Inzh.-Fiz. Zh.*, **37**, 5 (1979).

36. J. Chang, T. F. Lin, and C. H. Chien, *Int. J. Heat Mass Transfer*, **36**, 1315 (1993).
37. T. Fusegi, J. M. Hyun, and K. Kuwahara, *Int. J. Heat Mass Transfer*, **34**, 1559 (1991).
38. K. V. Pribytkova, S. I. Khudyaev, and E. A. Shtessel, in: *Heat and Mass Transfer-VI*, vol. 3, Minsk (1980), pp. 38–45.
39. E. K. Kalinin, G. A. Dreitser, and A. S. Neverov, in: *Heat and Mass Transfer in between Interaction Flows and Surfaces*, Moscow (1978), pp. 63–68.
40. G. H. Min and F. A. Kulacki, *Nucl. Sci. Des.*, **54**, 267 (1979).
41. K. Vafai and J. Ettefagh, *Int. J. Heat Mass Transfer*, **34**, 2555 (1991).
42. P. Wang, R. Kahawita, and D. L. Nguyen, *Int. J. Heat Mass Transfer*, **34**, 1429 (1991).
43. E. P. Kostogorov, E. A. Shtessel, and A. G. Merzhanov, in: *Heat and Mass Transfer-V*, vol. 1, Part 2, Minsk (1976), pp. 305–308.
44. G. D. Callahan and W. J. Marner, *Int. J. Heat Mass Transfer*, **19**, 165 (1976).
45. I. Pop, *Rev. Roun. Sci. Techn.- Mech. Appl.*, **13**, 41 (1968).
46. V. M. Soundalgekar, *Int. J. Heat Mass Transfer*, **15**, 1253 (1972).
47. G. C. Pande, *Ind. J. Pure Math.*, **6**, 819 (1975).
48. V. M. Soundalgekar and P. D. Wavre, *Int. J. Heat Mass Transfer*, **20**, 1363 (1977).
49. C. Y. Soong, P. Y. Tzeng, D. C. Chiang, and T. S. Sheu, *Int. J. Heat Mass Transfer*, **39**, 2869 (1996).
50. M. Okada and T. Suzuki, *Int. J. Heat Mass Transfer*, **40**, 3201 (1997).
51. T. D. Upton and D. W. Watt, *Int. J. Heat Mass Transfer*, **40**, 2679 (1997).
52. C. G. Jeevaraj and J. C. Patterson, *Int. J. Heat Mass Transfer*, **35**, 1573 (1992).
53. J. C. Patterson and J. Imberg, *J. Fluid Mech.*, **100**, 65 (1980).
54. M. Kazmierczak and Z. Chinoda, *Int. J. Heat Mass Transfer*, **35**, 1507 (1992).
55. W. Schopf and J. C. Patterson, *Int. J. Heat Mass Transfer*, **39**, 3497 (1996).
56. T. Nishimura, Y. Ogata, S. Sakura, and A. M. Morega, *Int. J. Heat Mass Transfer*, **42**, 1479 (1999).
57. J. Dosch and H. Beer, *Int. J. Heat Mass Transfer*, **35**, 1811 (1992).
58. Y.-M. Chen and J.-K. Liou, *Int. J. Heat Mass Transfer*, **40**, 711 (1997).
59. Yu. E. Karyakin, O. G. Martynenko, Yu. A. Sokovishin, and Kh. Z. Ustok, *Numerical Modeling of Nonstationary Natural Convection in Triangular Enclosures*, Minsk (1985).
60. V. M. Buznik, *Intensification of Heat Transfer in Ship Equipments*, Sudostroeniye Press, Leningrad (1969).
61. B. M. Galitseiskii, Yu. A. Ryzhov, and E. V. Yakush, *Heat and Hydrodynamics Processes in Oscillating Flows*, Mashinostroeniye Press, Moscow (1977).
62. R. G. Galiullin, V. B. Repin, and N. Kh. Khalitov, *Viscous Flows and Heat Exchange of Bodies in the Acoustic Field*, Kazanskii Universitet Press, Kazan (1978).
63. J. C. Dent, *Int. J. Heat Mass Transfer*, **11**, 605 (1968).
64. V. A. Demin, G. Z. Gershuni, and V. Verkholyantsev, *Int. J. Heat Mass Transfer*, **39**, 1979 (1996).
65. T. S. Sarma and S. P. Sukhatme, *Ind. J. Technol.*, **8**, 439 (1970).
66. P. P. Shilonosov, N. P. Ryazhentsev, and O. G. Shlyapkin, in: *Electrical Devices of Shock Action*, Novosibirsk (1969), pp. 39–46.

67. E. I. Udler, *Investigation of Free-Convective Heat Transfer of Electromagnetic Backward-Translatory Engines*, Abstract of Doctoral Dissertation (in Technical Sciences), Tomsk (1972).
68. K. K. Prasad and V. Ramanathan, *Int. J. Heat Mass Transfer*, **15**, 1213 (1972).
69. N. V. Selivanov, *Investigation of Heat Transfer between Liquid and Vertical Bounding Surfaces in Combined Free and Forced Convection Generated by Harmonic Oscillations of the Enclosure*, Abstract of Doctoral Dissertation (in Technical Sciences), Moscow (1979).
70. F. S. Dopkin and N. I. Kobasko, *Prikl. Mekh.*, **16**, 128 (1980).
71. W. S. Fu and W. J. Shieh, *Int. J. Heat Mass Transfer*, **35**, 1695 (1992).
72. R. Iwatsu, J. M. Hyun, and K. Kuwahara, *Int. J. Heat Mass Transfer*, **35**, 1069 (1992).
73. G. Z. Gershuni, A. K. Kolesnikov, J. C. Legros, and B. I. Myznikova, *Int. J. Heat Mass Transfer*, **42**, 547 (1999).
74. K. Ghorayeb, H. Khallouf, and A. Mojtabi, *Int. J. Heat Mass Transfer*, **42**, 629 (1999).
75. Y. Y. Jin and C. F. Chen, *Int. J. Heat Mass Transfer*, **40**, 1411 (1997).
76. B. V. Antohe and J. L.Lage, *Int. J. Heat Mass Transfer*, **39**, 1121 (1996).
77. G. Z. Gershuni, E. M. Zhuhovitski, and Yu. S. Yurkov, *Konvektivnye Techeniya* (Perm), No. 1, 73 (1979).
78. E. P. Dyban, M. V. Starodomskii, and V. A. Asmolovskii, in: *Heat and Mass Transfer-I*, part 1, Minsk (1972), pp. 346–350.
79. F. T. Ferguson and L. U. Lilleleht, *Int. J. Heat Mass Transfer*, **39**, (28951996).
80. C.-H. Cheng, J.-L. Hong, and W. Aung, *Int. J. Heat Mass Transfer*, **40**, 1825 (1997).
81. D.-Y. Lee, S.-J. Park, and S. T. Ro, *Int. J. Heat Mass Transfer*, **38**, 2529 (1995).
82. A. Gopinath and D. R. Harder, *Int. J. Heat Mass Transfer*, **43**, 505 (2000).
83. A. I. Leont'ev and V. M. Fomichev, *Mekh. Zhidkosti Gaza*, No. 5, 203 (1979).
84. D. V. Lyubimov, A. A. Cherepanov, T. P. Lyubimova, and B. Roux, *Int. J. Heat Mass Transfer*, **38**, 2089 (1995).

7

Heat Transfer by Mixed Convection

Calculations of heat transfer from high-temperature surfaces lead to the necessity of simultaneous allowance for the effects of different heat transfer mechanisms on the heat exchange characteristics. The role of radiative-convective heat transfer, when radiative, convective and conductive heat exchange occur simultaneously, is especially large. The complex heat transfer processes occur in this case in emission, absorption, and scattering of radiation by a liquid. However, under certain conditions the air is devoid of the indicated properties, and the effect of radiation should be taken into account only in the heat balance on the wall. If the wall has a constant temperature, the heat transfer for a nonabsorbing ambient is determined separately for radiating and convective components [7.1–7.3].

During simultaneous action of free and forced convection, the Reynolds number Re and the criteria describing the geometry of the boundaries and the flow orientation relative to the gravitational field are also essential in addition to Pr and Gr numbers. Depending on the correlation between viscous, buoyant, and inertial forces it is possible to distinguish different liquid movement regimes and heat transfer conditions. For the large Gr and small Re numbers the effect of forced convection on heat transfer can be neglected. For the total action of free and forced convection the largest practical interest represents a laminar regime. The viscous regime corresponds to natural convection with a disturbing effect of forced motion, and, conversely, for turbulent motion the free convection has a little effect on motion and is a small value in the total thermal balance [7.4].

7.1 Effect of Radiation on Free-Convective Heat Transfer

The heat flux q_{w} generated by a wall is transferred away by the radiating component q_r dependent on local temperature, and by the convective-conductive one q_t. The wall is a gray diffusive radiator with the emissivity ε, and for the radiative component we can use the Stefan–Boltzmann law:

$$q_{\mathrm{w}} = q_r + q_t = -\lambda \left(\frac{\partial T}{\partial n} \right)_0 + \varepsilon \sigma \left(T_{\mathrm{w}}^4 - T_\infty^4 \right) . \tag{7.1.1}$$

If the temperature T_{w} is assigned on the surface, the radiation has no effect on the liquid flow, since the liquid does not absorb. In the case of the assignment of a heat flux q_{w} on the surface, the wall temperature changes substantially and there occurs redistribution between the radiative and convective-conductive components of the heat flux. In the criterial dependence for heat transfer [7.1]

$$\mathrm{Nu} = f(\mathrm{Gr}^*, \mathrm{Pr}, \mathrm{Ki}, R) . \tag{7.1.2}$$

there appear the criteria Ki and R that establish correlations between heat fluxes on the surface.

The coupled radiation and laminar mixed convection in an absorbing and radiating gas mixture flowing past a vertical plate were investigated in [7.5]. Comparison of the results obtained for the cases with and without radiation under different conditions shows that liquid radiation increases the effect of buoyancy forces, increases the temperature, velocity, and conductive heat transfer near the wall and decreases the radiating heat flux on the wall. The study of ***boundary conditions at infinity*** shows that at the atmospheric pressure the radiation penetrates to the depth which is approximately equal to one equivalent meter in pure $\mathbf{H_2O}$ and which is an order of magnitude higher than the boundary layer thickness for mixed convection without liquid radiation.

Near the leading edge of a vertical plate at moderate temperature differences in a boundary layer the free-convective component of the heat flux plays the main part in heat transfer with the wall. In the upper part of the plate the boundary layer thickness increases, and the heat flux gradually decreases due to convection, whereas the contribution of the radiating component, on the contrary, increases. Since the latter does not depend on the boundary layer thickness and is determined by the surface temperature, the temperature surface tends to some constant value T_{w}. This value can be determined from the limiting boundary condition $(\partial T/\partial y) = 0$ for $x \to \infty$ [7.2, 7.3]:

$$T_{\mathrm{w}} = T_\infty \left(1 + R\right)^{1/4} .$$

Figure 7.1.1 presents the dependence of the local heat transfer coefficient for a ***vertical plate in air*** ($\mathrm{Pr} = 0.7$) on the longitudinal coordinate $\xi = \sigma \varepsilon T_\infty^3 x (\mathrm{Gr}_x^*/5)^{-1/5}/\lambda$. For $\xi = 0$ the curves converge into one point corresponding to heat transfer ***at a constant heat flux on a nonradiating surface*** (Fig. 7.1.1a), for large $\xi \mathrm{Nu}_x/\mathrm{Gr}_x^{1/4}$ they tend to the limiting value of heat transfer of a ***vertical plate of constant temperature*** (Fig. 7.1.1b) [7.6].

The asymptotic behavior of $\mathrm{Nu}_x/\mathrm{Gr}_x^{*1/5}$ for $\xi = 0$ is of another character (Fig. 7.1.1a). The experimental data coincide with numerical calculations for $R = 0$–1 with an error of 4–5%.

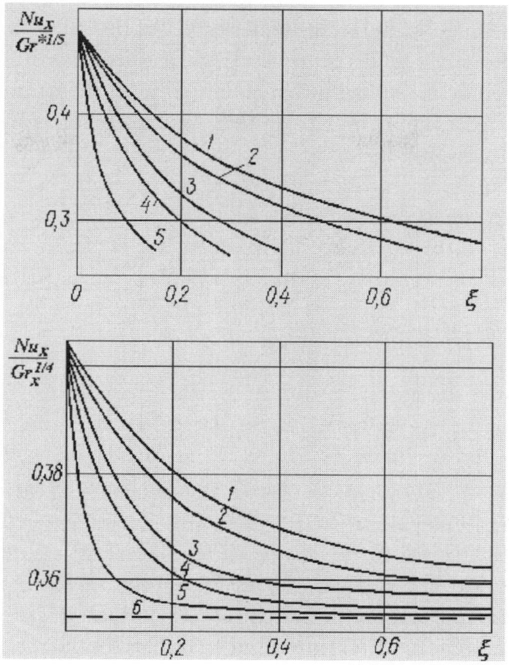

Fig. 7.1.1. Dependence of the local heat transfer $\mathrm{Nu}_x/\mathrm{Gr}_x^{*1/5}$ (*upper*) and $\mathrm{Nu}_x/\mathrm{Gr}_x^{1/4}$ (*lower*) on the longitudinal coordinate: (1) $R = 0$; (2) 1; (3) 5; (4) 10; (5) 50; (6) T_w = const

Detailed experimental data obtained in studying interaction of free convection and radiation ***for a single step mounted on a heated horizontal base*** obtained in [7.7] are correlated by empirical relations for both the total and radiative heat transfer coefficients:

for total heat transfer

$$\mathrm{Nu} = 0.287\,(1 + N_{r-c})^{2.816}\,\mathrm{Ra}_H^{0.277}\,(1 + \varepsilon_f)^{0.866}\ , \qquad (7.1.3)$$

for radiative heat transfer

$$\mathrm{Nu}_r = 660 N_{r-c}^{0.724}\mathrm{Ra}_H^{-0.143}\varepsilon_f^{0.264}\ , \qquad (7.1.4)$$

where ε_f is the hemispherical emissivity, d is the fin thickness, H is the fin height, σ is the Stefan–Bolzmann constant, and N_{r-c} is the radiation-conduction interaction parameter

$$N_{r-c} = \frac{2\sigma\varepsilon_f T_b^3 H^2}{\lambda d}\ . \qquad (7.1.5)$$

An experimental investigation of interaction of free convection and radiation ***in a horizontal fin array*** was carried out in [7.8]. The main conclusion drawn from that study is that radiation-convection interaction invalidates the additive approaches in which convection and radiation contributions are calculated independently assuming all surfaces to be isothermal and then adding this to obtain the total heat loss from the fin array. Separate correlations are given for the fin flats and the base.

Fin flat heat transfer coefficient:

$$\mathrm{Nu} = 0.791\,(1+N_{rc})^{-2.483}\,\mathrm{Ra}_s^{0.214}\,(1+\varepsilon_f)^{0.572}\,(S/b)^{0.468}\ . \tag{7.1.6}$$

Base heat transfer coefficient:

$$\mathrm{Nu} = 2.301\,(1+N_{rc})^{-0.682}\,\mathrm{Ra}_s^{0.118}\,(1+\varepsilon_f)^{-0.47} \times (S/b)^{0.635}\,(h/b)^{-0.086}\ . \tag{7.1.7}$$

For heat transfer rate of a ***vertical radiating cylinder of radius*** r_0 the parameter $S = 2\lambda/(r_0\varepsilon\sigma T_\infty^3)$, reflecting a relative effect of curvature and radiation on heat transfer, appears in the free convection equations. Figure 7.1.2 presents the curves of the dependences of heat transfer ***from the cylinder surface*** on the height [7.9]. For small ξ the heat transfer decreases due to radiation, but this is noticeable for $S < 1$. The curvature effect becomes dominating with increase of ξ and the heat transfer reaches the minimal value and starts to increase. The minimum points in Figs. 7.1.2a and 7.1.2b are different, as their location depends on the expression for local heat transfer, but in both cases they are displaced downstream with increase in the radiation parameter R. For $S = R = 1$ the interaction of curvature and radiation leads to the fact that $\mathrm{Nu}_x/\mathrm{Gr}_x^{*1/5}$ remains constant in a wide range of ξ. On a cylinder, like on a plate, the local heat transfer changes from $\mathrm{Nu}_x/\mathrm{Gr}_x^{1/4} = 0.403$ for $\xi = 0$ to 0.353 for $\xi \to \infty$.

The experimental data on heat transfer of a ***vertical radiating cylinder for constant heat flux on a surface*** are approximated by the calculations for a plate [7.10]:

$$l = 1\mathrm{m},\ d = 0.22\mathrm{m},\ R = 0.125 - 3.3,\ \xi = 0.01 - 70,\ S = 0.2 - 0.4\ ,$$

$$T_{\mathrm{av}} = T_{\mathrm{w}} - 0.38(T_{\mathrm{w}} - T_\infty),\ \mathrm{Gr}_x^* = 3\cdot 10^8 - 10^{11},\ \mathrm{Pr} = 0.71\ ,$$

$$\frac{\mathrm{Nu}_x}{\mathrm{Gr}_x^{*1/5}} R^{-7.4\cdot 10^{-3} R^{-0.51}} = 0.485 - 0.662 R^{1/5}\xi + 0.638 R^{2/5}\xi^2\ , \tag{7.1.8}$$

$$\frac{q_t}{q_{\mathrm{w}}} = -4.333 + 34.37\frac{\mathrm{Nu}_x}{\mathrm{Gr}_x^{*1/5}} - 93.22\left(\frac{\mathrm{Nu}_x}{\mathrm{Gr}_x^{*1/5}}\right)^2 + 92.84\left(\frac{\mathrm{Nu}_x}{\mathrm{Gr}_x^{*1/5}}\right)^3\ . \tag{7.1.9}$$

The numerical calculations and experiments on ***nonstationary free convection with radiation on a vertical flat surface*** show that the radiation

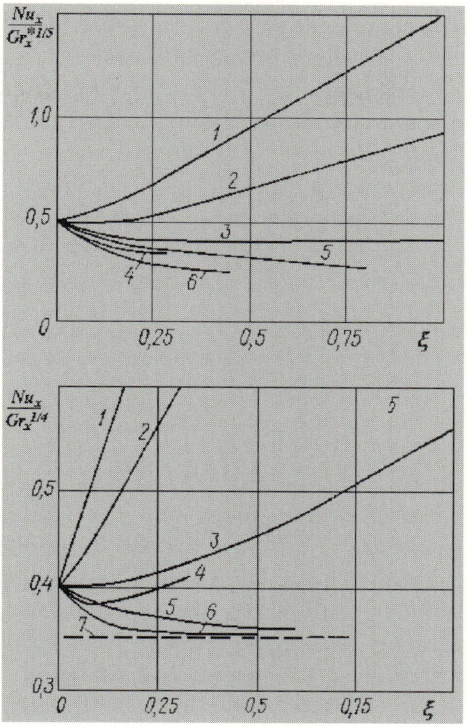

Fig. 7.1.2. Dependence of the local heat transfer $\mathrm{Nu}_x/\mathrm{Gr}_x^{*1/5}$ and $\mathrm{Nu}_x/\mathrm{Gr}_x^{1/4}$ on the longitudinal coordinate: (1) $S = 0$, $R = 0$; (2) 5, 1; (3) 1, 1; (4) 1, 10; (5) 10, 1; (6) 1, 5; (7) T_w = const

has a small effect on heat transfer processes for low heat fluxes [7.11]. For a small temperature difference at the initial moment the convective component of the heat flux increases with the temperature difference faster than the radiating one. The transition to a turbulent flow in nonstationary processes occurs a little earlier, than in steady ones.

The results of numerical calculations for the total heat transfer by radiation and free convection ***in isothermal channels of rectangular and triangular cross section*** (in the second case there is a equilateral cross section) are given in [7.12].

The numerical study of the effect of the ***number and dimensions of ventilating holes and their arrangement with respect to the entrance*** on radiative-convective cooling of a ***flat vertical channel in air with uniform heating of one of the lateral sides*** shows that the temperature of the heated walls in the presence of ventilating holes over the entire length upstream exceeds the corresponding values for unventilated channels [7.13]. The temperature of a wall with a single low-lying hole is smaller than the corresponding value for an unventilated channel only over the length downstream from the hole. For the channel heated with a constant heat flux the presence of holes usually leads to the decrease in the overall cooling intensity.

The results of numerical and experimental investigation of ***turbulent natural convection of air*** with allowance for thermal radiation ***in a vertical rectangular channel with one side heated*** show that at intermediate and high wall emissivities thermal radiation contributes significantly to the total heat transfer by natural convection, even at low heated wall temperatures [7.14]. Based on the experimental and numerical results, the following semi-empirical correlations were developed for describing the heat transfer data:

$$\mathrm{Nu} = 0.1\mathrm{Ra}^{1/3}\left\{1 + (1+2l)\,\frac{0.919R_s}{1+0.513R_s}\right\} , \tag{7.1.10}$$

where

$$R_s = \frac{1}{2+2l-\varepsilon\,\mathrm{Nu}_c}\,\frac{\mathrm{St}}{} , \quad \mathrm{Nu}_c = 0.1\mathrm{Ra}^{1/3} ,$$

ε is the dissipation rate of kinetic energy, $l = L/B$ (L and B are the depth and the width of the channel), and St is the Stefan number.

The numerical solution of a two-dimensional problem concerning the joint action of radiation and natural convection in ***radiating, absorbing, and isotropically scattering ambient medium in a cavity of square section*** was obtained in [7.15]. The radiation-conduction parameter changed from 0 to ∞, and the Rayleigh number – from 10^3 to 10^5.

7.2 Combined Free and Forced Convection

By the similarity theory methods it is shown that for combined free and forced convection the determining parameters are the numbers Gr, Pr and the complex $\mathrm{Gr}/\mathrm{Re}^2$. The values of the dimensionless heat transfer coefficient ***on a vertical flat surface with a constant wall temperature*** for $\mathrm{Pr} = 0.003-100$ are given in Fig. 7.2.1. The dashed lines refer to asymptotic values for free and forced convection. The upper curve corresponds to co-current flows, the lower – to counter-current ones. The end of the lower curve determines the point of separation of the boundary layer from the plate. The largest deviations from the asymptotic value occur at $\mathrm{Pr} = 0.003$ and make 25%. With increase in Pr this deviation decreases and, for example, for $\mathrm{Pr} = 100$ makes about 10%. For the co-current flows the equality of experimental and calculated data is within the accuracy of experiment, for the counter-current flows the experimental data are lower than the calculated ones [7.16].

The approximation dependence of the local value of heat transfer rate ***for co-current flows of free and forced convection*** ($\mathrm{Pr} = 0.7-10$), with an error of 1.5% is [7.17]

$$\mathrm{Nu}_x = C\left\{\exp\left[2.303\exp\left(2.303g\right)\right]\right\}^A\ \mathrm{Pr}^a\mathrm{Re}_x^{1/2} , \tag{7.2.1}$$

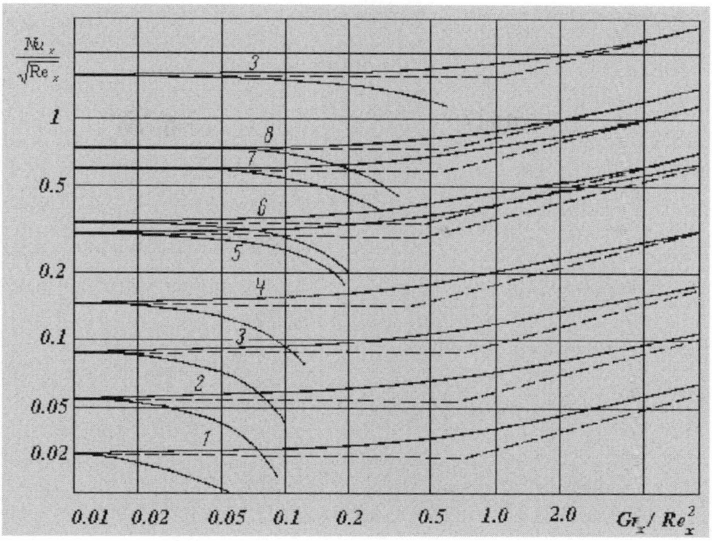

Fig. 7.2.1. Local heat transfer coefficient on a vertical surface of constant temperature: (1) Pr = 0.003; (2) 0.01; (3) 0.03; (4) 0.1; (5) 0.72; (6) 1; (7) 5; (8) 10; (9) 100

$\mathrm{Gr}_x/\mathrm{Re}_x^2 = 0 - 2$:

$$A = 0.2766, C = 0.3025 ,$$
$$a = (0.341 + 0.293\mathrm{Gr}_x/\mathrm{Re}_x)/(1 + \mathrm{Gr}_x/\mathrm{Re}_x) ,$$
$$g = -0.0576(\lg \mathrm{Gr}_x/\mathrm{Re}_x^2)^3 - 0.09779(\lg \mathrm{Gr}_x/\mathrm{Re}_x^2)^2 + 0.3902 \lg \mathrm{Gr}_x/\mathrm{Re}_x^2 - 0.1766 ,$$

$\mathrm{Gr}_x/\mathrm{Re}_x = 2 - 100$:

$$A = 0.32, C = 0.2784, a = 0.318(\mathrm{Gr}_x/\mathrm{Re}_x)^{-0.011} ,$$
$$g = -0.04984(\lg \mathrm{Gr}_x/\mathrm{Re}_x^2)^2 + 0.3374 \lg \mathrm{Gr}_x/\mathrm{Re}_x^2 - 0.163 .$$

The generalized dependence for calculating the average value of heat transfer by mixed convection ***for co-current flows for bodies of simple shape*** with an error of 10% is [7.18]

$$\mathrm{Re} = 15 - 10^4, \mathrm{Ra} = 2300 - 10^9, \mathrm{Pr} = 0 - \infty, T_{\mathrm{av}} = (T_{\mathrm{w}} + T_\infty)/2 ,$$

$$\left[\frac{(\mathrm{Nu} - \mathrm{Nu}_0) f_B(\mathrm{Pr})}{A_B \mathrm{Re}^{1/2} \mathrm{Pr}^{1/3}}\right]^3 = 1 + \left[\frac{A_C \mathrm{Ra}^{1/4} f_B(\mathrm{Pr})}{A_B \mathrm{Re}^{1/2} \mathrm{Pr}^{1/3} f_C(\mathrm{Pr})}\right]^3 , \qquad (7.2.2)$$

where

$$f_C(\mathrm{Pr}) = \left[1 + (C_C/\mathrm{Pr})^{9/16}\right]^{4/9} , \quad f_B(\mathrm{Pr}) = \left[1 + (C_B/\mathrm{Pr})^{2/3}\right]^{1/4} ,$$

Table 7.2.1.

Body Shape, Boundary Conditions	Determining Size	A_B	A_C	C_B	C_C	Nu_0
1. Vertical plate						
T_w = const, local values	x	0.339	0.503	0.0468	0.492	0
q_w = const, local values	x	0.464	0.563	0.0205	0.437	0
T_w = const, average values	L	0.677	0.670	0.0468	0.492	0
q_w = const, average values	$l/2$	0.656	0.669	0.0205	0.437	0
2. Horizontal cylinder						
T_w = const, average values	πd	1.08	0.690	0.412	0.559	1
q_w = const, average values	πd	–	0.694	0.442	–	–
3. Sphere						
T_w = const, average values	$\pi d/2$	0.69	0.659	–	–	
4. Approximate						
generalized expression	–	0.67*	0.67*	0.45**	0.45	–***
*, except for Nu_x; **, for vertical plate 0.035; ***, 0.5 for vertical plate; 1) for cylinder; π, for sphere.						

the coefficients A_C, A_B, C_C, and C_B and the characteristic dimensions for different bodies and boundary conditions are given in Table 7.2.1.

The local heat transfer value ***for turbulent mixed convection on a vertical surface with constant temperature*** is [7.19]

$$\mathrm{Nu}_x/\mathrm{Re}_x^{0.8} = 0.0288\,\mathrm{Pr}^{0.047}\,\Phi(\mathrm{Pr}, \mathrm{Gr}_x/\mathrm{Re}_x^2) \times \left[\left(\mathrm{Gr}_x/\mathrm{Re}_x^2\right) \Big/ \int_0^{\mathrm{Gr}_x/\mathrm{Re}_x^2} \Phi\left(\mathrm{Pr}, \xi\right) d\xi \right]^{0.2}, \tag{7.2.3}$$

where

$$\Phi\left(\mathrm{Pr}, \xi\right) = \left\{ \left[1 \pm \frac{0.6945\xi^{1/2}}{\left(2.14 + \mathrm{Pr}^{2/3}\right)^{1/2}} \right] \left[1 \pm \frac{1.845\xi^{1/2}}{\left(2.14 + \mathrm{Pr}^{2/3}\right)^{1/2}} \right]^3 \right\}^{1/4}.$$

The approximate expressions for parameters ***in the region of boundary layer separation in a counter-current flow***:

$$\left(\mathrm{Gr}_x/\mathrm{Re}_x^2\right)_{al} = \frac{2.14 + \mathrm{Pr}^{2/3}}{3.404} - \frac{2.14 + \mathrm{Pr}^{2/3}}{0.4823}.$$

The local heat transfer value ***for the variable volumetric expansion coefficient*** is [7.20]

$$\frac{\mathrm{Nu}_x}{\mathrm{Nu}_{x0}} = \left[1 - \frac{0.816\,\mathrm{Pr}^{-1/3}\,(1-\alpha_0)\,\mathrm{Gr}_x}{\alpha_0 \mathrm{Re}_x^2}\right]^{-1/2} \times \left[1 + \frac{2.335\,\mathrm{Pr}^{-1/3}\,(1-\alpha_0)\,\mathrm{Gr}_x}{\alpha_0 \mathrm{Re}_x^2}\right]^{-1}, \qquad (7.2.4)$$

$$\beta = \beta_\infty(1+c\theta), \mathrm{Gr}_x/\mathrm{Re}_x^2 \ll 1\,,$$

$$\alpha_0 = 1 - \beta_\infty\theta_\mathrm{w} - 0.5\beta_\infty\theta_\mathrm{w}(c\theta_\mathrm{w}) + 0.5(\beta_\infty\theta_\mathrm{w})^2 + 0.5(\beta_\infty\theta_\mathrm{w})^2(c\theta_\mathrm{w}) + 0.125(\beta_\infty\theta_\mathrm{w})^2(c\theta_\mathrm{w})^2\,.$$

The effect of the ***work of compression and viscous dissipation*** on mixed convection can be estimated from the data from Table 7.2.2 [7.21]. The heat transfer increases for the counter-current flow with increase in the criteria ε_1 and Ec.

Table 7.2.2.

$\mathrm{Gr}_x/\mathrm{Re}_x^2$	ε_1	$(\mathrm{Ec}_x\mathrm{Re}_x^2)/\mathrm{Gr}_x$	
		0.5	1
1	0.01	0.4378	0.4412
1	0.02	–	0.4430
5	0.05	0.5351	0.5411
−1	0.01	0.3050	0.3057
−1	0.02	–	0.3099
−2	0.02	0.1858	0.1936

The average value of the heat transfer ***downstream the separation point*** with an error of 6% is [7.22]

$$\mathrm{Pr} = 0.7, \mathrm{Gr}_l/\mathrm{Re}_l^2 = 0.004 - 1\,,$$

$$\mathrm{Nu}/\mathrm{Nu}_0 = \left[1 + 0.345\left(\frac{\mathrm{Gr}_l}{\mathrm{Re}_l^2}\right)^{1/4}\right]^{-1}. \qquad (7.2.5)$$

Figure 7.2.2 presents the data on heat transfer ***on a vertical surface with a constant heat flux*** [7.4].

In the presence of ***volumetric heat generation in the environment on the outer edge of the boundary layer*** the temperature difference changes according to the linear law $T_\infty = T_\mathrm{w} + Q_\nu x/(u_\infty \rho c_p)$. The local heat transfer value on a vertical flat surface, $\mathrm{Os}_x\mathrm{Gr}_x/\mathrm{Re}_x \ll 1$ is

$$\frac{\mathrm{Nu}_x}{\mathrm{Re}_x^{1/2}} = -\frac{1}{2}\left[\theta_0'(0) \pm \frac{\mathrm{Os}_x\mathrm{Gr}_x}{\mathrm{Re}_x}\theta_1'(0)\right]. \qquad (7.2.6)$$

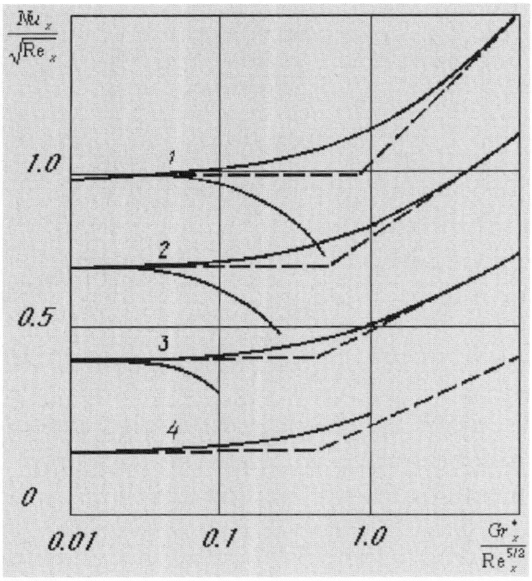

Fig. 7.2.2. Heat transfer coefficient on a vertical surface with a constant heat flux: (1) Pr = 10; (2) 3; (3) 0.7; (4) 0.1

Table 7.2.3 lists the values of $\theta_i'(0)$. Figure 7.2.3 gives the photos of the flame front cellular self-organization appearing due to a convective circulating gas flow above a horizontal permeable surface.

Table 7.2.3.

Pr	θ_0'	θ_1'
0.01	0.2219	−3.2157
0.1	0.7329	−0.4572
0.72	2.3124	−0.00915
1	2.8361	0.0398
5	7.9798	0.2912
10	12.59	0.3848

If the wall is insulated, it is possible to determine the influence ***of volumetric heat generation*** on its temperature, $\mathrm{Os}_x\mathrm{Gr}_x/\mathrm{Re}_x^3 \ll 1$ [7.4]:

$$T_\mathrm{w} - T_\infty = \frac{Q_\nu x}{u_\infty \rho c_p}\left[\theta_0(0) \pm \frac{\mathrm{Os}_x\mathrm{Gr}_x}{\mathrm{Re}_x^3}\theta_1(0)\right] . \tag{7.2.7}$$

The values of $\theta_i\prime(0)$ are given in Table 7.2.4.

Calculations by formula (7.2.7) coincide with an accuracy of 5% with experimental data on ***salt water***, through which the electric current passes.

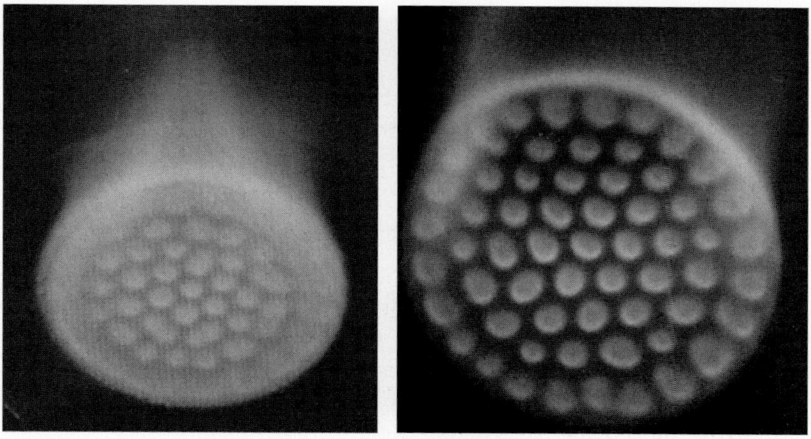

Fig. 7.2.3. Photos of the flame front cellular self-organization appearing due to a convective circulating gas flow above a horizontal permeable surface

Table 7.2.4.

Pr	$\theta_0(0)$	$\theta_1(0)$
0.01	1.1543	−4.4232
0.1	1.5205	−3.8985
0.72	2.3840	−3.1027
1	2.6141	−2.9137
5	4.2786	−1.8918
10	5.3542	−1.4806

Mixed convection ***on a vertical flat surface with a constant temperature for a counter-current flow. The external flow velocity changes according to the power law*** $u_\infty = Cx^m$. The approximation dependence for the local heat transfer value with an error of ±3% is [7.23]

$$m = 0 - 1/11, x/x_{al} = 0 - 0.99, \mathrm{Pr} = 0.7 - 1000 ,$$

$$\frac{\mathrm{Nu}_x}{\mathrm{Re}_x^{1/2}} \mathrm{Pr}^{1/3} = \sum_{i=0}^{4} a_i (x/x_{al})^i , \tag{7.2.8}$$

for Pr = 0.7–8

$$(\mathrm{Gr}_x/\mathrm{Re}_x^2)_{al} = 0.179(1 + m^{1/2})^2 (\mathrm{Pr}/8)^{0.215} ,$$

for Pr = 8–1000

$$(\mathrm{Gr}_x/\mathrm{Re}_x^2)_{al} = 0.3(1 + m^{1/2})^2 (\mathrm{Pr}/8)^{0.265} .$$

The values of the coefficients a_i are given in Table 7.2.5.

Table 7.2.5.

M	a_0	a_1	a_2	a_3	a_4
0	0.3358	0.005828	−0.3761	0.7035	−0.4707
1/35	0.3472	0.01417	−0.3676	0.6137	−0.3836
1/17	0.3614	−0.03801	−0.09022	0.1423	−0.1470
1/11	0.3808	−0.01710	−0.2795	0.5089	−0.3575

The mixed convection ***on a vertical flat surface with variable wall temperature*** $\theta_\text{w} = Ax^n$ ***and variable external flow velocity*** $u_\infty = Cx^m$ [7.4]. The numerical calculations on heat transfer $\text{Nu}_x/\text{Re}_x^{1/2}$ for $\text{Pr} = 0.7$ depending on $\text{Gr}_x/\text{Re}_x^2$ ***for a constant wall temperature*** $(n = 0, m = 0.5)$ and ***constant heat flux*** $(n = 0.2, m = 0.6)$ are given in Table 7.2.6.

Table 7.2.6.

$\text{Gr}_x/\text{Re}_x^2$	$n = 0$	$n = 0.2$
100	1.1349	1.2941
50	0.9626	1.0976
20	0.7807	0.8908
10	0.6733	0.7695
5	0.5892	0.6752
25/9	0.5345	–
2	–	0.5781
1	0.4713	0.5464
0.8	0.4820	0.5366
0.5	0.4368	0.5206
0.25	0.4325	0.5058
0.05	0.4196	0.4929
0	0.4162	0.4894
−0.05	0.4126	0.4858
−0.25	–	0.4707
−0.50	0.3736	0.4986
−0.80	0.3334	–
−0.90	0.3129	–
−0.94	0.3025	–
−0.95	0.2992	–
−1.0	–	0.3842
−1.1	–	0.3631
−1.15	–	0.3497
−1.1544	–	0.3484

Vertical plate with an unheated initial part of length l_0 (the coordinate x is measured from the beginning of the unheated part). The local heat transfer value for the laminar mixed convection ***in air*** is approximated by the dependence [7.24]

$$\mathrm{Pr} = 0.7, \mathrm{Re}_{l0} = 0 - 2 \cdot 10^4, \mathrm{Gr}_x/\mathrm{Re}_x^2 = 0.01 - 10\ ,$$

$$\left[\frac{\alpha_x\,(x+l_0)}{\lambda}\right]^3 = \left(0.295\,(\mathrm{Re}_x + \mathrm{Re}_{l0})^{1/2} \Big/ \left[1 - \left(\frac{l}{x_0 + l_0}\right)^{3/4}\right]^{1/2}\right)^3 + \left(0.354\mathrm{Gr}_x^{1/4}\right)^3 . \quad (7.2.9)$$

The effect of the buoyancy force on laminar forced convection ***along a heated vertical plate*** is amplified in the presence of the unheated initial part. For $\mathrm{Gr}_x/\mathrm{Re}_x^3 \cong 10^{-5}$ the contribution of the buoyancy force does not depend on the presence of the initial part.

A vertical wall moves up with a velocity $u_0 = C_1 x^m$ ***in an external flow with velocity*** $u_\infty = C_2 x^m$. ***The wall temperature changes along the direction of motion*** $\theta_\mathrm{w} = A x^{2m-1}$. Figure 7.2.4 gives the local heat transfer value ***for a co-current and counter-current flows*** for $\mathrm{Pr} = 0.733$, $m = 1/3$–1, $u_\infty/u_0 = 0.3$–0.8, $\mathrm{Gr}_x/\mathrm{Re}_{x0}^2 = 0$–$3.5$ [7.25].

The local heat transfer value ***on a vertical flat surface in the presence of injection and suction for co-current and counter-current flows*** is [7.26]

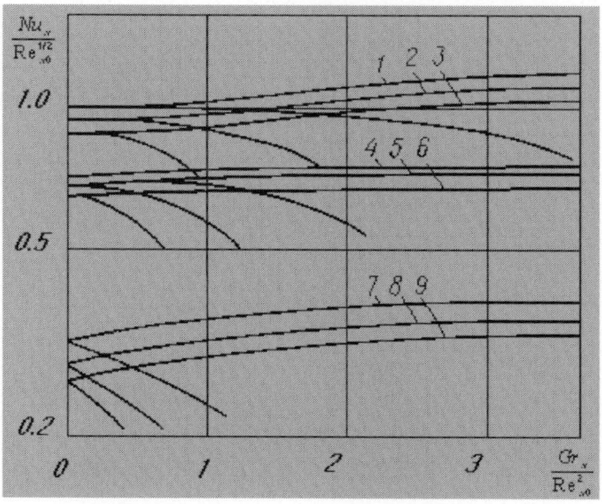

Fig. 7.2.4. Local heat transfer value on a moving surface: (1) $m = 1$, $u_x/u_0 = 0.8$; (2) 1, 0.5; (3) 1, 0.3; (4) 2/3, 0.8; (5) 2/3, 0.5; (6) 2/3, 0.3; (7) 1/3, 0.8; (8) 1/3, 0.5; (9) 1/3, 0.3

$$\mathrm{Nu}_x \left(\frac{\mathrm{Gr}_x}{\mathrm{Re}_x}\right)^{1/2} = \left(0.42 - 0.203\frac{\mathrm{Re}_{wx}}{\mathrm{Gr}_x^{1/4}}\right) \left(\frac{\mathrm{Gr}_x}{\mathrm{Re}_x^2}\right)^{-0.34+0.556\mathrm{Re}_{wx}/\mathrm{Gr}_x^{1/4}} . \tag{7.2.10}$$

In [7.27] the validity limits of the formulas for free and forced convection ***in the presence of uniform suction or injection*** are determined.

The analytical and numerical investigation of nonstationary mixed convection ***from a vertical surface in a porous surrounding medium saturated with a liquid in the case of a surface temperature change from*** T_∞ ***to*** T_w is carried out for the whole transient process from the initial stage after the change in the surface temperature to a fully developed boundary layer [7.28]. The cases of coinciding and opposite directions of free and forced flows are considered.

***On a vertical flat plate* 20(4 × 5) *copper modules* 1.27 × 1.27× 0.06 cm *are set with a distance between them* 0.06 cm**. The temperature of the module surfaces heated by an electric current is $T_\mathrm{w} = 24{-}80°\mathrm{C}$. The velocity of the external flow parallel to the plate is $u_\infty = 1.5{-}5$ m/sec. The experimental investigations of heat transfer for $\mathrm{Gr}_x = 7 \cdot 10^5{-}3 \cdot 10^6$ and $\mathrm{Re}_x = 1.5 \cdot 10^4{-}3.5 \cdot 10^5$ show that the value of the criterion Nu_x increases 2.5 times compared with a turbulent flow on a smooth plate [7.29].

The heat transfer by mixed convection ***on a horizontal surface*** is determined by the complex $\mathrm{Gr}_x/\mathrm{Re}_x^{5/2}$ ***for a constant wall temperature*** and by the value $\mathrm{Gr}_x^*/\mathrm{Re}_x^{5/23}$ ***for a constant heat flux*** [7.4]. Figure 7.2.5 presents the data on heat transfer ***on a horizontal surface with a constant wall***

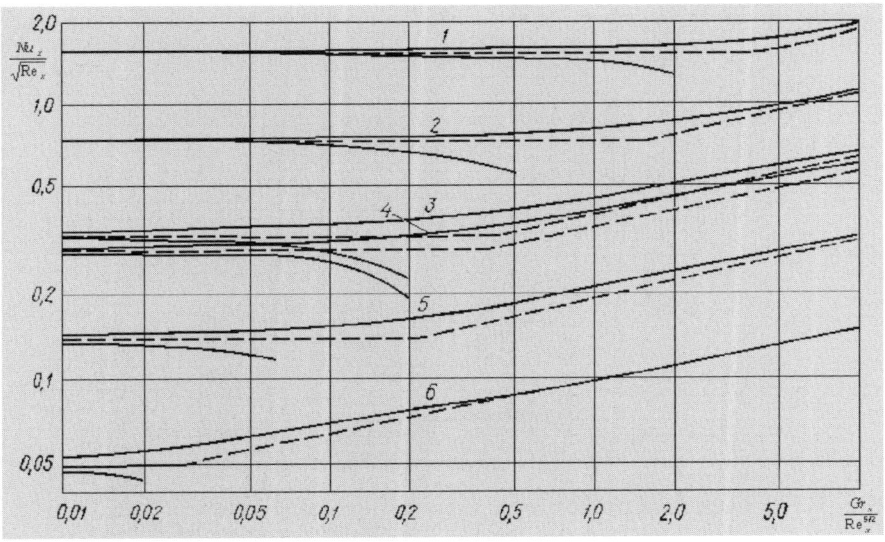

Fig. 7.2.5. Coefficient of heat transfer on a horizontal surface of constant temperature: (1) Pr = 100; (2) 10; (3) 1; (4) 0.72; (5) 0.1; (6) 0.01

Table 7.2.7.

Gr_x/Re_x^3	$\theta(\overline{x},0)$	Gr_x/Re_x^3	$\theta(\overline{x},0)$
0	2.46370	0.6	1.77946
0.1	2.14975	0.7	1.74716
0.2	2.01744	0.8	1.71622
0.3	1.93133	0.9	1.69132
0.4	1.86849	1.0	1.65827
0.5	1.81966		

temperature. The heat transfer ***for a constant heat flux*** is determined from the equation $Nu_x/Re_x^{1/2} = \theta^{-1}(Gr_x^*/Re_x^3)$. The values of the function $\theta(Gr_x^*/Re_x^3)$ for $Pr = 0.7$ are given in Table 7.2.7.

The ratio of the Nusselt numbers Nu_x ***for constant values of heat flux and temperature on the wall*** Nu_{xq}/Nu_{xt} is equal to 1.35 [7.30]. The value of the dimensionless heat transfer coefficient $Nu_x/Re_x^{1/2}$ ***on a horizontal surface for constant wall temperature difference*** θ_w ***and*** $u_\infty = Cx^{0.2}$ is given in Table 7.2.8 [7.4]. ***The horizontal surface moves with the velocity*** $u_0(Re_{x0} = u_0x/\nu)$ ***parallel to the wall in a motionless surrounding medium***.

Table 7.2.8.

Pr	$Gr_x/Re_x^{5/2}$	$Nu_x/Re_x^{1/2}$
1	0	0.405
1	0.05	0.415
1	0.10	0.424
1	0.50	0.476
1	−0.05	0.394
1	−0.10	0.370
1	−0.20	0.340
0.5	0.10	0.331
0.5	−0.10	0.234
5	0.10	0.745
5	−0.10	0.700

Table 7.2.9 lists the data on heat transfer for two Pr numbers [7.31].

On an inclined plate with a constant wall temperature the heat transfer for mixed convection is determined by the data for a vertical surface by introducing the parameter $Gr_{x\varphi}/Re_x^2$. As the calculations and experiments show, for angles $\varphi = 0{-}80°$ the error for heat transfer does not exceed 2% [7.4, 7.32]. The correction for the component at $\sin\varphi$ in the equations of motion can be determined by disturbance methods. For small deviation from a horizontal ($\varphi = 87{-}90°$), the heat transfer calculations should be carried out using the

Table 7.2.9.

$Gr_x/Re_{x0}^{5/2}$	$Nu_x/Re_x^{1/2}$	
	Pr = 0.7	Pr = 7
−0.05	–	1.38548
−0.02	–	1.38686
−0.01	–	1.38693
0	0.34924	1.38702
0.1	0.37951	1.38918
0.2	0.39625	1.39365
0.3	0.40880	1.39172
0.4	0.41909	1.39605
0.5	0.42773	1.39823
0.6	0.43556	1.40021
0.7	0.44250	1.40247
0.8	0.44874	1.40446
0.9	0.45457	1.40658
1	0.45994	1.40876

data for a horizontal surface replacing the principal parameter by the value $Gr_x \sin\varphi/Re_x^{5/2}$. The heat transfer of ***an inclined surface moving with the velocity*** u_0 ***in motionless surrounding medium*** for two Pr numbers is determined by the calculated data from Table 7.2.10 [7.33].

Table 7.2.10.

$Gr_{x\varphi}/Re_x^2$	$Nu_x/Re_x^{1/2}$		$Gr^*_{x\varphi}/Re_x^{5/2}$	$1/(Nu_x/Re_x^{1/2})$	
	Pr = 0.7	Pr = 7		Pr = 0.7	Pr = 7
Constant wall temperature			Constant heat flux		
5	0.59086	1.58510	5	–	0.43893
4.5	0.57893	1.56596	4.5	–	0.43982
4	0.56609	1.55334	4	1.29536	0.44072
3.5	0.55214	1.53641	3.5	1.31631	0.44165
3	0.53681	1.51864	3	1.34001	0.44259
2.5	0.51973	1.49999	2.5	1.36725	0.44355
2	0.50031	1.48026	2	1.39925	0.44453
1.5	0.47764	1.45938	1.5	1.43795	0.44554
1	0.45505	1.43712	1	1.48674	0.44656
0.5	0.41320	1.41322	0.5	1.55428	0.44761
0	0.34924	1.38703	0	1.66240	0.44878
−0.025	0.34339	1.38457	−0.05	1.68007	0.44880
−0.5	–	1.35907	−0.5	–	0.44979
−1	–	1.32776	−1	–	0.45092
−1.5	–	1.29249	−1.5	–	0.45207
−2	–	1.25173	−2	–	0.45326

The influence of the surface mass flux on mixed convection ***over horizontal plates with the boundary conditions*** $T_w = T_\infty + Ax^\lambda$ ***in saturated porous media*** was investigated in [7.34]. It is found that the heat transfer in the form of free, mixed or forced convection is enhanced by the withdrawal of fluid from the surface, while it is decreased due to injection of fluid. The heat transfer coefficient in terms of Nu number is given by

$$\frac{\mathrm{Nu}}{\mathrm{Re}^{1/3}} = \left[-\theta'(0)\right] , \tag{7.2.11}$$

where $-\theta'(0)$ are presented in Table 7.2.11.

Table 7.2.11. Selected Values of $-\theta'(0)$ for Mixed Convection over a Horizontal Plate

$\mathrm{Ra}/\mathrm{Pe}^{3/2}$	$f_w = -1$	$f_w = 0$	$f_w = 1$
	$\lambda = 1/2$		
0	0.6337	0.8862	1.2009
0.6	0.8037	1.0281	1.3123
1.0	0.8862	1.1020	1.3745
2.0	1.0450	1.2495	1.5041
5.0	1.3575	1.5503	1.7825
8.0	1.5724	1.7610	1.9836
15.0	1.9290	2.1137	2.3261
	$\lambda = 2$		
0	1.1258	1.5957	2.2117
0.6	1.4652	1.8627	2.4064
1.0	1.6309	2.0044	2.5182
2.0	1.9494	2.2889	2.7574
5.0	2.5716	2.8820	3.2827
8.0	2.9971	3.2927	3.6683
15.0	3.6971	3.9742	4.3308

The quantity f_w is the mass flux parameter for mixed convection, $\nu = ax^n$, $U_\infty = Bx^m$, α is the thermal diffusivity of the porous medium:

$$f_w = -\frac{2a}{(\alpha B)^{1/2}(1+m)} .$$

The linear theory of stability is used in [7.35] to analyze vortical instability of the mixed convection ***boundary layer*** in ***saturated porous medium near an inclined surface, the wall temperature is a power-law function of the distance, and the external flow is co-current and homogeneous***.

A theoretical model is formulated in [7.36], utilizing an integral technique, to describe the development of the thermal boundary layer. The closed-form solutions are obtained for $0.72 \le \mathrm{Pr} \le 10$ to predict the local and overage heat transfer coefficient for combined forced and natural convection ***from an isothermal vertical flat plate, for both co-current and counter-current flows***:

co-current flow

$$\frac{\mathrm{Nu}_x}{\mathrm{Re}_x^{1/2}} = 2\mathrm{Ri}_x^{1/2}\left\{-\left(\frac{\lambda_b}{3\lambda_a}\right)+\left[\left(\frac{\lambda_b}{3\lambda_a}\right)^2+\left(\frac{8}{3\lambda_a}\right)\left(\frac{\mathrm{Ri}_x}{\mathrm{Pr}}\right)\right]^{1/2}\right\}^{-1/2}, \tag{7.2.12}$$

counter-current flow; initial domination of natural convection

$$\frac{\mathrm{Nu}_x}{\mathrm{Re}_x^{1/2}} = 2\mathrm{Ri}_x^{1/2}\left\{\left(\frac{\lambda_b}{3\lambda_a}\right)+\left[\left(\frac{\lambda_b}{3\lambda_a}\right)^2+\left(\frac{8}{3\lambda_a}\right)\left(\frac{\mathrm{Ri}_x}{\mathrm{Pr}}\right)\right]^{1/2}\right\}^{-1/2}, \tag{7.2.13}$$

counter-current flow; initial domination of forced convection

$$\frac{\mathrm{Nu}_x}{\mathrm{Re}_x^{1/2}} = 2\mathrm{Ri}_x^{1/2}\left\{\left(\frac{\lambda_b}{3\lambda_a}\right)-\left[\left(\frac{\lambda_b}{3\lambda_a}\right)^2-\left(\frac{8}{3\lambda_a}\right)\left(\frac{\mathrm{Ri}_x}{\mathrm{Pr}}\right)\right]^{1/2}\right\}^{-1/2}, \tag{7.2.14}$$

where

$$\lambda = \delta_T(x)/\delta_V(x)\,.$$

For $\delta_T(x) < \delta_V(x)$:

$$\lambda_a(x) = \frac{1}{6\lambda^2}\left(\frac{\lambda}{15}-\frac{\lambda^2}{14}+\frac{9\lambda^3}{280}-\frac{\lambda^4}{180}\right),\quad \lambda_b(x) = \frac{2\lambda}{15}-\frac{3\lambda^3}{140}+\frac{\lambda^4}{180},$$

for $\delta_T(x) > \delta_V(x)$:

$$\lambda_a(x) = \frac{1}{6\lambda^2}\left(\frac{1}{20\lambda}-\frac{1}{30\lambda^2}+\frac{1}{140\lambda^3}-\frac{1}{504\lambda^4}\right),$$
$$\lambda_b(x) = \frac{3}{10}-\frac{3}{10\lambda}+\frac{2}{15\lambda^2}-\frac{3}{140\lambda^3}+\frac{1}{180\lambda^4}.$$

The cooling ***of an infinite moving plate of final thickness*** with allowance for the effect of a buoyancy force is investigated numerically in [7.37]. The numerical results determining the Biot numbers, the surface temperature, and the overall rate of heat transfer of a plate are presented for $0.7 \le \mathrm{Pr} \le 100$, $0.1 \le (\lambda\rho c_p)/(\lambda\rho c_p)_S$, $0 \le (\mathrm{Gr}_H/\mathrm{Re}_H^2)(\mathrm{Pe}/4) \le 10$.

The heat transfer intensification for a mixed-convective flow ***between parallel plates heated regularly from below*** is investigated in [7.38]. It was found that the occurrence of a secondary flow precedes the essential heat transfer intensification. The value of the Nusselt number for the location of the region of occurrence of the secondary flow is approximated by the dependence

$$\mathrm{Nu} = 0.3(\mathrm{Ra}^*)^{0.25}\,. \tag{7.2.15}$$

The experimental investigation of mixed convection ***between two horizontal plates for different temperatures*** was carried out in [7.39]. The laser-Doppler velocimeter was used to investigate the occurrence of mixed convection as transverse and longitudinal rolls in a channel of rectangular section. The experiments were carried out for the ranges of the Rayleigh Ra and Reynolds Re numbers $0.03 \le \mathrm{Re} \le 3.3$, $\mathrm{Ra} \le 4700$ using ***water*** as an investigated liquid.

The results of numerical calculations of the occurrence of the longitudinal-vortical instability of a laminar flow for mixed convection ***along an inclined and horizontal flat plates with surface temperature changing according to the law*** $T_\mathrm{w}(x) - T_\infty = Ax^n$ are given in [7.40]. The Prandtl number was equal to $\mathrm{Pr} = 0.7, 7, 100, 1000$. The ranges of the exponent and of the inclination angle change are $0.5 \le n \le 1.0, 0^\circ \le \theta \le 85^\circ$. It was found that for the fixed values of Prandtl number and inclination angle the instability decreases with increase of n. Moreover, for the specified values of n and Pr the critical value of $\mathrm{Gr}_x^*/\mathrm{Re}_x^{*3/2}$ increases with increase in the inclination angle with respective to the horizontal. The analysis has shown that the nonparallel flows, with allowance for the non-uniformity of amplitude functions in a flow have a stabilizing effect, unlike the parallel flow.

The results of numerical investigation of the initial stage of development of instability and transient oscillatory flow for mixed convection ***in a horizontal porous layer locally heated from below*** given in [7.41] show that for small Peclet numbers ($\mathrm{Pe} \le 1$), steady temperature fields and flow patterns are observed. For $\mathrm{Pe} > 1$, an oscillatory flow appears with complex interaction of forced and free-convective flows, the oscillation period for this flow depends on the Pe number ($\tau = 2.41\mathrm{Pe}^{-1.096}$). In the case of the domination of the forced flow the temperature and velocity fields become steady again. The experimental investigation of free and mixed convection is also carried out for three different dimensions of a heat source with the relative size of the heater (ratio of the heater length to its thickness) from 1.0 up to 5.0 [7.42]. For free convection the range of the Rayleigh numbers is $1 \le \mathrm{Ra} \le 1000$, for forced $0.1 \le \mathrm{Pe} \le 120$ for the same Ra values. The experimental and numerical data are generalized by the dependences given in Table 7.2.12.

The effect of buoyancy forces on forced convection in a laminar boundary layer ***on a semi-infinite inclined flat heated or cooled plate and also at different buoyancy force directions with respect to the forced flow*** was considered in works [7.43, 7.44]. A wide range of inclination angles (including cases of horizontal and vertical plates), various thermal boundary conditions and the effect of the Prandtl number (from 0 to ∞) are analyzed.

Table 7.2.12.

	Correlation
Similarity solution	Free convection: $\mathrm{Nu}_D = 1.288\mathrm{Ra}_D^{1/4}$
	Mixed convection: $\mathrm{Nu}_D/\mathrm{Pe}_D^{1/2} = [3.111 + 3.039(\mathrm{Ra}_D/\mathrm{Pe}_D^2)]^{1/4}$
Numerical results	Free convection: $\mathrm{Nu}_D = 0.520\mathrm{Ra}_D^{0.354}$
	Mixed convection: $\mathrm{Nu}_D/\mathrm{Pe}_D^{1/2} = [1.917 + 0.210(\mathrm{Ra}_D/\mathrm{Pe}_D^{3/2})]^{0.372}$
Experimental results	Free convection: $\mathrm{Nu}_D = 0.269\mathrm{Ra}_D^{0.451}$
	Mixed convection: $\mathrm{Nu}_D/\mathrm{Pe}_D^{1/2} = [1.274{+}0.079(\mathrm{Ra}_D/\mathrm{Pe}_D^{3/2})]^{0.506}$

It is shown that during action of opposite directed forces there can be both singular and regular flow observed. The results of calculations for two cases of boundary conditions (T_w = const q_w = const) are generalized by the dependences

***for* T_w = const:**

$$\mathrm{Nu}_x = \mathrm{Re}_x^{1/2}\left[(f_{1N})^{3.2} + \left[f_{3N}\left(\frac{\mathrm{Gr}_x}{\mathrm{Re}_x^2}\sin\alpha\right)^{1/4}\right]^{3.2} + \left[f_{2N}\left(\frac{\mathrm{Gr}_x}{\mathrm{Re}_x^{5/2}}\sin\alpha\right)^{1/5}\right]^{3.2}\right]^{1/3.2}, \tag{7.2.16}$$

***for* q_w = const:**

$$\mathrm{Nu}_x = \mathrm{Re}_x^{1/2}\left[(f_{1N}^*)^{3.4} + \left[f_{2N}^*\left(\frac{\mathrm{Gr}_x^*}{\mathrm{Re}_x^{5/2}}\sin\alpha\right)^{1/5}\right]^{3.4} + \left[f_{2N}^*\left(\frac{\mathrm{Gr}_x^*}{\mathrm{Re}_x^3}\sin\alpha\right)^{1/6}\right]^{3.4}\right]^{1/3.4}, \tag{7.2.17}$$

where

$$f_{1N} = 0.3320\ ,$$

$$f_{2N} = \left[\left(0.5755\ \mathrm{Pr}^{2/5}\right)^{-2.7} + \left(0.4562\ \mathrm{Pr}^{1/5}\right)^{-2.7}\right]^{-1/2.7},$$

$$f_{3N} = \left[\left(0.7609 \overset{2/5}{\mathrm{Pr}} \right)^{-2.8} + \left(0.6316 \overset{1/5}{\mathrm{Pr}} \right)^{-2.8} \right]^{-1/2.8},$$

$$f_{1N}^* = 0.3320 ,$$

$$f_{2N}^* = \left[\left(0.8309\,\mathrm{Pr}^{1/3} \right)^{-3.2} + \left(0.6231\,\mathrm{Pr}^{1/6} \right)^{-3.2} \right]^{-1/3.2},$$

$$f_{3N}^* = \left[\left(0.7609\,\mathrm{Pr}^{2/5} \right)^{-2.8} + \left(0.6316\,\mathrm{Pr}^{1/5} \right)^{-2.8} \right]^{-1/2.8}.$$

The coefficients of nonstationary heat transfer ***in a system of parallel concentric disks rotating in one direction with an internal heat generation*** representing a model of electric engine with unequal losses in the stator and rotor at start are experimentally determined [7.45]. The experimental data were obtained for different Taylor numbers corresponding to different heat loads. The effect of the rotation velocity of disks and the free convection in an ambient medium on the heat transfer rate in axial, radial, and tangential directions is investigated.

A vertical cylinder in a uniform longitudinal flow. In Table 7.2.13 the data are given on dimensionless heat transfer $\mathrm{Nu}_x/\mathrm{Re}_x^{1/2}$ ***for a constant wall temperature***, and Table 7.2.14 lists the dimensionless heat transfer value $\mathrm{Nu}_x/\mathrm{Nu}_x$ ***a for constant heat flux*** on a surface for $\mathrm{Pr} = 0.7$ [7.46].

Table 7.2.13.

$x/(r_0 \mathrm{Re}_x)$	$\mathrm{Gr}_x/\mathrm{Re}_x^2 = 0$	$\mathrm{Gr}_x/\mathrm{Re}_x^2 = 1$	$\mathrm{Gr}_x/\mathrm{Re}_x^2 = 2$
0	0.29270	0.41215	0.46630
0.0625	0.44098	0.53437	0.58698
0.25	0.55439	0.63645	0.68654
0.5625	0.65928	0.73236	0.78136
1	0.75710	0.82369	0.8203

The results of numerical calculation for the average Nusselt number for laminar mixed convection from an ***isothermal cylinder in a transverse external flow*** ($\mathrm{Pr} = 0.71$) at different attack angles γ are generalized by the dependence [7.47]

$$\frac{\mathrm{Nu}_m}{\mathrm{Nu}_{m0}} = \left[0.9826 + 0.4588 \lambda^{0.1235} \sin\gamma \right]^{0.3599}, \tag{7.2.18}$$

$$1 \le \mathrm{Re}_D \le 200, \quad 0 \le k \le 35, \quad -\pi \le \gamma \le \pi ,$$

where $k = \mathrm{Gr}_D/\mathrm{Re}_D^2$ (Re_D is the Reynolds number calculated on the basis of the cylinder diameter, Nu_m is the average Nusselt number for any attack angle, Nu_{m0} is the average Nusselt number for $\gamma = 0$, $\lambda = k^{0.2335} \mathrm{Re}_D^{0.06793}$).

Table 7.2.14.

$x/(r_0\mathrm{Re}_x)$	$\mathrm{Gr}^*_{r0}/\mathrm{Re}^{5/2}_{r0}=0$	$\mathrm{Gr}^*_{r0}/\mathrm{Re}^{5/2}_{r0}=0.5$	$\mathrm{Gr}^*_{r0}/\mathrm{Re}^{5/2}_{r0}=1$	$\mathrm{Gr}^*_{r0}/\mathrm{Re}^{5/2}_{r0}=2$
0	1	1	1	1
10^{-5}	1.0009	1.0009	1.0009	1.0009
10^{-4}	1.0082	1.0083	1.0083	1.0084
10^{-3}	1.0153	1.0159	1.0162	1.0163
10^{-2}	1.0798	1.0818	1.0822	1.0854
10^{-1}	1.3292	1.3374	1.3452	1.3982
0.5	1.7124	1.7650	1.8069	2.0103
1	1.9720	2.0781	2.1524	2.4615
2	2.3096	1.5096	2.6280	3.0624
5	5.9315	3.3522	3.5495	4.1976
7.5	3.2858	3.8480	4.0865	4.8453
10	3.5757	4.2612	4.5317	5.3786
25	4.7322	5.9402	6.3285	7.4955
50	5.9615	7.7801	8.2914	9.790

The results of numerical calculation of mixed convection near a ***heated rotating cylinder for small rotation velocities*** based on modeling by a method of large-scale vortices are generalized by the dependence [7.48]

$$\mathrm{Nu}=\mathrm{Nu}_{NC}-3.35\cdot 10^{-8}\mathrm{Ra}^*\sigma^{-0.783} \tag{7.2.19}$$

for $1\le\sigma\le\infty$ and $10^4\le\mathrm{Ra}^*\le 10^{12}$, where $\sigma=\mathrm{Gr}^*/\mathrm{Re}^2$, $\mathrm{Re}=\omega R^2/\nu$, Nu_{NC} – the Nusselt number for natural convection, which, according to [7.49], is obtained from the correlation

$$\mathrm{Nu}_{NC}=\left\{0.6+0.31x\left(\frac{\mathrm{Ra}^*}{\mathrm{Nu}_{NC}}\right)^{1/6}\right\}^{1/2}, \tag{7.2.20}$$

valid for $10^3\le\mathrm{Ra}^*\le 10^{12}$.

A vertical cylinder of constant temperature in a transverse flow. The generalizing dependence for calculating the average heat transfer coefficient with laminar free convection ***in air*** is [7.50]

$$\mathrm{Nu}^2=\left[0.57\left(\mathrm{Ra}_d\frac{d}{l}\right)^{1/4}\right]^2+\left\{\frac{0.609\mathrm{Re}_d^{1/2}\,\mathrm{Pr}^{1/3}}{\left[1+(0.412/\,\mathrm{Pr})^{2/3}\right]^{1/4}}\right\}^2. \tag{7.2.21}$$

The average value of heat transfer of a ***thin horizontal cylinder with a constant temperature in a uniform transverse flow***, $\mathrm{Pe}_d<\pi/5$, $(\mathrm{Ra}_d\mathrm{Pr})^3\ll\pi/5$, is given by [7.18]

$$\frac{4\pi}{\mathrm{Nu}}=-\ln\left[\left(\frac{Pe_d}{16.32}\right)^2+\left\{\frac{\mathrm{Nu}\,\mathrm{Ra}_d\,\mathrm{Pr}}{4.831\,(\mathrm{Pr}+9.4)^{1/2}}\right\}^{2/3}\right]. \tag{7.2.22}$$

Longitudinal flow around a horizontal cylinder with a constant temperature. The relative heat transfer value, $\mathrm{Gr}_{r0}/\mathrm{Re}_{r0}^2 \ll 1$, is [7.51]

$$\frac{\mathrm{Nu}_z}{\mathrm{Nu}_{z0}} = 1 + \frac{\mathrm{Gr}_{r0}}{\mathrm{Re}_{r0}^2}\frac{z}{r_0}\cos\varphi C(\mathrm{Pr}) , \tag{7.2.23}$$

where Nu_{z0} refers to forced convection. The values of $C(\mathrm{Pr})$ are given in Table 7.2.15.

Table 7.2.15.

Pr	0.01	1	10
$C(\mathrm{Pr})$	0.35352	0.39836	0.26304

The average heat transfer coefficient with an error of 7–10% ***for a transverse water-air flow around a horizontal cylinder*** is [7.52]:

$$\mathrm{Re}_d = 315 - 460, \mathrm{Gr}_d = 5.7\cdot 10^4 - 5\cdot 10^6, k = V_{air}/V_{water} = 0 - 2, T_{\mathrm{av}} = T_f ,$$

$$\begin{aligned}\mathrm{Nu} = 0.21\left(\mathrm{Re}_d + \mathrm{Gr}_d^{1/2}/2\right)^{0.62}\mathrm{Pr}_f^{0.38}\left(\mathrm{Pr}_f/\mathrm{Pr}_{\mathrm{w}}\right)^{0.25}\\ \times\left(1 + 8.473k - 3.97k^2\right) .\end{aligned} \tag{7.2.24}$$

The maximum value of heat transfer coefficient is $\alpha_{\max} = (5\text{–}6)\alpha_0$ for $k = 1\text{–}1.2$. A larger amount of air has a negative effect on the heat transfer coefficient. This is explained by transition of emulsion motion regime of a two-phase flow into a plug one.

The average heat transfer coefficient ***with formation hoarfrost from humid air in a transverse flow on a horizontal cylinder*** is [7.53]

$$Ar_d/\mathrm{Re}_d^2 = 0.1\text{–}10, T_{\mathrm{av}} = (T_{\mathrm{w}} + T_\infty)/2, \mathrm{Re}_d < 4600, Ar_d = 10^5\text{–}10^6 ,$$

$$\begin{aligned}\mathrm{Nu} &= 1.2\left(0.3737 + 0.37\,\mathrm{Re}_d^{0.5} + 0.057\,\mathrm{Re}_d^{0.667}\right)\\ &\times\left[0.74 + 0.07\left(1 - \lg\frac{\mathrm{Ar}}{\mathrm{Re}^2}\right)\right]\left[1 + \left(\frac{\mathrm{Ar}}{\mathrm{Re}^2}\right)^{0.5}\right]^{0.5} .\end{aligned} \tag{7.2.25}$$

The increase in convective heat transfer ***with formation hoarfrost*** in comparison with the heat transfer of ***dry air*** is explained by the effect of the surface roughness and by an increase in the effective heat transfer surface due to hoarfrost porosity.

The data of numerical calculations of the local value of heat transfer of a ***horizontal cylinder in a transverse air flow*** $\mathrm{Nu}_{\varphi d}/\mathrm{Re}_d^{1/2}$ ***on instantaneous temperature increase to*** θ_{w} for $\mathrm{Pr} = 0.7$ and $\mathrm{Gr}_d/\mathrm{Re}_d^2 = 5$ are given in Table 7.2.16 [7.54].

Table 7.2.16.

$2(U_\infty \tau / r_0)^{1/2}$	$\varphi_0 = 10$	$\varphi_0 = 20$	$\varphi_0 = 30$	$\varphi_0 = 40$	$\varphi_0 = 50$	$\varphi_0 = 60$
0.1	13.3678	13.3658	13.3658	13.3640	13.3620	13.3596
0.2	6.7010	6.6978	6.6978	6.6952	6.6920	6.6884
0.3	4.4864	4.4846	4.4820	4.4782	4.4736	4.4682
0.4	3.3860	3.3836	3.3798	3.3748	3.3686	3.3614
0.5	2.7318	2.7288	2.7238	2.7172	2.7092	2.6998
0.6	2.3012	2.2974	2.2912	2.2828	2.2728	2.2612
0.7	1.9988	1.9942	1.9866	1.9764	1.9642	1.9502
0.8	1.7772	1.7716	1.7626	1.7506	1.7358	1.7190
0.9	1.6100	1.6034	1.5930	1.5788	1.5614	1.5418
1	1.4812	1.4738	1.4618	1.4454	1.4256	1.4028
1.2	1.3020	1.2928	1.2780	1.2578	1.2326	1.2036
1.4	1.1914	1.1810	1.1638	1.1402	1.1106	1.0756
1.6	1.1240	1.1126	1.0940	1.0682	1.0354	0.9962
1.8	1.0844	1.0728	1.0532	1.0262	0.9914	0.9488
2	1.0630	1.0506	1.0308	1.0032	0.9676	0.9230
2.5	1.0442	1.0332	1.0124	0.9856	0.9524	0.9146
3	1.0420	1.0300	1.0104	0.9840	–	0.9216
∞	1.0414	1.0296	1.0096	0.9818	0.9464	0.9034

The local value of heat transfer of a ***horizontal cone with the angle*** 2γ ***in a longitudinal external flow with the velocity*** $u_\infty = Cx^m$ is [7.55]

$$\frac{\mathrm{Gr}_{l\varphi}}{\mathrm{Re}_l^2} \ll 1, \quad \frac{\mathrm{Nu}_z}{\mathrm{Nu}_{z0}} = 1 + \frac{\mathrm{Gr}_{l\varphi}}{\mathrm{Re}_l^2} \left(\frac{z}{l}\right)^{1-2m} C_k\,(\mathrm{Pr}) \ . \qquad (7.2.26)$$

The values of $C_k(\mathrm{Pr})$ are given in Table 7.2.17.

Table 7.2.17.

Pr	$\gamma = 10°$	$\gamma = 20°$	$\gamma = 30°$	$\gamma = 90°$
0.01	0.06823	0.05553	0.03844	0.01961
1	0.06927	0.05148	0.03114	0.00592
10	0.04790	0.03421	0.01934	0.00162

The coefficient of heat transfer by mixed convection of a ***vertical needle*** increases with decrease of the diameter [7.56].

Heat transfer ***in tube bundles with mixed convection. In the case of vertical arrangement of three tubes (staggered bundle) and four tubes (in-line bundle)*** numerical calculations for $s/d = 1.05$–1.55 show that heat transfer increases mainly with increase in the relative step between the tubes [7.57].

Staggered and in-line arrangement of horizontal tubes of six- and seven-tube bundle. The experimental data on heat transfer for a longitudinal ***water*** flow around the bundles are as follows [7.58]

$$s/d = 3, \mathrm{Re}_d = 100 - 2000, \mu_\mathrm{w}/\mu_f = 1.05 - 1.35, T_\mathrm{av} = T_f ,$$

constant tube temperature, $\mathrm{Ra}_d = 1.5 \cdot 10^7 - 5 \cdot 10^8$:

$$\mathrm{Nu} = 0.0025(\mathrm{Pe}_d d/l)^{1/3} \mathrm{Ra}_d^{0.42} (\mu_f/\mu_\mathrm{w})^{0.33} \tag{7.2.27}$$

constant heat flux, $\mathrm{Ra}_d^* = 5 \cdot 10^8 - 2 \cdot 10^{10}$:

$$\mathrm{Nu} = 0.0022(\mathrm{Pe}_d d/l)^{1/3} \mathrm{Ra}_d^{*0.32} (\mu_f/\mu_\mathrm{w})^{0.33} . \tag{7.2.28}$$

A numerical and experimental investigation of interaction of the Rayleigh-Benard convection and of a horizontal shear flow ***through lateral porous walls in a rectangular region*** carried out in [7.59] showed a strong dependence of flow structure on the channel length ***for silicon oil*** (Pr = 530) for $2300 < \mathrm{Ra} < 20000$ and $3 \cdot 10^{-5} < \mathrm{Re} < 0.1$. ***In a short channel*** for high Reynolds numbers and low Rayleigh numbers both longitudinal and transverse circulating vortices are observed at the same time in different regions of the container. ***In an infinitely long*** channel the simultaneous existence of two types of circulating flows is not observed, the vortical structure in this case is the same along the whole channel length. Transition from transverse to longitudinal vortices in an infinitely long channel occurs in the range $0.05 < \mathrm{Re} < 0.1$.

The analysis of laminar flow and heat transfer by mixed convection ***in a vertical round channel with an uniform wall temperature*** shows that the effect of buoyancy forces is the strongest for the case of an ascending flow in comparison with a descending one [7.60].

The numerical solution of the problem of mixed convection for the case of nonstationary laminar flow ***in a container filled with a porous medium saturated with the liquid which obeys the Darcy law, in the presence of internal heat generation*** was given in [7.61] ***Two vertical walls are adiabatically insulated, the horizontal ones are maintained at a constant temperature, and the upper surface moves with a constant velocity***. The numerical data are generalized by the dependence ($10^{-4} \le \mathrm{Da} \le 10^{-1}$)

$$\mathrm{Nu} = \frac{1.6064}{\left(1 + \mathrm{Da}^{-1}\right)^{0.1659}} (\mathrm{Ri})^{-0.1593} , \tag{7.2.29}$$

where $\mathrm{Ri} = \mathrm{Gr}/\mathrm{Re}^2$, Da is the Darcy number. The analysis of the numerical data demonstrates the strong dependence of the flow and heat transfer characteristics on the Ri number. At small Ri numbers the presence of internal heat generation has a strong effect on the temperature field, but exerts a little effect on the flow structure.

The numerical data given in [7.62] for a two-dimensional steady free, forced, and mixed convection ***in a vertical flat channel one of the walls of which is heated and the second one is maintained at a constant temperature or adiabatically insulated*** show the possibility of separation of the free-convective flow from an unheated wall at large Grashof numbers. The channel is heated from ***a heat source of finite size*** located on the wall. At large distances from the source reattachment of the separated flow is possible. The intensity of motion in circulating cells and their extent depend on the ratio of the Reynolds and Grashof numbers. As the Reynolds number increases, the circulating flows disappear, and for short channels there is the possibility of entrapping the flow escaping from them into the entrance. The numerical data ***for air*** are generalized by the dependences

for free convection

$$\mathrm{Nu}_x = 0.726\mathrm{Gr}_x^{0.17} , \tag{7.2.30}$$

$$\mathrm{Nu}_{NC} = 1.253\mathrm{Gr}^{0.152} , \tag{7.2.31}$$

for forced convection

$$\mathrm{Nu}_{FC} = 1.166\mathrm{Gr}^{0.393} , \tag{7.2.32}$$

for mixed convection

$$(\mathrm{Nu}_{MC})^n = (\mathrm{Nu}_{NC})^n + (\mathrm{Nu}_{FC})^n , \; n = 3 . \tag{7.2.33}$$

The numerical data for local values of viscous friction and heat transfer coefficients for mixed convection ***of air near two-dimensional or axisymmetric bodies of arbitrary shape*** are generalized by the dependences [7.63]

$$\begin{aligned} \mathrm{C}_f\mathrm{Re}_x^{1/2} &= 1.1416 - 0.4775R_x - 0.3013R_x\left(R_x - 1\right) \\ &\quad - 2.414R_x\left(R_x^2 - 1\right) + 2.017R_x\left(R_x^3 - 1\right) , \\ \mathrm{Nu}_x\mathrm{Re}_x^{-1/2} &= 0.297 - 0.0043R_x - 0.4359R_x\left(R_x - 1\right) \\ &\quad - 0.1578R_x\left(R_x^2 - 1\right) + 1.251R_x\left(R_x^3 - 1\right) , \end{aligned} \tag{7.2.34}$$

where $R_x = u_c/(u_c + u_a)$, (u_c is the velocity of a forced flow at the outer edge of the boundary layer and u_a is the pseudovelocity due to body forces).

The effect of the buoyancy-force inclination angle on the characteristics of flow and heat transfer in the case of mixed two-dimensional laminar convection ***in an inclined duct with a backward-facing step*** investigated in [7.64] shows that the reattachment length increases with increase in the inclination angle. The coordinate of the point on the heated wall, at which the maximum in the local Nusselt number distribution is attained, also increases with increase in the inclination angle almost by the linear law.

The results of numerical analysis of the effect of wall temperature, radii ratio, and the Re number on the characteristics of flow and heat transfer with mixed convection of an ***air-water mixture in vertical concentric channels with asymmetric heating*** are given in [7.65].

Comparison of experimental and numerical results for mixed convection ***in large chambers with partitions*** for co-current and counter-current forced and free-convective flows shows that the flow is predominantly free-convective and asymmetric [7.66]. The wall and entrance temperature was kept constant for uniform velocity distribution at the entrance. Two lines of two partitions were located asymmetrically relative to the vertical axis, the entrance was at the upper chamber wall, the exit at the base. The difference of the temperatures between the wall and the entrance changed from 1.0 to 3.3°C at the Prandtl number $\mathrm{Pr} = 0.7$, the Reynolds number at the entrance $\mathrm{Re} = 32$ and 235 and the Rayleigh number at the entrance $\mathrm{Ra} = 0\text{–}6.8 \cdot 10^7$. The temperature difference between the entrance and wall has a rather substantial effect on the flow structure in the chamber.

The experimental study of the effect of buoyancy forces on forced ***water*** flow ***in the entrance of a semicircular channel for constant axial heat supply and perimeter-uniform temperature*** showed that the local and average Nusselt numbers change strongly along the semicircle and increase with increase in the heat flux [7.67]. For average values of the Reynolds number from 400 to 1600 and high heating intensity ($\mathrm{Ra} \sim 4.6{\cdot}10^8$) an increase in the friction coefficient by a factor of more than two, and the heat transfer intensification more than four times are observed. The experimental data for the Nusselt average number in the case of mixed convection are generalized by the dependences

$$\mathrm{Ra}_M < 5 \cdot 10^7 \quad \frac{\mathrm{Nu}}{\mathrm{Nu}_0}\left(\frac{\mu_W}{\mu_M}\right)^{0.14} = 1 + 0.0267\,\mathrm{Ra}_M^{0.2522}\,, \tag{7.2.35}$$

$$\mathrm{Ra}_M > 5 \cdot 10^7 \quad \frac{\mathrm{Nu}}{\mathrm{Nu}_0}\left(\frac{\mu_W}{\mu_M}\right)^{0.14} = 0.2662\,(\mathrm{Re}_M\,\mathrm{Ra}_M)^{0.1052}\,, \tag{7.2.36}$$

where μ_M, Re_M, and Ra_M are the viscosity coefficient, the average Rayleigh and Reynolds numbers at the average liquid temperature in the region of fully developed flow, μ_W is the viscosity coefficient near the channel wall, and Nu_0 is the average Nusselt number for forced convection.

The flow and temperature field characteristics for mixed convection ***in a channel with rectangular partitions on one wall with periodically changing thermal boundary conditions*** are numerically obtained to model the process of cooling the chips in a system of multilayer boards [7.68]. The values of the local Nusselt number near the surfaces of the partitions and the temperature distribution on the surface of base plate are given for two typical cases for horizontal and vertical orientation of the channel. The Re number changed from 100 to 1500, the Gr number, from 0 to $2{\cdot}10^6$.

The results of numerical investigation of mixed convection ***in a square cavity, where the flow is caused by the motion of the upper surface and the buoyancy force due to the base heating given*** in [7.69] point to a strong effect of the Prandtle number on the two-dimensional laminar flow investigated. The model suggested describes the technological process of manufacturing ***a flat glass***, when the cooled and solidifying sheet is drawn from a melt. The numerical modeling was carried out in the ranges $100 \leq \mathrm{Re} \leq 2200$ and $0.01 < \mathrm{Pr} < 50$ for different Richardson numbers.

The results of a three-dimensional numerical analysis of flow and heat transfer characteristics ***in a horizontal annular channel with heated rotating internal circular cylinder*** are given in [7.70]. The presence of hydrodynamic instability leads to the formation of Taylor vortices in the annular channel between two concentric cylinders, one of which or both rotate. In the case of a horizontal arrangement for the aspect ratio from 1.0 to 10.0 and the radii ratio equal to 2.6 the effect of buoyancy and centrifugal forces (created by the heated rotating cylinder) leads to the fully developed spatial flow.

To determine the effect of buoyancy and inertial forces on the average equivalent heat conduction in mixed convection ***in a vertical rotating cylinder*** in [7.71] the criterion $\mathrm{Gr}(36\mathrm{Re}^{1.255})$ is suggested instead of $\mathrm{Gr}/\mathrm{Re}^2$ usually used for mixed convection. The average equivalent heat conduction in the vertical rotating cylinder on imposing the temperature difference between the upper and lower edges first decreases and then increases with increase in the rotational Reynolds number.

Mixed convection ***in a channel of rectangular section (the aspect ratio is equal to 2), with the bottom part being heated and the upper one cooled*** for $\mathrm{Ra} = 22200$ and $\mathrm{Re} = 18.75$, 36, 54 is investigated in [7.72] using the method of laser-Doppler anemometry. In the case of low Re values there was a symmetry distortion for stationary but spatially oscillating flows taking place at some distance from the leading edge of the part with variable heating. Further downstream the nonstationary regimes were observed even for $\mathrm{Re} = 18.75$. Numerical simulation was carried out for studying the effect of the channel inclination angle, the Prandtl number, and adiabatic and ideal conducting walls on the flow and heat transfer characteristics.

The fully developed mixed convection ***in horizontal channels with packings of spheres*** was analytically and experimentally investigated in [7.73]. The effects causing the deviation from the Darcy law and caused by flow inertia, thermal dispersion, and no-slip conditions on the boundary are considered. The analysis shows that the buoyancy force has a marked effect on the structure of a secondary flow and on the heat transfer rate for low Pe number, however, with increase in Pe this effect may decrease for the constant Re value. The Nusselt values in a fully developed region dependent on the Pe number and the ratio of channel and sphere diameters D_e/d for $\mathrm{Ra} = 10^5$ are given in Table 7.2.18.

The effect of mixed convection on the flow structure ***in large partitioned chambers with internal heat sources*** was investigated in [7.74]. The partitions and internal heat sources were placed symmetrically relative to the vertical axis, and the velocity and entrance and wall temperature were

Table 7.2.18. The Values of the Nusselt Number in the Region of a Fully Developed Flow

$D_e/d = 10$	Pe = 10	Nu = 22.001
	Pe = 100	Nu = 22.973
$D_e/d = 19$	Pe = 30	Nu = 31.423
	Pe = 100	Nu = 34.304

constant and homogeneous. Comparison of non-stationary experimental and two-dimensional numerical results for different chamber arrangements shows that the flow structure undergoes a number of bifurcations as the temperature of internal heat sources increases.

The results of numerical investigation of heat transfer by mixed convection for laminar two-dimensional flow ***in a vertical channel with a backward-facing step*** are given in [7.75]. The range of flow conditions at the entrance and wall temperature covers the region from a pure forced convective flow to the inlet degenerate convective flow. The flow caused by buoyancy forces causes a decrease in the reattachment length and pushes the recirculating region away from the heated wall.

The dynamic and thermal characteristics of laminar mixed convection ***in a radially rotating channel of porous material with uniform evaporation and constant wall temperature gradient*** was investigated in [7.76]. On the basis of the self-similar solution obtained the effect of rotation, evaporation, and heating of the wall on the velocity and temperature fields, surface friction, pressure difference, and heat transfer rate is investigated. Flow-reversal phenomena can be caused by the transpiration and buoyancy effects.

The experiments and three-dimensional numerical calculations with the purpose of investigation of mixed-convective heat transfer from four-row, in-line array of 12, square heat sources which are flush mounted in the lower wall of a horizontal rectangular channel, were carried out in [7.77]. Experimental data were obtained for heat transfer regimes for pure natural, mixed, forced laminar convection and also for a regime with transition to turbulence. The dependence of the row-average Nusselt number on the Reynolds number has a minimum, hence we can assume that due to the flow induced by the elevating force it is possible to intensify the heat transfer and to lower the pumping power due to the decrease in the flow velocity.

Measurement of the local heat transfer coefficients and the results of the experiments on flow visualization for orthogonal mixed convection in a partially closed cavity with uniformly heated lateral walls and adiabatic upper and lower walls are given in [7.78]. The experiments were carried out in an experimental module of small scale with different inlet and outlet configurations using ***water*** as a working fluid. The flow and heating conditions were chosen to simulate high values of the number Ra $= 10^{10}$–10^{11} and low and moderate values of the number Re $= 0$–9000 typical of building heat physics applications. With increase in the rate of flow through the cavity the heat transfer rate increases quickly in comparison with the case of pure free convection. For some flow conditions the local enhancement is as large as a factor of 7. The variation of the local heat transfer coefficient over the heated surface is found to be strongly dependent on the recirculation of portions of the forced flow within the enclosure as well as the impingement on or separation of flow from the side wall in some regions.

An experimental investigation of the problem of mixed convection ***in a packed bed of glass spheres on a horizontal isothermal surface*** was presented in [7.79]. The experimental results confirm the dependence of the growth of the thermal boundary layer and the variation of the local heat flux on the parameter $\mathrm{Ra}_x^{1/3}/\mathrm{Pe}_x^{1/2}$. This parameter was obtained from scaling analysis of the Darcy model and it describes the transition between forced and natural convection. The data for experimental dimensionless thermal boundary layer thickness is correlated by

$$\frac{\delta_T}{x}\mathrm{Pe}_x^{1/2} = 5.43\left(\frac{\mathrm{Ra}_x^{1/3}}{\mathrm{Pe}_x^{1/2}}\right)^{-0.34} . \tag{7.2.37}$$

The local Nusselt number for the mixed convection region is correlated by

$$\mathrm{Nu}_x/\mathrm{Pe}_x^{1/2} = 1.15\left(\mathrm{Ra}_x^{1/3}/\mathrm{Pe}_x^{1/2}\right)^{0.29} . \tag{7.2.38}$$

The results of experimental study results of the aspect ratio effects on longitudinal vortex flow in mixed convection ***of air in a bottom heated horizontal rectangular duct*** including combined flow visualization and temperature measurement were presented in [7.80]. Steady and unsteady longitudinal vortices were investigated at the Re number from 2.5 to 50, the Ra number from 3000 to 20.000, and the aspect ratio from 2 to 12. The results indicated that the spatial vortex flow structure in a low aspect ratio duct is similar to that in a high aspect ratio duct with the central portion of the wider duct around $x = A/2$ removed, but the unstable vortex flow at high Re and Ra can be stabilized when the aspect ratio is reduced. In a high aspect ratio duct ($A \geq 8$) reduction in the Re number causes the flow to change from steady to unsteady state with more frequent roll splitting and merging. While in a low aspect ratio duct ($A \leq 6$) at decreasing Re the roll cross-section exhibits cyclic expansion and contraction in time and the longitudinal rolls become snakelike and time periodic in the downstream portion of the duct. At a higher buoyancy-to-inertia ratio the snakelike motion of the rolls gets more intensive and is initiated at a shorter distance from the duct inlet. When the value of the aspect ratio is an odd number, it was found that there is still an even number of rolls induced in the duct. The transition from steady to unsteady state at increasing buoyancy-to-inertia ratio is subcritical in a high aspect ratio duct ($A \geq 8$). However, in a low aspect ratio duct ($A \leq 6$) there exists a finite range of the buoyancy-to-inertia ratio for the appearance of the time-periodic snakelike vortex flow.

The effect of Richardson number (Ri $= \mathrm{Gr}/\mathrm{Re}^2$) on the two-dimensional laminar mixed convective flow and heat transfer ***over in-line tube bundles of cylinders*** were numerically simulated for Re numbers of 50, 100, 150, Ri numbers of −1.0, −0.5, 0.0, +0.5, +0.1 and pitch to diameter ratios of 1.5 and 2.0 with a Pr number of 0.71. The results show a considerable effect of buoyancy over tube bundles [7.81].

Numerical results of investigation of transient mixed turbulent convective flows ***in an axisymmetric, cylindrical, chilled-water storage tank with stratification*** indicate that to ensure a stratified flow the Archimedes number should be greater that two (Ar > 2) and independent of the inlet Reynolds number [7.82]. The calculations involve injection of cold water through a slot in the base of an insulated tank, within which the fluid is initially at rest and at constant temperature. The inlet Re number ranges from 500 to 3000, Ar numbers range from 0.5 to 5.0.

Numerical results for turbulent mixed convection ***of air in a horizontal concentric annulus between a cooled outer cylinder and a heated, rotating, inner cylinder*** show that the mean Nusselt number increases with increase in Ra, but decreases with increase in Re or R (radius ratio) [7.83]. Ra numbers range from 10^7 to 10^{10}, Re numbers range from 0 to 10^5 and the radius ratio R from 2.6 to 10 for a constant Prandtl number of 0.7. When Ra $> 10^9$, a critical phenomena, the peak value of local Nusselt number occurring in the vicinity of thermal plume region is first discovered. For centrifugal configurations, it is found that rotation has caused significant reduction in the mean heat transfer and it generally increases the strength of the secondary flows. Table 7.2.19 shows that Nu values decrease with increasing R due to greater centrifugal effects.

Table 7.2.19. Mean Nu Number for Various Cases

Ra	10^7		10^8	10^9				10^{10}
R	2.6	5.0	2.6	2.6	3.5	5.0	10.0	2.6
Re = 0	10.286	9.953	21.823	42.084	40.932	38.179	36.231	86.432
Re = 10^2	10.129	9.578	20.826	41.937				
Re = 10^3	10.006	9.162	20.512	41.672				
Re = 10^4	8.826	7.583	19.340	41.328	39.541	35.621	30.074	85.478
Re = 10^5	6.091	5.211	15.832	40.803				

The experimental data for the average heat transfer coefficient with mixed convection are generalized by the empirical dependence constructed on the basis of the superposition method, modified for ***water*** ascending flow ***in vertical channels*** [7.84]

$$\frac{\mathrm{Nu_{up}}}{\mathrm{Nu_{df}}} = 1 - a \exp\left\{-0.8\left[\log\left(\frac{\mathrm{Bo}}{b}\right)\right]^2\right\} , \tag{7.2.39}$$

where

$$\mathrm{Nu}_{\mathrm{up}}^3 = \left|\mathrm{Nu}_{\mathrm{for}}^3 - \mathrm{Nu}_{\mathrm{nat}}^3\right| ,$$
$$\mathrm{Nu}_{\mathrm{df}}^2 = \mathrm{Nu}_{\mathrm{for}}^2 + \mathrm{Nu}_{\mathrm{nat}}^2 .$$

The coefficients a and b are determined from the experimental data

$$a = 0.36 + 0.00654\frac{L}{D}, \quad b = 869\left(\frac{L}{D}\right)^{-2.16} ,$$

$$\mathrm{Bo} = 8 \cdot 10^4 \frac{\mathrm{Gr}^*}{\mathrm{Re}^{3.425}\,\mathrm{Pr}^{0.8}} ,$$

L and D are the channel length and diameter, Nu_{up} and $\mathrm{Nu}_{df}\,\mathrm{Pe}_x^{1/2}$ are the Nusselt numbers for ascending and descending flows, respectively.

The stable laminar mixed convection ***in fully developed region of horizontal tubes with internal fins in the case of axial heat supply and circle-uniform temperature*** is investigated in [7.85]. It is found that the characteristics of fluid flow and heat transfer depend on the Gr and Pr numbers, the number of fins M and the fin relative height $H = l/r_0$ (l is the fin length, r_0 is the tube radius). The numerical investigation is carried out in the ranges $M = 4, 16$; $H = 0, 0.2, 0.5, 0.8$; $\mathrm{Pr} = 7$ and $\mathrm{Gr} = 0–2 \cdot 10^6$. The inner finning delays the threshold of appreciable effect of free convection and suppresses the increase of the friction coefficient and the Nusselt number compared with the case of smooth tubes. The values of the average Nu are given in Table 7.2.20.

Table 7.2.20.

M	H	Re_0	Nu
4	0.2	18.54	4.570
	0.5	36.21	7.763
	0.8	71.30	19.60
16	0.2	26.55	4.747
	0.5	129.2	8.326
	0.8	440.7	110.2

The effect of the rate of rotation and of the diverging angle on the temporal stability of a flow is studied experimentally in [7.86, 7.87] ***for air*** convection ***in a vertical, diverging cylinder heated from below and rotating steadily about its own axis***. Data are collected for the diverging angle varying from 0–30°, imposed temperature difference 5–15°C, and rotation rate of up to 346.1 rpm. The measured data show that the ranges of the rotation rate to stabilize the flow are wider for diverging cylinders than those for a right cylinder at $\varphi = 0°$(φ is the diverging angle between insulated wall and gravity). A power spectrum density analysis of the data for the unstable flow at high rotation rates indicates that the flow oscillation is mainly time periodic and dominated by single fundamental frequency. Moreover, the oscillation frequency varies nonmonotonically with the rotation rate.

The effects of Coriolis and buoyancy forces on the flow structure and on the flow properties ***in a rotating straight pipe*** is investigated in [7.88]. The flow properties, such as the shear stresses and the Nusselt number, depend on the values of the products ReRe_r (Re_r is the Reynolds number based on the angular velocity, Re is the Reynolds number based on the axial velocity) and ReRa, respectively.

Numerical data for three-dimensional mixed-convection laminar constant-property liquid flow ***through alternating two-pass square channel with radial rotation*** are presented in [7.89]. The emphasis is placed on rotating effects, including both the Coriolis force, and centrifugal buoyancy, on heat transfer in channels of different flow directions. The results reveal of calculations show that in buoyancy-affected rotating two-pass channel the local heat transfer is dependent on the flow direction, and the Coriolis-force effect is more significant for a radially converging flow than that for a radially expanding one. It is also found that the centrifugal buoyancy significantly enhances the peripherally averaged heat transfer in a channel with a radial inward flow, but relatively negligibly in a channel with a radial outward flow, which is analogous to the buoyancy-dependence of the cross-flow-intensity development. The prediction further demonstrates that, as long as the centrifugal buoyancy is sufficiently strong, the radial outward flow will separate from the leading surface, which largely deteriorates the wall heat transfer.

Experimental results for developing turbulent mixed convection ***in a horizontal circular tube with longitudinal and transverse strip inserts*** for $6500 \leq \mathrm{Re} \leq 19500$, $1.53 \cdot 10^{-4} \leq \mathrm{Gr}^*/\mathrm{Re}^2 \leq 2.31$ and $5.82 \cdot 10^4 \leq \mathrm{Gr}^* \leq 9.78 \cdot 10^7$ at different heat flux levels show that the buoyancy effect has a smaller influence for tubes with inserts than for bare tubes [7.90]. Moreover, the heat transfer enhancement was found to be 2–3 times those of bare tubes.

Experimental measurements and analysis of buoyancy-assisted mixed convection ***in a vertical square channel with asymmetric heating conditions*** are presented in [7.91]. Two opposite sides of the test section are heated by four heaters and the other two sides of the square channel are insulated. The Reynolds number is varied from 200 to 11200 and the buoyancy parameter, $\mathrm{Gr}/\mathrm{Re}^2$, is varied from 0.02 to 200. The local heat transfer coefficient increases with increasing buoyancy parameter.

Combined free and forced convection ***in a plane- parallel vertical channel*** was analyzed in the region of a fully developed flow taking into account the effect of viscous dissipation [7.92,7.93]. Two isothermal boundary surfaces were maintained either at equal or at different temperatures. The velocity field, the temperature field, and the Nusselt numbers were obtained by the method of successive perturbations which employs a perturbation parameter proportional to the Brinkman number. The Nusselt number can be defined at each boundary as

$$\mathrm{Nu}_- = 2R_T + 12\mathrm{Br}\ , \quad \mathrm{Nu}_+ = 2R_T - 12\mathrm{Br}\ , \tag{7.2.40}$$

where $R_T = (T_2 - T_1)/\Delta T$, $\Delta T = T_2 - T_1$, if $T_2 > T_1$, or $\Delta T = \nu^2/(c_p D^2)$, if $T_2 = T_1$, $D = 2L$ is the hydraulic diameter and $\mathrm{Br} = \mu U_0^2/(\lambda \Delta T)$ is the Brinkman number.

The effect of gas-liquid interface, variability of thermophysical properties, and ***film vaporization from a cooling of falling liquid film through on turbulent mixed convective heat and mass transfer in channel flow in an air-water system*** was numerically investigated in [7.94]. The results show that the heat transfer from the gas-liquid interface to the gas flow is predominantly determined by the evolvement of latent heat due to film evaporation. Moreover, the systems are best cooled by a liquid film when they have a higher temperature and small flow velocity at the entrance and also for systems characterized by high values of the Reynolds number for a gas flow.

Numerical calculations for a three-dimensional turbulent mixed convection and heat transfer ***in the entrance region of a curvilinear tube***, presented in [7.95], show that distribution of axial velocity and temperature of rotating cells of different extention are determined by the ratio $\sqrt{\mathrm{Gr}}/\mathrm{Dn}\sqrt{1+\mathrm{Pr}}$ (which is used to characterize the relative magnitude of buoyancy and centrifugal effect on the secondary flow in the curvilinear tube (Dean number $\mathrm{Dn} = \mathrm{Re}\delta^{1/2}$, $\delta = r/R_c$, r is the radius of the curvilinear tube, R_c is the radius of the coil). At high Gr numbers, the developing secondary flow field showed the existence of three vortices. The peripherally averaged Nusselt number and friction factor exhibit oscillatory behavior along the streamwise direction. The increase in the average Nu number and friction factor resulting from buoyancy was prominent at the entrance region of the tube, but gradually becomes weaker further downstream.

Mixed convection ***of air*** with $\mathrm{Pr} = 0.7$ ***between two horizontal concentric cylinders which are held at different uniform temperatures*** is numerically investigated in [7.96]. The forced flow is induced by the cooled outer cylinder slowly rotating with a constant angular velocity with its axis at the center of the annulus. Investigations were made for various combinations of Ra, Re, and σ (diameter of inner cylinder/gap width) in the range of $\mathrm{Ra} \leq 5 \cdot 10^4$, $\mathrm{Re} \leq 1500$ and $0.5 \leq \sigma \leq 5$. The flow patterns can be classified into three basic types according to the number of eddies: two-add, one-add and no-add flows.

The results of numerical examination the effects of wall transpiration on laminar mixed convection and heat transfer ***in the entrance region of horizontal rectangular ducts for both thermal boundary conditions of uniform heat flux (UHF) and uniform wall temperature (UWT)*** show that either wall injection or wall suction has a considerable effect on the flow structure and heat transfer [7.97]. The wall Reynolds number $\mathrm{Re_w}$ changes from −2 (suction) to 4 (injection) with Rayleigh numbers Ra ranging from 0 to $2 \cdot 10^5$ for the aspect ratios $\gamma = 0.2$, 0.5, 1, 2, and 5. Correlation equations for the longitudinal averaged Nu number for both boundary conditions in the entrance region of up to $z/\mathrm{Pr} = 0.1$ are

uniform heat flux (UHF)

$$\frac{\mathrm{Nu}}{\mathrm{Nu_0}} = a + b\exp\left(2.45\log Ra\right) + c\left(\log \mathrm{Ra}\right)^{15} + d\left(\log \mathrm{Ra}\right)^{7} \mathrm{Re_w}\,, \qquad (7.2.41)$$

uniform wall temperature (UWT)

$$\frac{\mathrm{Nu}}{\mathrm{Nu_0}} = a + b\exp\left(2.85\log \mathrm{Ra}\right) + c\left(\log \mathrm{Ra}\right)^{16.6} + d\left(\log \mathrm{Ra}\right)^{7} \mathrm{Re_w}\,, \qquad (7.2.42)$$

where the constants a, b, c, and d are listed in Table 7.2.21.

Table 7.2.21. Constants of the Correlating Equations of Average Nu Numbers

	UHF				UWT			
γ	a	$b{\cdot}10^7$	$c{\cdot}10^{12}$	$d{\cdot}10^7$	a	$b{\cdot}10^7$	$c{\cdot}10^{12}$	$d{\cdot}10^7$
0.2	0.9906	11.09	−4.596	2.148	0.9732	5.897	−1.516	3.85
0.5	0.9764	36.61	−17.51	2.514	0.9716	23.13	−7.131	8.037
1.0	0.9796	50.62	−24.68	2.349	1.014	25.03	−7.743	9.827
2.0	0.9799	55.88	−27.38	2.1	1.001	1.518	−4.693	4.682
5.0	0.9864	32.60	−15.59	1.518	0.9954	2.77	−0.781	1.011

The three-dimensional mixed convection fluid flow ***within a horizontal duct, the walls of which are assumed to be maintained at specified constant temperatures*** was investigated in [7.98]. The fluid flowing through the duct was assumed to be steady and laminar and the results presented were mainly those obtained when the parameter values were typically found in hydraulic fractures, a technique used in the oil industry to enhance production of an oil well. The results were obtained for Newtonian and power-law fluids, as well as Newtonian fluids with a temperature dependent viscosity.

The influence of buoyancy forces ($\mathrm{Gr} \sim 10^5$) on heat transfer and secondary flows ***in a helically coiled exchanger with a constant wall temperature*** for $\mathrm{Re} = 500$, $\mathrm{Pr} = 5$ and $\delta = 1/14$ was studied in [7.99]. It appeared that heat transfer was highly influenced by the secondary flow induced by centrifugal and buoyancy forces. For low Gr numbers a splitting phenomenon of the temperature field was observed due to large secondary velocities, resulting as separated areas of fluid. For high Gr numbers the fluid in the coiled tube becomes almost linearly stratified which results in small secondary velocities. A wavy behavior in the Nu number is observed for mean Gr numbers.

Three-dimensional interaction ***of an initially cylindrical vortex tube with a solid sphere in a stratified temperature field*** was investigated by solving the Navier–Stokes and energy equations in [7.100]. Particular attention was given to the effect exerted on the Nusselt number on the sphere surface by such phenomena as the vortex advection and temperature stratification. Transient Nu number patterns are entirely different from those in a uniform temperature field; temperature stratification influences the vortex impact nearly by a factor of three. The Nusselt number on the sphere surface can be represented as the sum of the Nu number for a uniform flow at a constant temperature equal to the average temperature of the flow and the Nu number for a stratified flow with fluctuations of temperature about the average temperature equal to zero. When the product of the vortex circulation and the gas-filled temperature gradient is positive, the Nusselt number average over the sphere surface increases monotonically with increase in the vorticity and with increase in the distance of the vortex from the flow symmetry axis; when this product is negative, the Nu number increases with decrease in vortex circulation or the vortex distance. Based on these findings, it can be shown that in spray combustion systems coupled vortex-temperature stratification could have a significant effect on convective heating of droplets and eventual evaporation.

A numerical investigation of mixed convection heat transfer ***from a circular cylinder with a heated core*** was presented in [7.101]. Numerical results are given for the radii ratio $r_c = 0.11$, Prandlt numbers $\mathrm{Pr} = 0.72$ and 5, Reynolds numbers $\mathrm{Re} \le 40$, buoyancy parameter $\mathrm{Gr}/\mathrm{Re}^2 \le 8$, and thermal conductivity ratios $k \le 400$. The heat transfer and the flow field were found to be primarily affected by both the buoyancy forces and the thermal conductivity ratio, but also depending on the inclination angle and on the Re and Pr numbers.

Mixed convection ***through a double-step sudden expansion into a duct with a uniform temperature of the walls*** was investigated in [7.102]. For buoyancy-assisted flow heat transfer accelerates the flow near the heated wall and flow reversal takes place in the centerline region as the Gr number becomes high enough. A pair of recirculating flows appears near the centerline as Gr becomes sufficiently large. When the duct is inclined to the vertical position, a symmetric pattern is obtained at $\varphi = 60°$. However, this critical angle may depend upon Re number. The heat transfer peak corresponds to the location where the wavy flow reaches the walls. The lowest rate of heat transfer is observed in the region of circulating cells. The overall heat transfer is enhanced by a wavy flow, compared with a horizontal duct.

The results of an experimental investigation of the local heat transfer ***in a vertical tube cooled by air*** for mixed turbulent convection under the conditions of coinciding directions of forced and free-convective flows for $\mathrm{Re} = 0.3 \cdot 10^4 – 5 \cdot 10^4$ were presented in [7.103]. The generalizing dependence for calculating the local heat transfer on the tube length is

$$\frac{\mathrm{Nu}}{\mathrm{Nu}_{FC}} = \frac{1}{0.98 + 0.54(\mathrm{Gr}/\mathrm{Re}^2)^{0.25}(x/d)^{0.975}} . \qquad (7.2.43)$$

The results of a numerical analysis of the effect of a buoyancy force on the flow pattern and heat transfer in ***an asymmetrically heated channel with a series of conductive fins*** were presented in [7.104]. The fins were attached to a relatively hotter wall at regular distances in the flow direction, so that in the developed area the periodic flow character prevails. It is found that within one module, along with a typical recirculating flow taking place at the back fin surface, a secondary recirculating vortex appears near the less heated wall for quite intense heating of the hotter wall. The strength and structure of the secondary vortex are determined by the Grashof and Reynolds numbers and the geometric parameters (distance between fins, fin height, and channel inclination angle).

A numerical study for the buoyancy effect on the fully developed laminar forced convection ***in an axially uniformly heated tubes with a concentric adiabatic plate insert*** was presented in [7.105]. An increase in the Ra number intensifies the secondary flow and collects colder fluid below the plate. The effect of Pr number on the flow and temperature patterns becomes obvious as the Ra number increases. Overall Nu numbers are given in Table 7.2.22.

Table 7.2.22. Overall Nu Numbers

φ	Pr	$\mathrm{Ra} = 10^3$	$\mathrm{Ra} = 10^4$	$\mathrm{Ra} = 10^5$	$\mathrm{Ra} = 10^6$
0°	0.7	4.276	4.466	6.072	9.416
	7.0	4.275	4.429	5.962	9.405
45°	0.7	4.277	4.524	6.197	9.314
	7.0	4.276	4.491	6.117	9.315
90°	0.7	4.278	4.564	6.126	9.351
	7.0	4.278	4.556	6.139	9.288

Experimental investigations were made in [7.106] to study the convective phenomena ***of an initially stratified salt-water solution due to bottom heating in a uniformly rotating cylindrical cavity***. Three types of global flow patterns initially appear depending on the effective Ra number and Ta number: stagnant flow regime, single mixed layer flow regime and multiple mixed layer flow regime.

The longitudinal and overall convective heat transfer ***of rotating spheres and spheroids*** with axis ratios of $0.50 \le \alpha \le 1.0$ and limited thermal diffusivity and, thus, non-uniform surface temperatures was experimentally investigated in a wind tunnel over the range of Reynolds numbers $1.1 \cdot 10^4 < \mathrm{Re}$

$< 5.2 \cdot 10^4$ [7.107]. The results indicate how the convective heat transfer of rotating particles varies with latitude, the rotation rate (equivalent to Strouhal numbers $0 \leq \mathrm{Sr} \leq 0.06$), and the position of rotation axis. Within experimental error, these factors do not influence the total heat transfer. However, they will be important in a two-component, three-phase system, where deposition of substances, chemical reactions including release of latent heat, radiation and other processes are specific for the location of particles and local temperature. The present study, directed towards the local transfer rates controlling hailstone growth, provides the basic method of how to approach such complex situations. The overall Nusselt number, determined by integrating latitudinal heat transfer over the entire particle surface, was parameterized as a function of Reynolds number and axis ratio:

$$\mathrm{Nu} = (0.236 + 0.094\alpha)\,\mathrm{Re}^{(0.0658-0.088\alpha)} \,, \tag{7.2.44}$$

$$\mathrm{Nu} = (55.6 - 13.5\alpha) + (0.00458 - 0.00218\alpha) \cdot 10^{-4}\mathrm{Re} \,. \tag{7.2.45}$$

Numerical simulation for a steady, developing $3D$ laminar mixed convection ***in axially uniformly heated horizontal tube with different longitudinal adiabatic inserts*** for a Prandtl number of 0.7 was performed in [7.108]. The average Nusselt number of each cross-section along the downstream distance decreases initially due to the growth of a thermal boundary layer and starts to rise as the secondary flow becomes more active. The effect of flow on the buoyancy force increases and decreases due to unstable stratification.

Combined experimental flow visualization and three-dimensional unsteady numerical simulation were carried out in [7.109–7.112] to investigate the formation of the traveling transverse and mixed vortex roll structures from the initially unidirectional, sub critical flow ***of air in a bottom heated horizontal flat duct*** when the Rayleigh number was raised to a supercritical level. In the study the Reynolds number was varied from 2.5 to 20 and the Rayleigh number from 1500 to 8000. At a low Reynolds number ($\mathrm{Re} \leq 5$) the transverse rolls are longer and stronger to expel all the longitudinal rolls out of the test section so that a pure and regular transverse vortex roll structure is formed. While at higher Reynolds number ($\mathrm{Re} \geq 7.5$) the transverse rolls are shorter and weaker and a few longitudinal rolls stay, all the time, near the duct sides, resulting in a mixed longitudinal and transverse roll structure.

Both forced and mixed convection flow and heat transfer ***in a convergent channel*** was studied experimentally in [7.113]. The parameter $\mathrm{Nu}/\mathrm{Re}^{0.4}$ for the convergent channel is correlated in terms of $\mathrm{Gr}/\mathrm{Re}^2$, and the results are written as follows.

For assisted convection and $\mathbf{Gr/Re^2 \leq 907}$:

$$100\log(\mathrm{Nu}/\mathrm{Re}^{0.4}) = -3.8 + 0.4124\log(\mathrm{Gr}/\mathrm{Re}^2) + 2.6234[\log(\mathrm{Gr}/\mathrm{Re}^2)]^2 \,. \tag{7.2.46}$$

***For opposite convection and* Gr/Re^2 $\leq$ 10:**

$$100\log(\mathrm{Nu}/\mathrm{Re}^{0.4}) = -4.52 + 1.314\log(\mathrm{Gr}/\mathrm{Re}^2) - 3.396[\log(\mathrm{Gr}/\mathrm{Re}^2)]^2 - 4.279[\log(\mathrm{Gr}/\mathrm{Re}^2)]^3 \,. \tag{7.2.47}$$

***For opposite convection and* 10 $\leq$ Gr/Re^2 $\leq$ 907:**

$$100\log(\mathrm{Nu}/\mathrm{Re}^{0.4}) = -17.03 + 9.834\log(\mathrm{Gr}/\mathrm{Re}^2) + 4.08[\log(\mathrm{Gr}/\mathrm{Re}^2)]^2 \,. \tag{7.2.48}$$

Numerical results for natural convection ***in an enclosure induced by a rotating cylinder*** show that total heat transfer depends on the direction of rotation [7.114]. For the counterclockwise rotation the contribution of cylinder to natural convection heat transfer is found to be substantial, when the value $\mathrm{Gr}/\mathrm{Re}^2$ is larger than 100; however, for the clockwise rotation this contribution is hardly found even when the value $\mathrm{Gr}/\mathrm{Re}^2$ is equal to 1.

Three-dimensional conjugate mixed convection ***in a rectangular channel*** was numerically investigated in [7.115]. The focus is on the effects of local buoyancy generated by a discrete heat source mounted on a thermally conducting board. Calculations covered $0.126 \leq 1260\mathrm{Re}$, $\mathrm{Gr} = 7.8789 \cdot 10^7$, $1 \leq k \leq 100$, and $\mathrm{Pr} = 0.71$, respectively. The flow field showed active interaction between the buoyancy-induced thermal plume and the approaching external flow in the mixed convection regime where the temperature distribution, heat flux, and heat transfer coefficients all strongly depend on such flow interaction. In addition, the Nusselt number on top of the heat source indicates significant three-dimensional effects. The mixed convection regime is determined as $1.81 \cdot 10^{-3} \leq \mathrm{Re}/\mathrm{Gr}^{1/2} < 2.80 \cdot 10^{-2}$. The numerical results for the Nusselt numbers are generalized by correlation

$$\mathrm{Nu} = \left(C_1 + C_2\mathrm{Re}/\mathrm{Gr}^{1/2}\right)^{c_3} , \tag{7.2.49}$$

where C_1, C_2 and C_3 are determined in Table 7.2.23 ($k = k_s/k_f$ is the thermal conductivity ratio; k_s, k_f are the thermal conductivities of the solid and fluid).

The averaged Nusselt number ***of a heated sphere rotating in a quiescent air*** was measured experimentally in [7.116] over the range $0 < \mathrm{Re}$

Table 7.2.23. Coefficients for a Correlation Equation

k	C_1	C_2	C_3
1	225.95	16025.70	0.4655
10	55.15	19375.20	0.4297
100	11.29	7148.34	0.4493

$<$ 33320 from free to forced convection regimes. The overall Nu number for correlating the measured data was found ($\mathrm{Gr} > 3 \cdot 10^5$, $\mathrm{Re} < 33320$, $\mathrm{Pr} = 0.71$)

$$\begin{aligned} \mathrm{Nu}^3 &= \mathrm{Nu}_{NC}^3 + \mathrm{Nu}_{FC}^3 , \\ \mathrm{Nu}_{NC} &= 2 + 0.392\,\mathrm{Gr}^{0.31} , \\ \mathrm{Nu}_{FC} &= 2 + 0.175\,\mathrm{Re}^{0.583} . \end{aligned} \tag{7.2.50}$$

The problem of laminar mixed convective heat transfer ***from a straight isothermal tube of elliptic cross-section placed in a uniform stream*** was investigated in [7.117]. The resulting velocity and thermal fields both were found to be either steady or quasi-steady depending on vortex shedding. The results also indicate that the increase of Gr for a given value Re tends to suppress vortex shedding. The averaged Nu number is found to increase considerably with increase in the ratio $\mathrm{Gr}/\mathrm{Re}^2$ (Table 7.2.24).

Hear transfer ***from a rotating sircular cylinder immersed in a spatially uniform, time-dependent convective environment*** was investigated numerically in [7.118]. The results show that vortex shedding is

Table 7.2.24. The Effect of Gr Number on the Average Nu Number (Axes Ratio $b/a = 0.6$, Angle of Inclination of the Tube is $\varphi = 30°$)

Re	$\mathrm{Gr}/\mathrm{Re}^2$	Nu
20	0.00	2.911
	1.95	3.028
	3.90	3.247
	5.86	3.445
	9.46	3.757
50	0.00	4.364
	1.95	4.581
	3.90	4.919
	5.86	5.204
	9.76	5.649
100	0.00	6.030
	1.95	6.380
	3.90	6.780
	5.86	7.120
	9.76	7.770
200	0.00	8.890
	1.00	9.283
	2.00	9.401
	3.00	9.573
	5.00	10.055
500	0.00	15.600
	1.00	16.230
	2.00	16.680
	3.00	16.800
	5.00	17.820

promoted by the cylinder rotation but vanishes when the buoyancy force is present. In opposing flows, the counter flow currents cause a large expansion of the streamlines and isotherms in the direction normal to the free stream velocity. These changes in the structure of the flow and the temperature field greatly modify the heat flux along the surface of the cylinder and consequently, the heat transfer rate is strongly dependent upon Re number, Gr number, rotation speed, and the gravity direction. In Table 7.2.25 steady state mean Nu numbers of mixed convection are listed.

Table 7.2.25. Mixed Convection Nu Numbers with Pr = 0.7

Re	Gr	Nu
1	4	1.09
5	100	1.89
20	1000	3.13
40	6400	4.59

Work [7.119] deals with mixed natural and forced turbulent convection heat transfer ***in vertical tubes***. Two correlations are suggested to predict Nu numbers for opposing and aiding mixed natural and forced convection and heat transfer:

$$\mathrm{Nu}_{-} = \sqrt{\mathrm{Nu}_{FC}^{2} + \mathrm{Nu}_{NC}^{2}}\,, \tag{7.2.51}$$

$$\mathrm{Nu}_{+} = \mathrm{Nu}_{-}\left\{1 - \left(1 - 2\frac{\mathrm{Nu}_{FCl}}{\mathrm{Nu}_{-}}\right) f(P)\right\}, \tag{7.2.52}$$

where

$$f(P) = \exp\left[-1.3\left(\frac{P}{1-(P)} + 0.5\right)^{2}\right], \quad P = \frac{\mathrm{Nu}_{NC} - \mathrm{Nu}_{FC}}{\mathrm{Nu}_{-}}\,,$$

$$\mathrm{Nu}_{NC} = 0.122 Ra^{0.333}\left(1 + \left(\frac{0.492}{\mathrm{Pr}}\right)^{9/16}\right)^{-16/27}. \tag{7.2.53}$$

For 2300 < Re < 10000

$$\mathrm{Nu}_{FC} = (1-\gamma)\,\mathrm{Nu}_{FCl} + \gamma \mathrm{Nu}_{FC}\,, \tag{7.2.54}$$

where $\gamma = (\mathrm{Re} - 2300)/(10000 - 2300)$,

$$\mathrm{Nu}_{FCl} = \left[3.66^{3} + 0.7^{3} + \left[1.615\sqrt{\mathrm{Gz}} - 0.7\right]^{3} + \left[\left(\frac{2}{1 + 22\,\mathrm{Pr}}\right)^{1/6}\sqrt{\mathrm{Gz}}\right]^{3}\right]^{1/3}, \tag{7.2.55}$$

where the Graetz number is $\mathrm{Gz} = \mathrm{RePr}d/L$.

For $\mathrm{Re} \geq 10000$

$$\mathrm{Nu}_{FC} = \frac{\mathrm{Re}\,\mathrm{Pr}\,\xi/8}{1 + 12.7\sqrt{\xi/8}\left(\mathrm{Pr}^{2/3} - 1\right)}\left(1 + \left(\frac{d}{L}\right)^{2/3}\right), \tag{7.2.56}$$

where $\xi = (1.8 \log(\mathrm{Re}) - 1.5)^{-2}$.

By the methods of the boundary layer theory the effect of radiation on free-convective flow of an ***optically dense viscous liquid along an isothermal truncated cone*** was investigated [7.120]. The results of calculation are given for the local Nusselt number Nu and the local viscous friction coefficient C_f depending on the dimensionless distance ξ, parameter $R_d = 4\sigma T_\infty^3/(k(a_r + \sigma_s))$ characterizing a relation between heat fluxes due to thermal conductivity and due to radiation and the parameter $H = T_\mathrm{w}/T_\infty$, where k is the thermal conductivity coefficient, a_r is the Rosseland mean extinction coefficient, C_f is the local viscous friction coefficient, $(C_f = 2\nu(\partial u/\partial y)_{y=0}/U_r^2)$, $U_r = \sqrt{g\beta\cos\gamma(T_\mathrm{w} - T_\infty)x^*}$, γ is half the cone angle, $x^* = x - x_0$, x is the coordinate along the flow, x_0 is the distance from the origin to the leading edge of the truncated cone. As ξ, R_d, and H increase, the local Nu number increases, and the local friction coefficient C_f decreases. For large values of R_d and H the C_f value in the range of ξ from 10^{-3} to 10^4 reaches its minimum. In the limiting cases the values of Nu and C_f correspond to the coefficients for an inclined plane ($\xi = 0$) and a full cone ($\xi = \infty$).

The heat transfer for mixed and forced convection ***in a uniformly heated bundle of seven cores*** with the ratio P/D equal to 1.38 (D is the diameter of cores, P is the step between the neighboring cores) was investigated experimentally. The Re number changed from 148 to 3800, Gr_q from $1.3 \cdot 10^5$ to $3 \cdot 10^6$ and Ri from 0.01 to 9. The transition from a forced laminar convection to a mixed one for the downstream occured at $\mathrm{Ri} = 0.1$, that is, an order of magnitude less than for an ascending flow. The dependences for the average Nusselt number in the case of the laminar mixed convection are [7.121]

for ascending flows:

$$\mathrm{Nu}_{C,L} = \left[\mathrm{Nu}_{F,L}^3 + \mathrm{Nu}_{N,L}^3\right]^{1/3}, \tag{7.2.57}$$

for descending flows:

$$\mathrm{Nu}_{C,L} = \left[\mathrm{Nu}_{F,L}^2 - \mathrm{Nu}_{N,L}^2\right]^{1/2}, \tag{7.2.58}$$

where $\mathrm{Nu}_{F,L}$ and $\mathrm{Nu}_{N,L}$ are the average Nu numbers for forced and free convection in laminar regime, $\mathrm{Ri} = \mathrm{Gr}_q/\mathrm{Re}^2$ is the Richardson number.

In work [7.122] unified similarity transformation was suggested which allowed one to obtain all possible solutions for free, forced, and mixed convection in ***different porous media*** is offered. For forming a modified Peckle number, the resulting flow velocity near the wall caused by the forced flow due to external and buoyancy forces was chosen as the characteristic velocity. The suggested technique allowed one to distinguish between three limiting flow regimes. The asymptotic dependences obtained for different flow regimes can be applied to a practical estimation of convective heat transfer in media obeying and not obeying the Darcy law.

References

1. O. G. Martynenko and Yu. A. Sokovishin, *Free-Convective Heat Transfer on a Vertical Surface (II Kind Boundary Conditions)*, Nauka i Tekhnika Press, Minsk (1978).
2. O. G. Martynenko and Yu. A. Sokovishin, in: *Future Energy Production Systems Heat and Mass Transfer*, vol. 1, New York (1976), pp. 131–138.
3. Yu. A. Sokovishin and M. V. Shapiro, in: *Heat and Mass Transfer Processes in Thermooptical Devices,* Minsk (1979), pp. 110–115.
4. O. G. Martynenko and Yu. A. Sokovishin, *Mixed Convection Heat Transfer*, Nauka i Tekhnika Press, Minsk (1975), pp. 1071–1079.
5. L. Zhang, A. Soufiani, J. P. Petit, and J. Taine, *Int. J. Heat Mass Transfer*, **33**, 319 (1990).
6. O. G. Martynenko, Yu. A. Sokovishin, and M. V. Shapiro, *Inzh.-Fiz. Zh.*, **32**, 1071 (1977).
7. V. R. Rao, C. Balaji, and S. P. Venkateshan, *Int. J. Heat Mass Transfer*, **40**, 2941 (1997).
8. V. Rao and S. P. Venkateshan, *Int. J. Heat Mass Transfer*, **39**, 779 (1996).
9. Yu. A. Sokovishin and M. V. Shapiro, *Inzh.-Fiz. Zh.*, **33**, 694 (1977).
10. T. Fujii, M. Fujii, H. Tanaka, and S. Fujiyama, *Trans. ASME, Ser. C, J. Heat Transfer*, **44**, 3832 (1978).
11. B. Sammakia, B. Gebhart, and Z. H. Qureshi, *Int. J. Heat Mass Transfer*, **23**, 571 (1980).
12. G. Yang, M. A. Ebadian, and A. Campo, *Int. J. Heat Mass Transfer*, **34**, 1319 (1991).
13. A. Moutsoglou, J. H. Rhee, and J. K. Won, *Int. J. Heat Mass Transfer*, **35**, 2855 (1992).
14. X. Cheng and U. Muller, *Int. J. Heat Mass Transfer*, **41**, 1681 (1998).
15. Z. Tan and J. R. Howell, *Int. J. Heat Mass Transfer*, **34**, 785 (1991).
16. O. G. Martynenko and Y. A. Sokovishin, *AIChE Symp. Ser.*, **74**, 352 (1978).
17. S. Tsuruno and I. Iguchi, *Sci. Eng. Repts Def. Acad.*, **14**, 207 (1976).
18. S. W. Churchill, *AIChE J.,* **23**, 10 (1977).
19. B. N. Kovalyev, *Investigation of Local Heat Transfer in Mixed Turbulent Convection on a Vertical Surface*, Abstract of Doctoral Dissertation (in Technical Sciences), Ashkhabad (1980).
20. R. S. R. Gorla and T. Stearns, *Lett. Heat Mass Transfer*, **2**, 429 (1975).
21. V. M. Soundalgekar, *Isr. J. Technol.*, **15**, 368 (1977).
22. J. K. Gloudemans and M. M. El-Wakil, in: *Future Energy Production Systems Heat and Mass Transfer*, vol. 1, New York (1976).
23. S. Tsuruno and I. Iguchi, *Trans. ASME, Ser. C, J. Heat Transfer*, **101**, 573 (1979).

24. R. M. Abdel-Wahed, E. M. Sparrow, and S. V. Patankar, *Trans. ASME, Ser. C, J. Heat Transfer*, **98**, 576 (1976).
25. V. M. Borovskii, V. A. Shelimanov, V. A. Kutishcheva, *Dokl. Akad. Nauk USSR, Ser. A*, No. 3, 212 (1979).
26. P. M. Brdlick, in: *Energy Conservation in Heating, Cooling and Ventilating Buildings*, vol. 1, Washington (1978).
27. S. Tsuruno and I. Iguchi, *Trans. ASME, Ser. C, J. Heat Transfer*, **102**, 168 (1980).
28. S. D. Harris, D. B. Ingham, and I. Pop, *Int. J. Heat Mass Transfer*, **42**, 357 (1999).
29. U. P. Hwang, *Lett. Heat Mass Transfer*, **6**, 459 (1979).
30. A. Mucoglua and T. S. Chen, *Proc. 6th Int. Heat Transfer Conf.*, Toronto (1978).
31. T. S. Chen and F. A. Strobel, *Trans. ASME, Ser. C, J. Heat Transfer*, **102**, 170 (1980).
32. A. Moutsoglou, S. K. L. Tzuoo, and T. S. Chen, *AIAA Paper*, No. 80-1525, 1 (1980).
33. A. Moutsoglou, T. S. Chen, *Trans. ASME, Ser. C, J. Heat Transfer*, **102**, 371 (1980).
34. F. C. Lai and F. A. Kulacki, *Int. J. Heat Mass Transfer*, **33**, 576 (1990).
35. J.-Y. Jang and K.-N. Lie, *Int. J. Heat Mass Transfer*, **35**, 2077 (1992).
36. C. J. Kobus and G. L. Wedekind, *Int. J. Heat Mass Transfer*, **39**, 2723 (1996).
37. S. L. Lee and J. S. Tsai, *Int. J. Heat Mass Transfer*, **33**, 457 (1990).
38. J. R. Maughan and F. P. Incropera, *Int. J. Heat Mass Transfer*, **33**, 555 (1990).
39. M. T. Ouazzani, J. K. Platten, and A. Mojtabi, *Int. J. Heat Mass Transfer*, **33**, 1417 (1990).
40. H. R. Lee, T. S. Chen, and B. F. Armaly, *Int. J. Heat Mass Transfer*, **35**, 1913 (1992).
41. F. C. Lai and F. A. Kulacki, *Int. J. Heat Mass Transfer*, **34**, 887 (1991).
42. F. C. Lai and F. A. Kulacki, *Int. J. Heat Mass Transfer*, **34**, 525 (1991).
43. G. Wickern, *Int. J. Heat Mass Transfer*, **34**, 1935 (1991).
44. G. Wickern, *Int. J. Heat Mass Transfer*, **34**, 1947 (1991).
45. B. Banerjee, K. V. C. Rao, and V. M. K. Sastri, *Int. J. Heat Mass Transfer*, **33**, 1177 (1990).
46. J. P. Narain, *Lett. Heat Mass Transfer*, **3**, 21 (1976).
47. B. A. K. Abu-Hijleh, *Int. J. Heat Mass Transfer*, **42**, 1383 (1999).
48. R. E. M. Morales, A. Balparda, and A. Silveira-Neto, *Int. J. Heat Mass Transfer*, **42**, 941 (1999).
49. S. W. Churchill and H. H. S. Chu, *Int. J. Heat Mass Transfer*, **18**, 1049 (1975).
50. P. H. Oosthuizen and R. Leung, *Trans. CSME*, **5**, 115 (1979).
51. L. S. Yao and I. Catton, *Trans. ASME, Ser. C, J. Heat Transfer*, **99**, 122 (1977).
52. V. F. Stepanchuk, N. F. Parfenova, V. F. Kravets, *Izv. VUZov, Energetika*, No. 1, 61 (1979).
53. L.V. Grudnikov, *Experimental Investigation of Heat and Mass Transfer in Parts of Pipe Heat Exchangers of Air Dryers Operating under Mixed Convection Conditions*, Abstract of Doctoral Dissertation (in Technical Sciences), Ashkhabad (1980).
54. N. Katagiri, I. Pop, *ZAMM*, **59**, 51 (1979).
55. L. S. Yao and I. Catton, in: *Proc. 6th Int. Heat Transfer Conf.*, vol. 1, Toronto (1978), pp. 13–18.

56. J. P. Narain and M. S. Uberoi, *Int. J. Heat Mass Transfer*, **16**, 1505 (1973).
57. J. W. Yang, *Proc. 6th Int. Heat Transfer Conf.*, vol. 1, Toronto (1978), pp. 49–54.
58. A.V. Fartushnov, *Investigation of Heat Transfer Process in a Horizontal Heat Exchanger in Combined Action of Free and Forced Convection*, Abstract of Doctoral Dissertation (in Technical Sciences), Tashkent (1979).
59. E. Schroder and K. Buhler, *Int. J. Heat Mass Transfer*, **38**, 1249 (1995).
60. A. Barletta, *Int. J. Heat Mass Transfer*, **42**, 2243 (1999).
61. K. M. Khanafer, A. J. Chamkha, *Int. J. Heat Mass Transfer*, **42**, 2465 (1999).
62. D. Elpidorou, V. Prasad, and V. Modi, *Int. J. Heat Mass Transfer*, **34**, 573 (1991).
63. M. R. Cameron, D. R. Jeng, and K. J. DeWitt, *Int. J. Heat Mass Transfer*, **34**, 582 (1991).
64. J. T. Lin, B. F. Armaly, and T. S. Chen, *Int. J. Heat Mass Transfer*, **34**, 1568 (1991).
65. W.-M. Yan and H.-C. Tsay, *Int. J. Heat Mass Transfer*, **34**, 1309 (1991).
66. K. L. Yerkes and A. Faghri, *Int. J. Heat Mass Transfer*, **34**, 1525 (1991).
67. Q. M. Lei and A. C. Trupp, *Int. J. Heat Mass Transfer*, **34**, 2361 (1991).
68. S. Y. Kim, H. J. Sung, and J. M. Hyun, *Int. J. Heat Mass Transfer*, **35**, 2941 (1992).
69. M. K. Moallemi and K. S. Jang, *Int. J. Heat Mass Transfer*, **35**, 1881 (1992).
70. L. Yang and B. Farouk, *Int. J. Heat Mass Transfer*, **35**, 1947 (1992).
71. Z.-Y. Guo and C.-M. Zhang, *Int. J. Heat Mass Transfer*, **35**, 1635 (1992).
72. T. A. Nyce, J. Ouazzani, A. Durand-Daubin, and F. Rosenberger, *Int. J. Heat Mass Transfer*, **5**, 1481 (1992).
73. F. C. Chou, C. J. Cheng, and W. Y. Lien, *Int. J. Heat Mass Transfer*, **35**, 1197 (1992).
74. K. L. Yerkes and A. Faghri, *Int. J. Heat Mass Transfer*, **35**, 1209 (1992).
75. J. T. Lin, B. F. Armaly, and T. S. Chen, *Int. J. Heat Mass Transfer*, **33**, 2121 (1990).
76. C. Y. Soong and G. J. Hwang, *Int. J. Heat Mass Transfer*, **33**, 1805 (1990).
77. H. V. Mahaney, F. P. Incropera, and S. Ramadhyani, *Int. J. Heat Mass Transfer*, **33**, 1233 (1990).
78. M. M. Rahman and V. P. Carey, *Int. J. Heat Mass Transfer*, **33**, 1307 (1990).
79. K. J. Renken and D. Poulikakos, *Int. J. Heat Mass Transfer*, **33**, 1370 (1990).
80. M. Y. Chang and T. F. Lin, *Int. J. Heat Mass Transfer*, **41**, 719 (1998).
81. Y. T. K. Gowda, P. A. A. Narayana, and K. N. Seetharamu, *Int. J. Heat Mass Transfer*, **41**, 1613 (1998).
82. R. E. Spall, *Int. J. Heat Mass Transfer*, **41**, 2003 (1998).
83. M. I. Char and Y.-H. Hsu, *Int. J. Heat Mass Transfer*, **41**, 1633 (1998).
84. G. P. Celata, F. D'Annibale, A. Chiaradia, and M. Cumo, *Int. J. Heat Mass Transfer*, **41**, 4037 (1998).
85. I. M. Rustum and H. M. Soliman, *Int. J. Heat Mass Transfer*, **33**, 1485 (1990).
86. Y. T. Ker and T. F. Lin, *Int. J. Heat Mass Transfer*, **41**, 4281 (1998).
87. Y. T. Ker, T. F. Lin, and Y. H. Li, *Int. J. Heat Mass Transfer*, **41**, 1445 (1998).
88. M. A. Petrakis, *Int. J. Heat Mass Transfer*, **41**, 4385 (1998).
89. J. J. Hwang and D. Y. Lai, *Int. J. Heat Mass Transfer*, **41**, 979 (1998).
90. S.-S. Hsieh, M.-H. Liu, and F.-Y. Wu, *Int. J. Heat Mass Transfer*, **41**, 1049 (1998).

91. X. Zhang and S. Dutta, *Int. J. Heat Mass Transfer*, **41**, 3255 (1998).
92. A. Barletta, *Int. J. Heat Mass Transfer*, **41**, 3501 (1998).
93. E. Zanchini, *Int. J. Heat Mass Transfer*, **41**, 3949 (1998).
94. W.-M. Yan, *Int. J. Heat Mass Transfer*, **41**, 3719 (1998).
95. L. J. Li, C. X. Lin, and M. A. Ebadian, *Int. J. Heat Mass Transfer*, **41**, 3793 (1998).
96. J.-S. Yoo, *Int. J. Heat Mass Transfer*, **41**, 293 (1998).
97. K.-T. Lee and W.-M. Yan, *Int. J. Heat Mass Transfer*, **41**, 411 (1998).
98. L. Elliott, D. B. Ingham, and J. D. Wood, *Int. J. Heat Mass Transfer*, **41**, 33 (1998).
99. J. J. M. Sillekens, C. C. M. Rindt, and A. A. V. Steenhoven, *Int. J. Heat Mass Transfer*, **41**, 61 (1998).
100. M. Masoudi and W. A. Sirignano, *Int. J. Heat Mass Transfer*, **41**, 2639 (1998).
101. M. Amaouche and A. Chennafi, *Int. J. Heat Mass Transfer*, **41**, 2419 (1998).
102. Y.-Y. Tsui and S-J. Shu, *Int. J. Heat Mass Transfer*, **41**, 2687 (1998).
103. J. V. Vilemas, P. S. Poskas, and V. E. Kaupas, *Int. J. Heat Mass Transfer*, **35**, 2421 (1992).
104. C.-H. Cheng, C. D. Luy, and W.-H. Huang, *Int. J. Heat Mass Transfer*, **35**, 2643 (1992).
105. J.-D. Chen and S.-S. Hsieh, *Int. J. Heat Mass Transfer*, **35**, 263 (1992).
106. J. Lee, S. H. Kang, and M. T. Hyun, *Int. J. Heat Mass Transfer*, **40**, 3387 (1997).
107. G. Zheng and R. List, *Int. J. Heat Mass Transfer*, **39**, 1815 (1996).
108. S.-S. Hsien and M.-Y. Wen, *Int. J. Heat Mass Transfer*, **39**, 299 (1996).
109. M. Y. Chang, C. H. Yu, and T. F. Lin, *Int. J. Heat Mass Transfer*, **40**, 1907 (1997).
110. M. Y. Chang, C. H. Yu, and T. F. Lin, *Int. J. Heat Mass Transfer*, **40**, 333 (1997).
111. M. Y. Chang, C. H. Yu, and T. F. Lin, *Int. J. Heat Mass Transfer*, **40**, 347 (1997).
112. M. Y. Chang, C. H. Yu, C. C. Huang, and T. F. Lin, *Int. J. Heat Mass Transfer*, **40**, 505 (1997).
113. T. M. Huang, C. Gau, and W. Aung, *Int. J. Heat Mass Transfer*, **38**, 2445 (1995).
114. W.-S. Fu, C. S. Cheng, and W.-J. Shieh, *Int. J. Heat Mass Transfer*, **37**, 1885 (1994).
115. C. Y. Choi and S. J. Kim, *Int. J. Heat Mass Transfer*, **39**, 1223 (1996).
116. S. M. Tieng and A. C. Yan, *Int. J. Heat Mass Transfer*, **36**, 599 (1993).
117. H. M. Badr, *Int. J. Heat Mass Transfer*, **37**, 2343 (1994).
118. H. D. Nguyen, S. Paik, and R. W. Duglass, *Int. J. Heat Mass Transfer*, **39**, 511 (1996).
119. T. Aicher and H. Martin, *Int. J. Heat Mass Transfer*, **40**, 3617 (1997).
120. K. A. Yih, *Int. J. Heat Mass Transfer*, **42**, 4299 (1999).
121. M. S. El-Genk, S. D. Bedrose, and D. V. Rao, *Int. J. Heat Mass Transfer*, **33**, 1287 (1990).
122. A. Nakayama and I. Pop, *Int. J. Heat Mass Transfer*, **34**, 357 (1991).

8

Heat Transfer in Media with Special Properties

8.1 Water at Extreme Density

In water at the temperature $T_m = 4.029325°\mathrm{C}(p = 0.1\,\mathrm{MPa})$ there is "free convection inversion", consisting in the change of fluid flow direction. The phenomenon of inversion is explained by parabolic change in the density with temperature [8.1]. If the dependence of density on pressure, is not taken into account then the additional parameter of convective heat transfer is the parametric criterion $P_{Ts} = (T_m - T_\infty)/(T_\mathrm{w} - T_\infty)$. When $P_{Ts} \leq 0$ the buoyancy force is positive in the whole region, and the motion is directed upwards. On the contrary, for $P_{Ts} \geq 0.5$ the movement is directed downwards [8.2]. In the range $P_{Ts} = 0{-}0.5$ the sign of the buoyancy force changes its sign in the boundary layer [8.3]. According to experimental and calculated data, the free convection inversion occurs at $T_\infty = 5.5{-}5.6°\mathrm{C}$ [8.2, 8.4]. The theoretical calculations for square-law density dependence on temperature in a pure water give the minimum of heat transfer at $T_\infty = (16 - T_\mathrm{w})/3$ [8.5].

Recently the following state equation is used to calculate heat transfer [8.1]:

$$\rho(T, s, p) = \rho_m(s, p)\left\{1 - \beta_s(s, p)\,[T - T_m(s, p)]^{q(s,p)}\right\} ,$$

according to which corresponding to two limiting cases are: $q = 1.894816$ – pure water at the pressure $p = 0.1\,\mathrm{MPa}$ and $q = 1.58295$ – pure water at $p = 100\,\mathrm{MPa}$ or salt water for $s = 0.4 \cdot 10^{-2}$.

The average and local heat transfer values ***on a vertical flat wall with constant temperature*** [8.1, 8.2]:

$$T_\mathrm{w} = 0°\mathrm{C}, T_\infty = 0 - 20°\mathrm{C}, p = 0.1 - 100\,\mathrm{MPa}, s = 0 - 0.4 \cdot 10^{-2} ,$$

$$\mathrm{Pr} = 7 - 13.6, \mathrm{Gr}_m \ll 10^9, T_\mathrm{av} = (T_\mathrm{w} + T_\infty)/2 ,$$

$$\mathrm{Nu}_x/\mathrm{Gr}_{xm}^{1/4} = -\theta'(0)/\sqrt{2}, \quad \mathrm{Nu} = 1.33\mathrm{Nu}_l . \qquad (8.1.1)$$

The values of $\theta'(0)$ are given in Table 8.1.1.

Table 8.1.1.

P_{Ts}	Pr = 8.6	Pr = 9.6	Pr = 10.6	Pr = 11.6	Pr = 12.6	Pr = 13.6
			$q = 1.894816$			
−16	−2.45117	−2.53026	−2.60322	−2.67105	−2.73451	−2.79421
−8	−2.10655	−2.17452	−2.23721	−2.29548	−2.35000	−2.40129
−4	−1.81651	−1.87508	−1.92911	−1.97935	−2.02634	−2.07052
−3	−1.71096	−1.76688	−1.81698	−1.86427	−1.90583	−1.95014
−2	−1.57626	−1.62704	−1.67387	−1.71742	−1.75815	−1.79647
−1	−1.38283	−1.42730	−1.46833	−1.50647	−1.54215	−1.57571
−0.5	−1.23420	−1.27380	−1.31033	−1.34429	−1.37606	−1.40596
0	−0.96210	−0.99263	−1.02079	−1.04696	−1.07144	−1.09447
0.5	−0.94073	−0.97181	−1.00047	−1.02713	−1.05207	−1.07533
1	−1.22606	−1.26589	−1.30264	−1.33680	−1.36877	−1.39883
2	−1.48633	−1.53443	−1.57881	−1.62006	−1.65867	−1.69497
3	−1.64551	−1.69872	−1.74780	−1.79342	−1.83611	−1.87626
4	−1.76423	−1.82124	−1.87383	−1.92273	−1.96848	−2.01150
8	−2.07606	−2.14310	−2.20496	−2.26246	−2.31626	−2.36685
16	−2.4337	−2.51193	−2.58441	−2.65178	−2.71483	−2.77410

P_{Ts}	$q = 1.859633$	$q = 1.727147$	$q = 1.58295$
		Pr = 11.6	
−16	−2.59421	−2.32197	−2.05462
−8	−2.24276	−2.05293	−1.86146
−4	−1.94515	−1.82000	−1.69011
−3	−1.83638	−1.73350	−1.62538
−2	−1.69715	−1.62159	−1.54065
−1	−1.49629	−1.45739	−1.41402
−0.5	−1.34098	−1.32770	−1.31164
0	−1.05222	−1.07289	−1.09707
0.5	−1.03156	−1.04794	−1.06494
1	−1.33368	−1.32118	−1.30604
2	−1.60458	−1.54633	−1.48209
3	−1.76927	−1.67972	−1.58479
4	−1.89167	−1.77756	−1.65843
8	−2.21174	−2.02891	−1.84389
16	−2.57623	−2.30834	−2.04495

$q(0,1) = 1.894816, q(0,100) = 1.859663, q(0,500) = 1.727147, q(0,1000) = 1.58295$

The Boussinesq approximation near the temperature of maximum density (Pr = 11.6) is applicable if the value $T_{av} = T_w - 0.69(T_w - T_\infty)/2$ is taken as the determining temperature. Experimental investigations of ***ice melting on a vertical surface show*** that the values of the heat transfer coefficient are 7% lower than the calculated curve (8.1.1) [8.6]. ***In the presence of melting*** there appears an additional parameter in the equations – the phase transition number k_φ. For the temperatures $T_\infty = 0\text{–}20°\text{C}$ the range of k_φ makes 0–0.25. In Table 8.1.2 the calculated data on heat transfer ***in the presence and absence of melting on a surface*** are compared ($T_w = 0$) [8.7]. For $T_\infty = 0\text{–}20°\text{C}$ a polynomial of the third order for density dependence on the temperature difference is used. Up to $T_\infty = 4.75°\text{C}$ the decrease in the heat transfer coefficient is insignificant and for $T_\infty = 20°\text{C}$ it makes 12%.

Table 8.1.2.

T_∞, °C	$\theta'(0)$ (without melting)	$\theta'(0)$ (ice melting on a surface)
0.2	−1.2818	−1.2800
0.6	−1.2894	−1.2841
1.6	−1.3207	−1.3066
2.6	−1.3935	−1.3698
3.6	−1.6833	−1.6452
4.0	−1.0676	–
4.4	−1.4594	−1.4215
4.6	−1.2368	−1.2047
6	−0.8190	−0.7772
8	−1.0204	−0.9646
10	−1.0564	−0.9868
12	−1.0673	−0.9849
14	−1.0696	−0.9752
16	−1.0680	−0.9622
18	−1.0644	−0.9478
20	−1.0597	−0.9328

A vertical surface with a constant heat flux on a wall. The local heat transfer value with an error of 1% for laminar free convection and the temperature of maximum density in a surrounding medium ($T_\infty = T_m$) [8.8]:

$$q = 1.5829 - 1.8948, \mathrm{Pr} = 8 - 13, P_{Ts} = 0 \,,$$
$$\mathrm{Nu}_x = 0.577\,\mathrm{Ra}_x^{*1/(4+q)} \,. \tag{8.1.2}$$

Horizontal flat wall with heated surface facing upwards. The local and average heat transfer values are [8.9]

$$T_w = 0°\text{C}, T_\infty = 0 - 20°\text{C}, p = 0.1 - 100\,\text{MPa}, s = 0 - 0.4 \cdot 10^{-2} \,,$$

$$\begin{aligned} \mathrm{Pr} = 10.6 - 12.6, \mathrm{Gr}_m < 10^9, T_{\mathrm{av}} = (T_{\mathrm{w}} + T_\infty)/2 \; , \\ \mathrm{Nu}_x/\mathrm{Gr}_{xm}^{1/5} = -\theta'(0)/5^{1/5} \; , \\ \mathrm{Nu} = 5/3 \; \mathrm{Nu}_l \; . \end{aligned} \tag{8.1.3}$$

For a horizontal flat plate with a constant wall temperature and constant heat flux on the surface the values of $\theta'(0)$ are given in Table 8.1.3, ***for a horizontal disk*** – in Table 8.1.4.

Table 8.1.3.

P_{Ts}	Pr = 10.6	Pr = 11.6	Pr = 12.6	Pr 4= 10.6	Pr = 11.6	Pr = 12.6
	$q = 1.894816$			$q = 1.58295$		
			$T_{\mathrm{w}} = \mathrm{const}$			
−2	−1.24856	−1.27408	−1.29786	−1.14721	−1.17067	−1.19253
−1	−1.11850	−1.14134	−1.16262	−1.06763	−1.08945	−1.10978
−0.5	−1.01298	−1.03364	−1.05282	−1.00026	−1.02069	−1.03972
0	−0.79006	−0.80604	−0.82092	−0.83668	−0.85365	−0.86946
0.5	−0.88131	−0.89952	−0.91647	−0.90388	−0.92253	−0.93990
1	−1.04749	−1.06900	−1.08904	−1.02238	−1.04336	−1.02689
2	−1.20876	−1.23354	−1.25661	−1.12318	−1.14620	−1.16763

q	$P_{Ts} = -16$	$P_{Ts} = -4$	$P_{Ts} = 0.08$	$P_{Ts} = 0.301$	$P_{Ts} = 0.1$
			Pr = 11.6		
1.894816	−1.82398	−1.61438	−1.43158	−0.70082	−0.74506
1.58295	−1.47908	−1.36600	−1.26312	–	−0.79159

q	$P_{Ts} = 0.4$	$P_{Ts} = 4$	$P_{Ts} = 8$	$P_{Ts} = 16$
1.894816	−0.83771	−1.40869	−1.60142	−1.81665
1.58295	−0.87169	−1.24991	−1.35885	−1.47520

Pr	$q = 1.894816$	$q = 1.58295$
	$q_{\mathrm{w}} = \mathrm{const},\ P_{Ts} = 0$	
10.6	−0.98448	−1.04569
11.6	−1.00503	−1.06648
12.6	−1.02325	−1.08584

For constant heat flux on a horizontal surface the wall temperature is determined from the equation

$$T_{\mathrm{w}} - T_\infty = \left[5q_c^2\nu^2x^2/(\lambda^5\left[-\theta'(0)\right]^5 g\beta_s)\right]^{1/(q+5)} .$$

Sphere covered with ice with the surface temperature $T_{\mathrm{w}} = 0°\mathrm{C}$. The local value of heat transfer ***from a sphere surface, on which ice melts***, changing substantially with water temperature (Fig. 8.1.1) [8.10].

The average value of heat transfer ***from the outer surface in the presence of melting*** [8.10, 8.11]:

Table 8.1.4.

P_{Ts}	Pr = 10.6	Pr = 11.6	Pr = 12.6	Pr = 10.6	Pr = 11.6	Pr = 12.6
	$q = 1.894816$			$q = 1.58295$		
	$T_w = \text{const}$					
−2	−1.52797	−1.55885	−1.58762	−1.40399	−1.43237	−1.45881
−1	−1.36871	−1.39635	−1.42209	−1.30653	−1.33293	−1.35753
−0.5	−1.23943	−1.26443	−1.28772	−1.22400	−1.24871	−1.27173
0	−0.96590	−0.98524	−1.00325	−1.02317	−1.04371	−1.06284
0.5	−1.07938	−1.01138	−1.12188	−1.10690	−1.12945	−1.15045
1	−1.28235	−1.30836	−1.33259	−1.25151	−1.27587	−1.30050
2	−1.47957	−1.50953	−1.53745	−1.37477	−1.40260	−1.42853

Pr = 11.6					
q	$P_{Ts} = -16$	$P_{Ts} = -8$	$P_{Ts} = -4$	$P_{Ts} = 0.08$	$P_{Ts} = 0.301$
1.894816	−2.23183	−1.97534	−1.75163	−0.98524	−0.85519
1.58295	−1.80981	−1.67144	−1.54552	–	–

q	$P_{Ts} = 0.4$	$P_{Ts} = 4$	$P_{Ts} = 8$	$P_{Ts} = 16$
1.894816	−0.91342	−1.72378	−1.95958	−2.22292
1.58295	–	−1.52946	−1.66274	−1.80510

Pr	$q = 1.894816$	$q = 1.58295$
	$q_w = \text{const},\ P_{Ts} = 0$	
10.6	−1.09789	−1.16168
11.6	−1.11963	−1.18473
12.6	−1.13988	−1.20620

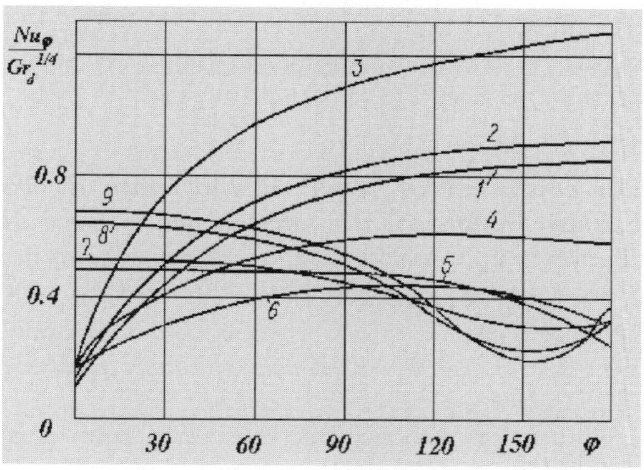

Fig. 8.1.1. Local coefficient of heat transfer from the surface of a sphere of diameter $d = 3.5\,\text{cm}$ covered with a layer of ice for $T_w = 0$: (1) $T_\infty = 2.69°\text{C}$; (2) 3.37; (3) 3.38; (4) 4.65; (5) 5.26; (6) 5.92; (7) 6.78; (8) 7.76; (9) 8.69

$$d = 3.5\ \text{cm}, T_\infty = 0 - 10°\text{C}, \text{Gr}_d = 0 - 10^5\ ,$$

$$\text{Nu} = 2 + 0.56\ \text{Ra}_d^{1/4}\ , \tag{8.1.4}$$

$$d = 5.08 - 10.16\ \text{cm}, T_\infty = 0 - 20°\text{C}, \text{Ra}_d = 1.7 \cdot 10^6 - 2.4 \cdot 10^8\ ,$$

$$\text{Nu} = 2 + 0.437\,\text{Ra}_d^{0.248}\ . \tag{8.1.5}$$

Near the density maximum $\text{Nu}_{lden}/\text{Nu}_{dsfer} = 1.14(l/d)^{3/4}$ and at the minimum value of $\text{Nu}_d T_\infty = 5.35 \pm 0.2°\text{C}$ [8.11].

The melting of a horizontal cylinder in water with an initial diameter d_0 ***and instantaneous radius d***. Figure 8.1.2 presents a change in the local heat transfer value ***along the cylinder perimeter***. The flow instability and the minimal rate of heat transfer is at $T_\infty = 6°\text{C}$ [8.12].

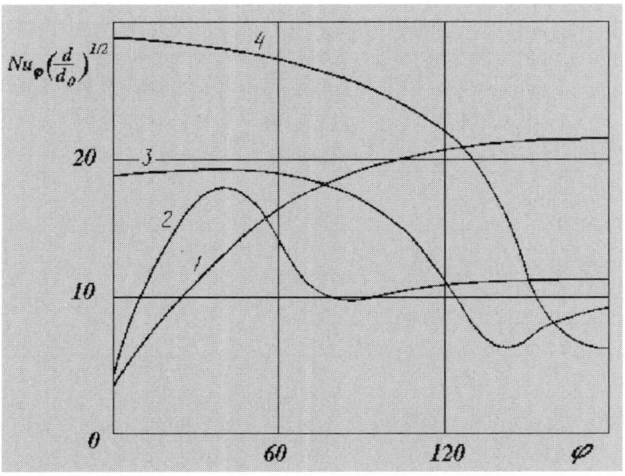

Fig. 8.1.2. Local heat transfer for melting of a horizontal cylinder: (1) $T_\infty = 4.6°\text{C}$, $\text{Ra}_d = 2 \cdot 10^4$, (2) $5.6°\text{C}$, $6.2 \cdot 10^4$; (3) $6°\text{C}$, $6.2 \cdot 10^4$; (4) $7°\text{C}$, $5 \cdot 10^5$

Rectangular container of height H and width B. The upper and lower surfaces are insulated, the vertical walls have different temperatures: $T_0 = 0°\text{C}$, $T_B = 0–12°\text{C}$. The average rate of heat transfer ***on a horizontal surface*** increases for $T_B < 4°\text{C}$ and $T_B > 8°\text{C}$. The minimal value of $\text{Nu}_{\text{av}B}$ is at $T_B = 8°\text{C}$ (Fig. 8.1.3). The value of $\text{Nu}_{\text{av}B}$ decreases with decrease of the width of the container B and is maximal at $B/H = 0.67–1.25$ [8.13].

The visual pictures of unsteady free-convective flow ***in a rectangular container with lateral cooling*** near the water inversion point during its supercooling and freezing are obtained experimentally [8.14]. Using the liquid crystals for temperature field visualization, the experimental data specifying the important role of free convection during water overcooling in the container are obtained. At the initial stage, alongside with the circulating flow with

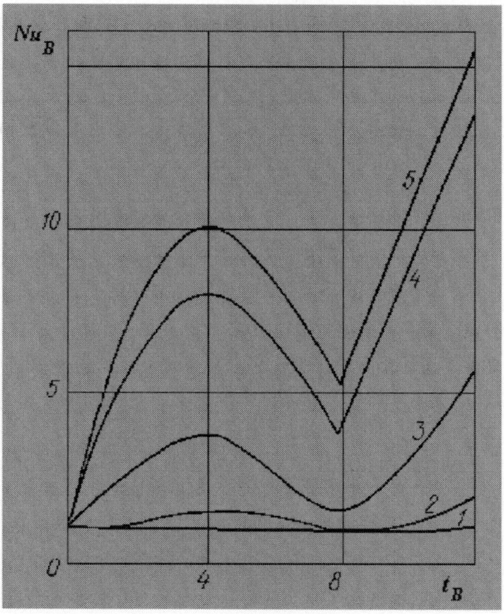

Fig. 8.1.3. Average value of heat transfer on a hot surface for $T_0 = 0°C$; $H = 100$ mm: (1) $B = 5$ mm; (2) 10; (3) 20; (4) 50; (5) 100

rotation clockwise, in the bottom corner near the cold wall a cell with opposite rotation is formed which moves upwards during the cooling of water. The interface between them corresponds approximately to the isotherm near 4°C. Further, ice formation in this area begins.

Melting of ***ice-porous media (glass beads) in a rectangular cell*** is investigated experimentally and numerically in [8.15] to determine an effect of natural convection and water density inversion in the melting region. It is shown that for small superheat across the liquid area the flow in a porous medium is weak, the interface is almost planar. For larger superheat the strength of natural-convective flow, the interface velocity depend on the supplied temperature difference and on the permeability of the porous medium. The presence of free convection leads to non-planar melting front form and to the increase of the melting rate.

Nonstationary natural-convective heat transfer for water near its density maximum ***in a rectangular cavity with opposite vertical walls maintained at different temperatures*** is investigated experimentally and theoretically in [8.16]. On the basis of the data obtained it is shown that the water density inversion has an essential effect on natural convection in the cavity, and the flow structure cannot be determined by classical models of natural convection with the use of the Boussinesq approximation.

An experimental investigation has been conducted in [8.17] with regard to the natural convection of water near the density extremum ***in rectangular enclosures with the vertical walls held at different temperatures.***

The water in the enclosure is initially set at a uniform temperature and then the temperatures of the cold and hot walls are suddenly changed to 0 and 8°C, respectively. At an initial temperature of 4°C, the convective flow field consists of a symmetric double-cell circulation forming a downward flow, called a sinking jet in the interior of the enclosure. For $\mathrm{Ra} > 9 \cdot 10^6$, the interior sinking jet shows an oscillatory behavior due to free-shear instability, which enhances the heat transfer rate between the vertical walls.

The beginning of free convection ***in a horizontal porous container*** is investigated by the method of linear stability analysis for $t = 3.98$°C appropriate to the maximal water density. The beginning of convective motion depends on the container aspect ratio (A), the Darcy number (Da), the inversion parameter (γ), and the hydrodynamic boundary conditions on horizontal walls of a porous layer. The results of the numerical solution show that free convection in such a system is possible provided that the upper steady layer occupies more than half the container depth [8.18].

Rectangular container with B/H = 2. The temperature of the upper wall is $T_H = 0$°C, ***of the lower one*** $-T_0 = 8$°C, ***the temperature on lateral surfaces changes according to the linear law.*** The isotherm $T_m = 4$°C passes through the container middle portion. Numerical calculations for a change in the square-law density are performed for $\mathrm{Pr} = 11.59$, $\mathrm{Gr}_H = 3.25 \cdot 10^3$–$1.6 \cdot 10^4$ [8.19]. The flow reached a steady regime when $\mathrm{Gr}_H < 4 \cdot 10^3$; when $\mathrm{Gr}_H > 5 \cdot 10^3$, it had self-generating oscillations. The critical value of the Gr_{H_K} number is 4400. The oscillations of the $\mathrm{Nu}_{\mathrm{av}}$ numbers corresponding to the lateral cavity boundaries are shifted in phase. The maximum or minimum of the heat flux through the right boundary corresponds to the constant flux through the left boundary vice versa. The heat fluxes ***through the upper and lower boundaries*** are in opposite phases. ***For a square cavity*** the convective motion begins at $\mathrm{Gr}_{H_K} = 3050$, for an elongated one – at $\mathrm{Gr}_{H_K} = 1800$. The rate of ice melting from below in spring in the case where under the ice there are conditions for the incipience convective motion for $\mathrm{Gr}_H = 1.6 \cdot 10^4$ is equal to 4.64 mm/h [8.20].

Horizontal sublayer of height H. The temperature of the base is T_0, ***of the upper horizontal wall is*** T_H. The average heat transfer value for ice melting from above and heating from below is [8.21]

$$T_H = 0,\ T_0 = 1\text{–}8^\circ\mathrm{C}, H = 120\,\mathrm{mm}, B \times L = 250 \times 250\,\mathrm{mm}, T_{\mathrm{av}} = (T_0 + T_H)/2\ ,$$

$$\text{for}\quad \mathrm{Ra}_H = 500\text{–}950 \qquad \mathrm{Nu} = 0.00369\,\mathrm{Ra}_H^{0.9}\ , \tag{8.1.6}$$

$$\text{for}\quad \mathrm{Ra}_H = 950\text{–}10^4 \qquad \mathrm{Nu} = 0.356\,\mathrm{Ra}_H^{0.24}\ , \tag{8.1.7}$$

$$\text{for}\quad \mathrm{Ra}_H = 10^4\text{–}1.5 \cdot 10^6 \qquad \mathrm{Nu} = 0.183\,\mathrm{Ra}_H^{0.31}\ . \tag{8.1.8}$$

Ice melting from below, heating from above. It is recommended to select the value $T_m - T_0$ for $T_m > 4$°C ($i = m$) and $T_H - T_0$ for $T_H \le 4$°C ($i = H$) as the characteristic temperature in terms of the Nu_i and Ra_{Hi} numbers.

For $T_0 = 0°C$, $T_H = 1 - 12°C$, $T_{av} = (T_m + T_0)/2, H < 25\,mm, B \times L =$ $250 \times 250\,mm$, and $Ra_{Hi} = 2 \cdot 10^2 - 10^5$ the data on the calculation of the average heat transfer coefficient are given in Fig. 8.1.4 [8.22]. At a temperature of the heated surface less than 4°C and higher than 8°C the earlier given dependences in horizontal sublayers can be used without taking into account the water density inversion.

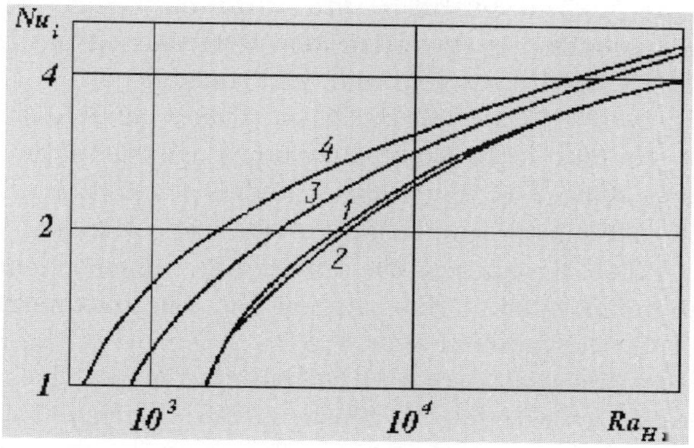

Fig. 8.1.4. Average heat transfer value in a horizontal water sublayer for heating from below $T_0 = 0°C$: (1) $T_H = 2°C$; (2) 4; (3) 5; (4) 8

Table 8.1.5 lists the values of Ra_{Hi_K} of transition from the thermal conductivity regime to free convection ***for a liquid limited by two rigid surfaces (A) and having a free upper surface (B)***. When ***the wall thermal conductivity*** is taken into account, the transition from thermal conductivity to convection ($T_H < 7°C$) is observed at $Ra_{H_K} = 1100$ [8.23].

Table 8.1.5.

T_H,°C	A	B
1	1636	1068
2	1644	1091
3	1649	1120
4	1649	1160
4.5	1173	–
5	892	665
6	619	513
7	545	532
8	565	548
9	576	575
10	573	572
12	572	–

A horizontal cylinder filled with water and cooled through lateral walls at a constant rate. For $Ra_d > 10^6$ the boundary layer and flow core are observed visually. Near the temperature of water maximum density there are three quasi-steady free convection regimes. During heating there occurs a change of the quasi-steady regimes [8.24]. In [8.25] a model of ***water cooling in a horizontal cylinder till ice formation inside a cavity*** is developed.

Horizontally located coaxial cylinders. The temperature of the inner cylinder is $T_{w1} = 0°C$ ***and of the outer cylinder*** – $T_{w2} = 1–15°C$. Numerical and experimental investigations were carried out in the following ranges [8.26, 8.27]: $d_2 = 65.5$, and 121.5 mm, $d_2/d_1 = 1.18–6.39, \delta = 5–51.2$ mm, $T_{w1} = 0°C$, $T_{w2} = 1–15°C$, and $Gr_\delta = 32–1.7{\cdot}10^5$. Figure 8.1.5 presents the dependence of the dimensionless heat transfer of the ***inner cylinder*** on the temperature of the outer one $[Nu = q_1\delta/(\pi d_1(T_{w2} - T_{w1})\lambda)]$. When $T_{w2} = 4–9°C$, in the gap there are two vortices caused by water density inversion. When the sizes of the vorticex are almost identical, one can expect the minimum value of the criterion Nu. The maximum value of Nu is at $T_{w2} = 4°C$, the minimal – at $T_{w2} = 6–7°C$ (Fig. 8.1.5). The average heat transfer value increases with increase of the gap thickness δ. In the range of temperatures without the density inversion effects ($T_{w2} < 4°C$ and $T_{w2} > 9°C$) the dependences obtained earlier for heat transfer ***in a cylindrical gap*** are applicable.

Using the approximate mathematical model of free convection ***on a vertical porous plate in the presence of injection and an external oscillating flow*** it is shown that the heat flux amplitude increases as the plate is cooled and decreases as the surface is heated.

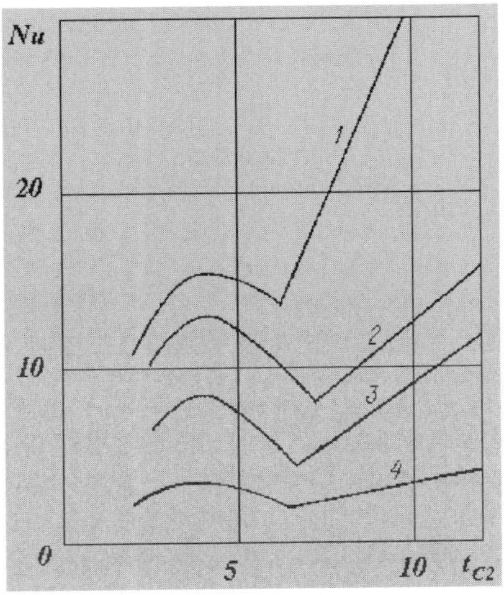

Fig. 8.1.5. Average heat transfer value in a gap between the cylinders ($T_{w1} = 0°C$, $T_{w2} = 12°C$): (1) $d_1/d_2 = 6.39, \delta = 51.2$; (2) 3.2, 41.7; (3) 3.44, 23.2; (4) 1.72, 17.7

The water thermostated at the temperature of maximum density is used ***in highly-sensitive measuring devices*** to suppress gravitational convection. To decrease temperature gradients in the investigated cavity, passive thermostabilization is used with the aid of multilayer shells [8.28].

The unsteady temperature distributions and heat transfer characteristics for free-convective flow in ***air-water layers inside a horizontal tube with a regularly decreasing wall temperature*** in the presence of density inversion are determined experimentally using a holographic interferometer [8.29]. The temperature field structure and flow regime depend substantially on the tube cooling rate. Using the track visualization method with aluminum powder three characteristic patterns of the dendritic ice growth and the growth of pure ice along the tube wall were revealed.

An apparatus for carrying out experiments on free convection in a cylindrical volume under strictly controlled boundary conditions is described in [8.30]. The ***water-deuterated water*** system is investigated in temperature ranges that include the investigated mixture density maximum and is compared with the data for pure water. The existence of the earlier observed anomalies of convective behavior in the given range shown as a typical plateau is confirmed, and their careful investigation is carried out to obtain new information about the maximum density of isotopic mixtures. For a sudden increase in the temperature at the boundary there appears a new effect of faster cooling near the maximum density.

The results of a theoretical and experimental study of the steady-state characteristics of a rectangular natural circulation loop with vertical heat transfer sections containing water near its density extreme are presented in [8.31]. Correlations for the Nu number on the basis of the predicted values have the following forms:

with density-inversion effect

$$\mathrm{Nu} = 0.0667\ \mathrm{Ra}^{0.279}\ , \tag{8.1.9}$$

without density-inversion effect

$$\mathrm{Nu} = 0.119\ \mathrm{Ra}^{0.251}\ . \tag{8.1.10}$$

Natural convection ***of a layered fluid system composed of two immiscible liquids***, ***silicon oil on top of water***, was studied numerically in [8.32]. Both flows are coupled hydrodynamically and thermally. Two counter-rotating natural convection roll cells of opposite vorticity develop when one side wall temperature in the density inversion fluid is above and the other below the density inversion temperature. At large Ra numbers the flow pattern in the density inversion layer becomes very complex. The two-roll cell pattern in the density inversion layer is impeding the total horizontal heat transfer. A vertical heat transport exists across the interface from the density inversion layer into the encapsulating upper layer. The moving interface between the immiscible liquids improves the heat transfer in each layer when compared to the cavity cases.

Transient natural convection ***in a water layer*** subjected to density inversion was studied numerically in [8.33]. The results illustrate that the temperature difference determining the position of the maximum density plane in the water layer can alter the flow field and heat transfer substantially. The significant effect of the aspect ratio on the transient natural convection was also investigated. The heat transfer is maximized in a square enclosure and is less intense at other aspect ratios.

The experimental data for the local and average Nusselt number for the value of parameter $A = 6.72 \cdot 10^{10}$ (where $A = gD^3/\lambda^2$) of a horizontal cylinder ***in water near the point of the density inversion*** are presented in Table 8.1.6 [8.34].

Table 8.1.6.

T_∞,°C	$\vartheta =$ 0°	30	60	90	120	150	180	Nu_D
2.00	16.61	16.40	15.75	14.37	13.04	10.77	3.03	13.66
2.75	16.86	16.65	15.98	14.91	13.20	10.96	3.05	13.86
3.50	16.44	16.23	15.57	14.51	12.85	10.71	3.03	13.53
4.60	14.09	13.89	13.29	12.28	10.84	8.89	2.90	11.54
5.70	5.67	5.20	8.85	10.09	10.32	10.30	10.25	8.80
5.80	5.61	5.07	10.15	11.41	11.94	12.22	12.31	9.98
6.00	4.88	7.20	11.38	12.55	13.25	13.65	13.78	11.14
7.00	3.54	11.96	15.14	16.40	17.50	18.13	18.36	15.01
8.00	3.44	14.27	17.63	19.04	20.29	21.18	21.18	17.38
10.00	3.76	16.45	20.12	22.62	24.12	25.03	25.13	20.83

8.2 Critical and Supercritical State of a Substance

Critical and supercritical parameters are characterized by a significant change in the thermodynamic and transport properties in a very narrow temperature range. As a result of the change of these properties the heat transfer coefficient near a critical point becomes very large, which allows achieving high heat fluxes for a small temperature difference. For a supercritical pressure the most significant change in the physical properties is observed near a pseudocritical point T_{pc} (the temperature corresponding to the maximum of isobaric heat capacity. Because of the high substance compressibility near the critical point, one can observe pulsational and turbulent regimes with substantially different mechanisms of heat transfer.

The experimental dependence with an error of ±7% for heat transfer of a ***platinum plate of constant temperature in carbonic acid near the critical and pseudo-critical points*** is [8.35]

$$l \times b \times \delta = 41 \times 4.6 \times 0.05\,\text{mm}, p/p_{cr} = 1 - 1.15, T/T_{cr} = 0.997 - 1.03\,,$$

$$T/T_{cr} = 0.98 - 1.3,\ \beta_p = 0.04 - 10\,\text{MPA}^{-1}, \text{Ar}_x = 10^6 - 2 \cdot 10^8\,,$$

$$\text{Nu}_x = 0.508\left[\text{Pr}^2/\left(0.952 + \text{Pr}\right)\right]^{1/4}\left[\text{Ar}_x e^{0.037(\text{Pr}-1)+0.027\beta_p \mathfrak{p}/p}\right]^{1/4}. \quad (8.2.1)$$

The heat flux ***from a vertical Ni-Cr plate of constant temperature near the critical point of carbonic acid*** is [8.36]

$$l \times b \times \delta = 63.5 \times 38.1 \times 0.05\,\text{mm}, \quad p = 7.59 - 8.97\,\text{MPa}\,,$$
$$T_\infty = 30.47 - 32.12°\text{C}, \quad q_\text{w} = (0.212 - 3.8)\,\text{kWt/m}^2\,,$$
$$\text{Pr}_0 = \mu c_p/\lambda_0, \text{Ra}_0 = \text{GrPr}_0, \quad \lambda = 5.31 \cdot 10^3\,\text{Wt/m} \cdot \text{grad}\,,$$
$$\text{Ra}_0\text{Pr}_0^{-1/2} = (3 - 200) \cdot 10^6, \text{Gr}_{l\text{w}} = 2 \cdot 10^7 - 10^8, \theta_\text{w} = 0.01 - 0.001°\text{C}\,,$$
$$\frac{ql}{\rho_{cr}\lambda_0\left(\partial T/\partial \rho\right)_p}\left[\left(T - T_{cr}\right)/T_{cr}\right]^{1/2} = 3.25 \cdot 10^{-9}\frac{\text{Ra}_{l0}}{\text{Pr}_0^{1/2}}\,. \quad (8.2.2)$$

The critical value of the number $\text{Gr}_{l,k} = 2 \cdot 10^7 - 10^8$ increases with increase of Pr_0.

Heat transfer of a vertical platinum sheet in water is given by [8.37]

$$l \times b = 152 \times 12.7\,\text{mm}, p = 23.25, 23.65\,\text{MPA}\,,$$
$$T_\text{w} = 375 - 376°\text{C}, \text{T}_\text{av} = 376 - 380°\text{C}\,,$$
$$\text{Nu} = 0.0872 Ar_l^{1/3}\,\text{Pr}^{0.247}\left[T_\infty/\left(T_\text{w} - T_\infty\right)\right]^{0.137}\,. \quad (8.2.3)$$

The average deviation of the experimental data from formula (8.2.3) is 15%.

Theoretical calculations of heat transfer ***on a plate of constant temperature for variable thermophysical characteristics in the region of supercritical parameters for water*** $p = 23.45$, 24.5, 25.5 MPa; ***for carbonic acid*** $p = 8.16$, 9.18 MPa yields [8.38]

$$T_\infty - T_{pc} = 0 - (\pm 5)°\text{C}, T_\text{w} - T_\infty = 0.25 - 16°\text{C}, T_\text{av} = (T_\text{w} + T_\infty)/2\,,$$

$$l^{-3/4}\text{Nu} = 0.56\left(l^{-3}\text{Ar}_l\,\text{Pr}\right)^{1/4}\,. \quad (8.2.4)$$

Comparison with experiments ***on water*** ($\text{Ar}_l\text{Pr} = 1.7 \cdot 10^7 - 1.3 \cdot 10^8$) and ***carbonic acid*** ($\text{Ar}_l\text{Pr} = 10^9 - 10^{12}$) testifies that the value of dispersion of experimental points is ±20%.

Numerical calculations of heat transfer ***of a vertical surface in helium for supercritical parameters*** give [8.39]

$$p = 0.608 - 2.53\,\text{MPa},\ T_\text{w} = 7 - 14\text{K},\ T_\infty = 6 - 11\text{K}\,,$$
$$T_\text{av} = (T_\text{w} + T_\infty)/2,\ \text{Ra}_l = 10^6 - 10^{11}\,,$$
$$\text{Nu} = 0.616\text{Ra}_l^{0.248}\,. \quad (8.2.5)$$

The generalization of experimental data on the heat transfer ***of a vertical flat plate in* He *under supercritical parameters*** with an error of ±20% is given by [8.40]

$$l \times b \times \delta = 30 \times 30 \times 3\,\text{mm}, p = (0.233 - 0.456)\text{MPa}, T_\infty = 4.5, 6 - 10\,\text{K}\,,$$
$$T_\text{w} - T_\infty = 0.07 - 23\,\text{K}, c_{p,\text{av}} = (i_\text{w} - i_\infty)/(T_\text{w} - T_\infty), \text{Ra}_{l\infty} = 6 \cdot 10^9 - 10^{14}\,,$$
$$\text{Nu} = 0.135\,\text{Ra}_{l\infty}^{1/3}(c_{p,av}/c_{p\infty})^n(\rho_w/\rho_\infty)^{0.15}\,, \qquad (8.2.6)$$

where $n = 0.5$ for $T_\infty \leq T_{pc}$, $n = 1$ for $T_\infty > T_{pc}$.

When $T_\infty = T_{pc}$, the divergence from experiments attains 30–35%, because of inexact determining of thermo physical characteristics.

Cooled vertical surface under supercritical pressure of carbonic acid. The calculations show that for $T_\infty = 40–70°\text{C}$, $T_\text{w} = 17°\text{C}–T_\infty$, $p = 8–10\,\text{MPa}$ the average heat transfer values increase with increase of temperature difference and are maximal at $T_\text{w} \cong T_{pc}$. The curves of the average heat transfer value depending on the temperature difference have no maximum, in contrast to a heated surface. It is explained by the dependence of the thermophysical characteristics of $\mathbf{CO_2}$ on temperature [8.41].

The average value of heat transfer of a ***vertical surface with a constant heat flux near the critical condition for carbonic acid*** $p/p_{pc} = 1.083–1.354$, ***for water*** $p/p_{pc} = 1.018–1.108$:

$$q_\text{w} = 1 - 50\,\text{kWt/m}^2, \text{Ra}_{l\infty} = 5 \cdot 10^6 - 5 \cdot 10^{10}, c_{p,\text{av}} = (i_\text{w} - i_\infty)/(T_\text{w} - T_\infty)\,,$$

carbonic acid, error ±10%

$$\text{Nu}_\infty = 0.48\,\text{Ra}_{l\infty}^{0.251}\left(c_{p,\text{av}}/c_{p\infty}\right)^{0.491}\left(\rho_\infty/\rho_\text{w}\right)^{0.573} \times \left(\mu_\infty/\mu_\text{w}\right)^{-0.378}\left(\lambda_\infty/\lambda_\text{w}\right)^{-0.584}\,, \qquad (8.2.7)$$

carbonic acid and water, error ±20%

$$\text{Nu}_\infty = 0.475\,\text{Ra}_{l\infty}^{0.25}\left(c_{p,\text{av}}/c_{p\infty}\right)^{0.431}\left(\rho_\infty/\rho_\text{w}\right)^{0.089} \times \left(\mu_\infty/\mu_\text{w}\right)^{0.201}\left(\lambda_\infty/\lambda_\text{w}\right)^{-0.675}\,. \qquad (8.2.8)$$

The numerical calculations of the local heat transfer value ***for laminar free convection in* He *and carbonic acid for supercritical parameters*** with dispersion of ±20% are approximated by the equation [8.42]

for T_w= **const**

for $\mathbf{CO_2}$ $T_\text{w} = 306\,W–316\,K$, $T_\infty = 303 - 310\,K$,
for **He** $T_\text{w} = 5–9\,K$, $T_\infty = 4 - 6\,K$,

for T_w= **const**

for $\mathbf{CO_2}$ $q_\text{w} = 5–10\,\text{kWt/m}^2$, $T_\infty = 293–310\,K$,
for **He** $q_\text{w} = 1–5\,\text{kWt/m}^2$, $T_\infty = 4–6\,K$, $\text{Ra}_{x,\infty} = 10^8–4 \cdot 10^9$,

$$\mathrm{Nu}_{x\infty} = 0.53\,(c_{p,\mathrm{av}}/c_{p\infty})^{0.65}\,(\rho_\infty/\rho_\mathrm{w})^{-0.4}\,\mathrm{Ra}_{x\infty}^{0.25}\,. \tag{8.2.9}$$

The calculations of heat transfer of a ***vertical flat plate for a constant wall temperature and constant wall heat flux for carbonic acid and*** **He** ***in a supercritical region*** show that the values of Nu_x for $T_\mathrm{w} = \mathrm{const}$ are by 10–20% smaller than the corresponding value of Nu_x for $q_\mathrm{w} = \mathrm{const}$.

Heat transfer of an ***inclined plate for supercritical parameters of carbonic acid*** [8.43]:

$$l = 100, 200\,\mathrm{mm},\ \varphi = (-90) - 60^\circ,\ \theta_\mathrm{w} = 0 - 20^\circ\mathrm{C}\,,$$

$$T/T_{cr}, = 0.95 - 1.11,\ p/p_{cr} = 0.86 - 1.19,\ \mathrm{Ra}_{l,cr} = \mathrm{Gr}_l\mathrm{Pr}_\mathrm{av}\,,$$

$$\mathrm{Pr}_\mathrm{av} = \mu c_{p,\mathrm{av}}/\lambda,\ \mathrm{Ra}_{l,\mathrm{av}} = 10^7 - 10^{14}\,,$$

$$\begin{aligned}\mathrm{Nu} &= 0.135\,\mathrm{Ra}_l^{1/3}\,(\mu_\mathrm{w}/\mu_\infty)^{1/4}\,(c_{p,\mathrm{av}}/c_{p\infty})^n \\ &\times (1 - 0.3\sin\varphi|\sin\varphi|)\,,\end{aligned} \tag{8.2.10}$$

$$n = 0.8 \text{ for } T_\mathrm{w} < T_{pc}, n = 0 \text{ for } T_\mathrm{w} > T_{pc}\,.$$

The root-mean-square deviation of experimental data from formula (8.2.10) is 9.5%. The initial unheated part has no significant effect on heat transfer of the plate and on the temperature distribution in a free convection turbulent regime.

Horizontal disk with the heated surface facing upwards. Free convection of **He** ***for near-critical and supercritical parameters*** in the fields of acceleration $g_t/g = 1{-}100$ and pressure $p = 0.15{-}0.6\,\mathrm{MPa}$. For heat transfer calculation formulas (8.1.10) with characteristic temperature $T_\mathrm{av} = T_\infty$ are applicable [8.42].

Numerical simulations performed in [8.44] to study turbulent convective heat transfer to fluids near the critical point ***in the entrance region of a vertical tube*** show a little higher value of turbulent diffusivity than the standard mixing length model. The buoyancy force parameters increase as the pressure approaches the critical pressure due to the steep density gradient with temperature that affects the characteristics of fluid flow and heat transfer in the developing region.

The effect of maximum density on laminar water ***pipe flow solidification*** was investigated experimentally and numerically in [8.45]. The experimental parameters cover the ranges of the dimensionless axial length $z/(r\mathrm{PrRe})$ from 0.008 to 0.32, the inlet water temperature – from 5 to 20°C, the wall temperature – from 12 to 0°C, and the Ra number – from 10^4 to $3 \cdot 10^6$. The numerical analysis employs the large Pr number assumption and considers a gradual variation of solid shell in the streamwise direction.

Deterioration in heat transfer at supercritical water-cooling ***in a vertical pipe*** was numerically analyzed in [8.46]. Heat transfer deterioration is attributable to two mechanisms depending on the flow rate. When the flow rate is large, viscosity increases locally near the wall by heating. This makes the viscous sublayer thicker and the Prandtl number smaller. Both effects reduce the heat transfer. When the flow rate is small, buoyancy force accelerates the flow velocity near the wall. This makes the flow velocity distribution flat, and generation of turbulence energy is reduced. This type of heat transfer deterioration appears at the boundary between forced and natural convection. As the heat flux increases above the deterioration heat flux, a violent oscillation of the wall temperature is observed.

The generalizing formula for free-convective heat transfer ***on a vertical cylinder***

$$\mathrm{Nu}_{x\infty} = C\,\mathrm{Ra}^n_{x\infty}\left(c_{p,\mathrm{av}}/c_{p\infty}\right)^n\left(\rho_p/\rho_\infty\right)^k \tag{8.2.11}$$

with a dispersion of ±20% describes the main experimental data [8.47]. In some cases the deviation of experimental points from the averaging curve attains 35%. The fluid pressure and heat flux on the wall changed so that the wall and liquid temperature covered a range of values from T_∞, $T_\mathrm{w} < T_{pc}$ to T_∞, $T_\mathrm{w} > T_{pc}$. For the fluid temperature $T_\infty < T_{pc}$ the local maximum in the wall temperature distribution is formed over the initial part, and for $T_\infty > T_{pc}$ the clearly expressed maximum is not observed on the initial tube part.

Carbonic acid* $\mathbf{CO_2}$**: $d = 8.18, 19.6$ mm, $l = 160$ mm, $p = 7.5$–10 MPa, $p/p_k = 1.02$–1.36, $T_\infty = 14.3$–$54°$C, $T_\infty/T_{cr} = 0.94$–1.07, $\theta_\mathrm{w} = 1.7$–$259°$, $q_\mathrm{w} = 3.5$–110 kBт/м^2, $\mathrm{Ra}_{x\infty} = 10^7$–$1.5 \cdot 10^{15}$ [8.47]. ***Laminar regime: $C = 0.53$, $n = 1/4$, $m = 0.65$, $k = 0.4$; ***turbulent regime***: $C = 0.135$, $n = 1/3$, $m = 0.75$, $k = 0.4$.

For small temperature differences, transition from laminar to turbulent flow occurs at $\mathrm{Ra}_{x\infty} = 2\cdot10^{10}$–$7\cdot10^{10}$. The transition origin for $T_\infty < T_{pc}$ with an error of ±50% can be estimated from the equation $\mathrm{Ra}_{x\infty}(c_{p,\mathrm{av}}/c_{p\infty})^{1.66} = 5\cdot10^{10}$.

Toluene, benzol, and ethylbenzol. A vertical stainless steel cylinder of diameter $d = 4$ mm and length $l = 110$ mm, $p = 4.3 - 9.5$ MPa, $T_\infty = 18 - 100°$C, $T_\mathrm{w} < 650°$C, $q_\mathrm{w} < 800$ kWt/m^2, $\theta_\mathrm{w} < 650°$C, $q_\mathrm{w} = 3.5$–110 kWt/m^2, $\mathrm{Ra}_{x\infty} = 2\cdot10^{10}$–$10^{12}$ [8.48]. ***Turbulent regime***: $C = 0.15$, $n = 1/3$, $m = 0.4$, $k = 0.2$.

The average value of heat transfer ***by turbulent convection of a vertical tube in the supercritical region*** of ethylbenzol and benzol with an error of ±20% is [8.49]

$$d = 4.05\,\mathrm{mm}, l = 100\,\mathrm{mm}, \mathrm{Gr}_l = 2\cdot10^9 - 10^{12}, T_\infty < T_{pc} \le T_\mathrm{w}, T_\mathrm{av} = T_\infty\,,$$
$$\mathrm{Nu} = 2.26\mathrm{Gr}_l^{1/3}\,\mathrm{Pr}\left(\mu_\mathrm{w}/\mu_\infty\right)^{1/3}\left(c_\mathrm{av}/c_\infty\right)^{1/3}\left(\rho_\infty/\rho_\mathrm{w}\right)^{1/3}$$
$$\times q_\mathrm{w}/\left[\left(T_\mathrm{w} - T_\infty\right)\rho_\infty c_{p\infty} v^*\right]\,, \tag{8.2.12}$$

where $\nu^* = \sqrt{\tau_\mathrm{w}/\rho_\mathrm{w}}$ is the dynamic velocity.

To determine $T_\mathrm{w}(x)$ for the given T_∞ and $q_\mathrm{w}(x)$, empirical formulas are suggested [8.50]. ***On a vertical cylinder*** with $d = 19.6$ mm and $l = 160$ mm ***for constant transport properties in carbonic acid***

$$\mathrm{Nu}_{x0} = 0.665\ \mathrm{Ra}_{x\infty}^{0.2}\ . \qquad (8.2.13)$$

The temperature distribution ***over the cylinder height for laminar free convection*** with an error of ±7% is determined from the equation

$$T_\infty < T_{pc} < T_\mathrm{w},\ \mathrm{Ra}_{x\infty} < 10^9,\ x = 5 - 20\,\mathrm{mm}\ ,$$

$$(i_\mathrm{w} - i_\infty)/(i_{0\mathrm{w}} - i_\infty) = (1.62 - 0.62x/l)(\beta\theta_{0\mathrm{w}})^{0.3}\ .$$

The average value of heat transfer ***from a surface of a vertical cylinder*** of $d = 4\,\mathrm{mm}$ ***in*** **He** can be determined from the formula (8.2.5) for $\mathrm{Ra}_{d\mathrm{av}} = 2 \cdot 10^7 - 10^9$, $p = 0.6 - 1.6\,\mathrm{MPa}$, $T_\infty = 4.2 - 7.2\,\mathrm{K}$ [8.39].

Heat transfer of a ***sphere in a cavity filled with*** **He** ***at supercritical parameters*** is described by [8.51]

$$T_\infty = 4.2 - 25\mathrm{K}, p = 0.3 - 3.5\,\mathrm{M\Pi a}, \theta_\mathrm{w} = 0.1 - 7\,\mathrm{K},\ d = 3.185\,\mathrm{cm}\ ,$$

$$\mathrm{Gr}_d = 7 \cdot 10^8 - 5 \cdot 10^{11}, T_\mathrm{av} = T_\infty\ ,$$

$$\mathrm{Nu} = C\,[T_\infty/\,(T_\mathrm{w} - T_\infty)]^a\,[(\rho_\mathrm{w} - \rho_\infty)\,/\rho_\infty]^b\,\mathrm{Pr}^d\ \mathrm{Ra}_d^e \qquad (8.2.14)$$

the values of the coefficients C, a, b, d, e are given in Table 8.2.1.

Table 8.2.1.

Gr_d	C	a	b	d	e
$7{\cdot}10^8 - 2.5 \cdot 10^{10}$	2.57	−0.55	−0.34	1.07	0.2
$2.5{\cdot}10^{10} - 1.25{\cdot}10^{11}$	0.154	0.14	0.25	0.28	0.3
$>1.25{\cdot}10^{11}$	0.072	0.32	0.37	0.16	0.13

Heat transfer of a ***horizontal cylinder at near-critical and supercritical parameters***.

Carbonic acid $\mathbf{CO_2}$, root-mean-square deviation 9.5% [8.43]:

$$d = 3 - 9.14\ \mathrm{mm}, p/p_{cr} = 0.88 - 1.19, T_\infty/T_{cr} = 0.95 - 1.11\ ,$$

$$\theta_\mathrm{w} = 0 - 20°, \mathrm{Ra}_{\mathrm{av},d} = 10^7 - 10^{14}, \mathrm{Pr}_\mathrm{av} = 0.87 - 20\ ,$$

$$\mathrm{Nu} = 0.135\ \mathrm{Ra}_d^{1/3}\,(\mu_\mathrm{w}/\mu_\infty)^{1/4}\,(c_{p,\mathrm{av}}/c_{p\infty})^n\ , \qquad (8.2.15)$$

where $n = 0.8$ for $T_\mathrm{w} < T_{pc}$ and $n = 0$ for $T_\mathrm{w} \geq T_{pc}$.

With an error of ±20% formula (8.2.15) is applicable for calculating heat transfer of ***cryogenic liquids*** of critical parameters, $\mathrm{Ra}_{d,\mathrm{av}} = 0.3 \cdot 10^6 - 10^{10}$ [8.52]:

nitrogen:

$$d = 2.8, 10.01\,\mathrm{mm}, l = 80\ \mathrm{mm}, p/p_{cr} = 1.175 - 2.92\ ,$$

$$T_\infty/T_{cr} = 0.87 - 1.19, q_\mathrm{w} = 3.15\ \mathrm{kWt/m^2}\ ,$$

argon:

$$d = 2.8\ \mathrm{mm}, p/p_{cr} = 1.175 - 2.04, T_\infty/T_{cr} = 0.87 - 1.19\ ,$$

$$q_\mathrm{w} = 0.565 - 11.5\ \mathrm{kWt/m^2}\ ,$$

air:

$$d = 2.8\ \mathrm{mm}, p/p_{cr} = 1.175 - 1.47, q_\mathrm{w} = 3.15\ \mathrm{kWt/m^2}\ .$$

Heat transfer of a ***horizontal wire in carbonic acid under critical and supercritical parameters*** [8.53]:

$$l = 67\ \mathrm{mm}, d = 0.05\text{–}0.3\,\mathrm{mm}, T_\infty = 10\text{–}50°\mathrm{C}, \mathrm{T_w} < 600°\mathrm{C}\ ,$$

$$p/p_\kappa = 1.002\text{–}1.287, \mathrm{Ra}_d = \frac{\mathrm{g}d^3}{\nu_\mathrm{av}\lambda_\mathrm{av}} \frac{2(\rho_\infty - \rho_\mathrm{av})}{\rho_\mathrm{av}} \frac{2(i_\mathrm{av} - i_\infty)}{T_\mathrm{w} - T_\infty} \rho_\mathrm{av}\ ,$$

$$b_\mathrm{av} = \frac{1}{T_\mathrm{w} - T_\infty} \int\limits_{T_\infty}^{T_\mathrm{w}} b(T)dT, \mathrm{Ra}_d = 1 - 6 \cdot 10^5\ ,$$

for $T_\infty < T_\mathrm{w} < T_{pc}, T_{pc} < T_\infty < T_\mathrm{w}$

$$\mathrm{Nu} = 0.35 + 0.25\mathrm{Ra}_d^{1/4} + 0.45\mathrm{Ra}_d^{1/8}\ , \tag{8.2.16}$$

for $T_\infty < T_\mathrm{w} < T_{pc} - 2\mathrm{K}, T_\mathrm{w} > T_\infty > T_{pc} + 2\mathrm{K}$

$$\begin{aligned} \mathrm{Nu} = {} & \left(0.35 + 0.25\mathrm{Ra}_d^{1/4} + 0.45\mathrm{Ra}_d^{1/8}\right) \\ & \times \left[1 + 4.5\left(\frac{T_\mathrm{w} - T_{pc}}{T_k}\right)^{1/2} e^{-4.5(T_\mathrm{w} - T_{pc})^{1/2}/T_k^{1/2}} \mathrm{th}\left(30\frac{T_{pc} - T_\infty}{T_\mathrm{w}}\right)\right] \\ & \times \left[1 - 0.3th\left(15\frac{p - p_k}{p_k}\right)\right]\ . \end{aligned} \tag{8.2.17}$$

In technical calculations the average values of the parameters can be determined on the basis of the temperature $T_\mathrm{av} = (T_\mathrm{w} + T_\infty)/2$. Then the error in calculation of heat transfer coefficient does not exceed ±10% [8.54].

The theoretical calculations of free convection of a ***horizontal cylinder at near-critical parameters in freons*** $\mathbf{CF_2Cl_2}$, $\mathbf{C_2F_5Cl}$ and $\mathbf{C_4F_8}$ yields [8.55]

$$T_\infty/T_{cr} = 1 - 1.021,\ p/p_{cr} = 1 - 1.73,\ T_{\rm av} = (T_{\rm w} + T_\infty)/2\ ,$$

for $\mathrm{Ra}_d = 10^8$–10^9

$$\mathrm{Nu}_{cp} = 0.6\ \mathrm{Ra}_d^{0.25} F\ , \tag{8.2.18}$$

for $\mathrm{Ra}_d = 10^9$–$2 \cdot 10^{11}$

$$\mathrm{Nu} = 0.15\ \mathrm{Ra}_d^{1/3} F\ , \tag{8.2.19}$$

$$F = 1 \quad \text{for } (\rho_{\rm w} - \rho_\infty)/\rho_{\rm av} < 0.2\ ,$$

$$F = \exp[0.25 - (\rho_{\rm av} - \rho_{\rm w})/(\rho_\infty - \rho_{\rm w})] \quad \text{for } (\rho_{\rm w} - \rho_\infty)/\rho_{\rm av} \geq 0.2\ .$$

The comparison with experiments ***on a tube*** ($d = 8$ mm) shows that the error of formula (8.2.18) attains ±20% [8.56].

The experimental data for heat transfer of a horizontal tube ***in*** **He** ***of supercritical parameters*** with an error of ±10% are generalized as follows [8.39]:

$$d = 4\,\mathrm{mm}, p = 0.102 - 0.51\,\mathrm{MPa},\ T_\infty = 5.4 - 12\,\mathrm{K}\ ,$$

$$T_{\rm av} = (T_{\rm w} + T_\infty)/2, \mathrm{Ra}_d = 10^5 - 10^9\ ,$$

$$\mathrm{Nu} = 0.615\ \mathrm{Ra}_d^{0.258}\ . \tag{8.2.20}$$

Heat transfer of a ***horizontal stainless steel wire in a dissociating*** $\mathbf{N_2O_4}$ [8.57]:

$$d = 1\,\mathrm{mm}, l = 57\,\mathrm{mm}, T_\infty = 12°\mathrm{C}, \theta_{\rm w} = 8 - 145°\mathrm{C}, \mathrm{p} = 0.1 - 16\,\mathrm{MPa}\ ,$$

$$q_{\rm w} < 3.5 \cdot 10^3\,\mathrm{kWt/m^2}, T_{\rm av} = (T_{\rm w} + T_\infty)/2, \mathrm{Ra}_d = 10^5 - 10^7\ ,$$

$$\mathrm{Nu} = 21.9\ \mathrm{Ra}_d^{0.26} (\mu_{\rm w}/\mu_\infty)^{0.04} (\lambda_\infty/\lambda_{\rm w})^{0.33} (c_{p\infty}/c_{p\rm w})^{0.6}\ . \tag{8.2.21}$$

Strong substance compressibility in the region of supercritical parameters can lead to the occurrence of some special features of free convection processes [8.58]. Study of the special convection regimes was carried out mainly on ***alumel wires*** of different lengths under a pressure of 7.85 MPa. For very high loadings, when the wire begins to shine in a dark field, the weak fluctuations of temperatures of an experimental part are observed. However, for laminar free convection there are no distinctions between convective flows of $\mathbf{CO_2}$ for under- and supercritical pressure. With the growth of thermal loadings $q_{\rm w} > 200\,\mathrm{kWt/m^2}$ the zone of laminar flow under the alumel wire becomes narrow enough, and the pulsations of the turbulent zone boundary begin.

The boundary of the ***turbulent zone*** approaches the ***wire*** rather slowly, and individual parts of the boundary move nonuniformly and reach the wire not simultaneously. Then the touch area is distributed along the whole length. The boundary pulsation frequency changes with the value of thermal loading from 0.5 to 10 Hz. On the bottom wire surface part the formations like the vapor bubbles are observed. These bubbles turbulize the boundary layer (such regime is sometimes referred to as ***bubble pseudoboiling***). For higher q_w alongside with normal $\mathbf{CO_2}$ temperature distribution the periodic disturbance θ_w is imposed.

The occurrence of pulsation and turbulent regimes is accompanied by the occurrence of a whistling noise, and in the pulsation regime the noise is modulated with the frequency of the turbulent zone boundary pulsations. In the turbulent regime the noise is uniform, its frequency attains some thousands hertz. With increase in thermal loading, the noise frequency and intensity increase.

The pulsation and turbulent regimes were observed at the temperature of $\mathbf{CO_2}$ less than pseudocritical and pressure above critical. With increase of pressure the special convection regimes occur at smaller thermal loadings. In processing experimental data for ***wire heat transfer*** corresponding to special regimes, the significant dispersion is observed. The unambiguous dependence between the criteria Nu and Ra_d for these points is not found [8.43, 8.58].

A vertical tube with ammonia and carbonic acid at near critical parameters. It is experimentally shown that in the mid part of such a pipe with liquids in critical condition the heat transfer is close to that in the case of a hard core with the outer diameter equal to the tube diameter and the effective thermal conductivity $\lambda_{ef}/\lambda_{Cu} = 4 \cdot 10^3$ [8.59]. For the above-critical temperatures the heat transfer quickly decreases.

The unsteady turbulent convection ***on a horizontal stainless steel sheet*** $l \times b \times h = 57 \times 2 \times 0.2\,\mathrm{mm}$ ***in a dissociating*** $\mathbf{N_2O_4}$ at $p = 12\,\mathrm{MPa}$, $T_\infty = 15°\mathrm{C}$. During the experiment there was a step change in heat generation from $Q_{w1} = 800\,\mathrm{kWt}$ to $Q_{w2} = 3500\,\mathrm{kWt}$. The generalized dependence $\mathrm{Nu}/\mathrm{Nu}_0 = f(k'_{T_1}, T_w/T_\infty)$ shows that for $k'_{T_1} < 2 \cdot 10^3$ the heat transfer process can be considered steady [8.60].

Experimental data for natural convective heat transfer ***from a heated horizontal cylinder to a fluid*** were presented in [8.61] for a wide range of bulk conditions. At near critical states the deviation of the dimensionless heat transfer coefficient from a Nusselt-type correlation was observed. The heat transfer from horizontal cylinder ***at a constant heat flux density on the surface*** immersed ***in a liquid near the critical point*** ($\mathbf{CClF_3}$, $P_{cr} = 3.96\,\mathrm{MPa}$, $\mathrm{T}_{cr} = 302.02\,\mathrm{K}$, $\rho_{cr} = 580\,\mathrm{kg/m^3}$) is given by

$$\mathrm{Nu} = 0.518(\mathrm{Ra}^*)^{0.25} , \tag{8.2.22}$$

where

$$\mathrm{Ra}^* = \mathrm{Ra}/\left[1 + (0.559/\mathrm{Pr})^{9/16}\right]^{16/9} .$$

8.3 Rarefied Gases and Evacuated Liquids

The coefficients of kinematic viscosity and thermal conductivity for gases depend on the average distance between molecule collisions l^*. When the density is so small that the average distance between molecule collisions is of the same order with the characteristic size, the gas cannot be considered as a complex medium. The gas rarefaction rate is determined by the Knudsen criterion $\mathrm{Kn} = l^*/l$. In a free-convective flow a discrete structure appears, when $\mathrm{Kn} > 0.02$. When the average distance between molecule collisions is rather small, the flow is referred to as the flow with slipping. The values $\mathrm{Kn} = 0.02-1$ correspond to this type of flow. If the average distance between molecule collisions is much greater than the body sizes, the free-molecular flow takes place. In a rarefied gas the thermal and hydrodynamical disturbances of the free-convective flow do not localize near the surface as the density decrease.

Heat transfer ***in a rarefied gas*** is determined by the dependence like

$$\mathrm{Nu} = f(\mathrm{Ra}, \mathrm{Pr}, \mathrm{Kn}, \mathrm{P}) .$$

With increase in rarefaction, the value of the Ra number decreases proportionally to the square of the density, and the effect of this criterion decreases. The effect of geometry becomes more significant. When $\mathrm{Kn} > 0.02$, there is a fast decrease of heat transfer due to the temperature step change near the heated surface.

In [8.62,8.63] ***in the presence of rarefaction*** the approximate method of calculating free-convective heat transfer is suggested. Assume that we obtained the criterial dependence of heat transfer of a continuous medium: $\mathrm{Nu} = f(\mathrm{Ra}, P)$. The values of $\lambda, a, \nu = \mu/\rho$ depend on pressure in the similarity criteria Nu, Ra. The dependence of density on pressure is determined directly from the state equation $\rho/\rho_0 = p/p_0$. The coefficients of thermal conductivity and dynamic viscosity are determined by the Knudsen criterion [8.62]:

$$\begin{aligned} \lambda &= \lambda_0/(1 + A\mathrm{Kn}) , \\ \mu &= \mu_0(1 + B\mathrm{Kn}) \\ l^* p &= l_0^* p_0 , \end{aligned} \tag{8.3.1}$$

for $A\mathrm{Kn}$ and $B\mathrm{Kn} \ll 1$.

Substituting dependence (8.3.1) into the similarity criteria, we get

$$\mathrm{Nu} = \frac{\alpha l}{\lambda_0}(1 + A\mathrm{Kn}) , \tag{8.3.2}$$

$$\mathrm{Ra} = \mathrm{Ra}_0 \left(\frac{p}{p_0}\right)^2 (1 + A\mathrm{Kn})(1 + B\mathrm{Kn}) ,$$

where $A = (9k - 5)(2 - b)/[bd(k + 1)]$, $B = 2/d$, $d = 1$ for the external and $d = 2$ for internal heat transfer problem; b is the accommodation coefficient.

The validity ranges for these results are estimated experimentally: $p = 1.33 \cdot 10^{-2} - 1.013 \cdot 10^5$ Pa [8.63].

The local value of heat transfer ***on a vertical flat plate for laminar free convection*** with an error of 8% is given by [8.63]

$$A\mathrm{Kn} \ll 1, B\mathrm{Kn} \ll 1, p = 1.33 \cdot 10^{-2} - 1.013 \cdot 10^5 \text{ Pa} ,$$

$$p_0 = 0.1013\,\text{MPa},\ \Pr = 0.7,\ \mathrm{Ra}_{x0}(p/p_0)^2 = 1 - 10^7 ,$$

$$\mathrm{Nu}_x = 0.4\ \mathrm{Ra}_x^{0.25} . \qquad (8.3.3)$$

Hence, for the local and average values we get $\mathrm{Nu} = \mathrm{Nu}_0(p/p_0)^{0.5}$. The numerical calculations of a self-similar problem ***in the presence of a temperature step change on a surface*** confirm dependence (8.3.3) within the limits of a specified accuracy [8.64, 8.65].

The experimental investigations of a ***thermal boundary layer on a vertical plate*** ($\mathrm{Ra}_l = 5 \cdot 10^7$) for the pressure decrease to $p = 0.67 \cdot 10^4$ Pa show that the boundary layer thickness increases, and the flow does not separate from the surface (there is no slipping and temperature discontinuity) [8.66]. If on change in the pressure we use expression (8.3.2) for the Ra_l number, then $\delta_T/\delta_{T0} = (p/p_0)^{1/2}$. Then, the determined thermal boundary layer thickness value is overestimated by 11% compared with experimental data (for the dispersion of the experimental points ±8%).

The method of simultaneous use of the continual transfer equations and generalized dependences (8.3.2) of the similarity criteria is recommended for calculating free-convective heat transfer ***on a vertical and horizontal cylinder and on a sphere*** [8.67]. The average heat transfer value on the sphere for the pressure decrease, the error ±5% [8.68]:

$$d = 0.05\,\text{m}, p = 10 - 10^5\,\text{Pa},\ \Pr = 0.7,\ \mathrm{Ra}_d = 10^{-3} - 5 \cdot 10^5 ,$$

$$\alpha_{\text{av}} = \alpha_{\text{av}0} = (p/p_0)^{0.3} . \qquad (8.3.4)$$

The decrease of the convective heat transfer under rarefaction leads to the increase of the effect of the body shape. ***The bodies of limited sizes of any convex shape*** have a limit nonzero value of Nu_l for free convection. If a still lower value is taken as the characteristic size, then the Nu_l criterion is between the values ***for sphere*** ($\mathrm{Nu}_d = 2$) and for ***circular unlimited cylinder*** ($\mathrm{Nu}_d \to 0$). The limit values of Nu_l for $\mathrm{Ra}_l < 10^{-5}$ and $A\mathrm{Kn}$, $B\mathrm{Kn} \ll 1$ ***and for bodies of simple shape*** [8.69]:

A cylinder of length l, diameter $d(l/d = 1\text{–}20$, error ±5%)

$$\mathrm{Nu}_d = 1.58\,(1/d)^{-0.32} , \qquad (8.3.5)$$

a plate of length l, width b

$$\begin{aligned} \mathrm{Nu}_b &= 2(l/b)^{-0.32} , \quad l/b = 1 - 20 , \\ \mathrm{Nu}_b &= \pi/\ln(4/b) , \quad b/l \ll 1 , \end{aligned} \qquad (8.3.6)$$

round plate of diameter d

$$\mathrm{Nu}_d = 2.55 \ ,$$

a cube with a side length b

$$\mathrm{Nu}_b = 1.36 \ .$$

The local and average value of free-convective heat transfer of ***a rarefied gas in a rectangular cavity of height H and width B*** with an error of 10% can be found from [8.63]

$$B/H = 0.025 - 0.175, T_0 = 25°\mathrm{C}, T_B = 45, 65, 85°\mathrm{C}, \mathrm{Pr} = 0.7 \ ,$$

$$\mathrm{Ra}_B = 1 - 10^5, \ \mathrm{Ra}_B = \mathrm{Ra}_{B0}(p/p_0)^2, \ \mathrm{Kn} \ll 1, T_{\mathrm{av}} = (T_{\mathrm{w}} + T_\infty)/2 \ ,$$

$$\text{for } x = 0 - x_k, \lambda_{ef,x}/\lambda = 2.111 lg, \ \mathrm{Ra}_B - 4.8, x/H - 4.96 \ ,$$

$$\text{for } x = x_k - 1, \lambda_{ef,x}/\lambda \cong 0.2, x_k/H = 1 \text{ for } \mathrm{Ra}_B = 1 - 10^3 \ ,$$

$$x_k/H = 0.77 \quad \text{for} \quad \mathrm{Ra}_B = 10^3 - 3.2 \cdot 10^4 \ ,$$

$$\lambda_{ef,\mathrm{av}}/\lambda = 1 + 0.024 \ \mathrm{Ra}_B^{1.4}/(10^4 + \mathrm{Ra}_B) \ . \tag{8.3.7}$$

A horizontal electrically heated tungsten wire of diameter d_1, ***length*** l_1 ***in a coaxial cylinder open from both ends of diameter*** d_2 ***and length*** l_2. The space between the cylinders is filled **He,** ***carbonic acid or air***. The experimentally investigated parameters: $d_1 = 0.025$–0.25 cm, $l_1 = 21$ cm, $d_2 = 1.9 - 8.9$ mm, $l_2 = 15$ cm, $d_2/d_1 = 39 - 356$, $p = 0.53$–$1.013 \cdot 10^5$ Pa, $T_{w2} = 16°$C, $\theta_{1,2} = 100$–$450°$C [8.70]:

free-molecular heat transfer regime:

$$p < 67 \, \mathrm{Pa}, d_2/d_1 = 39 - 356, \ \mathrm{Kn}_{d1} = 0.02 - 50 \ ,$$

$$\lambda_{ef}/\lambda = \left(1 + \frac{15}{2} \mathrm{Kn}_{d1} / \ln \frac{d_2}{d_1}\right)^{-1} \ . \tag{8.3.8}$$

For $p = 67 - 2 \cdot 10^4$ Pa the heat transfer is mainly determined by thermal conductivity. If ***an axis of the outer cylinder is inclined***, the minimum in the heat flux dependence on pressure is observed. The heat flux density decrease attains 5–13% for the inclination angle 5° and is reached at a pressure of $p = 2 \cdot 10^4$ Pa.

The free-convective regime begins to develop at $\mathrm{Gr}_{\sigma\kappa} = (2.4 \pm 0.2) \cdot 10^3$. The average heat transfer value ***in a region of thermal conductivity and free convection*** with an error of ±5% is determined from the equation

$$\mathrm{Ra}_{d1} = 10^{-4} - 1 \ ,$$

$$\mathrm{Nu} = 2.35 \ \mathrm{Ra}_{d1}^{0.12} \ . \tag{8.3.9}$$

The average values of heat transfer ***through rarefied gases (air, argon)*** were experimentally investigated for $d_2/d_1 = 1.9$–$9.4, T_{\mathrm{w2}}/T_{\mathrm{w1}} = 0.6$–$0.85$,

$\mathrm{Kn}_{d1} = 10^{-5}$–$10$, $\mathrm{Gr}_{d1} = 10^{-9}$–$5$ [8.71]. The calculated dependence in an intermediate heat transfer region is recommended:

$$\begin{gathered} \mathrm{Gr}_{d1} = 10^{-9} - 10^{-3},\ \mathrm{Kn}_{d1} = 0.02 - 10\ , \\ \lambda_{ef}/\lambda = A\left[2/(1 + T_{\mathrm{w}2}/T_{\mathrm{w}1})\right]^{1-A}\ , \\ A = \left[1 + \frac{19}{3}\frac{\mathrm{Kn}_{d1}}{\ln d_2/d_1}\right]^{-1}\ . \end{gathered} \tag{8.3.10}$$

The noticeable effect of natural convection is shown at $\mathrm{Gr}_{d1} = 10^{-3} - 10^4$.

In the case of free-convective heat transfer of heating steel surfaces ***in oil*** the heat transfer coefficient depends substantially on the wall material for $p < 670\,\mathrm{Pa}$. It takes place only for low pressure and is explained by the effect of surface phenomena on the heat transfer process [8.72]. ***On the interface heat transfer medium – wall*** it is necessary to consider the presence of three contact thermal resistances: of a surface film, of a layer of gas species adsorbed by the heated surface in the heat transfer medium, and of the thermal boundary layer. The surface film thickness and the thermal boundary layer depend on pressure insignificantly. The adsorption layer thickness depends on pressure substantially: it is small for $p < 670\,\mathrm{Pa}$ and reaches a constant value for $p \geq 10^5\,\mathrm{Pa}$. For $p \ll 10^5\,\mathrm{Pa}$ the adsorption layer thickness is so small that its thermal resistance can be neglected compared with the surface film thermal resistance. The latter depends on the wall material, and, hence, in ***vacuum conditions*** the heat transfer coefficient α must change with the change of the heating surface material.

The average heat transfer coefficient for ***liquid* Na *in a large volume*** is

$$p = 600, 1.7 \cdot 10^5\,\mathrm{Pa}, q = 1.4 \cdot 10^5 - 1.8 \cdot 10^6\,\mathrm{W/m^2}, T_\infty = 470 - 840^\circ\mathrm{C}\ ,$$

$$\alpha = 1060\ q^{0.2}\ . \tag{8.3.11}$$

In the pressure range $(0.3$–$1)\cdot 10^5\,\mathrm{Pa}$ the heating surface material has no effect on free-convective heat transfer of $\mathbf{CCl_4}$ ***in a large volume***.

The average heat transfer value of a ***steel horizontal disk*** of diameter $d = 70\,\mathrm{mm}$ ***in oil*** is [8.72]

$$\begin{gathered} \mathrm{Pr} = 8 - 2000,\ \mathrm{Gr}_d = 10^4 - 5 \cdot 10^9,\ T_{\mathrm{av}} = (T_{\mathrm{w}} + T_\infty)/2\ , \\ p = 160 - 800\,\mathrm{Pa} \quad \text{for}\ q_{\mathrm{w}} \leq 12\,\mathrm{kW/m^2}\ , \\ p = 800 - 1.013 \cdot 10^5\,\mathrm{Pa} \quad \text{for}\ q_{\mathrm{w}} \leq 41\,\mathrm{kW/m^2}\ , \\ \mathrm{Nu} = 2.3\mathrm{Gr}_d^{0.25}\,\mathrm{Pr}^{0.3+0.02\,\mathrm{Pr}^{-1/3}}\ . \end{gathered} \tag{8.3.12}$$

For $q_{\mathrm{w}} > 12\ \mathrm{kW/m^2}$, $p < 333$ Pa α increases with the pressure increase.

A horizontal steel helix $d = 0.7$–$0.8\,\mathrm{mm}$ with a coil diameter $D = 6\,\mathrm{mm}$ and a relative spacing $s/d = 1.16 - 5.5$. The average heat transfer values ***in vacuum conditions*** $p = 67 - 1.33 \cdot 10^3$ Pa are obtained from [8.73]

$$\mathrm{Nu} = C\mathrm{Gr}_d^n \,\mathrm{Pr}^{0.3+0.02\,\mathrm{Pr}^{-1/3}} (s/d)^{0.17} \ . \tag{8.3.13}$$

The constants C, n, wall material, heat transfer medium, and main parameters are given in Table 8.3.1.

Table 8.3.1.

Heat transfer medium	Pr	Gr_d	n	C		
				Steel	Ni	Steel Ni-Cr
vacuum oil	$69–3.2{\cdot}10^3$	10^{-4}–0.3	0.38	0.209	0.327	0.477
		0.3–0^2	0.28	0.182	0.301	0.382
engine oil	10–250	10^{-1}–0.35	0.62	0.342	0.884	0.884
		0.35–10^2	0.23	0.25	0.467	0.467

Dependences (8.3.12) and (8.3.13) refer to surfaces processed with purity class 4. On the walls with the processing purity class 7 the α values are about 2 times lower.

8.4 Convection Induced by Radiation

An electromagnetic radiation of optical and infrared frequencies traveling in a motionless gas or liquid can cause the medium heating due to absorption of apart of radiation energy. The heat equal to $k_a I$ is generated in a volume unit during a time unit, where I is the radiation intensity, k_a is the radiation absorption coefficient dependent on temperature, pressure, radiation spectrum, and medium composite. The medium heating causes its expansion, and the appearing buoyancy forces lead to the development of photo-absorption convection. The convection induced by radiation has a certain role in heat transfer in the atmosphere of planets, in an ocean, etc. [8.74] The interest to the photoabsorption convection is caused also by the development of laser engineering [8.75]. The beam of laser radiation has a limited width; hence, the medium heating by radiation is spatially non-uniform. This leads to the formation of the self-induced thermal lens by the laser beam. The medium refraction index n depends on temperature, and for the majority of media $dn/dT < 0$. The radiation laser beams defocus, and generally there occurs their deviation from the initial traveling direction. The thermal defocusing is characterized by low power limits and bounded the radiation intensity and power that can be effectively transferred through the medium [8.76–8.78].

Laminar photoabsorption convection in gas begins at [8.79]

$$I > \left[\lambda g/(k_a l \rho c_p),\ \lambda g^2/(\rho^2 \rho c_p^2)\right]_{\max} \ . \tag{8.4.1}$$

The main determining criteria are

$$\mathrm{Ra}_I = g\beta k_a I_0 l^5/(\lambda\nu a) \tag{8.4.2}$$

and Pr, and based on them it is possible to find the free-convective flow velocity u. The generalized dependence

$$\mathrm{Pe} = f\left(\mathrm{Ra}_I, \mathrm{Pr}\right) \tag{8.4.3}$$

is considered for the following stationary natural convection regimes [8.80, 8.81]:

weak convection (Re, Pe $\ll 1$)
for Pr $= O(1)$, Pr $\gg 1$, $\mathrm{Ra}_I = 0$–1,
for Pr $\ll 1$, $\mathrm{Ra}_I = 0$– Pr,

$$\mathrm{Pe} = C_1 \mathrm{Ra}_I \ , \tag{8.4.4}$$

where $C_1 \cong 10^{-5}$,

moderate convection
for Pr $\gg 1$, Re $\ll 1$, Pe $= 1$ – Pr, $\mathrm{Ra}_I = 1 - \mathrm{Pr}^2$,

$$\mathrm{Pe} = C_2\, \mathrm{Ra}_I^{0.5} \ , \tag{8.4.5}$$

for Pr $\ll 1$, Re $\gg 1$, Pe $\ll 1$, $\mathrm{Ra}_I = \mathrm{Pr}^2 - 1$,

$$\mathrm{Pe} = C_2\left(\mathrm{Ra}_I\,\mathrm{Pr}\right)^{0.5} \ , \tag{8.4.6}$$

where $C_2 \cong 10^{-2}$,

developed convection (Re, Pe $\gg 1$)
for Pr $\gg 1$, Pe $>$ Pr, $\mathrm{Ra}_I > \mathrm{Pr}^2$,
for Pr $= (1)$, $\mathrm{Ra}_I > 1$
for Pr $\ll 1$, $\mathrm{Ra}_I\mathrm{Pr} > 1$,

$$\mathrm{Pe} = C_3 \mathrm{Ra}_I^{1/3} \ . \tag{8.4.7}$$

where $C_3 \cong 0.2$.

The transition to turbulent convection occurs at $\mathrm{Re}_k = 3$ [8.79, 8.80, 8.82, 8.83]. The low Re_k value is connected with the presence of excesses in the velocity distribution due to the reversal streams and the absence of rigid boundaries stabilizing the flow. ***A limit of turbulence generation*** for the radiation intensity is $I_k \geq \nu\lambda a/(k_n g\beta l^5)$ [8.79].

The time of development for stationary velocity τ_u ***and temperature*** τ_T fields at free convection [8.78, 8.84]:

weak convection

$$\tau_u = \max(l^2/\nu, \tau_T), \tau_T = l^2/a \ , \tag{8.4.8}$$

moderate convection

$$\begin{aligned} &\mathrm{Pr} \gg 1, \tau_u = \max(l^2/\nu, \tau_T), \tau_T = l/u\ , \\ &\mathrm{Pr} \ll 1, \tau_u = \max(l/u, \tau_T), \tau_T = l^2/a\ , \end{aligned} \tag{8.4.9}$$

developed convection

$$\tau_u = \tau_T = l/u\ . \tag{8.4.10}$$

The nonstationary free convection takes place, when the time of interaction of radiation and environment $\tau \ll \tau_u$, τ_T. The heat generated at radiation absorption is localized within the limits of the beam. The ambient temperature nonuniformity is

$$\theta = k_a/(\rho c_p) \int_0^\tau I d\tau\ , \tag{8.4.11}$$

the photoabsorption convection velocity is [8.78, 8.79]

$$\vec{u} = \vec{g} \int_0^\tau \theta d\tau\ . \tag{8.4.12}$$

8.5 Biosystems

The main activity of green plants consists in photosynthesis and is carried out in the main plant bodies – leaves. The energy obtained by organisms goes to the performance of a biological work by them, however, its significant part turns to heat. The biosystem receives and gives back to the surrounding space a part of heat by free convection. For example, in a thermal balance of leaves this value attains 10–20 % [8.85, 8.86].

In investigation of convective heat transfer of plants many different and irregular leaf shapes are encountered.

In nature the leaf seldom remains motionless in a wind, and the flow is frequently disturbed by the presence of neighboring leaves and branches. The known formulae of convective heat transfer for simple geometrical shapes are applied to such irregular shapes as leaves in conditions approximating the nature.

The average heat transfer value of a coniferous tree leaf of width b and length l with accuracy of $\pm 20\%$ is determined depending on orientation as [8.87]

$$\mathrm{Ra}_b = 20 - 40^4, b = 0.44 - 1.18\,\mathrm{cm}, l = 4.19 - 8.02\,\mathrm{cm}, F = 1.7 - 8.2\,\mathrm{cm}^2\ ,$$

horizontal leaf arrangement

$$\mathrm{Nu} = 2.44\ \mathrm{Ra}_b^{0.09}\ , \tag{8.5.1}$$

leaf located vertically, the long axis is horizontal

$$\mathrm{Nu} = 2.65\ \mathrm{Ra}_b^{0.1}\ , \tag{8.5.2}$$

leaf located vertically, the long axis is vertical

$$\mathrm{Nu} = 2.37\ \mathrm{Ra}_b^{0.08}\ . \tag{8.5.3}$$

The leaves of leaf-bearing plants have considerably more diverse shapes. In [8.88] 10 idealized shapes are systematized. To apply the existing formulae of convective heat transfer such as (1.10.10), the universal characteristic size l_{ef} was introduced. The average characteristic length is determined by the average heat transfer coefficient on a flat plate using the correlation

$$l_{ef} = \int_0^B \alpha(x_B - x_H)dz \Big/ \int_0^B \alpha dz\ .$$

For free convection on a vertical flat surface from (1.10.10) $\alpha_{\mathrm{av}} \sim (x_B - x_H)^{-1/4}$, and for l_{ef} we finally have [8.88]

$$l_{\theta\phi} = \int_0^B (x_B - x_H)^{3/4} dz \Big/ \int_0^B (x_B - x_H)^{-1/4} dz\ .$$

For the simplicity of using the results by botanists and ecologists the numerical values of l_{ef}/l are given for 10 idealized shapes for various orientation of the axis relative to the gravity force [8.88]. The average value of heat transfer of ***a plant leaf*** with an accuracy of ±25% is determined from the equation

$$\mathrm{Gr}_{l,ef} = (1-5)\cdot 10^5, \vartheta = 0 - \pi/2\ ,$$
$$\mathrm{Nu} = 0.37(1 + 0.3 \sin\vartheta)\mathrm{Gr}_{l,ef}^{0.25}\ . \tag{8.5.4}$$

The average Nusselt number increases gradually as ***the leaf turns from the horizontal position to the vertical***.

The more exact results for the average heat transfer value can be obtained from (2.11.1) for the constant value of temperature difference and from (2.11.5) for any change in the wall temperature $\theta_{\mathrm{w}}(x,\ z)$. However, ***due to a casual orientation, curvature, and roughness of the leaves*** the heat transfer coefficient value obtained should be increased by 25% [8.89].

The presence of others leaves for the fixed temperatures difference leads to an increase of heat transfer by 15–20% [8.88]. For the account of the ***effect of the neighboring leaves*** on heat transfer the model consisting of a cavity $10 \times 10 \times 10\,\mathrm{cm}$ filled with vertically located elliptic plates is experimentally investigated. The heating of leaf models was made electrically by a heat flux decreasing exponentially to the base (simulating leaf heating by the sun). The leaf models were located as lines $z_1 \times z_2 \times z_3 = 5 \times 5 \times 15$. Beside this system, five identical systems were located. The average heat transfer of a ***vertical plate in a middle of the cavity filled with leaf models***, $l \times b = 8 \times 8\,\mathrm{cm}$, $d_э = 2/\sqrt{\pi}$ is given by [8.90]

for T_w $=$ ***const***

$$\mathrm{Nu} = 0.648\ \mathrm{Gr}_{d_e}^{0.214}\ , \tag{8.5.5}$$

for q_w $=$ ***const***

$$\mathrm{Nu} = 1.045\ \mathrm{Gr}_{de}^{0.184}\ . \tag{8.5.6}$$

The leaf fluctuations in the presence of a branch increase the heat transfer coefficient not more than by 6% [8.88].

For mixed convection of plant leaves the plenty of dimension formulae are recommended for different geometry, orientation, and heat transfer conditions [8.91, 8.92]. However, ***for calculating heat transfer by mixed convection*** it is possible to use the dependence [8.93]

$$\mathrm{Nu}^3 = \mathrm{Nu}_{ck}^3 + \mathrm{Nu}_{ek}^3\ . \tag{8.5.7}$$

For leaves of coniferous trees Nu_{ck} is defined by (8.5.1)–(8.5.3) and by the following dependences:

horizontal leaf arrangement ($\mathrm{Re}_b < 10^4$) [8.87]

$$\mathrm{Nu}_{ek} = 0.632\mathrm{Re}_b^{0.52}\ , \tag{8.5.8}$$

vertical arrangement ($\mathrm{Re}_b < 10^4$)

$$\mathrm{Nu}_{ek} = 0.689\mathrm{Re}_b^{0.52}\ . \tag{8.5.9}$$

For the leaves of leaf-bearing plants Nu_{ck} is defined by (8.5.4) and by dependence ($\mathrm{Re}_{l,ef} < 1800$) [8.88]

$$\mathrm{Nu}_{ek} = 0.6\mathrm{Re}_{l,ef}^{0.5}\ . \tag{8.5.10}$$

Such effect as wood fire is simulated by mixed convection above a horizontal surface in the presence of injection [8.94]. In a zone near the wood fire front the thermal veil takes place. Usually, in a nondisturbed near-ground atmosphere layer the logarithmic wind profile is supposed to take place. The approximation of numerical calculations of ***mixed convection in the near-ground atmosphere layer before the front of the upper fire for adiabatic temperature*** T_a ***behind the fire front and a heat flux from the surface*** q_w [8.94, 8.95]:

$$u_\infty = 3 - 8\ m/s, (p\nu)_F = 1 - 5\,\mathrm{kg/m^2}s\ ,$$

$$T_a = T_\infty + (T_F - T_\infty)\left(1 + \frac{15}{u_\infty^2}\frac{\mathrm{Re}_{\Delta x}}{\mathrm{Re}_{lx}}\right)^{-0.8} ,$$

$$q_\mathrm{w} = 7x_F^{-0.6}\rho_\infty c_{p\infty}\mathrm{Re}_x^{-0.2}\left(x/x_F\right)^{-0.6}\left(T_\mathrm{w}/T_\infty\right)^{0.39}\left(T_a - T_\mathrm{w}\right) , \qquad (8.5.11)$$

where $\mathrm{Re}_{\Delta x} = u_\infty(x - x_F)/\nu$; x_F is the width of combustion front; x is counted from the left combustion boundary.

These formulae with an accuracy of 30% approximate the results of numerical calculations at a distance of up to 90 m from the combustion front. The values of q_w and T_a decreases with increase of x, and for $x \to \infty$ we have $T_a = T_\infty$, $q_\mathrm{w} = 0$. The heat flux value increases with increase of T_F, $\rho\nu$ and x_F.

The average value of heat transfer ***from a river surface to the atmosphere*** ($u_\infty = 0.5$–$3.5\,\mathrm{m/s}$) is [8.96]

$$\alpha = 5.06 + 6.88u_\infty^{0.82} . \qquad (8.5.12)$$

Equation (8.5.12) can be applied for calculating the thermal balance of thermal electrical stations ***cooled by water***.

The characteristics of a plume generated by natural convection above heating pipes in greenhouse row crops are studied in [8.97]. This plume differs from the free plume mainly in two respects: owing to the particular geometry of the hotbed, a downward flow appears in the edges of the hotbed and interacts with the plume and owing to the presence of vegetation heat and momentum sinks appear in the hotbed crop. It shows that: the similarity hypothesis is relevant in this case while the entrainment hypothesis is not, the decrease of excess temperature with height is steeper than in a free plume, and the air velocity increases (decreases) with height at low (high) heights.

8.6 Solidifying Melt

A convective ***metal*** mixing in bath refers to one of the major parameters affecting the rate of heat transfer from overheated volumes ***to the solidification boundary and the solidification front motion velocity***, on redistributions of individual crystals, separation and distribution of nonmetal inclusions in an ingot and the arrangement of liquidating strips in it. In the initial period the convective flows move the crystals just formed in a narrow zone of two-phase state to the bottom part of the ingot. As the solidification boundary moves, the wide zone of two-phase state develops, and the convective mixing occurs along this zone near isoliquidus border.

The average velocity of a convective flow ***near a semi-infinite metal solidification front*** is [8.98]

$$u_x = 0.128\sqrt{g\beta H(T_\infty - T_m)} .$$

The velocity of a ***reversal flow*** of ***liquid steel*** on an ingot axis (directed upwards) is given by

$$u_\infty = u_x F_x/(F_{\rm ing} - F_x)\ .$$

For a prism of square section l and height $H/l = 5$ the viscous flow regime is realized at $\mathrm{Gr}_l \leq 10^4$. In this range the relative velocities in ***metal*** ($\mathrm{Pr} = 1.108$) and ***non-metal*** ($\mathrm{Pr} = 8.8$) melts almost coincide. The increase of the velocity occurs with increase of Gr_l under the linear law irrespective of the Pr number. For temperatures such area is observed in a little narrower range of the Gr_l numbers. ***On a vertical prism the symmetry axis on the upper horizon z = H and at the point of intermediate height z = H/2*** the velocity maximum is reached at $\mathrm{Fo_l} = 0.03$ for $\mathrm{Pr} = 0.108$ and at $\mathrm{Fo_l} = 0.005$ for $\mathrm{Pr} = 8.8$ [8.99]. The increase of the Grashof number leads to a change in the convective heat and mass transfer character. As the melt flows, the thermal centre is moved from the symmetry axis and is located near a lateral surface.

The maximum velocity of the thermal convection of the melt achievable in the region $l \times H = 0.92 \times 1.84\,\mathrm{m}$ on heating $\theta_\mathrm{w} = 100°\mathrm{C}$ ($\mathrm{Gr}_l = 10^{10}$, $\mathrm{Pr} = 0.224$) attains 0.43 m/s ***downstream***. ***In a region with overheating*** by 50°C and geometrical correlations $l \times H = 1.13 \times 2.56\,\mathrm{m}$ after 20 min of the solidification beginning the dimensionless overheating in the liquid area does not exceed 0.1°C. For the time of bottom region full solidification of 4 h the convection rate is noticeable after 1.5 h [8.100].

The conjugated problem of melt cooling in the bath of rectangular section $H \times B$ ***with wall thickness*** δ ***and uniform temperature at the initial moment. For carbonic steel*** ($\mathrm{Pr} = 0.108$) the interface between phases $\mathrm{Gr}_B > 10^8$ is close to straight lines parallel to coordinate axes, the curvature is observed only at the corners of the area. The temperature gradient at the initial time is concentrated near the crystallization front [8.99].

A rectangular region $H \times B$ ***with a motionless melt, the uniform initial temperature distribution*** $T_0 > T_{cr}$, ***and the uniform distribution of species*** c_0. At the time $\tau > 0$ the boundary temperature makes a step change to the crystallization temperature T_{cr}. The numerical investigation is carried out at $\mathrm{Gr} = \mathrm{Gr}_D = 0.2{\cdot}10^7$, $0.2{\cdot}10^8$ ***for steel with the carbon contents 0.01*** with the initial melt overheating $T_0 - T_{cr} = 1°\mathrm{C}$ [8.101]. After some time from the process beginning the direction of liquid core convective mixing turns to the opposite. At the removal of the initial temperature overheating of the ***liquid core*** (the time of existence of the initial overheating makes approximately 15–20% of the general solidification time) the temperature nonuniformity is smoothed. The thermo-concentration convection, unlike the thermal one, takes place in the liquid core of the crystallizing melt almost until the end of solidification.

The average value of heat transfer from a ***melt of potassium electrolyte to a vertical wall*** with an error of ±15% is [8.102]

$$T_p = 685 - 730°\text{C},\ T_{cr} = 650 - 660°\text{C},\ q_\text{w} = 4 - 34\ \text{kW/m}^2\ ,$$

$$\text{Gr}_l = 0.7 \cdot 10^9 - 4 \cdot 10^9,\ T_\text{av} = (T_p + T'_\text{w})/2\ ,$$

$$\text{Nu} = \left[1 + (T_{cr} - T'_\text{w})\,/T_{cr}\right]^2 = 0.00136\ \text{Pr}^{-T_p/T_\text{w}}\ \text{Gr}_l^{0.56}\ , \qquad (8.6.1)$$

where T_w is the wall temperature until the solid layer is formed; T'_w is the temperature on the border solid layer-melt after formation of a solid phase on the plate.

The effect of the value $[1 + (T_{cr} - T'_\text{w})/T_{cr}]$ is shown only in the region of a two-phase medium, when the wall temperature is lower than temperature of melt crystallization.

Vertical cylindrical bath H × D filled with tin up to the height h. Heat transfer ***in a melt heated from below*** [8.103]:

$$\begin{aligned} &\text{for}\quad \text{Ra}_h < 10^5 && \lambda_{ef}/\lambda = 1\ ,\\ &\text{for}\quad \text{Ra}_h = 10^5 - 10^7 && \lambda_{ef}/\lambda = 7.94 \cdot 10^{-3}\ \text{Ra}_h^{0.35}\ ,\\ &\text{for}\quad \text{Ra}_h = 10^7 - 10^8 && \lambda_{ef}/\lambda = 1.82 \cdot 10^{-9}\ \text{Ra}_h^{1.3}\ . \end{aligned} \qquad (8.6.2)$$

In a zone of the most intensive metal motion the vertical component of the local velocity reaches the values (2.6–5.3)$\cdot 10^{-3}$ m/sec, in zones of rather slow flows – (0.9–1.8)$\cdot 10^{-3}$ m/sec, and in stagnant zones the motion velocity does not exceed $0.6 \cdot 10^{-3}$ m/sec.

In glass production, free convection has a significant effect on the quality of the product [8.104]. The processes of production of monocrystal are associated with investigation of free convection in the presence of rotation [8.105]. The motion of the melt (especially of the phase interface) due to free convection decays as the crystal begins to rotate with a rather small velocity. The heat flux increases with increase of the angular velocity ω proportionally to $\sqrt{\omega}$. The heat flux value increases also with increase of the crystallization rate.

In zonal melting the coefficient of species distribution is the main characteristic of the purification process [8.106]. At the crystallization front a surplus (lack) of the component is created, which is transferred to the liquid phase by diffusion. Experiments on heat transfer at the ***crystallization front*** are processed with allowance for the Stefan flux and the zone motion velocity of $2 \cdot 10^{-5}$ m/sec ($\text{Ra}_l = 10^4$–10^6):

$$\text{Nu}_{cp} = 1.93\ \text{Ra}_l^{0.065}\ , \qquad (8.6.3)$$

where l is the length of the melt zone.

The mass transfer rate due to convection increases 3.5–4 times. The fluctuations of the temperatures cause a layer-by-layer crystallization and may cause deterioration of parification of an ingot. The maximum amplitude of temperature fluctuations in the zone of melting is equal to 1.2°C for the frequency of 6–8 periods per minute due to convective flows of vortical nature.

In aluminum electrolyzers the main part of heat of a melt of an ***industrial electrolyte*** is removed through the cathode. ***On the lateral bath surfaces*** during the start-up of aluminum electrolyzers a protection layer from solidification electrolyte melt is formed. The average value of the coefficient of heat transfer from the melt to the hardened slag is [8.107]

horizontal disk with the heated surface facing downwards

$$d = 30\,\mathrm{mm}, T_p = 965 - 1005°\mathrm{C}, T_g = 934.1 \pm 1°\mathrm{C}\ ,$$

$$T_p - T_g = 10 - 70°\mathrm{C}, q_\mathrm{w} = 6 - 55\,\mathrm{kWt/m}^2\ ,$$

$$\alpha_\mathrm{av} = 575 + 1.68\,(T_p - T_g)\ , \tag{8.6.4}$$

vertical cylinder

$$d = 29.6\,\mathrm{mm}, l = 125\,\mathrm{mm}, T_p = 972 - 1011°\mathrm{C}, T_\mathrm{g} = 935.6 \pm 0.6°\mathrm{C}\ ,$$

$$\alpha_\mathrm{av} = 353.6 + 7.635\,(T_p - T_\mathrm{g})\ . \tag{8.6.5}$$

In the absence of hardened slag the rate of heat transfer from the melt to the wall first increases with increase of the temperature difference θ_{pc} and then decreases.

Heat transfer ***from the electrolyte melt to the wall*** increases essentially with increase of the melt temperature (***for horizontal disk*** from 640 to 910 W/m^2·grad, and ***for a vertical cylinder*** from 700 to 970 W/m^2·grad).

One of the most important features of convection in weak force fields is a substantial increase of the temperature field local nonuniformities at a rather weak total effect of convection [8.108]. It is essential that the temperature stratification can reach the maximum value in a certain range of Ra numbers, while the average heat transfer decreases monotonically with decrease of Ra.

Alloy solidification in a two-dimensional rectangular cavity was analyzed in [8.109]. For the investigated alloy (***1% chromium-steel***) it is found that at the initial and final solidification stages the fluid flow due to compression is more intense than natural convection, while at the intermediate stage the natural convection dominates. Moreover, the fluid flow caused by compression in a transient zone is more intense than the flow due to natural convection, which specifies the necessity of the allowance for the compression effect in simulation with the purpose of determining the casting defects, connected with the flow.

The effect of the Prandtl number on natural convection is analyzed on the basis of numerical solution of a two-dimensional stationary flow in a square cavity for two different temperatures of the vertical wall both in the presence and absence of a solidification process. For a fixed Ra the characteristic momentum and heat transport depend insignificantly on Pr, and for Pr > 1 regardless of the presence of phase changes. For a monotonic decrease of the Pr number, starting with 1, its effect becomes more noticeable. In the presence of solidification the more intense convection caused by high Ra numbers in a combination with an effect of the Stefan number increases the effect of the Pr number on the stream function and the Nusselt number, but does not affect the global enthalpy patterns [8.110].

Melting of a ***pure phase change material (PCM) in a rectangular container heated from below*** was numerically simulated in [8.111]. Instability of a free convection flow during the melting process is discovered and flow patterns for a wide range of Ra numbers are presented.

Unidirectional solidification of ***a dilute alloy (tin-bismuth) in a two-dimensional rectangular cavity*** was investigated in [8.112]. Solute concentrations were found using a simplified model of binary alloy solidification similar to the enthalpy method. Results for the solidification of **Sn-0.5%Bi** alloy were presented.

Numerical results for the transient behavior and heat transfer during the melting ***of ice in non-Darcy porous media within a rectangular enclosure*** show that the solid-liquid interface becomes irregular due to the presence of porous media [8.113]. The nonlinerity, due to temperature-dependent density, of the molten liquid could be a significant contributor to temperature and flow field, position of interface, and capacity of heat transfer. As the Darcy number increases, the heat transfer is enhanced, the rate of ice melting increases, and the interface becomes more distorted.

The effect of anisotropic permeability of the two-phase zone on the solidification ***of a binary mixture of*** $\mathbf{NH_4Cl\text{-}H_2O}$ contained ***in a square cavity*** was simulated numerically in [8.114]. It was shown that the anisotropy effects were essential. Moreover, because of the temperature decrease due to the greater concentration and earlier development of thermal convection in the upper layer there is intense secondary melting. The interaction of phases through macroscopic solidification front increases with increase of the ratio of permeabilities.

The analysis of the process of pure substance melting ***in a square cavity with one of the vertical walls heated and the second maintained at the initial temperature*** shows that at the initial stage of heating the thermal conductivity is the dominating heat transfer mechanism; then, as the liquid phase appears, a convective flow develops leading to the heat flux nonuniformity at the melting front and to the deformation of the phase interface [8.115]. The cases of low (characteristic for ***metal*** melting) and high (referring to the melting of ***organic materials***) Pr values were considered ($\mathrm{Pr} \sim 10^2$).

A rectangular region $\boldsymbol{H/B} = 4$ ***filled with a melt;*** at the initial time it has a homogeneous structure. The temperature gradient is directed along the longitudinal axis, overloading – across the axis. The experiments and calculations were carried out under the following conditions: $\mathrm{Ra}_B < 350, g_t/g = 4 \cdot 10^{-3}$, $B = 1\,\mathrm{cm}$, $\mathrm{Pr} = 0.016$, $\mathrm{Sc} = 10$, $\varphi = 0°$ C [8.108].

In a regime of theoretical weightlessness ($\mathrm{Ra}_B = 0$) the motion is absent, and the species distribution is uniform. The effect of weak convection on the species distribution in ***germanium*** begins to exert its effect at

$\mathrm{Ra}_B = 10$. In the region a monovortical motion appears, which because of the smallness of Pr number, has no effect on the temperature field, but exerts its effect on the concentration field, as the Schmidt number is rather great (Sc = 10), which is characteristic for the majority of semi-conductor materials.

The solidification process ***of a binary solution of*** $\mathbf{Na_2CO_3}$ ***in a horizontal cylindrical annular channel*** is investigated numerically in [8.116]. The initial conditions were chosen to simplify the study of the opposing phenomena: thermal convection and convection in substance dissolution, which are accompanied by an intense increase in the two-phase zone. The calculated time scale of the diffusion layer evolution appeared much smaller than that in experiments.

An analytical model of the liquid freezing ***inside a vertical pipe*** was described in [8.117]. This provides an estimate of the freeze time and limiting temperature and can be applied to different pipe sizes, fluid temperatures, jacket lengths, cooling methods, and fluids. The study indicated that the limiting temperature is controlled by laminar natural convection in pipes less than 100–150 mm in diameter; in larger pipes the flow conditions become turbulent.

The effect of natural convection on solidification of a laminar fluid flow ***in the thermal entrance region of a horizontal isothermally cooled tube*** was investigated numerically in [8.118]. The theoretical solution assumes that the Prandtl number is large, and the variation of the liquid-solid interface is gradual in the axial direction. For smaller superheat ratio, the value of the local Rayleigh number decreases faster because of the rapid increase in the solid shell, which suppresses the effect of natural convection on the dimensionless radius of the circular interface. There are significant effects of superheat ratio and Ra on the Nusselt number and dimensionless heat transfer rate.

In a vertical cylinder, weak convective streams under the conditions of the lowered gravitation can cause a macroliquation along the sample radius. The effect of macrosegregation imposes certain restriction on the value of weak mass force admissible for a technological experiment. For reliable elimination of the by-effects of weak convection the mass force should be decreased to the value of $g_t/g \cong 10^{-5}$ [8.119].

A vertical cylinder of height H and radius r_0 ***is fully filled with n-hexadecane, n-octadecane, p-xilem or cyclohexan***. At the initial moment the wall and cylinder inner medium temperature are equal to T_{w2}. A cylindrical heater of small diameter with temperature T_{w1} is located coaxially inside. The change in the heat transfer coefficient in time in the presence of melting (the coordinate x is reckoned from the upper horizontal surface of the cylinder) [8.120]:

$$H = 70 - 140\,\mathrm{mm},\ r_0 = 50\,\mathrm{mm},\ \mathrm{Pr} = 8 - 50,\ \mathrm{Ra}_x = 3 \cdot 10^5 - 5 \cdot 10^8\ ,$$

$$\mathrm{Bi}_{r0}/k_f = 0.25 \cdot 10^{-2} - 1.25 \cdot 10^{-2}\ ,$$

$$\frac{\mathrm{Nu}_x}{\mathrm{Nu}'_x} = 1 + 0.43 \left(\frac{\mathrm{Bi}_{r0}}{k_f} \mathrm{Ra}_{r0}^{0.25} \right)^{1.7} e^{-4.05x/H} , \tag{8.6.6}$$

where

$$\mathrm{Nu}'_x = (0.4 - 0.6)\, \mathrm{Ra}_x^{0.24-0.27} . \tag{8.6.7}$$

The characteristics of fast metal solidification after laser melting were calculated numerically in [8.121]. The flow field and temperature distribution were investigated ***in a volume of fluid during fast_solidification*** of ***aluminum and steel*** at different laser powers (10^8 and 10^9 W/m^2) and interaction time values. The flow disappearance is observed soon after the laser beam is removed. In the case with ***aluminum*** for high power the volume of fluid remains very deep for a rather long time due to the formation of a large secondary vortex during melting. In spite of the different interface shape in the presence and absence of convection there is an insignificant distinction in solidification time for both metals.

Nonstationary solidification in a closed volume of fluid cooled from a side was investigated experimentally and theoretically in [8.122]. A special attention was given to the effect of convection at large Rayleigh numbers and to the coupled influence of convection and thermal conductivity on location of a freezing front. In the experiments carried out ***on paraffin*** at $\mathrm{Ra} \sim 10^{11}$ and $\mathrm{Ste} \sim 0.1$ the existence of a weak turbulent flow at early stages of the process is noticed. The range of parameters is extended using a theoretical model of a coupled process of thermal conductivity and convection with nonstationary stratification in the fluid volume. The effect of time, the Stefan number, and the ratio of coefficients of heat conductivity and thermal diffusivity on solidification rate was shown graphically.

On a three-phase model (solid body, liquid, mushy zone) the effect of the gravitational field was investigated analytically (under different gravitation conditions) on the ***binary alloy*** solidification [8.123].

A numerical analysis of the effect of natural convection on the heat transfer processes in the presence of phase change in view of simultaneous existence of solid, liquid and mushy zones in a wide range of the Prandtl, Rayleigh, and Stefan numbers showed that the size and strength of a convective cell and the arrangement and shape of phase interface depended on the correlation between the listed determining parameters and on the temperature range of the mushy zone existence [8.124].

An effect of natural convection on melting process was investigated analytically in [8.125]. A phase change material initially in the solid phase at its melting temperature was instantaneously exposed to a uniform higher temperature at the wall. It was supposed that the solid phase had the large density compared with the liquid and decreased as the metal melted. The average Nusselt number for the whole melting time was approximated by the dependence

$$10^4 \le \mathrm{Gr} \le 10^6, \quad 10 \le \mathrm{Pr} \le 100, \quad 0.5 \le \mathrm{Mt} \le 5.0 ,$$

$$\text{Md} \gg 1, \text{Ste} \ll 1, \text{Gr}/\text{Ar} \le 0.1, Sb < 0.75 \,.$$

$$\begin{aligned}\text{Nu} = 1.308\,\text{Gr}^{0.22}\,\text{Pr}^{0.08}\text{Mt}^{0.15} \\ \times \left(1 - 3.47\zeta + 4.69\zeta^2 - 1.68\zeta^3\right)\frac{dz}{dt}\,,\end{aligned} \tag{8.6.8}$$

where

$$\frac{dz}{dt} = 1.636 - 1.7\zeta + 1.425\zeta^2 - 0.93\zeta^3\,,$$

$$\zeta = \left(1.546t - 0.77t^2 + 0.135t^3\right)(1 + C)\,,$$

$$C = 2.94\,\text{Gr}^{0.13}\,\text{Pr}^{0.31}\text{Mt}^{0.13}\text{Md}\,(1 - \text{Sb})\,,$$

$$\text{Mt} = \left(\left[\frac{\lambda\Delta T}{l}\right]^3 \frac{\Delta\rho g D^3}{\rho_s^4\nu^5}\right)^{-0.25}, \quad \text{Md} = \left[\frac{\lambda\Delta T\nu}{l\Delta\rho g D^3}\right]^{-0.25},$$

Sb is the subcooling parameter.

Natural convection flow visualization experiments during melting and solidification ***of a hypoeutectic gallium five-weight percent indium*** were performed in [8.126] using a real time X-ray radioscopy technique. The experiments revealed a conductive-convective threshold for the onset of natural convection in the alloy. The threshold appeared to be caused by a substantial chemical stratification of indium in the melt.

Dynamic behavior of heat transfer, fluid flow, and interfaces ***in the vertical zone-melting crystal growth*** was studied numerically in [8.127]. Growth of a single crystal of $\mathbf{NaNO_3}$ in a computer-controlled transparent multizone furnace was simulated as an example. For unstable cases, periodically oscillatory flow in growth rate occurring at intermediate values of Ra number were observed. The upper flow cells beneath the feed front seem to be responsible for the instability, and this is consistent with the observation during crystal growth experiments. For stable cases, a steady state can be achieved smoothly.

The process of dissolution mass transport ***along a vertical soluble substrate submerged in a large pool of otherwise quiescent molten metal*** was studied theoretically in [8.128]. It was found that the Sherwood number depended strongly on the saturated concentration of the substrate at the moving dissolution front and the degree of saturation in the ambient pool. Based upon the numerical results, a correlation for the average Sherwood number is

$$\text{Sh} = (0.66 - 0.32a)\left(1 + 0.38C_s + 0.36C_s^2\right)\text{Sc}^{1/4}\text{Gr}_L^{1/4}\,, \tag{8.6.9}$$

where a is the ambient pool concentration, and C_s is the saturated concentration at the solid/liquid interface.

The experiments on freezing and melting were carried out in [8.129] to visualize the interaction between natural convection and phase change processes in porous media. It was found that the natural convection process was determined by the average Ra number and decays as the freezing process developed.

Occurrence and development of double-diffusive convection in the liquid phase during solidification of **NH_4Cl-H_2O** system ***in a confined cavity with lateral cooling*** were studied in [8.130]. Multiple point measurement of concentration show a step change in the vertical direction, and temperature visualizations reveal an S-shaped profile of isotherms, which indicate the existence of time-dependent horizontally-stacked roll cells separated by diffusive interfaces. The cells were generated in a sequential fashion, rather than simultaneously and the thickness of each cell increases with progression of solidification. The concentration in each cell was found to remain nearly constant, except for the initial development of the cells. Convection within each cell was largely controlled by the temperature field, and diffusion was dominant in the diffusive interface between two cells due to the solute field with a vertical concentration gradient. The fluid in the diffusive interface was initially stagnant, but the interface changed into a new cell under a certain condition.

References

1. B. Gebhart and J. C. Mollendorf, *Deep-Sea Res.*, **24**, 831 (1977).
2. B. Gebhart and J. C. Mollendorf, *J. Fluid Mech.*, **89**, 673 (1978).
3. P. Carey Van, B. Gebhart, and J. C. Mollendorf, *J. Fluid Mech.*, **97**, 279 (1980).
4. N. W. Wilson and B. D. Vyas, *Trans. ASME, Ser. C, J. Heat Transfer*, **101**, 313 (1979).
5. L. A. Aborin, *Inzh.-Fiz. Zh.*, **13**, 837 (1967).
6. M. S. Bendell and B. Gebhart, *Int. J. Heat Mass Transfer*, **19**, 1081 (1976).
7. C. R. Vanier, C. Tien, *Chem. Eng. Prog. Symp. Ser.*, **64**, 240 (1968).
8. Z. H. Qureshi and B. Gebhart, *NHT*, **2**, 467 (1979).
9. B. Gebhart, M. S. Bendell, and H. Shaukatullah, *Int. J. Heat Mass Transfer*, **22**, 137 (1979).
10. J. Schenk and F. A. M. Shenkels, *Appl. Sci. Res.*, **19**, 465 (1968).
11. C. R. Vanier and C. Tien, *AIChE J.*, **16**, 76 (1970).
12. S. Takeo, *Appl. Sci. Res.*, **32**, 429 (1970).
13. N. Seki, S. Fukusako, and H. Inaba, *Wu S*, **11**, 145 (1978).
14. T. Nishimura, M. Fujiwara, N. Horie, and H. Miyashita, *Int. J. Heat Mass Transfer*, **34**, 2663 (1991).
15. S. Chellaiah, R. Viskanta, *Int. J. Heat Mass Transfer*, **33**, 887 (1990).
16. S. L. Braga and R.Viskanta, *Int. J. Heat Mass Transfer*, **35**, 861 (1992).
17. T. Nishimura, Y. Hayashida. M. Mineoka, and A. Wake, *Int. J. Heat Mass Transfer*, **40**, 3449 (1997).
18. M. Mamou, L. Robillard, and P. Vasseur, *Int. J. Heat Mass Transfer*, **42**, 4487 (1999).
19. A. S. Blokhin, N. S. Blokhina, O. S. Makeeva, and Z. P. Startseva, *Vodnye Resursy*, No. 4, 154 (1977).
20. N. S. Blokhina, *Model of Convective Mixing in Fresh Basins (under Temperatures near* 4°C), Abstract of Doctoral Dissertation (in Technical Sciences), Moscow (1976).

21. M. Sugawara, S. Fukusako, and N. Seki, *Trans. JSME*, **40**, 3155 (1974).
22. N. Seki, S. Fukusako, and M. Suguwara, *Wu S*, **10**, 269 (1977).
23. G. P. Merker, P. Waas, J. Straub, and U. Grigull, *Wu S*, **9**, 99 (1976).
24. R. R. Gilpin, *Int. J. Heat Mass Transfer*, **18**, 1307 (1975).
25. K. S. Chen, M. Takeuchi, and R. R. Gilpin, *NHT*, **1**, 101 (1978).
26. N. Seki, S. Fukusako, and M. Nakaoka, *Trans. JSME*, **43**, 3064 (1977).
27. N. Seki, S. Fukusako, and M. Nakaoka, *Trans. ASME, Ser. C, J. Heat Transfer*, **97**, 64 (1975).
28. Y. L. Martisov, *Investigation of Liquid Flows in Closed Enclosure under Heat Disturbances*, Abstract of Doctoral Dissertation (in Technical Sciences), Minsk (1974).
29. S. Fukusako and M. Takahashi, *Int. J. Heat Mass Transfer*, **34**, 396 (1991).
30. C. Anselmi, M. De Paz, A. Marciano, M. Pilo, and G. Sonnino, *Int. J. Heat Mass Transfer*, **33**, 2519 (1990).
31. C. J. Ho, S. P. Chiou, and C. S. Hu, *Int. J. Heat Mass Transfer*, **40**, 3553 (1997).
32. J. N. Coster and K. Nguyen, *Int. J. Heat Mass Transfer*, **39**, 467 (1996).
33. W. Tong and G. N. Coster, *Int. J. Heat Mass Transfer*, **37**, 927 (1994).
34. P. Wang, R. Kahawita, and D. L. Nguen, *Int. J. Heat Mass Transfer*, 34, 1429 (1991).
35. H. Schmitt and H. Beer, *Brenst. – Warme-Kraft*, **29**, 15 (1977).
36. H. A. Simon, and E. R. G. Eckert, *Int. J. Heat Mass Transfer*, **6**, 681 (1963).
37. J. R. Larsen and R. J. Schoenhals, *Trans. ASME, Ser. C, J. Heat Transfer*, **88**, 407 (1966).
38. K. Nishikawa and T. Ito, *Int. J. Heat Mass Transfer*, **12**, 1449 (1969).
39. M. A. Hilal, *Cryogenics*, **18**, 545 (1978).
40. V. I. Deev, A. K. Kondratenko, V. I. Petrovichev, V. E. Kalinin, I. A. Kovalev, *Proc.* 6th *Int. Heat Transfer Conf.*, vol. 2, Toronto (1978).
41. S. P. Beschastnov, S. M. Kulkov, V. P. Petrov, *Teplofiz. Vys. Temp.*, No. 3, 651 (1978).
42. H. Ogata, W. Nakayama, *Cryogenics*, **17**, 461 (1977).
43. S. P. Beschastnov, *Free Convective Heat Transfer of Supercritical Liquids in Large Volume*, Abstract of Doctoral Dissertation (in Technical Sciences), Moscow (1974).
44. S. H. Lee and J. R. Howell, *Int. J. Heat Mass Transfer*, **41**, 1205 (1998).
45. C. W. Tsai, S. J. Yang, and G. J. Hwang, *Int. J. Heat Mass Transfer*, **41**, 4251 (1998).
46. S. Koshizuka, N. Takano, and Y. Oka, *Int. J. Heat Mass Transfer*, **38**, 3077 (1995).
47. V. S. Protopopov and G. K. Sharma, *Teplofiz. Vys. Temp.*, No. 4, 781 (1976).
48. A. M. Mamedov and F. I. Kolbaliev, *Teplofiz. Vys. Temp.*, No. 2, 436 (1977).
49. F. I. Kolbaliev, S. V. Suleimanov, *Izv. VUZov, Energetika*, No. 9, 116 (1980).
50. V. S. Protopopov, I. V. Kuraeva, N. E. Ferubko, and E. A. Osin, *Trudy MEI*, No. 405, 3 (1979).
51. M. A. Hilal, *Int. J. Heat Mass Transfer*, **23**, 697 (1980).
52. S. Uskenbaev, *Investigation of Free Convective Heat Transfer of Liquid Gases in Critical Region of Parameters*, Abstract of Doctoral Dissertation (in Technical Sciences), Leningrad (1971).
53. R. J. Neumann and E. W. P. Hahne, *Int. J. Heat Mass Transfer*, **23**, 1643 (1980).

54. E. N. Dubrovina, *Free Convective Heat Transfer from Compressed Gases to Thin Wires*, Abstract of Doctoral Dissertation (in Physics and Mathematics), Sverdlovsk (1971).
55. A. P. Solodov and D. Gorenflo, *Wu S*, **9**, 151 (1976).
56. K. Bier, H. R. Engelhorn, D. Gorenflo, and A. P. Solodov, *Wu S*, **9**, 193 (1976).
57. P. A. Maslichenko and N. Yu. Fedorov, *Izv. Akad. Nauk BSSR, Ser. Fiz.-Energ. Nauk*, No. 3, 61 (1977).
58. S. P. Beschastnov, P. L. Kirillov, V. P. Alabushev, *Teplofiz. Vys. Temp.*, No. 4, 775 (1976).
59. E. Schmidt, *Int. J. Heat Mass Transfer*, **1**, 92 (1960).
60. P. A. Maslichenko, N. Yu. Fedorov, *Izv. Akad. Nauk BSSR, Ser. Fiz.-Energ. Nauk*, No. 3, 114 (1979).
61. Y. G. Leonenko and V. A. Antipin, *Int. J. Heat Mass Transfer*, **40**, 1671 (1997).
62. G. N. Dulnev, R. S. Prasolov, and A. V. Sharkov, *Vopr. RE, Ser. TRTO*, No. 1, 15 (1971).
63. G. N. Dulnev, Yu. P. Zarichnyak, and A. V. Sharkov, *Int. J. Heat Mass Transfer*, **18**, 213 (1975).
64. A. N. Devoino, *Inzh.-Fiz. Zh.*, **4**, 70 (1961).
65. P. H. Oosthuizen, *Appl. Sci. Res.*, **16**, 121 (1966).
66. A. K. Rebrov and N. V. Mukhina, *Int. J. Heat Mass Transfer*, **9**, 819 (1966).
67. R. S. Prasolov, *Teplofiz. Vys. Temp.*, No. 3, 555 (1968).
68. P. A. Novikov and V. A. Solov'ev, in: *Common Questions of Heat and Mass Transfer*, Minsk (1966), pp. 25–30.
69. A. K. Rebrov, *Inzh.-Fiz. Zh.*, **11**, 703 (1966).
70. S. D. Probert, R. G. Broors, T. R. Thomas, and J. Maxwell, *Int. J. Heat Mass Transfer*, **10**, 135 (1967).
71. Yu. A. Koshmarov and A. E. Ivanov, *Izv. Akad. Nauk SSSR, Energetika Transport*, No. 4, 156 (1971).
72. A. V. Chechetkin, *High-Temperature Heat Transfer Media*, Energya Press, Moscow (1971).
73. A. Moutsoglou, S. K. L. Tzuoo, and T. S. Chen, *AIAA Paper*, No. 80-1525, 1 (1980).
74. A. V. Luikov and B. M. Berkovskii, *Convection and Heat Waves*, Energya Press, Moscow (1974).
75. O. G. Martynenko, P. M. Kolesnikov, and V. L. Kolpashchikov, *Introduction to the Theory of Convective Gas Lenses*, Nauka i Tekhnika Press, Minsk (1972).
76. D. C. Smith, *Proc. IEEE*, **65**, 1679 (1977).
77. V. V. Vorob'ev, *Izv. VUZov, Fizika*, No. 11, 61 (1977).
78. R. I. Soloukhin, O. G. Martynenko, and N. E. Galich, *Int. J. Heat Mass Transfer*, **23**, 1653 (1980).
79. N. E. Galich, *Zh. Tekh. Fiz.*, **50**, 1196 (1980).
80. B. P. Gerasimov, V. N. Gordienko, A. P. Sukhorukov, *Zh. Tekh. Fiz.*, **45**, 2485 (1975).
81. B. P. Gerasimov, V. N. Gordienko, A. P. Sukhorukov, *Inzh.-Fiz. Zh.*, **33**, 705 (1977).
82. R. A. Chodzko and S.-C. Lin, *Appl. Phys. Lett.*, **16**, 434 (1970).
83. B. P. Gerasimov, V. N. Gordienko, A. P. Sukhorukov, *Inzh.-Fiz. Zh.*, **36**, 331 (1979).
84. Yu. A. Sokovishin, A. N. Sinitsyn, and V. P. Mikhailov, in: *Modern Methods of Investigation of Clothes,* Leningrad (1973), pp. 139–140.

85. D. M. Gates, *Energy Exchange in Biosphere*, Harper-Row Publishers, New York (1962).
86. V. Larher, *Ecology of Plants*, Mir Press, Moscow (1978).
87. R. Sinclair, *Austral J. Biol. Sci.,* **23**, 309 (1970).
88. D. F. Parkhurst, P. R. Dunkan, D. M. Gates, and F. Kreith, *Trans. ASME, Ser. C, J. Heat Transfer*, **90**, 71 (1968).
89. P. H. Schnepp, *Boundary-Layer-Meteorol.*, **3**, 454 (1973).
90. M. Iqbal and J. A. Stoffers, *Proc.* 5th *Canad. Congr. Appl. Mech.,* Frederigton (1975), p. 557.
91. E. C. Tibbals, E. K. Carr, D. M. Gates, and F. Kreith, *Amer. J. Botany*, **51**, 529 (1964).
92. D. M. Gates, E. C. Tibbals, and F. Kreith, *Amer. J. Botany*, **52**, 66 (1965).
93. E. Ruckenstein and R. Rajagopalan, *Chem. Eng. Commun.*, **4**, 15 (1980).
94. A. M. Grishin, A. D. Grudin, and V. A. Kapustin, *Prikl. Mekh. Tekh. Fiz.*, No. 4, 57 (1980).
95. A. M. Grishin and A. D. Grudin, *Dokl. Akad. Nauk SSSR*, **253**, 549 (1980).
96. H. Glaser and E. Pruckner, *Proc.* 6th *Int. Heat Transfer Conf.*, vol. 3, Toronto (1978), pp. 49–54.
97. M. Aubinet and J. Deltour, *Int. J. Heat Mass Transfer*, **37**, 1795 (1994).
98. V. A. Efimov, *Casting and Crystallization of Steel*, Metalurgya Press, Moscow (1976).
99. B. I. Myznikova and E. L. Tarunin, in: *Mathematical Investigation Methods of Special Electrical Metallurgy Processes*, Kiev (1976), pp. 129–135.
100. G. M. Sevost'anov, *Numerical Investigation of Heat Convection in Solidifying Melt*, Abstract of Doctoral Dissertation (in Physics and Mathematics), Donetsk (1975).
101. P. F. Zavgorodnii, *Inzh.-Fiz. Zh.*, **35**, 155 (1978).
102. M. F. Bozhenko, *Investigation of Heat Transfer Intensity of Potassium Electrolyte Melts under in Free Convection and Heat Conduction of Materials Saturated by Melt*, Abstract of Doctoral Dissertation (in Technical Sciences), Kiev (1973).
103. L. A. Volokhonskii and L. A. Medvedovskaya, in: *Mathematical Methods in Investigation of Special Metallurgy Processes*, Kiev (1976), pp. 123–129.
104. M. G. Stepanenko, *Steklo Keramika*, No. 9, 17 (1955).
105. P. K. Konakov, G. E. Verevochkin, and L. A. Goryainov, *Heat and Mass Transfer in Manufacture of Monocrystalls*, Metalurgya Press, Moscow (1971).
106. N. A. Avdonin, in: *Heat and Mass Transfer*, vol. 4, Minsk (1968) pp. 311–323.
107. N. N. Urda, *Investigation of Heat Transfer Processes in Free Convection of Melt of Fluoric Salts, Thermal Conductivity of the Solid Phase and Materials Impregnated by Melt*, Abstract of Doctoral Dissertation (in Technical Sciences), Kiev (1974).
108. V. S. Avduevskii, I. V. Barmin, and S. D. Grishin, *Space Production Problems*, Mashinostroenie Press, Moscow (1980).
109. K. C. Chiang and H. L. Tsai, *Int. J. Heat Mass Transfer*, **35**, 1771 (1992).
110. W. Shyy and M.-H. Chen, *Int. J. Heat Mass Transfer*, **33**, 2565 (1990).
111. Z.-X. Gong and A. S. Mujumdar, *Int. J. Heat Mass Transfer*, **41**, 2573 (1998).
112. J. E. Simpson and S. V. Garimella, *Int. J. Heat Mass Transfer*, **41**, 2485 (1998).
113. W.-J. Chang and D.-F. Yang, *Int. J. Heat Mass Transfer*, **39** 2333 (1996).
114. H. Yoo and R. Viskanta, *Int. J. Heat Mass Transfer*, **35**, 2335 (1992).

115. D. Gobin and P. Le Quere, *Int. J. Heat Mass Transfer*, **42**, 357 (1999).
116. D. G. Neilson, F. P. Incropera, and W. D. Bennon, *Int. J. Heat Mass Transfer*, **33**, 367 (1990).
117. A. C. Keary and R. J. Bowen, *Int. J. Heat Mass Transfer*, **41**, 1129 (1998).
118. G. J. Hwang and C. W. Tsai, *Int. J. Heat Mass Transfer*, **38**, 2733 (1995).
119. I. T. Belyakov and Yu. D. Borisov, *Space Technology Fundamentals*, Mashinostroenie Press, Moscow (1980).
120. M. Bareiss and M. Beer, *LHMT*, **7**, 329 (1980).
121. B. Basu and A. W. Date, *Int. J. Heat Mass Transfer*, **35**, 1049 (1992).
122. Z. Zhang and A. Bejan, *Int. J. Heat Mass Transfer*, **33**, 661 (1990).
123. F. Vodak, R. Cerny, and P. Prikryl, *Int. J. Heat Mass Transfer*, **35**, 1787 (1992).
124. W. Shyy and M.-H. Chen, *Int. J. Heat Mass Transfer*, **33**, 2545 (1990).
125. S. K. Roy and S. Sengupta, *Int. J. Heat Mass Transfer*, **33**, 1135 (1990).
126. R. Derebail and J. N. Koster, *Int. J. Heat Mass Transfer*, **41**, 2537 (1998).
127. C. W. Lan and D. T. Yang, *Int. J. Heat Mass Transfer*, **41**, 4351 (1998).
128. S. W. Shiah, B .C. Yang, F. B. Cheung, and Y. C. Shih, *Int. J. Heat Mass Transfer*, **41**, 3759 (1998).
129. H. Lein and R. S. Tankin, *Int. J. Heat Mass Transfer*, **35**, 187 (1992).
130. T. Nishimura, T. Imoto, and H. Miyashita, *Int. J. Heat Mass Transfer*, **37**, 1455 (1994).

Printing: Krips bv, Meppel
Binding: Stürtz, Würzburg